Advances in Intelligent Systems and Computing

Volume 924

The series "Advances in Intelligent Systems and Computing" contains publications on theory, applications, and design methods of Intelligent Systems and Intelligent Computing. Virtually all disciplines such as engineering, natural sciences, computer and information science, ICT, economics, business, e-commerce, environment, healthcare, life science are covered. The list of topics spans all the areas of modern intelligent systems and computing such as: computational intelligence, soft computing including neural networks, fuzzy systems, evolutionary computing and the fusion of these paradigms, social intelligence, ambient intelligence, computational neuroscience, artificial life, virtual worlds and society, cognitive science and systems, Perception and Vision, DNA and immune based systems, self-organizing and adaptive systems, e-Learning and teaching, human-centered and human-centric computing, recommender systems, intelligent control, robotics and mechatronics including human-machine teaming, knowledge-based paradigms, learning paradigms, machine ethics, intelligent data analysis, knowledge management, intelligent agents, intelligent decision making and support, intelligent network security, trust management, interactive entertainment, Web intelligence and multimedia.

The publications within "Advances in Intelligent Systems and Computing" are primarily proceedings of important conferences, symposia and congresses. They cover significant recent developments in the field, both of a foundational and applicable character. An important characteristic feature of the series is the short publication time and world-wide distribution. This permits a rapid and broad dissemination of research results.

** Indexing: The books of this series are submitted to ISI Proceedings, EI-Compendex, DBLP, SCOPUS, Google Scholar and Springerlink **

More information about this series at http://www.springer.com/series/11156

Sanjiv K. Bhatia · Shailesh Tiwari ·
Krishn K. Mishra · Munesh C. Trivedi
Editors

Advances in Computer Communication and Computational Sciences

Proceedings of IC4S 2018

 Springer

Editors
Sanjiv K. Bhatia
Department of Mathematics
and Computer Science
University of Missouri–St. Louis
St. Louis, MO, USA

Krishn K. Mishra
Department of Computer Science
and Engineering
Motilal Nehru National Institute
of Technology
Allahabad, Uttar Pradesh, India

Shailesh Tiwari
CSED
ABES Engineering College
Ghaziabad, Uttar Pradesh, India

Munesh C. Trivedi
Department of Information Technology
Rajkiya Engineering College
Azamgarh, Uttar Pradesh, India

ISSN 2194-5357   ISSN 2194-5365   (electronic)
Advances in Intelligent Systems and Computing
ISBN 978-981-13-6860-8   ISBN 978-981-13-6861-5   (eBook)
https://doi.org/10.1007/978-981-13-6861-5

Library of Congress Control Number: 2019932704

This Springer imprint is published by the registered company Springer Nature Singapore Pte Ltd.
The registered company address is: 152 Beach Road, #21-01/04 Gateway East, Singapore 189721, Singapore

Preface

IC4S is a major multidisciplinary conference organized with the objective of bringing together researchers, developers, and practitioners from academia and industry working in all areas of computer and computational sciences. It is organized specifically to help the computer industry to derive the advances of next-generation computer and communication technology. Researchers invited to speak will present the latest developments and technical solutions.

Technological developments all over the world are dependent upon globalization of various research activities. Exchange of information and innovative ideas is necessary to accelerate the development of technology. Keeping this ideology in preference, the International Conference on Computer, Communication and Computational Sciences (IC4S 2018) was organized at Mandarin Hotel Bangkok, Bangkok, Thailand, during October 20–21, 2018.

This is the third time the International Conference on Computer, Communication and Computational Sciences has been organized with a foreseen objective of enhancing the research activities at a large scale. Technical Program Committee and Advisory Board of IC4S include eminent academicians, researchers, and practitioners from abroad as well as from all over the nation.

In this book, selected manuscripts have been subdivided into various tracks named—Intelligent Hardware and Software Design, Advanced Communications, Intelligent Computing Techniques, Web and Informatics, and Intelligent Image Processing. A sincere effort has been made to make it an immense source of knowledge for all, and this book includes 64 manuscripts. The selected manuscripts went through a rigorous review process and are revised by authors after incorporating the suggestions of the reviewers.

IC4S 2018 received around 350 submissions from around 550 authors of 15 different countries such as USA, Iceland, China, Saudi Arabia, South Africa, Taiwan, Malaysia, Indonesia, and Europe. Each submission went through the plagiarism check. On the basis of plagiarism report, each submission was rigorously reviewed by at least two reviewers with an average of 2.7 per reviewer. Even some submissions have more than two reviews. On the basis of these reviews,

64 high-quality papers were selected for publication in this proceedings volume, with an acceptance rate of 18.28%.

We are thankful to the keynote speakers—Prof. Raija Halonen, University of Oulu, Finland; Dr. K. K. Mishra, University of Missouri, St. Louis, USA; Mr. Aninda Bose, Senior Editor, Springer Nature, to enlighten the participants with their knowledge and insights. We are also thankful to delegates and the authors for their participation and their interest in IC4S 2018 as a platform to share their ideas and innovation. We are also thankful to Prof. Dr. Janusz Kacprzyk, Series Editor, AISC, Springer, for providing guidance and support. Also, we extend our heartfelt gratitude to the reviewers and Technical Program Committee members for showing their concern and efforts in the review process. We are indeed thankful to everyone directly or indirectly associated with the Organizing Committee of the conference, leading it toward the success.

Although utmost care has been taken in compilation and editing, a few errors may still occur. We request the participants to bear with such errors and lapses (if any). We wish you all the best.

Bangkok, Thailand   Organizing Committee
                                                                        IC4S 2018

About This Book

With the advent of technology, intelligent and soft computing techniques came into existence with a wide scope of implementation in engineering sciences. Nowadays, technology is changing with a speedy pace and innovative proposals that solve the engineering problems intelligently are gaining popularity and advantages over the conventional solutions to these problems. It is very important for the research community to track the latest advancements in the field of computer sciences. Keeping this ideology in preference, this book includes the insights that reflect the 'Advances in Computer and Computational Sciences' from upcoming researchers and leading academicians across the globe. It contains the high-quality peer-reviewed papers of 'International Conference on Computer, Communication and Computational Sciences' (IC4S 2018), held during October 20–21, 2018, at Mandarin Hotel Bangkok, Bangkok, Thailand. These papers are arranged in the form of chapters. The content of this book is divided into five broader tracks that cover a variety of topics. These tracks are: *Intelligent Hardware and Software Design, Advanced Communications, Intelligent Computing Technologies, Web and Informatics, and Intelligent Image Processing*. This book helps the prospective readers from computer and communication industry and academia to derive the immediate surroundings' developments in the field of communication and computer sciences and shape them into real-life applications.

Contents

About the Editors

Sanjiv K. Bhatia received his Ph.D. in Computer Science from the University of Nebraska, Lincoln, USA in 1991. He presently works as a Professor and Graduate Director (Computer Science) at the University of Missouri, St.Louis, USA. His primary areas of research include image databases, digital image processing, and computer vision. In addition to publishing over 40 articles in these areas, he has consulted extensively with industry for commercial and military applications of computer vision. He is an expert on system programming and has worked on real-time and embedded applications. He has taught a broad range of courses in computer science and was the recipient of the Chancellor's Award for Excellence in Teaching in 2015. He is a senior member of ACM.

Shailesh Tiwari currently works as a Professor at the Department of Computer Science and Engineering, ABES Engineering College, Ghaziabad, India. He is an alumnus of Motilal Nehru National Institute of Technology Allahabad, India. His primary areas of research are software testing, implementation of optimization algorithms and machine learning techniques in software engineering. He has authored more than 50 publications in international journals and the proceedings of leading international conferences. He also serves as an editor for various Scopus, SCI and E-SCI-indexed journals and has organized several international conferences under the banner of the IEEE and Springer. He is a senior member of the IEEE and a member of the IEEE Computer Society.

Krishn K. Mishra is currently working as an Assistant Professor at the Department of Computer Science and Engineering, Motilal Nehru National Institute of Technology Allahabad, India. He has also been a Visiting Faculty at the Department of Mathematics and Computer Science, University of Missouri, St. Louis, USA. His primary areas of research include evolutionary algorithms, optimization techniques and design, and analysis of algorithms. He has also authored more than 50 publications in international journals and the proceedings of leading international conferences. He currently serves as a program committee member of several conferences and an editor for various Scopus and SCI-indexed journals.

Munesh C. Trivedi is currently an Associate Professor at the Department of Information Technology, Rajkiya Engineering College, Azamgarh, India and also Associate Dean-UG Programs, Dr. APJ Abdul Kalam Technical University, Lucknow. He has published 20 textbooks and 95 research papers in various international journals and in the proceedings of leading international conferences. He has received young scientist and numerous other awards from different national and international forums. He currently serves on the review panel of the IEEE Computer Society, International Journal of Network Security, and Computer & Education (Elsevier). He is a member of the executive committee of the IEEE UP Section, IEEE India Council and also the IEEE Asia Pacific Region 10.

Part I
Intelligent Hardware and Software Design

Software Architecture Decision-Making Practices and Recommendations

Md. Monzur Morshed, Mahady Hasan and M. Rokonuzzaman

Abstract The role of Software Architect plays a vital role in entire software development process and becomes highly responsible when the software goes into deployment as business operations and automations are based on the software system. Since business operations are fully dependent on the software system, software architectural decision-making process becomes one of the major activities of a successful software projects. In this paper, we have depicted a series of empirical ways based on real life experiences so that software architects able to make accurate decision before going into software development. Regardless of architecture decision-making process, this paper may help software engineers, project managers, business owners and stakeholders in developing enterprise solutions.

Keywords Software · Architecture · Decision · Framework

1 Introduction

We are living in a fast growing business environment where most of the business operations are being handled by big software systems based on massive IT infrastructures. According to the famous Software Engineer and Academician Professor Barry Boehm, "*Marry your architecture in haste and you can repent in leisure*".

In the initial stage of software system development, a software architect should consider the following issues (not limited to):

Md. M. Morshed
TigerHATS, Independent University-Bangladesh, Dhaka, Bangladesh
e-mail: m.monzur@gmail.com; monzur@tigerhats.org

M. Hasan (✉)
Independent University-Bangladesh, Dhaka, Bangladesh
e-mail: mahady@iub.edu.bd

M. Rokonuzzaman
North South University, Dhaka, Bangladesh
e-mail: m.rokonuzzaman@northsouth.edu

© Springer Nature Singapore Pte Ltd. 2019
S. K. Bhatia et al. (eds.), *Advances in Computer Communication and Computational Sciences*, Advances in Intelligent Systems and Computing 924,
https://doi.org/10.1007/978-981-13-6861-5_1

(a) What type of software it will be?
(b) What type of business operation will be performed?
(c) Who are the end-users? Identify them.
(d) How many users will be using this software system?
(e) What should be the expected life span of the software?
(f) Will it be a cross platform application?
(g) What are the functional and non-functional requirements?
(h) Does the company have available resources such as human resources, funds?

It is also important to understand the organization's culture and business processes while architecting the software system. One might say that business analyst or software requirements engineer is responsible for such issues. But as a software architect is important not to ignore such issues because better understanding about the organization's work process and business operations will help him/her in architecting a cost effect software as several issues are involved here such as cost effectiveness, technical issues, business thinking, clients' satisfaction and user friendly system, effective use of resources, on time delivery and budgets, etc.

Typically software architecture decision-making process begins from software development to maintenance phase. Both the phases are equally important for a successful software project. As the most expensive phase of software development is maintenance [1], the software architecture decision making process also becomes very critical in this stage. According to Walker Royce, "for every $1 you spend on development, you will spend $2 on maintenance [2]". During software maintenance phase new change requests need to be continuously delivered and deployed on demand basis, any system anomaly will directly impact the end-users and stakeholders.

2 Related Works

Vliet et al. [3] outlined some research directions that may help to understand and improve the software architecture design process. They have discussed about two decision-making concepts like rational and bounded rational.

Dasanayake et al. [4] studied the current software architecture decision making practices in the industry using a case study conducted among professional software architects in three different companies in Europe. They have identified different software architecture decision-making practices followed by the software teams as well as their reasons for following them, the challenges associated with them, and the possible improvements from the software architects' point of view. They found out that improving software architecture knowledge management can address most of the identified challenges and would result in better software architecture decision-making.

Kazman et al. [5] discussed about economics-driven architecture which could produce the maximum amount of value, productivity, evolve ability, and reputation which should be the guide in making any architecture-related decisions.

Asundi et al. [6] developed a method known as CBAM (Cost Benefit Analysis Method) for performing economic modeling of software systems, centered on an analysis of their architecture. In the framework, they incorporate economic criteria like benefit and cost that are derived from the technical criteria like quality attribute responses.

3 SWOT Analysis in Architecture Decision Making

SWOT analysis is a strategic planning technique used to identify the strengths, weaknesses, opportunities, and threats related to business planning. In this paper, we have shown the use of SWOT analysis in Software Architecture-decision making process.

Strengths	Weaknesses
• Widely accepted	• Unstable platform
• Open source	• Unstable technology
• Cross-platform	• Too many bugs/defects
• Scalable	• Too much dependency
• Community support	• Lack of documentation
• Documentation	
Opportunities	Threats
• Competitive price	• Structural change leads to affect the core technology
• Available skilled human resource	• Third-party software dependency
• Popular technology	• Copyright issues
	• Unrealistic price
	• Security weakness

The lists we have shown may not accommodate all the critical issues associated in a software project. However, in future we will come up with a complementary checklist to serve the purpose of common software engineering issues.

4 Evaluation Criteria of Selecting Software Architectures

To select the right software architecture we are proposing the following evaluation criteria:

Table 1 Software development methodologies and project based complexity level

	Project size	Target group	Complexity level
Scrum	Small	S	1
Agile	Medium	S, M	2
Waterfall	Medium, large	M, L	3
Extreme programing	Medium	S, M	2
Microsoft solutions framework	Medium	S, M	1, 2
Test-driven development	Medium	S, M	1, 2
Spiral	Large	L	3

Table 2 Software architectural platforms and project based resource allocation

	Technology availability (%)	Cost effective (%)	Time (%)	Resources (%)	Skill (%)	Recomm.
Monolithic	80	90	90	70	80	S
Client-server	60	80	80	85	80	S
Distributed	80	80	90	85	90	M, L
SOA	80	70	90	90	90	M, L
Micro-services	90	70	95	90	90	M, L

Small enterprise	S
Medium enterprise	M
Large enterprise	L

Complexity level	Normal	1
	Average	2
	High	3

See Tables 1 and 2.

5 Economics-driven Software Architecture

By adopting proper software engineering methodology, software architecture and software product line, it becomes easier to reduce the software product cost which may increase in the number of software production. Reduction in software production cost enhances the product's market penetration and makes the product more

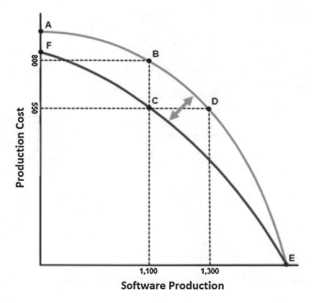

Fig. 1 Production cost versus software production

competitive among other players. It is important to remember that along with software features and functionality, software price also plays a vital role in today's global market. In the Fig. 1, we have depicted the relation between Production Cost versus Software Production.

6 Fishbone Model in Software Architecture Decision-Making

FishBone model also known as Cause and Effect Diagram created by Kaoru Ishikawa, a visualization tool is widely used for identifying the root cause of the problem [7]. This model can be used in software architecture decision making process to identify the root cause of the issues that usually occur during software system development and maintenance phase (Fig. 2).

7 Conclusion

It is one of the major tasks to identify the product's life span considering the existing technology, business, and market demand before developing a software system. As these days technology is evolving rapidly, it is a very challenging task to develop

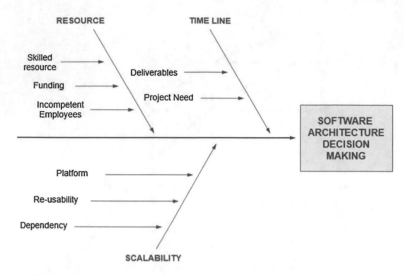

Fig. 2 Fishbone model in software architecture decision making-process

software system that converges with the Organization's vision. This is why it has become a vital task to identify proper software platform. Lack of proper planning, the project's future could be at risk and cause financial loss which can be tackled by selecting proper software development methodology, platform, strategy, and technology frameworks to enhance ROI. Adopting any of these may not be suitable for any project unless the need assessments are conducted properly.

In future, we are aiming to extend this work and develop a comprehensive check list that may help technology professionals, academicians and industry in identifying critical issues during project initiation, planning, and software development to maintenance stages. Furthermore, it would be our goal to incorporate the concept of software product lines and reusing the software component to ensure cost effective products development in a timely manner.

References

1. Martin, R.C.: Granularity. C++ Rep. **8**(10) (1996)
2. Walker, R.: Next-Generation Software Economics. The Rational Edge (2000)
3. Vliet, H., Tang, A.: Decision making in software architecture. J. Syst. Softw. (2016). https://doi.org/10.1016/j.jss.2016.01.017
4. Dasanayake, S., Markkula, J., Aaramaa, S., Oivo, M.: Software architecture decision-making practices and challenges: an industrial case study. In: 2015 24th Australasian Software Engineering Conference, Adelaide, SA, 2015, pp. 88–97 (2015)
5. Kazman, R., Bahsoon, R., Mistrik, I., Zhang, Y.: Economics-Driven Software Architecture: Introduction. https://doi.org/10.1016/b978-0-12-410464-8.00001-5

6. Asundi, J., Kazman, R., Klein, M.: Using Economic Considerations to Choose Among Architecture Design Alternatives (2001)
7. FishBone Diagram. https://whatis.techtarget.com/definition/fishbone-diagram

Abstraction in Modeling and Programming with Associations: Instantiation, Composition and Inheritance

Bent Bruun Kristensen

Abstract The association concept abstracts over collaboration between concurrent autonomous entities. Associations describe collaborations, i.e., requests and coordination of contributions from the entities. And the association concept also supports abstraction in the form of instantiation, composition and inheritance. Simple and expressive abstraction with associations is demonstrated by illustrative examples, schematic language presentations, brief characterizations of qualities and implementation experiments.

Keywords Language design · Abstraction with associations · Instantiation, composition and inheritance · Concurrent autonomous entities · Qualities

1 Introduction

The description of collaboration between autonomous entities is typically distributed among the participating entities in terms of their individual contributions, and the coordination is based on synchronization mechanisms. On the contrary, the association concept gives a unified description of collaboration—i.e., both requests and coordination of contributions from entities. The association abstracts over both structural and interaction aspects of the collaboration: It includes roles and a directive, where entities play the roles in interactions and the directive describes interaction sequence between the entities. In [1], the association concept is motivated and defined—and evaluated and related to similar modeling and programming concepts.

The aim of the present paper is to show the potential of abstraction in modeling and programming with instantiation, composition and inheritance of associations. The association offers instantiation, i.e., how to create an instance from an association category; composition, i.e., how to compose an association category from instances

B. B. Kristensen (✉)
University of Southern Denmark, Odense, Denmark
e-mail: bbk@mmmi.sdu.dk

© Springer Nature Singapore Pte Ltd. 2019
S. K. Bhatia et al. (eds.), *Advances in Computer Communication and Computational Sciences*, Advances in Intelligent Systems and Computing 924,
https://doi.org/10.1007/978-981-13-6861-5_2

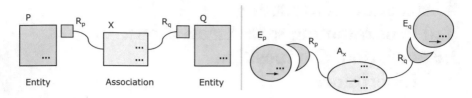

Fig. 1 Illustration of association and entity

of existing association categories; and inheritance, i.e., how to describe an association category by inheriting from an existing association category.

Section 2 is a brief introduction to the association concept, a presentation of a basic example, as well as related work and prerequisites for this kind of abstraction. Section 3 illustrates abstraction in terms of instantiation, composition and inheritance by giving illustrative but complex examples and schematic presentations and by characterizing the qualities of the abstractions. In Sect. 4, an experimental implementation of abstraction with associations is discussed.

2 Background

2.1 Associations and Autonomous Entities

Associations are inspired from a model for understanding, modeling and programming ambient systems [2]. The association concept offers an alternative way of describing interactions among concurrent autonomous *entities*. Association X and entities P and Q are illustrated in Fig. 1: Instances of entities (E_p and E_q) are autonomous (and therefore distinct from ordinary objects/classes [3]) and participate in instance A_x of association X. Because an entity is autonomous, only the entity itself controls the execution of its methods. An entity has an *action part* (illustrated by ...) in the form of a sequence of actions (similar to a class in SIMULA [4]). An entity also has a list of the actual contributions (i.e., methods to be executed) requested from associations in which an instance of the entity participates. An association requests an instance of an entity to contribute by executing a given method, but the instance itself chooses among the requests. Associations *integrate* activity (the directive) and role aspects. The directive (illustrated by ... ...) is an interaction sequence including requests to participating entities. The interactions of the directive are executed sequentially, whereas an entity executes its action part and list of requests interleaved. The description of a role (R_p and R_q) of an association includes a name of an entity category (like P or Q) that specifies which instances may play the role. The association concept combines activities [5] and roles [6] in one abstraction.

A textual version of associations and entities is illustrated in the *Association Language* in Fig. 2, i.e., association x with instance A_x and entities P and Q with

```
entity P {                association X {              X A_x = new X (...)
  void m_p (...) {...}        role R_p for P            ...
  ...                         role R_q for Q            P E_p = new P (...)
  action part {...}           ...                       ...
}                             directive {               Q E_q = new Q (...)
entity Q {                      ...                      ...
  void m_q (...) {...}          R_p::m_p (...)           E_p play A_x.R_p
  ...                           ...                      ...
  action part {...}            R_q::m_q (...)            E_q play A_x.R_q
}                               ...
                              }
                            }
```

Fig. 2 Textual version of illustration of association and entity

instances E_p and E_q. With respect to an autonomous instance E_p playing role E_q of A_x, we use the notation $R_p::m_p()$ for the request of an invocation of the method $mp()$ of E_p. But because entities are autonomous, the invocation is a request to E_p, i.e., E_p decides if and when it may execute $m_p()$ (interleaved with E_p's action part). With respect to the directive, different mechanisms for synchronization and communication among entities are possible: In [1], synchronization of autonomous entities is illustrated in rendezvous-based way [7]; in [8], interactions between agents are illustrated by asynchronous message-passing [9].

Concurrent object execution [10] implies that the objects execute in parallel and independently, but that these objects at various times synchronize and typically communicate to exchange data. Typically, the language mechanisms used to describe the synchronization are placed at the concurrent objects, i.e., in the action part of entities P and Q in Fig. 2. On the contrary, by associations, the synchronization mechanisms are placed outside the entities and shared by these entities, i.e., in the association x in Fig. 2. The association is superior with respect to modeling and programming collaboration because it supports our natural understanding of collaborations between autonomous entities and because the concept captures collaboration as a descriptive unit, c.f. association x in Fig. 2. The association abstraction x is formed by our conceptualization of collaborations and is essential for understanding, modeling and communication ("Without abstraction we only know that everything is different" [11]): The association is a language concept and not implementation-specific technology. As a shared plan for collaboration, the association makes it possible for entities precisely and understandably to explain their ongoing executions, i.e., in Fig. 2, when entity E_p executes method $m_p()$, this execution refers to a specific request, for example, $R_p::m_p()$ in association x.

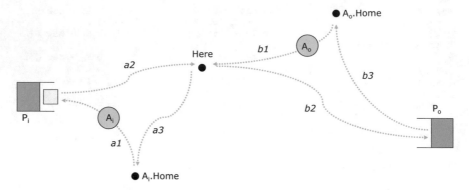

Fig. 3 Example with interacting AGVs

2.2 Example with Interacting AGVs[1]

The examples in the paper are inspired from a project about transportation systems [12]. The project investigates the process of moving boxes from a conveyor belt onto pallets and transporting these pallets. This scenario is inspired from the high bay area of the LEGO® factory with AGVs, no human intervention and only centralized control. A toy prototype inspired from this system (to bridge the gap between simulation and real physical applications) measures 1.5 by 2 m, with three mobile robots (LEGOBots), two input stations, two output stations, a conveyor belt and a station with empty pallets. The approach supports a fully distributed control for each LEGOBot. A LEGOBot is based on a LEGO® MindstormsTM RCX brick extended with a PDA and wireless LAN. Problems with combining and maintaining the basic technology (including LEGO® MindstormsTM, RCX, PDA, WLAN) have motivated the introduction of a virtual platform.

The basic example is illustrated in Fig. 3. The example includes two AGVs and two ports, and the objective is to describe how to organize the transport of packages by the AGVs between the ports. AGVs A_i and A_o bring packages from port p_i to port p_o (i for *input* and o for *output*). AGV A_i brings a package from port p_i to position Here, the AGVs shift the package from AGV A_i to AGV A_o, and AGV A_o brings the package from position Here to port P_o. The AGVs move along suitable paths: AGV A_i moves along *a1* followed by *a2* followed by *a3*; AGV A_o moves along *b1* followed by *b2* followed by *b3*. The AGVs must synchronize in order to shift the package from A_i to A_o. Furthermore, AGV A_i must synchronize with port P_i to pick up the package, and AGV A_o must synchronize with port P_o to put down the package.

The example presented in Fig. 3 is described in the Association Language in Fig. 4: AGV and Port are entity categories for AGVs and ports, respectively. Entity Port includes package x and Position P, whereas entity AGV includes

---

[1] AGV is the acronym for Automatic Guided Vehicle.

```
entity Port {            association Transport {      Transport T = new Transport()
   ...                      role Aᵢ for AGV
   Package x                role Aₒ for AGV            AGV A₁ = new AGV(...)
   Position p               role Pᵢ for Port           AGV A₂ = new AGV(...)
}                           role Pₒ for Port
                                                       Port P₁ = new Port(...)
entity AGV {             directive {                   Port P₂ = new Port(...)
   ...                      Position Here = f(...)
   Package x                (| Aᵢ::goto(Pᵢ.p)          A₁ play T.Aᵢ, A₂ play T.Aₒ
   Position Home               Pᵢ::x → Aᵢ::x
   void goto(...) {...}        Aᵢ::goto(Here)           P₁ play T.Pᵢ, P₂ play T.Pₒ
}                           ,  Aₒ::goto(Here)
                            |)
class Position {          Aᵢ::x → Aₒ::x
   ...                      (| Aᵢ::goto(Aᵢ.Home)
}                           ,  Aₒ::goto(Pₒ.p)
                               Aₒ::x → Pₒ::x
                               Aₒ::goto(Aₒ.Home)
                            |)
                         }
                       }
```

Fig. 4 Textual form of example with interacting AGVs

Package x, Position Home and method goto(...). Instances of AGV and Port exist at execution time, namely P_1, P_2 and A_1, A_2, respectively. The association category Transport is a description of the interactions of AGVs and ports. In the instance T of Transport, AGVs A_1, A_2 and Ports P_1, P_2 play roles A_i, A_o and P_i, P_o, respectively. The interactions are described by the association Transport on behalf of—and executed by—the participating AGVs and ports. The synchronization of AGVs and ports is described specifically by $P_i::x \rightarrow A_i::x$, $A_i::x \rightarrow A_o::x$ and $A_o::x \rightarrow P_o::x$.[2] The clause (| ..., ... |) means concurrent execution of the (two or more) parts included in the clause.[3] The example illustrates synchronization in time and place by "being at the right places before synchronization in time." In the directive of Transport, the meeting point is calculated as Position Here. Next the directive includes the sequence (| ... |) $A_i::x \rightarrow A_o::x$(| ... |) to be executed sequentially. In the first (| ... |) clause, the input AGV A_i moves to input Port P_i from which A_i picks up x and moves to Here and concurrently the output AGV A_o moves to Here. When both A_i and A_o are at Here, the package x is transferred from A_i to A_o by $A_i::x \rightarrow A_o::x$. In the last (| ... |) clause, the input AGV A_i moves to its Home and concurrently the output AGV A_o moves to output Port P_o where A_o puts down x and moves to its Home.

   The problem outlined in Fig. 3 is illustrative, and Fig. 4 offers an excellent solution. Variations of this problem are used in Sects. 3.2 and 3.3 for illustrating composition

---

[2]$S::x \rightarrow R::y$ describes synchronization between S and R where S is requested to transfer the value of x and R to receive this value in y. Entities S and R are synchronized exactly when the contents of x are transferred to y.

[3](|..., ..., ...|) describes concurrent execution of the parts included. The execution is completed when all parts have completed their execution.

and inheritance. The programs are created in order to illustrate the functionality of composition and inheritance but are not straightforward solutions to the problems.

2.3 Related Work and Prerequisites

In [1], the basic form of the association concept is compared to related work—in the present paper, the focus is on instantiation, composition and inheritance by means of associations: In [13], *relations* are introduced as abstractions between objects. However, the relation is an associative abstraction over structural aspects only, i.e., interaction aspects are not covered by the relation. Alternative approaches in [14–17] support the notion of relationships in object-oriented languages. But the purpose is to have structural relationships only—where the essence of the association is also to support interactions through the relationships: The alternative approaches do not support composition and inheritance of relationships—however, in [15], the term composition is used to establish something similar to transitivity of relations.

In general, programming languages contain mechanisms with relevant qualities for modeling time, programming time and runtime, e.g., [18–20]. Qualities may be discussed either for a specific language, i.e., the whole language or specific mechanisms of a language; or for languages in general or categories of structures in programming languages, e.g., control structures; or in the form of specific qualities (e.g., simplicity) where these may be interlinked or overlapping (e.g., simplicity and readability) with the inherent property of often being conflicting (e.g., readability and efficiency). We focus on the development process, i.e., we are concerned only about modeling and programming time, i.e., *development time*. And we focus on the association concept—where *abstraction* is a quality itself. In relation to abstraction by means of associations, we focus on qualities of language mechanisms for *instantiation*, *composition* and *inheritance*. For reasons of simplicity, we only include the qualities understandability, efficiency and flexibility—with the definition:

– *Understandability*: The clarity and the simplicity of the *designed* models and *described* programs. We see simplicity and readability as aspects of understandability.
– *Efficiency*: The effort needed to *develop* models and programs by using the mechanisms. In general, efficiency may be in conflict with understandability.
– *Flexibility*: The ability to easily *change* various aspects within the models or programs. In general, flexibility supports both efficiency and understandability.

To design inheritance for the association concept, action inheritance is essential. In general, action inheritance (e.g., methods in Java and Beta [21] or action sequences in classes in Simula and Beta) differs as exemplified by Java and Beta, as illustrated in Fig. 5 (the syntax is neither from Java nor from Beta): Methods x and x′ are declared in classes x and Y′, respectively. Class Y′ inherits from class Y, whereas method x′ inherits from method x. Action inheritance in Java is controlled by super.X(...) in x′, where x′ is declared as a usual method, and the names x′ and x must be

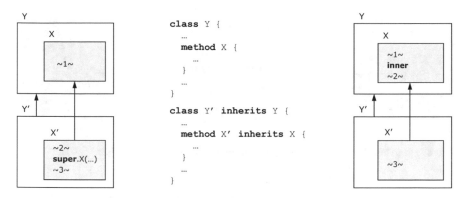

Fig. 5 Illustration of action inheritance in Java (left) and in Beta (right)

identical. Invocation of x (...) of an Y' object gives the sequence ~2~~1~~3~. The intention is that ~1~ is seen as a *preliminary* sequence to be combined as specified with *secondary* sequences ~2~ and ~3~. Action inheritance in Beta is controlled by inner in x and by declaring method x' inherits x {...}, where the name X' may be different from X. The invocation of x' (...) of an Y' object gives the sequence ~1~~3~~2~. The intention is that ~1~ and ~2~ are seen as general *before* and *after* sequences, respectively, whereas ~3~ is a special sequence *in between*, i.e., a conceptual understanding of action inheritance. Consequently, we use the approach from Beta (and Simula) because the intention of inheritance of associations is to support the development of conceptual models.

3   Abstraction

Abstraction in modeling and programming with associations include the following:

- Instantiation, i.e., an instance of an association is created from the association category,
- Composition, i.e., an association is described as a composition of instances of other associations, and
- Inheritance, i.e., an association is described as an extension of another association.

3.1   Instantiation

Instantiation refers to the creation of an instance from an association category. An association category is a description from which instances are created with unique identities, state and values according to this description. Instantiation supports a

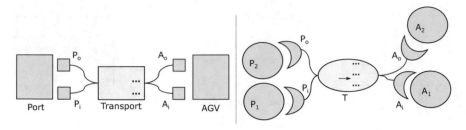

Fig. 6 Diagrams illustrating instantiation as example with interacting AGVs

number of qualities among which our focus is *efficiency* and *understandability*, i.e., we use a single description for creating several instances and our understanding is captured explicitly by the category.

Figure 1 gives a schematic illustration of instantiation of associations. Figure 1 (right) illustrates an instantiation of the model in Fig. 1 (left): A_x is an instantiation of association x, E_p and E_q are instances of entities P and Q, E_p and E_q play the roles R_p and R_q, and an arrow indicates the state of the execution of a directive/an action part. Instantiation of associations is similar to instantiation in object-oriented languages, e.g., Java and Beta, except for the properties role and directive.

In Fig. 6 (left), association Transport (with Ports P_i and P_o and AGVs A_i and A_o) is illustrated together with entity categories Port and AGV. This means that Transport is identified as an abstraction and described as a category. In Fig. 6 (right), instances of associations and entities are illustrated, namely instance T of Transport, instances P_1, P_2 of Port and A_1, A_2 of AGV, where ports and AGVs play appropriate roles of T. Figure 4 shows a textual version of the diagrams in Fig. 6.

For reasons of simplicity, we present instantiation of associations only in a very basic form. It is obvious to consider the inclusion of, for example, constructors also for instantiation of associations.

At development time, instantiation of associations especially supports the following:

- Understandability: The purpose of classification is mainly to understand: Without classification, very little is being modeled. By classifying concurrent execution including synchronization and communication by associations, we achieve some understanding and the associations express our understanding.
- Efficiency: Instantiation is time saving because no complete redesign and programming task is necessary—the classification by an association happens only once.

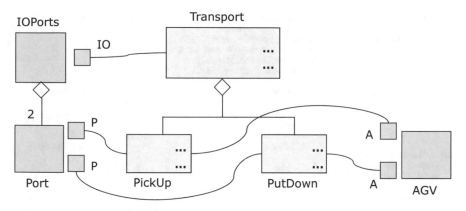

Fig. 7 Diagram illustrating composition of `Transport`

3.2 *Composition*

Composition refers to the description of an association, the *whole* association, by combining instances of other associations, the *part* associations. Composition supports a number of qualities among which our focus is *simplicity* and *understandability*, i.e., simple and smaller associations are developed in order for instances of these to be utilized as building blocks and combined in more complicated, but still understandable associations.

In Fig. 7, association `Transport` with role `IO` for `IOPorts` is composed of instances of associations `PickUp` and `PutDown`, each with roles `P` for `Port` and `A` for `AGV`. Entity `IOPorts` is a composition of two `Ports`. Consequently, the associations `PickUp` and `PutDown` are identified as categories where instances of these may be applied in, e.g., `Transport` that also is a category. When applied, the instances of `PickUp` and `PutDown` are initialized according to their purpose within `Transport`, i.e., how the AGVs and ports are used appropriately by `PickUp` and `PutDown`.

Fig. 8 is a textual version of the diagram in Fig. 7. As mentioned, `PickUp` and `PutDown` are part associations to *pick up* and *put down* a package: In `PickUp`, AGV A moves to `Position` P of `Port` P where the package x is transferred from P to A by P::x → A::x; in `PutDown`, the package x is transferred from AGV A to `Port` P by A::x → P::x after which A moves to its `Position` Home. Entity `IOPorts` is a composition of two ports P_i and P_o and is used for the `IO` role in `Transport` in order to specify the ports from which the transport starts and ends. The allocation of AGVs A_i and A_o is dynamic, e.g., by means of A_i = AGV.allocate(...), the input AGV A_i is allocated. Instances `PickUp` and `putdown` of `PickUp` and `PutDown`, respectively, are initialized appropriately concerning `Port` and AGV. In the remaining directive of `Transport`, the sequence (| ... |) A_i::x → A_o::x is executed sequentially. In the first (| ... |) clause, the execution of the `PickUp` instance is followed by the input AGV A_i moving to `Here` and concurrently the

```
association PickUp {          association Transport {         entity IOPorts {
  role P for Port               role IO for IOPorts            ...
  role A for AGV                PickUp pickup = new PickUp()      Port Pi
                                PutDown putdown = new PutDown()   Port Po
  directive {                   AGV Ai = AGV.allocate(…)       }
    A::goto(P.p)                AGV Ao = AGV.allocate(…)
    P::x → A::x                                                Port P1 = new Port(…)
  }                             directive {                    Port P2 = new Port(…)
}                                 Position Here = f(…)         Port PP =
                                  IO.Pi play pickup.P                new IOPorts(P1, P2)
                                  Ai play pickup.A
                                  IO.Po play putdown.P         Transport T =
                                  Ao play putdown.A                 new Transport()

association PutDown {             (| pickup                   PP play T.IO
  role P for Port                    Ai::goto(Here)
  role A for AGV                  ,   Ao::goto(Here)
                                  |)
  directive {                     Ai::x → Ao::x
    A::x → P::x                   (| Ai::goto(Ai.Home)
    A::goto(A.Home)              ,   Ao::goto(IO.Po.p)
  }                                  putdown
}                                 |)
                                }
                              }
```

Fig. 8 Textual form of composition of Transport

output AGV A_o moves to Here. When both A_i and A_o are at Here, the package x is transferred from A_i to A_o by A_i::x $\rightarrow$ A_o::x. In the last (| … |) clause, the input AGV A_i moves to its Home and concurrently the output AGV A_o moves to output Port P_o followed by the execution of the PutDown instance.

Figure 9 is a schematic illustration of the composition of associations. The *whole* association X is a composed association made up of *part* associations Y_1, …Y_n (a diamond denotes the composition). The part associations Y_1, …Y_n are available so that the description of (the properties of) X may use (the properties of) instantiations of Y_1, …Y_n, i.e., properties of instances of Y_i may be used for describing properties of X. In this sense, the composition of associations is similar to the composition in object-oriented languages, exemplified by, e.g., Java and Beta. The properties of associations, namely roles and directive, cause the difference: Role properties of instances of Y_i may be used for describing role properties of X, i.e., participants of the part associations are integrated into the description of the participants of the composed association. Similarly, the directive property of instances of Y_i may be used for describing the directive property of X, i.e., the directives of part associations may be used for executions in the directive of the composed association.

Figure 10 shows a textual version as a refinement of the schematic diagram in Fig. 9. Roles and directives of the whole associations may be composed as follows: A role, e.g., R_p or $R_p \cdot q_i$ of the composed association X is composed of roles of the part associations, e.g., $Y_i \cdot R_{pi}$ or $Y_i \cdot R_{qi}$, i.e., roles of instances Y_i are used in the application of the roles of the composed association, e.g., by R_p **play** $Y_i \cdot R_{pi}$ or by $R_p \cdot q_i$ **play** $y_i \cdot R_{qi}$. The directive of the composed association X utilizes directives of instances of part associations as exemplified by Y_i, i.e., directives of

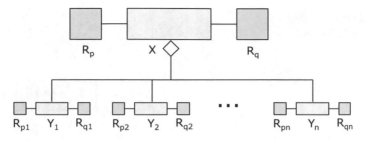

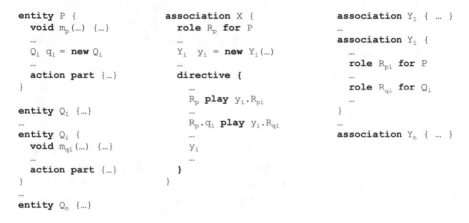

Fig. 9 Diagram illustrating composition

```
entity P {                    association X {                association Y₁ { ... }
  void mₚ(...) {...}            role Rₚ for P
                                                             ...
  ...                           ...                          association Yᵢ {
  Qᵢ qᵢ = new Qᵢ                Yᵢ  yᵢ = new Yᵢ(...)
                                                               ...
  ...                           ...                            role Rₚᵢ for P
  action part {...}             directive {
                                                               ...
}                                 ...                          role Rqᵢ for Qᵢ
                                  Rₚ play yᵢ.Rₚᵢ
                                                               ...
entity Q₁ {...}                   ...                        }
...                               Rₚ.qᵢ play yᵢ.Rqᵢ
entity Qᵢ {                       ...                        ...
  void mqᵢ(...) {...}             yᵢ                         association Yₙ { ... }
  ...                             ...
  action part {...}             }
}                             }

...
entity Qₙ {...}
```

Fig. 10 Schematic illustration of composition of associations

instances of the parts are included in the description of the directive of the composed association.

For reasons of simplicity, we present the composition of associations only in a very basic form. Composition is similar to aggregation in [22] where aggregation is used to model whole-part relationships between things. Composite aggregation means that the composite solely owns the part and that there is an existence and disposition dependency between the part and the composite. Shared aggregation implies that the part may be included simultaneously in several composite instances. It is obvious to consider the inclusion of the distinction between composite and shared aggregation also for the composition of associations. Furthermore, the distinction in [23] between the *is-component-for* relation and *has-part* relation (not necessarily the reverse relation) may be considered for the composition of associations.

At development time, the composition of associations especially supports

– Understandability: The ability to be able to see the whole association as constructed by the various part associations makes the form and content of the whole association more conceivable.

Fig. 11 Diagram illustrating
inheritance in general

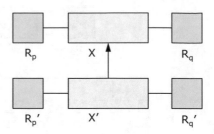

– Efficiency: The composition of a whole association from part associations makes
the development process simple and straightforward. In this sense, the efficiency
is based on the simplicity obtained through the divide-and-conquer principle for
problem solving in general.

3.3 Inheritance

Inheritance refers to the description of an association, the *specialized* association, by
supplying additional description to an existing association, the *general* associations.
Inheritance supports a number of qualities, among which our focus is *flexibility* and
understandability, i.e., basic and incomplete associations may be refined to more
specific and complete but still reasonably understandable associations.

Figure 11 is a schematic illustration of inheritance of associations in general: The
association X' is a *special* version of the *general* association X (an arrow denotes
the inheritance). The general association X is available so that the description of (the
properties of) X' may use (the properties of) X. In this sense, inheritance of associations is similar to inheritance in object-oriented languages, exemplified by, e.g., Java
and Beta. The properties of associations, namely roles and directive, cause the difference: The role properties, e.g., R_p of X, may be used in various ways for describing
specialized role properties Rp' of X', i.e., participants of the special association are
specialized participants of the general association. Similarly, the directive property
of X may be used for describing the specialized directive property of X', i.e., the
activity of the special association is the activity of the general association including
additional interaction sequences.

Figure 12 shows diagrams illustrating two variants of directive inheritance as a
refinement of the general inheritance from Fig. 11: The leftmost part is based on the
inner mechanism, whereas the rightmost part is based on virtual associations—also
shown in the textual form in Figs. 13 and 15, respectively.

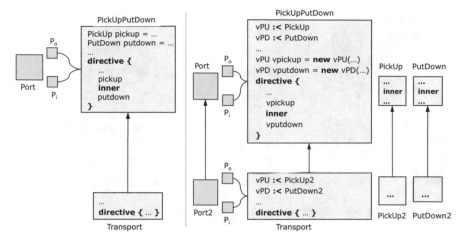

Fig. 12 Inheritance of `PickUpPutDown`: inner (left) and virtual associations (right)

```
association PickUpPutDown {          association Transport             Transport T =
  role Pᵢ for Port                     extends PickUpPutDown {              new Transport()
  role Pₒ for Port
  PickUp pickup = new PickUp()       directive {                         Port P₁ = new Port(...)
  PutDown putdown = new PutDown()      Position Here = f(...)            Port P₂ = new Port(...)
  AGV Aᵢ = AGV.allocate(...)           (| Aᵢ::goto(Here)
  AGV Aₒ = AGV.allocate(...)           ,  Aₒ::goto(Here)                 P₁ play T.Pᵢ
                                       |)                                P₂ play T.Pₒ
  directive {                          Aᵢ::x → Aₒ::x
    Pᵢ play pickup.P                   (| Aᵢ::goto(Aᵢ.Home)
    Aᵢ play pickup.A                   ,  Aₒ::goto(Pₒ.p)
    Pₒ play putdown.P                  |)
    Aₒ play putdown.A                }
    pickup                        }
    inner
    putdown
  }
}
```

Fig. 13 Textual form of inheritance of `PickUpPutDown`: inner

3.3.1 Inheritance by Means of Inner

In the illustration in Fig. 12 (left) based on **inner**, the general association `PickUpPutDown` is used by the specialized association `Transport`. `PickUpPutDown` uses instances `PickUp` and `putdown` of associations `PickUp` and `PutDown` (from Fig. 8) in order (before execution of **inner**) to pick up a package and (after execution of **inner**) to put down a package. The **directive**{...} of `Transport` describes the interactions to replace **inner**, i.e., to take place between these executions of `PickUp` and `putdown`.

In Fig. 13, the associations `PickUpPutDown` and `Transport` are given in textual form, corresponding to diagram Fig. 12 (left). As mentioned, `PickUpPutDown` uses instances of associations `PickUp` and `PutDown` and allocates appropriate

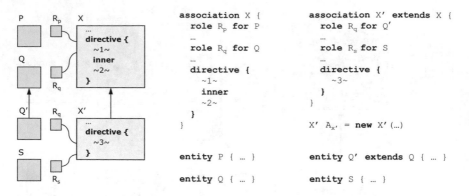

```
association X {                 association X' extends X {
   role R_p for P                  role R_q for Q'
   ...                             ...
   role R_q for Q                  role R_s for S
   ...                             ...
   directive {                     directive {
      ~1~                             ~3~
      inner                        }
      ~2~                        }
   }
}                                 X' A_x' = new X' (...)

entity P { ... }                  entity Q' extends Q { ... }

entity Q { ... }                  entity S { ... }
```

Fig. 14 Schematic illustration of inheritance of associations: **inner**

AGVs to be used. In the directive of `PickUpPutDown`, instances of `PickUp` and `PutDown` are initialized appropriately by ports P_i, P_o and AGVs A_i, A_o, i.e., A_i, P_i in instance `PickUp` and A_o, P_o in instance `putdown`. The remaining part of the directive has the form `PickUp` **inner** `putdown`: Before **inner**, input AGV picks up the package, and after **inner**, the output AGV puts down the package. In `Transport`, the contents of **inner** are added: The meeting point is calculated as `Here` after which the AGVs move to the meeting point, exchange the package and move to appropriate places by the sequential sequence $(|\ ...\ |)$ $A_i::x \rightarrow A_o::x$ $(|\ ...\ |)$. In the first $(|\ ...\ |)$ clause, the input AGV A_i moves to `Here` and concurrently the output AGV A_o moves to `Here`. When both A_i and A_o are at `Here`, the package x is transferred from A_i to A_o by $A_i::x \rightarrow A_o::x$. In the last $(|\ ...\ |)$ clause, the input AGV A_i moves to its `Home` and concurrently the output AGV A_o moves to `Port` P_o.

Figure 14 (left) shows a refinement of the schematic diagram in Fig. 11 based on **inner**, and Fig. 14 (right) gives a textual version: Roles of the general association X may be inherited by the special association X' as follows: Any role is inherited from the general association, e.g., R_p, and additional roles may be described for the specialized association, e.g., R_s. An existing role may be modified, i.e., role R_q from **role** R_q **for** Q in X may be specialized by replacing Q by Q' (where Q' inherits Q), i.e., in X' role R_q may be modified as **role** R_q **for** Q'.

The directive is always inherited from the general association and may be modified (the inherited directive may be extended by a supplementary description), i.e., inheritance of the directive is similar to the **inner** mechanism [4]: In Fig. 14 (right) in the directive of X, the notation ~1~ (indicating some initial general interaction sequence) is followed by **inner** that is followed by the notation ~2~ (indicating some concluding general interaction sequence). The meaning of **inner** is that the directive in any association X' inheriting from X takes the place of **inner** in instances of X': The supplementary description in X' is indicated by the notation ~3~. The modified directive of an instance of X' is the sequence ~1~~3~~2~. An alternative association X″ also inheriting from X may describe another directive to replace ~3~.

3.3.2   Inheritance by Means of Virtual Associations

In the illustration of Fig. 12 (right) based on virtual associations, the association PickUpPutDown is specialized to the association Transport. The intention of Transport is to move two packages (instead of one package) between the ports. Consequently, entity Port is specialized to Port2 in order also to include package y (entity AGV is specialized to AGV2 similarly). Associations PickUp and PutDown are both extended to include **inner** in order to allow the interactions not only to pick up and put down x but also to be specialized in order also to pick up and put down y. Associations PickUp2 and PutDown2 describe this inheritance from PickUp and PutDown. Consequently, the directive of association PickUpPutDown is described by means of instances of virtual versions of associations PickUp and PutDown, namely the virtual associations vPU and vPD, respectively. These associations are declared virtual by vPU:<PickUp and vPD:<PutDown, respectively. The actual instances vpickup and vputdown are declared accordingly, and the sequence vpickup **inner** vputdown specifies execution of vpickup, before execution of **inner**, before execution of vputdown. In Transport, the declarations of the virtual associations vPU and vPD are further specialized to be exactly of categories PickUp2 and PutDown2 in instances of Transport by the declarations vPU:<PickUp2 and vPD:<PutDown2, respectively. Consequently, **directive{…}** of Transport describes the interactions to be executed between the executions of these specialized versions of vpickup and vputdown.

In Fig. 15, the associations PickUpPutDown and Transport are given in textual form, corresponding to diagram Fig. 12 (right). Therefore, both Port and AGV are specialized to Ports2 and AGV2, respectively. Port, Port2 and AGV, AGV2 describe entities and are used in the associations to declare virtual entities. This means that the declarations vP:<Port and vA:<AGV in PickUp and PutDown are re-declared as vP:<Port2 and vA:<AGV2 in PickUp2 and PutDown2, respectively. Consequently, the instances of the entities declared by means of vP and vA are specialized to be declared by Port2 and AGV2, respectively. Another change of the purpose of the example is that only a single AGV2 picks up and puts down the packages instead of having two AGVs to meet in order to exchange a package. Consequently, only a single AGV2 is allocated in Transport, and this single AGV2 (named A2) is the only AGV used in the example. In order to be able to specialize the associations PickUp and PutDown to pick up and put down not only one package but also two packages, the declarations of the associations are (as mentioned) specialized by means of the **inner** mechanism. The associations PickUp2 and PutDown2 specialize PickUp and PutDown appropriately to also pick up and put down y.

```
association PickUp {              association PickUpPutDown {   association Transport
  role P for vP                     role Pᵢ for vP                 extends PickUpPutDown {
  role A for vA                     role Pₒ for vP
  vP :< Port, vA :< AGV                                            association InitT2
  directive {                       association InitPUPD {            extends initPUPD {
    A::goto(P.p)                      directive {                    directive {
    P::x → A::x                         Pᵢ play vpickup.P              A2 play vpickup.A
    inner                               Pₒ play vputdown.P            A2 play vputdown.A
  }                                     inner                       }
}                                     }                           }
association PickUp2                  }
    extends PickUp {                                              AGV2 A2 = AGV2.allocate(…)
  vP :< Port2, vA :< AGV2          vP :< Port                     vP :< Port2
  directive {                     vPU :< PickUp                   vPU :< PickUp2
    P::y → A::y                    vPD :< PutDown                 vPD :< PutDown2
  }                               vI :< InitPUPD                  vI :< InitT2
}
association PutDown {             vI vinit = new vI()             directive {
  role P for vP                   vPU vpickup = new vPU()           A2::goto(Pₒ.p)
  role A for vA                   vPD vputdown = new vPD()        }
  vP :< Port, vA :< AGV                                          }
  directive {                     directive {
    A::x → P::x                     vinit                        entity Port2 extends Port {
    inner                           vpickup                        Package y
    A::goto(A.Home)                 inner                        }
  }                                 vputdown                     entity AGV2 extends AGV {
}                                 }                                Package y
association PutDown2             }                                }
    extends PutDown {
  vP :< Port2, vA :< AGV2                                        Transport T = new Transport()
  directive {                                                    Port P₁ = new Port2(…)
    A::y → P::y                                                  Port P₂ = new Port2(…)
  }                                                              P₁ play T.Pᵢ, P₂ play T.Pₒ
}
```

Fig. 15 Textual form of inheritance of `PickUpPutDown`: virtual associations

In the association `PickUpPutDown`, the virtual associations `vPU` and `vPD` are declared as well as instances `vPickUp` and `vputdown` of these. In association `Transport`, the virtual associations `vPU` and `vPD` are specialized to `PickUp2` and `PutDown2`, implying that in an instance of `Transport`, the instances `vpickup` and `vputdown` actually are instances of `PickUp2` and `PutDown2`, respectively. The sequence `vpickup` **inner** `vputdown` in the directive of `PickUpPutDown` is utilized in `Transport` where its directive replaces **inner** in this sequence. Consequently, before **inner**, x and y are picked up by A2—and after **inner**, x and y are put down by A2. In between, the directive of `Transport`, namely `A2::goto(Po.p)`, replaces **inner**, meaning that in between A2 moves from the input `Port` to the output `Port`.

Furthermore, association `initPUPD` is a local association to `PickUpPutDown` used as an initialization in the beginning of the directive of `PickUpPutDown` to assign which ports to play which roles. In association `Transport`, the association `initPUPD` is specialized to `initT2` to also assign which AGVs play which roles, in order also to take care of this initialization in instances of `Transport`.

Figure 16 (left) shows a refinement of the schematic diagram in Fig. 11 based on virtual associations, and Fig. 16 (right) gives a textual version: Instance `iV` is declared in the general association `Y` to be an instance of virtual association `V` that is declared to

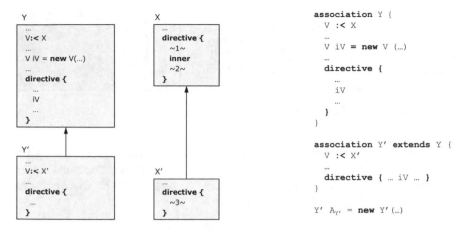

Fig. 16 Schematic illustration of inheritance of associations: virtual associations

be an association of category X (by virtual declaration V : <X). Instance iV becomes an instance of X' (a specialized association of X) when the virtual association V is further specialized to X' in the specialized association Y' (by the virtual re-declaration V : <X'). The effect is that the instance iV of category X in instances of Y retroactively becomes category X' in instances of Y' despite that instance iV actually is declared in Y. The virtual association is similar to a virtual class in Beta [24] and [25]: The virtual association inherited may be further specialized by succeeding re-declarations.

For reasons of simplicity, we present inheritance of associations only in a basic form. Multiple inheritance refers to a feature of some object-oriented programming languages in which a class can inherit behaviors and features from more than one superclass. Multiple inheritance for the association concept implies almost identical complications, and instead, we recommend to support multiple inheritance for associations similar to the approach based on part objects in [21].

3.3.3 Characterization: Inheritance of Associations

Inheritance, c.f. the previous sections, is seen as *structural parameterization* (by descriptions, not only by variables and values). The **inner** mechanism offers a fixed form, whereas virtual associations offer a generalized version.

At development time, inheritance of associations especially supports the following:

– Flexibility: Parameterization supports flexibility: Inheritance of associations is structural parameterization, i.e., at development time, inheritance enables description of additional associations by parameterizing existing associations.

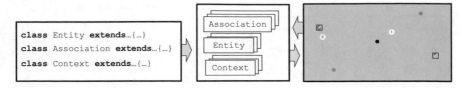

Fig. 17 Illustration of virtual environment

– Understandability: Given that associations are related by inheritance, the model or program becomes easier to comprehend due to the underlying conceptual model. Furthermore, the model or program typically becomes smaller but more complex and thus typically less understandable.

4 Experimental Implementation

The purpose of the experimental implementation is to test the informal definition of the Association Language, to experiment with the execution of example programs written in this language and to obtain experience with translator and runtime system. The implementation includes a translator and a virtual environment (implemented in Java). The translator transforms programs in the Association Language (mainly association and entity abstractions) to Java in terms of predefined abstract classes `Association` and `Entity`. The translated classes are included in the virtual environment, in which an interpreter produces an execution. The virtual environment is illustrated in Fig. 17: The rightmost part is a visualization of the system. The leftmost part is a logical framework with abstract classes `Entity`, `Association` and `Context`.[4] The interpreter system in the middle part maintains the repository of instances of entities and associations. When entities and activities are interpreted, the environment visualizes AGVs, ports and packages.

The implementation is experimental in the sense that neither completeness nor (translation and runtime) efficiency is important. The translation from the Association Language to Java is supported by a simple translation scheme [27]. The translation is incomplete in the sense that ordinary parts of the Java programming language (included in the Association Language) are not checked. Furthermore, the contextual analysis [28] is completed only when necessary, e.g., declarations and applications of associations and entities are matched, instantiated associations and entities are checked to be declared—by means of a simple symbol table [27]. Support of local associations and extensive analysis of declaration and re-declaration of virtual associations are incomplete parts of the translation. The interpreter part [28] is experimental in the sense that, e.g., the composition of associations is imple-

---

[4]An ambient system may also include contexts, i.e., universes in which entities and associations exist. In [26], contexts are named habitats and conceived as some kind of locality, providing to its inhabitants opportunities that allow its inhabitants to achieve their various goals.

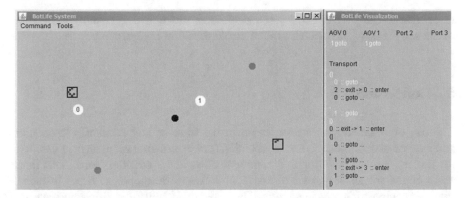

Fig. 18 Snapshot of BotLife System and BotLife Visualization

mented by copying the directive of a part association into the directive of a whole association (and by substituting variables according to the **... play ...** specifications). Similarly, the combination of directives by means of inheritance of associations is implemented by mixing the directive of the special association and the directive of the general association by replacing the **inner** clause appropriately.

The execution is presented by the BotLife System where the behavior of AGVs, ports and packages is illustrated, c.f. Fig. 18: The two white spots numbered 0 and 1 show the input AGV and the output AGV. The illustration matches the situation in Fig. 3: The input AGV moves from its Home toward the input Port and the output AGV moves from its Home to the meeting point Here. The Ports are loaded with packages indicated by small boxes. The input AGV picks up a package at the input Port and brings it to Here. The output AGV takes over the package, brings it to the output Port and puts down the package.

Figure 18 illustrates a general visualization of associations and entities exemplified by Transport and AGV 0, AGV 1, Port 2, Port 3, respectively. The internal situation (i.e., relation to the program description) in the BotLife Visualization (Fig. 18 right) matches the external situation in the BotLife System (Fig. 18 left). The column below each of the entities AGV 0, AGV 1, Port 2 and Port 3 shows the actual requested contributions from the entity. The method currently being executed by an entity is in white, whereas the remaining pending methods are in black (in the actual snapshot only a white method is present), and the number in front of the method refers to the (number of) requesting association (in this case 1 corresponding to Transport). The column below Transport shows its directive, i.e., its static operations. The clause (| ... |) is white because it is being executed. And because the operands of this clause are being executed concurrently, an operation in each of the two sequences of the clause is also in white, namely the operations corresponding to A_i::goto(P_i.p) and A_o::goto(Here). For reasons of simplicity of the implementation, the actual parameters are replaced by ... and the names of the AGVs and ports are replaced by a unique numbering of entities, namely 0, 1, 2 and 3, corresponding to AGV 0, AGV 1, Port 2 and Port 3, respectively. The `BotLife System` offers an instructive dynamic visualization of the semantics of the mechanisms of the Association Language.

5 Conclusions

The use of associations for abstraction in the form of instantiation, composition and inheritance supports our conceptualization of collaborations and is essential for understanding, modeling and communication about collaborations. Also these abstractions are simple and expressive at software development time: Instantiation and composition are characterized by understandability and efficiency, whereas inheritance is characterized by understandability and flexibility.

Future work includes additional experiments with software development based on instantiation, composition and inheritance for associations, as well as with efficiency of the implementation.

References

1. Kristensen, B.B.: Rendezvous-based collaboration between autonomous entities: centric versus associative. Concurr. Comput.: Pract. Exp. **25**(3), 289–308 (2013) (Wiley Press)
2. May, D.C.-M., Kristensen, B.B., Nowack, P.: Tangible objects: modeling in style. In: Proceedings of the Second International Conference on Generative Systems in the Electronic Arts (2001)
3. Lea, D.: Concurrent Programming in Java: Design Principles and Patterns, 2nd edn. Addison-Wesley (1999)
4. Dahl, O.-J., Myhrhaug, B., Nygaard, K.: SIMULA 67 Common Base Language (Editions 1968, 1970, 1972, 1984). Norwegian Computing Center, Oslo (1968)
5. Kristensen, B.B., May, D.C.-M.: Activities: abstractions for collective behavior. In: Proceedings of the European Conference on Object-Oriented Programming (1996)
6. Kristensen, B.B.: Object-oriented modeling with roles. In: Proceedings of the 2nd International Conference on Object-Oriented Information Systems (1995)
7. Hoare, C.A.R.: Communicating sequential processes. Commun. ACM (1978)
8. Kristensen, B.B.: Abstraction from collaboration between agents using asynchronous message-passing. In: 12th International Conference on Enterprise Information Systems, Portugal (2010)
9. Scott, M.L.: Programming Language Pragmatics. Morgan Kaufmann Publishers (2009)
10. Yonezawa, A., Tokoro, M.: Object-oriented Concurrent Programming. MIT Press Series in Computer Science (1986)
11. Booch, G.: Private Communication (2007)
12. Jensen, L.K., Kristensen, B.B., Demazeau, Y.: FLIP: prototyping multi-robot systems. J. Robot. Autonom. Syst. **53**, 230–243 (2005)
13. Rumbaugh, J.: Relations as semantic constructs in an object-oriented language. In: Proceedings of the Object-Oriented Systems, Languages and Applications Conference (1987)
14. Pearce, D.J., Noble, J.: Relationship aspects. In: Proceedings of the 5th International Conference on Aspect-Oriented Software Development (2006)
15. Østerbye, K.: Design of a class library for association relationships. In: Proceedings of the 2007 Symposium on Library-Centric Software Design (2007)

16. Nelson, S., Noble, J., Pearce, D.J.: Implementing first-class relationships in Java. In: Proceedings of the Workshop on Relationships and Associations in Object-Oriented Languages (RAOOL) (2008)
17. Østerbye, K.: Associations as a Language Construct. In: Michel, R. (eds.) Proceedings of TOOLS, vol. 29, pp. 224–235 (1999)
18. MacLennan, B.J.: Principles of Programming Languages Design, Evaluation, and Implementation, 3rd ed. Oxford University Press (1999)
19. Liskov, B., Guttag, J.: Program Development in Java: Abstraction, Specification and Object-Oriented Design. Addison-Wesley (2000)
20. Watt, D.A.: Programming Language Design Concepts. Wiley (2004)
21. Kristensen, B.B., Madsen, O.L., Møller-Pedersen, B.: The when, why and why not of the BETA programming language. In: Proceedings of the Third ACM SIGPLAN Conference on History of Programming Languages, California (1999)
22. Larman, C.: Applying UML and Patterns: An Introduction to Object-Oriented Analysis and Design and Iterative Development, 3rd edn. Prentice Hall (2004)
23. Østerbye, K.: Parts, wholes, and subclasses. In: Schmidt, B. (ed.) Proceedings of the 1990 European Simulation Multi-conference, pp. 259–263 (1990)
24. Madsen, O.L.: Semantic analysis of virtual classes and nested classes. In: Proceedings of Conference on Object-Oriented Programming, Systems, Languages and Application, Colorado (1999)
25. Madsen, O.L., Møller-Pedersen, B.: Virtual classes—a powerful mechanism in object-oriented programming. In: Proceedings of Conference on Object-Oriented Programming, Systems, Languages and Application, Louisiana (1999)
26. May, D.C.-M., Tang, O.: Designing for the Digitally Pervasive World. Ph.D. thesis, Maersk Mc-Kinney Moller Institute, University of Southern Denmark (2003)
27. Aho, A.V., Sethi, R., Ullman, J.D.: Compilers: Principles, Techniques, and Tools. Addison Wesley (1986)
28. Watt, D.A., Brown, D.F.: Programming Language Processors in Java: Compilers and Interpreters. Pearson Education (2000)

Ensuring Compliance with Sprint Requirements in SCRUM

Preventive Quality Assurance in SCRUM

Manuel Pastrana, Hugo Ordóñez, Ana Rojas and Armando Ordoñez

Abstract Compliance with the client requirements is a key factor for the success of projects. However, this compliance is not always easy to monitor and verify. SCRUM framework flexibility makes it possible to include a variety of techniques and good practices from other methods to help software development teams to achieve their goals. Some of these tools can track historical changes in the code (versioning), perform code inspection to identify bugs, vulnerabilities, code duplication, and code smells that can affect the software quality. This paper presents a preventive software quality assurance environment that aims at improving the quality of the development process.

Keywords SCRUM · Software development process · Software quality assurance · Agile development process

1 Introduction

Two out of three software projects require significant changes during their development, which negatively impacts the time and cost of the projects [1]. Software methodologies have been evolving to respond to these drawbacks [2]. Some propos-

M. Pastrana (✉) · A. Rojas
GrintTic, Universidad Antonio José Camacho, Cali, Colombia
e-mail: mapastrana@admon.uniajc.edu

A. Rojas
e-mail: amrojas@admon.uniajc.edu.co

H. Ordóñez
Research Laboratory in Development of Software Engineering, Universidad San Buenaventura, Cali, Colombia
e-mail: haordonez@usbcali.edu.co

A. Ordoñez
Intelligent Management Systems, University Foundation of Popayan, Popayán, Colombia
e-mail: jaordonez@unicauca.edu.co

© Springer Nature Singapore Pte Ltd. 2019
S. K. Bhatia et al. (eds.), *Advances in Computer Communication and Computational Sciences*, Advances in Intelligent Systems and Computing 924, https://doi.org/10.1007/978-981-13-6861-5_3

als have focused on improving the analysis stage [3], other on the representation of design decisions through formal UML diagrams [4], and other on representing the design in models in concurrent views oriented to different user groups [5, 6]. However, to improve the quality of the development processes, it is necessary to identify the required quality attributes at the design stage [7, 8].

In traditional methodologies, the stages are generally sequential and separated from each other, while in agile methodologies the process is different [9]. SCRUM, for example, proposes an iterative incremental model divided into sprints (periods of time from 1 to 4 weeks), where a set of functionalities described in user stories is developed [10]. Sprints planning make possible to identify the work to be done. Also, other quality factors may be considered to determine that a user story is finished, since here know as done by [10]. A user story is done if its functionality meets the acceptance criteria, correctly handles errors and exceptions messages, and its quality has been reviewed [11]. Problems in the estimation and follow-up of user stories are very frequent and may generate delays and increase in budgets [12].

Here, a strategy for agile software development that seeks to guarantee the completeness of sprints is presented. For this purpose, tools are used to automate the verification of compliance. The rest of the paper is organized as follows: Sect. 2 describes the motivating scenario. Section 3 describes the approach. Section 4 exposes the discussion of the results. Section 5 presents the background and related works. Finally, Sect. 6 concludes and exposed the future work.

2 Motivating Scenario

SCRUM framework was created in response to the problems of traditional development processes [11]. SCRUM proposes an incremental iterative cycle that allows controlling the result in partial deliveries [10]. In [12], several SCRUM projects were analyzed, finding that a large proportion of the problems are presented in the product backlog. The product backlog is the main artifact resulting from the requirements analysis [11]. This artifact is usually composed of user stories (although use cases or business process models can also be used [3]). In spite of there are no standard models for user stories [13], some models are widely accepted by the industry such as the one presented in [14] where the story definition and development activities are integrated. Also, in [15] a guiding phrase to construct user stories is proposed: I [as a role] + WANT [functionality] + SO THAT [reason] + FULFILLING [these criteria]. User stories should be written so that the user can understand and modify them in their own words.

During the planning phase, the user stories are organized according to their complexity and priority. In addition, the work teams set the duration of the sprint (1 to 4 weeks). In this phase, many of the problems may arise [12].

The flexibility of SCRUM makes it possible to integrate good practices to guarantee continuous improvement. In [10] and [12], it is proposed to include good practices taken from Extreme Programming or XP [14].

Among the good practices, it has been proposed that the source code belong to the whole team and be available at any time, and this is known as collective ownership of the code and can be achieved using a central repository that stores the progress of the project. Likewise, standardization of the source code is suggested to ensure that all team members can write and interpret the development in the same way, allowing the code to be easily maintained, modified, and scaled [14]. This may be verified through code inspection.

Pair programming is also useful when the teams are composed of members with different technical abilities. It is important to emphasize that the team must maintain a unit test environment to monitor the progress of the development. This is known as test-driven development and promotes the awareness of quality during the development.

3 Proposed Strategy

The good practices incorporated in the proposed strategy are reviewed below:

3.1 Collective Property of the Code

All team members must be able to know, understand, and modify the source code [14]. The purpose is twofold; first, in case any member of the team becomes unavailable, his activities may be distributed among other collaborators. Second, there is no one key member that always makes the deliveries to the client; on the contrary, all the team members may be key members.

To achieve this goal, it is very important to adopt a version control system, also known as VCS. These systems are central repositories that store the historical changes made to the code. Some widely used VCS are SVN, Git, GitLab, and Bitbucket.

3.2 Code Conventions

Code conventions are important for a number of reasons. All team members write code in the same way, generating order. This practice increases the maintainability, modifiability, scalability, and collective property of the code. Moreover, code conventions help to diminish bad practices. An example of code conventions is presented in [16]. However, team leaders can choose the standard that best suits their needs. Code inspection may be used to verify this practice. Inspections can be done manually or can be supported by tools such as Donar, GitLab, or Travis.

3.3 Pair Programming

In pair programming, two people work at the same time, in the same machine doing simultaneous things that aim at developing a user story [14]. This practice promotes a constant code inspection by the collaborator (who is not directly programming). This person compares the requirements and the code developed; also, the product design is continually reviewed. Moreover, in case of not having the same technical level of the programmer, the collaborator continuous improves his technical skills. This practice is widely used with new members of the team [12].

3.4 Continuous Integration

The continuous integration systems work together with version control systems. Each time the team generates changes in the code, the integration takes all the code from the repository and builds a deployment release (exe, war, ear, etc.), which is set it up in a server, and all the tests are performed. In case of failure, an email is sent. Although these environments may be expensive, they significantly speed up the development process, generating detailed reports that make possible to identify problems and take corrective measures [12]. Jenkins is a widely used tool for this practice. Even though, some version control systems also include this functionality such as Bitbucket and GitLab.

3.5 Test-Driven Development (TDD)

TDD starts by writing a test and building all the necessary code units to comply with the functional objective [12]. Tests may be written using frameworks such as JUnit (Java) or PHPUnit (PHP), Moreover, other frameworks, such as selenium, automate functional tests by automating the navigation in the graphic interface. TDD may result hard to adopt at first, but once the process is properly applied it fully guarantees that all user histories are done [12].

3.6 Sustainable Pace and Energized Work

A preventive quality environment is proposed that integrates all the good practices mentioned above. The entry point is a code repository that organizes the code, tracks changes, and ensures collective property. Besides, a continuous integration tool is installed and configured so that every time a change is made to the source code, the following steps are carried out:

1. Generation of the deployment version to detect syntactic errors, which are notified by email.
2. Execution of automated tests (unit tests, functional test, etc.).
3. Deployment release in the test server environment.

These steps ensure continuous integration of the project and the use of TDD. In case of not knowing TDD or wanting to improve the skills of the team, pair programming can be used.

Finally, along with continuous integration and the version systems, it is proposed to integrate automated code inspection. Code inspection tools ensure unification of code conventions and error detection and likewise improve the performance, modifiability, maintainability, and scalability of the developed product. Most code inspection tools allow identifying errors and duplicate code blocks. Moreover, by using TDD along with code inspection, it is possible to identify the test coverage.

The present approach based on preventive quality tools provides a sustainable development pace without reprocessing. Besides, this approach aims at ensuring compliance with the sprints requirements and reducing the occurrence of failures.

4   Discussion

Due to the wide variety of tools that exist to support the proposed strategy, the discussion focuses on the following topics:

- What infrastructure may support the proposed approach?
- Based on the infrastructure, which tools may be implemented?
- What is the impact of the use of the present approach on the projects?

4.1   Infrastructure that Supports the Proposal

At present, there are several applications such as GitLab and Bitbucket that include tools for versioning, continuous integration, and continuous deployment. These options may support the proposed strategy, and however, to guarantee the operation with several concurrent projects, some requirements must be fulfilled: first, due to the processing capacity required for the integration and continuous deployment; Second, RAM consumption for the versioning. Due to the above, the workload was distributed in three machines, one for each tool.

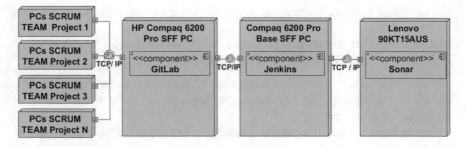

Fig. 1 Deployment diagram of the proposed approach

4.2 Selection of Tools

For the versioning, the selected tool is GitLab since it includes a robust versioning model, supports many users without restriction per project, and is free. Besides, the limit of projects is given by the disk capacity. Moreover, it can be installed on local servers, increasing the protection of the data, unlike other cloud-based tools such as Bitbucket or GitHub.

By its part, the continuous integration tool selected is Jenkins due to its multi-language support (Java, PHP, and .NET). Besides, it offers easy connection to Git repositories such as GitHub, GitLab, Gitolite, and Bitbucket. Additionally, Jenkins is designed to integrate several types of plug-ins and tools that guarantee a quality environment like the one proposed here. Regarding, dependency managers such as Maven and Gradel, the unit tests can be carried out prior to the generation of the release. Besides, by using selenium, it is possible to automate black-box functional tests on the built artifact.

Finally, the source code inspection tool selected is Sonar because of its transparent integration with Jenkins. Sonar enables team members to identify vulnerabilities based on the top ten of OWASP. Also errors, bugs, and bad practices (code smells) are identified. With Sonar, it is possible to identify code duplicity and the test coverage, even detecting if the tests have been carried out or not. Figure 1 depicts the selection of tools by means of a deployment diagram.

4.3 Results

The approach was applied in five projects (see Table 1). All projects were developed using SCRUM:

• Four of the five projects developed use Java.
• One project uses PHP.
• One java project is an android mobile development.
• Diverse visual frameworks were used in Java-based projects.

Table 1 Projects evaluated

Project	Description	PL	ES
APYSCC	System for the management of the pedagogical agreement between teachers and students, monitoring academic results for decision making	Java, PrimeFaces	10
CTSANDROID	System for the communication between students and teachers (forums, chats, and file sharing)	Java para Android	7
PSUNIAJC	System for the social projection office of a university	Java y Angular	12
SCRUM	System for the management of user stories, planning, and monitoring of SCRUM-based projects	PHP, Laravel	6
SIGEG	System that manages the information of the University graduates and their education needs	Java, Spring, Bootstrap	8

Programming language (PL), executed sprint (ES)

Below is a description of each of the projects mentioned in Table 1.

The first project was developed in PHP with 6 sprints of 2 weeks each. During the first quality measurement of sprint 1, a total of 5 bugs, 0 vulnerabilities, and 112 code smells (bad development practices) were identified. The second test showed 12 bugs, 0 vulnerabilities, and 126 code smells showing a decline in quality and the possibility of non-compliance with the sprint requirements. Due to the above, the development team decides to use pair programming to increase the review of errors. The final result of sprint 1 is presented in Table 2, no errors or vulnerabilities were identified, and code smells decreased to 125.

As can be seen from the above, the tests indicated that the sprint 1 could not reach the status of done, but once the correction plan was carried out, the sprint 1 was successfully accepted. During sprint 2 (see Table 2), the quality of the process increased, keeping errors and vulnerabilities at 0 and decreasing code smells (66). The reduction in bad practices is due to the fact that when errors were identified, the development team did not repeat them. During sprint 3, a new module was built, one vulnerability was obtained, and code smells were reduced to 40 (Table 2).

Despite having found one vulnerability, the tool suggested that the sprint could be approved because of its low impact. In sprints 4 and 5, both the vulnerabilities and the errors were 0, and the bad practices continued decreasing from 40 to 12 (see Fig. 2).

The second project called APYSCC was developed in Java, with 10 sprints of 2 weeks each. During the first quality measurement of sprint 1, a total of 15 bugs or errors, 8 vulnerabilities, and 140 code smells were identified. Once the team reviewed

Table 2 Results of the sprint compliance for project SCRUM

Project	SP 1			SP 2			SP 3			SP 4			SP 5		
	B	V	CS	B	V	CS	B	V	CS	B	V	CS	B	V	CS
SCRUM	0	0	125	0	0	66	0	1	40	0	0	33	0	0	12
APYSCC	0	1	106	0	0	87	0	0	52	0	0	16	0	0	4
CTSANDROID	0	29	412	0	0	25	0	0	0	0	0	0	0	0	0
PSUNIAJC	24	0	741	0	0	561	0	0	395	0	0	102	0	0	75
SIGEG	0	0	187	0	0	168	0	0	9	0	0	0	0	0	0

Sprint (SP), bug (B), vulnerability (V), code smells (CS)

the results of the first measurement, some coding issues were identified and corrected. The next measurement within the same sprint throws a total of 0 bugs or errors, 1 vulnerability and 106 code smells (see Table 2). This result showed the impact of the code refinement in the quality of the system.

During sprints 2 through 10, errors and vulnerabilities were maintained in 0 (see Fig. 3). Additionally, bad practices show a decreasing trend.

The Android project called CTSANDROID was developed in 7 sprints, of 2 weeks each. The first quality measurement done in sprint 1 showed 8 bugs, 45 vulnerabilities, and 396 code smells. The second revision (two days after the first) showed an alarming result of 8 bugs or errors, 50 vulnerabilities, and 472 code smells. This situation forced to take corrective measures to identify the most common bad practices.

During the rest of the sprint, pair programming was used, and the team managed to achieve 0 errors, 29 vulnerabilities, and 412 code smells (see Table 2). Figure 4 shows the results of the reviews of all sprints. In sprints 2–7, the errors remained at 0, the vulnerabilities in sprint 2 fall to 0, and a decreasing trend of bad practices can be seen until reaching the value of zero.

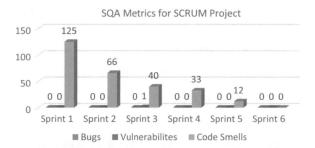

Fig. 2 SQA metrics for SCRUM project

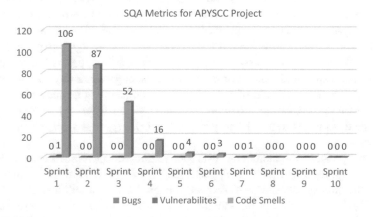

Fig. 3 SQA metrics for APYSCC project

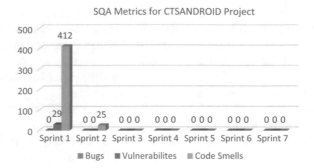

Fig. 4 SQA metrics for CTSANDROID project

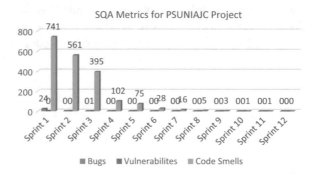

Fig. 5 SQA metrics for PSUNIAJC project

The project called PSUNIAJC was developed in Java and Angular and takes 12 sprints, each of 2 weeks. The first quality measurement done in sprint 1 shows 206 bugs, 61 vulnerabilities, and 996 code smells. Starting from there, it was decided to implement pair programming to increase the identification of errors. Thanks to this it was possible to reduce to 0 vulnerabilities, 24 bugs, and 741 code smells.

With these results, sprint 1 could not be accepted, and an agreement between the client and the team was necessary to define a new deadline, this is known as spike [10]. The team specified that the time to resolve the bugs was 2 days before the start of sprint 2. The first revision of sprint 2 indicates that the bugs and vulnerabilities reached a value of 0, besides the code smells go down from 741 to 690. Figure 5 shows the decreasing trend of bad practices and 0 bugs and vulnerabilities in the following sprints.

The last project evaluated under was SIGEG and was developed in Java, Spring, and Bootstrap. Its construction took 8 sprints, 2 weeks each. At the beginning of the project, the quality measurement showed 0 bugs, 12 vulnerabilities, and 431 code smells. By the end of the sprint, both the bugs and the vulnerabilities were 0 and the code smells were reduced to 187, which allow the sprint 1 to be accepted. The results of the remaining sprints are shown in Fig. 6. It can be observed that both bugs and

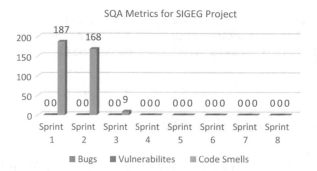

Fig. 6 SQA metrics for SIGEG project

vulnerabilities remained at 0, and in addition, the code smells were reduced to zero in the next 3 sprints.

5 Related Works

Agile methodologies arise in response to the problems presented by traditional methods [9]. These new models seek compliance with the client requirements, by using continuous delivery and constant feedback [17]. One of the widely disseminated frameworks is SCRUM [11]. SCRUM proposes an incremental iterative process [18] that unifies a management model and a controlled development process [17].

The fundamental artifact of the SCRUM is the product backlog [10]. This artifact enables the team to know how much work should be done, define the design of the solution, do the planning of development task, define the priority of these tasks, and define the size of the sprints. The team must address the incremental complexity of the project and react to problems that may arise along the way.

Due to the flexibility of SCRUM, it is possible to incorporate good practices from other methodologies to achieve better results [14]. For example, in [19], code versioning and continuous integration are implemented in SCRUM. However, it was evidenced that these two practices do not have a high impact on the quality of the product.

In [20], the code inspection is integrated. The authors divide the continuous delivery process in scenarios. The first is called commit stage, where the commit is made to versioning systems that deliver the changes to the continuous integrator, and the latter is responsible for compiling, executing unit tests, and defining if the integration is adequate. Once this task is completed, in the second scenario white-box and black-box tests are carried out. A third scenario appears to run all the automatic tests that should be executed like the inspection of code by some specialized tool. Subsequently, other scenario includes the manual tests that other scenarios have not covered. Finally, the last scenario is the deployment of the product in the production environment. This work invites to have a holistic perspective of the quality process. However, this work does not analyze compliance with the sprint requirements. Here, a gap was identified between the existing approaches and the proposal described in this paper.

6 Conclusions and Future Work

An approach based on good practices and automated tools to guarantee compliance with the sprint requirements is presented. This approach makes it possible to achieve sustainable development pace and management of possible errors as soon as possible.

Additionally, this environment of continuous improvement enables team members to identify errors using TDD and automated code inspection. Team members can learn from the identified errors and improve their yield.

Future work will be focused on including other tools for automated monitoring of goals and development task, tracking, and visualizing the quality of the development process not only for Java, .NET, and PHP but also mobile frameworks like IONIC.

References

1. Hastie, S., Wojewoda, S.: Standish Group 2015 Chaos Report - Q&A with Jennifer Lynch. http://www.infoq.com/articles/standish-chaos-2015, 2015. [Online]. Available: http://www.infoq.com/articles/standish-chaos-2015
2. Pressman, R.S., Ph, D.: Ingeniería del Software
3. Pastrana, M., Ordóñez, H., Ordonez, A., Thom, L.H., Merchan, L.: Optimization of the inception deck technique for eliciting requirements in SCRUM through business process models. In: International Conference on Business Process Management, vol. 4928, no. January, pp. 649–655 (2018)
4. Kruchten, P.: Planos Arquitectónicos: El Modelo de '4 + 1' Vistas de la La Arquitectura del Software. IEEE Softw. **12**(6), 1–16 (2006)
5. Pressman, R.S.: Software Engineering A Practitioner's Approach, vol. 33, p. 930. McGraw-Hill (2010)
6. Sommerville, I.: Ingeniera del Software. Septima (2005)
7. Larman, C., Vodde, B.: Scaling Agile Development—Large and Multisite Product Development with Large-Scale Scrum. CrossTalk, no. May–June, pp. 8–12 (2013)
8. Larman, C.: Applying UML and Patterns: An Introduction to Object-Oriented Analysis and Design and Iterative Development (2004)
9. Fowler, M., Highsmith, J.: The agile manifesto. Softw. Dev. **9**, 28–35 (2001)
10. Schwaber, K., Sutherland, J.: La Guía de Scrum TM La Guía Definitiva de Scrum: Las Reglas del Juego, pp. 1–21 (2017)
11. I. SCRUMstudy^{TM}, una marca de VMEdu: A Guide to the SCRUM Body of Knowledge (SBOK^{TM}GUIDE) 2016 Edition, 2016th ed. Phoenix, Arizona 85008 USA: SCRUMstudy^{TM}, una marca de VMEdu, Inc. (2016)
12. Kniberg, H.: Scrum and XP desde las trincheras (2007)
13. IEEE 1998, "IEEE Std 830-1998": Especificación Requisitos Softw. según el estándar IEEE 830, p. 19 (2000)

14. Beck, K.: Extreme Programming Explained: Embrace Change, no. c. 1999
15. Cohn, M.: User Stories Applied: For Agile Software Development. Addison Wesley Signature Series, vol. 1, no. 0 (2004)
16. Hommel, S., Microsystems, S., Molpeceres, A.: Java ^{TM} (2001)
17. Hanssen, G.K., Haugset, B., Stålhane, T., Myklebust, T., Kulbrandstad, I.: Quality assurance in scrum applied to safety critical software. Lecture Notes in Business Information Processing, vol. 251, pp. 92–103 (2016)
18. Pressman, R.S.: Software Engineering A Practitioner's Approach, 7th edn. In: Pressman, R.S. (2009)
19. Deshpande, A., Riehle, D.: Continuous integration in open source software development. Open Source Dev. Communities Qual. **275**, 273–280 (2008)
20. Humble, J., Farley, D.: Continuous Delivery: Reliable Software Releases Through Build, Test, and Deployment Automation (2010)

Remote Collaborative Live Coding in SuperCollider-Based Environments via Open Sound Control Proxy

Poonna Yospanya

Abstract Networked live coding as a form of musical ensemble performance is generally limited to having performers coding in the same venue, using the same programming environment. There is an inherent latency issue with any networked musical performance, although to a lesser extent in live coding. Different performers may also be familiar with different live coding languages and environments, thus making collaboration options more limited. We propose a proxy system that enables different SuperCollider-based live coding environments to collaborate over the Internet via Open Sound Control protocol. The proxy intercepts the communication between a SuperCollider server and its clients at the protocol level and relays it to other participating peer servers. We address the network latency issue with latency-compensated synchronization, allowing each participant to have their performance played back similarly on all participating machines regardless of their physical location and distance from each other.

Keywords Live coding · SuperCollider · Open Sound Control

1 Introduction

Live coding is an act of writing code while the program is running. The code is executed on-the-fly, and the result is immediately produced on the running program. It is most prominently featured in performing arts, such as musical performance or audiovisual installation arts, although uses in other areas, including teaching programming and conference presentation, are also gaining popularity.

In the context of musical performance, the performer (or live coder) creates musical pieces by writing code in a programming language supported or provided by a live coding environment. While the music is playing as a result of the code execution, the

P. Yospanya (✉)
Faculty of Engineering at Sriracha, Kasetsart University, Sriracha Campus,
Chon Buri 20230, Thailand
e-mail: poonna@eng.src.ku.ac.th

© Springer Nature Singapore Pte Ltd. 2019
S. K. Bhatia et al. (eds.), *Advances in Computer Communication
and Computational Sciences*, Advances in Intelligent Systems and Computing 924,
https://doi.org/10.1007/978-981-13-6861-5_4

code can be changed and manipulated to shape the music being played. Compared to traditional musical performance, live coding has advantages of music being highly flexible and malleable with potential beyond that of real instruments, although there are downsides, such as risks of programming errors, and coding not being the most immediate mode of affecting sounds [5].

Networked live coding, where live coders collaborate in real time either co-located or remotely, has been present since the inception of live coding and poses challenges in facilitating collaboration and coordinating shared programming practices [10]. Many live coding environments have been developed with networking capability to facilitate the collaboration [4, 7, 9, 12, 13, 17, 18]. Barbosa [3] suggested two classification criteria for networked musical performances based on interaction (*synchronous/asynchronous*) and location (*co-located/remote*). For our purpose, we focus only on the synchronous mode of interaction where live coders perform together in real time.

Most of the work in this area require performers to be co-located, where network latency is minimal, thereby having much lesser issue with time synchronization among different performers' machines. In cases where there is a central server handling the code execution, there is no need for time synchronization at all. However, when performers remotely collaborate from different geo-locations, a number of issues need to be taken into consideration. Sound rendering has to be done at all participating venues, with proper synchronization so that the rendering results are exactly the same. Higher network latency, jitters, and loss rate also need to be carefully handled.

Several tools have been designed to address some or all of these issues. OSC-groups [4] allows peer-to-peer message exchange among nodes with "NAT hole punching" to accommodate nodes that are behind NAT routers, but does not provide a synchronization mechanism. OSCthulhu [13] takes a different approach based on a technique from multiplayer online games where object update messages are multicasted to all clients, and the server also periodically multicasts a server sync message to invalidate and replace all the client object states. This method keeps the states consistent throughout the whole system. However, it demands a specific object-based approach to music coding. EspGrid [18] is a decentralized system where each client sends messages to other clients directly. Exchanged messages are synchronized to the measured network latency between each corresponding node pair. EspGrid is not specifically designed for remote collaboration. Instead, it is based on the idea of having multiple participants, each controlling one or more applications, sharing resources, and exchanging messages and code in the same performance environment. Another interesting approach taken by SuperCopair [7] and Troop [9] is code sharing, where performers are writing and modifying the same code using shared buffers. Depending on the ensemble's preference, this can be a feature or a limitation.

Our work, SCProxy, takes a centralized approach where a master node handles all the distribution of messages to all participating nodes. Round-trip latencies are accounted for in message exchange. The main feature is that we leverage the existing SuperCollider ecosystem which has been widely used in live coding. There are many third-party programming clients that use a SuperCollider server as their audio engine.

Most of these clients, including the *sclang* client which is included with SuperCollider itself, do not directly support remote collaborative live coding. By providing a thin layer between these clients and SuperCollider servers, and taking control of message exchange, SCProxy augments existing systems with the capability to perform together remotely using multiple different clients.

2 SuperCollider Ecosystem

SuperCollider [12] is a text-based programming environment originally designed for real-time audio synthesis and algorithmic composition. It has since been widely used in acoustic research, interactive programming, and live coding. Other audio and music programming environments that are also widely used include Max/MSP [19], Pure Data (Pd) [20], and Reaktor, but they are for the most part graphical environments which are not as well suited for use in live coding.

SuperCollider environment consists of two components: a server, *scsynth*, which provides the synthesis engine that handles all the audio processing tasks, and a client, which hosts an interpreter for the programming language *sclang*. The client communicates with the server via Open Sound Control (OSC) protocol [22] (see Fig. 1). This mode of communication allows third-party clients to be used instead of *sclang* to provide alternative programming languages and environments. For this reason, there have been many live coding environments that are built on top of the SuperCollider server to take advantage of its vastly capable synthesis engine and large library of unit generators. The list includes, but not limited to, Sonic Pi [1], Overtone [2], FoxDot [8], Tidal [14, 15], ixi lang [11], Skoar [6], Steno [16], and Xi [21]. These live coding environments replace the *sclang* client with their own clients and programming languages, some as a domain-specific language embedded in an existing programming language, while others being a new language specifically designed for a particular way of music programming.

The *scsynth* server is also multiclient and network-capable. It supports OSC communication with multiple clients over either UDP datagrams or TCP streams. Therefore, networked live coding is possible natively. However, this is generally limited to

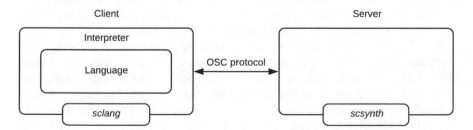

Fig. 1 Structure of the SuperCollider application. Adapted from "Client versus Server," *SuperCollider 3.9.3 Help*. Retrieved September 21, 2018, from http://doc.sccode.org/Guides/ClientVsServer.html

co-located performance, as there is only one central server producing all the sounds, thus requiring all the performers to be in the same venue.

The separation of the server and the client allows us to enhance the capability of SuperCollider in networked live coding to cover remote collaboration and enable heterogeneous mix of different clients. We introduce a proxy-based approach to handle these requirements with our tool, *SCProxy*,[1] which acts as a middleman between clients and servers. We describe the architecture and its working mechanism in the next section.

3 SCProxy Architecture

SCProxy adopts a master–participant strategy as its architecture. An SCProxy instance works as the intermediary layer between a client and a SuperCollider server on a machine, and intercepts and relays messages from the client through the master SCProxy instance to other SCProxy instances on other machines, which in turn distribute the messages to their respective server on the same machine that they are running. Messages are modified by the proxy such that all the machines play back the same sound in a synchronized manner and without one client interfering another. Figure 2 shows an example of a three-machine ensemble.

3.1 OSC Message Relay

Communication with the SuperCollider server can only be done using Open Sound Control (OSC) protocol. OSC is a standardized protocol for communication among musical instruments, multimedia devices, and computers over a network. An OSC packet can be a single *OSC message* or an *OSC bundle*, which can in turn contain multiple OSC messages or other OSC bundles.

An OSC message consists of an *OSC address pattern* followed by an *OSC type tag string* followed by zero or more *OSC arguments*. OSC address patterns and arguments are application-specific and determine the request or the response one party wants to make on another party. An OSC bundle additionally contains an *OSC time tag*, which determines when the bundle should be executed.

In Fig. 2, we have Machine A designated as the master node. Machine B and Machine C join node A to form an ensemble, establishing communication channels called *proxy trunks* to the master node. Each node has a client, a SuperCollider server, and a proxy instance. The client connects to the server through the proxy instance. For each client in the ensemble, either local or remote, the proxy creates a corresponding client stub between the proxy and the server to keep each

---

[1] Source code available at https://github.com/Poonna/SCProxy.

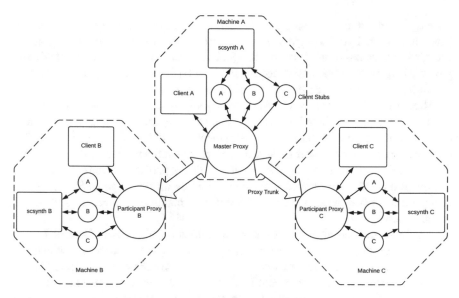

Fig. 2 Architecture of SCProxy is shown with three participants, one serving as the master proxy distributing time-synchronized OSC bundles to all connected proxies

client–server communication pair distinct, as messages sent from the server do not generally contain identification of target clients.

Event messages sent from the client are intercepted and converted to bundles before sending to the server in the same node. The bundles are also relayed to the master node (unless they originate in the master node themselves) through the proxy trunk with client tagging to identify originating clients. The master node then distributes the bundles to all other nodes through their corresponding proxy trunks. Proxy instances at the receiving ends subsequently relay the bundles to their local server through the respective client stubs.

It should be noted that server responses from remote nodes have to be discarded. Otherwise, clients will receive multiple duplicated responses, thereby defeating the transparency property of the proxy.

3.2　Event Synchronization

One of the most important issues in remote collaborative live coding is the latency and jitters, which are more serious than in the case of a local network. We handle this issue with beat-quantized event synchronization. As a node joins the master node, it follows a handshake procedure which exchanges local system times and establishes a round-trip latency estimate between the master node and the joining node. The master node takes two nodes with the highest round-trip latency values t_1 and t_2, and estimates the maximum end-to-end latency $t_{max} = t_1/2 + t_2/2$. Minimum tempo

bpm_{min} is computed by $bpm_{min} = \frac{60000}{ct_{max}}$ where $c \geq 1$ is a safety factor to account for jitters and internal processing delay. As the participant's local system time T_p and the master local system time T_m are exchanged, the time offset between that particular participating node to the master node is taken to be $T_p + t_p/2 - T_m$ where t_p is the round-trip latency between the participant and the master.

These parameters are used for event synchronization. Client messages that cause events to happen are delayed to the next beat, skipping the incoming beat. SCProxy turns messages into OSC bundles and tags them with the time of the next beat. The bundles are sent to the master node, who will re-adjust the time tag with appropriate time offsets before distributing the bundles to all the participant nodes. In this way, the musical events will happen at the same time on all nodes, including the nodes that generate the events, and sequential events will maintain their temporal distances on all nodes.

A globally synchronized metronome is provided on all nodes and can be optionally turned on to provide cues. Beat synchronization can be turned off, which will delay events for a duration equal to ct_{max}. Measure synchronization option is also available which will delay events until the next measure, unless the events are issued at a duration less than ct_{max} before the next measure, where the events will be postponed for another measure. With the current limitation of the system, however, all participants have to manually set the tempo on their programming environments to the same value.

3.3 Namespace Separation

Although a SuperCollider server accepts multiple clients, it has a single namespace which contains nodes, groups, bus channels, and buffers. Each instance of these resources has a unique ID to refer to it. When a client creates a node, the client also assigns an ID to it (unless it is a one-off node in which case the client lets the SuperCollider server assigns an arbitrary ID from the pool of reserved IDs). Another client may come along and assign the same ID to another newly created node, which results in an error.

The *sclang* client has a mechanism to separate namespaces using its resource allocator classes, so that multiple *sclang* clients are assigned distinct sets of IDs to use. However, few, if any, non-*sclang* clients provide such mechanism, and even if they provide one, the mechanism is likely different. This situation makes different third-party clients difficult to interoperate in the same session. There is a great risk of having unexpected results or conflicts arise during the performance.

SCProxy provides an object ID translation mechanism such that IDs assigned in one client will never clash with those assigned in another client. Each ID sent from a client is remapped to an unused ID by the proxy. This process is transparent, and the clients are not aware of the IDs to which their own IDs are remapped. This allows different types of clients to play together regardless of their resource allocation strategies.

4 Conclusion

We have introduced SCProxy, an OSC-based proxy for remote collaborative live coding which connects heterogeneous SuperCollider clients together in a quasi-transparent way. The proxy is closely tied to SuperCollider server OSC commands, allowing us to achieve tighter and more seamless collaboration. By providing mechanisms for OSC message distribution, and latency-compensated, beat-quantized event synchronization, we no longer require performers to be co-located. Sound is reproduced the same way at the originating node and at other nodes. With namespace separation by ID remapping, we also enable performers to use programming environments that are different from each other, as long as they use SuperCollider server as the backend.

However, due to the beat-quantized nature of the proxy, effects of executing commands are no longer immediate. Live coding in this mode requires anticipation of incoming beats and, as such, there is a need for performers to adjust their live coding practice to suite.

5 Future Work

Although namespace separation makes collaboration among heterogeneous environments possible without conflicts, it is also a limitation. A better approach might be to enable different environments to share resources and synthesizer definitions and gain access to ones that are created by other environments. SuperCollider server does not directly provide a means to do so. SCProxy will need to maintain a global directory of all available resources and synthesizer definitions by intercepting all resource creation and destruction commands, and provide an API for the client to gain knowledge about those resources. Also, the client needs to have a way to create new objects that reference existing resources that are created by other clients. We are considering implementing it in Overtone, as it is one of the most comprehensive third-party SuperCollider clients and live coding environments.

References

1. Aaron, S.: Sonic Pi performance in education, technology and art. Int. J. Perform. Arts Digital Media **12**(2), 171–178 (2016)
2. Aaron, S., Blackwell, A.F.: From Sonic Pi to Overtone: creative musical experiences with domain-specific and functional languages. In: Proceedings of the First ACM SIGPLAN Workshop on Functional Art, Music, Modeling and Design, pp. 35–46. ACM Press, New York (2013)
3. Barbosa, Á.: Displaced soundscapes: a survey of network systems for music and sonic art creation. Leonardo Music J. **13**, 53–59 (2003)

4. Bencina, R.: OSCgroups: peer-to-peer Internet OSC multicast without the pain. Accessed 27 August 2018. http://www.rossbencina.com/code/oscgroups
5. Collins, N., McLean, A., Rohrhuber, J., Ward, A.: Live coding in laptop performance. Organised Sound **8**(03), 321–330 (2003)
6. Cornelisse, L.: Skoar. Accessed 27 August 2018. https://github.com/sofakid/Skoarcery
7. de Carvalho Jr. A.D., Lee, S.W., Essl, G.: SuperCopair: collaborative live coding on Super-Collider through the cloud. In: McLean, A., Magnusson, T., Ng, K., Knotts, S., Armitage, J. (eds.) Proceedings of the First International Conference on Live Coding, ICSRiM, University of Leeds, pp. 152–158 (2015)
8. Kirkbride, R.: FoxDot: live coding with Python and SuperCollider. In: Proceedings of the International Conference on Live Interfaces, pp. 194–198 (2016)
9. Kirkbride, R.: Troop: a collaborative tool for live coding. In: Proceedings of the 14th Sound and Music Computing Conference, pp. 104–109 (2017)
10. Lee, S.W., Essl, G.: Models and opportunities for networked live coding. In: Proceedings of The Live Coding and Collaboration Symposium (2014)
11. Magnusson, T.: The IXI Lang: a SuperCollider parasite for live coding. In: Proceedings of International Computer Music Conference, pp. 503–506. Michigan Publishing (2011)
12. McCartney, J.: Rethinking the computer music language: SuperCollider. Comput. Music J. **26**(4), 61–68 (2002)
13. McKinney, C.: Oscthulhu: applying video game state-based synchronization to network computer music. In: Proceedings of the 2012 International Computer Music Conference, Ljubljana, Slovenia, pp. 309–314 (2012)
14. McLean, A.: Making programming languages to dance to: live coding with Tidal. In: Proceedings of the 2nd ACM SIGPLAN International Workshop on Functional Art, Music, Modeling and Design - FARM '14, pp. 63–70. ACM Press, New York (2014)
15. McLean, A., Wiggins, G.: Tidal pattern language for the live coding of music. In: Proceedings of the 7th Sound and Music Computing Conference (2010)
16. Musikinformatik: Steno: concatenative little meta language for live coding. Accessed 27 August 2018. https://github.com/musikinformatik/Steno
17. Narveson, J., Trueman, D.: LANdini: a networking utility for wireless LAN-based laptop ensembles. In: Proceedings of the Sound and Music Computing Conference, Stockholm, Sweden, pp. 309–316 (2013)
18. Ogborn, D.: EspGrid: a protocol for participatory electronic ensemble performance. In: Audio Engineering Society 133rd Convention, San Francisco, CA (2012)
19. Puckette, M.S.: Combining event and signal processing in the max graphical programming environment. Comput. Music J. **15**(3), 68–77 (1991)
20. Puckette, M.S.: Pure Data. In: Proceedings: International Computer Music Conference 1997, The International Computer Music Association, Thessaloniki, Hellas, 25–30 September 1997, pp. 224–227 (1997)
21. Silvani, D.E.: XI: a domain-specific language for live coding musical patterns in Ruby. Accessed 27 August 2018. https://github.com/xi-livecode/xi
22. Wright, M.: Open Sound Control: an enabling technology for musical networking. Organised Sound **10**(03), 193 (2005)

A Collaborative Model for Customer Retention on User Service Experience

Pushpa Singh and Vishwas Agrawal

Abstract The traditional brick-and-mortar economy that we are accustomed to is transforming into a realm that is fully digitized. To enable such an age with modern businesses, differential services through multiple providers must be utilized. This compels service providers to collaborate with one another to maximize their payoffs. The paper suggests a model based on machine learning algorithms to tag loyal customers and provides them better quality of service (QoS) experiences to retain them in a system, thereby affecting the overall payoffs of the e-business as retention of loyal customers proves to be less expensive as compared to acquiring new customers.

Keywords E-commerce · Loyalty · K-nearest neighbor (KNN) · Customer relationship management (CRM) · Machine learning · Quality of service (QoS)

1 Introduction

Nowadays, businesses are in the form of e-businesses or m-businesses. The dominant features of e-commerce can be identified in 5C sectors such as commerce, collaboration, communication, connection, and computation [1]. Not only is every product just a click away, but we have numerous sellers to choose from in the form of websites or mobile applications. This has led to an extremely fierce competition, forcing service providers to ensure that the consumers receive the highest possible standards of quality. Hence, utilizing the existing resources along with improving the customer experience has become an important factor for enterprises to guarantee their sustained growth. In e-commerce, the majority of the revenue is generated

P. Singh
Accurate Institute of Management and Technology, Greater Noida, India
e-mail: puspha.gla@gmail.com

V. Agrawal (✉)
Cochin University of Science & Technology, Kochi, India
e-mail: vishwas283@gmail.com

© Springer Nature Singapore Pte Ltd. 2019
S. K. Bhatia et al. (eds.), *Advances in Computer Communication and Computational Sciences*, Advances in Intelligent Systems and Computing 924, https://doi.org/10.1007/978-981-13-6861-5_5

through the online transactions that take place on the respective platform. Revenues can grow substantially each year if new customers can be added without losing existing customers. Therefore, it is essential for every business to retain as many customers as possible. Since it is practically not possible for any enterprise to retain each and every customer, focus must be given to the customers, which prove to be loyal to their service providers. The objective of customer relationship management (CRM) is to improve the customer service relationship and assist in customer retention. Loyalty attitudes of e-commerce services are very important for the long-term advancements of many businesses.

The quality of service (QoS) experience of a customer is crucial for any service provider to retain its existing customers. In the current scenario, adding a new customer can be a challenge, as it needs new marketing strategies and schemes. In such a competitive market, it is important and advisable to retain an existing loyal customer [2, 3]. Customer loyalty is a subjective parameter that can be categorized according to the service provider's perception and may differ from one service provider to the other as for e-commerce, the buying frequency, online payment, and period of the transaction of a customer will be the major parameter to decide its loyalty toward the service provider. Customer retention is the intention of a customer to stay with the current service provider, whereas customer loyalty is defined as maintaining the business relationship with the current service provider.

A very primal question is: 'Why should a customer be loyal to service providers?' It is only when a customer is content with the services offered to him by the provider that the customer will develop a sense of trust and reliance toward that provider, and once such a feeling is forged, the consumer will then feel confident enough to rely on the provider for any need that may arise.

At present, it has become extremely challenging to continue the business without a mobile phone and Internet. E-commerce is the result of using communication facilities, computers, and programs to permit instant interaction between any two or more resources on the network [4] and hence requires better network QoS for successful transaction. There is, therefore, an obvious indication of collaboration between a telecommunication enterprise and e-commerce establishment. CRM applications help organizations assess customer loyalty and profitability based on measures such as repeat purchases, amount spent, and longevity. While a number of companies are adopting CRM technology as a way of managing customer relationships, it becomes a daunting task to educate the user about the proper and effective use of CRM. Businesses which adopt artificial intelligence and machine learning are showing cost benefits, sharp increases in customer satisfaction and revenue coupled with exponential growth in overall performance.

2 Related Works

Efficient CRM raises customer satisfaction and retention rates [5, 6] and resulted in higher revenues and lower operational costs. Reference [7] discussed strategies and barriers for successful implementation of CRM in e-business and m-business. CRM is implemented successfully in large business enterprises because they have the significant financial strength to operate. There are various key factors of success or failure during CRM implementations [8]. Reference [9] identified poor planning, lack of clear objectives, etc., as the key factors for CRM failures. For small enterprises, CRM is still a costly solution. Reference [10] discussed why small and medium enterprises (SMEs) do not implement CRM system and suggested to use open-source CRM system in SMEs. The same solution was offered by [11]. CRM is one of the many ways to identify a loyal customer, customer brand commitment, acceptance, and buying, three different aspects are discussed for understanding customer loyalty [12]. Reference [13] discussed the concern of telecom corporations to integrate with underlying electronic services or e-services. E-services are intangible in nature, and its rapid use has raised the need to define standards and means to assess and assure quality [14].

There may be several combinations of metrics or attributes with different possible values that can be utilized in the computation of customer loyalty. Various utility functions can be used to compute the weight of each attribute from (0–1) [15]. These computations carry some complexity in the case of several possible combinations according to the choice of providers and need a quick solution. Machine learning (ML) provides the ability to learn without being explicitly programmed utilizing mathematical models, heuristic learning, knowledge acquisitions, and decision trees for decision making [16]. Reference [17] proposed machine learning methods for challenging problems such as a customer churning prediction in the telecommunications industry. By taking advantage of machine learning techniques with the CRM database, it is possible to find the best customers [18]. Machine learning is providing solutions in a vast area of research, including engineering, to the manufacturing environment [19].

This paper aims to identify the loyal customer based on the Econometrics (payoff) toward the service provider and render them better QoS through discounts, promotional benefits to gain the trust of the customer and subsequently increase the revenue of the service provider. The paper utilizes a machine learning algorithm to predict a loyal customer based on the parameters defined by an user, which is a novel approach toward suggesting ways to retain the customer in a system to improve growth of the commercial and communication sectors.

3 Proposed Framework

Companies in e-commerce already realize that acquiring new customers is not enough for lasting success and efforts need to be made in order to identify customer loyalty. It is a very costly affair to implement CRM for small-scale enterprises. Machine learning provides a simple solution to compute the loyalty of customers. Loyalty of customer proposed by the authors is based on the following parameters: customer

Table 1 Linguistic term for loyalty parameters

S.No.	Linguistic term	Range
1.	Very high (VH)	≥0.9
2.	High (H)	0.7–0.9
3.	Medium (M)	0.7–0.4
4.	Low (L)	<0.4

lifetime (T), revenue (R_v), and data usage (U). If data usage measured in GB is higher, it means that the person is utilizing the service extensively and is a potential case.

Online shopping is a major part of e-commerce or m-commerce applications. In online shopping, customers can access the portal through heterogeneous devices such as computers, laptops, tablet, and smartphones. Customer can opt for various electronic payment systems (EPS) in order to complete the online transaction. Credit card, debit card, and net banking are the various popular payment options available to directly buy goods or services from a seller over the Internet using a Web browser. In fact, mobile payment services with their increasing popularity are presently under the phase of transition toward a future payment system [20]. In both types of payment system EPS and mobile payment system (MPS) [21], the mobile phone plays an important role for online payment.

In this paper, we intend to propose a version with added parameters F and C_p, where F is the frequency of online shopping and C_p is the amount customer paid for the product and services. Let (P_i, C_i) where i = 1, 2, ..., n be data points such that P_i denotes parameter value and C_i denotes the target class for P_i. It is assumed that there are two classes: loyal and not loyal. Label L belongs to the category of loyal customers, and label NL belongs to the customers who are not loyal. The type of customer (cust_type) is measured on the basis of parameters U, T, R_v, F, and C_p. U, T, R_v, F, and C_p are normalized between 0 and 1 according to (1)

$$N(P_i) = \frac{P_a}{P_{max}} \quad \text{where } P_a \le P_{max} \tag{1}$$

where P_a is actual value collected by the user equipment and P_{max} is the maximum expected value of that parameter. Maximum expected value can be set by different service providers. After normalizing each of the loyalty parameters in the range of 0–1, a linguistic term is assigned, as given in Table 1. The customer can have several combinations of parameter values 0–1. To simplify this task, normalized value of the parameter is converted to the associated linguistic variables such as 'very high', 'high', 'medium,' and 'low' as depicted in Table 1.

A data set (DS) consisting of labeled instances to predict the type of customer is used. The data set is designed as a comma-separated value (CSV) file. A CSV file is a human-readable text file where each line has a number of fields, separated by commas or some other delimiter. Proposed CSV file represents cust_type = 'Loyal' on the basis of the following two rules:

Table 2 Few records in data set to predict cust_type

U	T	R_v	F	C_p	Label
VH	VH	VH	VH	VH	Loyal
H	H	H	H	H	Loyal
H	M	H	H	VH	Loyal
VH	M	VH	VH	H	Loyal
M	H	M	M	M	Not loyal
H	H	H	M	M	Not loyal
M	M	M	M	H	Not loyal
L	L	L	L	L	Not loyal

Rule-1: If the value of each parameter is either 'high' or 'very high.'
Rule-2: If the values of F and Cp parameters are either 'very high' or 'high.'

All other cases are representative of cust_type = 'Not loyal'.

To model the linguistic terms as numeric variables, the set or range of values must be defined through which linguistic terms can be associated. The data of around forty-five users were taken to predict cust_type on the basis of the above rule. The data set in terms of a linguistic variable is shown in Table 2. The numeric values associated with the linguistic variable of a data set are given in the range between 0 and 1 in the CSV file.

Let x be a point for which the label is unknown and the problem is to determine the class. K-nearest neighbor (KNN) approach is used to predict cust_type directly from the training data set. KNN is used for classification, estimation, and prediction algorithm in the area of data mining [22]. The following parameter value was used for KNN classifier:

KNeighboursClassifier(algorithm = 'auto', leaf_size = 30, metric = 'minkowski', metric_params = None, n_jobs = 1, n_neighbours = 5, p = 2, weights = 'uniform')

Algorithm = 'auto' refers to the most appropriate algorithm based on the values passed to KNN fit method. Other choices of algorithms are 'ball_tree', 'kd_tree', and 'brute'. Parameter leaf_size can affect the speed of the construction and query, as well as the memory required to store the tree and default value is 30. P is a power parameter for the Minkowski metric. The default metric is Minkowski, and with p = 2 is equivalent to the standard Euclidean metric. The metric_params is an additional keyword parameter for the metric function. The parameter n_jobs shows the number of parallel jobs to run for neighbor search. Parameter n_neighbors is the number of neighbors for queries and by the default value of this parameter is 5. If n_neighbors = 1, then the object is simply assigned to the class of that single nearest neighbor. Weights = 'uniform' means all points in each neighborhood are weighted equally. Except for weight and the metric parameter, all of the other parameters are optional.

In scikit-learn, a random split into training and test sets can be quickly computed with the train_test_split () helper function with state parameters as shown in Eq. 2. This function splits arrays or matrices into the random train and test subsets. That means that every time it is run without specifying random_state, a different result is obtained, this is expected behavior.

$$\text{train\_test\_split}(X, y, \text{test\_size} = 0.3, \text{random\_state} = \text{seed}) \qquad (2)$$

where X represents a list of features set involved in cust_type, i.e., P (U, T, R_v, F, C_p), y is the target set such as {'Loyal', 'Not Loyal'}, and random_state is the seed value used by random number generator for random sampling.

Predictions are made for a new instance P (U, T, R_v, F, C_p) by searching through the entire training set for the K-most similar instances (the neighbors) and summarizing the output variable for those K instances. Pseudocode for cust_type prediction is represented in Algorithm 1.

Algorithm 1: Pseudocode for cust_type Prediction

1. Reading the comma-separated values (CSV) file by using read_csv() function, which has a data set for training and testing
2. Splitting CSV file into training and testing sets by using the Python function: Train_test_split()
   X_train, X_test, y_train, y_test = Train_test_split(X, y, test_size = 0.3, random_state = seed)
3. Predict the target label from test sample by following function:

   knn = KNeighborsClassifier (n_neighbors = K)
   y_pred = knn.predict (X_test)

4. Find the KNN accuracy metrics by using:

   accuracy_score (y_test, y_pred)

5. Set the maximum value of each parameter in the data set and normalized the each attribute value of (0–1), by (1).
6. After getting the normalized value of each sample,
   P (U_i, T_i, R_{vi}, F_i, C_p), predict the category of cust_type by:

   (i) sample = np.array ([[U_i, T_i, R_{vi}, F_i, C_p]]) (np is instance of library NumPy library)
   (ii) knn.predict (sample)

7. Prediction category is 'loyal' or 'not loyal'

Table 3 Value of K and testing accuracy

K	2	4	6	8	10	12	14
Accuracy	0.928	0.928	0.928	1	1	0.625	0.928

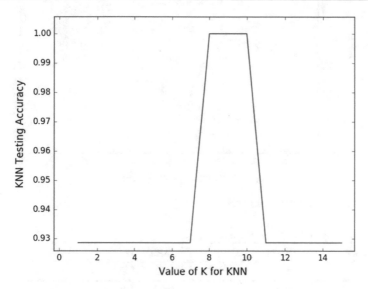

Fig. 1 KNN testing accuracy for different value of K

4 Results and Discussion

Algorithm 1 is implemented in the Python language. Scikit-learn is a machine learning library that is imported in Python. Scikit-learn has features of classification, regression, K-means, and clustering algorithm and also is designed to interoperate with the Python numerical and scientific libraries NumPy and SciPy [23]. Pandas library is used to import CSV file in Python. The read_csv method of the Pandas library is used to read a CSV file as a data set. Here, data set is split at 70% for training purpose and 30% for testing purpose by taking test_size $= 0.3$ and random_state $= 7$. Maximum accuracy 1 or 100% occurs at K $= 8$ and 10, and then, accuracy reduces as given in Table 3 and is plotted as shown in Fig. 1.

Further, the algorithm was modified in accordance with different classifiers to form a basis of comparison. Classification and regression tree (CART) also referred to as decision trees and support vector machine (SVM) are the two other classifiers that were simulated. The accuracy of these classifiers depends on the test_size and random_state parameters described in Algorithm 1. For simplicity, the test_size was taken to be equal to 0.3 for each simulation run. Simulation results of the models are represented in Fig. 2.

The results were collected at a particular instant of time by varying the random_state as stated in Table 4. On an average, KNN proves to be the most accurate,

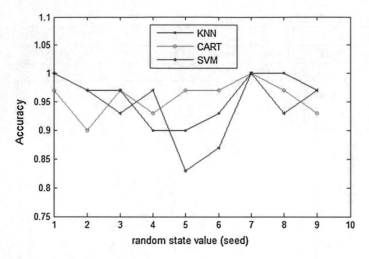

Fig. 2 Accuracy of different classifiers

Table 4 Accuracy of different classifiers

Random_state	KNN	CART	SVM
1	1	0.97	1
2	0.97	0.9	0.97
3	0.97	0.97	0.93
4	0.9	0.93	0.97
5	0.9	0.97	0.83
6	0.93	0.97	0.87
7	1	1	1
8	1	0.97	0.93
9	0.97	0.93	0.97
Average	0.96	0.956667	0.941111

closely followed by CART. CART proves to be too sensitive if small changes are made in the training data set and tends to overfits in the model.

5 Scope and Future Work

Proposed work shows a significant scope of e-collaboration where telecom industry and e-commerce-based companies can increment their revenues and improve customer retention. Online shopping in the future will be a shopping experience where companies utilize data-driven insights to improve their marketing and promotional schemes, thereby forging a deeper relationship with a loyal customer. They can increase conversion rates by enabling collaborative idea sharing and conversation to their Web sites to move shoppers from consideration to purchase and telecom industry can provide better network QoS in terms of bandwidth, packet loss, delay, etc. [24] to loyal customers during e-transaction.

The proposed framework can be extended to many services related sectors where the retention of the customer is important and can be utilized to expand operations and boost e-business growth. This work can find application in the future with the fifth generation (5G) of mobile communication where user retention is gaining importance. Further, the classifiers can be simulated for different test sizes to produce new insights.

6 Conclusion

The proposed work establishes the role of machine learning algorithms in marking customer data into different sets based on the parameters defined. This information is then utilized to retain the user. The results of the simulation plotted in Fig. 1 reveal that KNN classifier is better suited as compared to other classifiers such as CART and SVM. The model predicts cust_type, which can be utilized in e-collaboration to enhance providers' payoff as well as the user experience. The application of the work can be extended to any domain where retention of customer is essential for the business to sustain and flourish.

References

1. Zwass, V.: Electronic commerce and organizational innovation: aspects and opportunities. Int. J. Electronic Commer. **7**(3), 7–37 (2003)
2. Jain, S.C.: CRM shifts the paradigm. J. Strateg. Mark. **13**(4), 275–291 (2005)
3. Sanchez, J.G.: Customer relationship marketing: building customer relationships for enduring profits in a wired economy. Research Paper. Atthapholj. net (2003)
4. IPC E-Business and Supply Chain Committee (2000). The Myths of E-commerce (2000). http://www.ipc.org/4.0_Knowledge/4.1_Standards/E-CommerceWhitePaper.pdf. Accessed 12 Feb 2018
5. Arab, F., Selamat, H., Ibrahim, S., Zamani, M.: A survey of success factors for CRM. In: Proceedings of the World Congress on Engineering and Computer Science, vol. 2, pp. 20–22 (2010, October)
6. Chen, I.J., Popovich, K.: Understanding customer relationship management (CRM) people, process and technology. Bus. Process Manage. J. **9**(5), 672–688 (2003)
7. Nguyen, T.H., Sherif, J.S., Newby, M.: Strategies for successful CRM implementation. Inf. Manage. Comput. Secur. **15**(2), 102–115 (2007)
8. Wu, J.: Customer relationship management in practice: a case study of hi-tech company from China. In: International Conference on Service Systems and Service Management, June 30–July 2, 2008, pp. 1–6. IEEE Computer Society (2008)
9. Foss, B., Stone, M., Ekinci, Y.: What makes for CRM system success-or failure? J. Database Mark. Customer Strategy Manage. **15**(2), 68–78 (2008)

10. Tereso, M., Bernardino, J. Open source business intelligence tools for SMEs. In: 2011 6th Iberian Conference on Information Systems and Technologies (CISTI), pp. 1–4. IEEE (2011, June)
11. Sampaio, D., Bernardino, J.: Open source CRM systems for SMEs. In: Proceedings of the 2014 International C* Conference on Computer Science & Software Engineering, p. 22. ACM (2014, August)
12. Uncles, M.D., Dowling, G.R., Hammond, K.: Customer loyalty and customer loyalty programs. J. Consum. Market. **20**(4), 294–316 (2003)
13. Chou, T.H., Lee, Y.M.: Integrating E-services with a telecommunication e-commerce using service-oriented architecture. JSW **3**(9), 60–67 (2008)
14. Batagan, L., Pocovnicu, A., Capisizu, S.: E-service quality management. J. Appl. Quant. Methods **4**(3), 372–381 (2009)
15. Wu, Q., Li, W., Wang, R., Yu, P.: An access network selection mechanism for heterogeneous wireless environments. J. Comput. Inf. Syst. **9**(5), 1799–1807 (2013)
16. Mishra, A., Mishra, D.: Customer relationship management: implementation process perspective. Acta Polytech. Hung. **6**(4), 83–99 (2009)
17. Vafeiadis, T., Diamantaras, K.I., Sarigiannidis, G., Chatzisavvas, K.C.: A comparison of machine learning techniques for customer churn prediction. Simul. Model. Pract. Theory **55**, 1–9 (2015)
18. Dullaghan, C., Rozaki, E.: Integration of Machine Learning Techniques to Evaluate Dynamic Customer Segmentation Analysis for Mobile Customers (2017) arXiv preprint arXiv:1702. 02215
19. Wuest, T., Weimer, D., Irgens, C., Thoben, K.D.: Machine learning in manufacturing: advantages, challenges, and applications. Prod. Manuf. Res. **4**(1), 23–45 (2016)
20. Bezovski, Z.: The future of the mobile payment as electronic payment system. Eur. J. Bus. Manage. **8**(8), 127–132 (2016)
21. Carr, M.: Mobile payment systems and services: an introduction. In: Mobile Payment Forum, vol. 1, p. 12 (2007)
22. Han, J., Pei, J., Kamber, M.: Data Mining: Concepts and Techniques. Elsevier (2011)
23. Pedregosa, F., Varoquaux, G., Gramfort, A., Michel, V., Thirion, B., Grisel, O., … Vanderplas, J.: Scikit-learn: machine learning in Python. J. Mach. Learn. Res. **12**(Oct), 2825–2830 (2011)
24. De Gouveia, F.C., Magedanz, T.: Quality of service in telecommunication networks. Telecommun. Syst. Technol. **2**(5), 77 (2009)

Smart Services for Smart Cities: New Delhi Versus Jaipur

Devesh Kumar Srivastava

Abstract By following previously conducted studies by various researchers, private corporations, and public offices around the globe, people have found out various loopholes not only regarding the ground aspects of a failing system but also the somewhat manipulated officials who generally short come in fulfilling their duties toward a continuous development and modifications to our cities. Even though, every annual budget highlights the capital investment made for the growth of various services fabricated for a city, I am still unable to provide an impact for the public, which results in an overall failure of our master plan to provide for a better living of the public. In this paper, I focused on the facts and hurdles that an administration faces while trying to deliver smart services to develop a smart city. In that framework, I included six of the most essential domains that have already been introduced in two major smart cities in India which can effectively contribute in the enhancement of currently developing models of smart city which were used by those two cities earlier. Altogether, I elaborated how each component has proved its worth in either one or both the cities and integrate it into a new developed model. Using the devices which used the concepts of things based on Internet, I formed conclusion regarding how a city can be upgraded into a smart city taking the lessons from two of the already declared smart cities, Jaipur (JAI) and New Delhi (DEL).

Keywords Smart system · Jaipur (JAI) · New Delhi (DEL) · Wi-fi · Internet of Things (IoT)

1 Introduction

The definition of the word "smart" essentially states that the something is now being done in a more intelligent or clever way. For a city, it could be smart design, smart utilities, smart housing, smart mobility, smart technology, smart surroundings, and

D. K. Srivastava (✉)
SCIT Manipal University, Jaipur, India
e-mail: devesh988@yahoo.com

© Springer Nature Singapore Pte Ltd. 2019
S. K. Bhatia et al. (eds.), *Advances in Computer Communication
and Computational Sciences*, Advances in Intelligent Systems and Computing 924,
https://doi.org/10.1007/978-981-13-6861-5_6

smart disposal of solid or liquid based, etc., for its people. It also includes smart employment and better economic opportunities. Smart living opportunities in terms of quality affordable housing, cost efficiency and sustainable community infrastructure, and smart service infrastructure include quality water supply, regular sanitation, 24×7 power supply, convenient health care, enhanced security, high-speed internet connectivity, and smart governance which mean a faster and more effective delivery of services. In its basic form, a smart city is the one which successfully provides a basic standard of living, sustainability, and economic development for all those people who commute, work, and live in it in a more efficient and effective way. But the above challenges beyond any doubt have proved that the current existing system is unsustainable and outnumbered by the needs of the increasing population in the cities. Therefore, I am observing an increase in numbers of daily calls from our own citizens for major changes in the existing framework of commute system, healthcare, sewage, water treatment, solid waste management, etc. [1]. For example, sewage system on the road pavements should be able enough to survive a heavy rainfall or the healthcare system should be proactive and predictive in diagnostic process [2].

The recent advancements in information and communication technologies (ICT) could play a vital role in curbing these problems. For example, wearing smart bands and watches could help patients monitor their health 24/7, even when they are asleep or using quad copters and droids to monitor the traffic conditions and to observe any obsolete behavior by drivers such as running a signal, speeding, and reckless driving. [3]. As I am observing a growing trend of urbanization worldwide, it is expected that by 2050, somewhat 6.3 billion people, more than 70% of the population will move to major metropolitan cities, from a study published by UN urbanization perspective 2011 revision [4]. Every major metropolitan city is a well-defined ecosystem of a set of subsystems such as food and agriculture, transit and transportation, water supply and treatment, residential and industrial complexes, electricity distribution, and mobile communication. [5]. Such cities in India are New Delhi, Jaipur, Mumbai, Pune, Kolkata, etc. [6]. These cities are already termed as smart cities by the Indian government as they have proposed there framework during the smart India movement under the prime minister and got the funding to implement it. But even after becoming a smart city, officials have wondered that what they could have done differently in order to make the solutions more efficient and cost effective [7, 8]. Every framework proposed by the different cities was customized according to the size and the population of the city and how much resources it had on its disposal. But, I believe in a more wholesome and teamwork approaches rather than individually trying to make it to the top. Hence I decided to review the models of two of the most influential and successful models of all, New Delhi and Jaipur, and observe which six features were most effective and essential for them. When I observed the comparison between these two cities and hence proved the worth of that feature. It tolled us that the feature is relevant and it must be added to other frameworks proposed by smart cities.

2 Literature Survey

I first reviewed the committee that is going to implement these changes, overseas, and maintains every prospect of a smart city. I looked at the reinventing government: How the entrepreneurial spirit is transforming the public sector, by David Osborne and Ted Gaebler published in 1992. It suggested that the governments of the city should believe in changing their whole hierarchy model that they have used for ages of conducting business and suggests a new framework that can help governments conduct businesses more effectively [9]. The author proposed a new format consisting of ten principles which delivered more service-oriented approach and delivered impeccable quality of services. Those ten principles involved the elimination of whole commanding and controlling, preventative planning, taking bribes, the concept of decentralization, thinking of clients as valuable customers, reducing rules, audits and reviews of agencies, introducing competition, and more agile approach to work with clients [10]. The author called it lynchpin; the framework aimed to introduce alternatives to their insider's program creation and recommends delivery of smart services by developing other public and private relationships. These public–private partnerships, which is being very old, are also a featuring suggestion of President Bill Clinton's National Performance Review, which was a famous effort to inject efficient features to improve the federal government [11, 12]. It also helped in defining the triple helix model to define the smart city [13]. It is the base of other projects like Intel cities, which again recommends relationships between public and private sector [14, 11]. In the beginning of the year of 2013, Pardo and Nam proposed a model consisting of three core dimensions [15]. They are institutions, people, and technology. This model has vastly helped committees strategize smart cities. Pardo and Nam view the concept of a smart city as the one which primarily involves improving the existing city services and supports development of economy driven by the efforts of local government. The conclusion in the framework highlighted a reiterated concept which was found in all existing models of smart city. They all consist of distinct social parts and their importance in the conclusions of planning of smart cities. It also consists of linking them with technology and their power to change the community and economy. Aggregating this, a smart city has a purpose of resolving all the problems of urban areas (shortages, traffic, congestion, environmental challenges, sanitation shortcomings, service unavailability, etc.) with the help of IoT technologies and data analysis [6]. The final aim is to rejuvenate a city's social and economic imbalances via efficient analysis of data collected from various IoT-enabled devices, and eventually making everyone's life healthy and the environment sustainable toward a better tomorrow for the forthcoming generations.

3   Research Framework

Developing the components of the framework based compositely on the aforementioned literature, the article highlights various key general sectors and their sub-
dimensions of implementation of the smart city [5]. These sectors are dubbed as the
groundwork of the development process and recommend a heterogeneous view with
the current accent and future scope of a smart city. After this session of explained
work, some of the sectors were rephrased for clarification purposes with respect to the
citations. For example, a previous version of the framework proposed for the smart
city cited a sector that was used to measure the overall percentage of development
of the city by quantifying the total count of the tech firms or startups in that city's
industrial sector [10]. A smart city must contain a plan to sustainable development
for the future; highlighting our commitment to the betterment of our citizens should
I find another greener or smarter solution to our daily activities. It will inspire other
cities to adopt this approach who are on their way of becoming a smart city.

3.1   Core Services of a Smart City

1. Sustainable Infrastructure: Management and governing of a smart city and draft-
   ing new policies to keep up with development and feasibility of adding them to
   our existing framework. Making use of the ICT technology to collect, analyze,
   and draw conclusions as if I must update existing solution into smarter one. Every
   change must also be accountable and transparent in nature.
2. Material Infrastructure: Updating the legacy urban transportation system, preserving water, energy in our homes, betterment of sanitation facilities around the
   city, and cleaner and power saving technology for public facilities like improving
   waste management, sewage, drainage, etc.
3. Social Infrastructure: Infrastructure related to healthcare, education, and entertainment around the city. Improving public places like shopping complexes,
   movie theaters, children parks, and gardens.
4. Economic Infrastructure: Improving the quality of the factors related to economy
   of the city like its export processing zones (seaports, airports etc.) warehousing
   and trading freight terminals, industrial parks, IT hubs, business parks, logistics
   hubs, other trade centers, financial centers, service centers, etc.

To consume the benefits after implementation, of the services of smart city as
shown in the Fig. 1, I put them into my practices. I must have a concrete plan in
place to draw benefits from a conceptual framework that we created for improving
a city into a smart city. For this to happen, I came up with a list of core elements
of a proposed smart city. These elements tell indicate the factors that are absolutely
mandate if I am planning to update the whole city's infrastructure.

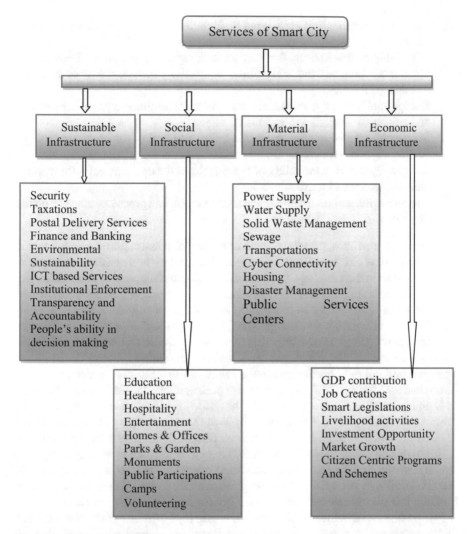

Fig. 1 Services of smart city

3.2 Core Infrastructure Elements of a Smart City

1. Appropriate supply of water, which includes treatment and recycling.
2. Assuring 24 × 7 electricity, a majority of which must be generated through renewable resources power generation alternatives like solar power, hydropower, wind energy, tidal energy etc.
3. Effectively and efficiently mobilizing public transport, which must be available for everyone.

4. Sanitation facilities available to public including sewage and solid waste treatment.
5. Affordable or free housing for the people living in the quagmire of poverty.
6. Constant and robust cyber and cellular connectivity in at least 95% of the effective area of the city.
7. Exceptional form of governance, additionally promoting e-governance services and citizen participation in various decisions.
8. Sustainable use of natural resources, using and opting for green technology, healthcare, and education.
9. Adequately assuring the safety of the citizens of the city especially for women, children, and the elderly.
10. Appropriation and transparency across the system and processes whether public or private.

However, sometimes, the total growth of the smart infrastructure might not match the total growth of citizens of the city. Even after developing and updating the city with smart solutions in the most impacted sectors, I might miss the parts or fields of the city that does not seem important at the time of action, but when I analyze the whole infrastructure, I realized that they were a critical part of our way toward a complete smart city. Population boom, traffic congestions, inadequate and unnecessary infrastructure, limited resources, increasing migration, inadequate discipline and supervision, rapidly growing mobility, improper urbanization, conventional energy crisis, global warming, and natural phenomena are the issues of urbanization. Contrary to all the issues that are highlighted above, I must also portray all the salient features of a smart city, no matter its size or population. Our basic idea is to build a smart city to finally improve the quality of the life of the citizens in the city, bringing smart alternatives to everyone and personalize the service according to their usage to best serve them.

If a state decides to finally initiate their journey on the path of making every city, which contents them to a smart city, they must begin with at least one of the city that they believe has a potential of becoming a smart city in the shortest possible time. Usually, this city is selected by the state government on various selection criteria which are discussed below. Furthermore, it is safe to say that generally states choose their capitals as the city that must be transformed into a smart city. Reason behind this is that capital cities are bound to have airports, bus ports, and multiple railway stations ensuring the greatest connectivity in the state to the outside world. Most of the resources are directed to the capital city for usage and further redistribution. Major MNCs always prefer the capital city to set up their sites and other point of presences.

3.3 Selection Criteria

1. The process of selection of the city is modularly based on the idea of cooperative federalism and competitive federalism.

2. A very open and generalized formula is applied which is the equal weight age, which must be 50:50 to the effective urban population of that union territory or state and the count of all statutory towns in that union territory or state.
3. Based on the calculations performed on this formula, every union territory or state must select at least one nominated city for the upgradation process.

4 Data Analysis

I drew publicly available data and conducted interviews from those city officials, which have first-hand information about these services, wherever possible, and hence had an assessment process to evaluate the type of service and how mature it was. Keeping in mind our goals, I conducted the cross-case analysis by citing the contrast between the services as shown in Fig. 2 provided by the two cities New Delhi and Jaipur highlighting their smart city approaches. Participatory information services consist of one user or customer and one information service provider which helps and provides an answer to all the queries fired by user. Other type of service being examined is a unidirectional information service which helps the data from an IoT-enabled device to flow from "low" side to "high" side in order to provide best security across the paradigm of the service [16] as discussed.

4.1 Digital Accessibility: Freedom

The cities taken in consideration, DEL and JAI have a pretty firm idea and realize the potential of a unified platform that not only allows virtually everyone that can contribute to its potential growth but also invite various multinational corporations to give their valuable contribution toward a platform that is vital for the growth of smart services which can be provided by the government and the third-party vendors. For developing a platform that supports all major third-party vendors' interfaces, I need open application programming interfaces (APIs). I studied the data provided by the Delhi and Rajasthan governments which uses a range of IoT-enabled devices and found out that Delhi uses more number of open APIs as Delhi itself uses more

Fig. 2 Division of smart city services in DEL and JAI

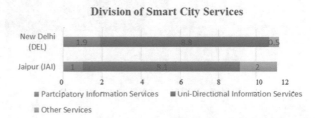

Division of Smart City Services

number of IoT devices ranging from PDAs to our mobile phones. These devices have the ability to connect to various existing legacy devices and with more devices using these open APIs, a larger number of devices can get connected and programmed to become an aspect of a smart service.

4.2 Solutions: People and Communities

Out of the smart services that both cities provide for the betterment of their infrastructure, Jaipur emerged as the city which took a larger number of initiatives with 63% in transportation system which includes apps that helps user to book their choice of flights, busses, trains, cabs, and shuttles. Nearly 59% in the travel and tourism services, websites provide tailor made packages for singleton, couple, and family trips for traveling locally, nationally, and internationally. This was also followed by cultural recreation at 52%. Internet applications help us to book different types of social and cultural events like plays, shows, music and food festivals, movies, and live performances to restaurant.

4.3 Collaborations: Management and Organization

Teamwork done with these corporations can help us to build new ways to reach to the public through the power of the Internet, or can help us in improving our own loopholes and areas that I was lacking in. In Fig. 3, I can clearly see that Delhi proves that it has successfully created strong and effective direct partnerships with private sectors with 24% and has changed the way it used to provide its smart services. Jaipur still leads in the outsourcing development and managerial processes with 92%.

Fig. 3 Comparison of partnerships in smart services

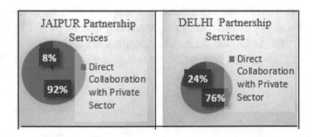

Fig. 4 Comparison of
impact of smart services on
the environment

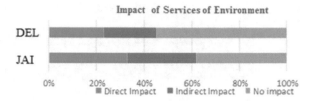

4.4 Modern Sustenance: Future

This sector aims completely takes aim at the needs of the hour but keeping in mind the needs of the future. If a service provided to the user consumes too much power or harms the environment or nature after a detailed audit, it should be terminated because our belief has always been to protect and rejuvenate the environment in any way I could. Optimizing our services in a way such that it uses the utmost of renewable resources and affect the environment in the most minimalistic way possible. I finally found out that even though Delhi uses more services than Jaipur, it also does it in a way that minimizes the affect and reduces the carbon footprint of that service on the environment and hence holding its pristine state. One of the ways to do this is to introduce the role of intelligence in our services using Internet-oriented services and inbuilt hardware services like GPS. JAI leads the way with 32% of the advanced intelligence-oriented services which would, say, consume less power when the service is left idle or provide us with better options which leads to a less environment disturbance as shown by Fig. 4.

4.5 Smart City Governance

As I have stated before, even after completing the tedious task of successfully upgrading a city into a smart city, if I fail to maintain and timely modify and update the services and the city itself, the aim creating of a smart city in the first place will be defeated. I can manually analyze the data provided by the IoT-enabled devices and update their services accordingly to best serve the needs of the customer as discussed in the sector of discussed in dedicated solutions. Another way to do this is to introduce the role of intelligence in our services using Internet-oriented services and inbuilt hardware services like GPS. JAI leads the way with 32% of the advanced intelligence-oriented services which would, say, consume less power when the service is left idle or to provide us with better options which leads to a less environment disturbance as shown in Fig. 5.

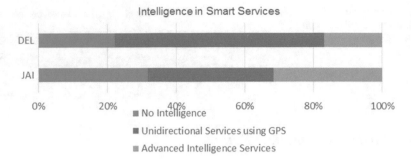

Fig. 5 Comparison of intelligence of smart services

Fig. 6 Comparison of integration of systems

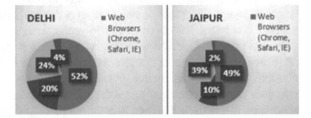

4.6 Systems Technology

All the devices that can access the services provided by the smart city like Del and JAI are called as IoT devices, which means that through the availability of Internet, a device can ubiquitously access the services from either government ex- registration for charity marathons, registrations for polio vaccination camps, or blood donation camps to private services like finding the best restaurant or movies. These services are now accessible over multiple device platforms like Apple's iOS, Google's Android, Microsoft's Windows Phone, etc. in forms of apps and web apps. Devices may include smart TVs, smart glasses, smart phones, laptops, and tablets etc. Portable devices can virtually access these services from anywhere. No surprises that DEL, the capital of the country, leads JAI on both the number of IoT devices and the services used on those devices. Reason is transparently simple, population of DEL is far more than JAI and because of the city being the political capital, and it has always been a trade hub for various multinational tech companies and has helped them to set their roots in as shown in Fig. 6.

The leading single device platform on portable devices in both the cities is Android with 24% in DEL and a staggering 39% in JAI, cutting close to the dominant field traditional computer's web browsers which are running Windows, Macintosh, and Linux on them. This makes our fundamental clear that people from both the cities are already using and benefitting from smart services provided by their governments and other private companies. This has proved to be useful for all the citizens and the government, as open data is no longer a dream in these cities and hence, smart services have become a successful reality.

5   Conclusions and Future Work

After conducting an extensive study on the role of IoT in the development of a smart city, the result was somewhat predictable and expected. It is imminent with respect to today's technology to use the services of IoT-enabled devices to collect the data, analyze it, and draw results to perform a successful modification of the services that end clients are using. Other recommended topics for future work comprises of the questions like how smart cities and its services can be developed and analyzed through this framework. How other smart city initiatives and local small entrepreneurs with their companies and startups can reinforce innovation in this traditional and orthodox system. Studies should also begin to explore other scales of measurements for every single one of the six parameters and define evolutionary versions of these parameters after they validated as profitable to both the governments and the citizens of that city. The framework provided can be further confirmed by applying the same on an existing framework of a potential smart city. The framework, upon its completion, must ensure a cleaner, less cost-oriented and greener method to build and maintain smart city services.

References

1. Lee, J.H., Hancock, M.G., Hu, M.C.: Towards an effective framework for building smart cities: lessons from Seoul and San Francisco. Technol. Forecast. Soc. Change. Elsevier Journal, TFS-17832 (2013)
2. van Ark, B., Broersma, L., den Hertog, P.: Service Innovation, Performance and Policy: A Review. Research Series, vol. 6. Ministry of Economic Affairs, The Hague (2003)
3. Bakıcı, T., Almirall, E., Wareham, J.: A smart city initiative: the case of Barcelona. J. Knowl. Econ. (2012). http://dx.doi.org/10.1007/s13132012-0084-9. Available at http://www.springerlink.com/content/9318pq8q61r06345/
4. Duggar, J.: The core of smart city must be smart governance. Forrester Research Inc, Cambridge, MA (2011)
5. Kapoor, E., Kemerer, C.F.: Network externalities in microcomputer software: an econometric analysis of the spreadsheet market. Manage. Sci. **42**(12), 1627–1647 (1996)
6. Mitra, A., Mehta, B.: Cities as the engine of growth: evidence from India. J. Urban Plann. Dev. **137**(2), 171–183 (2011)
7. Downs, A.: Smart growth: why we discuss it more than we do it. J. Am. Plann. Assoc. **71**(4), 367–378 (2005)
8. S. S. Division: US Census Bureau The 2012 Statistical Abstract: Historical Statistics. [Online]. Available: http://www.census.gov/compendia/statab/hist_stats.html. Accessed: 27 May 2012
9. United Nations Department of Economics and Social Affairs, Population Division, Population Estimates and Projections Section: UN: World Urbanization Prospects, 2011 Revision. United Nations Department of Economics and Social Affairs, 25 Apr 2012. [Online]. Available: http://esa.un.org/unpd/wup/Country-Profiles/country-profiles_1.htm. Accessed: 27 May 2012

10. Goodsell, C.T.: Reinvent government or rediscover it? Public Adm. Rev. **53**(1), 85–87 (1993)
11. Reinventing the Business of Government: An Interview with Change Catalyst David Osborne. Harvard Business Review. [Online]. Available: http://hbr.org/1994/05/reinventing-the-business-ofgovernment-an-interview-with-change-catalyst-david-osborne/ar/pr. Accessed: 28 May 2012
12. Allwinkle, S., Cruickshank, P.: Creating smart-er cities: an overview. J. Urban Technol. **18**(2), 1–16 (2011)
13. Leydesdorff, L., Deakin, M.: The triple-helix model of smart cities: a neo-evolutionary perspective. J. Urban Technol. **18**(2), 53–63 (2011)
14. Deakin, M., Lombardi, P., Cooper, I.: The Intel Cities community of practice: the capacity building, co-design, evaluation, and monitoring of e-government services. J. Urban Technol. **18**(2), 17–38 (2011)
15. Nam, T., Pardo, T.A.: Conceptualizing smart city with dimensions of technology, people, and institutions. In: Proceedings of the 12th Annual International Digital Government Research Conference: Digital Government Innovation in Challenging Times, pp. 282–291 (2011)
16. Putnam, R.D.: Bowling alone: America's declining social capital. In: The City Reader, pp. 120–128 (1995)
17. Winters, J.V.: Why are smart cities growing? Who moves and who stays*. J. Reg. Sci. **51**(2), 253–270 (2011)
18. Ratti, C.: The social nexus. Sci. Am. **305**(3), 42–48 (2011)
19. Cruickshank, P.: SCRAN: the network. J. Urban Technol. **18**(2), 83–97 (2011)

Global Terrorism Predictive—Analysis

Sandeep Chaurasia, Vinayak Warikoo and Shanu Khan

Abstract In the recent view of increasing number and lethality of terrorist attacks, it has become important for us to recognize a strategic vision to help prepare and prevent such events from happening. This paper includes descriptive and predictive analyses of Global Terrorism Database which reveal vital information about the trends of such events and help identify the perpetrators of any such future terrorist activities. The descriptive phase covers elucidation of the dataset to identify useful features for forecasting and predictive phase involves data manipulation and compares the performance of various multi-class classification and regression algorithms like decision trees, random forest, etc., on the dataset. Python with Scikit-learn library was used for the experimentation purpose.

Keywords Global terrorism database · Descriptive analysis · Predictive analysis · Classification and regression

1 Introduction

In the recent scenario there is an increase in the number of terrorist activity and lethality of terrorist attacks, it has become important for us as citizens to recognize a strategic vision, prepare, and if possible prevent such events. Each step taken into this direction enables us to tackle such adverse situations better and bring justice.

Golabal Terrorism Database [GTD] is an open—source database maintained by US Department of Homeland Security which includes all statistical information about the terrorist events around the world from 1970 to 2017. This database can be used to detect terrorist group and analyze frequency of the attacks [1, 2]. The purpose is to develop a supervised predictive model [3] around this database where it can be trained

S. Chaurasia (✉) · V. Warikoo
Manipal University Jaipur, Jaipur, India
e-mail: chaurasia.sandeep@gmail.com

S. Khan
Indian Institute of Technology Roorkee, Roorkee, India

© Springer Nature Singapore Pte Ltd. 2019   77
S. K. Bhatia et al. (eds.), *Advances in Computer Communication and Computational Sciences*, Advances in Intelligent Systems and Computing 924,
https://doi.org/10.1007/978-981-13-6861-5_7

to make predictions about future terrorist activities and provide vital information for counter-measures. The interest in predictive analytics for counter-terrorism agenda can be traced back to the reaction to the tragic events of 9/11 attacks in 2001 [4]. The attack itself was a concrete evidence of the fact that unlikely and unfortunate events with disastrous effect on society could happen and that pre-existing social structure system is unpredictable and complex.

Many data analytic organizations began mining and profiling the data but the sheer vastness of the uncategorized data has posed a challenge and most recently, a dependable source of database was released by the US Department of Homeland Security in partnership with University of Maryland, Global Terrorism Database [5].

2 Literature Review

2.1 Survey

The literature on models of predictive analysis used to identify or make predictions on the terrorism group is divided between technically practical literature and theoretically ideological literature.

The technical literature consists of massive number of reports directed toward intelligent service practice. This literature is pivotal to understand the possibilities extended by pattern-based data-mining and classification [6]. The theoretical literature shares similar trust in the technical possibilities but many practitioners discover that it is not possible to predict terrorism established on outliner algorithms as the noise is great for the recorded database and there are too many feature variables to classify or form any pattern.

This fact is just seen as another social difficulty that can be addressed technically by the new avenues presented by big data. It is widely known that prevention and prediction are now possible. Now that there are new methods of predictive analytics, society's prevailing profiling of terrorist groups and elimination of terrorist attacks can be more targeted, just and effective.

2.2 Previous Shortcomings

Earlier works in this field have led to development of models with moderate range of predictive accuracy [2] which means none of them are qualified to be deployed as a successful model for forecasting any group responsible for such events.

This can be traced back to methods adopted to develop such models which fail to address the use of outliner algorithms [7], as all types of descriptive data is potentially relevant, resulting in a huge quantity of variables making it improbable for the model to converge on one ideal fit model.

2.3 Distinctive Methodology

Running big stock of unconventional data can be time consuming and counterproductive for the desired result to be accomplished. The data need to be categorized in nature and superflous or erronous data need to be removed. Proper visualization of data is required to find out the key factors which can help train the model better and a distinctive categorization of the data is required to better train multiple models to help achieve high accuracy.

2.4 Result Significance

A predictive accuracy of anything above eighty-five percent would be a huge step of success in the direction of eliminating terrorism and deploying a real-time model which can be fed required metadata to forecast future terrorist activities.

3 Methodology

Main aim of this research is to infer from the Global Terrorism Database and obtain vital demographics of the data provided to develop a predict model which can learn and utilize the dataset fed to make forecast on identifying future terrorist groups and casualties when provided real-time data of the recent event.

To streamline and maintain the order of the work, the task at hand is divided into three phases.

3.1 Experimental Setup

The Global Terrorism Database tool database kit was acquired from their official Web site after verification which includes the dataset and a codebook guide for understanding the dataset.

Primarily Python 2.7 was used as a programming language for the development of models alongside several open-source libraries. Pandas library was used to import dataset into a data structure for the editing and modification of the dataset. Scikit-learn library was used for the machine learning aspect of the model development. Plotly, an open-source client-based framework was used to obtain all the desired graphs and plots.

3.2 Descriptive Analysis

This gives a clearer picture of what kind of database it is and summarizes it in a meaningful way. In this particular study, we have a large amount of data to simplify and understand. Each descriptive statistics reduces lot of data into a simple summary and breaks down hard to understand trends or important features [8]. Several graph plots were obtained identifying the most active terrorist groups, most incidents by a group, and methods of active terrorist groups.

3.3 Data Manipulation and Selection

Data manipulation and selection include all steps of data transformation, formatting, and structuring. Based on the criteria of selection of important features, tasks such as updating, adding/removing, sorting, selection, merging, shifting, and aggregation. of data were undertaken.

A total of one thirty-six features were reduced to nineteen features providing vital information about the terrorist event keeping the prediction model performance in mind. Doubtful attacks were removed and unknown or missing data were filled with median data of the same group to help better categorize the terrorist activity. Categorical data like attack type, weapon type, and target type were one hot encoded to be made suitable for the predictive classification and regression algorithms.

The data were divided into 12 regions geographically to maintain continuity and obtain specialized models. Top five terrorist groups were identified from every region and rest were labeled in "Others" category.

3.4 Predictive Analysis

The first step is to design a predictive model using supervised learning algorithms based on classifiers and regression. Then the dataset is divided at random into training and test set for the predictive model. The training set is utilized for the machine learning algorithms to train themselves for the features provided to identify the corresponding labels. The test set is used to score the accuracy of the trained model to forecast the label in regard to the feature information provided.

The predictive analysis is broken into constituents' models based on different machine learning predictive algorithms.

3.4.1 K-Nearest Neighbors

It is a simple machine learning algorithm which stores all normalized cases and classifies new cases based on measure of similarity [8]. The similarity is measured on basis of distance function (Euclidean) and number of neighbors specified. KNN is based on learning by analogy and known for simplicity and applicability supporting multiple data structures. A drawback of KNN classifiers is assigning equal weight to all the attributes which may cause irregular results when there are many irrelevant attributes in the data [9]. KNN also incurs expensive computational cost when a number of potential neighbors are great, therefore require efficient indexing technique.

3.4.2 Logistic Regression

It is a type of classification algorithm involving a linear discriminant, and it does not try to predict the value of a numerical variable given set of input data [10]. It instead outputs the probability that the given input point belongs to certain label. It is a machine learning algorithm used for predicting binary output. For our multi-label (multiple group name) data, N number of distinct logistic regression model are built to provide probabilities for the N number of groups [11].

3.4.3 Decision Trees

These are a non-parametric supervised learning structure used for classification. The objective is to develop a model that predicts the value of target variable by learning basic decision trends or rules derived from the data features. A decision tree is a decision support tool that uses a tree-like graph or model of decision and their possible consequences [10]. It assigns a class label to unseen record and explains why the decision was made in a classification rule [8]. Decision trees have some weakness when it comes to small number of instances for large variety of different classes as it brings out higher error rate in the classification.

3.4.4 Random Forest

It is an ensemble-based classifier which uses several decision tree models to predict the result. It is an estimator that fits a number of decision tree classifiers on various sub-samples of the dataset and uses averaging to improve the predictive accuracy [8] and control over-fitting.

3.4.5   XGBoost

It is an extreme gradient boosting approach to build new models which account for predicting residuals or errors of older models and then added together to make final prediction [12]. It is so named because it uses gradient descent algorithm to minimize the loss while adding new models.

4   Result

On the given dataset, five different classifiers were applied to obtain the results. Scikit python library was used with Anaconda IDE on i5 core processors. Thus, the following results are obtained for the mentioned regions (Table 1).

It is observed from the above table that among the five classifiers, random forest and the XG boost dominate over the other classifier. For American continent, random forest gave the better accuracy, whereas for the Asian continent, XG boost provides the better accuracy.

Refer Fig. 1 for the prediction accuracy; best accuracy was considered on Y-axis and twelve regions were considered on X-axis. It has been observed that for the later Asian and African countries obtained accuracy was much higher.

Table 1 Performance accuracy of all models regionwise on the test dataset

Regions	Model accuracy in (%)				
	KNN	Logistic regression	Decision tree	Random forest	XGBoost
North America	76.34	64.15	78.49	80.864	80.28
South America	73.90	49.10	78.19	79.18	76.63
Western Europe	75.36	72.06	80.40	81.52	81.41
Central America and Caribbean	82.05	80.46	87.57	89.06	90.39
Sub-Saharan Africa	75.97	56.74	93.5	94.76	94.77
Middle East and North Africa	79.87	47.65	89.37	90.58	90.98
East Asia	60	65	70	70	70
Eastern Europe	69.38	65.30	77.55	81.63	77.55
Australia and Oceania	28.57	64.28	85.71	57.14	85.71
South East Asia	71.74	66.70	72.30	78.81	80.27
South Asia	77.32	60.57	87.57	89.05	89.11
Central Asia	36.36	72.72	100	100	100

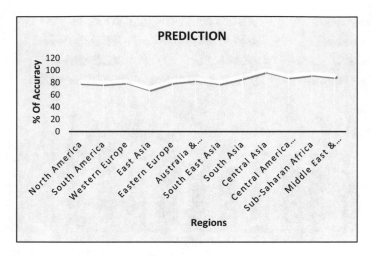

Fig. 1 Prediction on various regions

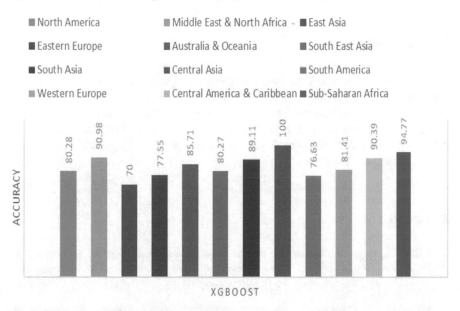

Fig. 2 XG boost classifier performance on regions

Based on the experimentations, it is observed that random forest classifier and XG boost classifier perform better. So, a comparative graph has been plotted as Figs. 2 and 3 for the mentioned regions.

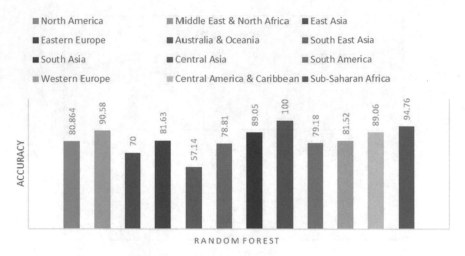

Fig. 3 Random forest classifier performance on regions

5 Conclusions

The selection of features from the data as well as the strategy to divide the dataset into different regions to improve the classification result has proven to be successful. The desired accuracy can be obtained by fine tuning the aforementioned models. search and study works need to be done in mathematical and statistical domain in order to better understand the intricacies of the machine learning models to improve the performance of the same.

References

1. LaFree, G., Dugan, L.: Introducing the global terrorism database. Polit. Violence Terrorism **19**, 181–204 (2007). Zarri, G.P.: Semantic web and knowledge representation. In: Proceedings of the 13th International Workshop on Database and Expert System Applications (DEXA'02), pp. 1529–4188 (2002)
2. LaFree, G.: The global terrorism database: accomplishments and challenges. Perspect. Terrorism **4**(1) (2010)
3. Al Hasan, M., Chaoji, V., Salem, S., Zaki, M.: Link prediction using supervised learning. In: SDM06: Workshop on Link Analysis, Counter Terrorism and Security, Rensselaer Polytechnic Institute, NY (2006)
4. Munk, T.B.: Why anti-terror algorithms don't work. First Monday **22**(9) (2017)
5. National Consortium for the Study of Terrorism and Responses to Terrorism (START): Global Terrorism Database [Data file] (2017). Retrieved from https://www.start.umd.edu/gtd
6. Taipale, K.A.: Data mining and domestic security: connecting the dots to make sense of data. Columbia Sci. Tech. Law Rev. **5**, 1–83 (2013)
7. Malathi, A., Dr. Santhosh Baboo, S.: Evolving data mining algorithms on the prevailing crime trend—an intelligent crime prediction model. Int. J. Sci. Eng. Res. **2**(6) (2016)

8. Hongbo, D.: Data Mining Techniques and Applications: An Introduction. Cengage Learning, Boston, (2010)
9. Xiao, X., Ding, H.: Enhancement of K-nearest neighbor algorithm based on weighted entropy of attribute value. In: 2012 5th International Conference on Bio-Medical Engineering and Informatics (BMEI) (2012)
10. Cerri, R., et al.: An extensive evaluation of decision tree-based hierarchical multilabel classification methods and performance measures. Comput. Intell. **31**(1), 1–46 (2015)
11. Logistic Regression Module. http://scikit-learn.org
12. Chen, T., Guestrin, C.: Xgboost: a scalable tree boosting system. In: Proceedings of the 22nd ACM SIGKDD International Conference on Knowledge Discovery and Data Mining. ACM (2016)

Part II
Advanced Communications

Effect of Constraint Variation on the Efficiency of Protocols Under Realistic Environment

Prem Chand Vashist, Ashish Tripathi and Pushpa Choudhary

Abstract The performance of mobile ad hoc networks hinges on various factors such as bandwidth, low-power/selfish nodes, transmission range, and environment of network deployment. This paper is devoted to learn the consequence of such parameters on network efficiency in idealistic and realistic (where a node may face obstacles in the network) situation. To observe these effects, a simulator in MATLAB has been developed which uses mentioned factors in deliberation. The reproduction results have shown notable effect on POR value, hops count, and total length of the path with respect to the divergence in the taken factors.

Keywords MANET · Hop count · Performance evaluation · Environmental conditions · Bandwidth · Residual power · Transmission range · Probability of reachability

1 Introduction

Mobile ad hoc networks are rapidly deployable without any dependency on external support. Due to the absence of outer infrastructure and decentralized network the routing [1–3], liabilities are the collective responsibility of all the devices/nodes contributing to the system. Several routing protocols have been established by keeping these characteristics in consideration. The efficiency of such protocols hinges upon different features such as transmission range, power of nodes, and bandwidth. Another feature which touches the efficiency [2] of MANET is the existence of

P. C. Vashist · A. Tripathi (✉) · P. Choudhary
Department of Information Technology, G. L. Bajaj Institute of Technology and Management,
Greater Noida, India
e-mail: ashish.mnnit44@gmail.com

P. C. Vashist
e-mail: pcvashist@gmail.com

P. Choudhary
e-mail: pushpak2728@gmail.com

© Springer Nature Singapore Pte Ltd. 2019
S. K. Bhatia et al. (eds.), *Advances in Computer Communication
and Computational Sciences*, Advances in Intelligent Systems and Computing 924,
https://doi.org/10.1007/978-981-13-6861-5_8

obstacles in the network or merely we can say that it is the situation of MANET deployment in realistic environment, as it may contain random obstacles like ponds, hilly regions, or areas without any hosts. Consequently, this situation will result in hindering or obstructing message relay. Many hindrances in the atmosphere hamper the nodes program and concrete broadcast routes among the nodes.

In this paper, we have considered the environmental conditions (idealistic and realistic) to estimate the recital of protocols used in routing along with three constraints [2–6] bandwidth, transmission range, and power. For this purpose, a square type of obstacle to create a real environment and a constantly growing proportion of low-bandwidth/transmission range/power nodes are inserted into the network by evolving an emulator in MATLAB. The emulator allocates nodes arbitrarily in the recreation area and using Dijkstra's shortest path routing strategy, it selects all available pairs of source and destinations to establish route between them.

The paper has been prepared in the following given divisions: Sect. 2 gives the allied efforts. System factors and performance metric taken are mentioned in Sect. 3. Section 4 gives result and its analysis. Interpretation from result and conclusion has been specified in Sects. 5 and 6, respectively.

2 Allied Efforts

The allied efforts taken by different researchers have been discussed in this section; some papers that deliberate the influence of hurdles and effect of selfish nodes in the network are mentioned here.

The study is executed on realistic reproduction area by Chang et al. [7] enlighten the existence of hindrances, such as highlands, ponds, structures, or areas without any hosts in the region, generally causes delay or hindrance in message relay. This problem was solved by geo-casting protocol.

Various techniques have been applied by Koshti et al. [8] to identify selfish nodes in MANET.

Chand et al. [9] have proposed a routing strategy that includes residual battery power of nodes as an important parameter of interest to manage the nodes.

In paper of Ahmed et al. [10], an arrangement for finding selfish nodes is given.

A new routing protocol named optimized link state routing (OLSR) has proposed by Thiagarajan et al. [11] which applies periodic route metric technique to reduce the control overhead for route maintenance and also the contention problem.

Rajesh et al. [12] have proposed a novel cluster-based algorithm with minimized energy consumption for the evaluation of the performance of routing algorithms in highly dense MANET architecture.

A security enhanced zone routing protocol (SEZRP) has been proposed by Rajput et al. [13] that provide security to the routing packets in zone routing protocol (ZRP) and prevent the attacks that most frequently occurred in an efficient manner. They have used MAC and key pre-distribution technique to maintain the confidentiality of the message and to reduce the overhead at run time.

Arya et al. [14] have introduced a new version of AODV. To provide security for the routing packet in AODV, an authentication technique has been used to prevent different attacks like black hole attacks, attacks related to changes in routing information, etc. They have used the key pre-distribution technique to minimize the overhead due to key distribution at execution time.

Sharma et al. [15] have discussed the problem of congestion control in MANET. They have discussed different congestion control techniques in their paper.

Probability-based Random Early Detection (P-RED) has been proposed by Sharma et al. [16]. In this paper, they have proposed Probability-based Random Early Detection (P-RED). P-RED is a modified version of Random Early Detection (RED) to enhance the fairness in allocation of gateway.

The study discussed in this section gives the consequence of limitations and plans to survive in idealistic conditions, and however, no one has designed the usefulness of the routing protocols with different restraints in realistic environment, i.e., in occurrence of hindrances. Thus, a simulator using MATLAB with capability of providing idealistic and realistic environment [17–20] and facility to insert controlled fraction of selfish nodes in the network. The numerous concert aspects their reproduction results and inspection are deliberated in the coming sections.

3 System Factors and Performance Metric Taken

The basis of selecting these factors is purely based upon realistic conditions, as we must have an area where experimentation is to be done, and similarly, on the same basis mobility model, number of iterations, packet size, etc. are being chosen.

This section describes brief information of experimental work performed to achieve the desired results, and the setup parameters and performance metric [19–22] are mentioned in below subsections.

3.1 System Factors

Various factors to set up the system and simulation process have been taken which are mentioned below; similarly, Table 1 gives the explanation of numerous factors used in the reproduction progression.

   i. **Area**: The amount of unit is squared into which the testing has taken place.
  ii. **Mobility model**: the arrangement of node movement.
 iii. **Count of nodes**: The sum of overall nodes used in experimentation.
  iv. **Number of iterations**: It is the total count of attempts used in simulation process.
   v. **Number of packets used**: 40.
  vi. **Packet Size**: It gives length of packet in terms of bytes.

Table 1 System factors

System factors	Values
Area	2250,000 m^2
Count of nodes	40
Tr. range	300 m
Mobility model	Random walk
Node movement rate	2.5 m/s
Number of repetitions	25
Shape of obstacle	Square
Packet time	1 s
Packet magnitude	512 bytes
Count of packet sent per second	40
Fraction of nodes having any one of the constraints taken for investigation	It starts from 0%, 10%, and goes up to 100%

vii. **Transmission range**: the energy requisite for accurate functioning of neighboring nodes at some defined distance.

viii. **Node movement**: the rate of displacement of nodes.

ix. **Fraction of low-bandwidth nodes**: It is the number of nodes having any one of the mentioned constraint taken for experimentation for this paper.

x. **Shape of obstacle**: In case of network deployment in realistic environment, various obstacles in environment restrict the nodes movement and operational communication tracks among nodes. To feel this experience, a square type of obstacle in the network has been developed.

3.2 Performance Metrics

Different factors taken for assessing the routing protocol performance are mentioned below.

i. **Hop count**: The count of midway nodes exists from starting point to ending point.

ii. **Reachability**: The fraction of total packets effectively acknowledged by the ending point to the packets directed by starting point.

iii. **Path length**: the precise count of nodes from starting point to ending point.

4 Result and Analysis

The explanation of outcomes of emulator and graphical examination of each category has been given in this section.

4.1 Low Bandwidth and Hop Count

It is shown in Fig. 1 that with the gentle increase in the fraction of low-bandwidth [21–25] nodes started at 0% till 100%, the count of hops declines. Thus, paths which need additional nodes are not developing in the network. Basically, lengthier paths were not being shaped on the evolution of nodes having low bandwidth.

4.2 Reachability and Low Bandwidth

The gentle increase in the fraction of low-bandwidth [19–24] nodes started at 0% till 100%, with respect to the growth in the fraction of nodes with low bandwidth, the probability of reachability has started to decrease, and it extends near to zero with 100% low-bandwidth nodes. The graphical analysis is shown in Fig. 2.

4.3 Low Bandwidth and Length of the Path

The precise count of nodes from starting point to ending point is called as path length. Nodes having low bandwidth also distress the length of the path [25–28] of a definite route. Effect and analysis of such nodes are shown in Fig. 3.

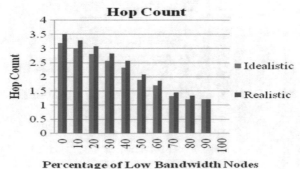

Fig. 1 Hop count in the presence of low bandwidth

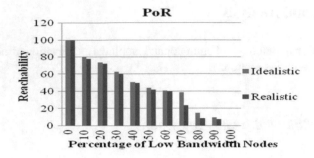

Fig. 2 Effect measurement on reachability in the presence of low bandwidth

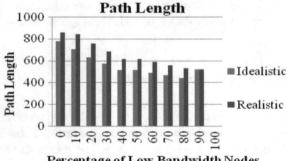

Fig. 3 Effect of nodes with low bandwidth on path length

4.4 Transmission Range and Hop Count

This factor [28–30] affects the hop count in reverse order, i.e., increment in former factor decreases later. Similarly, graphical analysis for the same is shown in Fig. 4.

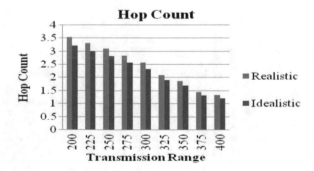

Fig. 4 Effect on hop count with low transmission range

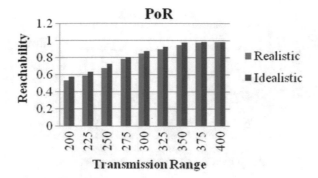

Fig. 5 Effect of low transmission range on reachability

4.5 Transmission Range and Reachability

In network, the POR [29–31] is directly proportional to the transmission range of nodes. It is shown in Fig. 5, as we increase value of POR, the transmission range also increases.

4.6 Transmission Range and Path Length

Path length of a route from home to endpoint is affected by transmission range of a network [24, 25]. This effect is presented in Fig. 6.

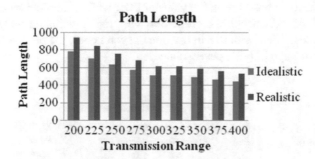

Fig. 6 Low transmission range effect on path length

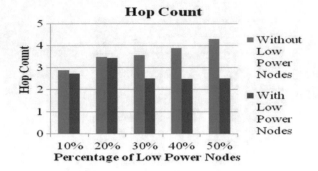

Fig. 7 Hop count in the presence of low-power nodes

4.7 Hop Count with Low-Power Nodes

As anticipated, the number of hop count declines with growth in deliberation of nodes with low power [21–25]. The graphical analysis and demography of both types of nodes are shown in Fig. 7. Thus, with rise in fraction of nodes with low power, the route with large quantity of midway node becomes unreliable.

4.8 Effect of Power on Reachability

Here, value of reachability in appearance and nonappearance of nodes with low power is shown in Fig. 8. Similarly, mentioned below is the gist of graphical analysis.

i.   Less disparity is observed in the value of POR while using idle type nodes.
ii. The value of POR declines considerably with rise in deliberation level of low-power nodes. Thus, with rise in fraction of such nodes, the probabilities of data packets drop-age [29–31] get escalated.

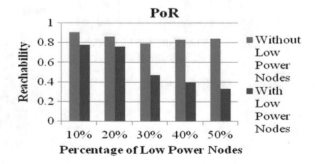

Fig. 8 Impact of nodes with low power on reachability

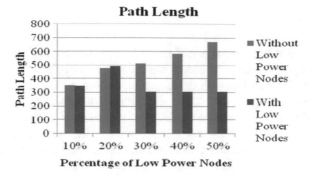

Fig. 9 Low-power nodes effect on path length

4.9   Nodes with Low Power and Path Length

The graphical assessment of path length in the presence and absence of low-power nodes is given in Fig. 9. Consequently, it is observed that path length declines [25–29] with the evolution of low-power nodes.

5   Interpretation

Mentioned below in Table 2 is the interpretation of three performance factors, in correspondence with bandwidth transmission and power restraint. The thorough examination completes to below-mentioned points.

Table 2 Inference from result and analysis

	With bandwidth constraint	With transmission range constraint	With power constraint
Hop count	Number of hop count declines with the increase in low-bandwidth nodes	The count of hopes declines as we raise transmission range	Hop count declines with rise in the fraction of nodes with low power
POR	The value of POR decreases with increase in low-bandwidth nodes	The transmission range and POR are directly proportional to each other. Thus, one factor rises with the rise in another	The value of POR decreases with increase in low-power nodes
Path length	Due to less number of intermediate nodes, path length also going to decrease	With the rise in transmission range route/path length declines	Due to less number of intermediate nodes, path length also going to decrease

6   Conclusion

This paper includes the environmental conditions to assess the efficiency of routing protocols along with three constraints bandwidth, transmission range and power. Simulation result shows that with the growth of fraction of low transmission range, low-power and low-bandwidth nodes, the number of hop count and path length declines. Similarly, the value of probability of reachability (POR) also depreciates considerably with the rise in fraction of such nodes in the system. Thus, it has been concluded that a routing protocol under realistic environment has to perform on many fronts along with power, transmission, and bandwidth constraint.

References

1. Iwata, A., Chiang, C.C., Pei, G., Gerla, M., Chen, T.W.: Scalable routing strategies for ad hoc wireless networks. IEEE J. Sel. Areas Commun. **17**(8), 1369–1379 (1999)
2. Abusalah, L., Khokhar, A.A., Guizani, M.: A survey of secure mobile ad hoc routing protocols. IEEE Commun. Surv. Tutor. **10**(1–4), 78–93 (2008)
3. Patel, N.R., Kumar, S.: Energy conscious DSR in MANET. In: 2012 2nd IEEE International Conference on Parallel Distributed and Grid Computing (PDGC), 2012 Dec 6, pp. 784–789. IEEE (2012)
4. Ovalle-Martínez, F.J., Stojmenović, I., García-Nocetti, F., Solano-González, J.: Finding minimum transmission radii for preserving connectivity and constructing minimal spanning trees in ad hoc and sensor networks. J. Parallel Distributed Comput. **65**(2), 132–141 (2005)
5. Ingelrest, F., Simpol-Ryl, D., Stojjmenovic, I.: Optimal transmission radius for energy efficient broadcasting protocol in ad-hoc and sensor networks. IEEE Trans. Parallel Distributed Syst. **17**(6) (2006)
6. Samuel, H., Zhuang, W., Preiss, B.: DTN based dominating set routing for MANET in heterogeneous wireless networking. Mob. Netw. Appl. **14**(2), 154–164 (2009)
7. Chang, C.Y., Chang, C.T., Tu, S.C.: Obstacle-free geo casting protocols for single/multidestination short message services in ad hoc networks. Wirel. Netw. **9**(2), 143–155 (2003)
8. Koshti, D., Kamoji, S.: Comparative study of techniques used for detection of selfish nodes in mobile ad hoc networks. Int. J. Soft Comput. Eng. (IJSCE) ISSN, 2231–2307 (2011)
9. Chand, P., Soni, M.K.: A novel routing scheme for mobile ad hoc network. Int. J. Comput. Netw. Inf. Secur. **5**(4), 17 (2013)
10. Ahmed, A., Bakar, K.A., Channa, M.I., Haseeb, K., Khan, A.W.: TERP: a trust and energy aware routing protocol for wireless sensor network. IEEE Sens. J. **15**(12), 6962–6972 (2015)
11. Thiagarajan, R., Moorthi, M.: Efficient routing protocols for mobile ad hoc network. In: 2017 Third International Conference on Advances in Electrical, Electronics, Information, Communication and Bio-Informatics (AEEICB), 2017 Feb 27, pp. 427–431. IEEE (2017)
12. Rajesh, M.V., Gireendranath, T.V., Murthy, J.V.: A novel energy efficient cluster based routing protocol for highly dense MANET architecture. Int. J. Comput. Intell. Res. **13**(5), 719–744 (2017)
13. Rajput, S.S., Trivedi, M.C.: Securing zone routing protocol in MANET using authentication technique. In: 2014 International Conference on Computational Intelligence and Communication Networks (CICN), 2014 Nov 14, pp. 872–877. IEEE (2014)
14. Arya, K.V., Rajput, S.S.: Securing AODV routing protocol in MANET using NMAC with HBKS technique. In: IEEE International Conference on SPIN, 2014 Feb 20, pp. 281–285 (2014)

15. Sharma, N., Gupta, A., Rajput, S.S., Yadav, V.K.: Congestion control techniques in MANET: a survey. In: 2016 Second International Conference on Computational Intelligence & Communication Technology (CICT), 2016 Feb 12, pp. 280–282. IEEE (2016)
16. Sharma, N., Rajput, S.S., Dwivedi, A.K., Shrimali, M.: P-RED: probability based random early detection algorithm for queue management in MANET. In: Advances in Computer and Computational Sciences, pp. 637–643. Springer, Singapore (2018)
17. Chen, T.W., Gerla, M.: Global state routing: a new routing scheme for ad-hoc wireless networks. In: 1998 IEEE International Conference on Communications, ICC 98, Conference Record, 1998 June 7, vol. 1, pp. 171–175. IEEE (1998)
18. Jardosh, A., Belding-Royer, E.M., Almeroth, K.C., Suri, S.: Towards realistic mobility models for mobile ad hoc networks. In: Proceedings of the 9th Annual International Conference on Mobile Computing and Networking, 2003 Sept 14, pp. 217–229. ACM (2003)
19. Sun, B., Gui, C., Liu, P.: Energy entropy multipath routing optimization algorithm in MANET based on GA. In: 2010 IEEE Fifth International Conference on Bio-Inspired Computing: Theories and Applications (BIC-TA), 2010 Sept 23, pp. 943–947. IEEE (2010)
20. Divecha, B., Abraham, A., Grosan, C., Sanyal, S.: Impact of node mobility on MANET routing protocols models. JDIM 5(1), 19–23 (2007)
21. Varaprasad, G., Wahidabanu, R.S.: New power-aware multicast algorithm for mobile ad hoc networks. IEEE Potentials 32(2), 32–35 (2013)
22. Sengul, C., Kravets, R.: Conserving energy with on-demand topology management. In: 2005 IEEE International Conference on Mobile Adhoc and Sensor Systems Conference, 2005 Nov 7, p. 10. IEEE (2005)
23. Sheriff, I., Belding-Royer, E.: Multipath selection in multi-radio mesh networks. In: 3rd International Conference on Broadband Communications, Networks and Systems, 2006, BROADNETS 2006, 2006 Oct 1, pp. 1–11. IEEE (2006)
24. Dorsey, J., Siewiorek, D.: 802.11 Power Management Extensions to Monarch ns. Carnegie-Mellon Univ Pittsburgh PA School of Computer Science, 2004 Dec
25. Huang, T.C., Chung, W.J., Huang, C.C.: A revised AODV protocol with energy management for real-time/non-real-time services in mobile ad hoc network. In: 10th IEEE International Conference on High Performance Computing and Communications, HPCC'08, 2008 Sept 25, pp. 440–446. IEEE (2008)
26. Chen, T., Han, T., Katoen, J.P., Mereacre, A.: Reachability probabilities in Markovian timed automata. In: 2011 50th IEEE Conference on Decision and Control and European Control Conference (CDC-ECC), 2011 Dec 12, pp. 7075–7080. IEEE (2011)
27. Demir, C., Comaniciu, C.: An auction based AODV protocol for mobile ad hoc networks with selfish nodes. In: IEEE International Conference on Communications, ICC'07, 2007 June 24, pp. 3351–3356. IEEE (2007)
28. Chen, L., Heinzelman, W.B.: QoS-aware routing based on bandwidth estimation for mobile ad hoc networks. IEEE J. Sel. Areas Commun. 23(3), 561–572 (2005)
29. Viswanath, K., Obraczka, K., Tsudik, G.: Exploring mesh and tree-based multicast. Routing protocols for MANETs. IEEE Trans. Mob. Comput. 5(1), 28–42 (2006)
30. Sengul, C., Kravets, R.: Conserving energy with on-demand topology management. In: IEEE International Conference on Mobile Adhoc and Sensor Systems Conference, 2005 Nov 7, pp. 10. IEEE (2005)
31. Lin, C.R., Wang, G.Y.: A duplicate address resolution protocol in mobile ad hoc networks. J. Commun. Netw. 7(4), 525–536 (2005)

Performance Evaluation of OLSR-MD Routing Protocol for MANETS

Rachna Jain and Indu Kashyap

Abstract Mobile Adhoc Networks (MANETS) are gaining popularity because of interconnected networks. Routing is a key issue which needs to be addressed for efficient forwarding of packets from source to destination. OLSR (Optimized Link State Routing) is a proactive or table driven routing protocol in MANETS which works on the principal of link sensing. These mobile nodes are battery operated and due to limited battery early death of node results in network partitioning. Energy becomes an important parameter to be conserved so design of energy efficient routing protocol is the need of hour for extending network lifetime. In this work, OLSR routing protocol is implemented according to modified Dijakstra's algorithm which works on residual energy of nodes. Proposed energy aware OLSR-MD is compared with conventional OLSR. Extensive simulations were performed using NS-2 Simulator, and simulation results show improved network parameters such as higher throughput, more PDR (Packet Delivery Ratio) and lesser End to End delay.

Keywords OLSR · MANETS · PDR · Throughput

1 Introduction

Wireless is the technology which helps people to access data from any corner of the world. These wireless networks are broadly divided into two categories. Infrastructure-based networks communicate with the help of access point, whereas infrastructure less networks or Adhoc Networks communicate directly without any fixed infrastructure [1]. Mobile Adhoc Networks (MANETS) gain ample interest due to their wide range of applications in military areas, search and rescue operations, habitat monitoring, etc. These networks consists of various mobile nodes communicating through wireless medium with limited resources such as battery power, bandwidth, memory, etc. and without any fixed backbone structure [2, 3].

R. Jain (✉) · I. Kashyap
Manav Rachna International Institute of Research and Studies, Faridabad, Haryana, India
e-mail: Rachna_19802000@yahoo.com

© Springer Nature Singapore Pte Ltd. 2019
S. K. Bhatia et al. (eds.), *Advances in Computer Communication and Computational Sciences*, Advances in Intelligent Systems and Computing 924, https://doi.org/10.1007/978-981-13-6861-5_9

Router is served through base station to route the information in case of MANETS. Nodes which are in communication range of each other can communicate directly. Optimized path has to be found in order to support time constrained communication such as video conferencing, voice, etc. Multipath routing protocols are proposed in order to obtain node disjoint and link disjoint path using modified Dijakstra's algorithm. Cost functions are defined to obtain node disjoint and link disjoint multiple paths. But performance of this protocol is reduced drastically if any of the nodes is having low energy levels. Shortest path is always not the best choice, and sometimes, there is a trade-off between different QoS (Quality of Service) parameters. Energy efficiency of links is an important parameter to be studied for enhancing network life time.

Section 2 discusses related work, and Sect. 3 discusses proposed work. Section 4 shows the network scenario during simulation as well as simulation results of performance parameters such as throughput, end to end delay, packet delivery ratio (PDR). Sect. 5 comprises conclusion of the work.

2 Related Work

Jabbar et al. [4] in 2014 has proposed multipath battery aware OLSR (MBA-OLSR) which is based on OLSRv2 and its multipath links between source and destination. In this work, author has considered remaining battery power of nodes to calculate initial cost of links between sender and destination. Simulations are done using EXata Simulator, and results show better network lifetime. Same author in 2015 [5] has suggested queue aware multicriteria-based OLSR in multi hop networks. Again in 2017, same author [6, 7] has suggested energy and mobility aware OLSR (EMA-OLSR). Since mobility of nodes plays an important role, so he has suggested MCNR (Multi Criteria Node Rank) metric to compute rank of node which considers willingness of node to become MPR (Multi Point Relay) in OLSR.

Natarajan and Rajendran in 2014 [8] proposed AOLSR which is hybrid advanced OLSR by modifying Dijakstra algorithm to compute multiple paths in both dense and sparse networks from source to destination. Whereas Dhanalakshmi in 2015 [9] proposed Energy Conserving Advanced Optimised Link State Routing (ECAO) model by calculating energy costs of all nodes and compared results with OLSR and AOLSR. Author has also implemented modified Dijakstra's algorithm to compute multipath and also checked for link failure for further analysis.

Moussaoui [10] devised stability of nodes (SND) and fidelity of nodes (FND) parameters to elect MPR (Multi Point Relays) in case of OLSR. The author has proved via simulation that his selected method provides better QOS (Quality of Service) parameters than traditional method in OLSR for MPR selection which considers Expected Transmission Time metric (ETX). The limitation of using ETX is overestimating link delivery ratio when packet size is too large.

Gerharz et al. [11] discussed method to select link with high probability in different scenarios under different mobility conditions. He devised that random way point and Guass Markov scenarios give similar results. In random way point model, large simulation area results in links of longer duration. The author has concluded that stability of link depends on the age of link. Then, in 2003 [12], same author demonstrated statistical methods to compute path stability.

Singal [13] has introduced multi-objective function to improve OLSR routing protocol. He has considered three routing objective, i.e., to minimize end to end delay by **applying local queuing delay** model which is affected by congestion. Another routing objective includes to minimize residual battery energy in order to enhance network energy lifetime and to maximize PDR (Packet Delivery Ratio). Survey paper by Moussaoui [14] has discussed all the techniques to discuss link stable routing protocol in MANETS.

Wu et al. [15] gave different analytical models for investigating effects of node movement on wireless links. He has used two state Markov-model for finding the link life which is a strong tool to study link dynamics. But limitation of this approach is that it considered only link breakage which is not successful for OLSR where link breakage as well as formation of link both can initiate any event. CCDF (complimentary cumulative distribution function) of collision at MAC layer is shown for different scenarios. Wu [16] suggested distributed Q learning to consider both link stability and bandwidth efficiency to better version of AODV protocol.

So authors proposed different metrics for link selection such as selecting the oldest link, selecting the youngest link, selecting the link with maximum expected residual lifetime, selecting the link with maximum 'persistent probability' or selecting the link with lowest failure probability.

The author further extended to the concept of **Path Stability**. It has been emphasized on avoiding instable links or minimizing the use of instable links. But there is reduced capacity due to increase in path length. So, stability of link can be estimated by following methods: received signal strength measurement [17], pilot signals as used in ABR (associativity-based routing) [18], relative speed between two nodes forming the link [19], link duration distribution or link lifetime [9], remaining battery power of nodes [4–7].

3 Proposed Work

In multipath battery aware MBA-OLSR [4] link cost function is defined on residual energy of battery as:

$$Eij = \left(\frac{Ei}{Ei + MB} + \frac{Ej}{Ej + MB} \right) \tag{1}$$

where MB stands for maximum battery capacity, Ei and Ej are residual energies of node i, and node j. Eij stands for ratio of residual battery energies of node i and node j.

$$L_{Cij} = 1 + k * \left(1/e^{Eij}\right) \tag{2}$$

where L_{Cij} = Quality of Link
k = Stability Factor

Dijakstra's Algorithm [20] is most famous algorithm to calculate shortest distance between source node to destination node. Generally shortest path is found from source node to all other nodes of the network creating shortest path tree. Initial cost of link is calculated considering residual energy of both the nodes. Link with highest residual battery energy will have minimum value of link cost as computed by Eq. (2). Status flag (s_flag) for all the nodes is checked. Initially s_flag of source node is set to zero. Multiple paths are found using modified Dijakstra's Algorithm, and s_flag is updated to 1. Path selection is carried out on the basis of link cost function which results in optimized path with highest value of residual energies of the nodes amongst obtained multiple paths.

Modified Dijakstra's Algorithm

Find Multiple paths P in graph G from source s to destination d, no. of paths, cost functionfC, group of arcs G_a, group of vertices Gv
MultiPath Dijakstra(s,d,G,P)
fC1 = fC//Initial Cost function
G1 = G//Initial Graph
For i = 1 to P do
Sourcetree(i) = Dijakstra(Gi,s)
P(i) = Getpath(Sourcetree(i),d)
For all arcs a or a^{-1} in Gi
fC_{i+1} = fp(fC_i (a))
For all vertices v in Gi
fC_{i+1} = fe(fC_i (a))
else
fC_{i+1} = fC_i (a)
end if
end for
return (P_1,P_2,....P_N)

4 Experimental Analysis

See Fig. 1.

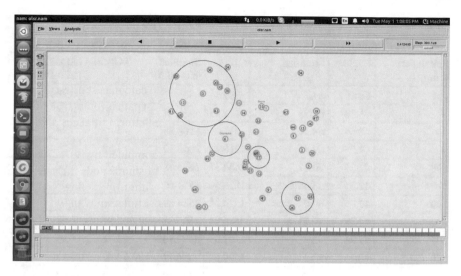

Fig. 1 Network scenario during simulation

5 Simulation Scenario and Results

In this work, network parameters are listed in Table 1, whereas Fig. 1 shows network scenario during the simulation.

Table 1 Network parameters

Network parameters	Values
Channel type	Wireless channel
Radio propagation model	Two ray ground
Mobility model	Random way point
MAC type	IEEE 802.11 a
Interface queue type	Drop tail
Antenna model	Omni directional
Agent	UDP
Application	CBR
Topology	Dynamic
Speed of nodes	0.01 m/s
Maximum packet in ifq	50
Number of nodes	20
Simulation area (X)	870
Simulation area (Y)	870
Simulation time	20, 40, 60, …400 s
Initial energy	100 J per node

Table 2 Comparison of different QoS parameters for OLSR and modified MD-OLSR

Packet delivery ratio (%)			Throughput (Kbps)		End to end delay (s)	
Simulation time (s)	OLSR	Modified MD-OLSR	OLSR	Modified MD-OLSR	OLSR	Modified MD-OLSR
20	51.88	57.26	379.64	426.12	0.95	0.89
40	41.32	65.75	411.45	955.48	0.90	0.85
60	42.65	68.56	812.49	1508.86	1.82	0.99
80	44.35	69.85	1216.26	2063.19	1.73	1.00
100	42.35	70.37	1430.37	2604.48	1.98	1.02
120	42.57	68.66	1529.64	2554.40	1.95	1.04
140	42.95	69.67	1529.64	2554.89	1.96	1.01
180	44.57	70.21	1540.86	2665.81	1.96	1.01
200	47.24	72.21	1501.83	2777.70	1.18	0.98
260	48.24	73.31	1529.64	2554.79	1.97	1.11
320	49.31	73.39	1649.64	2554.89	1.98	1.08
360	49.46	74.84	1629.51	2854.47	1.88	1.06
400	49.65	79.85	2659.43	3699.39	1.84	1.01

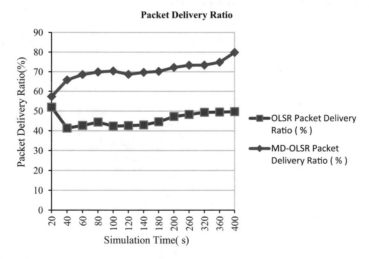

Fig. 2 Packet delivery ratio

Table 2 indicates that there is improvement in different QoS parameters such as higher values of PDR and throughput, whereas end to end delay is reduced considerably.

Figures 2 and 3 show that there is considerable improvement in throughput as well as PDR, whereas Fig. 4 clearly shows there is reduction in end to end delay in case of proposed MD-OLSR in comparison with conventional OLSR.

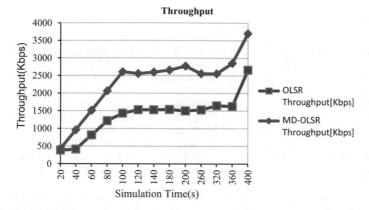

Fig. 3 Throughput of the network

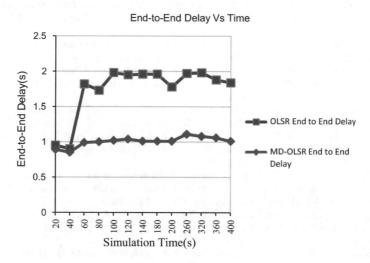

Fig. 4 End to end delay

6 Conclusion and Future Scope

In this work, network performance parameters such as throughput, PDR, and end to end delay of proposed MD-OLSR have been found better than conventional OLSR. This work can be further extended by considering different network speed scenarios as well as varying pause time of nodes. MPR (multipoint relay) node is mainly responsible for forwarding of control packets in OLSR, this routing protocol can be further improvised by selecting energy efficient MPR node.

References

1. Conti, M., Giordano, S.: Multihop ad hoc networking: the evolutionary path. Mobile Ad Hoc Networking: Cutting Edge Directions, Second Edition **11**, 1–33 (2013)
2. Chlamtac, I., Conti, M., Liu, J.J.N.: Mobile ad hoc networking: imperatives and challenges. Ad Hoc Netw. **1**(1), 13–64 (2003)
3. Clausen, T., Dearlove, C., Jacquet, P, Herberg, U.: The optimized link state routing protocol version 2, Internet Engineering Task Force (IETF), No. RFC 7181 (Standards Track) (2014)
4. Jabbar, W.A., Ismail, M., Nordin, R.: Performance evaluation of MBA-OLSR routing protocol for MANETs. *Journal of computer networks and communications* (2014)
5. Jabbar, W.A., Ismail, M., Nordin, R.: Multi-criteria based multipath OLSR for battery and queue-aware routing in multi-hop ad hoc wireless networks. Wirel. Netw. **21**(4), 1309–1326 (2015)
6. Jabbar, W.A., Ismail, M., Nordin, R., Ramli, R.M.: EMA-MPR: energy and mobility-aware multi-point relay selection mechanism for multipath OLSRv2. In:IEEE 13th Malaysia International Conference on Communications, (MICC), pp. 1–6. IEEE, November 2017
7. Jabbar, W.A., Ismail, M., Nordin, R.: Energy and mobility conscious multipath routing scheme for route stability and load balancing in MANETs. Simul. Model. Pract. Theory **77**, 245–271 (2017)
8. Natarajan and Rajendran: AOLSR: hybrid ad hoc routing protocol based on a modified Dijkstra's algorithm. EURASIP J. Wirel. Commun. Netw. **2014**, 90 (2014)
9. Dhanalakshmi, N., Alli, P.: Efficient energy conservation in MANET using energy conserving advanced optimised link state routing model. Int. J. Parallel, Emergent Distrib. Syst. https://doi.org/10.1080/17445760.2015.1103850 (2015)
10. Moussaoui, A., Semchedine, F., Boukerram, A.: A link-state QoS routing protocol based on link stability for Mobile Ad hoc Networks. J. Netw. Comput. Appl. **39**, 117–125 (2014)
11. Gerharz, M., de Waal, C., Frank, M., Martini, P.: Link stability in mobile wireless ad hoc networks. In: 27th Annual IEEE Conference on Local Computer Networks, 2002. Proceedings. LCN 2002, pp. 30–39. IEEE, Nov 2002
12. Gerharz, M., de Waal, C., Martini, P., James, P.: Strategies for finding stable paths in mobile wireless ad hoc networks. In: 28th Annual IEEE International Conference on Local Computer Networks, 2003. LCN'03. Proceedings, pp. 130–139. Oct 2003
13. Singal, G., Laxmi, V., Gaur, M.S., Todi, S., Rao, V., Zemmari, A.: MCLSPM: multi-constraints link stable multicast routing protocol in adhoc networks. In: Wireless Days (WD), pp. 1–6, Mar 2016
14. Moussaoui, A., Boukeream, A.: A survey of routing protocols based on link-stability in mobile ad hoc networks. J. Netw. Comput. Appl. **47**, 1–10 (2015)
15. Wu, X., Sadjadpour, H.R., Garcia-Luna-Aceves, J.J., Xu, H.: A hybrid view of mobility in MANETs: analytical models and simulation study. Comput. Commun. **31**(16), 3810–3821 (2008)
16. Wu, C., Kumekawa, K. and Kato, T.: A MANET protocol considering link stability and bandwidth efficiency. In: International Conference on Ultra Modern Telecommunications and Workshops, 2009, ICUMT'09, pp. 1–8. IEEE, Oct 2009
17. Baboo, S.S., Narasimhan, B.: Genetic algorithm based congestion aware routing protocol (GA-CARP) for mobile ad hoc networks. Proced. Technol. **4**, 177–181 (2012)
18. Toh, C.K.: Associativity-based routing for ad hoc mobile networks. Wirel. Pers. Commun. **4**(2), 103–139 (1997)
19. Song, Q., Ning, Z., Wang, S., Jamalipour, A.: Link stability estimation based on link connectivity changes in mobile ad-hoc networks. J. Netw. Comput. Appl. **35**(6), 2051–2058 (2012)
20. Meghanathan, N.: Graph Theory Algorithms for Mobile Ad hoc Networks. Informatica **36**(2), 185–200 (2012)

Privacy Attack Modeling and Risk Assessment Method for Name Data Networking

Vishwa Pratap Singh and R. L. Ujjwal

Abstract Assessment of security threat for Internet has become a major concern because of the advancement and expansion of IT in recent years, and it is equally important for future Internet architectures. Name data networking architecture is designed from scratch and is immune to most of the security attacks that are common in IP-based networks. The newly added features give rise to new security attacks which can compromise user privacy, data confidentiality, and integrity. Formal mathematical modeling is essential for proper assessments and mitigation from the security attacks. Security attack modeling provides insight of network vulnerabilities and helps in identifying the areas, which have to be kept on priority. This paper presents an attack tree-based privacy attack modeling and risk assessment technique for NDN. First attack tree is constructed in a top-down approach to find out possible attacks and threats that compromise user privacy from the attacker's point of view and then presented risk assessment technique to ascertain degree of threat that an attack imposes on user privacy.

Keywords NDN · Risk assessment · Attack tree · Attack modeling

1 Introduction

Name data networking (NDN) [1] is NSF-sponsored project for developing future Internet architecture. It is based on content-centric approach where data is on top priority instead of its location. Content in NDN is identified by application generated variable length names, and the routing is performed on the basis of names only. It uses hourglass architecture of thin waste line. Two new layers, security and name data, are added into previous Internet architecture. There are only two types of packets

V. P. Singh (✉) · R. L. Ujjwal
Guru Gobind Singh Indraprastha University, New Delhi, India
e-mail: vishwa.iiit@gmail.com

R. L. Ujjwal
e-mail: ujjwal@ipu.ac.in

© Springer Nature Singapore Pte Ltd. 2019
S. K. Bhatia et al. (eds.), *Advances in Computer Communication and Computational Sciences*, Advances in Intelligent Systems and Computing 924,
https://doi.org/10.1007/978-981-13-6861-5_10

in NDN—interest packet and data packet [2]. NDN uses pull mechanism where user requests data by sending interest packet that contains name of the requested content. Producer returns data in the form of data packets. Every piece of data is encrypted for providing security to data and user. NDN is provided with data caching facility which is one of the most important characteristics of NDN. Intermediate routers in NDN have capability to cache data packets that pass through it. Router checks whether it has cached copy of content in its content store (CS) rather than going to producer every time, and if copy is present there, then it is returned to user using incoming interface. NDN is designed to overcome the shortcomings present in our current Internet architecture. Security was not considered while designing our present Internet architecture, and it is provided later as patch which leads to various vulnerabilities.

NDN considers security from the beginning and resolves most of the security issues, but its new features and architecture impose some new security threats [3] due to caching, naming, and forwarding mechanisms [4]. User privacy, denial of service attack (DoS), and distributed denial of service attack (DDoS) [5] are the major concerns for NDN. The existing research (in area of NDN security) on NDN is mainly focused on preventing and mitigating DDoS attacks (cache pollution [6], content poisoning [7], Interest flooding) and privacy attacks (name privacy, cache enumeration, timing attack, conversation cloning). The idea was to introduce techniques for user privacy protection and mitigating DoS. To analyze the impact of security attacks, using expected as well as unexpected threats, it is necessary to have proper mathematical modeling of attacks and evaluate attacks for risk assessments. An all-around characterized assessment and evaluation enable NDN administrator to recognize most critical privacy attacks strategies and threats that enable them to take preventive measures. To best of our knowledge, there is no research paper available on privacy attack modeling and assessment based on cost analysis as well as quantitative risk analysis for NDN.

In this paper, we have modeled privacy security attacks in NDN by using attack tree model [8], performed assessment of security threats, and proposed risk assessment methods for preserving user privacy. Rest of the paper is organized as follows. Section 2 discusses attack tree modeling and its uses in NDN. Section 3 proposed NDN privacy attack modeling using attack tree, and in Sect. 4, we have described our risk assessment method. Finally, the conclusion is prepared in Sect. 5.

2 Attack Tree and Its Application to NDN Privacy Attack Assessment

Attack tree [9] is a meticulous way to systematically and structurally analyze a system for all possible security attacks. Attack tree can be represented in graphical as well as in a mathematical way. Root node in attack tree defined a global goal of attackers, and children nodes refine the attack and define how that can be achieved by attacker also known as intermediate goals. Leaf nodes are possible attacks that cannot be refined further and also represent attacks. Attack tree systematically defines the number a way a system can be attacked. An attack tree also defines a attack suite, which is a collection of all possible attacks on a system for achieving a particular goal.

Definition: An attack tree is a rooted tree represent by a three-tuple set $(V, \rightarrow, n_o)$.

- V is the finite set of nodes representing sub-goals and states followed by the attacker in a sequence to successfully attack a system to achieve global goal.
- n_o is a special node(root node) $\in V$; every node in V is reachable from n_o
- Leaf nodes are attacking nodes that leads to achieve sub-goals (internal nodes) or global goal.
- Leaf nodes $\cap$ Internal nodes $= \emptyset$
- $\rightarrow$ represents an acyclic relation between nodes. It represents state transition from child to parent.

It is convenient for a system administrator to analyze system for possible threats using attack trees. After successful construction of attack tree for a system, it can be used for analyzing different security attributes like cost. Root node in attack tree, also known as goal, is decomposed into concrete security attacks, and these attacks are decomposed further till they cannot be refined further. Nodes of attack tree can be decomposed by *OR* and *AND* relations. In *AND* relationship, accomplishment of parent requires all of its children to be successful $P_{AND} = \prod_{j=1}^{n} P_j$ and in *OR* relation any one of the child needs to be successful.

3　Attack Tree Model for Privacy Attacks in NDN

3.1　System Model

We have considered three types of nodes in NDN system (Fig. 1): (i) producer, (ii) consumer nodes, and (iii) intermediate routers. Content is produced, initially stored, and provided by producer nodes. Producer nodes can produce various types of content, and it may be organized in the form of prefix according to various applications; every content has a unique name and can be accessed by names only. Consumer nodes are the nodes that send request for the content (where content is being utilized). Intermediate routers have caching capability according to predefined algorithms. There are three types of data structures present in router: content store (CS), pending interest table (PIT), and forwarding information base (FIB). CS caches content and PIT is used to store the interfaces of the request that needs to be fulfilled. FIB is populated by routing algorithms, and it stores the forwarding information [10]. Content can be requested by consumer in form of *I_packets*(*I_packets* are not encrypted), data can be returned in form of *D_packets*, all data packets are encrypted and content signatures, and key information is attached with D_packets. Attacker nodes (malicious user nodes) can be connected to the same edge router on which legitimate user nodes are connected.

3.2 Building Attack Tree

We have set "getting usage information" as goal in attack tree, which is denoted by G.
There are two types of nodes

- G_j where $j = 1$–13, representing sub-goals.
- A_i where $I = 1$–16 representing attacks.

There are two types of gates AND gate and OR gate which represent AND and OR relationship. We have constructed attack tree in a top-down approach with divide and conquer manner. Usage pattern can be identified in three ways, by getting I_names or D_names, by getting publisher information, and by identifying content cached at router. They are also known as intermediate causes of getting usage pattern, and they are denoted by G1, G7, and G13. The objective can be achieved by any of the following paths to achieve goal successfully. Starting with right most, further names can be of two types—simple names and complicated names, so G1 is decomposed into G2 and G6. Taking left-most sub-tree, G2 can be achieved by getting $I\_packets$ which is denoted by G3. There are two possibilities of getting $I\_packets$:

- Attacker is connected to the same router as the consumer ($G4$)

Attacker is connected to different routers or farther from consumer (G5).

For G4 attacker, only need to sniff Interests sent by the consumer to the router (A1) and G5 can be achieved by monitoring attack (A2). G3 can be achieved by sniffing $I\_packets$ (A3) as well as knowledge of previous history (A4).

There is AND relationship between A3 and A4. A3 and A4 need to be successful to get names info by complicated names. Goal G can also be achieved by getting publisher information from each data packets denoted by sub goal $G7$. It can be further decomposed in two sub goals publisher information (G8) and key information (G9) (Fig. 2).

Goal G8 can be achieved by sniffing data packets (A5). The attacker can also get the publisher's information from the publisher's security key, that can be extracted by launching A6, A7 or A8 attacks first is A6 sniffing data packet and get signature and then in A7 attacker has to map key locator to publisher and in third A8 needs to know $P.Frequency$ [11]. Identify content cached at router: It is another method to find what a legitimate user is requesting. As NDN cached data previously requested by the user for fulfilling similar requests from different users. This can be done by object discovery [12] at router or timing information for getting data packet. It is based on the theory that if a user requests for data, then it is also cached at the router, and when attacker requests for same data, then it will take less time in comparison to the time when it is not cached at router. By finding what data is cached at router,

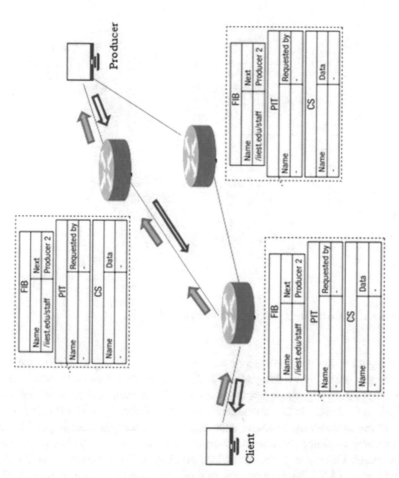

Fig. 1 Name data network architecture (with router data structure)

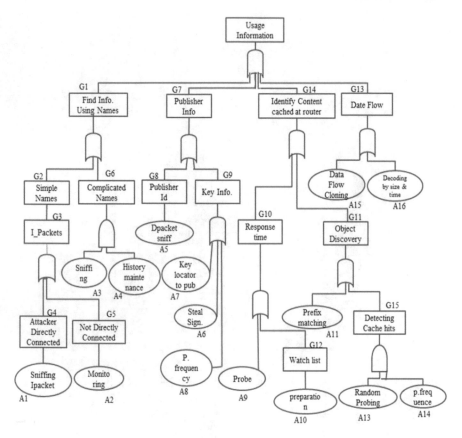

Fig. 2 Attack tree model for privacy attack assessment in NDN

the attacker gets to know about recent requests of users. The goal (*G10*) can be achieved in two ways: by probing router [13] with random data or by watch list attack, attacker needs to prepare a watch list (*A10*) for particular data he is looking for and then sends request to know whether it is cached or not. For goal G11, attacker can perform prefix matching to know what data a particular producer is providing. Object discovery at router can also be done by detecting cache hits. For detecting cache hits, attacker needs to perform random probing (*A13*) and needs to calculate probing frequency (*A14*). Right-most sub-tree of attack tree is used to define attack where attacker tries to get encrypted or plain text data before the user or with the user. It can be done by data flow cloning (A15), and for that, attacker also needs to perform object discovery attack. By data flow cloning attack [14], attacker can replicate entire data flow in the machine. Then, attacker can predict future requests of the victim and can request future data packets in the same way as legitimate user. As the attacker gets the data in encrypted format, he needs to decode it using packet size and time, etc., for extracting further knowledge represented by (*A16*).

4 Risk Assessment

For assessment, we have given quantitative values for attacks (Ai). These values are based on cost (C_a), probability to be discovered (P_a), time (T_a), technical difficulty (T_a) [15], and values assigned to these attributes in Table 2 are based on Table 1. The assignment of values to each node can be different for different scenarios and can be more precisely calculated according to application. We have calculated the total probability to reach root node (attack goal) from each leaf node (attack node). Based on multiple attributes, we need to find out occurrence probability for each node, which is also known as attacker utility [16]. This can be done by multi-attribute utility theory [17].

We have used additive decomposition, which can be used for taking decision under certainty. According to additive decomposition, there is a function

$$\mu_1 : x_1 \rightarrow R \text{ and } \mu(x_1 \ldots x_n) = \sum_{i=1}^{n} (k_i u_i x_i) \tag{1}$$

We have used the following formula to calculate utility for each leaf node

$$P_a = w_1 \times u_1(C_a) + w_2 \times u_2(T_a) + w_3 \times u_3(P_a) + w_4 \times u_4(I_a) + w_5 \times u_5(T_a) \tag{2}$$

where $u_1(C_a), u_2(T_a), u_3(P_a), u_4(I_a), u_5(T_a)$ are   utility   functions   of $(C_a), (T_a), (P_a), (I_a), \text{and}(T_a)$

Weight $w_1 + w_2 + w_3 + w_4 + w_5 = 1$, we have use $w_1 = w_2 = w_3 = w_4 = w_5 = 1/5$

As all the attributes are inversely proportional to their utility value, we have supposed that

$$u_1(C_a) = u_2(T_a) = u_3(P_a) = u_4(I_a) = u_5(T_a) = d/x \tag{3}$$

We have chosen value of d as 0.2 for assuring that occurrence probability lies in between 0 and 1. We have populated occurrence probability (P_a) values in Table 2 using Eqs. (2) and (3).

We have used binary decision diagram [18] to calculate total probability for reaching global goal (G). We have converted attack into BDD, and after that, counting rule is applied.

4.1 Attack Scenarios

Attack scenario is a path that an attacker can follow. For reaching a global goal from leaf nodes, all sub-goals have to be achieved successfully. It is easy to calculate

Table 1 Standard used for grading

Attribute	Description	Value
Cost	How much cost resources required, software, educational expenses, development	Graded in level 1–5, 5 for maximum, and 1 for minimum
Probability to be discovered	Chances that attacker is discovered	Unlikely: below 10%. Low: between 11% and 25%. Medium: between 26% and 75%. High: more than 76%. Certain: close to 100%. Optional: specific percentage value [15]
Time	Time needed for performing the attack	Very fast: performed in an instance Fast: attack can be performed in a moment Medium: attacker needs to wait for some time (hours) Slow: requires a lot of patience (take days to complete) Very slow: takes weeks to complete
Technical difficulty	Special skills and special resources required	Very difficult (1): require highly specialized skill, applications, and equipment Difficult(2), Medium (3): moderate knowledge and application can work well Simple (4): should be aware of system and well as application equipment required Very simple (5): no specialize equipment or application required, person with average knowledge can conduct this procedure
Impact	Consequences (profit) from attacker point of view	Very low (1): very low reward, get very few information. Low(2): Do not have very high impact in achieving goal. Moderate (3): moderate information retrieval or moderate achievement of goal. High (4): high amount of information retrieval. Extreme(5): reward is very high in term or resources and near to the goal

Table 2 Attribute values for leaf nodes

Attack nodes	Attribute					Occurrence probability (P_a)
	Cost (C_a)	Technical difficulty (T_a)	Discovering probability (D_a)	Impact (I_a)	Time (T_a)	
A_1	2	1	1	3	1	0.153
A_2	3	2	2	2	2	0.093
A_3	2	1	1	3	1	0.153
A_4	3	2	2	2	2	0.093
A_5	2	1	1	3	1	0.153
A_6	2	1	1	2	2	0.140
A_7	3	2	2	2	2	0.093
A_8	1	3	2	2	1	0.133
A_9	1	1	5	3	5	0.546
A_{10}	3	3	2	2	2	0.086
A_{11}	5	3	5	2	5	0.057
A_{13}	4	1	5	2	5	0.086
A_{14}	1	3	1	2	1	0.153
A_{15}	2	3	3	3	3	0.073
A_{16}	4	5	1	5	5	0.074

the occurrence probability if attack scenario is known. Admin can identify which attack path is most likely to be followed by attacker by calculating and comparing occurrence probability for attack scenarios. If an attack scenario is represented by $AS_i = (Ai_1, Ai_2, \ldots Ai_n)$, then the probability of attack scenario can be defined as:

$$P(AS_i) = P(Ai_1) * P(Ai_2) \ldots P(Ai_n)) \tag{4}$$

Following are attack scenarios identified by Boolean algebra method:

$$G_1 = G_2 + G_6 \Rightarrow G_3 + (A_3 * A_4) \Rightarrow G_4 + G_5 + (A_3 * A_4) \Rightarrow A_1 + A_2 + (A_3 * A_4) \tag{5}$$

$$G_7 = G_8 + G_9 \Rightarrow A_5 + (A_6 + A_7 + A_8) \tag{6}$$

$$G_{14} = G_{10} + G_{11} \Rightarrow (A_9 + G_{12}) + (A_{11} + G_{15}) \Rightarrow (A_9 + A_{10}) + (A_{11}) + (A_{13} + A_{14}) \tag{7}$$

$$G_{13} = (A_{15} + A_{16}) \tag{8}$$

All attack scenarios to achieve global goal can be identified by formula of G. All attack scenarios are $AS_1 = \{A_1, A_3, A_4\}$, $AS_2 = \{A_2, A_3, A_4\}$, $AS_3 = \{A_5, A_6, A_7, A_8\}$, $AS_4 = \{A_6, A_{10}, A_{11}, A_{13}, A_{14}\}$, and $AS_5 = \{A_{15}, A_{16}\}$. From Table 2, we can identify that the attacker most likely follows AS_1 scenario. So administrator needs to have eyes on this path and need to take measures for securing this path first.

5 Conclusion

The two main issues related to security of NDN are DoS and user privacy. In this paper, we have presented an attack tree-based privacy attack modeling for NDN. We have identified possible attacks that can compromise users privacy and modeled them using attack tree. We have also presented a quantitative assessment method for finding possible attack paths that an attacker can follow. We have used multi-attribute utility theory and additive decomposition. Proper attack assessment can help in identifying the path an attacker is likely to follow. Prior knowledge of attack paths can help network administrator to keep eyes on threats which should be considered first and to take preventive measures. To best of our knowledge, this is the first paper to present attack modeling and risk assessment for privacy attack in NDN. This paper intends to demonstrate the privacy attack assessment technique for NDN and to identify accurate attack path.

References

1. Zhang, L., Estrin, D., Burke, J., Jacobson, V., Thornton, J.D., Smetters, D.K., Zhang, B., Tsudik, G., Massey, D., Papadopoulos, C.: Named data networking (ndn) project. Relatório Técnico NDN-0001, Xerox Palo Alto Research Center-PARC **15**, 158 (2010)
2. Zhang, L., Afanasyev, A., Burke, J., Jacobson, V., Crowley, P., Papadopoulos, C., Wang, L., Zhang, B.: Named data networking. ACM SIGCOMM Comput. Commun. Rev. **44**(3), 66–73 (2014)
3. Chen, S., Mizero, F.: A survey on security in named data networking. arXiv preprint arXiv: 1512.04127 (2015)
4. Yi, C., Afanasyev, A., Moiseenko, I., Wang, L., Zhang, B., Zhang, L.: A case for stateful forwarding plane. Comput. Commun. **36**(7), 779–791 (2013)
5. Dai, H.: Mitigate ddos attacks in ndn by interest traceback. In: 2013 IEEE Conference on Computer Communications Workshops (INFOCOM WKSHPS). IEEE (2013)
6. Conti, M., Gasti, P., Teoli, M.: A lightweight mechanism for detection of cache pollution attacks in named data networking. Comput. Netw. **57**(16), 3178–3191 (2013)
7. Wu, D., Xu, Z., Chen, B., Zhang, Y.: What if routers are malicious? mitigating content poisoning attack in ndn. In: Trustcom/BigDataSE/I200B SPA, 2016 IEEE 2016 Aug 23 (pp. 481–488). IEEE
8. Mauw, S., Oostdijk, M.: Foundations of attack trees. In: International Conference on Information Security and Cryptology. Springer, Berlin, Heidelberg (2005)
9. Schneier, B.: Attack trees. Dr. Dobb's J. **24**(12), 21–29 (1999)
10. Amadeo, M., Campolo, C., Molinaro, A.: Forwarding strategies in named data wireless ad hoc networks: design and evaluation. J. Netw. Comput. Appl. **1**(50), 148–158 (2015)
11. Singh, A.K., Mohan, A.: Computation of probability coefficients using binary decision diagram and their application in test vector generation. Int. J. Comput. Sci. Eng. **3**(1), 33–40 (2009)
12. Hand, R.S.: Toward an active network security architecture

13. Tourani, R., Mick, T., Misra, S., Panwar, G.: Security, privacy, and access control in information-centric networking: a survey. arXiv preprint arXiv:1603.03409, 10 Mar 2016
14. Vasilakos, A.: Information centric network: research challenges and opportunities. J. Netw. Comput. Appl. **52**, 1–10 (2015)
15. Bagnato, A.: Attribute decoration of attack–defense trees. Int. J. Secure Softw. Eng. **3**(2), 1–35 (2012)
16. Hu, N., Steenkiste, P.: Evaluation and characterization of available bandwidth probing techniques. IEEE J. Sel. Areas Commun. **21**(6), 879–894 (2003)
17. Mateo, San Cristóbal, J.R.: Multi-attribute utility theory. In: Multi Criteria Analysis in the Renewable Energy Industry. Springer, London, pp. 63–72 (2012)
18. Akers, S.B.: Binary decision diagrams. IEEE Trans. Comput. **6**, 509 (1978)

Analysis of Classification Methods Based on Radio Frequency Fingerprint for Zigbee Devices

Jijun Wang, Ling Zhuang, Weihua Cheng, Chao Xu, Xiaohu Wu
and Zheying Zhang

Abstract Radio frequency fingerprint is an inherent characteristic of wireless communication devices which can be extracted from communication signals and be applied in wireless device identification for communication system security. This paper selects different characteristics of RF fingerprints and compares the identification accuracy of Zigbee devices with five classification algorithms, including support vector machine, bagging, neural network, naive Bayes, and random forest algorithms. The experimental research shows that the highest identification accuracy reaches approximately 100% by using multi-features of frequency offset, IQ offset, and circle offset based on the neural network algorithm under high SNR. With the reduction in SNR, the identification accuracy based on bagging algorithm with multi-features of frequency offset and IQ offset is the highest. The performance of support vector machine algorithm is the most stable.

Keywords Radio frequency fingerprint · Machine learning · Feature
classification · Access authentication

1   Introduction

Due to the openness of wireless networks, it is more vulnerable to large-scale malicious attacks than traditional wired networks which lead to social concentration of its security. Traditional network security is bit-level oriented, achieving integrity and confidentiality of data and identity authentication through cryptographic mechanism-based security protocols. However, the actual wireless network security protocols

J. Wang (✉) · W. Cheng · C. Xu · X. Wu
Jiangsu Electric Power Information Technology Co., Ltd., Nanjing, China
e-mail: 220170820@seu.edu.cn

L. Zhuang
State Grid Jiangsu Electric Power Co., Ltd., Jiangsu, China

Z. Zhang
School of Cyber Science and Engineering, Southeast University, Nanjing, China

© Springer Nature Singapore Pte Ltd. 2019   121
S. K. Bhatia et al. (eds.), *Advances in Computer Communication
and Computational Sciences*, Advances in Intelligent Systems and Computing 924,
https://doi.org/10.1007/978-981-13-6861-5_11

usually exist flaws [1]. In the past decade, extraction and identification of wireless fingerprints for wireless communication devices aroused broad interest as radio frequency fingerprint recognition (RFFR) can identify and authenticate wireless devices through the transient or steady-state part of RF signal to avoid impersonate or tempering of network nodes. Tolerance of hardware circuit components, consisting mainly of manufacturing tolerance and drift tolerance, which is hardware foundation, ensures the uniqueness of RF fingerprint [2] of every device and makes RFFR possible in information security. A general process of wireless device authentication by RFF extraction and recognition consists of RFF collection, preprocessing, transform of RFF, RFF feature extraction, RFF classification, and authentication of wireless devices [3]. RFF classification, whose accuracy rate determines the device recognition performance, is a typical pattern recognition problem, divided into two steps of training and recognition. In training phase, register RFF of legal devices into the library, and bind the RFF with its corresponding device ID. In identification phase, extracting RFF from communication signal of the device is to be identified, classifying them with employment of legal RFF in database to authenticate whether it is the device it claims.

Although current researches propose improvement methods in extraction and classification, there is a lack of comprehensively comparison of several common classification methods. In addition, there is no agreement in classification characteristics. In view of existing problems, this paper mainly studies in following aspects:

(1) This paper studies different RFF characteristics including frequency offset, IQ offset, circle offset, differential constellation trace figure (DCTF), transient signal starting point (TS), and transient signal end point (TE) to classify devices and combine them as multi-features to improve classification accuracy.

(2) Based on obtained RFF characteristics of Zigbee wireless devices, this paper compares the classification and recognition accuracy of different machine learning classifiers, including Bayesian classification, support vector machine algorithm, neural network algorithm, bagging algorithm and random forest algorithm.

(3) By optimizing the classification accuracy of RF fingerprints under different signal-to-noise ratios (SNR), this paper proposes the optimal classification algorithm corresponding to SNR and propose the most stable algorithm. This paper is organized as follows. In Sect. 2, we look back on proposed RFF classification methods. In Sect. 3, we discuss our experimental setup and RF fingerprint characteristics we employ in device classification. In Sect. 4, we discuss common classification methods including SVM algorithm, bagging algorithm, neural network algorithm, navies Bayes algorithm, and random forest algorithm and give the classification accuracy rates of device recognition, respectively. In Sect. 5, we conclude our experimental results to give optimal classification method and correspondent RF fingerprint characteristics under different SNR.

2 Related Work

Methods of RF characteristic extraction are mainly based on transient response and steady-state response. RFFR of transient signal extracts RF fingerprint of the device according to the on/off transient signal which requires extremely accurate starting point detection and transient signal interception [4]. With lower accuracy requirements for equipments and less change caused by channel, RFFR of steady-state signal extracting from preamble before data segment form becomes a research hotspot. After extraction of RFF, processing is required for purer RFF characteristic. Wang [5] uses the DBSCAN algorithm to eliminate extracted RFF data of lager errors or deviations to build a high-quality fingerprint database. Klein and Temple [4] shows that wavelet domain (WD) fingerprints with DT-CWT (complex wavelet transform) features emerged as the superior alternative for all scenarios at SNRs below 20 dB while achieving performance gains of up to 8 dB at 80% classification accuracy employing Fisher-based multiple discriminant analysis (MDA) with maximum likelihood (ML) classification. Ureten and Serinken [6] reduced the dimension B of the feature vector by using PCA, and a probabilistic neural network (PNN) is used as a classifier. The PNN provides a general solution to pattern classification problems by following an approach developed in statistics called Bayesian classification. Peng et al. [7] presents constellation trace figure as an RFF characteristic to identify wireless communication devices. After extracting RFF features, classification algorithms including SVM, neural network [8], naive Bayes [9], and random forest [10] algorithm are applied to identify if the corresponding device is authenticated.

3 Experimental Setup and Classification Feature

3.1 Experimental Setup

The experimental platform setup is shown in Fig. 1, and 54 Zigbee wireless communication devices are used as transmitters to send certain signals to a receiver. By adding different power of noise, we collect different RFF features from the signals to classify devices.

3.2 Classification Feature

Classification features extracted from RF signals are RF fingerprints of communication devices which are unique for each device. The selection of different classification characteristics affects the classification accuracy a lot. This paper selects six characteristics as RF fingerprints of 54 Zigbee wireless devices to classify, including frequency offset (FO), IQ offset (IQO), differential constellation trace

Fig. 1 Experimental platform setup

Table 1 Average frequency offset of wireless devices	Device number	1	11	21
	FO/kHz	66.1544	30.4994	25.0556

figure (DCTF), circle offset (CO), transient signal starting point (TS), and transient signal end point (TE).

Frequency offset Frequency offset refers to the systematic frequency offset, which refers to the offset between the average of signal frequency and the receiver's frequency of 2.505 GHz in this experiment. Different wireless devices have different frequency offsets as shown in Table 1.

IQ offset IQ offset is the signal offset from the origin on the complex plane coordinate system after quadrature modulation. The IQ offset used in this experiment is the deviation value normalized to 1 in the unit circle. Figure 2 shows the IQ offset of different wireless devices. IQ offset is 100 times of itself for convenience of illustration.

Differential constellation trace figure Since the carrier frequency of the receiver and transmitter may have a certain deviation, it is necessary to perform differential processing on received signal at a fixed interval to obtain constellation trace figure, that is, a differential constellation trace figure (DCTF) [7] as shown in Fig. 3.

Fig. 2 IQ offset of different wireless

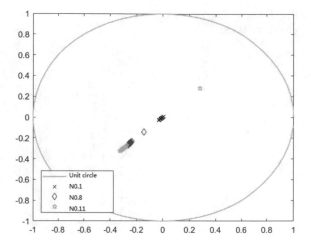

Fig. 3 Differential constellation trace figure devices of different wireless devices

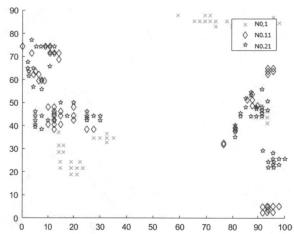

Circle offset Due to the phase deviation and frequency deviation of the demodulated signal, the rotation of DCTF results in a distortion in constellation trace figure, called circle offset. Although the curves in Fig. 4 are cluttered, the circle offset of different devices is significantly different.

Transient signal starting point The feature of transient signal starting point (TS) primarily reflects the amplitude envelope and phase information of the electromagnetic waves emitted by the wireless device when it is turned on, as shown in Fig. 5. The feature of transient signal end point (TE) primarily reflects the amplitude envelope and phase information of the electromagnetic waves emitted by the wireless device when it is turned off, and the feature figure of TE is similar to Fig. 5.

Fig. 4 Circle offset of
different wireless

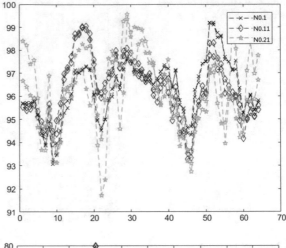

Fig. 5 Transient signal
starting point devices of
different wireless devices

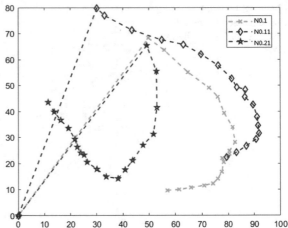

4 RF Fingerprint Classification Algorithm and Identification Accuracy

This paper selects different characteristics of RF fingerprints and compares the classification accuracy of Zigbee devices with five classification algorithms, including support vector machine, bagging, neural network, naive Bayes, and random forest algorithm. For the first 3 algorithms, this paper uses MATLAB toolbox to train 30 sets of RFF data for 54 Zigbee devices, including 6 single features and 4 multi-features, and then use 15 remaining sets of data to test the classification accuracy. For naive Bayes algorithm, classification prior probability is computed with 30 sets of RFF data known and then tested the accuracy of trained naive Bayes classifier with remaining RFF data. Then, random forest algorithm randomly selects several features as the decision method in sample features. Then, random forest algorithm randomly selects several features as the decision method in sample features. Naive Bayes classification is also used as a weak classifier in random forest algorithm.

4.1 SVM Algorithm

Support vector machine (SVM) algorithm is a supervised learning model to analyze data and identify patterns for classification and regression analysis. We apply SVM to high dimensionality characteristics for classification. Unlike logistic regression, SVM outputs category which is decided by $w^T x + b$ instead of probability. An important innovation in SVM is kernel trick which concluded the form of dot products between samples in many machine learning algorithms. The linear function in SVM can be written as

$$\omega^T x + b = \sum_{i=1}^{m} a_i x^T x^{(i)} + b \tag{1}$$

where $x^{(i)}$ is ith training sample, and a is coefficient vector. x can be substituted for the output of characteristic function $\phi(x)$, which can be expressed by

$$\phi(x) = \sum_{i} a_i \cdot k(x, x^{(i)}) + b \tag{2}$$

Since our RFF characteristics belong to high-dimensional spaces, $k(x, x^{(1)})$ is the kernel function whose basic function is transforming two vectors in low-dimensional space into the inner product value of vector in a high-dimensional space. The kernel function is equivalent to preprocessing all the inputs with $\phi(x)$ and then learning the linear model in the new transformation space [7]. Table 2 shows the classification accuracy of SVM algorithm under different SNR.

Experimental results show that single-feature classification based on circle offset is more stable and efficient under high SNR while single-feature classification based on frequency offset performs best under low SNR. On the whole, multi-feature classification based on the frequency offset and IQ offset has a high performance under both high and low SNR.

Table 2 Classification accuracy of SVM and bagging algorithm

Classification feature	Classification accuracy %							
	SVM				Bagging			
	30 dB	20 dB	10 dB	0 dB	30 dB	20 dB	10 dB	0 dB
FO	55.6	55.7	55.4	55.1	87.6	87.6	87.2	86.1
IQO	93.2	79.5	43.6	11.0	92.0	77.1	39.2	7.5
DCTF	80.5	81.0	67.2	16.4	83.9	82.50	68.3	12.1
TS	35.2	10.4	1.1	2.1	31.0	8.5	1.6	1.5
TE	20.1	4.5	1.7	2.1	20.1	4.5	1.7	1.5
CO	96.0	92.7	64.6	11.0	96.0	90.7	45.0	4.6
FO, IQO	95.7	94.8	87.0	69.0	95.4	94.8	92.50	79.0
IQO, CO	95.9	93.3	64.8	12.8	96.4	93.5	58.2	9.4
FO, CO	96.1	94.8	75.2	24.5	96.3	95.1	74.1	27.7
FO, IQO, CO	95.9	94.7	74.2	26.4	96.0	95.9	80.2	37.7

4.2 Bagging Algorithm

Bagging algorithm improves the accuracy of learning algorithm by constructing a series of predictive functions and combining them into a predictive function. Bagging votes for decision by drawing the predictive function sequences, respectively, from training several different models separately. Compared to weak classifier, bagging has a repetitive process that increases classification accuracy because different models seldom produce the exact same error on the test sets. Table 2 shows the classification accuracy of bagging algorithm under different SNR. Experimental result shows that three-feature classification based on frequency offset, IQ offset, and circle offset is more stable and efficient under high SNR while single-feature classification based on frequency offset performs best under low SNR. On the whole, multi-feature classification based on the frequency offset and IQ offset has a high performance under both high and low SNR.

4.3 Neural Network Algorithm

BP (error back-propagation network) neural network algorithm is one of the most widely used and successful neural networks. The basic idea is that the learning process consists of two processes: the forward propagation of the signal and the back-propagation of the error. While forward propagation, samples are transmitted from the input layer to output layer, processed through the hidden layer. If the actual output does not match the expected output in output layer, it turns to error back-propagation. The error back-propagation is to invert the output error to input layer

Table 3 Classification accuracy of BP neural network and naive Bayes algorithm

Classification feature	Classification accuracy (%)							
	BP neural network				Naive Bayes algorithm			
	30 dB	20 dB	10 dB	0 dB	30 dB	20 dB	10 dB	0 dB
FO	53.1	55.4	51.6	49.6	53.1	55.4	51.6	49.6
IQO	85.5	84.0	78.8	20.4	76.2	55.3	19.8	4.8
DCTF	86.3	86.4	76.5	20.5	59.6	67.2	64.1	20.5
TS	42.7	17.2	2.7	1.4	16.5	5.6	2.6	0
TE	26.0	5.1	3.3	2.5	1.6	3.5	2.2	2.0
CO	99.4	98.5	69.5	3.9	81.2	73.7	23.7	6.2
FO,IQO	91.8	88.8	87.6	40.2	94.6	91.7	77.4	45.1
IQO,CO	100.0	99.3	94.4	43.1	92.1	80.1	33.7	6.9
FO,CO	100.0	99.9	70.7	54.0	90.7	87.4	72.8	52.7
FO,IQO.CO	100.0	98.5	81.4	45,4	97.2	95.9	82.6	43.6

through the hidden layers and distribute the error to all the units of each layer to obtain the error signal of each layer unit which can be used for correction of weight of each unit. The process of adjusting weight of each unit is repeated which is the learning and training process of the network. This process continues until the network output error is reduced to an acceptable level or to a preset number of learning processes.

Table 3 shows the classification accuracy of BP neural network algorithm under different SNR. Experimental result shows that under high SNR, three-feature classification based on frequency offset, IQ offset, and circle offset, two-feature classification based on IQ offset and circle offset, and two-feature classification based on frequency offset and circle offset all have a good classification accuracy. Under low SNR, two-feature classification based on frequency offset and circle offset performs well.

4.4 Naive Bayes Algorithm

The naive Bayes classifier is the simplest probability classifier which has good classification accuracy and the ability to adapt to new samples [7]. Bayes formula is expressed as

$$p(y = j|x = i) = \frac{p(x = i|y = j) \cdot p(y = j)}{p(x = i)} \tag{3}$$

where x, y are RFF characteristic and classification number, respectively, for test sample, i, j are RFF characteristic i and classification j in training. When $x = i$, compare $y = j|x = i$, $j = 1, 2...n$, n is number of devices. The max $y = j|x = i$ represents the most possible classification. For 54 devices with 45 sets of RFF data of each device, this paper calculates the classification prior probability with 30 sets of RFF data known and then uses naive Bayes classifier to judge the posterior probability of classification with remaining RFF data. Table 3 shows the classification accuracy of naive Bayes classifier under different SNR. Experimental result shows that three-feature classification based on frequency offset, IQ offset, and circle offset is more stable and efficient under high SNR while two-feature classification based on frequency offset and circle offset performs well under low SNR.

4.5 *Random Forest Algorithm*

Random forest classifier consists of multiple decision trees. The classifier first samples in sample space and repeat sampling if necessary [7]. Then, random forest algorithm randomly selects several features as the decision method in sample features. This experiment uses naive Bayes in 4.4 as a weak classifier. Since the number of samples is large enough, instead of repeated sampling, all samples are used as the input of random forest algorithm. Previous experiments show that classification accuracy based on TS and TE is non-ideal, so random forest algorithm classification experiment is divided into two groups: One's classification characteristics include TS and TE, one without to determine whether the two low-resolution features will affect the final classification accuracy when using random forest algorithm. Figure 6 shows the classification accuracy of random forest algorithm. The experimental results show that TE and TS features have little effect on the classification accuracy, and random forest algorithm has good performance under high SNR.

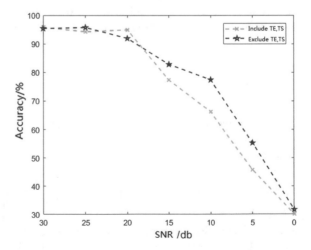

Fig. 6 Classification accuracy of random forest algorithm

5   Results and Conclusion

5.1   *Experimental Results*

This paper applies 5 classification algorithms to 6 single RF fingerprint characteristics and 4 multi-features of 54 Zigbee wireless devices. We optimize the classification accuracy to figure out the optimal algorithm with optimal RF fingerprint characteristic under different SNR, as shown in Fig. 7.

Neural network algorithm is the optimal classification method under high SNR, whose classification accuracy is close to 100%. With the reduction in SNR, classification method based on bagging algorithm performs better. SVM, random forest, and naive Bayesian classifier work only under high SNR. In general, neural network algorithm is selected under high SNR, and bagging algorithm is selected under low SNR. The performance of SVM algorithm shows stability under all SNR.

5.2   *Conclusion*

This paper studies classification accuracy of different RFF in wireless devices under different SNR with five classification methods, including SVM, bagging, neural network, naive Bayes, and random forest algorithms. The experimental results show that RFF feature, SNR, classification algorithm are crucial factors which affect RFF identification accuracy. Neural network algorithm is the optimal classification method when SNR is between 10 and 30 dB. From 0 to 10 dB of SNR, classification method based on bagging algorithm performs the best. The performance of SVM algorithm

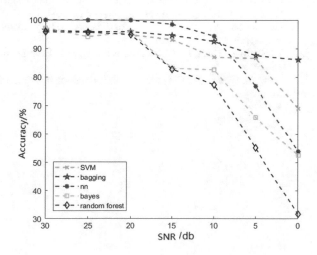

Fig. 7 Comparison of optimal classification forest algorithm accuracy of five classification algorithms

Table 4 Optimal classification method under different SNR

SNR	30 dB	20 dB	10 dB	0 dB
Classification algorithm	NN	NN	NN	Bagging
RF fingerprint	IQO,CO	FO,CO	ICO,CO	FO
Accuracy (%)	100	99.9	99.4	86.1

NN represents the neural network algorithm

shows stability under all SNR. The selection of classification algorithms and RF fingerprint characteristic is shown in Table 4.

References

1. Yuan, H.L.: Research on physical-layer authentication of wireless network based on rf fingerprinting, Ph.D. dissertation, Southeast University (2011)
2. Yuan, H., Hu, A.: Fountainhead and uniqueness of rf fingerprint. J. SE Univ. **39**(2), 230233 (2009)
3. Yu, J., Hu, A., Zhu, C., Peng, L., Jiang, Y.: Rf fingerprinting extraction and identification of wireless communication devices. J. Cryptologic Res. (2016)
4. Klein, R.W., Temple, M.A., Mendenhall, M.J., Reising, D.R.: Sensitivity analysis of burst detection and rf fingerprinting classification performance. In: IEEE International Conference on Communications, p. 15 (2009)
5. Wang, F.: Research on radio fingerprints localization techniques based on cluster analysis. Beijing University of Posts and Telecommunications (2014)
6. Ureten, O., Serinken, N.: Wireless security through RF fingerprinting. Can. J. Electr. Comput. Eng. **32**(1), 27–33 (2007)
7. Peng, L., Hu, A., Jiang, Y., Yan Y., Zhu, C.: A differential constellation trace figure based device identification method for zigbee nodes. In: International Conference on Wireless Communications and Signal Processing, pp. 1–6 (2016)
8. Goodfellow, I., Bengio, Y., Courville, A.: Deep learning. MIT Press. http://www.deeplearningbook.org (2016)
9. Dupre, M.J., Tipler, F.J.: New axioms for rigorous bayesian probability. Bayesian Anal. **1**(1), 18 (2009)
10. Liaw, A., Wiener, M.: Classification and regression by random forest. R News **23**(3), 23 (2013)

To Identify Visible or Non-visible-Based Vehicular Ad Hoc Networks Using Proposed BBICR Technique

Kallam Suresh, Patan Rizwan, Balusamy Balamurugan,
M. Rajasekharababu and S. Sreeji

Abstract In vehicular ad-hoc network design, the border node can be select based on one-hop neighbor data using a minimum neighbor based distance concept. Where the different existing approach and various protocols are examined for nodes are located nearest neighbor position lists are follows a distributed network based strategy. Thus, determine which vehicle/nodes share the least number of common neighbors. In this proposed paper, nodes which satisfy present state are typically outermost from next forwarding node of border side of intercommunication system model with border-based routing making hybridization, minimizing end-to-end delay and improving average throughput with our hybrid protocol that is BBICR, using MATLAB 2014Ra version.

Keywords VANET · BBR · ICS · Packet forwarding · Neighbor node

1 Introduction

VANET stands for the vehicular ad hoc network. The new scenario of VANET is mobile ad hoc network (MANET) that is a self-framing network and exceptionally adaptable, which has a few characteristics like dynamic topologies, less transmission capacity and energy consumption, and it can work without the need of any concentrated control. In this manner, VANET [1] is a wireless communication among vehicle to vehicle for different area vehicles to the roadside framework in light of wireless local area network (WLAN) innovation. Every hub in an ad hoc network goes about as both an information terminal and a switch. The hubs in the network at that point utilize the wireless medium to speak with different hubs in their radio range. The advantage of utilizing ad hoc networks is it is conceivable

K. Suresh (✉) · P. Rizwan · B. Balamurugan · S. Sreeji
SCSE, Galgotias University, Noida, India
e-mail: sureshkallam@gmail.com

M. Rajasekharababu
VIT University, Vellore, Tamilnadu, India

© Springer Nature Singapore Pte Ltd. 2019
S. K. Bhatia et al. (eds.), *Advances in Computer Communication and Computational Sciences*, Advances in Intelligent Systems and Computing 924,
https://doi.org/10.1007/978-981-13-6861-5_12

133

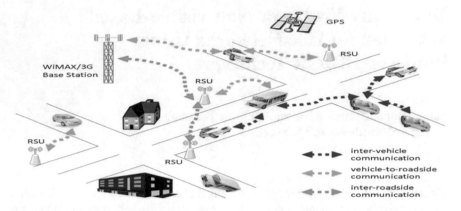

Fig. 1 Vehicular ad hoc network

to convey these networks in areas where it is not doable to introduce the required framework. Another significance of ad hoc networks is that it could be immediately conveyed with no administrator association. The administration of a substantial scale vehicular network would be a troublesome assignment. These reasons add to the ad hoc networks being connected to vehicular environments.

Figure 1 [1] shows a vehicular ad hoc network with all its working. The moving vehicles are communicating with each other as well as the roadside units (RSUs). Different roadside units are also communicating with each other as well as the infrastructure.

1.1 Components of VANETs

There are three components of VANET:

(i) Position of Onboard Unit (OBU)

The OBU is prepared on a vehicle between vehicle communications for different areas or communications between the vehicle and the roadside units that identified for a radio wire are equipped in an OBU to such an extent that the vehicle communications with various sides or the roadside units can be applied.

(ii) Position of Roadside Unit (RSU)

The RSU is distributed along the roads. The principal function of the roadside units is to sidestep the messages among the vehicles and trust authority.

(iii) Major Trust Authority (MTA)

The MTA is a server, which is overseen by a specialist co-op or the legislature. The function of a trust authority is to keep up the administration, to keep the records of every vehicle or to issue the testament for every vehicle.

1.2 Mobility of Border Node-Based Routing (BBR)

In this area, we display a border routing protocol in part associated different states. The proposed routing is intended for sending messages from random node to other nodes (unicast) or from one node to every single other node (broadcast). The general plan objectives are to improve the broadcast conduct for low node thickness and high-range mobility.

2 Motivation

Different VANETs mobility trace generators advanced for the very aim of testing VANETs routing protocols in simulation. Realistic mobility traces collected from vehicles are discussed below:

In [2], the routing in mobile ad hoc networks was described. Routing in mobile ad hoc networks is quite a challenging task since the nodes move arbitrarily and have limited power back, topology changes and lack of other resources. In other routing protocols [3], there is a part of routing overhead because of the flooding of the messages [4].

There are numerous strategies for finding the portable nodes like radiolocation procedures, global positioning system (GPS), the bat framework, the cricket compass framework and radar. This work has made the blend of the area following instrument that has been designed in the genuine MANET test bed with geo-throwing ability. Every portable node (MN) has run the area following system and put away the situating table at the nodes itself in a record. The information was in succession and began with the MAC address, IP address and the separation of each neighbor from MN.

In [5], a protocol named GeoTORA was proposed, as it was gotten from the temporally ordered routing algorithm (TORA) (unicast) routing protocol. Flooding is additionally joined in GeoTORA; however, it is restricted to nodes inside a little area. This integration of TORA and flooding can fundamentally diminish the overhead of geocast delivery, while keeping up sensibly high exactness. TORA is one of the groups of connection inversion algorithms for routing in ad hoc networks. For every conceivable goal in the ad hoc network, TORA keeps up a goal arranged in the directed acyclic graph (DAG).

In [6], a protocol was suggested that expands the existing geocast protocols by supporting a packet delivery framework with adequate flooding. In basic flooding, there are repetitive retransmissions of geocast messages, expanding network traffic, possibly bringing about broadcast storms. The paper proposed an answer to this issue as takes after. Instead of having, all nodes partake in the packet transmission, just the nodes that fulfill the following two conditions: Those that have a transmission go that is bigger with respect to the normal transmission run, and those that can cover the areas that are yet revealed. In this way, this approach diminishes the quantity of transmissions. To stop a node tolerating copy packets, one of the kind succession numbers is related to every packet, which is contrasted and already recorded (source, grouping) sets.

In [7], vehicular delay-tolerant network (VDTN) is a Delay-Tolerant Network (DTN) based architecture concept for transit networks, where vehicles movement and their bundle relaying service is opportunistically exploited to enable non-real time applications, under environments prone to connectivity disruptions, network partitions and potentially long delays. VDTN expects offbeat, package-situated communication and a store-convey and store-forward routing paradigm. A routing protocol for VDTNs should make the best utilization of the tight assets accessible in network nodes to make a multi-jump way that exists after some time. A VDTN routing protocol, called GeoSpray, was proposed, which takes routing choices based on geographical area information and consolidates a hybrid approach between different duplicate and single-duplicate plans. In the first place, it begins with a numerous duplicate plan, spreading a set number of package duplicates, keeping in mind the end goal to abuse elective ways.

In [8], GeoCross, a basic, yet novel, occasion-driven geographic routing protocol is suggested that expels cross-interfaces progressively to abstain from routing circles in urban vehicular ad hoc networks (VANETs). GeoCross misuses the regular planar component of urban maps without falling back on bulky planarization. Its component of dynamic circle identification makes GeoCross reasonable for profoundly versatile VANET. We have likewise demonstrated that storing (GeoCross + Cache) gives a similar high packet delivery ratio and yet utilizes less bounces.

3   System Model

3.1   Improved Edge–Border Node-Based Routing (EBBR) Protocol in VANETS

In our proposed edge count, measurement for border nodes, which take after by the framework scope detecting separation in BBR protocol, can endure network segment because of low node thickness and high node mobility. The execution of route nodes and BBR are assessed with Geographica and Traffic Information (GTI) based mobility. The simulation comes about to demonstrate that under country network conditions, a restricted flooding protocol, for example BBR, performs well and offers the advantage of not depending on an area benefit required by different protocols proposed for VANETs.

Improved novel proposed system will combine the probabilistic and detachment plots remembering the true objective to manhandle the benefits of the two strategies. The proposed system is designed with help of mathematical model for fundamental arrangement. It upgrades the Density Aware Probabilistic Flooding (DAPF) for eliminating the topology issues in the territory networks. To find the near by the nodes based on the following procedure.

A host will rebroadcast, flooding the messages with the probability $p = f\text{mode2}(n)$:

$$F\text{mode} = K/N \tag{1}$$

The parameter to accomplish the reachability is k which is a productivity of the broadcast. The $f$mode2 picked in light of the fact that has been naturally the ideal likelihood of broadcast is the converse of the local thickness. Moreover, we have watched the accuracy of this presumption.

3.2 Probabilistic-Based Flooding Border Retransmission

The existing models are inconvenience to be locally constant. In actuality, each node of a given region gets, communicates and chooses based on the probability as demonstrated by a consistent or from the nearby thickness. This resembles an extra zone scope plot [5], yet that arrangement is mainly position-based. The main establishment is divided between two hubs with help of full duplex. It can be taken into evaluation by differentiating their neighbor nodes and Z_{src} (Source region for zone a_n) and Z_{dest} (destination zone for zone a_n) as far the intersection point between Z_{src} (Source region for zone b) and Z_{dest} (destination territory for zone b) and in c zone all fringe hub from source to objective underneath in 3 condition:

$$Z_a = Z_{\text{src}} \cap Z_{\text{dest}} : \text{ the correspondence region anchored just by src,} \dots \tag{2}$$

$$Z_b = Z_{\text{src}} \cap Z_{\text{dest}} : \text{ the correspondence region anchored just by dest,} \dots \tag{3}$$

$$Z_c = Z_{\text{src}} \cap Z_{\text{dest}} : \text{ the correspondence region anchored by two of src and dest} \dots \tag{4}$$

The nodes cannot calculate the zones Z_a, Z_b and Z_c without arranging workplaces in conditions 1, 2 and 3. In any case, these zones can be depicted in the form of mobiles in each node. Each node can be represented by the following format (src, dest (neighbor nodes Nd), inside Z_c (number of hubs)). This is the uniform case for src and dest (Fig. 2).

$$\mu = N_b/(N_b + N_c) \tag{5}$$

It identifies all its neighbors in the message.

a. **Density-Based Border Node Retransmission-Based Probabilistic Flooding Neighbor Elimination**

Fig. 2 src and dest positions
in the radio areas

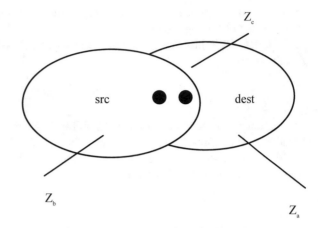

In a couple of conditions, a social occasion can be surveyed by the maximal estimation of the proportion $Z_{src \cap dest}/Z_{src}$ which identifies the circumstance when the partition among src and dest is proportionate to the communication length.

The $M = 1\ 3 + \sqrt{3}\ 2\pi \sim\ = 0$, and a couple of center points are numerical values. In the most negative situations, a portion of the system can happen paying little mind to whether the missed hubs are reachable. An answer presented in [6] depends on a neighbor end plan: Each vehicle checks if each one of the nearby nodes has gotten the imparted message. Each node in the modes 3 and 4 sends message to the nearby node which transmits the own neighbor node list.

4 Proposed Work

In this paper, we proposed a novel algorithm to broadcast the neighboring vehicles and finding the mutual broadcast between the router nodes.

1. protocol BBICR()
2. {
3. IF messages can receives for initial iteration
4. find the Broadcast ID from initial message
5. THEN
6. Create initial execution BTbid in the Broadcast Table for roadside distance.
7. Generate a list Lbid with all the IDs with help of neighbor distance table
8. Fetch route with avg density of vehicle in minimum distance grid or area;
9. Set parameter state=1
10. While (overlap==1)
11. Calculate grid sensing ratio GSR;
12. Vehicle overlap optimal angle with 90(degree towards left or right)
13. if(network is constant)

Roadside distance	Technique name (end-to-end delay)	
	BBICR	BBR
5	0.28	0.41
10	0.15	0.61
15	0.089	0.50
20	0.05	0.42
25	0.02	0.4
30	0.01	0.35

Table 1 Performance table for roadside distance of the proposed approach with the existing algorithm

14. overlap=min 3 vehicle at once;
15. End
16. End
17. calculate GSR;
18. while(GSR>=predefined threshold vehicle count)
19. calculate priority;
20. if(priority is highest for void road)
21. state=0;
22. send state information to its neighboring Vehicle
23. Else
24. calculate GSR(Grid sensing ratio) from vehicle to next upcoming vehicle;
25. End if
26. FOR EACH
27. Vehicle id included same message
28. DO
29. IF id in Lbid
30. remove id from Lbid
31. END IF
32. END FOR
33. }

4.1 Simulation Results

In simulation results, VANET has been implemented using MATLAB. The performance of the proposed approach is evaluated and compared with an earlier work, including intelligence into a VANET and finding the unique grid to improve the safety that makes use of V2V and V2R communication (Figs. 3 and 4; Table 1).

Figure 5 shows the improvement of traffic safety and comfort of driving at different routes with a fixed threshold and minimized delay, traffic intensity, locating vehicles from one-side road for the proposed protocol (Fig. 6; Tables 2 and 3).

Table 2 Performance table for radio range of the proposed approach with the existing algorithm

Radio range	Technique name (PDR)	
	BBICR	BBR
100	7.0	7.1
200	6.2	6.0
300	5.6	4.8
400	5.1	3.8
500	4.8	4.8
600	4.6	4.6
700	4.4	3.1
800	4.1	3.7
900	4.08	3.91

5 Conclusion

VANET simulation bunching of neighbors will not rebroadcast the message. In this way, a few nodes will not be reached. In the most pessimistic scenarios, a segment of the network can happen regardless of whether the missed nodes are reachable. An answer is based on a neighbor disposal plot: Every node checks if every one of the neighbors has gotten the broadcast message. The packet delivery ratio, throughput and end-to-end delay significantly improved; however, in BBICR, the transmission range of the source is divided into grid segments with radio range. It means that there is almost only one vehicle in a division of network area.

Fig. 3 VANET simulation with congestion and collision problem

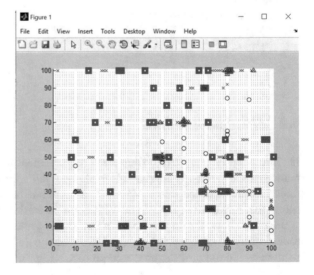

Fig. 4 End-to-end delay with roadside distance for proposed hybrid protocol

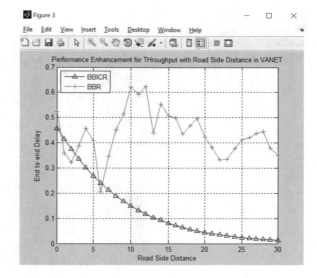

Fig. 5 VANET for packet delivery ratio at different radio ranges at minimum distance

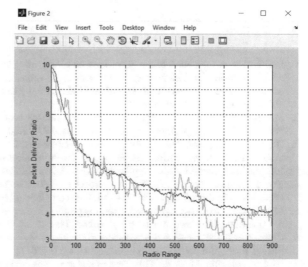

Table 3 Performance table for roadside distance-based throughput ratio of the proposed approach with the existing algorithm

Roadside distance	Technique name (throughput ratio)	
	BBICR	BBR
0	0.58	0.50
5	0.74	0.65
10	0.85	0.70
15	0.91	0.73
20	0.95	0.71
25	0.98	0.63
30	0.991	0.5

Fig. 6 VANET for throughput ratio at different roadside distances using hybrid approach

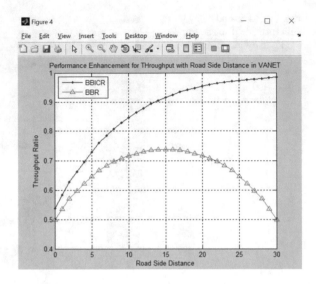

References

1. Latiff, L. et al.: Location-based geocasting and forwarding (LGF) routing protocol in mobile ad hoc network. In: Proceedings of the Advanced Industrial Conference on Telecommunications/Service Assurance with Partial and Intermittent Resources Conference/ELearning on Telecommunications Workshop, IEEE
2. Soares, V.N.G.J. et al.: GeoSpray: a geographic routing protocol for vehicular delay-tolerant networks. Fusion. https://doi.org/10.1016/j.inffus.2011.11.003 (2011)
3. Hughes, L., Maghsoudlou, A.: An efficient coverage-based flooding scheme for geocasting in mobile ad hoc networks. In: Proceedings of the 20th International Conference on Advanced Information Networking and Applications (AINA'06), IEEE (2006)
4. Ko, Y., Vaidya, N.H.: GeoTORA: a protocol for geocasting in mobile ad hoc networks. In: Proceedings of the 8th International Conference on Network Protocols (2000)
5. Lee, K. et al.: GeoCross: a geographic routing protocol in the presence of loops in urban scenarios. Ad Hoc Netw. https://doi.org/10.1016/j.adhoc.2009.12.005 (2010)
6. Chen, Y. et al.: Mobicast routing protocol with carry-and-forward in vehicular ad hoc networks. Fusion. https://doi.org/10.1016/j.inffus.2011.11.003 (2011)
7. Kihl, M., Sichitiu, M., Joshi, H.P.: Design and evaluation of two geocast protocols for vehicular ad-hoc networks. Swedish Governmental Agency for Innovation Systems (Vinnova), 28 Feb 2008
8. Kihl, M., et al.: Reliable geographical multicast routing in vehicular ad-hoc networks. J. Softw. **9**(3), 684–688 (2014)

Energy Efficient Improved SEP for Routing in Wireless Sensor Networks

Deepak Kumar Sharma, Siddhant Bagga and Rishabh Rastogi

Abstract A network created by multiple sensor nodes disseminated in an area to detect numerous physical conditions of region is called wireless sensor network. These sensor nodes are used to gather information from a terrain inaccessible to humans such as battlefields, inside a nuclear reactor, volcanic hotspots, and inside a human body. The nodes can be used to gather data about physical parameters such as weather or other parameters like the radioactivity and radiation density in a region. The data which is collected is sent to a sink called the base station where this data is used for various estimations and calculations. The data gets communicated from the nodes to the sink (base station), either by a multi-hop technique or via direct transmission. The routing protocol based on cluster formation for heterogeneous networks proposed in this paper is a modification of the stable election protocol (SEP), called energy efficient improved SEP (EISEP). It is based on optimizing the probability of selection of the cluster heads of the clusters taking into account various energy factors.The proposed protocol shows improvement in the network lifetime, period of stability, and the throughput of the wireless network in comparison to SEP and LEACH protocol.

Keywords SEP · LEACH · WSN · Clustering · Energy efficient SEP

1 Introduction

A standard wireless sensor network [1] (WSN) is a specific type of wireless ad hoc network where the small-sized sensor nodes have significantly more infrastructure

D. K. Sharma · S. Bagga (✉) · R. Rastogi
Division of Information Technology, Netaji Subhas Institute of Technology, New Delhi, India
e-mail: siddhantbagga1@gmail.com

D. K. Sharma
e-mail: dk.sharma1982@yahoo.co.in

R. Rastogi
e-mail: rishabhrastogi1509@gmail.com

© Springer Nature Singapore Pte Ltd. 2019
S. K. Bhatia et al. (eds.), *Advances in Computer Communication and Computational Sciences*, Advances in Intelligent Systems and Computing 924,
https://doi.org/10.1007/978-981-13-6861-5_13

than the nodes in the ad hoc network. The WSNs involve wireless communication of data and minimal computational capability requirement from the nodes which are involved in data collection. The wireless sensor network is used to sense or gather data relating to a physical object, environmental circumstances, or changes in their state. These include changes in temperature, weather, humidity, air pressure, etc. A particular sensor node in the wireless sensor network comprises of a small-sized sensor, memory to save data temporarily, microcontroller, radio transceiver for communication, and a battery for power supply.

Various factors need to be taken in consideration while designing a routing protocol viz. limited power supply of the battery of the sensor nodes because of their small size, lifetime of the sensor nodes, throughput achieved by the network, period of stability of the network, and the average energy in the network, etc. In order to improve on the above factors, clustering [2]-based routing algorithms were designed. Cluster-based routing algorithms were introduced because direct transmission of the node with the base station resulted in a very high rate of loss of energy. In clustering algorithm, the nodes form a group or cluster and each such group or cluster has a cluster head. This cluster head receives the information sensed by the sensor nodes in that cluster and then compresses it to send these packets of data to the base station via direct transmission or using multi-hop. In this paper, we propose a new routing protocol based on clustering which is an improvement of the conventional SEP protocol. The probability of the selection of the cluster head has been optimized taking into consideration various energy factors. The threshold computation has also been modified taking into account the requisite energy considerations. This protocol has significantly better results than the conventional routing protocols like LEACH and SEP considering the network lifetime, stability period, and the throughput of the network.

2 Related Work

2.1 LEACH (Low-Energy Adaptive Clustering Hierarchy)

LEACH [3, 4] is a conventional routing protocol which involves clustering. LEACH protocol was proposed by W. R. Heinzelman. LEACH is developed for both homogeneous and heterogeneous wireless sensor networks in which the wireless sensor network is broken down into various clusters with one cluster head per cluster which is chosen by a threshold formula which ensures randomness and uniqueness in the cluster head chosen for every group of nodes (cluster). All nodes transfer data to the cluster head which eventually channels it to the base station. The term heterogeneous means that the network contains sensor nodes with different energies initially whereas in homogeneous networks, all the nodes have equal energy in the beginning. In two-level heterogeneous network, only two energy levels are considered where in advance nodes (sensor nodes with higher level of energy) have energy equal to $(1 + a)$ times the initial energy of the normal nodes (sensor nodes with lower level of energy), where a specifies the amount of energy advance nodes have more than that of normal nodes. The threshold is calculated and then, for each node, a random number between 0 and 1 is generated which determines if the sensor will become a cluster head or not. If the chosen random number is smaller than the value of the threshold function then the sensor is chosen as the cluster head in that particular round, else the process is continued for the rest of the sensor nodes.

The threshold formula for LEACH protocol is calculated as

$$T(S_i) = \begin{cases} \frac{P_i}{1-P_i*\left(R \bmod \left(\frac{1}{P_i}\right)\right)}, & S \in N' \\ 0 & \text{Otherwise} \end{cases} \tag{1}$$

P_i is the fraction of nodes which can become the cluster heads in a particular round. It can also be defined as the chances of a node to be chosen as the cluster head. As a matter of fact, this probability is decided prior to the execution of the transmission within network. R refers to the present round number. N' is the group of nodes which are capable of becoming the cluster heads. A particular node becoming the head of the cluster also depends upon the fact that when was the last time the node was chosen as the cluster head. It is because after becoming the cluster head, a node cannot become the cluster head for certain period of time called the epoch $(=1/P_i)$. This election of cluster head is termed as cluster head selection phase. After selection of the cluster head, there is a cluster setup phase in which every sensor intimates the corresponding cluster head to which it will transfer its data and become a part of its cluster. The energy dissipation model followed is the radio model in which if the separation between the source and the sink is greater than do then data are sent using a multi-hop technique and the energy dissipation is proportional to the distance between the source and sink powered to 4, while when the distance is less than do, direct transmission of data takes place and energy used is directly related to the square of the distance between the source and sink of the data. The following formulas are used to calculate energy dissipation

$$d_0 = \sqrt{\frac{E_{\text{fs}}}{E_{\text{mp}}}} \tag{2}$$

where E_{fs} is the energy dissipation factor involved in first step or direct transmission and E_{mp} is the energy dissipation factor involved in multi-hop transmission.

$$E_{\text{diss}}(b, \text{distance}) = \begin{cases} \left((\text{ETX} + \text{EDA}) * (b) + E_{\text{mp}} * b * (\text{distance})^4\right) \text{for distance} > d_0 \\ \left((\text{ETX} + \text{EDA}) * (b) + E_{\text{fs}} * b * (\text{distance})^2\right) \text{for distance} \leq d_0 \end{cases} \tag{3}$$

where *ETX* is the energy required in transmitting from a node, *EDA* is the energy used to aggregate data in a node, b is the size of the message that has to be sent to the information sink or cluster head, and *distance* represents the distance from the source and the sink for the message.

2.2 SEP (Stable Election Protocol)

SEP [5, 6] protocol is an enhancement of LEACH protocol. The SEP protocol is better than the low-energy adaptive clustering hierarchy (LEACH) protocol as chances of a node of being chosen as a cluster head is different for different energy level nodes in the network. The optimal fraction of the total nodes that must be chosen as cluster heads in a particular round is defined as P_{opt} and the probabilities of nodes of different energy levels (normal and advance) becoming a cluster head can be calculated as:

$$P_{nrm} = \frac{P_{opt}}{1 + \alpha m} \tag{4}$$

$$P_{adv} = \frac{P_{opt} * (1 + \alpha)}{1 + \alpha m} \tag{5}$$

α is the additional initial energy factor for advanced nodes; and m is the percentage of advanced nodes in the entire sensor network. The threshold formula for SEP protocol is calculated as

$$T(S_{nrm}) = \begin{cases} \frac{P_{nrm}}{1 - P_{nrm} * \left(r \bmod \left(\frac{1}{P_{nrm}}\right)\right)}, & S_{nrm} \in G' \\ 0, & \text{Otherwise} \end{cases} \tag{6}$$

$$T(S_{adv}) = \begin{cases} \frac{P_{adv}}{1 - P_{adv} * \left(r \bmod \left(\frac{1}{P_{adv}}\right)\right)}, & S_{nrm} \in G'' \\ 0 & \text{Otherwise} \end{cases} \tag{7}$$

where G' and G'' are the group that contains the normal nodes and advance nodes, respectively, which have not become cluster heads yet in the current epochs. The formula for an epoch is given as

$$(E_{nrm}) = \frac{1}{P_{nrm}} \tag{8}$$

$$(E_{adv}) = \frac{1}{P_{adv}} \tag{9}$$

$E_{adv} < E_{nrm}$ since $P_{adv} > P_{nrm}$ which implies that advance nodes will become cluster heads more frequently.

3 Proposed Protocol

This protocol is a modification of SEP. In SEP, there is no consideration of any forms of power (energy) of nodes in the selection of a cluster head in a round. In our proposed protocol, we take into account the initial energy of node, current energy in a node, as well as mean energy in the network in computing the chances of a node to become a cluster head. Moreover, the current energy of a node and the mean energy in the network are also taken into account while computing the threshold. This ensures that the nodes with more energy relative to their initial energies and relative to the average energy of the network have the greatest chances of being chosen as the cluster head. The modified formulas for calculating the probabilities of nodes are as follows:

$$P_{\text{nrm}} = \frac{P_{\text{opt}}}{1 + \alpha m} * \frac{E_c}{E_i} * \frac{E_c}{E_{\text{an}}} \tag{10}$$

$$P_{\text{adv}} = \frac{P_{\text{opt}} * (1 + \alpha)}{1 + \alpha m} * \frac{E_c}{E_i} * \frac{E_c}{E_{\text{an}}} \tag{11}$$

E_c is the current energy of the node, E_i is the initial energy of the node, and E_{an} is the average energy of the network. The threshold calculation is modified as follows:

$$T(S_{\text{nrm}}) = \begin{cases} \frac{P_{\text{nrm}}}{1 - P_{\text{nrm}} * \left(r \bmod \left(\frac{1}{P_{\text{nrm}}} \right) \right)} * \frac{E_c}{E_{\text{an}}}, & S_{\text{nrm}} \in G' \\ 0, & \text{Otherwise} \end{cases} \tag{12}$$

$$T(S_{\text{adv}}) = \begin{cases} \frac{P_{\text{adv}}}{1 - P_{\text{adv}} * \left(r \bmod \left(\frac{1}{P_{\text{adv}}} \right) \right)} * \frac{E_c}{E_{\text{an}}}, & S_{\text{nrm}} \in G'' \\ 0, & \text{Otherwise} \end{cases} \tag{13}$$

G' and G'' are the sets of normal and advance nodes, respectively, which can become the cluster heads.

The above modification has very significant improvement in the lifetime of the network because it ensures that nodes with high ratios of current energy and initial energy and a higher ratio of current energy and mean energy in the network will have the most chances of becoming the cluster head. Since cluster head is involved in the maximum energy dissipation, it is very much understandable to consider the factors of energies in calculation for the chances of a node to become a cluster head and threshold so that nodes with the high energies are selected as cluster heads more often.

Table 1 Simulation setup parameters

Parameters	Description	Value
P_i	Desired fraction of cluster heads in a particular round	0.1
E_0	Initial energy of normal nodes	0.5 J
ETX	Energy dissipated in transmitting data	50×10^{-9} J/bit
ERX	Energy dissipated in receiving data	50×10^{-9} J/bit
E_{fs}	Amplifier energy dissipated in transmission when $d \leq d_0$ (d is the distance between nodes) $d_0 = \sqrt{E_{fs}/E_{mp}}$	10×10^{-12} J/bit/m^2
E_{mp}	Amplifier energy used in transmission when $d > d_0$	1.3×10^{-15} J/bit/m^4
EDA	Data aggregation energy	10×10^{-9} J
r_{max}	Maximum number of rounds	15000
m	Fraction of advance nodes	0.1
α	The advance nodes have α times more energy than normal nodes	1
M	The nodes are deployed in a square field of 100 * 100	100
b	Message size	4000 bits

4 Simulation Setup

For the implementation of the proposed protocol and for comparing the proposed protocols with the conventional protocols, MATLAB is used. In our execution, a square field of network of length and width 100 units is considered and 100 sensor nodes are randomly deployed in the network field. The base station (sink) has been fixed at the centre (50,50) of the field. The simulation setup parameters have been been described as follows (Table 1).

4.1 Performance Evaluation Parameters

Stability Period The time gap between the time when the network starts and till the first node dies in the network.

Throughput The amount of data transmitted successfully per unit time in the network.

Network Lifetime The time period from the commencement of the transmission in the network till the last node dies.

Average Energy It is the mean of the energies of all the sensor nodes present in the network.

There are 15,000 rounds executed for each routing protocol and all the parameters have been computed and compared with the proposed protocol through which a conclusion has been drawn proving the advantages of the proposed protocol over SEP and LEACH.

5 Simulation Results

The above graphs in Figs. 1 and 2 show that the network lifetime of the EISEP is maximum taking in consideration the number of nodes dying as the number of rounds proceed further and the number of nodes alive. The round number at which the last node dies is the maximum for the EISEP. This is because of the energy considerations in both the probability and the threshold calculation in EISEP. The current energy (also called residual energy) is taken into account relative to both the average energy of the network and the initial energy of that particular node. This surely ensures that the nodes with higher energies relative to the average energy of the network and with high ratios current energies of nodes to their initial energies have much more possibilities of becoming the cluster heads.

Figure 3 exhibits that the throughput is maximum for EISEP. Throughput is actually determined by the number of packets sent to the base station per unit time. The slope of the graph of EISEP is maximum in comparison to the other protocols. This is correlated with the previous result since the number of alive nodes in EISEP are significantly greater than the other protocols, the number of packets transmitted to the base station would also be higher as more number of nodes would be involved in transmission.

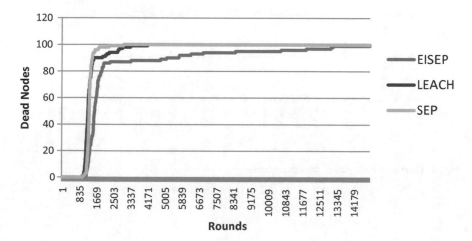

Fig. 1 Graph of number of nodes dying with the rounds

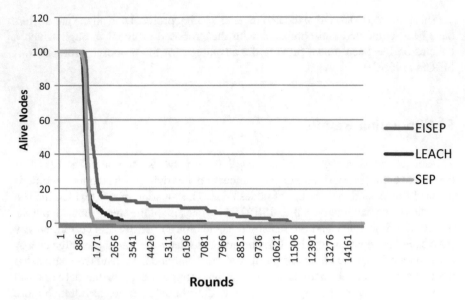

Fig. 2 Graph of number of nodes alive with the rounds

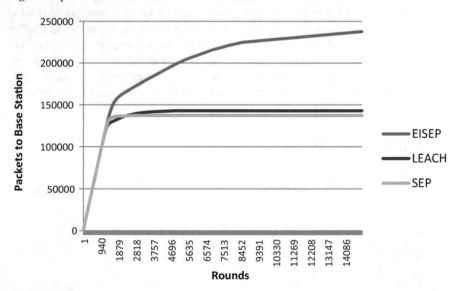

Fig. 3 Packets to base station with the rounds

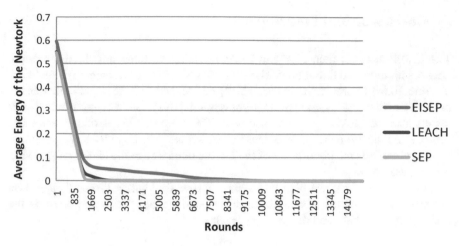

Fig. 4 Average energy of the network with the rounds

Table 2 Quantitative comparison of protocols

	First node dead	Last node dead	Maximum packets sent to base station
LEACH	1003	3233	142,893
SEP	1087	2700	137,248
EISEP	1203	13,990	237,761

Table 3 Qualitative comparison of protocols

	Stability period	Last node dead	Maximum packets sent to base station
LEACH	Low	Moderate	Moderate
SEP	Moderate	Moderate	Moderate
EISEP	20% more than LEACH 10.6% more than SEP	332% more than LEACH 418% more than SEP	66.3% more than LEACH 73.23% more than SEP

Figure 4 shows the graph of average energy of the network of various protocols with the rounds. The initial average energy of all the protocols is similar because all the protocols are implemented as two-level heterogeneity energy levels. However, as the rounds proceed, the average energy of the EISEP is more than the average energy of the other protocols. This is also related to the network lifetime of the routing protocol. Since the nodes last for a longer period of time, the average energy of the network would certainly be higher for EISEP than the rest of the protocols. Quantitative analysis and qualitative analysis of protocols has been shown in Tables 2 and 3 respectively.

6   Conclusion and Future Work

The EISEP is better than SEP and LEACH taking into account stability period, network lifetime, and throughput. The EISEP has 418% improvement in the network lifetime, 10.6% improvement in the stability period, and 73.23% improvement in the packets sent to the base station over the conventional SEP. It shows significantly better results than the LEACH protocol also with 20% more stability period, 332% more lifetime, and 66.3% more packets sent to base station. The significant improvement of results exhibits the importance of the factors of energy in the choice of selection of the cluster head.

In our future work, we plan to further improve the routing protocols based on clustering for WSN using some latest meta-heuristic algorithms and dividing the nodes into zones based on their energy levels.

References

1. Hammoudeh, M., Newman, R.: Adaptive routing in wireless sensor networks: Qos optimisation for enhanced application performance. Inf. Fusion **22**, 315 (2015)
2. Jiang, C., Xiang, M., Shi, W.: Overview of cluster-based routing protocols in wireless sensor networks. In: International Conference on Electric Information and Control Engineering (2011)
3. Heinzelman, W.B., Chandrakasan, A.P., Balakrishan, H.: An application-specific protocol architecture for wireless microsensor networks. IEEE Trans. Wirel. Commun. **1**(4), 660–670 (2002)
4. Thapa, R., Singh, H., Sharma, A.: A comparative analysis of LEACH and SEP using NS2. In: 8th International Conference on Computing, Communication and Networking Technologies (ICCCNT), July 2017
5. Smaragdakis, G.: SEP: a stable election protocol for clustered heterogeneous wireless sensor networks. OpenBU (2004)
6. Sharma, T., Kumar, B., Tomar, G.: Performance comparison of LEACH, SEP and DEEC protocol in wireless sensor network. In: Conference: Proceedings of the International Conference on Advances in Computer Science and Electronics Engineering, January 2012

Part III
Intelligent Computing Technologies

SVM Hyperparameter Optimization Using a Genetic Algorithm for Rub-Impact Fault Diagnosis

Alexander Prosvirin, Bach Phi Duong and Jong-Myon Kim

Abstract Differentiating data samples within various classes is a typical example of machine learning applications. The support vector machine (SVM) is a robust and useful tool capable of resolving the problems of data classification with high accuracy. One of the most reliable kernels that are widely used with SVM to map features into high-dimensional feature space and perform classification is a Gaussian radial basis function (RBF kernel). However, the classification performance of SVM with RBF kernel strongly depends on two hyperparameter values of the trained model. This paper proposes the use of genetic algorithm (GA) as an optimization technique to select proper hyperparameter values for the SVM classifier. GA is known as an intelligent optimization approach based on heuristics that allows finding a relatively good solution to the optimization problem, especially when the exhaustive search of optimal values can be computationally expensive or may not deliver a proper result. In the experimental part of this manuscript, the classification accuracies of the trained one-against-all multiclass SVM (OAA-MCSVM) classifier with hyperparameter values adjusted by the conventional exhaustive grid search optimization algorithm and the proposed one were evaluated and compared using five datasets containing mechanical rub-impact faults of various intensity levels.

Keywords Fault diagnosis · Frequency-domain features · Genetic algorithm · Rub-impact fault · Support vector machines · Optimization

A. Prosvirin (✉) · B. P. Duong · J.-M. Kim
School of Electrical, Electronics and Computer Engineering,
University of Ulsan, Ulsan 44610, South Korea
e-mail: a.prosvirin@hotmail.com

B. P. Duong
e-mail: duongbachphi@gmail.com

J.-M. Kim
e-mail: jmkim07@ulsan.ac.kr

© Springer Nature Singapore Pte Ltd. 2019
S. K. Bhatia et al. (eds.), *Advances in Computer Communication
and Computational Sciences*, Advances in Intelligent Systems and Computing 924,
https://doi.org/10.1007/978-981-13-6861-5_14

1 Introduction

Support vector machine (SVM), first introduced by Vapnik [1], is a powerful machine learning technique that is capable of solving both the problems of regression and classification [2, 3]. SVM has a strong mathematical background and obtaining the solution of SVM requires finding the solution of a well-studied quadratic programming optimization problem [4]. SVM approach supports various kernel functions that are used to map the features extracted from the original data samples into a high-dimensional feature space, where those features can be well-separable. Usually, the choice of a kernel function depends on the knowledge about data structure and distribution. However, the most widely used kernel function for SVM classifier is a Gaussian radial basis kernel function (RBF kernel) due to its easy mathematical formulation and several pros comparing to the other kernel functions [5].

The formulations of SVM when using Gaussian RBF kernel involve two hyper-parameters that should be appropriately assigned (optimized) to enhance the classification accuracy of the trained model. Specifically, these parameters are the penalty coefficient C, which determines the tolerance of classifier to classification errors, and the coefficient of RBF kernel σ which governs the training process. Those two parameter settings may have a profound effect on the resulting performance of the model. The basic idea is to find the combination of these parameters that minimize the generalization error of the trained SVM model. The most commonly used approach for hyperparameter selection is to assign some ranges of values for both parameters, and then apply an exhaustive grid search [3] over the parameter space to find the best setting. Unfortunately, even moderately high-resolution searches using exhaustive approaches can result in a large number of evaluations. Also, grid search does not guarantee that the best-selected values will enhance the classification abilities of SVM model because its capabilities are limited by the initial setup of the grid, i.e., the ranges of values included in it.

Recently, the development of evolutionary algorithms brings new approaches to solve the problems of optimization, such as genetic algorithm (GA). GA is a meta-heuristic approach that is generally used to produce high-quality solutions in mathematical optimization problems based on natural selection, which is inspired by biological evaluation [6]. This study proposes the use of GA for the optimization of hyperparameters for OAA-MCSVM classifier with RBF kernel. GA is capable of providing an improved parametrization of SVM's RBF kernel function and penalty coefficient, enhancing model's classification performance even for complex problems. To investigate the robustness of the proposed method, this paper compares the classification accuracies achieved by OAA-MCSVM models optimized by GA and conventional grid search method while used for resolving a real and complex industrial problem—the diagnosis (classification) of rub-impact faults [7] with various intensities.

The remaining parts of this paper are organized as follows. Section 2 briefly introduces the necessary background about SVM classifier in its standard form and the conventional grid search optimization method. Section 3 contains the necessary information about genetic algorithms and shows how they are utilized in the proposed optimization approach. The experimental evaluation of the GA optimization for OAA-MCSVM model is presented in Sect. 4. Finally, Sect. 5 concludes this paper with a summarization of obtained results.

2 Necessary Background

2.1 Support Vector Machine

SVM is a useful and robust method for data classification, insensitive to the dimensionality of feature vector [8]. The basic underlying idea of this binary classification method is to separate the instances belonging to various classes with a decision surface, maximizing the margin between those classes. For instance, let us consider a given training set of sample-label pairs $(x_i, y_i), i = 1, \ldots, l$ where $x_i \in R^n$ and $y \in \{1, -1\}^l$. SVM classifier requires to solve the following quadratic programming optimization problem:

$$\min_{w,b,\xi} \left(\frac{1}{2} w^T w + C \sum_{i=1}^{l} \xi_i \right)$$

$$\text{subject to } y_i \left(w^T \phi(x_i) + b \right) \geq 1 - \xi_i, \xi_i \geq 0. \tag{1}$$

Here, training vectors x_i are mapped into a higher dimensional space by the mapping function ϕ and ξ_i is a slack variable. SVM aims to find a separating hyperplane with the maximal margin in this high-dimensional space that can separate well the features corresponding to different classes. C is the penalty coefficient of slack variable. It controls the trade-off between the misclassification errors and generalization of the model. High values of the penalty coefficient will strictly penalize misclassified instances, and therefore the resulting hyperplane will be the one that firmly avoids classification errors, even sacrificing generalization capabilities of the classifier. On the other hand, low values only lightly penalize misclassifications, and the result might be an erroneous separation of samples corresponding to different classes.

As was mentioned in the introduction section, one of the most commonly used kernels for SVM classifier is Gaussian RBF kernel, and its formulation is defined as follows:

$$K(x_i, x_j) = \exp(-\gamma \|x_i - x_j\|^2), \ \gamma > 0, \tag{2}$$

where $\gamma = 1/2\sigma^2$ is a kernel parameter. This kernel nonlinearly maps samples into a higher dimensional space and can handle the cases when the relation between the class labels and attributes is nonlinear. The value of σ kernel parameter in this formulation is the second subject of optimization.

2.2   Cross-Validation and Conventional Grid Search

When utilizing SVM model with RBF kernel, it is important to identify suitable pairs of hyperparameters $(C; \sigma)$ to achieve a good classification performance on unknown data. To evaluate the classification accuracy of a trained model, a common strategy is separating the dataset into two parts: one subset is used for training the classifier, and the rest of data is considered as an unknown data and used for performance evaluation. Classification accuracy achieved on the "unknown" subset allows more precisely generalize the performance on classifying an independent, previously "unseen" data samples. Another variant of this technique is called cross-validation. In k-fold cross-validation, the dataset available for training is first should be split into k sets of equal size. Sequentially, each subset is evaluated by the classifier trained on the remaining $k-1$ subsets. Therefore, each data sample in the training set is evaluated once, so the cross-validation accuracy is the percentage of samples that are correctly classified. For SVM model tuning, grid search on C and σ parameters using cross-validation is a common practice used to optimize SVM hyperparameters [3]. Various pairs of $(C; \sigma)$ values for each class represented in training data set should be evaluated, and the ones with the best cross-validation accuracy are then chosen for final training of each binary SVM classifier that are the parts of complete OAA-MCSVM model for multiclass classification.

3   Proposed Optimization Methodology

3.1   Outline of the Proposed Method

Unlikely the exhaustive grid search algorithm which finds pairs of hyperparameters from predetermined ranges for each binary classifier in OAA-MCSVM scheme, the main idea of the proposed GA-based optimization approach is to find the only one optimal combination of hyperparameters that allows classifier to differentiate data samples with the accuracy better than a traditional exhaustive approach. In other words, in this paper, GA is proposed to find an optimal pair of hyperparameters for the classification problem based on the data selected for the training of OAA-MCSVM model. The GA principles are based on Darwin's idea of natural selection. In here, the solutions to the optimization problem are represented as abstract "individuals" in the population. Each solution from the population must be validated using fitness function, where fitness value expresses survivability of the solution, i.e., the probability of becoming a member of the next generation and creating an offspring with identical or better characteristics by passing the genetic information using evolutionary mechanisms such as selection, reproduction, and variation. Within GA framework, reproduction and variation procedures are usually performed by the operators of gene mutation and crossover of chromosomes. The crossover operator combines the characteristics of two different solutions to produce the offspring.

Also, the proper encoding scheme of a real-world optimization problem into a GA representation and the quality of the designed fitness function have a powerful influence on GA performance [9]. The GA capability of the heuristic global search for an optimal solution can be efficiently used to solve the problem of optimal SVM hyperparameter selection. The main difference behind the concepts of the proposed and conventional exhaustive approach is that during the exploration of problem space, GA finds one optimal solution that can be efficiently used for training all individual binary SVM models corresponded to classes represented in the training dataset. At the same time, the referenced grid search optimization method finds one specific pair of hyperparameters from the preassigned ranges of values that can be used only for the particular binary SVM trained to differentiate one specific class from all others.

3.2 Design of Genetic Algorithm Architecture

Encoding and Initial Population Creation. The hyperparameters C and σ were simultaneously encoded to create a chromosome. The initial population was created using the same ranges of values C and σ as it has been specified for the grid search approach: $C = 2^{-5}, 2^{-3}, \ldots, 2^{15}$, $\sigma = 2^{-15}, 2^{-13}, \ldots, 2^3$ [3]. Due to the form of the initial representation of hyperparameters, in this paper, the encoding procedure was applied to the values of power of each parameter. Therefore, the 8-bit ones' complement encoding was used to encode the signed integer values of powers into a binary representation to construct a chromosome which was a bit string, as shown in Fig. 1. Each power of parameters was encoded in a range from $-(2^7 - 1)$ to $+(2^7 - 1)$. The positive integers were represented as themselves, whereas the negative numbers were expressed as the ones' complement of their absolute values. This property allowed encoding to be both simple to implement and able to handle higher precision arithmetic efficiently.

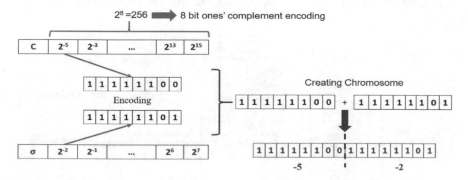

Fig. 1 Encoding scheme

After encoding, both parameters created arrays each of 8-bit length. The chromosome was then constructed by concatenating those arrays into a one 16-bit length string. All the chromosomes were initialized by choosing random values of C and σ from the initial ranges, encoding them, and finally, combining the obtained 8-bit arrays into 16-bit strings.

Fitness Function Design. Finding an optimal pair of hyperparameters by GA is an iterative process. Each generation the whole set of chromosomes in the population should be evaluated by fitness function to select the "fittest" solutions of the current generation. In this paper, OAA-MCSVM classifier directly participated in the quality evaluation process of the chromosomes. First, the chromosome was chosen from the population set. Then, the chromosome was decoded to obtain the real-valued representations of hyperparameter values. Next, these values were involved in the training of each binary SVM model that were the parts of the complete OAA-MCSVM classifier. Finally, the quality of hyperparameters was evaluated using three-fold cross-validation scheme. Specifically, after each iteration of the cross-validation process, the average classification accuracy (ACA) was computed based on the accuracies of every single individual SVM included in the whole multiclass classification scheme. The complete fitness function of the particular chromosome was the averaged classification accuracy, demonstrated by OAA-MCSVM during the threefold cross-validation process. The detailed process of the fitness function computation is presented in Fig. 2.

The Remaining GA Operators. Whereas the fitness function is utilized to select the solutions for reproduction, the reproduction is performed using the mutation and crossover operators. As a selection operator, a tournament selection with a tournament size of three was chosen in this paper. The crossover operator in this study was implemented as a two-point crossover. Specifically, the new child was created by combining the first part of a first parent, the second part of a second parent, and the third part from the first parent, respectively. Since each gene of a chromosome was represented as a binary number due to the proposed encoding scheme, the mutation of binary genes was performed by flipping 1–0 and vice versa in this study. Usually, it

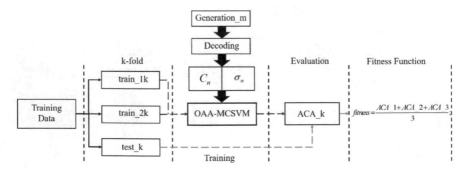

Fig. 2 Process of fitness function computation

is recommended to keep a high rate for crossover operator to enhance the exploration properties, whereas the low mutation rate is recommended to support exploitation of the current situation. Based on a series of experiments, it was observed that the rate of crossover equal to 0.8 and the mutation rate for one gene equal to 0.1 led to the best optimization results in this study.

4 Experimental Results

4.1 Data Collection

Figure 3 shows the test rig with a blade rotor used to acquire experimental rub-impact fault data for evaluating a GA-based OAA-MCSVM optimization approach. For collecting experimental data, rubbing fault was simulated by causing interaction between the rotor blades and rubbing device. Various intensities of rubbing faults were created by adjusting seven different distances between the rubbing device and rotor blades in this study. Specifically, the indicator of distance adjusting equipment showed the distances equal to 2.5, 3, 3.5, 4, 5, 6, and 7 mm during the experiment. The distance of "7 mm" in the indicator corresponded to the position of rubbing device when no interaction with the blades of the rotor was observed. The values from "6" to "3 mm" corresponded to slight rubbing conditions with different intensity levels, and "2.5 mm" was the distance when severe rubbing process occurred in the system. The rubbing fault was detected by the trace of paint on the rubbing device. The signal was collected under constant rotational speed equal to 2400 RPM with a sampling frequency of 65.5 kHz. Four vibration sensors capturing the displacement values of the rotor in various directions were used to collect the information in this experiment.

As the result of the experiment, five datasets containing seven groups of rubbing signals were collected. Each dataset comprised 1260 samples, whereas each group of signals consisted of 180 data instances which were labeled from "7" to "2.5 mm". To investigate the classification performance of the optimized OAA-MCSVM, in

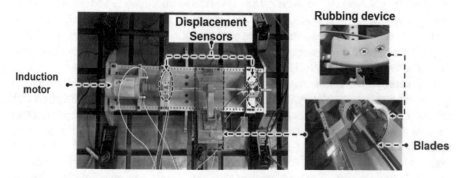

Fig. 3 Experimental testbed for simulating rub-impact faults

this study, each acquired dataset was randomly separated into training and testing subsets. Specifically, each training subset contained 840 data samples and the subset for testing consisted of 420 remaining instances. Each data sample was a 1-s length.

4.2 Extracted Features

The motivation for choosing the datasets containing a complex mechanical fault such as rubbing fault for evaluating GA-based optimization approach for OAA-MCSVM hyperparameters is not occasional and can be explained as follows. As we know, the rub-impact fault is a complex nonlinear and non-stationary mechanical fault that causes many transients appearing in a collected signal. These characteristics make it difficult to apply traditional frequency-domain signal analysis approaches that are based on fast Fourier transform (FFT) for feature extraction [10, 11]. Thus, it was decided to extract three simple frequency-domain statistical feature parameters from FFT frequency spectrum of the acquired vibration signal [12], with prior knowledge that these features cannot represent rubbing fault conditions well. Thus, these low-quality features for rubbing faults can be able to cause a challenge for individual SVM models while distinguishing instances within various classes. Hence, the differences in classification accuracies of those models when optimized by the proposed and conventional approaches can be more observable. Note that four vibration sensors were used to acquire data in this study. Therefore, the total length of the feature vector corresponding to each data instance was equal to 12 feature parameters.

4.3 Results and Discussion

To investigate the capability of the proposed GA-based optimization method to find an optimal pair of hyperparameters for SVM that allows classifier to achieve better accuracy than the classifier tuned by the conventional exhaustive grid search (referred to as GSearch) approach, the classification accuracies achieved by SVM models optimized using both of the above methods were compared in this experiment. Classification of data samples was accomplished by means of OAA-MCSVM. To perform feature extraction, GA implementation, GSearch implementation, and classification routines, MATLAB software was utilized in this study.

In this study, the classifier optimized by the proposed method used only one pair of hyperparameters delivered by GA, whereas the classifier optimized by GSearch was trained in a traditional manner and utilized seven different pairs of hyperparameters selected for each signal group presented in the dataset. That is, SVM itself is a binary classifier which is suitable to classify the samples within two different classes. Its modification OAA-MCSVM for multiclass classification consists of a set of individual SVMs where each of them is trained to classify only one specific class against all the remaining classes. Thus, GSearch provided the pairs of hyperparameters that can be best applied only to classify the specific classes. On the other hand, the search operation performed by GA resulted to one optimal pair of hyperparameters that can be applied to classify every class in the presented dataset and allowed OAA-MCSVM to achieve a better classification performance in comparison with its counterpart. This became possible due to the well-designed fitness function for chromosome evaluation that reflected the classification accuracies of every single individual SVM classifier within OAA-MCSVM structure, trained with a pair of hyperparameters delivered by GA.

The classification accuracies for each class were estimated by true positive rate (TPR) evaluation metric [13], whereas the final classification accuracy was computed as a sum of all true positive samples divided by the total number of signal instances presented in the specific testing dataset. The experimental results are tabulated in Table 1.

Table 1 shows that the classification accuracies achieved by OAA-MCSVM optimized with GA for each dataset were higher than those demonstrated by the model tuned using GSearch approach. In the result, GA appeared to be capable of finding one "relatively good" solution for all the classes presented in each dataset. The smallest accuracies and the most significant drops of TPR values for both of the methods were observed in the third dataset. After examining this dataset more precisely, it was concluded that this behavior was mostly caused by the appearance of additional noise in the experimental environment during the data collection.

Overall, the experimental results demonstrated that the optimal solutions found by GA were suitable for training OAA-MCSVM classifier to resolve complex classification problems. The performance achieved by the classifier optimized by GA outperformed the one optimized by the conventional scheme when solving the problem of diagnosing rub-impact faults of various intensities using not-informative low-quality features. The examination of TPRs allows concluding that GA-optimized OAA-MCSVM classifier yielded with relatively good solutions for all the individual classes maximizing the overall classification accuracy.

Table 1 Experimental results

Data set	Method	TPR (%)							Acc.
		2.5 mm	3 mm	3.5 mm	4 mm	5 mm	6 mm	7 mm	(%)
1	Proposed	100	95	95	75	98.3	100	100	94.7
	GSearch	100	93.3	93.3	68.3	96.6	96.6	100	92.6
2	Proposed	100	88.3	95	88.3	95	96.6	100	94.7
	GSearch	100	93.3	76.6	90	96.6	93.3	98.3	92.6
3	Proposed	100	53.3	70	95	95	96.6	100	87.1
	GSearch	100	35	51.6	93.3	93.6	88.3	95	79.2
4	Proposed	100	86.6	98.3	90	91.6	96.6	98.3	94.5
	GSearch	100	85	93.6	80	83.3	95	93.3	89.7
5	Proposed	100	100	83.3	70	91.6	98.3	100	92
	GSearch	100	100	80	46.6	88.3	93.3	96.6	86.4

5   Conclusion

This paper proposed an intelligent framework for OAA-MCSVM hyperparameter values optimization using GA. For a search process of optimal hyperparameters, the traditional genetic algorithm operators, such as selection, crossover, and mutation were employed in this study. 8-bit ones' complementary encoding was applied to express signed real-valued solution variables as chromosomes, which were represented as bit strings each of 16-bit length. To apply selection operator, specific handcrafted fitness function was introduced in this study. In the experimental part, the low-quality frequency-domain features were extracted from the vibration signals corresponded to various rubbing fault intensity levels to create a challenge for OAA-MCSVM classifier. Those features were used as an input data for OAA-MCSVM, hyperparameter values of which were optimized by GA and the conventional exhaustive grid search. The experimental results demonstrated that GA-based hyperparameter selection for OAA-MCSVM was able to improve the classification abilities of the model even when it was trained on low-quality features, i.e., frequency-domain features extracted by FFT for rubbing fault diagnosis. The performance achieved by the classifier optimized using GA outperformed the exhaustive grid search optimization method for all testing datasets presented in this study.

Acknowledgments This work was supported by the Korea Institute of Energy Technology Evaluation and Planning (KETEP) and the Ministry of Trade, Industry and Energy (MOTIE) of the Republic of Korea (No. 20181510102160, No. 20162220100050, No. 20161120100350, and No. 20172510102130). It was also funded in part by the Leading Human Resource Training Program of Regional Neo Industry through the National Research Foundation of Korea (NRF) funded by the Ministry of Science, ICT and future Planning (NRF-2016H1D5A1910564), and in part by the Basic Science Research Program through the National Research Foundation of Korea (NRF) funded by the Ministry of Education (2016R1D1A3B03931927).

References

1. Vapnik, V.N.: An overview of statistical learning theory. IEEE Trans. Neural Networks **10**, 988–999 (1999)
2. Gunn, S.R.: Support vector machines for classification and regression 66
3. Hsu, C.-W., Chang, C.-C., Lin, C.-J.: A practical guide to support vector classification 16
4. Boser, B.E., Guyon, I.M., Vapnik, V.N.: A training algorithm for optimal margin classifiers 9
5. Zhao, X., Zhao, K.: Study of a new online least squares support vector machine algorithm in gas prediction. Presented at the July 2008
6. Whitley, D.: A genetic algorithm tutorial. Stat. Comput. **4**, 65 (1994)
7. Rubio, E., Jáuregui, J.C.: Time-frequency analysis for rotor-rubbing diagnosis. In: Advances in Vibration Analysis Research. InTech (2011)
8. Wu, X., Kumar, V., Ross Quinlan, J., Ghosh, J., Yang, Q., Motoda, H., McLachlan, G.J., Ng, A., Liu, B., Yu, P.S., Zhou, Z.-H., Steinbach, M., Hand, D.J., Steinberg, D.: Top 10 algorithms in data mining. Knowl. Inf. Syst. **14**, 1–37 (2008)
9. Beasley, D.: An overview of genetic algorithms: part 1. Fundamentals 16

10. Xiang, L., Tang, G., Hu, A.: Analysis of rotor rubbing fault signal based on Hilbert-Huang transform. In: International Conference on Measuring Technology and Mechatronics Automation (2009)
11. Zhao, Y., Liu, E., Zhu, J., Zhang, B., Wang, J., Tian, H.: Rub-impact fault diagnosis of rotating machinery based on hilbert-huang transform 5
12. Rauber, T.W., de Assis Boldt, F., Varejao, F.M.: Heterogeneous feature models and feature selection applied to bearing fault diagnosis. IEEE Trans. Industr. Electron. **62**, 637–646 (2015)
13. Nguyen, P.H., Kim, J.-M.: Multifault diagnosis of rolling element bearings using a wavelet kurtogram and vector median-based feature analysis. Shock Vibr. **2015**, 1–14 (2015)

Adaptive Credit Card Fraud Detection Techniques Based on Feature Selection Method

Ajeet Singh and Anurag Jain

Abstract Credit card fraud is a crucial issue that has been faced by cardholder and card issuing companies for decades. Credit card frauds are performed at two levels, application-level frauds and transaction-level frauds. This paper focus on credit cards fraud detection at application level using features selection methods. In this paper, J48 decision tree, AdaBoost, Random Forest, Naive Bayes, and PART machine learning techniques have been used for detection of financial frauds of a credit card and the performance of these techniques are compared on the basis of the five parameters namely sensitivity, specificity, precision, recall, MCC, and accuracy. A German credit dataset is used to evaluate these machines learning techniques efficiency based on filter and wrapper features selection method. The experiment outcomes show that the prediction accuracy of J48 and PART has been increased after applying filter and wrapper methods. Finally, precision and sensitivity of J48, AdaBoost, and the random forest have been enhanced.

Keywords Credit card fraud · Fraud detection · Feature selection · Machine learning technique

1 Introduction

Financial fraud (FF) is a growingly serious problem, a big concern of organizations and requires attention. Financial frauds may be carried out by many organizations such as government and private [1]. The banking fraud, insurance fraud, corporate fraud, and telecommunication fraud are called as financial fraud. Credit card frauds are the core part of banking frauds. Credit card fraud (CCF) happens when somebody

A. Singh (✉) · A. Jain
University School of Information, Communication and Technology, Guru Gobind Singh
Indraprastha University, Delhi, India
e-mail: ajeetsinghiet@gmail.com

A. Jain
e-mail: anurag@ipu.ac.in

© Springer Nature Singapore Pte Ltd. 2019
S. K. Bhatia et al. (eds.), *Advances in Computer Communication
and Computational Sciences*, Advances in Intelligent Systems and Computing 924,
https://doi.org/10.1007/978-981-13-6861-5_15

Table 1 Year-wise financial
loss due to credit and debit
card frauds

Year	Amount (in billion)
2012	$11.27
2013	$13.70
2014	$18.11
2015	$21.84
2016	$24.71
2017	$27.69

uses credit card information without knowledge of cardholder for personal reason. Therefore, the universal CCFs are application fraud, phishing, counterfeiting, identity theft fraud, and skimming fraud.

According to Mieke et al. [2], the definition of fraud is "Fraud always involves one or more persons who, with intent, act secretly to deprive another of something of value, for their own enrichment." Billions of dollars have been lost due to financial fraud in the globe each year. Bank of America has to pay 16.5 billion USD to settle down financial fraud cases [3]. Another example is that of 17,504 cases of government and private banks fraud reported between 2013 and 2017 in India. According to RBI Data, the total loss has reached 10,12,89,35,216.40 US Dollar (Rs. 66,066 Crore) [Source: RBI data via TOI]. Fraudulent always struggles to find out the vulnerability of the e-commerce system to get credit card details. Vulnerability means the weakness of the cyber-physical system [4].

The professionals and researchers need more concentration to decrease financial loss. With the purpose of mitigating the financial fraud problems, there is need of optimizing a fraud detection technique to "maximize correct predictions and minimize incorrect predictions." FF detection is essential to prevent and detect the financial losses. Hence, financial frauds have to be detected within a feasible time predict so that fraudsters can be prevented from carrying out unauthorized activities [5]. Table 1 depict the financial loss effect from the plastic card frauds, it is based on the Nilson Report (https://nilsonreport.com).

Feature selection is the most important task in data mining applications to reduce irrelevant, redundant features from the real dataset and it improves learning performance [14]. The accuracy of most of the classifiers has increased after applying feature selection methods; we know that irrelevant features contain noisy data that affect the classification accuracy negatively. Feature selection method is used to decrease the impact of credit card frauds and achieve good fraud detection rate.

The purpose of this study is to propose an advanced machine learning model to detect credit card fraud based on feature selection approach. The objective of this study is to provide a comparative study of five machine learning techniques (J48, AdaBootM1, Random Forest, Naive Bayes, and PART) and German credit dataset is used to compare machine learning techniques statistics [6, 10, 11].

The paper is designed as follows: Sect. 2 explains various feature selection methods. In Sect. 3, machine learning (ML) techniques are discussed. In Sect. 4, the research methodology is discussed. The setup of the experiment is explained in Sect. 5, and five machine learning results are compared in Sect. 6. Section 7 provides a conclusion followed by some discussion.

2 Feature Selection Methods

Feature selection (FS) is called an attribute selection or variable subset selection method. Feature selection is one of the most widespread and important techniques for decrease feature dimensionality without compromising the performance of methodology. Those features do not give the useful information called as irrelevant features and those features do not offer more information called as redundant features. The feature selection method (FSM) can reduce the data and the computational complexity.

The major goal of the FSM is to improve learning performance to enhance accuracy and comprehensibility. The most universal feature selection methods are based on the theory discussed in [7, 8].

Filter Method: This method is also called an open-loop method and used in data preprocessing phase. The selections of the best subset of the features are independent of any learning algorithms used. Filter method evaluates the dataset on the basis of the correlation between the attributes. The advantages of this method are computationally much more efficient and better generalization ability than the wrapper and embedded method.

In this method; AttributeEval, Gain RatioAttributeEval, Information gain, Chi-square test, Correlation coefficient, Linear discriminate analysis, and Variance threshold techniques with a search method are used to find the correlation coefficients between features.

Wrapper method: Wrapper method is a closed-loop method to evaluate problem on the basis of the learning algorithm for subset evaluation. Therefore, the performance of the wrapper method is better. Search methods of wrapper method are forward feature selection, backward feature elimination, recursive feature elimination, etc.

Embedded: This method is a combined advantage of both filter and wrapper method which embed the feature selection with the model learning in the guide feature evaluation. Examples of this method are LASSO and RIDGE regression.

3 Machine Learning Techniques

This section discussed the most frequently used machine learning techniques (MLTs) in various domains and these different MLTs are employed on German dataset to detect the credit card fraud at the application level with help of feature selection method.

J48: Ross Quinlan developed a J48 algorithm which is used to produce a decision tree for classification. This classifier is called an optimized implementation of the C4.5 algorithm and an extension of the ID3 algorithm [9]. The outcomes of J48 come in the form of the decision tree. In a J48 classifier, features are used for decision trees pruning, missing values, continuous features value ranges, and derivation of rules, etc.

AdaBoost: Adaptive Boosting (AdaBoost) is a boosting and machine learning meta-algorithm. This technique merges the weak classifier algorithm into a strong classifier form [10]. This classifier is used in conjunction with various types of learning algorithms ("weak learners") to enhance their performance. The result of the weak learners is combined by a weighted sum which shows the concluding result of the boosted classifier. This classifier is used for minimizing the error of learning methods that produce a slightly enhanced performance. An advanced version of the AdaBoost algorithm is called AdaBoostM1.

Random Forest (RF): Random forest is also an ensemble learning techniques of the decision tree [6]. This classifier is used for classification, regression tasks, and feature engineering. The fundamental concept of ensemble methods is that a collection of "weak learners" can come together to form a "strong learner." It grows numerous decision trees and each individual decision tree is a "weak learner," whereas all the decision trees took together are a "strong learner." Each tree provides a classification output or "vote" for a class. This method is fast, and efficiently handles large databases and unbalanced with thousands of features.

Projective Adaptive Resonance Theory (PART): This is an efficient and accurate machine learning technique for deriving learning rules by generating partial decision trees again and again. The key idea of the PART classifier is based on a C4.5 algorithm and RIPPER learning scheme.

PART classifier can learn a very good rule set without using a global optimization by learning one rule at a time. The PART classifier has used learning rules, inferring techniques and separate & conquer technique to detect credit card frauds. PART overcomes the huge slow performance of the C4.5 algorithm on pathological datasets by not using post-processing steps [11].

Naive Bayes (NB): NB is a supervised machine learning (SML) algorithm that can learn and represent probabilistic knowledge. This classifier is a statistical approach based on Bayes theorem with the strong or naïve assumption for classification, NB opts the decision based on the highest probability [12].

Bayes rule is used in order to calculate the probability of every transaction when given the data examples. NB algorithm is based on two simplifying hypotheses. The first hypothesis states that the data features used in the prediction are conditionally independent while given the class label and second hypothesis states that there is no hidden feature or latent feature that could affect the process of prediction.

4   Research Methodology

Credit card fraud (CCF) detection is a binary classification and transactions predicated and classified as a fraudulent transaction and legitimate transaction. This study has carried out a variety of research activates; the first one is the collection of the dataset from the UCI repository. The second one is a data preprocessing to remove redundant data in a dataset. In the third phase of the credit card fraud detection, the evaluation metrics are chosen to measuring the performance of the classifiers using machine learning techniques on German credit card dataset discussed and finally, the impact of the FS method has been evaluated and results of the classifications approaches are compared.

4.1   German Dataset

German credit dataset has been used in this study to classify the transaction in either of the categories. The dataset is available in the UCI repository [13] and it is highly imbalanced. The dataset contains 1000 instances (loan applicants) and 21 features (with 7 numerical, 14 categorical/nominal) where each instance describes the credit status of an individual that can be a good or bad, entry in the dataset represents a person who takes a credit from a bank. Every person is classified as good or bad credit risk according to a set of features. Table 2 presents a brief summary of the dataset features, data type, and nature of features.

4.2   Data Preprocessing

Data preprocessing is an important work in data mining and machine learning applications which is used to eradicate irrelevant, redundant features, and improves learning performance. Data preprocessing methods are namely data cleaning, integration, transformation, and reduction. These techniques can be used to preprocess the unbalanced dataset. The aim of data pre-processing is to enhance the performance of the classifiers by reducing training time, and computation time of classifiers. In this study, the credit card dataset contains 21 features and identified that no missing value, redundant features are found.

4.3   Parameters Matrix

In this study, the following standard evaluation metrics [14, 15] are used to evaluate the results of five learning techniques with and without feature selection methods.

Table 2 Features of German credit card dataset

S. No.	Features	Type	Description
1	Status of existing checking account	Nominal	'<0', '$0 \leq X < 200$', '≥ 200' salary assignments for at least 1 year, and 'no checking'
2	Duration in month	Numerical	Credit duration minimum is 4 months and maximum 72 months
3	Credit_history	Nominal	Applicant history: no credits/all paid, all paid, existing paid, delayed previously, etc
4	Purpose	Nominal	The purpose of credits for education, vacation, other
5	Credit amount	Numerical	Minimum 250 and maximum 18,424 amount
6	Savings _status	Nominal	Savings account/bonds. '100', '$100 \leq X < 500$', '$500 \leq X < 1000$', '≥ 1000', 'no known savings'
7	Employment	Nominal	Applicant is employed ('<1', '$1 \leq X < 4$', '$4 \leq X < 7$', '≥ 7' ($X = $ Year)) or unemployed
8	Installments commitment	Numerical	Numbers of installments are 1, 2, 3, and 4
9	Personal status	Nominal	The applicant is male or female, and single, married, divorced
10	Other _parties	Nominal	Meansco-applicant, guarantor and none [none, "co applicant", guarantor]
11	Residence since	Numerical	The current address of person leaving since
12	Property magnitude	Nominal	Means real estate, life insurance, car, no known property
13	Age	Numerical	Age of Person for apply credit card
14	Other payment plans	Nominal	Bank, stores, and none
15	Housing	Nominal	A Person has own, rent, and for a free house
16	Existing credits	Numerical	Current credit limit
17	Job	Nominal	The applicant is an unemployed/unskilled non-resident, unskilled resident, skilled employee/official, etc.
18	Num dependents	Numerical	Number of family members dependent on the applicant
19	Own telephone	Nominal	Have own telephone or not
20	Foreign worker	Nominal	The applicant is a foreign worker or not
21	Class	Nominal	It means good and bad credit

Here, positives (P) means a number of fraud applications and negatives (N) means a number of the non-fraud applications.

True Positive (TP): The number of fraud application correctly predicted fraud application. **True Negative (TN)**: The number of non-fraud application correctly predicated non-fraud application.

False Positive (FP): Number of the non-fraud application incorrectly predicated fraud application. **False Negative (FN)**: Number of fraud application incorrectly predicated non-fraud application.

Precision or Hit rate: This measure gives the accuracy in cases classified as positive.

$$\text{Precision} = \frac{TP}{(TP + FP)} \tag{1}$$

Recall (sensitivity or true positive rate or fraud catching rate or fraud detection rate): This measure gives the accuracy on positive (fraudulent) cases classification.

$$\text{Recall (Sensitivity)} = \frac{TP}{(TP + FN)} \tag{2}$$

Specificity: It gives the accuracy on the non-fraud cases classification.

$$\text{Specificity} = \frac{TP}{(FN + TN)} \tag{3}$$

False Positive Rate: False positive rate is known as false alarm rate. In fraud detection, the legal application is classified as fraud.

$$FPR = \frac{FP}{(FP + TN)} \tag{4}$$

Mathews Correlation Coefficient: MCC is used to measure the quality of binary classifications problem. The MCC is mainly used for the unbalanced datasets because its evaluation consists of true positive, true negative, false positive, and false negative. The MCC value is usually between -1 and $+1$. A $+1$ value represents excellent classification while a -1 value represents the total distinction between classification and observation.

$$MCC = \frac{(Tp * Tn) - (Fp * Fn)}{\sqrt{(Tp + Fp)(Tp + Fn)(Tn + Fp) + (Tn + Fn)}} \tag{5}$$

Accuracy: The number of correct predictions of positive application divided the total number of predictions.

$$\text{Accuracy} = \frac{(\text{TP} + \text{TN})}{(\text{TP} + \text{TN} + \text{FP} + \text{FN})} \tag{6}$$

5 Experiments Setup

The number of fraud cases is generally a very small as compared with the total number of the cases. All experiments focused on 10 folds cross-validation approach because the dataset is imbalanced. This is used for moderately sized samples in the process of training and testing the different classification data mining techniques.

The machine learning tool has been used to perform the experiment on the dataset to classify fraud and non-fraud applications. Filter method (Information gain) and wrapper methods are employed to select the best features before passing a dataset to classifiers. Experimental outcomes of five classifiers are tabulated in Tables 3, 4, and 5 without feature selection and with feature selection respectively.

Information gain approach has been applied in the original dataset to identify the correlation coefficient between the 21 features and four features removed from the original dataset because these features have no correlation with good attributes. Table 4 shows the result of five classifiers after deployed on the remaining 17 features.

The wrappersubseteval method has been applied to the original dataset for evaluates attribute sets by using a learning scheme. The outcome of this method indicated that the prediction accuracy of classifiers has improved when 5, 4, and 6 features are selected.

Table 3 Experimental outcomes of five classifiers without feature selection method

Parameter	Classifiers				
	J48	AdaBoost M1	Random forest	Naive Bayes	Part
Sensitivity	0.840	0.877	0.917	0.864	0.801
Specificity	0.390	0.270	0.407	0.497	0.47
Precision	0.763	0.737	0.783	0.800	0.779
Recall	0.840	0.877	0.917	0.864	0.801
MCC	0.251	0.181	0.386	0.385	0.277
Accuracy (%)	70.5	69.5	76.4	75.4	70.2
Instances	1000	1000	1000	1000	1000
Attributes	21	21	21	21	21

Table 4 Experimental outcomes of five classifiers after filter method

Parameter	Classifiers: filter method				
	J48	AdaBoost M1	Random Forest	Naive Bayes	Part
Sensitivity	0.841	0.883	0.910	0.861	0.803
Specificity	0.453	0.257	0.423	0.477	0.47
Precision	0.782	0.735	0.786	0.793	0.781
Recall	0.841	0.883	0.910	0.861	0.803
MCC	0.313	0.175	0.389	0.363	0.282
Accuracy	72.5	69.6	76.4	75.4	70.4
Instances	1000	1000	1000	1000	1000
Selected attributes	17	17	17	17	17

Table 5 Experimental outcomes of five classifiers after wrapper method and evaluation metrics. Some features are removed from the original dataset by wrapper method

Parameter	Classifiers: wrapper method				
	J48	AdaBoost M1	Random forest	Naive Bayes	Part
Sensitivity	0.919	0.893	0.860	0.897	0.854
Specificity	0.343	0.400	0.457	0.407	0.403
Precision	0.765	0.776	0.787	0.779	0.770
Recall	0.919	0.893	0.860	0.897	0.854
MCC	0.327	0.339	0.342	0.352	0.284
Accuracy	74.6	74.5	73.9	75.0	71.9
Instances	1000	1000	1000	1000	1000
Selected attributes	05	05	04	05	06

6 Results Analysis

This section presents the comparison between five classifiers which applied to the dataset to detect fraud applications [16]. The five classifiers' performance is compared based on feature selection methods with help of evaluation metrics (such as specificity, precision, recall, MCC, and accuracy). These techniques are trained and investigated by using 10-fold cross-validation approaches. Each technique runs ten times and average results are shown in the figures (from 1a, b and 2a, b, c). According to Fig. 1a, the performance of J48, AdaBoost, Naive Bayes, and PART classifiers has increased after applying filter and wrapper methods but in the case of random forest classifier, performance has decreased. From Fig. 2b, predication accuracy of J48 (from 70.5 to 72.5 and 74.6) AdaBoost (from 69.5 to 69.6 and 74.5), and PART (from 70.2 to 70.4, and 71.9) has increased by the filter and the wrapper method.

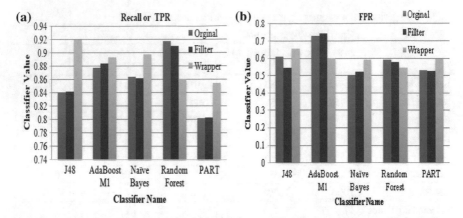

Fig. 1 **a** Comparison of recall of different classifier applied on a dataset with and without feature selection methods. **b** Comparison of FPR of different classifier applied on a dataset with and without feature selection method. J48, Naïve Bayes, and PART classifiers are most affected by wrapper method

And also precision of J48 (from 0.76 to 0.782, 0.765), AdaBoost (from 0.737 to 0.735, 0.776), and random forest (from 0.783 to 0.786, 0.787) have enhanced.

Finally, The J48 technique is most affected and showed superior performance across the evaluation metrics used for feature selection method. It reached the highest value for specificity and precision (that is 1) for the two data distributions (see Fig. 1a, b).

7 Conclusion and Future Work

In this study, credit card fraud (CCF) has been classified into two categories application-level frauds, and transaction-level frauds. The German dataset has been used to detect application-level frauds through machine learning techniques (MLTs) with help of feature selection methods. The use of MLTs is benefited from filter and wrapper methods for selecting very high correlated features that are effective for decreasing the runtime, incorrect prediction, and increasing the prediction accuracy of classifiers.

This study compares and analyzes the performance of five MLTs with feature selection method to detect CCF at the application level. The most important parameters (prediction accuracy, MCC, sensitivity, specificity, precision, and recall) are considered for performance evaluation.

It is observed that the prediction accuracy of J48 classifier and PART classifier has increased from 70.5% to 74.6% & 70.2% to 71.9% using information gain method

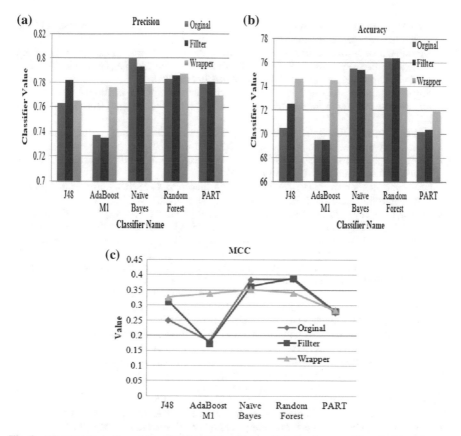

Fig. 2 **a** Comparison of precision of different classifier applied on a dataset with and without feature selection method. Based on Fig. 2a, J48 and random forest are the most affected classifiers because of used both feature selection methods. **b** Presented an accuracy comparison of five classifiers between two feature selection methods and original dataset. **c** Comparison of Mathews correlation coefficient of five classifiers with and without feature selection methods

and wrappersubseteval method. The precision of J48 classifier AdaBoost classifier and random forest classifier are also improved (78.2%, 77.6%, and 78.7%).

Acknowledgements Sincere thanks to the University Grants Commission (UGC), Delhi, India for providing fellowship to work on a research problem. We also thank USICT, Guru Gobind Singh Indraprastha University, Delhi, India to research ambiance for carrying out research.

References

1. Phua, C., Gayler, R., Lee, V., Smith-Miles, K.: A comprehensive survey of data mining based fraud detection research (2005)

2. Jans, M., Lybaert, N., Vanhoof, K.: A framework for internal fraud risk reduction at IT integrating business processes: the IFR framework (2010)
3. Department of Justice, Office of Public Affairs. Bank of America to pay $16.65 Billion in Historic Justice Department Settlement for Financial Fraud Leading up to and During the Financial Crisis. August 21, 2014. Last accessed 10 Mar 2018
4. Singh, A., Jain A.: Study of cyber attacks on cyber-physical system. In: 3rd International Conference on Advances in Internet of Things and Connected Technologies (ICIoTCT), Springer 26, 27 Mar 2018
5. Bhattacharyya, S., Jha, S., Tharakunnel, K., Westland, J.C.: Data mining for credit card fraud: a comparative study. Decis. Support Syst. **50**(3), 602–613 (2011)
6. Seeja, KR., Zareapoor, M.: FraudMiner: a novel credit card fraud detection model based on frequent itemset mining. Sci. World J. (2014)
7. Sheikhpour, R., Sarram, M.A., Gharaghani, S., Chahooki, M.A.Z.: A survey on semi-supervised feature selection methods. J. Pattern Recog. **64**, 141–158 (2017)
8. Panday, P.: Analysis of machine learning techniques (J48 and AdaBoost) for classification. In: 2016 1st India International Conference on Information Processing (IICIP), 12–14 Aug 2016
9. Karabulut, E.M., Özel, S.A., Ibrikci, T.: A comparative study on the effect of feature selection on classification accuracy. Procedia Technol. **1**, 323–327 (2012)
10. Randhawa, K., Loo, C.K., Seera, M., Lim, C.P., Nandi, A.K.: Credit card fraud detection using AdaBoost and majority voting. IEEE Access **6**, 14277–14284 (2018)
11. Awoyemi, J.O., Adetunmbi, A.O., Oluwadare, S.A.: Credit card fraud detection using machine learning techniques: a comparative analysis. In: 2017 International Conference on Computing Networking and Informatics (ICCNI), pp. 1–9. IEEE (2017)
12. Mishra, M. K., Dash, R.: A comparative study of chebyshev functional link artificial neural network, multi-layer perceptron and decision tree for credit card fraud detection. In: 2014 International Conference on Information Technology (ICIT), pp. 228–233. IEEE (2014)
13. UCI Repository. https://archive.ics.uci.edu/ml/. Last accessed 3 Mar 2018
14. Fadaei Noghani, F., Moattar, M.H.: Ensemble classification and extended feature selection for credit card fraud detection. J. AI Data Min. **5**(2), 235–243 (2017)
15. Xuan, S., Liu, G., Li, Z.: Random forest for credit card fraud detection. In: 2018 IEEE 15th International Conference on Networking, Sensing, and Control, 27–29 Mar 2018
16. Stolfo, S.J., Fan, D.W., Lee, W., Prodromidis, A.L.: Credit card fraud detection using metalearning: issues and initial results (1999)

Emousic: Emotion and Activity-Based Music Player Using Machine Learning

Pranav Sarda, Sushmita Halasawade, Anuja Padmawar
and Jagannath Aghav

Abstract In this paper, we propose a new way of personalized music playlist generation. The mood is statistically inferred from various data sources primarily: audio,
image, text, and sensors. Human's mood is identified from facial expression and
speech tones. Physical activities can be detected by sensors that humans usually
carry in form of cellphones. The state-of-the-art data science techniques now make
it computationally feasible to identify the actions based on very large datasets. The
program learns from the data. Machine learning helps in classifying and predicting
results using trained information. Using such techniques, applications can recognize
or predict mood, activities for benefit to user. Emousic is a real-time mood and activity recognition use case. It is a smart music player that keeps learning your listening
habits and plays the song preferred by your past habits and mood, activities, etc.
It is a personalized playlist generator.

Keywords Data science · Emotion/mood and activity recognition · Music
analysis · Music playlist generator · Classification

1 Introduction

Smart gadgets people carry everyday with them can obtain lot of data. Data from
fitness band, emotion detection from face captured by smartphone, and activities
detected by the sensors, further can be used for various applications. One of the

P. Sarda (✉) · S. Halasawade · A. Padmawar · J. Aghav
College of Engineering Pune, Wellsley Road, Pune, Maharashtra, India
e-mail: sardapranav1@gmail.com

S. Halasawade
e-mail: sushmita.halasawade@gmail.com

A. Padmawar
e-mail: anujapadmawar.ap@gmail.com

J. Aghav
e-mail: jagannath.aghav@gmail.com

© Springer Nature Singapore Pte Ltd. 2019   179
S. K. Bhatia et al. (eds.), *Advances in Computer Communication
and Computational Sciences*, Advances in Intelligent Systems and Computing 924,
https://doi.org/10.1007/978-981-13-6861-5_16

very precise and straight approaches to detect mood is using human facial expressions. Most of the time, emotion is revealed by face itself. Industries have their trained model for the emotion extraction and are now delivering customer services as frameworks providing API with help of cloud-based services. We used Microsoft's cognitive services and Google's own activity recognition API in our approach for faster implementation. Machine learning can be used for classifying music into set of particular emotions. Once, all these data are present, user can be studied about his/her preferences and habits of listening by time, mood, activity, etc. Training on this data can generate better playlist for future listening. Objective behind this work is to let daily factors get considered for better music recommendation.

2 Literature Survey

2.1 On Emotion Recognition

The mood is statistically inferred from various data sources primarily: audio, image, text, and sensors. In paper [1], author used boosted tree classifier for emotion extraction from short video sequence using audio and video, classified them into seven emotion categories. For audio feature extraction, openSMILE [2] toolkit is used. This model gives better accuracy for three emotions viz. angry, happy and neutral. In [3], Author et al. proposed an approach for analyzing the extracted facial features, with artificial neural network (ANN) used to classify those into six emotions viz. anger, happy, sad, disgust, surprise, and fear. Gabor Wavelets and Markov random fields are used in [4, 5], respectively.

Paper [6] is about emotion detection in voice from voice mail messages. Three different types of training sets viz. PhoneShell messages, CallHome corpus, and Oasis database were used. Feature vectors are trained using hidden Markov models (HMM) emotion wise. They have also used Gaussian mixture models and Zwickers model for loudness. Text independent method is presented in [7] for emotion recognition from speech. Hidden Markov model is proposed for the classification of speech emotions into six categories as disgust, fear, anger, joy, and sad. In paper [8], author proposed GMVAR model as the statistical classifier for modeling the temporal structure of the data which is useful for speech emotion recognition. In [9], the author designed oneclassinoneneural (OCON) network [10] (i.e., for each of eight emotions different sub-neural networks) for emotion recognition. Paper [11] presented emotion extraction on textual data. Author proposed the emotion detector algorithm which gives weight to each emotion word with the help of traversing and parsing the emotional ontology. Emotional class with the highest weightage is the final emotion of the corresponding text data. In [12], separate mixture model (SMM) is used to find the similarity between input sentence and EARs. Maximum probability shows the emotional state of the input sentence as happy, unhappy, or neutral.

Unsupervised emotion detection using semantic and syntactic relations is explored in [13].

In [14], author discussed a way to recognize the emotion of the user by using built-in sensors of mobile phone. They created a soft keyboard which uses data provided by these sensors to find the user's emotional state. The soft Keyboard uses the multiresponse linear regression. Another approach of using mobile data statistics and activity is explained in [15].

2.2   Emotion in Music and Emotion-Based Music Player

Music plays vital role in our life; different kinds of music or songs have different impacts on our lives. People like to listen music according to their mood, but how do we get if particular song belongs to particular category of emotion? Just like detecting human emotion, there are systems developed for music emotion recognition. There are various approaches for this, and one of them is discussed in [16]. Generally, the songs are classified according to their metadata like title, singer, etc., but to classify music into emotion set, music signal analysis is considered. Problem is approached using multilabel classification, also it is possible for one music to belong to various classes. SVM is the classifier used, for feature extraction MARSYAS is used with feature vector having 30 dimensions. Simple confusion matrix is used to check the accuracy, precision, and recall. Same topic is explored in [17].

Paper [18] is about mood-based music player application's design. Overall system works with two main modules, one that extracts emotion and other one is music audio feature extraction module. Output from these two modules is considered further in Emotion-Audio recognition module. For emotion extraction, facial image is processed. Image is given as input after converting it to binary format. Viola& Jones object detection framework is used for detecting parts of face. According to them, facial points of eyes and mouth depict the emotions accurately. Support vector machine (SVM) is used as classifier. In audio feature extraction module, music audio signal is categorized into eight types of mood pair, e.g., sadanger using auditory toolbox. Once emotion is obtained, songs that maps with it from classified dataset in second module are taken. Randomizer generates the playlist.

In [19], there is another similar approach discussed about mood extraction and mapping it with assumed mood wise classified dataset of songs. Here, facial detection method is used again as their crucial factor but instead of shapes of eyes and lips, points on face are taken. They used support vector machines (SVM) as their classification method. These surveys helped us how researches have achieved their own way of emotion recognition in human and music, determining activity recognition as well, this motivated us to integrate such small factors to create a new way of generating playlist in music apps which in turn ultimately considers outputs of such small emotion recognition modules and calculates result based on it. These small modules have their own working machine learning-based model.

3 Problems in Current Systems

Though there exists few similar music players, they utilized machine learning in facial emotion recognition or emotion wise classification of songs. These two aims can be achieved with different number of ways. Our goal is not to build the same thing again, rather use outputs of different components for better playlist generation. Current solutions consist of typically capturing face and mapping the mood recognized with songs and generates the playlist. Some people like to listen songs preferred to their activities like while reading, walking, running, while using social media, etc. Hence, considering factors like user's activities, preference, timing, etc., Emousic is the solution.

4 Proposed Solution

4.1 Random Forest Classifier

Random forest is the supervised classification algorithm. It can be used in both classification and regression problems. Random forest builds many classification trees by selecting best feature among a random subset of features. This process brings randomness, which results in a better model. In the classification process, each tree in the forest gives votes for that class. The forest chooses the classification having most of the votes. The general technique of bootstrap aggregating or bagging is applied in training algorithm for random forests. There is direct relation between the number of trees in the forest and the result we get, as larger the number of trees, better the accuracy. The advantages of random forest algorithm are those it can handle the missing values, avoid the problem of overfitting and it can be modeled for categorical values.

4.2 Algorithm of Emousic's Classifier

```
begin
      1.   Fetch the dataset.
      2.   Remove duplicate entries if present.
      3.   If same set of independent variable has multiple
             dependent class then categorise them in one class.
      4.   Divide the dataset into test set and training set
      5.   Apply Random forest classifier to train the dataset.
      6.   Take the inputs from data frame
      7.   Predict the result as song by giving input to 5.
   end
```

4.3 Architecture

As shown in Fig. 1, system involves three modules that work independently alto-gether. Output from this modules acts as input to server which has trained model, model predicts the song and returns back to phone. Modules are explained below.

Getting session

This is simple function that detects if it is morning, afternoon, evening, or night based on time and stores it into "session" variable.

Module—1: Mobile Activity Detection

This module looks into application usage statistics of android and gets the current fore ground app being used by user other than music player itself. Since we have considered few predefined categories as Social media, browsing, chatting, gaming, reading and if none of these, then nothing. Say for, e.g., user just opened player after/while using WhatsApp, it will be stored as "chatting" in the "Mact" variable.

Module—2: Physical Activity Detection

This module detects physical activity from the data collected from smartphone sensors. Google has already trained their model and provide service in the form of API. These activities are classified into categories like driving, walking, running, still, and bicycling. It stores them into "Pact" variable.

Module—3: Emotion Recognition System

This module takes input as user's selfie and passes it to cognitive services API provided by Microsoft for emotional analysis. We used this API to build our project in short time; however, one can implement his own model for facial emotion recognition

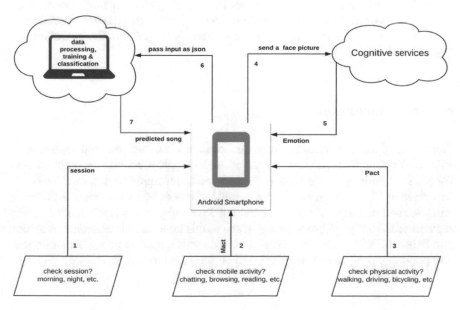

Fig. 1 Architecture

and use it as a module. This API classifies emotions into eight emotion categories viz. happy, sad, angry, surprise, neutral, contempt, disgust, and fear.

Output of these three modules and sessions are stored into JSON object and passed on to server. Server runs a Python script which accepts this JSON objects fields into variables and passes it to R script as command line arguments to run classifier script. R script then computes and predicts the best suitable song and prints out song name in form of JSON. This STDOUT is collected by Python script and sent back to android smartphone. Cursor points to the directory of emotion wise classified songs and plays the song returned as JSON. Till this, it is completely automated, then user can take full control of application.

5 Dataset

Since this approach is used for personalization which varies from user to user, dataset varies too. We used one smartphone and dumped output of emotion, physical activity, mobile activity, and session and songs listened by phone owner periodically in the phone itself. Later, this database in csv format was used in server side for training. There are four independent variables, i.e., mood, physical activity(pact), mobile activity(mact), session, and one dependent variable, i.e., song. Each independent variable is categorically defined. Mood contains happy, sad, fear, anger, contempt, disgust, neutral, and surprise (output by Microsoft's cognitive services). Pact has values as driving, walking, running, still, and bicycling (output by Google's activity recognition API). Mact contains social media, chatting, gaming, reading, and nothing. Session is daytime as morning, afternoon, evening, and night. Experimentation was performed on 101 songs. These songs are manually tagged into their emotion sets (Fig. 2).

6 Experimentation

Above-explained architecture was implemented in the android application. Microsoft's cognitive services for mood and Google's activity recognition API for physical activity recognition were used for faster implementation. It works as explained in Fig. 4. All modules run parallelly, output of each module is stored in variables and this input is passed in form of JSON object to webservice for getting recommended song. Webservice then receives this input, it runs classifier prediction and builds a JSON containing song name and sends it back to phone. Emousic then autoplays this song and shows playlist of all songs mapped to its emotion (Fig. 3).

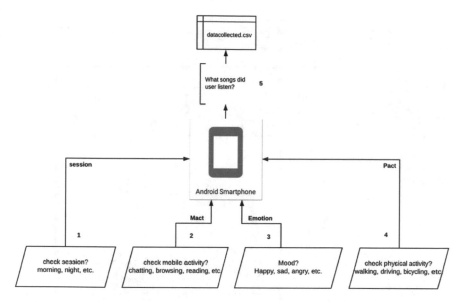

Fig. 2 Data collection model

Fig. 3 Emousic app demo. *Source* A screenshot taken on our local setup

7 Results

Parameters	J48 algorithm (%)	Random forest (%)
Correctly classified instances	50.86	96.33
Incorrectly classified instances	49.14	3.67

We trained the dataset which had approximately 1000 entries in training set and 100 entries in the test set. Weka tool is used to find the accuracy of different machine learning algorithms used to train the model. From the above table, it is observed

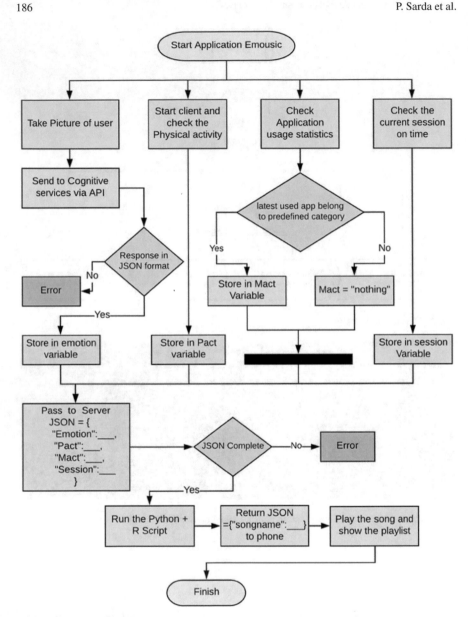

Fig. 4 Flowchart of Emousic

that random forest algorithm gives better accuracy than J48 algorithm with the mean absolute error of 0.0066 and 0.014, respectively. The accuracy of random forest algorithm can be increased by increasing the number of trees in the forest.

8 Conclusion and Future Scope

This paper presents Emousic, a new way of personalizing songs playlist by using machine learning techniques. Our solution works and gives better user preferable playlist. Due to the large number of classes, the performance of random forest classifier is better than decision tree algorithms. Since the experiment was performed on a small dataset and limited number of features, it still can be improved by adding more features like age, weather, etc. More number of attributes will improvise decision making and prediction of song. Each user has their own preferences about what kind of song is to be played for corresponding mood for e.g., some users listen sad songs when they are sad while some may prefer happy songs to change their mood.

Collecting this data from every user can help us build better user-specific radio application. Implementing this prototype in current music applications can provide better music experience to user.

References

1. Day, M.: Emotion recognition with boosted tree classifiers. In: ICMI 2013 Proceedings of the 2013 ACM International Conference on Multimodal Interaction, pp. 531–534. https://doi.org/10.1145/2522848.2531740
2. Eyben, F., Wllmer, M., Schuller, B.: Opensmile the munich versatile and fast opensource audio feature extractor. In: Proceedings of ACM Multimedia, pp. 1459–1462
3. Londhe, R.R., Pawar, V.P.: Analysis of facial expression using LBP and artificial neural network. Int. J. Comput. Appl. **44**(21), 975–8887 (2012)
4. Lyon, M., Akamatsu, S.: Coding facial expression with gabor wavelets. IEEE Conference on Automatic Face and Gesture Recognition, Mar 2000
5. Maglogiannis, Ilias, Vouyioukas, Demosthenes, Aggelopoulos, Chris: Face detection and recognition of natural human emotion using Markov random fields. Pers. Ubiquit. Comput. **13**, 95–101 (2009)
6. Inanoglu Z, Caneel R (2005) Emotive alert: HMM-based emotion detection in voicemail messages. In: Appeared in Intelligent user Interfaces (IUI 05), 2005, San Diego, California, USA, MIT Media Lab Technical Report No. 585, Jan 2005
7. Nwe, T.L., Foo, S.W., De Silva, L.C.: Speech emotion recognition using hidden Markov models. Speech Commun. J. **41**(4), 603–623 (2003)
8. El Ayadi, M.M.H., Kamel, M.S., Karray, F.: Speech emotion recognition using Gaussian mixture vector autoregressive models. In: IEEE International Conference on Acoustics, Speech and Signal Processing, 2007. ICASSP 2007
9. Nicholson, J., Takahashi, K., Nakatsu, R.: Emotion recognition in speech using neural networks. Neural Comput. Appl. **9**, 290–296 (2000). ISSN 1433-3058
10. Markel, J.M., Gray, A.H.: Linear Prediction of Speech. Springer, New York (1976)

11. Shivhare, S,N., Khethawat, S.: Emotion detection from text. CoRR, volume abs/1205.4944 (2012)
12. Wu, ChungHsien, Chuang, ZeJing, Lin, YuChung: Emotion recognition from text using semantic labels and separable mixture models. ACM Trans. Asian Lang. Inf. Process. **5**, 165–183 (2006)
13. Agrawal, A., An, A.: Unsupervised emotion detection from text using semantic and syntactic relations. In: 2012 IEEE/WIC/ACM International Conferences on Web Intelligence and Intelligent Agent Technology, Macau, pp. 346–353 (2012)
14. Aloul, Z.F., Shapsough, S., Hesham, A., ElKhorzaty, Y.: Emotion recognition using mobile phones. Comput. Electr. Eng. **60**, 113 (2017)
15. Rachuri, K.K., Musolesi, M., Mascolo, C., Rentfrow, P.J., Longworth, C., Aucinas, A. Emotion-Sense: a mobile phones based adaptive plat-form for experimental social psychology research, pp. 281–290. https://doi.org/10.1145/1864349.1864393 (2010)
16. Li, T., Ogihara, M.: Detecting emotion in music. In: ISMIR International Conference on Music Information Retrieval (2003)
17. Mahajan, N., Mahajan, H.: Detecting emotion in music. Int. J. Electr. Electron. Res. **2**(2), 56–60 (2014). ISSN 2348-6988
18. Kabani, H., Khan, S., Khan, O., Tadvi, S.: Emotion based musicplayer. Int. J. Eng. Res. Gen. Sci. **3**(1), 2091 (2015)
19. Patel, A.R., Vollal, A., Kadam, P.B., Yadav, S., Samant, R.M.: MoodyPlayer: a mood based music player. Int. J. Comput. Appl. **141**(4), 0975–8887 (2016)

Extra-Tree Classifier with Metaheuristics Approach for Email Classification

Aakanksha Sharaff and Harshil Gupta

Abstract It is very normal for any user to receive hundreds of emails every day. Almost 93% of them are spam messages which include mainly advertisements from the industries like software, phishing, gambling, stocks, electronics, pharmaceutical, loan, and malware attempts etc. Spams messages not only waste user's time but also eats up user valuable space. In this paper, a nature inspired metaheuristics technique has been used for email classification which emphasizes on reducing false-positive problem of treating spam messages as ham. It uses metaheuristics-based feature selection methods and employs extra-tree classifier to classify emails into spam and ham. The proposed model has accuracy of 95.5%, specificity of 93.7%, and F_1-score of 96.3%, which is clearly a major improvement over the previous researches which have been conducted in this field using decision trees. The comparative analysis of extra-tree classifiers with other classifiers like decision trees and random forest has also been studied.

Keywords Ham and spam detection · Feature selection · Extra tree · Binary particle swarm optimization

1 Introduction

Nowadays, email has become a major source of communication. Email is used as a personal as well as professional medium of communication and can be considered as official document of text by the users. As the usage of email has been increasing day by day, spam messages need to be detected. Different approaches have been adopted to stop the spam messages. Scientists/researchers are continuously working

A. Sharaff (✉) · H. Gupta
Department of Computer Science and Engineering, National Institute of Technology Raipur,
Raipur 492001, Chhattisgarh, India
e-mail: asharaff.cs@nitrr.ac.in

H. Gupta
e-mail: hghasty21@gmail.com

© Springer Nature Singapore Pte Ltd. 2019
S. K. Bhatia et al. (eds.), *Advances in Computer Communication
and Computational Sciences*, Advances in Intelligent Systems and Computing 924,
https://doi.org/10.1007/978-981-13-6861-5_17

189

on different feature selection method for text as well as image-based spam email. Feature selection plays an important role to achieve performance, efficiency, simplification, and accuracy. If better features are obtained, then detecting spam messages will become quite easier. Spam filtering is a very important issue in any communication media. It can be either email or Short Message Service (SMS). Many studies and researches are continued to automatically classify emails into spam and ham. These studies use several approaches including Neural Network, Evolutionary Computation, Machine Learning techniques, etc. In this paper, an aggregate system based on metaheuristics and machine learning approach has been proposed. In this work, feature selection process based on metaheuristics approach has been used to achieve better efficiency. The main objective of this research was to reduce the slow execution time. There are some flaws mentioned below in the literature which can be solved by using the proposed approach.

1. The first problem is the slow execution, which can be reduced by using binary particle swarm optimization method.
2. The second problem is lack of generality, which can be corrected by using a validation technique namely K-fold cross-validation.

2   Related Work

Some recent research work on email classification technique has been discussed in this section. Idris et al. classify emails based on Negative Selection Algorithm (NSA) and Particle Swarm Optimization (PSO) to improve the traditional random generation detectors in real values of negative selection algorithm and optimize the generated detectors in a spam space [1]. Brezočnik identified features by using a binary version of Particle Swarm Optimization method and used the method for classification task. Here the author has compared different classification algorithm on different particle dataset using SVM and Naïve Bayes in which Naïve Bayes outperforms infrequently [2]. Chakraborty also used a Binary version of Particle Swarm Optimization (BPSO)-based feature selection technique with a fuzzy fitness function based on feature subsets [3]. The concept behind this technique is to reduce the discrepancy in the class and maximize the discrepancy among the classes. The efficacy of binary PSO is compared with many different approaches like principle component analysis (PCA), chi-square, and empirical outputs show that the binary particle swarm optimization-based feature selection algorithm could achieve more classification accuracy, and it takes less processing time than these algorithms [3]. Wang et al. proposed a novel Document Term Frequency combined Feature Selection Method (DTFS) method to decrease the size of feature space and to improve performance of email classification. In this work, Fuzzy Support Vector Machine learning (FSVM) classification technique has been used with other algorithm to classify emails [4]. Zhang et al. proposed a mutation operator BPSO feature selection method for email classification using decision trees. The approach gave substantial results but the use of decision trees meant the cut point selected among targeted nodes which are not completely random, and hence, the result was not more efficient [5]. Sharaff and Nagwani proposed email categorization approach by identifying categorical terms based on latent Dirichlet allocation (LDA) [6]. Aski and Sourati developed a classification process to extract different features of email by using machine learning algorithms [7]. Some general descriptive features, normalization, and semantic indexing have also been explored to detect malicious messages in email and Short Message Service (SMS) [8, 9]. Machine learning algorithms have also been explored to classify emails for spam detection and multiclass categorization of emails [10]. Polat and Gunes proposed a hybrid method combining C4.5 decision tree classifier and one a gains tall approach for tackling multiclass classification problem but the problem with this approach was C4.5 algorithm consumes lot of processing power and take more memory space [11]. Proença et al. introduced a probabilistic fuzzy approach over stochastic gradient descent for classification. They have used metaheuristics approach to optimize the fuzzy system [12]. Dong et al. gave a hybrid model using GA with granular information for selecting features. However, just like PSO approach, this approach also suffers from premature convergence problem and therefore it is also less efficient than BPSO approach for feature selection [13]. Wei et al. proposed a BPSO-SVM method for feature selection. Although useful for feature selection, SVM is less efficient than tree-based classifiers for email classification [14].

3 Methodology and Proposed Model

Classification generally starts with removing redundancy in terms of attributes. The techniques used for removing these redundancies are feature selection and feature extraction. In this paper, feature selection has been chosen over feature extraction for getting greater efficiency.

In this work, extra-tree classifier has been selected because of its explicit meaning, simple properties, and easily conversion to "if–then" rules. The extra-tree method has been chosen because of its randomizing property for numerical inputs. This idea is very useful in problems involving huge number of numerical features. It leads often to increased accuracy in this kind of situation.

The concept of cost matrix and confusion matrix has been employed to categorize emails into different possible combinations of spam and ham. In the confusion matrix, the columns represent actual class and rows represent the prediction class instances. With the help of confusion matrix, many performances-related parameters have been calculated. These are accuracy, precision, recall or sensitivity, specificity, and F_1-score. Accuracy is defined as ratio of right predictions and all the predictions [5]. Precision is defined as ratio of true positives and true positives plus false positives [5]. Recall is defined as the ratio of true positives and true positives plus false negatives [5]. Specificity is defined as the ratio of true negative and true negative plus false positive. F_1-score is the harmonic mean of precision and recall.

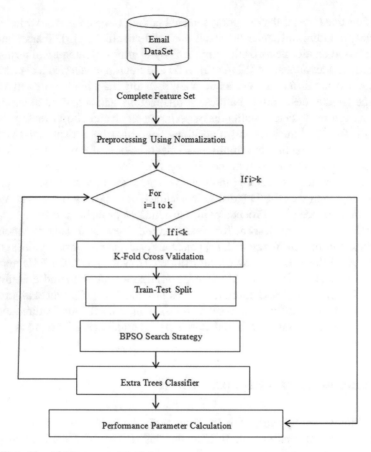

Fig. 1 Flowchart of the proposed model

For classification purpose, firstly data have been divided into k folds where k is taken as 15 and in each of the iterations, one of the folds has been used for testing purposes and the remaining folds have been used for training purposes. The purpose of doing so is to overcome the problem of overfitting and under fitting. The training samples thus created were then employed for feature selection purpose. The objective of this paper would be met when the problem of slow execution will be solved with estimating the cost factor. The cost function chosen as objective function is evaluated by using the equation given by Zhang et al. [5]:

$$j = \frac{(FN + \alpha * FP)}{\alpha + 1} \tag{1}$$

j is the output, FP is false positives, FN is false negatives, alpha is the weight parameter. The step-by-step procedure is explained as follows and shown in Fig. 1:

Step1:   (Dataset Acquisition) A complete dataset has to be collected from Spambase
          Dataset. Read the dataset and apply preprocessing techniques with normal-
          ization.
Step2:   (Select relevant features) Metaheuristics Feature Selection techniques like
          PSO, BPSO, and GA have been applied in order to get the relevant feature.
Step3:   (Classify emails into ham and spam) Classification algorithms such as extra
          tree, decision tree, and random forest are used to categorize the emails into
          ham message or spam messages.
Step4:   (Validation) The classifier is trained using the validation set. A k-fold cross-
          validation technique has been used for the testing purpose.
Step5:   (Performance parameter computation) Compute various parameters like
          accuracy, precision, recall, specificity, F_1-score with the help of confusion
          matrix.

4 Experimental Analysis and Result Discussion

A complete dataset of 4601 emails (from the UCI library) has been used for experimental analysis. These emails have three different types of features. The first one
constitutes 48 most common words frequency. The second one constitutes frequency
of six different types of symbols and characters. The third one is feature related to
capital run length. In this way, each email constitutes a 57 dimensional feature character. Among these, capital run length-related features have been chosen as the most
important feature as it is very much known that most of the spams contain capital run
length-related features. To classify emails, all the features are not necessary as some
features are correlated and some features are redundant. Therefore, feature selection
can be used to remove these kinds of features. Table 1 summarizes the result of various combinations of classifier and metaheuristics technique, namely PSO, BPSO,
and GA for email classification.

   Table 1 shows that extra-tree classifier with BPSO search strategy outperforms all
other combination of classifiers. The close competitors of BPSO are PSO and GA
but BPSO wins over them as a feature selection method as firstly it is less susceptible
to premature convergence problem and secondly unlike PSO, being discrete, it either
considers a feature completely or completely neglects it.

   With the accuracy of 95.5% (shown in Fig. 2), precision of 95.9% (shown in
Fig. 3), recall of 96.6% (shown in Fig. 4), specificity of 93.7% (shown in Fig. 5), and
F_1-score of 96.3% (shown in Fig. 6), this method is clearly a major improvement
over the previous researches which have been conducted in this field. The other two
tree classifiers, random forest and decision trees, also perform well. In terms of time
complexity, random forest classifier takes a lot of time especially with BPSO just
like extra-tree classifier but the output is also better than the other combinations of
feature selection and classifier method used. The value of weight parameter α was
chosen as 7 because this gave the optimal mean cost value. Other values of weight
parameter did not give the optimal value.

Table 1 Comparative study of various classifiers using different metaheuristics approach

S. No.	Classifier	Metaheuristics FS method	Accuracy	Precision	Recall	Specificity	F_1-score	Mean cost
1	Random forest	BPSO	0.954	0.945	0.965	0.913	0.955	0.069
2	Random forest	PSO	0.930	0.937	0.948	0.902	0.942	0.028
3	Random forest	GA	0.915	0.928	0.933	0.888	0.930	–
4	Extra tree	BPSO	0.955	0.959	0.966	0.937	0.963	0.054
5	Extra tree	PSO	0.935	0.942	0.945	0.941	0.953	0.011
6	Extra tree	GA	0.910	0.928	0.924	0.889	0.925	–
7	Decision tree	BPSO	0.921	0.946	0.922	0.920	0.934	0.078
8	Decision tree	PSO	0.908	0.928	0.920	0.890	0.924	0.028
9	Decision tree	GA	0.903	0.919	0.921	0.875	0.920	–

Fig. 2 Performance
comparison of each classifier
w.r.t. accuracy

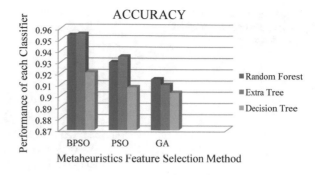

Fig. 3 Performance
comparison of each classifier
w.r.t. precision

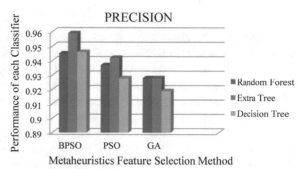

Fig. 4 Performance
comparison of each classifier
w.r.t. recall

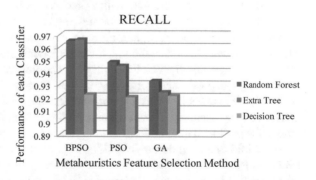

Fig. 5 Performance
comparison of each classifier
w.r.t. specificity

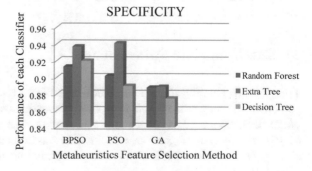

Fig. 6 Performance comparison of each classifier w.r.t. F_1-score

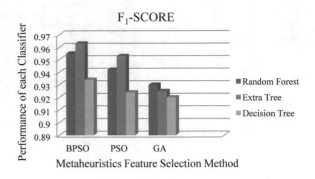

As different weights were assigned to spam going into ham and ham going to spam, this method also takes care of the relative importance of different types of possibilities of spam and ham. Generally, if a spam goes to ham, it is not much harmful for the user. Other than time consumption, it would not have other deadly effect. At the worst, user would be required to delete the spam. But if ham goes to spam, the user may never get to know about it. It might result in the loss of some of the most vital information concerning the user. If this happens, this would be a great loss for the user. That's why the latter has been assigned more weight as compared to the former in this paper. The classification has been aimed at reducing the number of ham going to spam as far as possible. With a precision of 95.9% with extra-tree classifier and BPSO method, the amount of ham going to spam is greatly reduced. Because of improving nature of spam, adding more common words might prove beneficial. The use of newer methods which includes statistical measure to further increase the efficiency of the classifier can be employed. Training the classifier with new dataset periodically can prove useful in increasing the accuracy of the classifier greatly. Using ensemble of classifiers for classification can also be one of the methods that can aid in achieving more accuracy. There are some problems with the proposed BPSO method. Due to the random nature of BPSO, it might not always produce the optimum result. Using K-fold cross-validation, classifier might not get trained accurately for some folds. As BPSO requires lot of time, lot of processing is required. In future work, it will be useful to try fast subset strategies to lessen the processing cost.

5 Conclusion and Future Scope

This paper has employed metaheuristics-based feature selection techniques to classify emails. The experiment and result showed that the best performance was obtained when BPSO was used as subset strategy and extra-tree classifier was chosen as the classifier model. Although extra-tree classifier was the best classification model, decision tree and random forest also produced optimal result when applied along with

BPSO. The methodology employed in this paper can be useful for solving some other problems like text classification, image classification, and web page classification.

Future work includes improving the time complexity of the method given in the paper and to gather more data to train the classifier and use an ensemble of classifiers for training for better results.

References

1. Idris, I., Selamat, A., Nguyen, N.T., Omatu, S., Krejcar, O., Kuca, K., Penhaker, M.: A combined negative selection algorithm–particle swarm optimization for an email spam detection system. Eng. Appl. Artif. Intell. **39**, 33–44 (2015)
2. Brezočnik, L.: Feature selection for classification using particle swarm optimization. In: 17th International Conference on Smart Technologies, IEEE EUROCON 2017, pp. 966–971. IEEE (2017)
3. Chakraborty, B.: Feature subset selection by particle swarm optimization with fuzzy fitness function. In: 3rd International Conference on Intelligent System and Knowledge Engineering, 2008, ISKE 2008, vol. 1, pp. 1038–1042. IEEE (2008)
4. Wang, Y., Liu, Y., Feng, L., Zhu, X.: Novel feature selection method based on harmony search for email classification. Knowl.-Based Syst. **73**, 311–323 (2015)
5. Zhang, Y., Wang, S., Phillips, P., Ji, G.: Binary PSO with mutation operator for feature selection using decision tree applied to spam detection. Knowl.-Based Syst. **64**, 22–31 (2014)
6. Sharaff, A., Nagwani, N.K.: Identifying categorical terms based on latent Dirichlet allocation for email categorization. In: Emerging Technologies in Data Mining and Information Security, pp. 431–437. Springer, Singapore (2019)
7. Aski, A.S., Sourati, N.K.: Proposed efficient algorithm to filter spam using machine learning techniques. Pac. Sci. Rev. A: Nat. Sci. Eng. **18**(2), 145–149 (2016)
8. Cohen, A., Nissim, N., Elovici, Y.: Novel set of general descriptive features for enhanced detection of malicious emails using machine learning methods. Expert. Syst. Appl. (2018)
9. Almeida, T.A., Silva, T.P., Santos, I., Hidalgo, J.M.G.: Text normalization and semantic indexing to enhance instant messaging and SMS spam filtering. Knowl.-Based Syst. **108**, 25–32 (2016)
10. Proença, H.M., Vieira, S.M., Kaymak, U., Almeida, R.J., Sousa, J.M.: Optimizing probabilistic fuzzy systems for classification using metaheuristics. In: 2016 IEEE International Conference on Fuzzy Systems (FUZZ-IEEE), pp. 1635–1641. IEEE (2016)
11. Sharaff, A., Nagwani, N.K., Dhadse, A.: Comparative study of classification algorithms for spam email detection. In: Emerging Research in Computing, Information, Communication and Applications, pp. 237–244. Springer, New Delhi (2016)
12. Dong, H., Li, T., Ding, R., Sun, J.: A novel hybrid genetic algorithm with granular information for feature selection and optimization. Appl. Soft Comput. **65**, 33–46 (2018)
13. Polat, K., Güneş, S.: A novel hybrid intelligent method based on C4. 5 decision tree classifier and one-against-all approach for multi-class classification problems. Expert Syst. Appl. **36**(2), 1587–1592 (2009)
14. Wei, J., Zhang, R., Yu, Z., Hu, R., Tang, J., Gui, C., Yuan, Y.: A BPSO-SVM algorithm based on memory renewal and enhanced mutation mechanisms for feature selection. Appl. Soft Comput. **58**, 176–192 (2017)

A Parametric Method for Knowledge Measure of Intuitionistic Fuzzy Sets

Zhen-hua Zhang, Shen-guo Yuan, Chao Ma, Jin-hui Xu and Jing Zhang

Abstract Entropy is an important tool in measuring the uncertainty research of fuzzy system, in which many achievements have been obtained. Due to the hesitant degree, intuitionistic fuzzy sets (IFS) is more complex than the traditional fuzzy sets in ambiguity and uncertainty. Therefore, there are many challenges in the research of intuitionistic fuzzy entropy and intuitionistic fuzzy knowledge measure, especially the lack of deep research on the axiom system, unified theory, and method. Hence, this paper firstly studies the Szmidt and Kacprzyk's axiom system, which is composed of order, non-negative boundedness, and symmetry. Meanwhile, we put forward some necessary and sufficient conditions and necessary conditions for order property. Simultaneously, the order property of some traditional and classical operators is proved by using the necessary and sufficient conditions. Then, a simple and convenient knowledge measure with parameter is proposed based on the order condition. Finally, based on intuitionistic fuzzy set, an experiment to test the order property is proposed. The simulation demonstrates that the results of the presented method are widely similar to that of the traditional algorithm under different parame-

Z. Zhang (✉)
School of Mathematics and Statistics, Guangdong University of Foreign Studies,
Guangzhou 510006, China
e-mail: zhangzhenhua@gdufs.edu.cn

Department of Informatics, University of Leicester, Leicester LE1 7RH, UK

S. Yuan
School of Economics and Statistics, Guangzhou University, Guangzhou 510006, China

C. Ma
Department of Computer Science and Technology,
University of Cambridge, Cambridge CB3 0FD, UK

J. Xu
School of Mathematical and Statistical Sciences, Arizona State University,
Tempe, AZ 85281, USA

J. Zhang (✉)
Education Technology Center, Guangdong University of Foreign Studies,
Guangzhou 510006, China
e-mail: zhangjing325@126.com

© Springer Nature Singapore Pte Ltd. 2019   199
S. K. Bhatia et al. (eds.), *Advances in Computer Communication
and Computational Sciences*, Advances in Intelligent Systems and Computing 924,
https://doi.org/10.1007/978-981-13-6861-5_18

ter values, and in some special values, the results of this parametric method are more accurate than those of most classic methods.

Keywords Intuitionistic fuzzy sets · Knowledge measure · Entropy · Order property

1 Introduction

In 1965, Zadeh [1–3] presented fuzzy sets (FS), which is generalized to intuitionistic fuzzy sets (IFS) in 1986 by Atanassov [4–9] due to the introduction of the hesitation degree. Therefore, the research of IFS is more complex than that of FS. As the tools measuring the degree of order and disorder in fuzzy sets, fuzzy knowledge measure and fuzzy entropy are both important. De Luca and Termini [10] introduced a classic axiom system of fuzzy entropy according to Shannon's entropy function in 1972, Yager [11] presented some fuzzy entropy functions in terms of distance formula. Hence, some researchers have begun to study fuzzy entropy [12] and intuitionistic fuzzy entropy [13–18], and extended them to interval value fuzzy sets (IVVS) [19–22] and interval valued intuitionistic fuzzy sets (IVIFS) [23–25]. Some researchers studied entropy and proposed the knowledge measure of IFS [13–18, 23–30]. During the past 20 years, a series of methods measuring entropy and knowledge are introduced. However, the existing research focuses on the construction of entropy and knowledge measure, and few on the study of the axiom system and the rule system of these information measures. This has contributed to the lack of theoretical basis and the specific rules of judgment, and hence, people are confused on how to judge the performance of all the models of entropy and knowledge measure of IFS. For the study on the axioms of entropy of IFS, Szmidt and Kacprzyk [17, 18] introduced a standard axiom system with a relatively wide applications [17, 18, 26, 27, 30, 31]: order, symmetry, and non-negative boundedness. In all the relevant literatures, most researchers only introduced their entropy formulas according to Szmidt and Kacprzyk's axioms system. Based on this axiom system, a dual axiom system of knowledge measure was established by Guo et al. [27–30]. In terms of the order property in Szmidt and Kacprzyk's axioms system, Guo [30] put forward a generalized order model and its knowledge measure formulas.

Due to the extensiveness of the application of Szmidt and Kacprzyk's axiom system [17, 18], this paper studied this axiom deeply. We first presented some equivalent conditions and necessary conditions of the order property based on the Szmidt and Kacprzyk's axiom system. Hence, we analyze the differences among some classic operators of information measure. And further, a new construction method is brought about and proved theoretically that the proposed formula satisfies all the conditions of the presented axiom system. In theory, we proved that the operators with the order property in the axiom system will be better than those without. Finally, an experimental case is introduced, and simulation results show that the performance of the knowledge measure with parameters is better than that of many classical operators, and the functions with the order property will be more effective and accurate than those without.

The rest of the paper is organized as follows. Section 2 introduces the concept of IFS and Sect. 3 contains the knowledge measure of IFS and its axiom system. Section 4 presents a parametric knowledge measure model. Section 5 focuses on the experiments and result analysis. Section 6 concludes the article.

2 Intuitionistic Fuzzy Sets

Definition 1 An IFS A in a finite set X is an object with the form:

$$A = \{\langle x, t_A(x), f_A(x)\rangle | x \in X\}$$

$$t_A(x) : X \rightarrow [0, 1], x \rightarrow t_A(x); f_A(x) : X \rightarrow [0, 1],$$
$$x \rightarrow f_A(x), t_A(x) + f_A(x) \in [0, 1].$$

$t_A(x)$ and $f_A(x)$ are the degree of membership and non-membership, respectively. $\pi_A(x) = 1 - t_A(x) - f_A(x) \in [0, 1]$, and $\pi_A(x)$ is the degree of hesitancy.

Definition 2 Let A and B be two IFSs, then we have:

(1) $A = B$ if and only if $t_A(x) = t_B(x)$ and $f_A(x) = f_B(x)$.
(2) $A \subseteq B$ if $t_A(x) \leq t_B(x)$ and $f_A(x) \geq f_B(x)$.
(3) $A^C = \{\langle x, t_{A^C}(x), f_{A^C}(x)\rangle | x \in X\} = \{\langle x, f_A(x), t_A(x)\rangle | x \in X\}$.

3 Intuitionistic Fuzzy Knowledge Measure

Entropy and knowledge measure are both basic tools for characterizing the state of matter. The function of entropy is to measure the degree of disorder, uncertainty, and irregularity of a system, while that of knowledge measure is to estimate the degree of order, certainty, and regularity of a system. Hence, the knowledge measure of IFS A can be defined by:

$$K(A) = 1 - E(A) \tag{1}$$

As a tool for information measure, the intuitionistic fuzzy entropy is defined as follows [13–18, 23–26]:

Definition 3 For IFS A, $E(A) = \text{Entropy}(t_A, f_A, \pi_A)$ is the entropy of A with properties:

(EP1) $t_A(x) = f_A(x) = 0 \Rightarrow E(A) = 1$.
(EP2) $E(A) = 0 \Leftrightarrow A$ is a crisp set.
(EP3) $E(A) = E(A^C)$.
(EP4) For another IFS B, $E(B) = \text{Entropy}(t_B, f_B, \pi_B)$ denotes the entropy of B, and we have: If $t_B(x) \le f_B(x)$ and $A \subseteq B$, $E(A) \le E(B)$; If $t_B(x) \ge f_B(x)$ and $B \subseteq A$, $E(A) \le E(B)$.

Knowledge measure of IFS can be defined as follows [26–30]:

Definition 4 A is an IFS, $K(A) = \text{Knowledge}(t_A, f_A, \pi_A)$ is an intuitionistic fuzzy knowledge measure of A if $K(A)$ have properties:

(KP1) $t_A(x) = f_A(x) = 0 \Rightarrow K(A) = 0$.
(KP2) $K(A) = 1 \Leftrightarrow A$ is a crisp set.
(KP3) $K(A) = K(A^C)$.
(KP4) If B is also an IFS, and $K(B) = \text{Knowledge}(t_B, f_B, \pi_B)$ is the knowledge measure of B, then we have: If $t_B(x) \le f_B(x)$ and $A \subseteq B$, $K(A) \ge K(B)$; If $t_B(x) \ge f_B(x)$ and $B \subseteq A$, $K(A) \ge K(B)$.

KP1 and KP2 denote the property of non-negative boundedness, KP3 the property of symmetry, and KP4 the property of order.

In terms of KP4, we obtain the following equivalent condition:

(KP4 I) If $t_A(x) \le t_B(x) \le f_B(x) \le f_A(x)$ or $t_A(x) \ge t_B(x) \ge f_B(x) \ge f_A(x)$, then we have $K(A) \ge K(B)$.

Proof $A \subseteq B \Leftrightarrow \begin{cases} t_A(x) \le t_B(x) \\ f_B(x) \le f_A(x) \end{cases}$, and $t_B(x) \le f_B(x)$, then $t_A(x) \le t_B(x) \le f_B(x) \le f_A(x)$. And if $t_A(x) \le t_B(x) \le f_B(x) \le f_A(x)$, $\Rightarrow t_B(x) \le f_B(x) \& t_A(x) \le t_B(x) \& f_B(x) \le f_A(x)$,

and then we get $t_B(x) \le f_B(x)$ and $A \subseteq B$.

Similarly, $B \subseteq A \Leftrightarrow \begin{cases} t_A(x) \ge t_B(x) \\ f_B(x) \ge f_A(x) \end{cases}$, and $t_B(x) \ge f_B(x)$, then $t_A(x) \ge t_B(x) \ge f_B(x) \ge f_A(x)$.

And if $t_A(x) \ge t_B(x) \ge f_B(x) \ge f_A(x)$, $\Rightarrow t_B(x) \ge f_B(x) \& t_A(x) \ge t_B(x) \& f_B(x) \ge f_A(x)$,

and then we have $t_B(x) \ge f_B(x)$ and $B \subseteq A$.

KP4 is equivalent to KP4I, thus we get $K(A) \ge K(B)$.

Based on KP4 I, condition KP4 II can be achieved.

(KP4 II) If $t(x) \le f(x)$, and $K = \text{Knowledge}(t, f, \pi)$ is a monotonous and non-increasing function for t and a monotonous and non-decreasing function for f, then $K(A) \ge K(B)$; If $t(x) \ge f(x)$, and $K = \text{Knowledge}(t, f, \pi)$ is a monotonous and non-decreasing function for t and a monotonous and non-increasing function for f, then $K(A) \ge K(B)$.

Proof According to KP4 I, for two IFSs A and B, we obtain:

$$t_A(x) \le t_B(x) \le f_B(x) \le f_A(x) \Rightarrow K(A) \ge K(B)$$

Which means

If $t(x) \le f(x), t_A(x) \le t_B(x) \& f_B(x) \le f_A(x) \Rightarrow K(A) \ge K(B)$, then we conclude that $K = \text{Knowledge}(t, f, \pi)$ is a monotonous and non-increasing function for t and a monotonous and non-decreasing function for f.

If $t(x) \ge f(x), t_A(x) \ge t_B(x) \& f_B(x) \ge f_A(x) \Rightarrow K(A) \ge K(B)$, then we conclude that $K = \text{Knowledge}(t, f, \pi)$ is a monotonous and non-decreasing function for t and a monotonous and non-increasing function for f.

From KP4 II, if $t(x) \le f(x)$ and $K = \text{Knowledge}(t, f, \pi)$ is a monotonous and non-increasing function for t and a monotonous and non-decreasing function for f, then when $A \subseteq B$, we can infer:

$$t_A(x) \le t_B(x) \le f_B(x) \le f_A(x) \Rightarrow K(A) \ge K(B).$$

Similarly, if $t(x) \ge f(x)$ and $K = \text{Knowledge}(t, f, \pi)$ is a monotonous and non-decreasing function for t and is a monotonous and non-increasing function for f, then when $A \subseteq B$, we can also infer:

$$t_A(x) \ge t_B(x) \ge f_B(x) \ge f_A(x) \Rightarrow K(A) \ge K(B)$$

Thus, KP4 II is equivalent to KP4 I.

If $K = \text{Knowledge}(t, f, \pi)$ is a continuous and derivative function of t and f, according to KP4 II, then we get KP4 III, which is also equivalent to KP4.

(KP4 III) $K = \text{Knowledge}(t, f, \pi)$ is a continuous and derivative function of t and f. If $t(x) \le f(x)$, $\frac{\partial K}{\partial t} \le 0$, $\frac{\partial K}{\partial f} \ge 0$, then $K(A) \ge K(B)$; If $t(x) \ge f(x)$, $\frac{\partial K}{\partial t} \ge 0 \& \frac{\partial K}{\partial f} \le 0$, then $K(A) \ge K(B)$.

From KP4 III, it is easy for us to obtain KP4 IV, a necessary condition of $K(A) \ge K(B)$.

(KP4 IV) $K = \text{Knowledge}(t, f, \pi)$ is a continuous and derivative function of t and f, and then we have $\frac{\partial K}{\partial t} \frac{\partial K}{\partial f} \le 0$.

Lemma 1 *For each* $j \in \{1, 2, \ldots, J\} \subseteq N$, N *is the Natural number set.*
$K_j(A) = Knowledge_j(t_{j,A}(x), f_{j,A}(x), \pi_{j,A}(x))$ *is a knowledge measure of IFS A, and for* $w_j \in [0, 1]$, $\sum_{j=1}^{J} w_j = 1$, *we have:*

$$K(A) = \sum_{j=1}^{J} w_j K_j(A) = \sum_{j=1}^{J} w_j \text{Knowledge}_j \left(t_{j,A}(x), f_{j,A}(x), \pi_{j,A}(x) \right) \qquad (2)$$

And then K(A) is also a knowledge measure of IFS A.

Proof From the definition of the knowledge measure of IFS, $K_j(A)$ will satisfy KP1-KP4. Hence, we have:

(KP1) $\forall j \in \{1, 2, \ldots, J\}, t_{j,A}(x) = f_{j,A}(x) = 0 \Rightarrow$

$$K_j(A) = 0 \Rightarrow K(A) = \sum_{j=1}^{J} w_j K_j(A) = 0.$$

(KP2) for each $j \in \{1, 2, \ldots, J\} \subseteq N$,

$$K_j(A) = \text{Knowledge}_j (t_{j,A}(x), f_{j,A}(x), \pi_{j,A}(x)) = 1 \Leftrightarrow$$

$t_{j,A}(x) = 0 \,\&\, f_{j,A}(x) = 1$ or $t_{j,A}(x) = 1 \,\&\, f_{j,A}(x) = 0$.
$\therefore K(A) = 1 \Leftrightarrow \forall j \in \{1, 2, \ldots, J\}, t_{j,A}(x) = 0 \,\&\, f_{j,A}(x) = 1$ or $t_{j,A}(x) = 1 \,\&\, f_{j,A}(x) = 0$.

$$K(A) = \sum_{j=1}^{J} w_j K_j(A) = 1 \Leftrightarrow \forall j \in \{1, 2, \ldots, J\}, \quad K_j(A) = 1.$$

(KP3) $K(A) = \sum_{j=1}^{J} w_j K_j(A) = \sum_{j=1}^{J} w_j K_j(A^C) = K(A^C)$.

(KP4) For another IFS B, $K(B) = \sum_{j=1}^{J} w_j K_j(B)$ denotes the knowledge measure of B, and then for each $j \in \{1, 2, \ldots, J\} \subseteq N$, we have:

$$t_{j,A}(x) \leq t_{j,B}(x) \leq f_{j,B}(x) \leq f_{j,A}(x) \text{ or}$$
$$t_{j,A}(x) \geq t_{j,B}(x) \geq f_{j,B}(x) \geq f_{j,A}(x) \Rightarrow K_j(A) \geq K_j(B),$$

$$\Rightarrow K(A) = \sum_{j=1}^{J} w_j K_j(A) \leq \sum_{j=1}^{J} w_j K_j(B) = K(B).$$

From the process of the proof above, we conclude that $K(A)$ is also a knowledge measure of IFS A, which means that the weighted average function composed of some intuitionistic fuzzy knowledge measures satisfying boundedness, symmetry, and order must have the same properties. Lemma 2 provides a good idea for finding the original intuitionistic fuzzy knowledge measure satisfying condition KP1-KP4, and the weighted average of these original intuitionistic fuzzy knowledge measures is a formula that satisfies these properties.

Some classic formulas of knowledge measure will be introduced as follows:

$$K_{BB}(A) = 1 - \frac{1}{n} \sum_{i=1}^{n} \pi_A(x_i) \tag{3}$$

$$K_{SK1}(A) = 1 - \frac{1}{n} \sum_{i=1}^{n} \left(\frac{\min\{t_A(x_i), f_A(x_i)\} + \pi_A(x_i)}{\max\{t_A(x_i), f_A(x_i)\} + \pi_A(x_i)} \right) \qquad (4)$$

$$K_{SKB}(A) = 1 - \frac{1}{2n} \sum_{i=1}^{n} \left(\frac{\min\{t_A(x_i), f_A(x_i)\} + \pi_A(x_i)}{\max\{t_A(x_i), f_A(x_i)\} + \pi_A(x_i)} + \pi_A(x_i) \right) \qquad (5)$$

$$K_G(A) = 1 - \frac{1}{2n} \sum_{i=1}^{n} (1 - |t_A(x_i) - f_A(x_i)|)(1 + \pi_A(x_i)) \qquad (6)$$

$$K_{HC}^{\alpha}(A) = \begin{cases} 1 - \sum_{i=1}^{n} \frac{1 - t_A(x_i)^{\alpha} - f_A(x_i)^{\alpha} - \pi_A(x_i)^{\alpha}}{(\alpha - 1)n}, \alpha \neq 1 (\alpha > 0) \\ 1 + \frac{1}{n} \sum_{i=1}^{n} \begin{array}{l} (t_A(x_i) \log(t_A(x_i)) + f_A(x_i) \log(f_A(x_i)) \\ + \pi_A(x_i) \log(\pi_A(x_i))), \alpha = 1 \end{array} \end{cases} \qquad (7)$$

$$K_R^{\beta}(A) = 1 - \frac{1}{n} \sum_{i=1}^{n} \frac{\log(t_A(x_i)^{\beta} + f_A(x_i)^{\beta} + \pi_A(x_i)^{\beta})}{1 - \beta}, \quad 0 < \beta < 1 \qquad (8)$$

$$K_H^p(A, B) = 1 - \frac{1}{n} \sum_{i=1}^{n} (\max\{|t_A(x_i) - t_B(x_i)|^p, |f_A(x_i)$$

$$- f_B(x_i)|^p, |\pi_A(x_i) - \pi_B(x_i)|^p\} \qquad (9)$$

$$K_{SK2}^P(A) = \frac{1}{n} \sum_{i=1}^{n} \left(\max\{|t_A(x_i)|^p, |f_A(x_i)|^p, (1 - \pi_A(x_i))^p\} \right) \qquad (10)$$

$K_{BB}(A)$ is from an entropy $E_{BB}(A)$ (1996, Bustince and Burillo) [13], and $K_{SK1}(A)$ from an entropy $E_{SK1}(A)$ (2001, Szmidt and Kacprzyk) [16]. Szmidt and Kacprzyk [17] introduced the Hausdorff distance $D_H^p(A)$ in 2011, and hence according to $D_H^p(A)$, in 2017, Das et al. [31] proposed $K_{SK2}^P(A)$. And in 2014, Szmidt et al. [26] presented $K_{SKB}(A)$, which is derived from the entropy $E_{SKB}(A)$. Obviously, $K_{SKB}(A)$ is the average of $K_{SK}(A)$ and $K_{BB}(A)$. In accordance with the Szmidt and Kacprzyk's axiom system of knowledge measure of IFS, in 2016 Guo put forward $K_G(A)$ [30]. Moreover, $K_{HC}^{\alpha}(A)$ and $K_R^{\beta}(A)$ are both derived from $E_{HC}^{\alpha}(A)$ and $E_R^{\beta}(A)$ (Huang and Yang [15]). The formulas above can be proved easily to satisfy the property of non-negative boundedness and symmetry. For the order property, we have Lemma 2. It is easy to prove Lemma 2, which is shown in the appendix.

Lemma 2 $K_{SK1}(A)$ *and* $K_G(A)$ *meet the order property KP4 III, while* $K_{BB}(A)$, $K_{SKB}(A)$, $K_{HC}^{\alpha}(A)$, $K_R^{\beta}(A)$, *and* $K_{SK2}^P(A)$ *don't meet KP4 III.*

4 Parametric Knowledge Measure of IFS

According to **KP4 II**, a reasonable measurement tool that conforms to Szmidt and Kacprzyk's axioms must satisfy: if $t \leq f$ then $K = \text{Knowledge}(t, f, \pi)$ is a monotonous and non-increasing function for t and a monotonous and non-decreasing function for f, otherwise $K = \text{Knowledge}(t, f, \pi)$ is a monotonous and non-decreasing function for t and a monotonous and non-increasing function for f. Hence, t and f change in the opposite direction for knowledge measure. When constructing a knowledge measure, the monotonic direction of membership degree is opposite with that of non-membership degree.

The smaller the gap between membership degree and non-membership degree is, the higher the degree of confusion will be. For the most extreme cases, $|\mu(x) - \nu(x)| = 0$, we have Entropy $= 1$ and Knowledge $= 0$.

The greater the difference between membership degree and non-membership degree is, the lower the degree of confusion and the greater the amount of information will be. For the most extreme cases, when the absolute difference between membership degree and non-membership degree reaches a maximum value of 1, the degree of clarity is 1 and the degree of confusion is 0, and we have Entropy $= 0$ and Knowledge $= 1$. Therefore, the knowledge measure should be changed in the same direction from the absolute difference between membership degree and non-membership degree.

Based on the analysis above and the definition of knowledge measure, for IFS A, the following formula is a knowledge measure:

$$K_p(A) = \frac{1}{n} \sum_{i=1}^{n} \left| (t_A(x_i))^p - (f_A(x_i))^p \right|, \quad p > 0. \tag{11}$$

Next, $K_p = |t^p - f^p|$ can be proved to meet all properties of Definition 4.

Proof For each $p > 0$,

  (EP1) $t_A(x) = f_A(x) = 0 \Rightarrow K(A) = 0$.
  (EP2) $K_p = 1 \Leftrightarrow t_A(x) = 0 \,\&\, f_A(x) = 1$ or $t_A(x) = 1 \,\&\, f_A(x) = 0$.
  (EP3) $K_p(A) = |t^p - f^p| = |f^p - t^p| = K_p(A^C)$.
  (EP4)

$$K_p = |t^p - f^p| = \begin{cases} f^p - t^p, t \le f \Rightarrow \frac{\partial K_p}{\partial t} = -pt^{p-1} \le 0, \frac{\partial K_p}{\partial f} = pf^{p-1} \ge 0, \\ t^p - f^p, t \ge f \Rightarrow \frac{\partial K_p}{\partial t} = pt^{p-1} \ge 0, \frac{\partial K_p}{\partial f} = -pf^{p-1} \le 0. \end{cases}$$

Therefore, for $p > 0$, $K_p(A)$ is a knowledge measure of IFS A.

5 Experimental Example and Result Analysis

Example 1 Taking into account eight IFSs,

$$A_1 = \{\langle x, 1, 0 \rangle\}, \quad A_2 = \{\langle x, 0.9, 0.1 \rangle\}, \quad A_3 = \{\langle x, 0.8, 0.1 \rangle\},$$
$$A_4 = \{\langle x, 0.8, 0.2 \rangle\}, \quad A_5 = \{\langle x, 0.7, 0.3 \rangle\}, \quad A_6 = \{\langle x, 0.6, 0.4 \rangle\},$$
$$A_7 = \{\langle x, 0.5, 0.4 \rangle\}, \quad A_8 = \{\langle x, 0.5, 0.5 \rangle\}.$$

Table 1 Comparison of experimental results of $E(A_i)$, ($i = 1, 2, 3, 4, 5, 6, 7, 8$)

Function	A_1	A_2	A_3	A_4	A_5	A_6	A_7	A_8	Right or wrong	Number of wrong	Accuracy (%)
K_{SK1}	1	0.889	0.778	0.75	0.571	0.333	0.167	0	Right	0	100
K_G	1	0.9	0.835	0.82	0.37	0.6	0.505	0.5	Right	0	100
K_{SK2}	1	1	0.81	1	1	1	0.81	1	Wrong	8	0
K_{SKB}	1	0.944	0.839	0.875	0.786	0.667	0.533	0.5	Wrong	1	87.5
K_{BB}	1	1	0.9	1	1	1	0.9	1	Wrong	8	0
$K_{HY}1$	1	0.859	0.722	0.783	0.735	0.708	0.590	0.699	Wrong	2	75
$K_r0.5$	1	0.796	0.632	0.745	0.717	0.703	0.562	0.699	Wrong	2	75
$K_{0.25}$	1	0.412	0.383	0.277	0.175	0.085	0.046	0	Right	0	100
$K_{0.333}$	1	0.501	0.464	0.343	0.218	0.107	0.057	0	Right	0	100
$K_{0.5}$	1	0.632	0.578	0.447	0.289	0.142	0.075	0	Right	0	100
K_1	1	0.8	0.7	0.6	0.4	0.2	0.1	0	Right	0	100
K_2	1	0.8	0.63	0.6	0.4	0.2	0.09	0	Right	0	100
CK_p	1	0.629	0.551	0.453	0.296	0.147	0.073	0	Right	0	100

$$CK_p(A) = \frac{(K_{0.25}(A) + K_{0.333}(A) + K_{0.5}(A) + K_1(A) + K_2(A))}{5} \tag{12}$$

where $CK_p(A)$ is a combination model being composed of $K_p(A)$.

It is clear that

$$
\begin{aligned}
f_{A_1}(x) = 0 &< f_{A_2}(x) = f_{A_3}(x) < f_{A_4}(x) < f_{A_5}(x) < f_{A_6}(x) \\
&= f_{A_7}(x) < f_{A_8}(x) = t_{A_8}(x) = t_{A_7}(x) < t_{A_6}(x) < t_{A_5}(x) < t_{A_4}(x) \\
&= t_{A_3}(x) < t_{A_2}(x) < t_{A_1}(x) = 1.
\end{aligned}
$$

$\therefore K_{A_1}(x) > K_{A_2}(x) > K_{A_3}(x) > K_{A_4}(x) > K_{A_5}(x) > K_{A_6}(x) > K_{A_7}(x) > K_{A_8}(x).$

$$\text{Accuracy} = \frac{\text{Number (Entropies\_with\_Right\_Order\_in\_}A_i)}{\text{Number}(A_i)} \tag{13}$$

where variable accuracy means the proportion of correctly ranked values in all knowledge measure values of IFS. Obviously, the higher the precision accuracy, the better the operator. From formulas (3, 4, 5, 6, 7, 8, 10, 11, 12), we get Table 1. Results illustrate that for the knowledge measures with the order property, such as K_G, K_{SK1}, and K_p, the order of the results is completely correct. However, the order of the results for the knowledge measures without the order property, such as K_{SKB}, K_{BB}, $K_{HY}1$, $K_r 0.5$, and K_{SK2}, does not meet the property KP4I. From Table 1, K_G, K_{SK1}, and K_p will be better than the others.

6 Conclusions

Based on the definition of knowledge measure of IFS in judging the degree of disorder and order of the system, this paper proposes an axiom system in terms of the property of order, and puts forward a series of equivalent conditions of the order judgments. On the basis of the axiom system theory, this paper presents a parametric model of knowledge measure and illustrates the validity of the measure in theory and in practice. Furthermore, this paper also applies the proposed formula, along with some classical formulas of intuitionistic fuzzy knowledge measure, to verify a conclusion: In most of the formulas of intuitionistic fuzzy knowledge measure, the accuracy of those operators with the order property will be higher than those without. In the future research, we can adopt the distance satisfying the order property [32] to create new knowledge measure.

Acknowledgements This paper is funded by the National statistical research key projects (No. 2016LZ18), Soft Science Project (No. 2015A070704051) and Natural Science Projects (No. 2016A030313688, No. 2016A030310105, No. 2018A030313470) and Quality engineering and teaching reform project (No. 125-XCQ16268) of Guangdong Province.

References

1. Zadeh, L.A.: Fuzzy sets. Inf. Control **8**, 338–353 (1965)
2. Zadeh, L.A.: Fuzzy sets and systems. In: Proceedings of the Symposium on Systems Theory, pp. 29–37. Polytechnic Press, Brooklyn (1965)
3. Zadeh, L.A.: The concept of a linguistic variable and its application to approximate reasoning-I. Inf. Sci. **8**(3), 199–249 (1975)
4. Atanassov, K.: Intuitionistic fuzzy sets. Fuzzy Sets Syst. **20**(1), 87–96 (1986)
5. Atanassov, K.: Intuitionistic Fuzzy Sets Theory and Applications, pp. 1–2. Springer, Berlin (1999)
6. Atanassov, K.: More on intuitionistic fuzzy sets. Fuzzy Sets Syst. **51**(1), 117–118 (1989)
7. Atanassov, K.: New operations defined over the intuitionistic fuzzy sets. Fuzzy Sets Syst. **61**(2), 137–142 (1994)
8. Atanassov, K.: On intuitionistic fuzzy sets theory. Stud. Fuzziness Soft Comput. **283**(1), 1–321 (2012)
9. Atanassov, K., Gargov, G.: Interval valued intuitionistic fuzzy sets. Fuzzy Sets Syst. **31**(3), 343–349 (1989)
10. De Luca, A., Termini, S.: A definition of non-probabilistic entropy in the setting of fuzzy sets theory. Inf. Control **20**(4), 301–312 (1972)
11. Yager, R.R.: On the measure of fuzziness and negation, part 1: membership in the unit interval. Int. J. Gen. Syst. **5**(4), 221–229 (1979)
12. Higashi, M., Klir, G.: On measures of fuzziness and fuzzy complements. Int. J. Gen. Syst. **8**(3), 169–180 (1982)
13. Bustince, H., Burillo, P.: Entropy on intuitionistic fuzzy sets and on interval-valued fuzzy sets. Fuzzy Sets Syst. **78**(3), 305–316 (1996)
14. Bustince, H., Barrenechea, E., Pagola, M., et al.: Generalized Atanassov's intuitionistic fuzzy index: construction of Atanassov's fuzzy entropy from fuzzy implication operators. Int. J. Uncertain., Fuzziness Knowl.-Based Syst. **19**(1), 51–69 (2011)
15. Hung, W.L., Yang, M.S.: Fuzzy entropy on intuitionistic fuzzy sets. Int. J. Intell. Syst. **21**(4), 443–451 (2006)
16. Szmidt, E., Kacprzyk, J.: Entropy for intuitionistic fuzzy sets. Fuzzy Sets Syst. **118**(3), 467–477 (2001)
17. Szmidt, E., Kacprzyk, J.: Intuitionistic fuzzy sets-two and three term representations in the context of a Hausdorff distance. Acta Univ. Matthiae Belii **19**, 53–62 (2011)
18. Szmidt, E., Kacprzyk, J.: Some problems with entropy measures for the Atanassov intuitionistic fuzzy sets. Lect. Notes Comput. Sci. **4578**, 291–297 (2007)
19. Zhang, H.Y., Zhang, W.X., Mei, C.L.: Entropy of interval-valued fuzzy sets based on distance and its relationship with similarity measure. Knowl.-Based Syst. **22**(6), 449–454 (2009)
20. Barrenechea, E., Bustince, H., Pagola, M., et al.: Construction of interval-valued fuzzy entropy invariant by translations and scalings. Soft. Comput. **14**(9), 945–952 (2010)
21. Vlachos, I.K., Sergiadis, G.D.: Subsethood, entropy and cardinality for interval-valued fuzzy sets—an algebraic derivation. Fuzzy Sets Syst. **158**(12), 1384–1396 (2007)
22. Zeng, W.Y., Li, H.X.: Relationship between similarity measure and entropy of interval-valued fuzzy sets. Fuzzy Sets Syst. **157**(11), 1477–1484 (2006)
23. Farhadinia, B.: A theoretical development on the entropy of interval valued fuzzy sets based on the intuitionistic distance and its relationship with similarity measure. Knowl.-Based Syst. **39**(2), 79–84 (2013)
24. Wei, C.P., Wang, P., Zhang, Y.Z.: Entropy, similarity measure of interval-valued intuitionistic fuzzy sets and their applications. Inf. Sci. **181**(19), 4273–4286 (2011)
25. Liu, X.D., Zheng, S.H., Xiong, F.L.: Entropy and subsethood for general interval-valued intuitionistic fuzzy sets. Lect. Notes Artif. Intell. **3613**(7), 42–52 (2005)
26. Szmidt, E., Kacprzyk, J., Bujnowski, P.: How to measure the amount of knowledge conveyed by Atanassov's intuitionistic fuzzy sets. Inf. Sci. **257**(1), 276–285 (2014)

27. Montes, I., Pal, N.R., Janis, V., et al.: Divergence measures for intuitionistic fuzzy sets. IEEE Trans. Fuzzy Syst. **23**(2), 444–456 (2015)
28. Pal, N.R., Bustince, H., Pagola, M., et al.: Uncertainties with Atanassov's intuitionistic fuzzy sets: fuzziness and lack of knowledge. Inf. Sci. **228**(4), 61–74 (2013)
29. Zhang, Q.S., Jiang, S.Y., Jia, B.G., et al.: Some information measures for interval-valued intuitionistic fuzzy sets. Inf. Sci. **180**(24), 5130–5145 (2010)
30. Guo, K.H.: Knowledge measure for Atanassov's intuitionistic fuzzy sets. IEEE Trans. Fuzzy Syst. **24**(5), 1072–1078 (2016)
31. Das, S., Guha, D., Mesiar, R.: Information measures in the intuitionistic fuzzy framework and their relationships. IEEE Trans. Fuzzy Syst. **26**(3), 1626–1637 (2018)
32. Zhang, Z.H., Wang, M., Hu, Y., et al.: A dynamic interval-valued intuitionistic fuzzy sets applied to pattern recognition. Math. Probl. Eng. **2013**(6), 1–16 (2013)

Hybrid PPSO Algorithm for Scheduling Complex Applications in IoT

Komal Middha and Amandeep Verma

Abstract The Internet of Things (IoT) is boosting revolution in almost every aspect of our lives. It provides networking to connect things, applications, people, data with the help of Internet. It is widespread across multiple domains extending its roots from civil to defense sectors. Although it has deepened its roots, it has certain shortcomings associated with it such as limited storage space, limited processing capability, scheduling complex applications. Large complex applications are normally represented by workflows. A lot of workflow scheduling algorithms are prevailing but somehow each one is having certain issues associated with them. In this paper, we have presented a new workflow scheduling algorithm, i.e., PPSO which is a hybrid combination of heuristic technique, i.e., Predict Earliest Finish Time (PEFT) and meta-heuristic technique, i.e., Particle Swarm Optimization (PSO). The proposed approach is analyzed for different workflows on WorkflowSim simulator. The overall outcomes validates that it outperforms better than existing algorithms.

Keywords Cloud computing · Internet of Things · Particle swarm optimization · Predict earliest finish time · Workflow scheduling

1 Introduction

Cloud Computing has developed as an overwhelming stage to understand different logical applications. It employs "pay-per-utilize" demonstrate, which implies clients are charged by asset utilization [1]. It offers multiple advantages such as versatility, adaptability, cost-effective, etc. [2]. Likewise, IoT is another new wave in the field of worldwide industry. It is characterized as a system of physical gadgets having detecting and system abilities that empowers the gadgets to store and trade information yet

K. Middha (✉) · A. Verma
U.I.E.T., Panjab University, Chandigarh, India
e-mail: mail2komalmiddha@gmail.com

A. Verma
e-mail: amandeepverma@pu.ac.in

© Springer Nature Singapore Pte Ltd. 2019
S. K. Bhatia et al. (eds.), *Advances in Computer Communication
and Computational Sciences*, Advances in Intelligent Systems and Computing 924,
https://doi.org/10.1007/978-981-13-6861-5_19

with constrained capacity limit. As cloud offers boundless capacity and handling, IoT and cloud computing are amalgamated together. This blend is known as Cloud of Things [3]. Both of these applications can take the upsides of each other, and the blending of these applications has offered beginning to different new large complex applications such as smart city, smart home, health care, etc. These vast complex applications are generally represented as workflows.

Workflows constitute a typical model for portraying an extensive variety of logical applications in dispersed frameworks. These generally represent the large complex application that belongs to different area of science [4]. Workflow scheduling is a procedure of mapping between subordinate tasks on the accessible assets to such an extent that workflow application can finish its execution within the client's predetermined Quality of Service (QoS) limitations, for example, deadline, budget [5]. It plays an eminent role in workflow management. Appropriate scheduling can have huge effect on the execution of framework [6]. There are two primary stages for workflow scheduling: the first is the asset provisioning stage in which the processing assets that will be utilized to run the errands are chosen and after this, a schedule produced and each task is mapped onto the most appropriate asset [7]. Initially, workflow scheduling has been studied only for grid and cluster computing but with the expanding interest for process automation in the cloud, particularly for distributed application, the examination on cloud workflow scheduling procedure is turning into a remarkable issue not just in the territory of cloud workflow frameworks yet additionally broad cloud computing [8]. There is a huge difference between workflow scheduling for grid and cloud computing. The main objective of scheduling in grid is to reduce the execution time only and neglecting the costs of hired resources while scheduling in cloud considers both the execution time and the costs of hired resources [9].

A lot of workflow scheduling algorithm exists, and these are continuously evolving which yields better and better results than the existing algorithms. So, to further optimize the performance of scheduling algorithm, we have proposed a new hybrid algorithm, i.e., PPSO. The proposed approach uses meta-heuristic technique, i.e., Particle Swarm Optimization (PSO) and heuristic technique, i.e., Predict Earliest Finish Time (PEFT) to generate a new hybrid algorithm. PSO is the most widely used optimization technique as it outperforms as compared to other existing optimization techniques [10, 11], and PEFT is one of the list heuristic algorithm also as better results in terms of efficiency and scheduling length ratio [12]. This novel approach tends to minimize the execution time as well as the cost. Finally, the simulation is done of the proposed work using the WorkflowSim simulator and the analysis validates the result.

The rest of the paper is divided into different sections. Section 2 reviews the prior work done. Section 3 defines scheduling problem formulation followed by Sect. 4 defining the baseline PEFT algorithm. The next section discusses about the proposed heuristic, i.e., PPSO. Section 6 evaluates the proposed heuristic results, and the last section concludes the whole work.

2   Related Work

It is well known that Cloud Computing has broadened, dynamic, and adaptable nature, which offers multiple benefit to the users. But, sometimes these natures constitute challenge in developing scheduling algorithm [13]. The process of assignment of resources to heterogeneous tasks is referred as scheduling. There are multiple factors based on which scheduling is done such as maximizing the CPU utilization and throughput, while minimizing the turnaround time, waiting time, and response time [14]. Scheduling algorithms are divided into two major classes, namely static and dynamic algorithm [15]. In static scheduling, all data about chores, for example, execution and correspondence costs for each undertaking and the association with different chores is known in advance. This is utilized to detail the schedule in progress. In case of dynamic scheduling, these data are not accessible and scheduling choices are made at runtime [16]. Subsequently, the objective of dynamic calendar is to produce a schedule within the schedule overhead point of confinement along with the internal failure concerns [17].

In this segment, we looked into current investigations which utilize the diverse heuristic, meta-heuristic, and hybrid algorithms for task scheduling. Heuristic algorithms are further classified as list scheduling heuristics, clustering heuristics, guided random search methods, and task duplication heuristics [18]. Among these, most widely used technique for scheduling is list scheduling since they produce great quality schedules with less intricacy [17]. The primary idea of the list scheduling calculation is that it sort tasks based on their priorities and the planning result is acquired when all task finish their execution [19]. The traditional list scheduling algorithm incorporates Mapping Heuristics (MP) [20], Dynamic Level Scheduling (DLS) [21], Modified Critical Path (MCP) [22], Critical Path on Processor (CPOP) [23], Heterogeneous Earliest Finish Time (HEFT) [24] etc.

Topcuoglu et al. [24] proposed two task scheduling HEFT and CPOP algorithms for heterogeneous computing. This approach minimizes the execution time and meets high-performance results for limited number of processors. They have also compared the performance of proposed algorithms with various other algorithms such as DLS, MH. Different metrics have been used for comparison, namely schedule length ratio, speedup, running time of algorithm and based on the comparison outcome it has been proved that HEFT has better performance among these and it is one of the acceptable solution for heterogeneous systems.

To further optimize the performance of HEFT algorithm, many variants of HEFT have been proposed. Lin and Lu [25–28] improve the performance based on different criteria. Lin et al. [25] proposed workflow scheduling algorithm called as Scalable Heterogeneous Earliest Finish Time (SHEFT) algorithm. According to this paper, even in spite of the fact that the quantity of allotted assets to a work process can be flexibly scalable but there exist some planning issues. To settle this issue, SHEFT is proposed for distributed computing. Bala et al. [26] proposed an improved HEFT algorithm. Initially, HEFT do scheduling on the basis of earliest finish time, while does not overlooks other factors. But this proposed scheduling incorporates various new factors for coordinating right assets for specific task to be scheduled such as storage, data transmission rate. This work demonstrates that it reduces the waiting time, execution time, as well as turnaround time. Dubey et al. [27] modified HEFT to reduce the make span as well as to disseminate the workload among processors in an efficient way. Verma and Kaushal [28] presented Budget and Deadline constrained Heuristic in the light of HEFT, i.e., BHEFT. In this work, they demonstrate that the proposed algorithm can reduce the execution cost along with making make span as good as other existing scheduling algorithm.

Rini et al. [29] have presented variants of Particle Swarm Optimization (PSO) algorithm. In this paper, they reviewed 46 studies related to the PSO of the time period between 2002 and 2010 along with the advantages, disadvantages, variants, and applications based on PSO. Li et al. [30] have researched an improved PSO algorithm. They combine PSO with Ant Colony Optimization (ACO) algorithm, which improves its performance in terms of search speed and accuracy of results. Verma and Kaushal [31] presented Multi-Objective Particle Swarm Optimization (MOPSO) algorithm to produce an ideal solution for task offloading.

Arabnejad et al. [12] gave a list scheduling algorithm known as Predict Earliest Finish Time (PEFT). In this paper, it has been proved that PEFT has better outcomes than existing list scheduling algorithms, namely HEFT, Lookahead, and many more in terms of efficiency and scheduling length ratio. Kaur and Kalra [32] presented a modified and improved version of PEFT algorithm by merging it with genetic algorithm. As it is proved in previous year research work [10, 31, 33, 34], PSO has better performance than genetic algorithm. In this paper, we presented another variant of PEFT by merging it with PSO and called it as PPSO. In [32], they have not considered user defined constraint but here, we have considered them also.

3 Problem Assumptions

3.1 Workflow Application Model

A workflow application model is configured as a Directed Acyclic Graph (DAG) with $G(T, E)$, consisting of set of T number of tasks $(t_1, t_2, t_3 \ldots, t_n)$ and E is the set of edges, which denotes the reliance between the tasks. In layman language we can say, workflow defines parent–child task relationship graphically and the rules followed in a workflow is that, no child task can start execution until its parent task completes its execution and the resources needed by the task are also accessible at the virtual machine, that is going to accomplish the task [34, 35]. For example, if $(t_i, t_j) \in T$, it implies that t_i is the parent of t_j and t_j is the child of t_i. A node with no parent is called as entry node and a node with no child is called as exit node [36]. An illustration of DAG, consisting of ten tasks, is shown in Fig. 1. The Infrastructure as a Service (IaaS) provides many services such as security, networking, etc. and along with services it provides wide range of VM types. The performance of VM

Fig. 1 Sample DAG

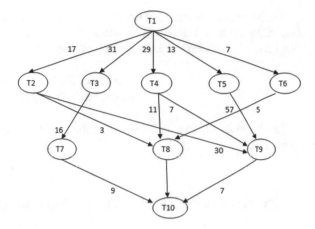

depends on its processing speed as well as on the cost per unit time. To measure execution time (ET) and transfer cost (TT), we have the following formulas:

$$\mathrm{ET}(t_i, p_j) = \frac{(\text{length})t_j}{\mathrm{PS}(r_j) * (1 - (\mathrm{PD}\,(p_j)))} \tag{1}$$

$$\mathrm{TT}(t_i, t_j) = \frac{\mathrm{Data}(t_i, \text{out})}{\beta} \tag{2}$$

where $\mathrm{Data}(t_i, \text{out})$ is the output data size, β is the average bandwidth. When t_i, t_j are executed on same resources, then $\mathrm{TT}(t_i, t_j)$ becomes equal to zero.

3.2 Problem Definition

A schedule is often defined as a metrics of an arrangement of resources (R), a task to asset mapping (M), the Total Execution Cost (TEC) and the Total Execution Time (TET), i.e., $S = (R, M, \mathrm{TEC}, \mathrm{TET})$ [7]. TEC and TET are calculated using (3) and (4), respectively

$$\mathrm{TEC} = \sum_{j=1}^{|P|} C(p_j) * \left(\frac{\mathrm{LFT}(p_j) - \mathrm{LST}(p_j)}{T} \right) \tag{3}$$

$$\mathrm{TFT} = \max\{\mathrm{FT}(t_i) : t_i \in T\} \tag{4}$$

where $\mathrm{LST}(p_j)$ is the least start time, i.e., when the resource is enrolled by the processor, $\mathrm{LFT}(p_j)$ is the least end time, i.e., when the resource is released by the processor, T is the unit of time, and $\mathrm{FT}(t_i)$ is the finish time of task t_i.

Each workflow scheduling heuristics have distinct targets but this work emphasis on finding a schedule that will minimize the total execution cost within the specified deadline. The defined objective can be outlined as:

$$\begin{aligned} &\text{minimize TEC} \\ &\text{subject to TET} \le D \end{aligned} \tag{5}$$

In the next section, a brief description of baseline algorithm, i.e., PEFT is explained.

4 Predict Earliest Finish Time Algorithm (PEFT)

PEFT [12] is categorized under list scheduling algorithm and encompasses two phases, namely prioritizing phase and resource selection phase. The various terms used in this algorithm are explained below:

Optimistic Cost Table (OCT). It is usually represented in the form of matrix where, rows signify the tasks, and columns signify the processors. It is calculated as (6) using backward approach, i.e., from exit task to entry task.

$$\text{OCT}(t_i, p_j) = \max_{t_j \in \text{succ}(t_i)} \left[\min_{r_w \in R} \{ \text{OCT}(t_j, p_w) + w(t_j, p_w) + c(t_i, t_j) \} \right]$$
$$c(t_i, t_j) = 0, \quad \text{if } p_w = p_j \tag{6}$$

where $c(t_i, t_j)$ is the average transfer cost of task t_i to task t_j and $w(t_j, p_w)$ is the execution time of task t_i on processor p_w.

rank$_{\text{OCT}}$. The first phase of PEFT is dependent on rank$_{\text{OCT}}$, which means that for assigning preference to each task, rank$_{\text{OCT}}$ is calculated. It is calculated as (7) and the computed values are sorted in decreasing order. Higher the value of rank$_{\text{OCT}}$, higher the priority of task.

$$\text{rank}_{\text{OCT}}(t_j) = \frac{\sum_{j=1}^{P} \text{OCT}(t_i, r_j)}{P} \tag{7}$$

Optimistic Earliest Finish Time (O_{EFT}). The second phase of PEFT is dependent on O_{EFT}, which means that it is computed for assigning processor to a task. It is calculated using Eq. (8)

$$O_{\text{EFT}}(t_i, p_j) = \text{EFT}(t_i, p_j) + \text{OCT}(t_i, p_j) \tag{8}$$

PEFT Algorithm

(1) Compute the OCT matrix as defined in Eq. (6).
(2) Calculate $rank_{OCT}$ for each matrix as given in Eq. (7) and sort them in descending order.
(3) Repeat step 4–6, until the list is empty.
(4) for all the processor (p_i) do.
(5) Calculate EFT (n_i, p_j) value using insertion based scheduling policy for the top most task [15].
(6) Using Eq. (8), calculate $O_{EFT}(n_i, p_j)$.
(7) Allocate task to the processor with minimum O_{EFT}.
(8) Return the optimal schedule.

5 Proposed PPSO Algorithm

The proposed algorithm PPSO is a composite combination of two algorithms namely, PEFT and PSO. Here, PEFT forms the initial base. The different steps followed are shown with the help of flow chart in Fig. 2. And a brief introduction is explained below:

Initialization. This is the most crucial step in any algorithm as, the whole performance and results both depends on input. Here, initialization is done in such a way that the first set of swarm is obtained using schedule of PEFT algorithm and the rest of the swarm are chosen randomly.

Fitness Value Computation. The computation of fitness value is directly dependent on the objective function. In this algorithm, the main motive is to minimize the execution cost as well as time. The swarm with minimum cost as well as the one with minimum time consumption will be opted, as it is stated in Eq. 5.

Parameters. The two main parameters used are p_{best} and g_{best}, which are responsible for optimizing the performance of the algorithm. The p_{best} defines the best solution attained till now, and g_{best} value determines the global best value achieved by the swarm. After selecting the parameters, the position and velocity for each swarm is calculated.

PPSO Algorithm

(1) Initialize the first set of swarm using schedule obtained from PEFT algorithm and the remaining are initialized randomly.
(2) Calculate the fitness value for each particle position as stated in Eq. 5.
(3) Compare the calculated fitness value with p_{best}, if the current value is better than p_{best}, then update the value of particle with the current value.

Fig. 2 Flowchart for
proposed heuristic

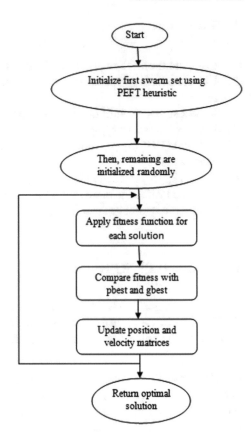

(4) Also compare the fitness value with g_{best}, if the current value is better than g_{best}, then update the value of particle with the current value.
(5) Assign best of all p_{best} and g_{best}.
(6) Update particle position and velocity as stated in [37].
(7) Repeat steps 2–5 until optimal solution is obtained.

6 Experimental Analysis

This section highlights the simulation results. The simulation is done using the Work-flowSim simulator consisting of various components such as workflow mapper, work-flow scheduler, workflow engine, etc., and the details about this simulator is explained in [38]. Different workflow structure is used for the simulating the results such as Montage, Inspiral, Epigenomics, Inspiral, CyberShake. The Directed Acyclic Graph in XML (DAX) configuration format for all the above-mentioned structure is defined in config file of the simulator. Figure 3 shows some sample scientific workflows.

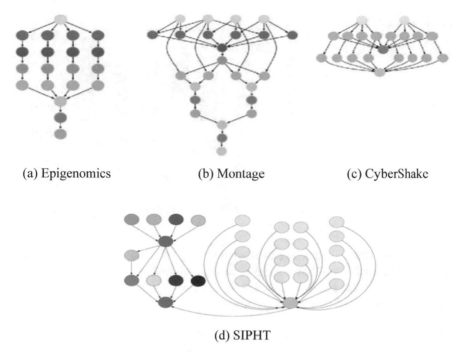

(a) Epigenomics   (b) Montage   (c) CyberShake

(d) SIPHT

Fig. 3 Structure of scientific workflows [39]

6.1 Experiment Configuration

For our experiment, we have considered four virtual machines in a single data center, each having different processing speed ranging from 1000 to 4000 Million Instruction Per Second (MIPS) and also, we have specified cost for each one of them that varies from 2 to 10, such that the ratio between slow and fast machine is 1:5. The cost is specified in dollar ($). The virtual machine with the minimum speed will have minimum cost, and the virtual machine with the maximum speed will have maximum cost. We have included two factors for performance comparison, namely deadline and cost. The deadline is evaluated as (9), by executing all tasks on fast and slow machines while excluding the data transfer cost and then subtracting the time incurred from both the machines. Then, the resultant is added with PEFT makespan time.

$$\text{Deadline} = \alpha * \{(M_{\text{slow}} - M_{\text{fasr}} + M_{\text{PEFT}}\}: \alpha \in \{0.5, 3\} \tag{9}$$

where α specify the deadline factor, that varies from 0.5 to 3 and M_{slow} and M_{fasr} specify the makespan of slowest and fastest schedule. M_{PEFT} is the makespan of PEFT.

Fig. 4 Execution time of
HEFT and PEFT

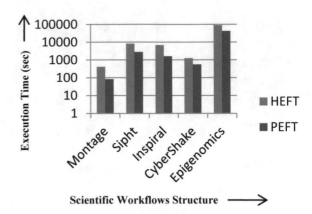

The cost for final schedule is evaluated as (10) by subtracting the finish time from start time of the task, and the resultant value is multiplied by cost of virtual machine, to which a particular task is assigned. The total cost is calculated as (11), which is the summation of cost of the final schedule.

$$EC_{i,j} = ET_{i,j} * Cost_j \tag{10}$$

$$Total\ Cost = \sum_{i=1}^{n} EC_{i,j} \tag{11}$$

where $EC_{i,j}$ denotes the cost of the final schedule, $ET_{i,j}$ indicate the execution time of the task on particular virtual machine to which it is allocated, and $Cost_j$ denotes the cost of virtual machine.

6.2 Experiment Results

For the experiment validation, initially we have compared the PEFT and HEFT execution time for different workflows structure with 100 tasks. As it has already been proved in previous research work, that PEFT has better performance than HEFT. Here, also it is proved that execution time of HEFT is more than PEFT as shown in Fig. 4.

Then, we have compared the performance of PPSO with PSO and HPSO. HPSO is hybrid combination of HEFT and PSO. As PSO has random solutions, all three algorithms are executed 10 times. Initially, we have scrutinized these algorithms in terms of average execution time and average execution cost for a Montage workflow consisting of 25 tasks, 50 tasks and then we compared it for workflow consisting of 1000 tasks. Figure 5 shows the comparison among these in terms of average execution time, and Fig. 6 demonstrates the comparison in terms of average execution cost.

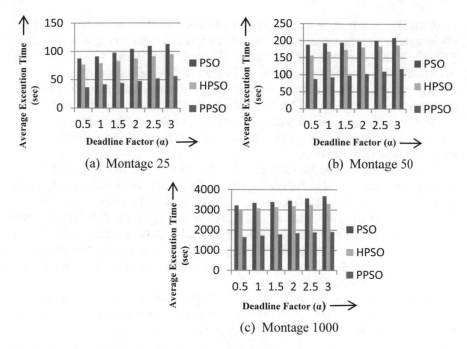

Fig. 5 Average execution time of scientific workflows

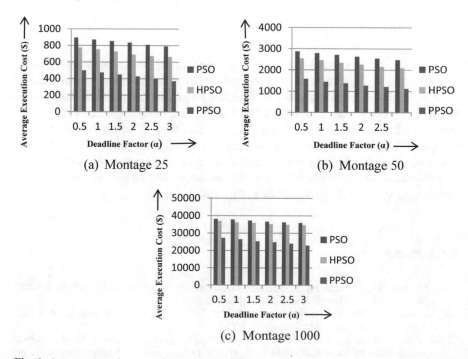

Fig. 6 Average execution cost of scientific workflows

From Fig. 5, it is clear that PPSO outperforms than PSO and HPSO. All these algorithms schedule each workflow within the specified deadline. For Montage 25, PPSO is 55.29% better than PSO and 48.6% better than HPSO. For Montage 50, the result of PPSO is 43.3 and 52.7% better than HPSO and PSO, respectively. The average execution time for Montage 1000 is more, as it has more number of tasks to execute than Montage 25 and Montage 50. It is 40.6% better than PSO and 38.6% better than HPSO. Hence, it depicts a direct relation between the deadline factor (α) and the average execution time, i.e., as the deadline factor (α) increases the average execution time also increases.

Similarly, from Fig. 6, it is clear that PPSO outperforms than PSO and HPSO in terms of execution cost also. Here, it depicts an inverse relation between the deadline factor (α) and the average execution cost, i.e., as the deadline factor (α) increases the average execution cost also decreases. For Montage 25, PPSO is 54.6% better than PSO and 42.3% better than HPSO. For Montage 50, the result of PPSO is 41.8 and 36.2% better than HPSO and PSO, respectively. The average execution cost for Montage 1000 considerably is more than other two structure of Montage. It is 28.9% better than PSO and 33.3% better than HPSO.

From the results, we can conclude that PPSO gives better performance in both the cases either in terms of average execution time or in average execution cost. Hence, it can be used to schedule large complex applications.

7 Conclusion

Workflow scheduling is a very eminent topic for research, and many research work has been done in this filed. But, each proposed work had certain limitations associated with it. This paper presents a new methodology to solve this problem, i.e., PPSO. It employs a combination of heuristic technique, i.e., Predict Earliest Finish Time (PEFT) and meta-heuristic technique, i.e., Particle Swarm Optimization (PSO) and compared the performance of proposed algorithm PPSO with the PSO and HPSO. The main objective is to optimize scheduling along with certain constraints such as minimum cost and time. Subsequently, we would like to amplify our work by introducing other parameters such as energy, reliability, etc.

References

1. Al-Roomi, M., Al-Ebrahim, S., Buqrais, S., Ahmad, I.: Cloud computing pricing models: a survey. Int. J. Grid Distrib. Comput. 6(5), 93–106 (2013)
2. Kavitha, K.: Study on cloud computing models and its benefits, challenges. Int. J. Innov. Res. Comput. Commun. Eng. 2(1), 2423–2431 (2014)
3. Aazam, M., Khan, I., Alsaffar, A.A.: Cloud of things: integrating internet of things and cloud computing and the issues involved. In: Proceeding of 11th International Bhurban Conference on

Applied Sciences and Technology (IBCAST), Pakistan (2014). https://doi.org/10.1109/ibcast.2014.6778179

4. Gil, Y., Deelman, E., Ellisman, M., Moreau, L., Myres, J.: Examining the challenges of scientific workflows. IEEE Comput. **40**(12), 26–34 (2007)

5. Taylor, I., Deelaman, E., Gannon, D., Shields, M.: Workflow for e-Science: Scientific Workflows for Grid, 1st edn. Springer, Berlin (2007)

6. Zhang, Y., Mandal, A., Koebill, C., Cooper, K.: Combined fault tolerance and scheduling techniques for workflow applications on computational grid. In: 9th IEEE/ACM International Symposium on Clustering and Grid, pp. 244–251 (2009)

7. Rodriguez, M.A., Buyya, R.: Deadline based resource provisioning and scheduling algorithm for scientific workflows on clouds. IEEE Trans. Cloud Comput. **2**(2), 222–235 (2014)

8. Bardsiri, A.K., Hashemi, S.M.: A review of workflow scheduling in cloud computing environment. Int. J. Comput. Sci. Manag. Res. (IJCSMR) **1**(3) (2012)

9. Rahman, M., Hassan, R., Buyya, R.: Adaptive workflow scheduling for dynamic grid and cloud computing environment. Concurr. Comput.: Pract. Exp. **25**(13), 1816–1842 (2013). https://doi.org/10.1002/cpi.3003

10. Wahab, M.N.A., Meziani, S.N., Tyabi, A.A.: A comprehensive review of swarm optimization algorithms. PLoS J. (2015). https://doi.org/10.1371/journal.pone.0122827

11. Kachivichyanukul, V.: Comparison of three evolutionary algorithms: GA, PSO and DE. Ind. Eng. Manag. Syst. **11**(3), 215–223

12. Arabnejad, H., Barbosa, J.G.: List scheduling algorithm for heterogeneous systems by an optimistic cost table. IEEE Trans. Parallel Distrib. Syst. **25**(3), 682–694 (2014)

13. Zhang, Q., Cheng, L., Boutaba, R.: Cloud computing: state-of-the-art and research challenges. J. Internet Serv. Appl. **1**(1), 7–18 (2010). https://doi.org/10.1007/s13174-010-0007-6

14. Dave, Y.P., Shelat, A.S., Patel, D.S., Jhaveri, R.H.: Various job scheduling algorithms in cloud computing: a survey. In: International Conference on Information Communication and Embedded Systems (ICICESS). IEEE, Chennai, India

15. Arya, L.K., Verma, A.: Workflow scheduling algorithm in cloud environment—a survey. In: 2014 Recent Advances in Engineering and Computational Sciences (RAECS), Chandigarh, India (2014). https://doi.org/10.1109/races.2014.6799514

16. Kwok, Y.K., Ahmad, I.: Dynamic critical path scheduling: an effective technique for allocating task graph to multiprocessors. IEEE Trans. Parallel Distrib. Syst. **7**(5), 506–521 (1996)

17. Illavarsan, E., Thambiduraj, P.: Low complexity performance effective task scheduling algorithm for heterogeneous computing environment. J. Comput. Sci. **3**(2), 94–103 (2007). https://doi.org/10.3844/jessp.2007.94.103

18. Sharma, N., Tyagi, S., Atri, S.: A survey on heuristic approach on task scheduling in cloud computing. Int. J. Adv. Res. Comput. Sci. **8**(3), 260–274 (2002)

19. Li, K.: Analysis of the list scheduling algorithm for precedence constrained parallel tasks. J. Comb. Optim. **3**, 73–88 (1999)

20. El-Rewini, H., Lewis, T.G.: Scheduling parallel program tasks onto arbitrary target machines. J. Parallel Distrib. Comput. **9**(2), 138–153 (1990). https://doi.org/10.1016/0743-7315(90)90042-n

21. Sih, G.C., Lee, E.A.: A compile-time scheduling heuristic for interconnection-constrained heterogeneous processor architecture. IEEE Trans. Parallel Distrib. Syst. **4**(2), 175–187 (1993). https://doi.org/10.1109/71.207593

22. Wu, M.Y., Gajski, D.D.: Hypertool: a programming aid for message passing. IEEE Trans. Parallel Distrib. Syst. **1**(3), 330–343 (1990)

23. Topcuoglu, H., Hariri, S., Wu, M.: Task scheduling algorithms for heterogeneous processors. In: Proceeding of 8th Heterogeneous Computing Workshop (HCS), USA, pp. 3–14 (1999). https://doi.org/10.1109/hcw.1999.765092

24. Topcuoglu, H., Hariri, S., Wu, M.Y.: Performance effective and low complexity task scheduling for heterogeneous computing. IEEE Trans. Parallel Distrib. Syst. **13**(3), 260–274 (2002). https://doi.org/10.1109/71.80160

25. Lin, C., Lu, S.: Scheduling scientific workflows elastically for cloud computing. In: International Conference on Cloud Computing (CLOUD), pp. 746–747. IEEE, Washington (2011)
26. Bala, R., Singh, G.: An improved heft algorithm using multi-criterion resource factors. Int. J. Comput. Sci. Inf. Technol. (IJCSIT) **5**(6), 6958–6963 (2014)
27. Dubey, K., Kumar, M., Sharma, S.C.: Modified heft algorithm for task scheduling in cloud environment. In: 6th International Conference on Smart Computing and Communications (ICSCC), Kurukshetra, India, pp. 725–732 (2017)
28. Verma, A., Kaushal, S.: Cost-time efficient scheduling plan for executing workflows in the cloud. J. Grid Comput. **13**(4), 495–506 (2015)
29. Rini, D.P., Shamsuddin, S.M., Yuchaniz, S.S.: Particle swarm optimization technique, system and challenges. Int. J. Comput. Appl. (IJCA) **14**(1), 19–27
30. Li, D., Shi, H., Liu, J., Tan, S., Liu, C., Xie, Y.: Research on improved particle swarm optimization algorithm based on ant-colony-optimization algorithm. In: 29th Chinese Control and Decision Conference (CCDC), China, pp. 853–858 (2017)
31. Verma, A., Kaushal, S., Sangaiah, A.K.: Computational intelligence based heuristic approach for maximizing energy efficiency in internet of things. In: Intelligent Decision Support Systems for Sustainable Computing, vol. 705, pp. 53–76. Springer, Berlin (2017). https://doi.org/10.1007/978-3-319-53153-3_4
32. Kaur, G., Kalra, M.: Deadline constrained scheduling of scientific workflows on cloud using hybrid genetic algorithm. In: IEEE Conference, Noida, pp. 281–285 (2017). https://doi.org/10.1109/confluence.2017.7943162
33. Li, Z., Liu, X., Duan, X.: Comparative research on particle swarm optimization and genetic algorithm. Comput. Inf. Sci. (CCSE) **3**(1), 120–127 (2010)
34. Verma, A., Kaushal, S.: Deadline and budget distribution based cost-time optimization workflow scheduling algorithm for cloud. In: IJCA Proceeding of international conference on Recent Advances and Future Trends in IT, Patiala, India, pp. 1–4 (2012)
35. Smanchat, S., Viriyapant, K.: Taxonomies of workflow scheduling problem and techniques in the cloud. Futur. Gener. Comput. Syst. **52**, 1–2 (2015)
36. Kwok, Y.K., Ahmad, I.: Benchmarking and comparison of the task graph scheduling algorithms. J. Parallel Distrib. Comput. **65**(5), 656–665 (2005)
37. Bai, Q.: Analysis of particle swarm optimization algorithm. Comput. Inf. Sci. **3**(1), 180–184 (2010). https://doi.org/10.5539/cis.v3n1p180
38. Chen, W., Deelman, E.: WorkflowSim: a toolkit for simulating scientific workflows in distributed environments. In: IEEE 8th International Conference on E-Science, USA, pp. 1–8 (2012). https://doi.org/10.1190/escience.2012.6404430
39. Sharma, V., Kumar, R.: A survey of energy aware scientific workflows execution techniques in cloud. Int. J. Innov. Res. Comput. Commun. Eng. (IJIRCCE) **3**(10), 10336–10343 (2015). https://doi.org/10.15680/ijircce.2015.0310176

Skyline Probabilities with Range Query on *Uncertain Dimensions*

Nurul Husna Mohd Saad, Hamidah Ibrahim, Fatimah Sidi and Razali Yaakob

Abstract To obtain a different perspective, a user often would query information in a range of search rather than a fixed search. However, the issue arises when attempting to compute skyline on data that consist of both exact and continuous values. Thus, we propose an algorithm named *SkyQUD-T* for computing skyline with range query issued on *uncertain dimensions* by eliminating objects that do not satisfy a given query range before advance processing is performed on the surviving objects. Our behaviour analysis of *SkyQUD-T* on real and synthetic datasets shows that *SkyQUD-T* is able to support skyline on data with *uncertain dimensions* in a given range query.

1 Introduction

In the attempt towards achieving an intuitive information systems, skyline queries have become one of the most popular and frequently used preference queries. It is especially found to be important on uncertain data where the data are lacked of certainty and preciseness. It is irrefutable that computing skyline probabilities on uncertain data is more complex than processing skylines on conventional data. It becomes even more intricate particularly when a search region is being queried on the uncertain data and skyline objects are expected to be reported within the search region. With the notion that in a dimension where an object may be represented by a continuous range, when taking into consideration a range query on the underlying

N. H. M. Saad (✉) · H. Ibrahim · F. Sidi · R. Yaakob
Universiti Putra Malaysia, Kuala Lumpur, Malaysia
e-mail: nhusna.saad@gmail.com

H. Ibrahim
e-mail: hamidah.ibrahim@upm.edu.my

F. Sidi
e-mail: fatimah@upm.edu.my

R. Yaakob
e-mail: razaliy@upm.edu.my

© Springer Nature Singapore Pte Ltd. 2019
S. K. Bhatia et al. (eds.), *Advances in Computer Communication
and Computational Sciences*, Advances in Intelligent Systems and Computing 924,
https://doi.org/10.1007/978-981-13-6861-5_20

dimension, we have to consider the many possible ways that an object may overlap with the range query and other objects. Granted that range queries are not some new issues faced by the database community, especially when analysing skyline objects within a range query. Cases in point are the works by [1–3, 5–8, 12], where they introduced several algorithms to process skyline with range queries on traditional data where there are no uncertainties involved. To the best of this author's knowledge, regardless of having a various studies that are concentrating on skyline query and its variant, thus far there is no work accomplishes on range skyline query regarding data with uncertainty.

In this paper, we shall concern ourselves with range queries that are being queried on dimensions, where objects are presented as exact values and continuous ranges, and finding skyline objects that satisfy a query range. To compute skyline with regard to a given range query would be straightforward on traditional datasets where the data are certain and complete. However, the issue exists when the datasets comprises of continuous range as well as exact values. We term this type of uncertainty in data as *uncertain dimensions*. Therefore, it is essential to propose an approach to determine skyline objects that satisfy a given query range. Hence, we summarise our contributions thusly:

1. We demonstrate the challenge of processing skyline with range query on *uncertain dimensions*.
2. We introduce and define the concept of *satisfy* to determine skyline on objects with *uncertain dimensions* that meet the requirements of a given range query.
3. We propose an algorithm, named *SkyQUD-T*, that will filter objects with *uncertain dimensions* that τ-*satisfy* a given range query. The algorithm would then determine skyline on the filtered objects.

This paper is an extension of the works in [10, 11]. We further extend the work in [10] to incorporate range query with skyline query. We also further extend the work in [11] by revising the three cases of possible ways a pair of objects may overlap with a range query to better include all possible relations in order to compute the dominance probability of an object, further refine the method proposed in the *SkyQUD-T* algorithm and perform additional experimentation to analyse the behaviour of the *SkyQUD-T* algorithm.

The rest of the paper is organised thusly. In Sect. 2, we review the related works. Then, in Sect. 3 we define concepts that are relevant to finding skyline probabilities with range query and formally define the problem in this paper. In Sect. 4, we define three major cases that would most likely occur between a pair of objects with uncertain dimensions that overlap with the range query and propose *SkyQUD-T* to compute the probabilities of skyline objects on *uncertain dimensions*. We evaluate the behavioural analysis performed on *SkyQUD-T* in Sect. 5, while in Sect. 6 we conclude the paper.

2 Related Work

The term constrained skyline query was coined by [7] in which the query would report *preferred* objects in the constrained query and proposed *Branch and Bound Skyline* (*BBS*) algorithm to process constrained skyline query. Conversely, [3] tackled the challenge of computing interval skyline queries on data presented as time series. Each time series is regarded as a point in the space, and each timestamp is regarded as a dimension. They propose two interesting algorithms, named *On-the-Fly* (*OTF*) and *View-Materialisation* (*VM*).

Then, [12] proposed *multi-directional scanning* (*MDS*) algorithm to cater the issue of dynamic skyline computation with respect to range queries, in which dynamic skyline is performed if no data points exist within the range query. Working on the same context, [8] introduced the concept of objects that is represented by a set of *range attributes* and *feature attributes*. The *range attributes* are coordinates in the *d*-dimensional *range space*, while the *feature attributes* are real-value attributes in the *t*-dimensional *feature space*. They proposed an algorithm named *range skyline query* (*RSQ*) to filter out based on *range attributes* all objects that are not contained in the *range space*, and then compute skyline points based on the *feature space* with respect to the *range space*.

To the best of our knowledge, regardless of having a variety of studies that are focusing on skyline query and its variant, there is no work thus far that has been done on skyline with range query with respect to uncertain data. The works in [3, 7, 8, 12] concentrated on applying an appropriate indexing methods to process and compute skyline objects which are contained in a given query range, and these objects are presented as exact values only.

3 Preliminaries

To begin, we formally define the concept of *uncertain dimensions* and derive several concepts that are relevant to finding skyline probabilities on objects with *uncertain dimensions* with respect to a range query.

Definition 1 (*Continuous range*) Let $[w_{lb} : w_{ub}]$ indicates the continuous range of w in such a way that $w_{lb} \leq w \leq w_{ub}$ and $\{w_{lb}, w_{ub} \in \mathbb{R} | 0 \leq w_{lb} \leq w_{ub}\}$.

Definition 2 (*Uncertain dimensions*) Given a d-dimensional datasets $\mathbb{D} = (D_1, D_2, \ldots, D_d)$ with n objects. A dimension $D_i \in \mathbb{D}$, $(1 \leq i \leq d)$, is considered as uncertain (indicated as $\mathbb{U}(D_i)$) if the dimension has at least one continuous range of value $a_j = [a_{lb} : a_{ub}]_j \in \mathbb{U}(D_i)$, $(1 \leq j \leq n)$; else, the dimension is said to be certain dimension. The range values of an object have a function ($f(v)$) defined on its range $[a_{lb} : a_{ub}]_j$, in which a_{lb} indicates the lower-bound and a_{ub} indicates the upper-bound of object a.

From this definition, it follows that when dimensions have at least one continuous range of value, then the dimensions are considered as *uncertain dimensions*.

Following this, the objective is to compute and report skyline on objects when skyline query is issued within a range query. Therefore, we define the term range query in our work as follows:

Definition 3 (*Range query*) A range query $[q_{lb} : q_{ub}]_k$ denotes a range of search being queried on dimension k, and $q_{[lb]k} < q_{[ub]k}$.

Given a query with a specific range of values $[q_{lb} : q_{ub}]_k$, a range query on a set of d-dimensions objects $\mathbb{O}$ would return all objects that are contained within the query, that is

$$\{v \in \mathbb{O} \mid q_{[lb]k} \leq v_k \leq q_{[ub]k}\}$$

Accordingly, a skyline issued within a range query would return the skyline of objects that are contained within the query, or in other words, all objects that are contained within the query and are not dominated by any other objects. Let us look at the example in Fig. 1, in which a user wish to search for an apartment with a rent that is within the range [500 : 900]. Thus, assuming that lesser value is much preferred, the skyline apartments will be apartments a, b and h, since these apartments are the preferred apartments within user's budget.

Now, let us look at another example. Suppose there exist *uncertain dimensions* in the apartment listing database as illustrated in Fig. 2, therefore, the rental price of the apartments is recorded as either fixed value or within some range. To narrow down the search, a user would query for a list of apartments that is between the rent budget from \$250 to \$450. To filter for skyline objects, obviously, for every object that does not overlap at all with the range query, we can definitely rule them out, while for objects that definitely falls in the boundary of the range query, we can definitely accept these objects. Nonetheless, how do we deal with objects that overlap and crossover the boundaries the range query? Do we directly accept these objects regardless if the objects mostly lie out of the range query? Or do we absolutely

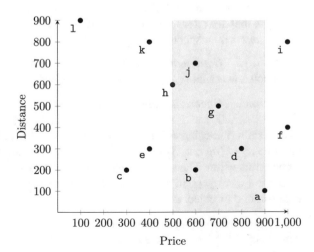

Fig. 1 Range query on traditional datasets

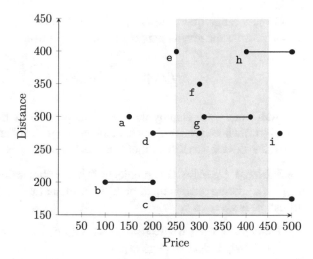

Fig. 2 Range query on *uncertain dimension*

reject these objects? How can we defend that these objects *satisfy* the range query? Consequently, by using existing skyline algorithms can skyline be efficiently reported when there exists objects with *uncertain dimension* presented as exact values and continuous ranges, and having the continuous ranges overlap with the range query? Figure 2 demonstrates the above discussion.

Problem Definition 1 Given a range $[q_{lb} : q_{ub}]_k$ queried on dimensions $\mathbb{U}(D_k)$ of a d-dimensional datasets $\mathbb{D} = (D_1, D_2, \ldots, D_d)$, with the uncertain dimensions defined as $\mathbb{U}(D_k)$, $(1 \leq k \leq d)$. Find skyline objects of $\mathbb{D}$ that satisfy range query $[q_{lb} : q_{ub}]_k$.

4 Probabilities of Skyline Objects in an *Uncertain Dimension* with Regard to the Range Query

Throughout this paper, it is assumed that a d-dimensional datasets $\mathbb{D}$ has the structure as illustrated in Fig. 3. We know that all probability statements about object v can be answered in terms of its pdf $f(v)$ as follows [9]:

$$Pr(v_{lb} \leq v \leq v_{ub}) = \int_{v_{lb}}^{v_{ub}} f(v)\, dv \qquad (1)$$

| | C | | | | UC | | |
	D_1	D_2	$\cdots$	D_α	$D_{\alpha+1}$	$D_{\alpha+2}$	$\cdots$	D_d
v	6	4	$\cdots$	3	$[220 : 230]$	95	$\cdots$	$[230 : 270]$
w	5	5	$\cdots$	4	88	$[70 : 95]$	$\cdots$	$[210 : 255]$

Fig. 3 Structure of *uncertain dimensions* datasets

Given a range query $[q_{lb} : q_{ub}]$, to determine that an object with a continuous range *satisfies* the given range query, we obtain from (1) that

$$Pr\,(q_{lb} \leq v \leq q_{ub}) = \int_{q_{lb}}^{q_{ub}} f\,(v)\,dv \tag{2}$$

Additionally, we employ the use of the probability threshold τ in order to filter out objects whose probabilities to be in the range query are small enough. Therefore, we define the term *satisfy* in this paper as below:

Definition 4 *(Satisfy)* Given objects $\mathbb{O}$ in a d-dimensional datasets $\mathbb{D}$, and a range $[q_{lb} : q_{ub}]_k$ being queried on dimension k, where $(\alpha + 1 \leq k \leq d)$, and a probability threshold is given as τ.

1. An object $r \in \mathbb{O}$, where $r = (r_1, \ldots, r_\alpha, r_{\alpha+1}, \ldots, r_d)$, is considered to satisfy $[q_{lb} : q_{ub}]_k$ if $q_{[lb]k} \leq r_k \leq q_{[ub]k}$.
2. An object $v \in \mathbb{O}$, where $v = (v_1, \ldots, v_\alpha, [v_{lb} : v_{ub}]_{\alpha+1}, \ldots, [v_{lb} : v_{ub}]_d)$, is considered to satisfy $[q_{lb} : q_{ub}]_k$ if $Pr(q_{[lb]k} \leq v_k \leq q_{[ub]k}) \geq \tau$.

Theorem 1 *Given a range $[q_{lb} : q_{ub}]_k$ queried on a set of objects $\mathbb{O}$ and three probability threshold values τ_0, τ and τ_1. Let $S_0, S_\tau, S_1 \in \mathbb{O}$ be a set of objects that have probabilities to satisfy the range query of at least τ_0, τ and τ_1, respectively. If $\tau_0 < \tau < \tau_1$ then $S_1 \subseteq S_\tau \subseteq S_0$.*

Proof We prove by contradiction. Assume that $\tau_0 = 0$, $\tau = 0.5$ and $\tau_1 = 1$. Suppose there exists an object v such that $Pr\,(q_{[lb]k} \leq v_k \leq q_{[ub]k}) \geq \tau$, thus $v \in S_\tau$ but $v \notin S_0$. As $v \notin S_0$, then it must mean $Pr(q_{[lb]k} \leq v_k \leq q_{[ub]k}) < \tau_0$. However, $\tau_0 < \tau$, therefore it would mean $Pr\,(q_{[lb]k} \leq v_k \leq q_{[ub]k}) < \tau_0 < \tau$ which leads to a contradiction since $v \in S_\tau$. The inequality if $\tau < \tau_1$ thus $S_1 \subseteq S_\tau$ is established in exactly the same manner as the above inequality. $\qquad\square$

Henceforth, we say that an object τ-*satisfy* a range query if its probability to *satisfy* the range query is at least τ. Theorem 1 indicates that if $\tau_0 < \tau$, hence the number of objects that τ-*satisfy* $[q_{lb} : q_{ub}]_k$ is smaller than the number of objects that τ_0-*satisfy* $[q_{lb} : q_{ub}]_k$. Therefore, by having parameter τ to be a user defined parameter, users are able to manage the size of objects that τ-*satisfy* their range query.

Thus far, we have concerned ourselves with determining objects that *satisfy* $[q_{lb} : q_{ub}]_k$. However, we desire to determine skyline objects on a set of objects $\mathbb{O}$ with *uncertain dimensions* that τ-*satisfy* the range query. For two continuous range objects v and w, where $v = [v_{lb} : v_{ub}]$ and $w = [w_{lb} : w_{ub}]$, respectively, given that it is assumed minimum values are preferred, we may compute the probability of object w to dominate another object v based on the following relations which follow the probability theory [9].

Definition 5 *(Relations between two range values)*
Disjoint: If $v_{lb} \geq w_{ub}$

$$Pr\,(w < v) = 1 \tag{3}$$

Disjoint-inverse: If $v_{ub} \leq w_{lb}$

$$Pr\,(w < v) = 0 \tag{4}$$

Overlap: If $w_{lb} \leq v_{lb} \leq w_{ub} \leq v_{ub}$

$$Pr\,(w < v) = 1 - \frac{1}{2} \int_{v_{lb}}^{w_{ub}} f\,(w)\,dw \int_{v_{lb}}^{w_{ub}} f\,(v)\,dv \tag{5}$$

Overlap-inverse: If $v_{lb} \leq w_{lb} \leq v_{ub} \leq w_{ub}$

$$Pr\,(w < v) = \frac{1}{2} \int_{w_{lb}}^{v_{ub}} f\,(w)\,dw \int_{w_{lb}}^{v_{ub}} f\,(v)\,dv \tag{6}$$

Contain: If $w_{lb} \leq v_{lb} \leq v_{ub} \leq w_{ub}$

$$Pr\,(w < v) = \int_{w_{lb}}^{v_{lb}} f\,(w)\,dw + \frac{1}{2} \int_{v_{lb}}^{v_{ub}} f\,(w)\,dw \tag{7}$$

Contain-inverse: If $v_{lb} \leq w_{lb} \leq w_{ub} \leq v_{ub}$

$$Pr\,(w < v) = \int_{w_{ub}}^{v_{ub}} f\,(v)\,dv + \frac{1}{2} \int_{w_{lb}}^{w_{ub}} f\,(v)\,dv \tag{8}$$

Equals: If $w_{lb} = v_{lb}$ and $w_{ub} = v_{ub}$

$$Pr\,(w < v) = \frac{1}{2} \int_{w_{lb}}^{w_{ub}} f\,(w)\,dw \int_{v_{lb}}^{v_{ub}} f\,(v)\,dv \tag{9}$$

On the other hand, to adapt the probability of objects with an exact value v and a continuous range $[w_{lb} : w_{ub}]$ to dominate one another, the relations that were previously defined in Definition 5 are remodel as follows:

Definition 6 (*Relations between range and exact values*)
Disjoint: If $v \geq w_{ub}$

$$Pr\,(w < v) = 1 \tag{10}$$

$$Pr\,(v < w) = 0 \tag{11}$$

Disjoint-inverse: If $v \leq w_{lb}$

$$Pr\,(w < v) = 0 \tag{12}$$

$$Pr\,(v < w) = 1 \tag{13}$$

Contain: If $w_{lb} \leq v \leq w_{ub}$

$$Pr\,(w < v) = \int_{w_{lb}}^{v-\varepsilon} f\,(w)\,dw \tag{14}$$

$$Pr\,(v < w) = \int_{v+\varepsilon}^{w_{ub}} f\,(w)\,dw \tag{15}$$

where ε symbolise a correction value that is introduced around a particular object and that the correction value is as small as possible. Following the above discussions, we define the range skyline query on $\mathbb{O}$ to be objects that τ-*satisfy* a query range and are not worse than any other objects that τ-*satisfy* the query range as well. That is,

$$\left\{v \in \mathbb{O} \mid \nexists w \in \mathbb{O},\, Pr\left(q_{[lb]k} \leq v_k, w_k \leq q_{[ub]k}\right) \geq \tau \wedge w \prec v\right\}$$

Range skyline query on *uncertain dimensions* has the following property.

Property 1 (Skyline with range query on uncertain dimensions) *Suppose a range* $[q_{lb} : q_{ub}]_k$ *is queried on dimension* $\mathbb{U}\,(D_k)$ *of a d-dimensional datasets* $\mathbb{D}$ *with objects* $\mathbb{O}$. *Given a probability threshold* τ, *an object* $v \in \mathbb{O}$ *is said to* τ-*satisfy* $[q_{lb} : q_{ub}]_k$ *and is a skyline object with regard to* $[q_{lb} : q_{ub}]_k$ *if* v *is not dominated by any other objects that* τ-*satisfy* $[q_{lb} : q_{ub}]_k$.

Proof We prove by contradiction. Suppose that v τ-*satisfies* the given range query and v is not worse than any other objects, yet v is not in the set of skyline objects that τ-*satisfy* the range query. Consequently, there must be another object $w \in \mathbb{O}$ such that w τ-*satisfy* the given range query and that w is better than v in such a way that:

1. $\forall i \in \{1,\,2,\ldots,\alpha\},\, w_i \leq v_i$ and $\exists j \in \{1,\,2,\ldots,\alpha\},\, w_j < v_j$,
2. $\forall h \in \{\alpha+1, \alpha+2,\ldots,\, d\},\, (1 - Pr\,(w_h < v_h)) < \tau$, and
3. $\exists k \in \{\alpha+1, \alpha+2,\ldots,d\},\, k \neq h,\, \left(1 - Pr\left(w_k <_{[q_{lb}:q_{ub}]_k} v_k\right)\right) < \tau$.

In other words, v is dominated by w, which leads to a contradiction. $\square$

Below, we define all the possible cases that would most likely occur between a pair of objects with *uncertain dimensions* that overlaps with the range query. For the purpose of discussion and ease of description for each of the cases defined below, it is assumed that in a d-dimensional datasets $\mathbb{D}$ with objects $\mathbb{O}$ has the structure as such $\mathbb{D} = \{\mathbb{C},\, \mathbb{UC}\}$, where $\mathbb{C} = \{D_2, D_3,\ldots, D_d\}$ and $\mathbb{UC} = \{D_1\}$. Suppose that the range $[q_{lb} : q_{ub}]_1$ is being queried on $\mathbb{U}(D_1)$. Given $\{r, v, w\} \in \mathbb{O}$, where $r = (r_1, r_2,\ldots, r_d)$, $v = ([v_{lb} : v_{ub}]_1, v_2,\ldots, v_d)$ and $w = ([w_{lb} : w_{ub}]_1, w_2,\ldots, w_d)$. We consider three cases.

Case 1: Objects Fully Lie Within the Range Query

In this case, when objects r, v and w clearly have their values within the range query $[q_{lb} : q_{ub}]_1$ so that $q_{[lb]1} \leq r_1 \leq q_{[ub]1}$, $q_{[lb]1} \leq v_{[lb]1} < v_{[ub]1} \leq q_{[ub]1}$ and $q_{[lb]1} \leq w_{[lb]1} < w_{[ub]1} \leq q_{[ub]1}$, respectively, as demonstrated in Fig. 4a, it easily follows all the dominance probabilities that have been deliberated in Definition 5 and Definition 6 in order to determine skyline objects.

Fig. 4 Objects that overlap with range query

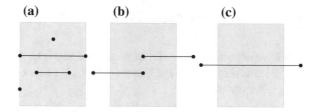

(a)   **(b)**   **(c)**

Case 2: Objects with One Endpoint Within the Range Query

In this case, when objects v or w overlap with range query $[q_{lb} : q_{ub}]_1$ so that $q_{[lb]1} < v_{[lb]1} < q_{[ub]1} < v_{[ub]1}$ or $w_{[lb]1} < q_{[lb]1} < w_{[ub]1} < q_{[ub]1}$, respectively, as exhibited in Fig. 4b. Based on Definitions 5 and 6, we know there are several different possible relations that would occur between a pair of objects with continuous ranges, for instance (v, w), or a pair of objects with an exact value and a continuous range, for instance (r, w), and each relation has its own way to compute the dominance probability. Nonetheless, with the range query thrown into the mix, we have to consider these dominance probabilities with regard to the range query.

The dominance probability of an object such as v or r in $\mathbb{U}(D_1)$ with respect to a range query is essentially the probability that v or r is preferable than another object w in the range query in addition to the probability that w occurs outside of the range query (written $Pr\left(v <_{[q_{lb}:q_{ub}]_1} w\right)$). This probability can be computed following the relations previously defined in Definitions 5 and 6 as follows:

Disjoint

Given objects v and w that are disjointed in $\mathbb{U}(D_1)$ so that $w_{[lb]1} \geq v_{[ub]1}$. Suppose that v and w overlap with range query $[q_{lb} : q_{ub}]_1$ so that $v_{[lb]1} \leq q_{[lb]1} < v_{[ub]1} \leq w_{[lb]1} < q_{[ub]1} \leq w_{[ub]1}$. We obtain:

$$Pr(v <_{[q_{lb}:q_{ub}]_1} w) = \int_{q_{[lb]1}}^{v_{[ub]1}} f(v)dv \qquad (16)$$

Given objects r and w that are disjointed in $\mathbb{U}(D_1)$ so that $r_1 \geq w_{[ub]1}$. Suppose that r and w overlap with range query $[q_{lb} : q_{ub}]_1$ so that $w_{[lb]1} \leq q_{[lb]1} < w_{[ub]1} \leq r_1 \leq q_{[ub]1}$. We obtain:

$$Pr(r <_{[q_{lb}:q_{ub}]_1} w) = \int_{w_{[lb]1}}^{q_{[lb]1}} f(v)dv \qquad (17)$$

Disjoint-Inverse

Given objects v and w that are disjointed in $\mathbb{U}(D_1)$ so that $w_{[ub]1} \leq v_{[lb]1}$. Suppose that v and w overlap with range query $[q_{lb} : q_{ub}]_1$ so that $w_{[lb]1} \leq q_{[lb]1} < w_{[ub]1} \leq v_{[lb]1} < q_{[ub]1} \leq v_{[ub]1}$. We obtain:

$$Pr(v <_{[q_{lb}:q_{ub}]_1} w) = \int_{w_{[lb]1}}^{q_{[lb]1}} f(w)dw \int_{v_{[lb]1}}^{q_{[ub]1}} f(v)dv \qquad (18)$$

Given objects r and w that are disjointed in $\mathbb{U}(D_1)$ so that $r_1 \leq w_{[lb]1}$. Suppose that r and w overlap with range query $[q_{lb} : q_{ub}]_1$ so that $q_{[lb]1} \leq r_1 < w_{[lb]1} < q_{[ub]1} \leq w_{[ub]1}$. We obtain:

$$Pr(r <_{[q_{lb}:q_{ub}]_1} w) = 1 \qquad (19)$$

Overlap

Given objects w and v that are overlapped in $\mathbb{U}(D_1)$ so that $v_{[lb]1} \leq w_{[lb]1} \leq v_{[ub]1} \leq w_{[ub]1}$. Suppose that v and w overlap with range query $[q_{lb} : q_{ub}]_1$ so that $v_{[lb]1} < q_{[lb]1} < w_{[lb]1} < v_{[ub]1} < q_{[ub]1} < w_{[ub]1}$. We obtain:

$$Pr(v <_{[q_{lb}:q_{ub}]_1} w) = \int_{q_{[lb]1}}^{w_{[lb]1}} f(v)dv + \frac{1}{2}\left(\int_{w_{[lb]1}}^{v_{[ub]1}} f(v)dv \int_{w_{[lb]1}}^{v_{[ub]1}} f(w)dw \right)$$
$$+ \left(\int_{w_{[lb]1}}^{v_{[ub]1}} f(v)dv \int_{v_{[ub]1}}^{w_{[ub]1}} f(w)dw \right) \qquad (20)$$

Overlap-Inverse

Given objects w and v that are overlapped in $\mathbb{U}(D_1)$ so that $w_{[lb]1} \leq v_{[lb]1} \leq w_{[ub]1} \leq v_{[ub]1}$. Suppose that v and w overlap with range query $[q_{lb} : q_{ub}]_1$ so that $w_{[lb]1} < q_{[lb]1} < v_{[lb]1} < w_{[ub]1} < q_{[ub]1} < v_{[ub]1}$. We obtain:

$$Pr(v <_{[q_{lb}:q_{ub}]_1} w) = \left(\int_{w_{[lb]1}}^{q_{[lb]1}} f(w)dw \int_{v_{[lb]1}}^{q_{[ub]1}} f(v)dv \right)$$
$$+ \frac{1}{2}\left(\int_{v_{[lb]1}}^{w_{[ub]1}} f(w)dw \int_{v_{[lb]1}}^{w_{[ub]1}} f(v)dv \right) \qquad (21)$$

Contain

Given object w that is contained within object v in $\mathbb{U}(D_1)$ so that $v_{[lb]1} \leq w_{[lb]1} \leq w_{[ub]1} \leq v_{[ub]1}$. Suppose that v and w overlap with range query $[q_{lb} : q_{ub}]_1$ so that $v_{[lb]1} < w_{[lb]1} < q_{[lb]1} < w_{[ub]1} < v_{[ub]1} < q_{[ub]1}$. We obtain:

$$Pr(v <_{[q_{lb}:q_{ub}]_1} w) = \left(\int_{w_{[lb]1}}^{q_{[lb]1}} f(w)dw \int_{q_{[lb]1}}^{v_{[ub]1}} f(v)dv \right)$$
$$+ \frac{1}{2}\left(\int_{q_{[lb]1}}^{w_{[ub]1}} f(w)dw \int_{q_{[lb]1}}^{w_{[ub]1}} f(v)dv \right) \qquad (22)$$

Given object r that is contained within object w in $\mathbb{U}(D_1)$ so that $w_{[lb]1} \leq r_1 \leq w_{[ub]1}$. Suppose that r and w overlap with range query $[q_{lb} : q_{ub}]_1$ so that $w_{[lb]1} < q_{[lb]1} < r_1 < w_{[ub]1} \leq q_{[ub]1}$. We obtain:

$$Pr(r <_{[q_{lb}:q_{ub}]_1} w) = \int_{w_{[lb]1}}^{q_{[lb]1}} f(w)dw + \int_{r_1+\varepsilon}^{w_{[ub]1}} f(w)dw \qquad (23)$$

Contain-Inverse

Given object v that is contained within object w in $\mathbb{U}(D_1)$ so that $w_{[lb]1} \leq v_{[lb]1} \leq v_{[ub]1} \leq w_{[ub]1}$. Suppose that v and w overlap with range query $[q_{lb} : q_{ub}]_1$ so that $w_{[lb]1} < v_{[lb]1} < q_{[lb]1} < v_{[ub]1} < w_{[ub]1} < q_{[ub]1}$. We obtain:

$$Pr(v <_{[q_{lb}:q_{ub}]_1} w) = \left(\int_{w_{[lb]1}}^{q_{[lb]1}} f(w)dw \int_{q_{[lb]1}}^{v_{[ub]1}} f(v)dv \right)$$
$$+ \frac{1}{2}\left(\int_{q_{[lb]1}}^{v_{[ub]1}} f(w)dw \int_{q_{[lb]1}}^{v_{[ub]1}} f(v)dv \right) + \left(\int_{v_{[ub]1}}^{w_{[ub]1}} f(w)dw \int_{q_{[lb]1}}^{v_{[ub]1}} f(v)dv \right)$$
$$(24)$$

Equals

Given objects v and w that are equals in $\mathbb{U}(D_1)$ so that $v_{[lb]1} = w_{[lb]1}$ and $v_{[ub]1} = w_{[ub]1}$. Suppose that v and w overlap with range query $[q_{lb} : q_{ub}]_1$ so that $v_{[lb]1} = w_{[lb]1} < q_{[lb]1} < v_{[ub]1} = w_{[ub]1} \leq q_{[ub]1}$. We obtain:

$$Pr(v <_{[q_{lb}:q_{ub}]_1} w) = \frac{1}{2}\left(\int_{q_{[lb]1}}^{v_{[ub]1}} f(v)dv \int_{q_{[lb]1}}^{w_{[ub]1}} f(w)dw \right)$$
$$+ \left(\int_{q_{[lb]1}}^{v_{[ub]1}} f(v)dv \int_{w_{[lb]1}}^{q_{[lb]1}} f(w)dw \right) \qquad (25)$$

Case 3: Objects with All Endpoints Out of the Range query

In this case, when object such as v overlaps with the range query $[q_{lb} : q_{ub}]_1$ so that $v_{[lb]1} < q_{[lb]1} < q_{[ub]1} < v_{[ub]1}$, as exhibited in Fig. 4c. Following from the preceding case, there are several different possible relations that would occur between a pair of objects with continuous ranges, for instance (v, w), or a pair of objects with an exact value and a continuous range, for instance (r, w), and based on Definitions 5 and 6, each relation has its own way to compute the dominance probability with regard to the range query.

Note that the *Disjoint* or *Disjoint-inverse* relations will not be considered in Case 3 since in this case, both objects are assumed to overlap with the range query, yet with both endpoints of the objects lie outside of the range query. Now, suppose that the *Disjoint* or *Disjoint-inverse* relations do occur in Case 3, then it would mean that one object overlaps with the range query, while another object totally lies outside of the range query, and that object does not *satisfy* the range query. This contradicts the aim of this section, as in this section we desire to compute only the probabilities of objects that do *satisfy* the range query, hence the reasoning.

Overlap

Given objects w and v that are overlapped in $\mathbb{U}(D_1)$ so that $v_{[lb]1} \leq w_{[lb]1} \leq v_{[ub]1} \leq w_{[ub]1}$. Suppose that v and w overlap with range query $[q_{lb} : q_{ub}]_1$ so that $v_{[lb]1} < q_{[lb]1} < w_{[lb]1} < q_{[ub]1} < v_{[ub]1} < w_{[ub]1}$. We obtain:

$$Pr(v <_{[q_{lb}:q_{ub}]_1} w) = \int_{q_{[lb]1}}^{w_{[lb]1}} f(v)dv + \frac{1}{2} \int_{w_{[lb]1}}^{q_{[ub]1}} f(v)dv \int_{w_{[lb]1}}^{q_{[ub]1}} f(w)dw$$
$$+ \int_{w_{[lb]1}}^{q_{[ub]1}} f(v)dv \int_{q_{[ub]1}}^{w_{[ub]1}} f(w)dw \tag{26}$$

Overlap-Inverse

Given objects w and v that are overlapped in $\mathbb{U}(D_1)$ so that $w_{[lb]1} \leq v_{[lb]1} \leq w_{[ub]1} \leq v_{[ub]1}$. Suppose that v and w overlap with range query $[q_{lb} : q_{ub}]_1$ so that $w_{[lb]1} < v_{[lb]1} < q_{[lb]1} < w_{[ub]1} < q_{[ub]1} < v_{[ub]1}$. We obtain:

$$Pr(v <_{[q_{lb}:q_{ub}]_1} w) = \left(\int_{w_{[lb]1}}^{q_{[lb]1}} f(w)dw \int_{q_{[lb]1}}^{q_{[ub]1}} f(v)dv \right)$$
$$+ \frac{1}{2} \left(\int_{q_{[lb]1}}^{w_{[ub]1}} f(w)dw \int_{q_{[lb]1}}^{w_{[ub]1}} f(v)dv \right) \tag{27}$$

Contain

Given object w that is contained within object v in $\mathbb{U}(D_1)$ so that $v_{[lb]1} \leq w_{[lb]1} \leq w_{[ub]1} \leq v_{[ub]1}$. Suppose that v and w overlap with range query $[q_{lb} : q_{ub}]_1$ so that $v_{[lb]1} < q_{[lb]1} < w_{[lb]1} < w_{[ub]1} < q_{[ub]1} < v_{[ub]1}$. We obtain:

$$Pr(v <_{[q_{lb}:q_{ub}]_1} w) = \int_{q_{[lb]1}}^{w_{[lb]1}} f(v)dv + \frac{1}{2} \int_{w_{[lb]1}}^{w_{[ub]1}} f(v)dv \tag{28}$$

Given object r that is contained within object w in $\mathbb{U}(D_1)$ so that $w_{[lb]1} \leq r_1 \leq w_{[ub]1}$. Suppose that r and w overlap with range query $[q_{lb} : q_{ub}]_1$ so that $w_{[lb]1} < q_{[lb]1} \leq r_1 \leq q_{[ub]1} < w_{[ub]1}$. We obtain:

$$Pr(r <_{[q_{lb}:q_{ub}]_1} w) = \int_{w_{[lb]1}}^{q_{[lb]1}} f(w)dw + \int_{r_1+\varepsilon}^{w_{[ub]1}} f(w)dw \tag{29}$$

Contain-Inverse

Given object v that is contained within object w in $\mathbb{U}(D_1)$ so that $w_{[lb]1} \leq v_{[lb]1} \leq v_{[ub]1} \leq w_{[ub]1}$. Suppose that v and w overlap with range query $[q_{lb} : q_{ub}]_1$ so that $w_{[lb]1} < q_{[lb]1} < v_{[lb]1} < v_{[ub]1} < q_{[ub]1} < w_{[ub]1}$. We obtain:

$$Pr(v <_{[q_{lb}:q_{ub}]_1} w) = \int_{w_{[lb]1}}^{q_{[lb]1}} f(w)dw + \frac{1}{2} \int_{v_{[lb]1}}^{v_{[ub]1}} f(w)dw + \int_{v_{[ub]1}}^{w_{[ub]1}} f(w)dw \tag{30}$$

Equals

Given objects v and w that are equals in $\mathbb{U}(D_1)$ so that $v_{[lb]1} = w_{[lb]1}$ and $v_{[ub]1} = w_{[ub]1}$. Suppose that v and w overlap with range query $[q_{lb} : q_{ub}]_1$ so that $v_{[lb]1} = w_{[lb]1} < q_{[lb]1} < q_{[ub]1} < v_{[ub]1} = w_{[ub]1}$. We obtain:

$$Pr(v <_{[q_{lb}:q_{ub}]_1} w) = \left(\int_{w_{[lb]1}}^{q_{[lb]1}} f(w)dw \int_{q_{[lb]1}}^{q_{[ub]1}} f(v)dv \right)$$
$$+ \frac{1}{2} \left(\int_{q_{[lb]1}}^{q_{[ub]1}} f(w)dw \int_{q_{[lb]1}}^{q_{[ub]1}} f(v)dv \right) + \left(\int_{q_{[ub]1}}^{w_{[ub]1}} f(w)dw \int_{q_{[lb]1}}^{q_{[ub]1}} f(v)dv \right) \tag{31}$$

Algorithm 1 presents the general outline of *SkyQUD-T*. As established in Theorem 1, we say that the *SkyQUD-T* algorithm is able to provide an approximation to the optimal answer. This is because the *SkyQUD-T* algorithm used probabilistic inference to determine skyline objects that *satisfy* a given range query. Although it requires the evaluation of all objects in the datasets to *satisfy* the range query, it has assisted in reducing the number of pairwise comparisons in skyline analysis, as all objects whose probabilities to *satisfy* the range query that is below a threshold value will be eliminated earlier.

Algorithm 1 *SkyQUD-T*

Input: S: a d-dimensional datasets with $\mathbb{U}(D_i)$, $1 \leq i \leq d$, τ: threshold, $[q_{lb} : q_{ub}]$: range query, z: dimension of range query
Output: Sky: a set of skyline objects that τ-*satisfy* $[q_{lb} : q_{ub}]_z$
1: Initialise Sky: Skyline
2: **for** each object $s \in S$ **do**
3:    **if** $q_{[lb]z} \leq s_z \leq q_{[ub]z}$ **or**
         $q_{[lb]z} \leq s_{[lb]z} \leq s_{[ub]z} \leq q_{[ub]z}$ **then**
4:       add s to S'
5:    **else if** $s_{[lb]z} \leq q_{[lb]z} \leq s_{[ub]z} \leq q_{[ub]z}$ **or**
         $q_{[lb]z} \leq s_{[lb]z} \leq q_{[ub]z} \leq s_{[ub]z}$ **or**
         $s_{[lb]z} \leq q_{[lb]z} \leq q_{[ub]z} \leq s_{[ub]z}$ **then**
6:       calculate $Pr(q_{[lb]z} \leq s_z \leq q_{[ub]z})$
7:       **if** $Pr(q_{[lb]z} \leq s_z \leq q_{[ub]z}) \geq \tau$ **then**
8:          add s to S'
9:       **end if**
10:    **end if**
11: **end for**
12: **for** each object $s \in S'$ **do**
13:    **for** each candidate of skyline objects $sc \in Sky$ **do**
14:       apply dominance probability w.r.t. $[q_{lb} : q_{ub}]_z$ between s and sc as deliberated in either Case 1, Case 2, or Case 3
15:       **if** object s is not dominated **then**
16:          add s into Sky
17:       **end if**
18:    **end for**
19: **end for**
20: **return** Sky

---

5 Behaviour Analysis for *SkyQUD-T*

We report the experiment results of computing probabilities of skyline objects within range queries on *uncertain dimensions* in this section. Due to the lack of research works in skyline query processing that is able to accommodate range query on *uncertain dimensions*, therefore as a basis of comparison we have consider the following two naïve algorithms: (1) *SkyQUD-LA* algorithm, where it is basically the *SkyQUD-T* algorithm. Instead of computing the probability of objects to be anywhere within the range query as implemented by *SkyQUD-T* algorithm, *SkyQUD-LA* will utilise a technique of simply accepting all objects that overlap with the border of the range that is being queried, regardless if the objects do not fully lie inside the range query, and (2) *SkyQUD-SR* algorithm, where again, it is basically the *SkyQUD-T* algorithm, yet *SkyQUD-SR* will utilise a technique of strictly rejecting all objects that do not fully lie in the range query.

Algorithms 2 and 3 elucidate the pruning method in Algorithm 1 (line 2–11) that are implemented in the *SkyQUD-LA* and *SkyQUD-SR* algorithms, respectively. We discuss and analyse the behaviour of the proposed algorithm *SkyQUD-T* in comparison to the aforementioned naïve algorithms when given a range $[q_{lb} : q_{ub}]$ that is being queried on an *uncertain dimension*. All experiments were executed on a PC running Ubuntu Linux operating system, with processor Intel Core i5-3470 3.20 GHz, and main memory 7.8 GB.

Algorithm 2 *SkyQUD-LA*

1: **for** each object $s \in S$ **do**
2:     **if** $q_{[lb]z} \leq s_z \leq q_{[ub]z}$ **or**
         $q_{[lb]z} \leq s_{[lb]z} < s_{[ub]z} \leq q_{[ub]z}$ **or**
         $s_{[lb]z} < q_{[lb]z} < s_{[ub]z} \leq q_{[ub]z}$ **or**
         $q_{[lb]z} \leq s_{[lb]z} < q_{[ub]z} < s_{[ub]z}$ **or**
         $s_{[lb]z} < q_{[lb]z} < q_{[ub]z} < s_{[ub]z}$ **then**
3:       add s to S'
4:     **end if**
5: **end for**
6: **return** S'

Algorithm 3 *SkyQUD-SR*

1: **for** each object $s \in S$ **do**
2:     **if** $q_{[lb]z} \leq s_z \leq q_{[ub]z}$ **or**
         $q_{[lb]z} \leq s_{[lb]z} < s_{[ub]z} \leq q_{[ub]z}$ **then**
3:       add s to S'
4:     **end if**
5: **end for**
6: **return** S'

Following [4], n objects of d-dimensional is generated for synthetic datasets, in which d is set to 3 and n is set to 100 k. The distribution of values for each dimension follows uniform random variable, in which the values are generated from 1 to 10,000. To establish the concept of *uncertain dimensions*, we have made dimension (D_1) into an *uncertain dimension* $(\mathbb{U}(D_1))$. Meanwhile, the NBA datasets contains the statistics

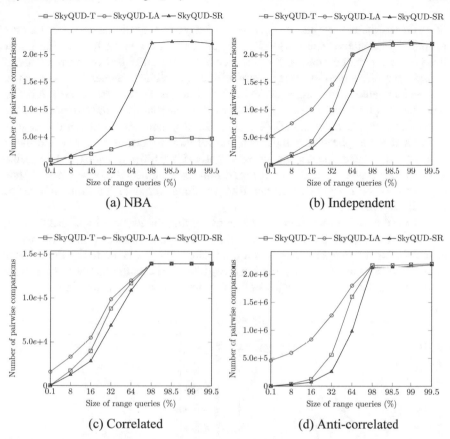

Fig. 5 Effect of γ on number of pairwise comparisons ($\tau = 0.5$)

of 21,961 NBA players for 16 attributes in basketball matches from year 1946 to 2009. Nevertheless, the NBA records are originally presented as exact values, and thus, following [4], we explicitly introduce another dimension to represent the *uncertain dimension* in the datasets. The size of range queries (γ) is uniformly generated as percentage of the volume of the datasets following the benchmark in [7]. For each size of the range queries, 50 range queries are generated and we report the average result.

Figure 5 presents the number of pairwise comparisons for the *SkyQUD-T*, *SkyQUD-SR* and *SkyQUD-LA* algorithms with regard to the size of range queries. The results clearly show that the number of pairwise comparisons performed in *SkyQUD-T* is always between the numbers of pairwise comparisons performed in *SkyQUD-SR* and *SkyQUD-LA*. Intuitively, the list of objects that *satisfies* the range queries reported by *SkyQUD-LA* will always have the highest number since the algorithm will always directly accept all objects that overlap with the border of the range that is being queried, regardless if the objects do not fully lie inside the range queries.

Meanwhile, the list of objects that *satisfies* the range queries reported by *SkyQUD-SR* will always have the lowest number as the algorithm will always absolutely reject all objects that do not fully lie in the range queries. Conversely, the number of objects that *satisfies* the range queries produced by *SkyQUD-T* would be dependent on the probability threshold value (τ), which in this instance τ is set to 0.5. The larger the size of the range queries, the more chances that an object overlaps with and *satisfies* the range queries. This in turn will lead to higher chances for an object to not be dominated in the range queries, and thus causes the increase in the number of pairwise comparisons between objects. It can be seen that when the size of the range queries is covering more than 98% of the datasets, there are not much differences in the number of pairwise comparisons for all three algorithms. This would mean that when the size of the range queries is approaching 100% (i.e. the regular skyline without any range queries), all three algorithms report almost the same list of objects that *satisfies* the range queries, therefore, resulting in nearly the same number of pairwise comparisons.

Figures 6 and 7 demonstrate the effect of varying the probability threshold τ on the number of pairwise comparisons performed within each algorithm with regard to the size of the range queries. From the figures, when τ is set to 0.0 the number of pairwise comparisons performed in *SkyQUD-T* reflects the number of pairwise comparisons executed in *SkyQUD-LA*, and the number of pairwise comparisons

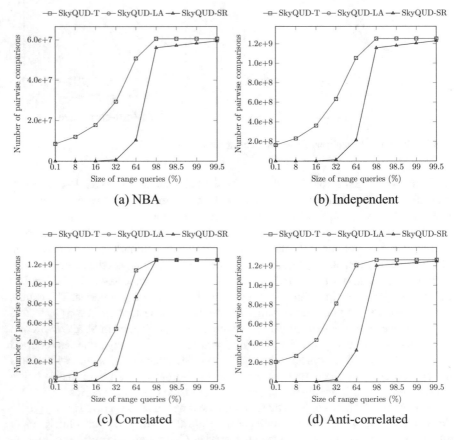

Fig. 6 Number of pairwise comparisons when $\tau = 0.0$

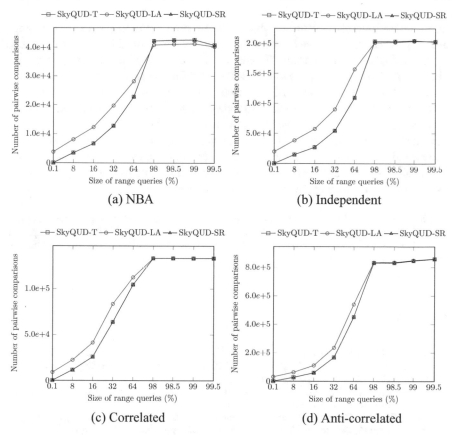

Fig. 7 Number of pairwise comparisons when $\tau = 1.0$

executed in *SkyQUD-T* clearly echoes the number of pairwise comparisons executed in *SkyQUD-SR* when τ is set to 1.0. This is due to when $\tau = 0.0$, *SkyQUD-T* would report all objects that overlap with the range queries, which follows the idea of *SkyQUD-LA*. Conversely, when $\tau = 1.0$, *SkyQUD-T* would only report objects that are clearly contained within the range queries, which follows the idea of *SkyQUD-SR*. The observation here is that all objects that are returned by *SkyQUD-T* will be returned by *SkyQUD-SR* and *SkyQUD-LA* too, and intuitively, all skyline objects returned by *SkyQUD-T* will be returned by *SkyQUD-SR* and *SkyQUD-LA* too. This would mean that *SkyQUD-T* will never return a skyline object that is returned neither by *SkyQUD-SR* nor *SkyQUD-LA*.

6 Conclusion

We employ the use of the probability threshold τ to filter out objects whose probabilities to be in the range query are small enough, and accordingly, we define the term of *satisfy* in this paper to be objects whose probability to *satisfy* a given range query is at least τ. We define three major cases that would most likely occur between a pair of objects with uncertain dimensions that overlap with the range query. We propose *SkyQUD-T* which aims at processing skyline queries that are within a range of search on data with *uncertain dimensions*. Due to lack of research works in skyline query processing that is able to accommodate range query on *uncertain dimensions*, we have thus consider two naïve algorithms as a basis of comparison to *SkyQUD-T*. The analysis clearly shows that the performance of *SkyQUD-T* is always between the aforementioned naïve algorithms. In future, we intend to extend our work to incorporate an efficient and effective index structure that is able to cater uncertainties in data as well as answering skyline with range query.

Acknowledgements This research was supported by MOSTI under the FRGS grant scheme (Grant no. 08-01-16-1853FR). All opinions, findings, conclusions and recommendations in this paper are those of the authors and do not necessarily reflect the views of the funding agencies. We thank the anonymous reviewers for their comments.

References

1. Chester, S., Mostensen, M.L., Assent, I.: On the suitability of skyline queries for data exploration. In: Workshop Proceedings of the International Conference on Extending Database Technology (EDBT), pp. 161–166 (2014)
2. He, G., Chen, L., Zeng, C., Zheng, Q., Zhou, G.: Probabilistic Skyline Queries on Uncertain Time Series. Neurocomputing **191**, 224–237 (2016)
3. Jiang, B., Pei, J.: Online interval skyline queries on time series. In: Proceedings of the 25th International Conference on Data Engineering (ICDE), pp. 1036–1047 (2009)
4. Khalefa, M.E., Mokbel, M.F., Levandoski, J.J.: Skyline query processing for uncertain data. In: Proceedings of the 19th ACM International Conference on Information and Knowledge Management (CIKM), pp. 1293–1296 (2010)
5. Lin, X., Xu, J., Hu, H.: Range-based skyline queries in mobile environments. IEEE Trans. Knowl. Data Eng. **25**(4), 835–849 (2013)
6. Mortensen, M.L., Chester, S., Assent, I., Magnani, M.: Efficient caching for constrained skyline queries. In: Proceedings of the International Conference on Extending Database Technology (EDBT), pp. 1–12 (2015)
7. Papadias, D., Tao, Y., Fu, G., Seeger, B.: Progressive skyline computation in database systems. ACM Trans. Database Syst. **30**(1), 41–82 (2005)
8. Rahul, S., Janardan, R.: Algorithms for range-skyline queries. In: Proceedings of the 20th International Conference on Advances in Geographic Information Systems (SIGSPATIAL), pp. 526–529 (2012)
9. Ross, S.M.: Introduction to Probability Models, 8th edn, American Press, Milwaukee (2003)
10. Saad, N.H.M., Ibrahim, H., Sidi, F., Yaakob, R.: Non-index based skyline analysis on high dimensional data with uncertain dimensions. In: The International Baltic Conference on Databases and Information Systems (DB&IS) (2018)
11. Saad, N.H.M., Ibrahim, H., Sidi, F., Yaakob, R.: Computing range skyline query on uncertain dimension. In: Proceedings of the International Conference on Database and Expert Systems Applications (DEXA), pp. 377–388 (2016)
12. Wang, W.-C., Wang, E.T., Chen, A.L.P.: Dynamic skylines considering range queries. In: Proceedings of the 16th International Conference on Database Systems for Advanced Applications, pp. 235–250 (2011)

KDD-Based Decision Making: A Conceptual Framework Model for Maternal Health and Child Immunization Databases

Sourabh Shastri and Vibhakar Mansotra

Abstract This paper focuses on the issues apposite to the use of maternal health and child immunization data and throws light on how the KDD (Knowledge Discovery in Databases) process makes use of maternal health and child immunization data for model building and decision making at various levels in healthcare sector. Data mining techniques and algorithms play a critical and vital role in these types of KDD systems, and the idea of using the same in such models and systems is to build an automated tool for identifying and spreading important healthcare information and knowledge. The implementation of these types of developed models is possibly the much-needed intervention that improves the usability of maternal health and child immunization data as it is used for better decision making by healthcare professionals to develop plans and policies to help the society and to achieve the better outcomes for an effective healthcare management. For these motives, a conceptual framework model based on KDD was designed during the present study to discover knowledge from maternal health and child immunization databases.

Keywords Maternal health · Child immunization · KDD · Data mining · Decision making

1 Introduction

Health care is one of the most significant applications of KDD (Knowledge Discovery in Databases) and still requires more growth and progress for the reason that it is the requirement of all individuals owing to which healthcare generates enormous extent of data that is complicated to handle and analyze by conventional approaches. KDD is an automatic, exploratory analysis, and modeling of large data repositories. It is

S. Shastri (✉) · V. Mansotra
Department of Computer Science and IT, University of Jammu, Jammu, India
e-mail: sourabhshastri@gmail.com

V. Mansotra
e-mail: vibhakar20@yahoo.co.in

© Springer Nature Singapore Pte Ltd. 2019
S. K. Bhatia et al. (eds.), *Advances in Computer Communication and Computational Sciences*, Advances in Intelligent Systems and Computing 924,
https://doi.org/10.1007/978-981-13-6861-5_21

243

an organized process of identifying valid, novel, useful, and understandable patterns from large and complex datasets along with data mining techniques and algorithms can transform these repositories of data into knowledge for decision making. Poor health among mothers and children has remained an immense problem for a long time in our country, thereby, pressing upon the need for more efficient actions to improve the health of women and their newborns. Thus, maternal health and child immunization domain is a potential area where knowledge discovery process along with data mining techniques and algorithms can be applied to discover knowledge for making decisions.

In India, maternal health care and child immunization comes under the aegis of Union Ministry of Health and Family Welfare (MoHFW), Government of India which facilitates the system's overall functioning at national and state levels. The main objective of MoHFW is to supervise the effective and systematic management by the states and local authorities [1]. India's public healthcare system follows centralized planning and policy making along with decentralized implementation [2]. The central government encourages the states to focus on precise health objectives and priorities [3]. There are primarily three levels of public healthcare organizations, viz. national, state, and district. The key task at the national level is to set goals for national health policy, whereas at the state level, reviewing of implementation of different national level programs is done, thereby, bringing them into action at the district level.

The reasons for the adoption and preference of KDD over other techniques like SEMMA and CRISP-DM is that the latter techniques can be viewed as an implementation of the KDD process itself, thereby, making the process a fundamental platform for analysis of huge repositories of data. Furthermore, the steps involved in KDD are simple and easy to work on and thus are formative of other offshoots. In addition to it, KDD can be applied in all general datasets for the analysis and prediction of data, thence, covering a wider range but SEMMA and CRISP-DM are specific decision making approaches.

2   Role of KDD in Decision Making: Significance of Study

A decision is always required to overcome problems and challenges originating from the constraints of the system [4]. Decision making is the ability of a decision maker to choose the best solution among the various alternatives to solve a particular problem to achieve some objectives. In any organization or system, mostly three types of decisions can be taken, viz. strategic, tactical, and operational [5]. There are various approaches to support decision making process in any organization. One of the most famous approaches that are used in the organizations is decision support system (DSS). A DSS is an information system that serves the management, operations, and planning levels of an organization, company, or business and solves various problems to make better decisions [6]. A well-developed DSS can help decision makers in compiling lot of contents from many sources and make decisions by solving problems. Although DSSs are primarily used in context of decision making in organizations, but in recent times it has been combined with various analytical tools including data warehouses, online analytical processing, data mining, knowledge discovery, and many others [2].

In this paper, we have designed a conceptual framework model by integrating the KDD process with decision making. The data mining is the core sub-process of KDD that applies different techniques and algorithms on the databases for developing predictive models to extract proficient and informative patterns. With the assistance of KDD process, various patterns and knowledge in the datasets of maternal health and child immunization can be discovered that are previously unknown and can be used to make better and efficient decisions for effective healthcare management. The main purpose of integrating decision making with KDD process is to strengthen and improve DSS through the KDD models developed especially where huge quantity of historical data is available for the purpose of knowledge discovery.

The main objective to undertake the present study is to facilitate and aid maternal and child health services to be offered in the fastest, most accurate, highest quality, and in the most responsive way in the state of Jammu and Kashmir in particular and in India generally, as healthcare professionals need access to the most accurate and latest knowledge to use this through decision support systems. To proceed for the aim, KDD and data mining techniques can be used to extract large amounts of hidden, valuable, usable knowledge within the data to further provide strategic decision support by developing decision making models based on the analysis of data. Although decision support system with knowledge discovery and data mining techniques can be employed for better decision making in different domains but their utilization in health care, in particular can definitely assist the policy makers for better policy formulation and interventions. Thus, keeping into view the relevance of KDD, present study was aimed to develop a KDD-based conceptual framework model for knowledge discovery to transform maternal health and child immunization data from public healthcare databases into knowledge in a systematic manner for an efficient healthcare management. This conceptual framework model shall assist healthcare researchers to identify and exploit new knowledge and insights into the form of trends, patterns, rules, and associations from large volumes of health indicators of maternal health and child immunization datasets in the future that can be further used for the purpose of decision making.

In addition to it, no work has been undertaken in the study area as regards KDD and data mining techniques in case of diverse maternal health care and child immunization indicators. Thus, the present piece of work is the first of its kind from the study area and thus carries a great technical implementation after the results analysis for different trends, patterns, rules, and associations are shared with diverse government agencies related to health sector of the state of Jammu and Kashmir, viz. Department of Health and Medical Education and Department of Social welfare (for the overall welfare of women and children).

3   Review of Literature

Some of worth mentioning works which helped in the current study are mentioned in this section. The idea of designing this conceptual framework model has been taken from the research work of Sharma and Mansotra [2] who designed a conceptual framework model DM-PHCS based on CRISP-DM methodology having two stages, viz. data mining and decision making. The DM-PHCS model was proposed for public healthcare system, and they illustrated how data mining can assist decision making at different levels of health industry in India. Mehta et al. [7] investigated data mining techniques for decision support system of maternal health care and submitted that the chances of high risk maternal patients can be predicted by using data mining for timely detection and providing quality care. Gupta et al. [8] applied random decision tree classification technique on public maternal health data of Uttar Pradesh state for the year 2015–16. Shastri et al. [9] developed a tool in Java NetBeans for classifying child immunization data of Jammu and Kashmir state by applying Naïve Bayes algorithm. Jindal et al. [10] introduced decision support system called DMDSS (data mining decision support system) based on CRISP-DM to create a new approach for problem solving. Chung et al. [11] focused on the use of advanced computational methods and visual tools for the development of DSSs for analysis of complex mental health systems to improve system knowledge and evidence-informed policy planning.

Khan et al. [12] designed a unified theoretical framework for data mining by fabricating clustering, classification, and visualization by means of composite functions. Cifci and Hussain [13] explained data mining as a decision support tool that facilitates health authorities to access the most accurate and latest information by utilizing the available data and support decision making process to manage healthcare institutions more efficiently. Huang et al. [14] designed safety decision making framework to make more intelligent and felicitous safety decisions by integrating and applying big data analytics, tools, and techniques. Alves and Cota [15] elaborated that visualization together with advanced algorithms allows extracting useful information from a huge volume of data that play a key role in decision making.

A number of researchers have fabricated decision support system with knowledge discovery and data mining techniques for better decision making in different domains including health care that assists the policy makers for better policy formulation and interventions. Better provisioning of primary health services can be measured by the level of health outcomes such as infant mortality rate (IMR), under 5 mortality rate (U5MR), and maternal mortality rate (MMR). These health outcomes are directly related to health indicators such as child immunization, antenatal care, postnatal care, and number of institutional deliveries which form a part of the primary health services. Therefore, the use of KDD as a decision support tool in the delivery of maternal health and child immunization services at all levels will help health professionals to make the most optimal decisions and better health policy formulation. The KDD and data mining experts can use the data available on HMIS portal for model building and extracting knowledge to assist the healthcare professionals for optimal decision making and better health policy formulation.

4 Proposed Conceptual Framework Model

The main objective of fabricating a conceptual framework model is to make possible the use of KDD and data mining techniques in decision making process for handling and managing maternal health and child immunization data available on Health Management Information System (HMIS) portal facilitated under the National Health Mission (NHM) scheme of Ministry of Health and Family Welfare, Government of India for decision making. A KDD reference model consists of various stages including data selection, data preprocessing, data transformation, data mining, interpretation, and knowledge representation [16].

The KDD reference model has been modified according to the requirements of the proposed model, i.e., KDD-based model for managing maternal health and child immunization databases (KDD-MHCI) framework (Fig. 1). Initially, KDD-MHCI is used to build a number of KDD models based on maternal health care and child immunization data of past several years for analysis and afterward, health administrators, and professionals are required to participate during the final decision making process without going deep into model building process.

As data mining is the core component of KDD process so this theoretical framework also incorporates various data mining techniques like classification, regression, association, clustering, time series, etc. Depending on the MHCI data, the KDD-MHCI uses various data mining techniques accordingly to develop the models and then subsequently use these models for decision making process.

The proposed conceptual framework model KDD-MHCI follows KDD research methodology, and the various stages of KDD-MHCI including KDD stages are discussed below in detail.

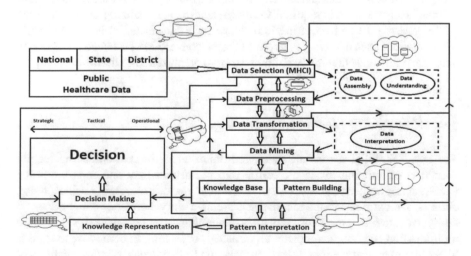

Fig. 1 Proposed conceptual framework model (KDD-MHCI)

4.1 Data Selection

The Ministry of Health and Family Welfare, Government of India has provided pub-lic healthcare data on a web-based monitoring system, namely HMIS that contains facility wise data related to maternal health, child immunization, family planning, patient services, laboratory testing, and stock position in the form of standard and analytical reports [17]. The data selection stage of KDD-MHCI framework has been used to select only the maternal health and child immunization data from the pub-lic healthcare data available on the supra mentioned portal and stored it into data assembly component for further analysis.

4.2 Data Understanding and Assembly

The data understanding includes the important issues regarding data, viz. data design, range, size, completeness, availability, and confidentiality. As the data collected in data assembly component particularly pertains to maternal healthcare and child immunization, it is necessary to understand the various health indicators that are the part of collected data. The maternal health data contains health indicators related to antenatal care services, deliveries, cesarean deliveries, pregnancy outcome, com-plicated pregnancies, weight of new-born, postnatal care, medical termination of pregnancy, etc., whereas child immunization data contain health indicators related to various vaccinations received by children including BCG, DPT, Pentavalent, OPV, Hepatitis, Measles, etc. In this stage, there is a possibility of redefining of goals, there-fore, understanding of domain and data becomes a pre-requisite for the KDD process. Domain understanding is required for understanding and realizing the goals of the KDD process [18] and acquiring knowledge regarding maternal health and child immunization. Data assembly is required for the conversion of selected unorganized maternal health and child immunization data into organized form for further analysis.

4.3 Data Preprocessing

The health facilities across all states and union territories are uploading facility wise data on monthly basis on HMIS web portal that may encompass missing values, noise, outliers, and extremes. The various reasons of missing values, noise, outliers, and extremes in the real datasets are data entry, human or computer error, faulty instruments, etc. Thus, raw data pertaining to maternal health and child immunization may be incomplete, inconsistent, or lacking in certain trends and is likely to contain many errors. Data preprocessing is used to enhance the reliability of data by resolving such errors and issues. It prepares the data for data transformation stage by performing cleaning on data such as handling missing values and remove noise, outliers, and extremes [18].

4.4 Data Transformation

In this stage, the generation of better data for the data mining stage is prepared by transforming data into appropriate forms. The maternal health and child immunization data for national, state, and district levels can be transformed into proper formats as per the requirement. The transformation methods may include dimension reduction and attribute transformation [18]. Dimension reduction is used to reduce the number of input health indicators of maternal health and child immunization under consideration and attribute transformation modifies the data by replacing a selected health indicator by one or more new health indicators. This stage is one of the most critical steps for the success of the whole KDD process where the data present is converted to a form which is more usable and is versatile to be used on a new program or system that is totally different from the previous one. This step also improves the sturdiness of KDD process and consolidates the data into a form which is appropriate for mining by performing summary or aggregation operations. The data transformation stage also includes data interpretation for making sense of the transformed data for the next stage of data mining.

4.5 Data Mining

In this stage, the KDD experts use various techniques of data mining, viz. association, clustering, time series, classification, etc. depending upon the requirement to find hidden patterns or knowledge in the transformed data. In association method, hidden patterns are discovered in the MHCI dataset which can be represented by association rules. In clustering technique, similarities are found out in the MHCI dataset so that they can be grouped together. These groups or clusters consist of similar objects in themselves but different to objects of another group. The time series analysis can be used to forecast the future values of the health indicators in MHCI data for the different districts, states, or whole nation. The classification technique is used to accurately predict the target class for each case in the dataset. For example, in MHCI data, classification model can be used to classify the districts into high priority districts where there are lack of immunization facilities and nonhigh priority districts where the immunization facilities are adequate.

4.6 Pattern Building and Knowledge Base

In pattern building stage, a data mining algorithm that is to be selected by the KDD experts in the previous stage is finally applied to the transformed MHCI data for pattern or model building. The algorithm depends on the level and type of MHCI data on which analysis is to be carried out. Different algorithms can be used on the same dataset which provides flexibility. As the KDD is an iterative process, the experts can develop patterns or models any number of times by applying different algorithms. The knowledge base serves as a knowledge repository of models or patterns extracted from transformed MHCI data [4] and stores all models or patterns obtained from previously used iterations. The patterns and models thus obtained are studied and then interpreted as discussed in the next section.

4.7 Pattern Interpretation

This stage is used to estimate the quality of the patterns or model discovered in pattern building stage by applying the various algorithms on transformed MHCI data. The data mining algorithms might be applied several times until a satisfactory outcome is reached. The models are developed according to the management levels, viz. national, state, and district for which the decisions are to be taken. The performance of these models can be evaluated by using various measures such as accuracy, recall, precision, F-measure, kappa statistic, ROC curves, etc. and finally after evaluation, the selected models can be visualized by various knowledge representation methods for decision making phase to carry out various tasks and programs.

4.8 Knowledge Representation

Knowledge representation is used to visualize the knowledge obtained from the models and patterns to health professionals and decision makers in terms of tables, trees, graphs, charts, matrices, rule graphs for further decision making at national, state, or district levels. The knowledge thus obtained from the models or patterns can be incorporated into decision making system for making strategic, tactical or operational decisions, policies, and planning. The knowledge becomes active in such a sense that we may make changes to the traditional system and measure the better effects. In fact, the success of this step depends upon the effective and efficient use of all the previous stages of KDD-MHCI framework. So, in decision making phase, the knowledge obtained from the model can be used to sort out various problems in maternal health and child immunization domain of public health care.

4.9 Decision Making

The decision making phase of KDD-MHCI uses the available MHCI data and knowledge obtained from proficient and informative patterns or models. The health-care professionals, administrators, policy, and decision makers have to participate in this phase of KDD-MHCI for a crucial responsibility of decision making at various levels, viz. national, state, or district levels. Depending upon the level, the different kinds of decision, viz. strategic, tactical, and operational are taken for the better health outcome in maternal and child immunization domain. The strategic decisions refer to the direction in which a program should head and are taken at the national level. It defines the mission and vision of the whole program related to maternal health and child immunization and the decision making to fulfill this mission into design. These decisions are of long-term nature and taken by higher authorities. A tactical decision divides a strategic decision into guidelines to be implemented at the state or regional levels. Operational level decisions concern how tactical decisions are implemented and are taken at the district level. These include blueprints for the implementation of tactical decisions, routine decisions, planning, and emergency level situations. These decisions should be consistent with strategic and tactical decisions for their smooth functioning.

The proposed KDD-MHCI model is showing high levels of integration with decision making for optimum results at various levels. The proposed framework shall assist the policy makers and healthcare administrators for taking decisions in the domain of maternal health and child immunization that are totally based on the present and past data of all the districts and states of the country available on the HMIS portal. With the use of KDD-MHCI framework, it can be easily discovered that which districts or states of the country require more attention in terms of maternal health and child immunization services, thereby, further assisting the healthcare decision makers for formulating the health policies to give more attention and services to the areas that require more attention as compared to other ones for speedily diagnose the diseases. This proposed approach for taking decisions in the field of maternal health and child immunization is better than the traditional approaches and methods of framing health policies because this approach is based on the evidence of discovering knowledge from the actual data of maternal health and child immunization and it assists in a better way for framing new health policies for the country.

5 Conclusion

KDD-MHCI is basically a KDD-based decision making theoretical framework which encourages decision making by providing evidence based knowledge in the form of patterns, relationships, and rules obtained by applying various data mining algorithms. The whole framework is divided mainly into two phases, viz. KDD and decision making that allows the healthcare administrators to make decisions for future programs at decision making phase with a minimum basic understanding of KDD process. On the other hand, the KDD process that requires expertise is hidden from the decision makers and performed by KDD experts. The KDD experts have responsibility to perform all the stages of KDD process including the use of various data mining algorithms for discovering patterns and models for knowledge representation. Therefore, the healthcare professionals and administrators have the final responsibility of decision making based on the derived proficient knowledge. Depending upon the availability of MHCI data at different levels, the framework can be applied accordingly for improved decision making.

6 Future Scope

In the present study, a KDD-based conceptual framework model called as KDD-MHCI has been proposed that can be used to build models in the field of maternal health and child immunization for better decision making at different levels, viz. national, state, and district. In the future, a KDD-MHCI framework based tool shall also be developed in order to build models state wise and for the whole country by applying various data mining algorithms on maternal health and child immunization data for better decision making. In addition to it, the predictions resulted out of the present work shall be discussed with government stakeholders in the state for the technical implementation of the study.

References

1. Ministry of Health and Family Welfare. [Online]. Available: https://mohfw.gov.in. Accessed 10 July 2018
2. Sharma, A., Mansotra, V.: Data mining based decision making: a conceptual model for public healthcare system. In: Proceedings of IEEE 3rd International Conference on Computing for Sustainable Global Development (INDIACom), pp. 1226–1230 (2016)
3. Priya, R., Chikersal, A.: Developing a public health cadre in 21st century India: addressing gaps in technical, administrative and social dimensions of public health services. Indian J. Public Health 57(4), 219–224 (2013)
4. Szeghegyi, A.: Investigation of decision-making process by the use of knowledge based system. In: Proceedings of 5th International Conference on Management, Enterprise and Benchmarking, pp. 209–222 (2007)
5. Murtola, L.M., Laine, H.L., Salantera, S.: Information systems in hospitals: a review article from a nursing management perspective. Int. J. Netw. Virtual Organ. 13(1), 81–100 (2013)
6. Kumar, A., Saurav, S.: Supply Chain Management Strategies and Risk Assessment in Retail Environments. IGI Global, Hershey (2017)
7. Mehta, R., Bhatt, N., Ganatra, A.: A survey on data mining technologies for decision support system of maternal care domain. Int. J. Comput. Appl. 138(10), 20–24 (2016)

8. Gupta, S., Singh, S.N., Jain, P.K.: Data mining on maternal healthcare data for decision support. In: Proceedings of 5th International Conference on Computing for Sustainable Global Development (INDIACom), 2017, pp. 4989–4993 (2018)
9. Shastri, S., et al.: Development of a data mining based model for classification of child immunization data. Int. J. Comput. Eng. Res. **8**(6), 41–49 (2018)
10. Jindal, K., Sharma, M., Sharma, B.K.: Data mining to support decision process in decision support system. Int. J. Emerg. Technol. Adv. Eng. **4**(1), 41–46 (2014)
11. Chung, Y., et al.: Use of the self-organizing map network (SOMNet) as a decision support system for regional mental health planning. Health Res. Policy Syst. **16**(35), 1–17 (2018)
12. Khan, D.M., Mohamudally, N., Babajee, D.K.R.: A unified theoretical framework for data mining. Inf. Technol. Quant. Manag. **17**, 104–113 (2013)
13. Cifci, M.A., Hussain, S.: Data mining usage and applications in health services. Int. J. Inform. Vis. **2**(4), 225–231 (2018)
14. Huang, L., et al.: Big-data-driven safety decision-making: a conceptual framework and its influencing factors. Saf. Sci. **109**, 46–56 (2018)
15. Alves, C.M.O., Cota, M.P.: Visualization on decision support systems models: literature overview. In: Proceedings of Springer World Conference on Information Systems and Technologies (WorldCIST'18), pp. 732–740 (2018)
16. Rupnik, R., Kukar, M.: Decision support system to support decision processes with data mining. J. Inf. Organ. Sci. **31**(1), 217–232 (2007)
17. Health Management Information System [Online]. Available: https://nrhm-mis.nic.in. Accessed 28 July 2018
18. Maimon, O., Rokach, L.: Introduction to Knowledge Discovery and Data Mining. Springer, Boston (2009)

An Efficient Detection of Malware by Naive Bayes Classifier Using GPGPU

Sanjay K. Sahay and Mayank Chaudhari

Abstract Due to continuous increase in the number of malware (according to AV-Test institute total $\sim 8 \times 10^8$ malware are already known, and every day they register $\sim 2.5 \times 10^4$ malware) and files in the computational devices, it is very important to design a system which not only effectively but can also efficiently detect the new or previously unseen malware to prevent/minimize the damages. Therefore, this paper presents a novel group-wise approach for the efficient detection of malware by parallelizing the classification using the power of GPGPU and shown that by using the Naive Bayes classifier, the detection speedup can be boosted up to 200x. The investigation also shows that the classification time increases significantly with the number of features.

Keywords Malware detection · GPGPU · Machine learning · Computer security

1 Introduction

The ubiquity of the Internet has engendered the prevalence of information sharing among networked users and organizations, and in todays information era, most of the computing devices are connected to the Internet, which has rendered possible countless invasions of privacy/security worldwide from the malware (**mal**icious soft**ware**). In 1970, the first virus was created [2], and since then malware are not only continuously evolving with high complexity to evade the available detection techniques, but also the new variants of malware are increasing exponentially; as a consequence, malware defense is becoming a difficult task to protect the computational devices from it. The use of malware for espionage, sophisticated cyber attacks, and other crimes motivated to develop an advanced method to combat the threats/attacks from

S. K. Sahay (✉) · M. Chaudhari
Department of CS & IS, BITS, Pilani, Sancoale, Goa Campus, India
e-mail: ssahay@goa.bits-pilani.ac.in

M. Chaudhari
e-mail: h20160014@goa.bits-pilani.ac.in

© Springer Nature Singapore Pte Ltd. 2019
S. K. Bhatia et al. (eds.), *Advances in Computer Communication and Computational Sciences*, Advances in Intelligent Systems and Computing 924, https://doi.org/10.1007/978-981-13-6861-5_22

it [3, 9, 18, 21]. However, due to the exponential increase in the number of malware (according to AV-Test institute total $\sim 8 \times 10^8$ malware are already known, and every day they register $\sim 2.5 \times 10^4$ malware [11])), anti-malware industries not only facing major challenges to check the potential malicious content but to detect the malware efficiently. The reason behind these high volumes of malware is basically in the advancement of second-generation malware which can create millions of its variants by using different obfuscation techniques [20]. The malware attack/threat is not only limited to individual boundaries, but they are highly skilled state-funded hackers writing customized malicious payloads to disrupt political, industrial working, and military espionage [5, 17, 24]. The most high-profile, subversive incident was a series of intrusions against the Democratic Party in the US presidential election [5].

In 2017, McAfee has more than 780 million malware samples in their database, and in the third quarter of 2017, there was a 10% increase in the number of the new malware; in addition, in the same quarter, they have observed a 60% increase in new mobile malware in the Android devices which are mainly due to increase in Android screen locking ransomware [14]. The Symantec 2017 Internet Security Threat report indicates that there were 357 million new malware variants [5]. The recent Internet Security Threat Report from Symantec shows an increase of 88% in overall malware variants [6]. Hence, if adequate advancement in anti-malware technique is not achieved, consequences at this scale at which new malware are being developed can create fatal effects, and the results will be more severe then past. In this recently, various machine learning techniques have been proposed by authors [1, 4, 23, 25], which can enhance the capabilities of traditional malware detection system viz. signature matching technique, but with the use of a complex machine learning, the detection time increases. Therefore, understanding the exponential increase in the number of malware released every year and files in the computational device, it is very important to design a system which not only effectively but can also efficiently detect the new or previously unseen malware to prevent/minimize the damages. Hence in this paper, we present a novel efficient group-wise static malware analysis approach for the efficient detection of malware by parallelizing the classification using the power of general-purpose graphics processing unit (GPGPU) and shown that by using the Naive Bayes classifier, the detection speedup can be boosted up to 200x. Accordingly, Sect. 2 briefly discusses the related work done in this field. Section 3 describes the data preprocessing and how features are selected for the classification. Section 4 contains the experimental and the result analysis of our approach. Finally, we summarize our conclusion in Sect. 5.

2 Related Work

With the evolution of complex second-generation malware which can generate millions of its variant, the detection techniques have also been made significant progress from the early day traditional signature matching to deep learning techniques to improve the detection accuracy [8]. In this, recently, Ashu et al. showed that group-wise classification of Windows malware in the range of 5 KB, the detection accuracy can be achieved up to 97.95% [21]. Similarly, they have also shown that on an average 97.15% detection of Android malware can be achieved by permission-based group-wise detection system [22]. However, understanding the exponential growth of malware and the number of file in our system, it is equally important to focus on the design of an efficient malware detection system. In this, Ciprian Pungila and Viorel Negru in 2012 has proposed an efficient memory compression model for virus signature matching using GPGPU [15]. They were able to achieve 22 less memory utilization and 38 times higher bandwidth compared to their single-core implementation. In 2014, Che-Lun Hung et al. proposed a GPU-based botnet detection technique [10]. They implemented the network traffic reduction on GPU and were able to achieve eight times performance over CPU-based traffic reduction. For Android devices, Manel Abdellatif et al. in 2015 have designed and implemented a host-based parallel anti-malware based on mobile GPU [13]. Their implementation was three times faster than the serial implementation on CPU. In 2016, to accelerate the statistical detection of zero-day malware, Igor Korkin et al. have proposed a technique using CUDA-enabled GPU hardware [12]. In their work, they used GPU mainly for achieving speedup in memory forensic task. Recently, Radu Velea et al. have proposed a CPU/GPU-based hybrid approach to accelerate pattern matching of the malware [26]. In their work, they found that the hybrid approach takes half of the time compared to the CPU implementation only and consumes 25% less power.

3 Data Preprocessing and Feature Selection

For the experimental analysis, we downloaded 11,355 malware from the malicia-project dataset (one of author possess the dataset [23])) and collected 2967 benign programs (also verified from virustotal.com) from different windows operating systems. It has been observed that 97.18% malware in the Malicia dataset is below 500 KB [21]. Therefore, we took both the samples (i.e., malware and benign programs) which are below 500 KB, and left with 11,305 malware and 2360 benign executables for the analysis. Also, the investigation by Ashu et al. shows that the variation in the size of the malware generated by malware kits viz. NGVCK, PS-MPC, and G2 do not vary by more than 5 KB range. Therefore, for efficient and effective classification, we partitioned the datasets in 100 group each of 5 KB size.

We selected the opcodes of the executable as a feature for the classification of malware because the difference in the opcode occurrence between the malware and benign executable differ in large [19, 21]. Therefore, the prominent features i.e., opcodes from the dataset which can differentiate the malware from benign programs are obtained as given by the Ashu et al. [21, 23] i.e., by normalizing the opcodes occurrence difference between the malware and benign executables for all the formed groups independently, and finally top k-features (opcodes) are selected from each group separately for the efficient and effective detection of the malware.

4   Our Approach and Experimental Analysis

A schematic of the experimental analysis of our approach is shown in the Fig. 1. For the purpose, we randomly split the dataset (containing only opcode occurrence of every executable) for the training and testing of the malicious and benign dataset separately (a conservative side as per the suggested norms to ensure optimal performance [7]) in the ratio of 2:1 and used Naive Bayes classifier (as the paper focusses on the efficient detection of malware, not on the accuracy; therefore, for simplicity, we selected the Naive Bayes classifier) which assumes strong class independence between different attributes under consideration, i.e., if the given set of (opcodes in

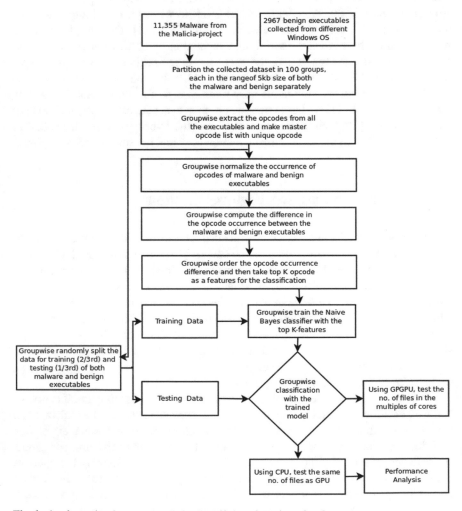

Fig. 1 A schematic of our approach for the efficient detection of malware

our case) $A = a_1, a_2, a_3, \ldots, a_n$, then the Naive Bayes model computes posterior probability for target class C (malware/benign) and can be represented as [23]

$$P(C|a_1, a_2, a_3, \ldots, a_n) = \frac{P(a_1, a_2, a_3, \ldots, a_n|C)}{P(a_1, a_2, a_3, \ldots, a_n)}$$

where $P(C|a_1, a_2, a_3, \ldots, a_n)$ is the posterior probability of an executable sample of belonging to class C. Hence, one can calculate the posterior probabilities for the test executable, and if the malware class probability is higher, then it is classified as malware otherwise it is labeled as benign.

The experiment has been conducted in Intel i7-7700HQ quad-core processor with a base frequency of 2.8 GHz, 8 GB RAM, Pascal architecture (GP107)-based Nvidia 1050Ti GPU with 768 CUDA cores distributed across 6 SMP and 4GB GPU DRAM operating on a base speed of 1291 MHz.

To improve the detection efficiency, we trained the model for all the groups independently with top k-features obtained from each group, except the group which has less than six files either in malware or benign. We find that 5, 8, 61, 65, 97, 98, and 100th group have less then six files in either category (benign or malware). For the actual implementation if any group have less than the minimum set number of file, then that group file can be classified/tested with the next group trained model.

First we investigated the classification time taken by the CPU by selecting the top 20, 40, 80, 100, 160, and 200 features (Figs. 2 and 3) from each group and the number of file in multiple of 768 (i.e., number of cores in the GPU), and the results obtained are shown in Fig. 2. Next, we find the time taken by GPU after distributing

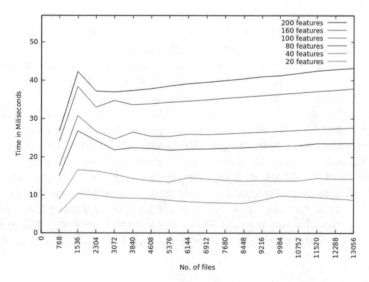

Fig. 2 Time taken by the CPU to classify the files in the multiple of 768 files by varying the number features

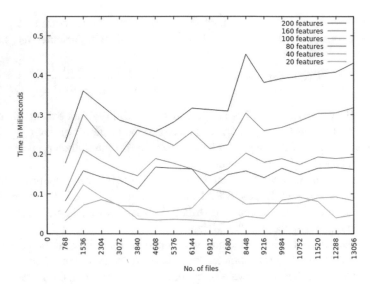

Fig. 3 Time taken by the GPU to classify the files in the multiple of 768 files by varying the number features

our trained model among all the 768 cores such that trained model of the particular group, the corresponding test file, and top k-features shall be in same core so that parallelization of the tasks in GPU can be optimally used. Then, we observed the time taken by the GPU by giving the file in the multiple of 768 for the testing/classification and the results obtained are shown in Fig. 3.

The analysis shows that the classification/testing time is also dependent on the number of features, and with the increase in the number of files in the multiple of number cores of GPU, the CPU proportionally take more time than GPU (Figs. 2 and 3). Therefore, experimented with various sets of features and found that the detection accuracy improves by increasing the number of features till top 200 features, after that there is no significant change in accuracy, and remains around ∼87%. Therefore, we investigated the speedup with top 200 features (speedup can be written as, Sp = Tc/Tp, where Tc is the time taken to execute the sequential program and Tp is the time taken to execute the program in parallel, i.e., in GPU, with P number of cores [16]) and almost all the dataset (as our focus is on the efficiency not on the effectiveness of the classification) for the improvement in the performance due the parallelization of the task using GPU can be achieved from the given system, and the obtained result is shown in Fig. 4. We observed that the parallel implementation for the classification of malware using GPU by Naive Bayes algorithm is able to achieve speedup up to 200 (not taking in account of the overhead involved in the processes).

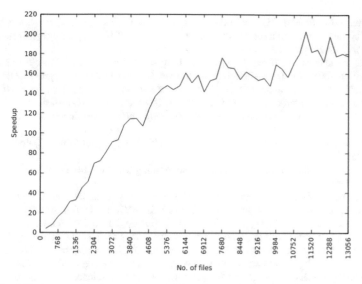

Fig. 4 Classification speedup due to parallelization of the task using the GPU

5 Conclusion

We present a novel group-wise approach (to the best of our knowledge, this is the first paper that group-wise classifies the malware using the power of GPGPU) for the efficient detection of new or previously unseen malware by parallelizing the classification using the power of GPGPU and shown that by using the Naive Bayes classifier, the classification speedup can be boosted up to 200x (not taking in account of the overhead involved in the processes). However, one has to study the performance using the classical random forest classifier and deep learning methods for the efficient and high-accuracy classification. Also, the trade-off between the efficiency and accuracy has to be investigated in-depth, i.e., optimal features selection (as classification time significantly increases with the number of features) to train the model after appropriately grouping the input data for the detection of malware, and in this direction, work is in progress.

References

1. Allix, K., Bissyandé, T.F., Jérome, Q., Klein, J., State, R., Le Traon, Y.: Large-scale machine learning-based malware detection: confronting the "10-fold cross validation" scheme with reality. In: Proceedings of the 4th ACM Conference on Data and Application Security and Privacy. CODASPY '14, ACM, New York, NY, USA, pp. 163–166 (2014). http://doi.acm.org/10.1145/2557547.2557587
2. Bilar, D.: Opcodes as predictor for malware. Int. J. Electron. Secur. Digit. Forensic 1(2), 156–168 (2007). http://dx.doi.org/10.1504/IJESDF.2007.016865

3. Bowen, B.M., Prabhu, P.V., Kemerlis, V.P., Sidiroglou, S., Stolfo, S.J., Keromytis, A.D.: Methods, systems, and media for detecting covert malware (2018). http://www.freepatentsonline.com/9971891.html
4. Canto, J., Dacier, M., Kirda, E., Leita, C.: Large scale malware collection: lessons learned. In: SRDS 2008, 27th International Symposium on Reliable Distributed Systems, October 6–8, 2008, Napoli, Italy. Napoli, ITALY (2008). http://www.eurecom.fr/publication/2648
5. Corporation, S.: Internet Security Threat Report. Technical report ((Date last accessed 31-May-2018)) (2017). https://www.symantec.com/content/dam/symantec/docs/reports/istr-22-2017-en.pdf
6. Corporation, S.: Internet Security Threat Report. Technical report (2018) (Date last accessed 31-May-2018)
7. Guyon, I.: A scaling law for the validation-set training-set size ratio. In: AT & T Bell Laboratories (1997)
8. Huang, A., Al-Dujaili, A., Hemberg, E., O'Reilly, U.: Adversarial deep learning for robust detection of binary encoded malware. CoRR (2018). arXiv:abs/1801.02950
9. Huda, S., Islam, R., Abawajy, J., Yearwood, J., Hassan, M.M., Fortino, G.: A hybrid-multi filter-wrapper framework to identify run-time behaviour for fast malware detection. Future Gener. Comput. Syst. **83**, 193–207 (2018). https://doi.org/10.1016/j.future.2017.12.037
10. Hung, C.L., Wang, H.H.: Parallel botnet detection system by using GPU. In: 2014 IEEE/ACIS 13th International Conference on Computer and Information Science (ICIS), pp. 65–70 (2014)
11. institute, A.T.: Malware statistics (2018). https://www.av-test.org/en/statistics/malware/ (Online; accessed 10-June-2018)
12. Korkin, I., Nesterow, I.: Acceleration of statistical detection of zero-day malware in the memory dump using CUDA-enabled GPU hardware. CoRR (2016). arXiv:abs/1606.04662
13. Manel Abdellatif, C.T., Hamou-Lhadj, A., Dagenais, M.: On the use of mobile GPU for accelerating malware detection using trace analysis. In: 2015 IEEE 34th Symposium on Reliable Distributed Systems Workshop (SRDSW), pp. 38. Montreal, QC, Canada (2016)
14. McAfee: McAfee Labs Threats Report. Technical report (2017)
15. Pungila, C., Negru, V.: A highly-efficient memory-compression approach for GPU-accelerated virus signature matching. In: International Conference on Information Security (ISC 2012), pp. 354–369 (2012)
16. Quinn, M.J.: Parallel Computing: Theory and Practice, pp. 80–83 (2002)
17. Lee, R.M., Assante, M.J., Conway, T.: Analysis of the cyber attack on the Ukrainian power grid. Technical report, E-ISAC group SANS (2016)
18. Ronen, R., Radu, M., Feuerstein, C., Yom-Tov, E., Ahmadi, M.: Microsoft malware classification challenge. CoRR (2018). arXiv:abs/1802.10135
19. Sahay, S.K., Sharma, A.: Grouping the executables to detect malwares with high accuracy. Procedia Comput. Sci. **78**(C), 667–674 (Mar 2016), https://doi.org/10.1016/j.procs.2016.02.115
20. Sharma, A., Sahay, S.K.: Evolution and detection of polymorphic and metamorphic malwares: a survey. Int. J. Comput. Appl. **90**(2), 7–11 (2014)
21. Sharma, A., Sahay, S.K.: An effective approach for classification of advanced malware with high accuracy. Int. J. Secur. Appl. **10**(4), 249–266 (2016)
22. Sharma, A., Sahay, S.K.: Group-wise classification approach to improve android malicious apps detection accuracy. Int. J. Netw. Secur. (2018)
23. Sharma, A., Sahay, S.K., Kumar, A.: Improving the detection accuracy of unknown malware by partitioning the executables in groups. In: Proceedings 9th ICACCT, 2015 Advances in Intelligent System and Computing , p. 421. Springer (2016)
24. Stone, R.: A call to cyber arms. Science **339**(6123), 1026–1027 (2013)
25. Ucci, D., Aniello, L., Baldoni, R.: Survey on the usage of machine learning techniques for malware analysis. CoRR (2017). arXiv:abs/1710.08189
26. Velea, R., Dragan, S.: CPU/GPU hybrid detection for malware signatures. In: 2017 International Conference on Computer and Applications (ICCA), pp. 85–89 (2017)

Content-Based Audio Classification and Retrieval Using Segmentation, Feature Extraction and Neural Network Approach

Nilesh M. Patil and Milind U. Nemade

Abstract The volume of audio data is increasing tremendously daily on public networks like Internet. This increases the difficulty in accessing those audio data. Hence, there is a need of efficient indexing and annotation mechanisms. Non-stationarity and discontinuity present in the audio signal rise the difficulty in segmentation and classification of audio signals. The other challenging task is to extract and select the optimal features in audio signal. The application areas of audio classification and retrieval system include speaker recognition, gender classification, music genre classification, environment sound classification, etc. This paper proposes a machine learning- and neural network-based approach which performs audio pre-processing, segmentation, feature extraction, classification and retrieval of audio signal from the dataset. We have proposed novel approach of classification and retrieval using FPNN by combining fuzzy logic and PNN characteristics. We found that FPNN classifier gives better accuracy, F1-score and Kappa coefficient values compared to SVM, k-NN and PNN classifiers.

Keywords SVM · k-NN · PNN · FPNN · Accuracy · Recall · Precision

1 Introduction

In this paper, an approach for classification of audio signals using machine learning and neural network and retrieval of audio signal is proposed. In audio signal

N. M. Patil (✉)
Computer Engineering, Pacific Academy of Higher Education and Research University,
Udaipur, India
e-mail: nileshdeep@gmail.com

Fr. CRCE, Mumbai, India

M. U. Nemade
Electronics Engineering Department, K. J. Somaiya Institute of Engineering
and Information Technology, Sion, Mumbai, India
e-mail: mnemade@somaiya.edu

© Springer Nature Singapore Pte Ltd. 2019
S. K. Bhatia et al. (eds.), *Advances in Computer Communication
and Computational Sciences*, Advances in Intelligent Systems and Computing 924,
https://doi.org/10.1007/978-981-13-6861-5_23

processing applications, segmentation and classification play a vital role. Audio segmentation [1] is an essential pre-processing step widely used in various applications like audio archive management, surveillance, medical applications, entertainment industry, etc. However, the traditional segmentation techniques like decoder-based segmentation approach which only place the edges at the silence locations and metric-based approach which empirically sets threshold value are easy and quite simple. Hence, there is a need to perform segmentation in an efficient manner. The feature extraction [2] process gives meaningful information about the audio signal and helps in classifying those audio signals. Feature extraction can be done in time domain, frequency domain and coefficient domain. But larger the number of audio signals in the dataset, more is the computational complexity of traditional feature extraction approaches. To index the audio signals in the dataset, audio classification is the popular approach used. For audio classification techniques, we need to consider both the single similarity measure and the perceptual similarity of audio signals.

In our proposed work, we have used averaging filter to filter the audio signals and reduce the Gaussian noise. Pitch extraction method is used for segmentation of filtered audio signal. In this approach, we have computed ZCR, STE, spectral flux and spectral skewness. Segmentation process divides the audio signal into voiced and unvoiced segments. In time domain, we have extracted features like root mean square (RMS), ZCR and silence ratio. In frequency domain, the features like bandwidth, spectrogram, spectral centroid and pitch are extracted. In coefficient domain, we have considered mel-based frequency cepstral coefficients (MFCCs) and linear predictive coding (LPC). Classification method with multiple labels and hierarchy is used to classify the audio signal as music signal, speech signal or the environment sound. In the proposed system, fuzzy probabilistic neural network (FPNN) provides better classification accuracy as compared to SVM, k-NN and PNN classifiers. Also, the Kappa coefficient and Dice coefficient values for FPNN are better than the other classifiers mentioned in this research.

2 Related Work

Muthumari and Mala [3] used PNN classifier to classify the audio signals from GTZAN dataset and achieved an accuracy of 96.2%. Nagavi et al. [4] achieved accuracy of 85% in time duration of 2–3 min using sort–merge technique for classification based on acoustic features of the audio signal. In [5], authors highlighted weighted MFCC (WMFCC) on GTZAN dataset and obtained precision value of 96.40% and better recall value. Al-Maathidi and Li [6] obtained 78.1% accuracy using MFCC and supervised neural network to classify audio signal into music, speech and other noise. Murthy and Koolagudi [7] used MFCC and artificial neural network (ANN) to classify Telugu audio clips into vocal with accuracy of 85.89% and non-vocal with accuracy of 88.52%. Zhang et al. [8] applied spectral decomposition technique on audio signals to achieve an accuracy of 84.1%. In [9], authors performed classification using correlation-intensive fuzzy c-means (CIFCM) algorithm and SVM classifier on compressed audio signals achieving an accuracy of 89.53%. In [10], authors performed classification on park, restaurant and tube station audio signals using SVM classifier and achieved an accuracy of 73%. Dhanalakshmi et al. [11] used AANN and GMM for classification of environment sounds, machine noise, etc., to achieve an accuracy of 93.1% for AANN and 92.9% for GMM. Riley et al. [12] performed vector quantization on dataset of 4000 songs with 90% accuracy. Park [13] had used centroid neural network (CNN) for classification of 2663 audio signals achieving accuracy 75.62%. Zahid and Hussain [14] applied bagged SVM and ANN on GTZAN dataset to obtain 98% classification accuracy. Mahana and Singh [15] performed classification of audio signals using k-NN and SVM with accuracy of 74.6 and 90%, respectively. Tzanetakis and Cook [16] classified music signals into ten genres using GMM and k-NN with 61% accuracy. Miotto and Lanckriet [17] used Dirichlet mixture model (DMM) and SVM to classify audio signals from CAL500 dataset and achieved precision of 0.475, recall of 0.235 and f-score of 0.285. In [18], authors applied fuzzy c-means algorithm on GTZAN dataset for classifying music signals with an accuracy of 84.17%. Dhabarde and Deshpande [19] used local discriminant bases to classify audio signals as artificial sounds, natural sounds, instrumental music and speech with an accuracy of 95, 93, 97 and 95%, respectively. Baniya et al. [20] used timbral texture, rhythmic content features, MFCC and extreme learning machine (ELM) on GTZAN dataset and achieved accuracy of 85.15%. Kesavan Namboothiri and Anju [21] applied dynamic time warping (DTW) method with SVM classifier to classify audio signals from MARSYAS web to obtain an accuracy of 96.2%. Rong [22] used SVM with Gaussian kernel and obtained a classification accuracy of 87.6% for general sounds and 86.3% for audio scenes. Kaur and Mohan [23] used MFCC feature vectors to classify music signal using SVM and back-propagation neural network (BPNN) classifiers achieving an accuracy of 83 and 93%, respectively. Hirvonen [24] used MFCC features with zero-phase component analysis to classify speech and music signals obtaining an accuracy of 95%. Singh et al. [25] applied fuzzy logic and knowledge-base filtering on audio signals from CMUSphinx4 library with classification accuracy of 70%.

From the above literature survey, we conclude that the accuracy of classification of audio signals is computed mostly by MFCC feature vectors along with one more classifier like SVM, ANN, GMM, etc. The experiment is carried on music signals and speech signals. The classification accuracy varies from a minimum value of 73% to a maximum value of 98%. But hardly any work is done using PNN. We propose a classifier which combines PNN characteristics and also fuzzy logic for multi-label and multi-level classification of audio signals.

3 Proposed Approach

In this section, we explain the proposed system shown in Fig. 1. For creating a dataset of input audio signals, we have taken into consideration GTZAN dataset from MARSYAS web [26], which consists of 1000 music signals with 10 different genres and 64 speech signals. Each audio track in GTZAN dataset is 16-bit, 30-s-long and 22,050 Hz mono file in .au format. For environment sound classification, we have taken 200 audio clips classified in ten different classes in .wav format each of 5 s long from ESC-50 dataset [27].

Moving average filter is used for filtering the audio signal. In our proposed system, we have taken 16 samples at a time for averaging in the sliding window [28]. Average filter normalizes the amplitude and reduces the gaussian noise present in the audio signal.

Figures 2 and 3 show the plot of 60.au signal and signal after filtering, respectively.

The segmentation process separates voiced and unvoiced segments in the audio signal. For segmentation, we have used pitch extraction method wherein we had used ZCR and STE as time domain features and spectral flux and spectral skewness as frequency domain features.

A. **Zero-Crossing Rate**

Zero-crossing rate represents sign-changes rate in the signal [29]. Figure 4 shows the zero crossings in the filtered audio signal. A zero-crossing rate is calculated as follows:

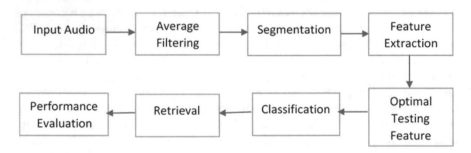

Fig. 1 Block diagram of proposed system

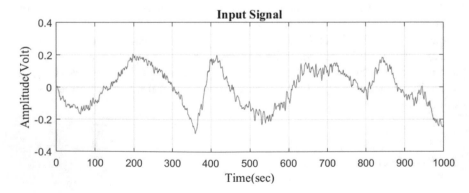

Fig. 2 Input audio signal 60.au (genre: Blues)

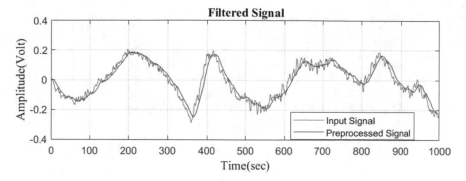

Fig. 3 Filtered audio signal

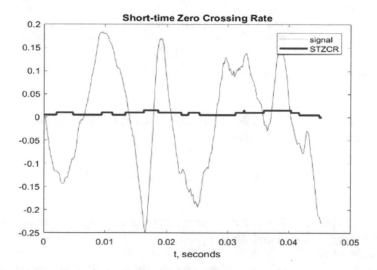

Fig. 4 Zero crossing in filtered signal

$$Z_n = \sum_{m=-\infty}^{\infty} |\text{sgn}[y(m)] - \text{sgn}[y(m-1)]| w(n-m) \qquad (1)$$

where

$$\text{sgn}[y(m)] = \begin{cases} 1, & \text{if } y(m) \geq 0 \\ -1, & \text{if } y(m) < 0 \end{cases} \qquad (2)$$

$$w(n) = \begin{cases} \frac{1}{2N}, & \text{if } 0 \leq n \leq N-1 \\ 0, & \text{otherwise} \end{cases} \qquad (3)$$

and $y(m)$ is the time domain signal for frame m.

B. **Short-Time Energy**

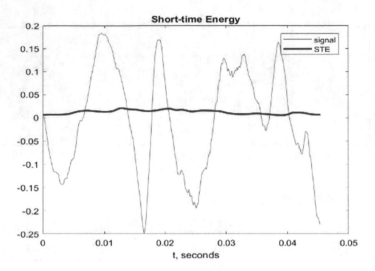

Fig. 5 Short-time energy of filtered signal

The energy of the signal is the representation of amplitude variations [29]. Short-time energy can be defined as:

$$E_n = \sum_{m=-\infty}^{\infty} [x(m)w(n-m)]^2 \tag{4}$$

In our proposed system, we used rectangular window. Figure 5 shows the short-time energy in the filtered audio signal.

C. **Spectral Flux**

The average variation value of spectrum between two adjacent frames in a given clip is called spectral flux (SF) [30]. The spectral flux is calculated as follows:

$$SF = \frac{1}{(N-1)(k-1)} \sum_{n=1}^{N-1} \sum_{k=1}^{k-1} [\log A(n,k) - \log A(n-1,k)]^2 \tag{5}$$

where A(n, k) is the discrete Fourier transform of the nth frame of input signal.

$$A(n,k) = \left| \sum_{m=-\infty}^{\infty} x(m)w(nL-m)e^{j\frac{2\pi}{L}Km} \right| \tag{6}$$

$x(m)$ is the original audio data, $w(m)$ is the window function, L is the window length, k is the order of discrete Fourier transform (DFT), and N is the total number of frames. Figure 6 shows the plot of spectral flux of the pre-processed audio signal.

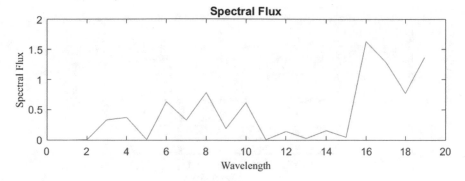

Fig. 6 Spectral flux

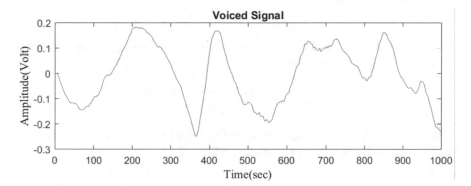

Fig. 7 Voiced segment of filtered audio signal

D. **Spectral Skewness**

Skewness represents the pitch in the audio signal. Skewness indicates more energy on the higher and lower parts of the spectrum [31].

Figure 7 shows the voiced segment of the filtered audio input signal.

The next step is the feature extraction. In time domain, we have estimated root mean square (RMS), zero-crossing rate and pitch saliency ratio using autocorrelation.

A. **RMS**

The RMS represents the square root of the average power of the audio signal for a given period of time. It is calculated as follows:

$$\text{RMS}_j = \sqrt{\frac{1}{N} \sum_{m=1}^{N} x_j^2(m)} \tag{7}$$

where $x_j(m)$ for $(m = 1, 2, …, N)$ is jth frame of windowed audio signal of length N.

B. **Pitch silence ratio**

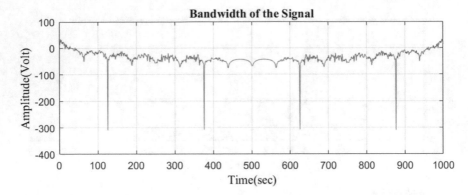

Fig. 8 Bandwidth of the pre-processed signal

It is the ratio of silent frames (determined by a preset threshold) and the entire frames. It is calculated as follows:

$$SR = \frac{\text{Number of silence frames}}{\text{Total number of frames}} \tag{8}$$

In frequency domain, we have estimated features like bandwidth, spectrogram, frequency centroid, spectral centroid and pitch.

A. **Bandwidth**

It represents the frequency up to which the signal holds information. Figure 8 shows the bandwidth plot of the pre-processed audio signal. It is calculated using the formula:

$$B_j = \sqrt{\frac{\int_0^{\omega_0} (\omega - \omega_c)|X_j(\omega)|^2 d\omega}{\int_0^{\omega_0} |X_j(\omega)|^2 d\omega}} \tag{9}$$

B. **Spectrogram**

Spectrogram splits the signal into overlapping segments, windows each segment with the hamming window and forms the output with their zero-padded, N points discrete Fourier transforms. Figure 9 shows the spectrogram of the pre-processed audio signal.

C. **Frequency Centroid**

It deals with the brightness in the signal and is calculated using the formula:

$$\omega_c j = \frac{\int_0^{\omega_0} \omega|X_j(\omega)|^2 d\omega}{\int_0^{\omega_0} |X_j(\omega)|^2 d\omega} \tag{10}$$

D. **Spectral Centroid**

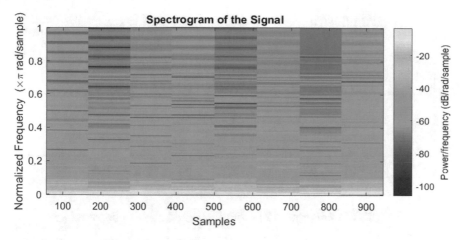

Fig. 9 Spectrogram of audio signal

Centroid deals with the sound sharpness, i.e. high-frequency components of the spectrum. The spectral centroid is calculated using the formula:

$$C_r = \frac{\sum_{k=1}^{N/2} f[k]|X_r[k]|}{\sum_{k=1}^{N/2} |X_r[k]|} \qquad (11)$$

where $f[k]$ is the frequency at bin k.

E. **Pitch**

Pitch refers to the fundamental period of a human speech waveform.

F. **Salience of Pitch**

The ratio of first pitch (peak) value and the zero-lag value of the autocorrelation is called as pitch saliency ratio. It is defined by the function $\frac{\emptyset_j(P)}{\emptyset(0)}$.

$$\emptyset_j(P) = \sum_{m=-\infty}^{\infty} x_j(m)x_j(m-P), \emptyset(0) = \sum_{m=-\infty}^{\infty} x^2(m)^2 \qquad (12)$$

In coefficient domain, we have estimated mel-frequency cepstral coefficients (MFCCs) and linear predictive coding (LPC) coefficients.

A. **MFCC**

MFCCs are the coefficients obtained in the mel-frequency cepstrum. These are computed from the FFT power coefficients. We adopt the first 12 orders of coefficients, out of which first three are used in building fuzzy inference system (FIS) for classification. At the end, we use sinusoidal lifter to suppress higher cepstral coefficients. This achieves better recognition performance for clean speech. Figure 10 shows the plot of MFCCs before liftering. The relation between the frequency and the mel scale is expressed as follows:

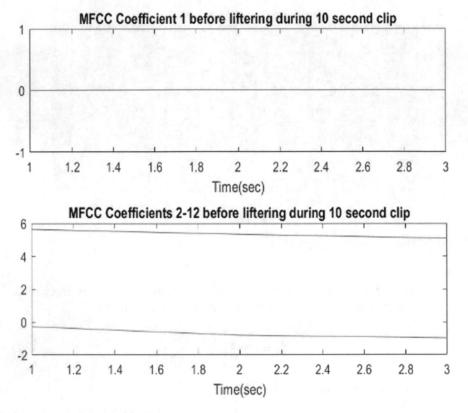

Fig. 10 MFCCs before liftering

$$m = 2595 \log_{10}\left(1 + \frac{f}{700}\right) = 1127 \ln\left(1 + \frac{f}{700}\right) \tag{13}$$

B. **LPC**

The LPC coefficients are a short-time measure of the speech signal. The plot of 1000 LPC coefficients of voiced segment using autocorrelation method is shown in Fig. 11. The values range from −1.0209 to +1.0000.

Figure 12 summarizes the values of the features extracted for the classification purpose.

The next step in the proposed system is classification process. In this process, we have classified the audio signals into three main classes, viz. music, speech and environment sound. Further, the music signals are classified into ten genres as in [26]. Speech signals are further categorized into the gender male or female and also get the age of the speaker [32]. Environment sounds are also categorized into ten classes as Chirping Birds, Church Bell Ring, Clapping, Crying Babies, Door-Wood-Knock, Frog, Glass Breaking, Keyboard Typing, Train and Water Drops. For classification, we have designed FIS taking into consideration features like mean of spectrogram,

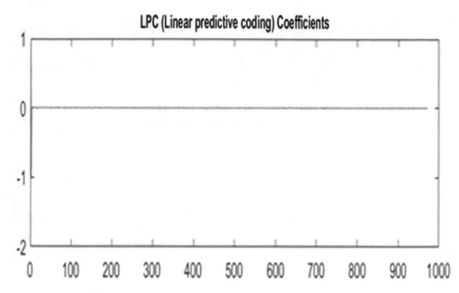

Fig. 11 LPC coefficients of voiced segment

	Extracted features
Mean of Spectral Flux	0.4590
Spectral Skewness	-0.1041
Root Mean Square	0.1130
Zero Cross Rate	0.0044
Mean of Bandwidth	-12.5062
Mean of Spectrogram	0.0058
Frequency Centroid	15.0718
Pitch	22.0500
MFCC-1	-9.5174
MFCC-2	-0.0807
MFCC-3	1.1055
Mean of LPC	6.1936e-05
Standard Deviation of Pitch Saliency ratio	0
Mean of Pitch Saliency ratio	0

Fig. 12 Features extracted

Fig. 13 SVM hyperplane

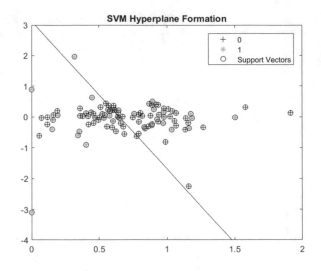

mean and standard deviation of pitch saliency ratio, mean of pitch, mean of zero crossing, mean of first MFCCs, mean of third MFCCs for fuzzy logic rules. We have considered four different classifiers for classification of audio signals. They are SVM, k-NN, PNN and FPNN.

A. **SVM**

SVM is a supervised machine learning algorithm that classifies data by finding the linear decision boundary (hyperplane) separating all the datapoints of one class from those of the other class as shown in Fig. 13. In the training phase, SVM receives some extracted speech features as input. In the testing phase, SVM constructs a hyperplane which can be used for classification. We have adopted linear kernel function of SVM in classification. In Fig. 13, 0 indicates true positive and 1 indicates true negative.

B. **k-NN**

k-NN categorizes objects based on the classes of their nearest neighbours in the dataset. We have used Euclidean distance to find the nearest neighbour. We have taken value of k as 3. Figure 14 shows the k-NN classification.

C. **PNN**

PNN is feed-forward neural network. It is widely used in classification and pattern recognition problems. A PNN is a multi-layered feed-forward network with four layers as shown in Fig. 15. Figure 16 shows how PNN divides the input space into three classes.

D. **FPNN**

In FPNN, the network is a combination of probabilistic and generalized regression neural networks and is capable of data classification on the basis of fuzzy decision on the membership of a particular observation to a certain class. Fuzzy probabilistic neural network (FPNN) is shown in Fig. 17.

Fig. 14 k-NN classification

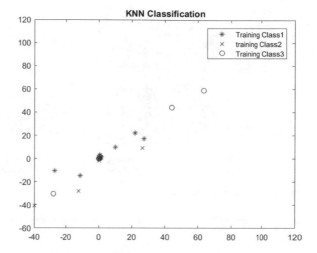

Fig. 15 Architecture of PNN classifier

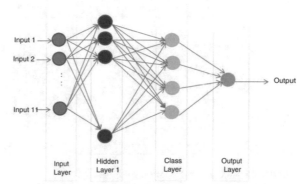

Fig. 16 PNN classification into three classes

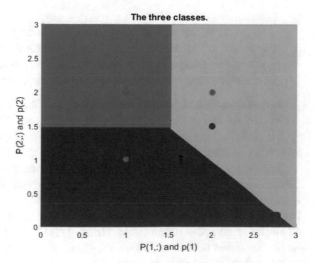

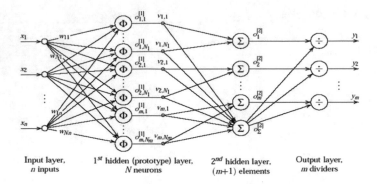

Fig. 17 Architecture of FPNN

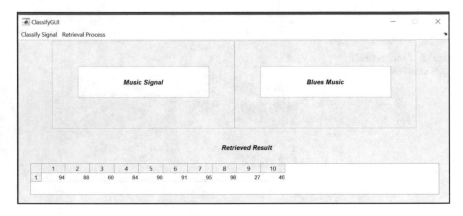

Fig. 18 Classification GUI for music signal

The output of the classification process is shown in Fig. 18. The input signal (numbered 60 in dataset in .au format) is classified as Music Signal and the genre of that music signal is classified as Blues. Similarly, the input signal can be classified into Speech Signal giving its gender and age as shown in Fig. 19.

In case of speech signal, we have considered 1026.au from dataset as input signal which is classified as Female Voice with age between 10 and 25.

In the similar manner, we classified 1140.wav signal as Environment Sound Signal which is classified as Clapping as shown in Fig. 20.

The next step is the retrieval process. Here, we have calculated the Euclidean distance among the feature vectors and compared the input audio signal with the audio files in the dataset. This step is basically to perform optimization to get the ten best audio files that closely match the feature vectors of the input audio file and also to get the input audio file in those retrieved result. We had taken 60.au as the input audio file in this paper and in Fig. 18, lower part denotes the retrieved result which contains the number of input audio file.

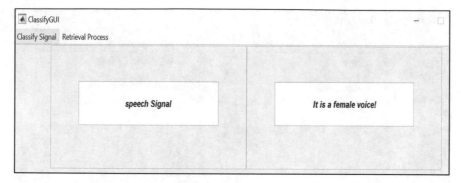

Fig. 19 Classification GUI for speech signal

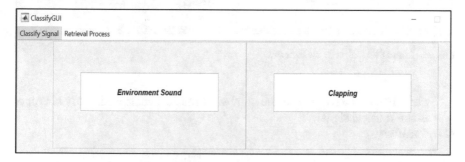

Fig. 20 Classification GUI for environment sound

4 Performance Analysis

In this section, we illustrate various performance parameters for different classifiers. Figure 21 shows the confusion matrix of different classifiers. We have evaluated metrics such as precision, recall, sensitivity, specificity, Jaccard's coefficient, Dice coefficient, Kappa coefficient and accuracy.

A. **Precision**

$$\text{Precision} = \frac{TP}{(TP + FP)} \quad (14)$$

where true positive (TP) is the number of audio signals correctly classified as music, speech or environment sound and false positive (FP) is the number of audio signals incorrectly classified, for example music signal classified as speech signal.

B. **Recall**

$$\text{Recall} = \frac{TP}{(TP + FN)} \quad (15)$$

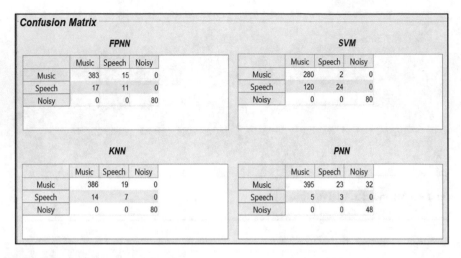

Fig. 21 Confusion matrix of classifiers

where false negative (FN) is the number of music signals incorrectly classified as non-musical signal.

C. **Sensitivity**

$$\text{Sensitivity}(\%) = \frac{\text{TP}}{\text{TP} + \text{FN}} \times 100 \tag{16}$$

D. **Specificity**

$$\text{Specificity}(\%) = \frac{\text{TN}}{\text{TN} + \text{FP}} \times 100 \tag{17}$$

where true negative (TN) is the number of audio signals incorrectly classified as music, speech or environment sound.

E. **Jaccard's Coefficient**
It is a measure of similarity for the two sets of data, with a range from 0 to 100%.

$$J(A, B) = \frac{|A \cap B|}{|A \cup B|} \tag{18}$$

F. **Dice Coefficient**
The harmonic mean of precision and recall is called the Dice coefficient. It is also called F1-score. Higher value of F1-score indicates improved efficiency in segmentation and classification.

$$\text{Dice Coefficient(F1-Score)} = \frac{2 \times \text{Recall} \times \text{Precision}}{\text{Recall} + \text{Precision}} \tag{19}$$

	FPNN	SVM	KNN	PNN
TP	383	280	386	395
TN	91	104	87	51
FP	17	120	14	5
FN	15	2	19	55
Sensitivity	96.2312	99.2908	95.3086	87.7778
Specificity	84.2593	46.4286	86.1386	91.0714
Precision	95.7500	70	96.5000	98.7500
Recall	96.2312	99.2908	95.3086	87.7778
Jaccard Coefficient	92.2892	69.6517	92.1241	86.8132
Dice Coefficient	95.9900	82.1114	95.9006	92.9412
Kappa Coefficient	0.8104	0.4834	0.7996	0.5669
Accuracy	93.6759	75.8893	93.4783	88.1423

Fig. 22 Performance metrics of classifiers

G. **Kappa Coefficient**

The Kappa statistic is a metric that is based on the difference between the observed accuracy and the expected accuracy that the samples randomly would be expected to reveal [33]. It is calculated as

$$k = \frac{O_a - E_a}{1 - E_a} \tag{20}$$

where O_a is the observed accuracy and E_a is the expected accuracy.

H. **Accuracy**

$$\text{Accuracy}(\%) = \frac{\text{Number of correctly classified results}}{\text{Total number of classified results}} \times 100 \tag{21}$$

Figure 22 shows the performance metric values obtained after executing the system for input audio signal 60.au.

5 Conclusions

In this paper, we had presented an approach for classification and retrieval of preprocessed audio signals. We had proposed a novel approach for classification using FPNN and compared it with classifiers like SVM, k-NN and PNN. We had used GTZAN dataset for music and speech signals and ESC-50 dataset for environment sounds. The music signals are classified into ten genres. For speech signals, we identified the speaker gender and their age. Environment sounds are also classified into ten labels as Chirping Birds, Church Bell Ring, Clapping, Crying Babies, Door-Wood-Knock, Frog, Glass Breaking, Keyboard Typing, Train and Water Drops. Of the total 1264 audio signals, we had used 60% of samples as training set and 40% samples as testing set. We found that FPNN provides higher classification accuracy of 93.6759%. The Jaccard, F1-score and Kappa coefficient values are also high for FPNN classifier.

References

1. Castán, D., Tavarez, D., et al.: Albayzín-2014 evaluation: audio segmentation and classification in broadcast news domains. EURASIP J. Audio Speech Music Process. **33**, 1–15 (2015)
2. Ludeña-Choez, J., Gallardo-Antolín, A.: Feature extraction based on the high-pass filtering of audio signals for acoustic event classification. J. Comput. Speech Lang. **30**(1), 32–42 (2015)
3. Muthumari, A., Mala, K.: An efficient approach for segmentation, feature extraction and classification of audio signals. J. Circuits Syst. **7**, 255–279 (2016)
4. Nagavi, T.C., Anusha, S.B., Monisha, P., Poornima, S.P.: Content based audio retrieval with MFCC feature extraction, clustering and sort-merge techniques. In: Proceedings of IEEE 4th International Conference on Computing, Communications and Networking Technologies, July 2013, pp. 1–6
5. Christopher Praveen Kumar, R., Suguna, S., Becky Elfreda, J.: Audio retrieval based on cepstral feature. Int. J. Comput. Appl. **107**(17), 28–33 (2014). ISSN: 0975-8887
6. Al-Maathidi, M.M., Li, F.F.: NNET based audio content classification and indexing system. Int. J. Digit. Inf. Wirel. Commun. (IJDIWC) **2**(4), 335–347 (2012). ISSN: 2225-658X
7. Srinivasa Murthy, Y., Koolagudi, S.G.: Classification of vocal and non-vocal regions from audio songs using spectral features and pitch variations. In: Proceedings of IEEE 28th Canadian Conference on Electrical and Computer Engineering (CCECE), Halifax, May 2015, pp. 1271–1276
8. Zhang, X., Su, Z., Lin, P., He, Q., Yang, J.: An audio feature extraction scheme based on spectral decomposition. In: Proceedings of IEEE International Conference on Audio, Language and Image Processing (ICALIP), Shanghai, July 2014, pp. 730–733
9. Haque, M.A., Kim, J.M.: An enhanced fuzzy C-means algorithm for audio segmentation and classification. Int. J. Multimed. Tools Appl. **63**(2), 485–500 (2013)
10. Geiger, J.T., Schuller, B., Rigoll, G.: Large-scale audio feature extraction and SVM for acoustic scene classification. In: Proceedings of IEEE Workshop on Applications of Signal Processing to Audio and Acoustics (WASPAA), New Paltz, Oct 2013, pp. 1–4
11. Dhanalakshmi, P., Palanivel, S., Ramalingam, V.: Classification of audio signals using AANN and GMM. Appl. Soft Comput. **11**(1), 716–723 (2011)
12. Riley, M., Heinen, E., Ghosh, J.: A text retrieval approach to content-based audio retrieval. In: Proceedings of ISMIR 9th International Conference on Music Information Retrieval, Sept 2008, pp. 295–300
13. Park, D.-C.: Content-based retrieval of audio data using a Centroid Neural Network. In: Proceedings of IEEE International Symposium on Signal Processing and Information Technology (ISSPIT), South Korea, Dec 2010, pp. 394–398
14. Zahid, S., Hussain, F., Rashid, M., Yousaf, M.H., Habib, H.A.: Optimized audio classification and segmentation algorithm by using ensemble methods. Math. Problems Eng. **2015**, 1–11 (2015). Article ID 209814
15. Mahana, Poonam, Singh, Gurbhej: Comparative analysis of machine learning algorithms for audio signals classification. Int. J. Comput. Sci. Netw. Secur. (IJCSNS) **15**(6), 49–55 (2015)

16. Tzanetakis, G., Cook, P.: Musical genre classification of audio signals. IEEE Trans. Speech Audio Process. **10**(5) (2002)
17. Miotto, R., Lanckriet, G.: A generative context model for semantic music annotation and retrieval. IEEE Trans. Audio Speech Lang. Process. **20**(4), 1096–1108 (2012)
18. Haque, Mohammad A., Kim, Jong-Myon: An analysis of content-based classification of audio signals using a fuzzy c-means algorithm. J. Multimed. Tools Appl. **63**(1), 77–92 (2013)
19. Dhabarde, S.V., Deshpande, P.S.: Feature extraction and classification of audio signal using local discriminant bases. Int. J. Ind. Electron. Electr. Eng. **3**(5), 51–54 (2015). ISSN: 2347-6982
20. Baniya, B.K., Ghimire, D., Lee, J.: Automatic music genre classification using timbral texture and rhythmic content features. ICACT Trans. Adv. Commun. Technol. (TACT) **3**(3), 434–443 (2014)
21. Kesavan Namboothiri, T., Anju, L.: Efficient audio retrieval using SVM and DTW techniques. Int. J. Emerg. Technol. Comput. Sci. Electron. (IJETCSE) **23**(2) (2016)
22. Rong, F.: Audio classification method based on machine learning. In: IEEE Proceedings of International Conference on Intelligent Transportation, Big Data & Smart City, pp. 81–84 (2016)
23. Kour, G., Mehan, N.: Music genre classification using MFCC, SVM and BPNN. Int. J. Comput. Appl. **112**(6) (2015)
24. Hirvonen, T.: Speech/music classification of short audio segments. In: IEEE Proceedings of International Symposium on Multimedia, pp. 135–138 (2014)
25. Singh, M., Tiwary, U.S., Siddiqui, T.J.: A speech retrieval system based on fuzzy logic and knowledge-base filtering. In: IEEE Proceedings of International Conference on Multimedia, Signal Processing and Communication Technologies (IMPACT), Nov 2013, pp. 46–50
26. GTZAN Dataset: http://marsyasweb.appspot.com/download/data_sets/
27. ESC-50 Dataset: https://github.com/karoldvl/ESC-50
28. Smith, S.W.: The Scientist and Engineer's Guide to Digital Signal Processing, pp. 277–284
29. Sunitha, R.: Separation of unvoiced and voiced speech using zero crossing rate and short time energy. Int. J. Adv. Comput. Electron. Technol. (IJACET) **4**(1), 6–9 (2017). ISSN: 2394-3416
30. Thiruvengatanadhan, R., Dhanalakshmi, P., Suresh Kumar, P.: Speech/music classification using SVM. Int. J. Comput. Appl. **65**(6), 36–41 (2013). ISSN: 0975-8887
31. Radha Krishna, S., Rajeswara Rao, R.: SVM based emotion recognition using spectral features and PCA. Int. J. Pure Appl. Math. **114**(9), 227–235 (2017). ISSN: 1314-3395
32. https://xpertsvision.wordpress.com/2015/12/04/gender-recognition-by-voice-analysis/
33. http://shodhganga.inflibnet.ac.in/bitstream/10603/150477/12/12_chapter%204.pdf

Information Extraction from Natural Language Using Universal Networking Language

Aloke Kumar Saha, M. F. Mridha, Jahir Ibna Rafiq and Jugal K. Das

Abstract Contemporaneous research has strongly indicated that most of the data on the Internet are unstructured data due to the phenomenon that during the input, processing of data and collection and storage of data by almost all the entities involved do not keep the data in a format that complies with a certain structure; this scenario has a domino effect on retrieving information should there be any inquiry. A part and parcel of semantic web area is data extraction and crucial for linking question and answer in the web. Should a question is pitched, it requires semantic analysis of data—both, structured and unstructured, map each part of the answer to the relevance of the question. Information extraction entails a crucial area of natural language processing and without the proper application of data acquisition from really large data set, for instance billions of alphanumeric words—the required data are hardly ever on the receiving end. The practical application, however, certainly needs answers that are succinct, correct and to the point; often times, the readers would skim-read through each answer as they themselves have to decide on which is more accurate to their question. This poses a unique challenge, a scenario where the question is incomplete; the answer is hidden under layers of data, and to make the query even more complex, researchers add the languages that are available. For English, a lot of researches have been conducted and due to the exceptional amount of usage among all the entities alike, English language has passed the initial issues and has been producing nearly ninety-nine percent accurate data. That is not the case for Bengali semantic analysis, and deriving meaningful information has been a challenge. This paper proposes a decisive algorithm to acquire meaningful and relevant data from

A. K. Saha (✉) · M. F. Mridha · J. I. Rafiq
Department of Computer Science and Engineering, University of Asia Pacific, Dhaka, Bangladesh
e-mail: aloke@uap-bd.edu

M. F. Mridha
e-mail: Firoz@uap-bd.edu

J. I. Rafiq
e-mail: Jahir@uap-bd.edu

J. K. Das
Department of Computer Science and Engineering, Jahangirnagar University, Dhaka, Bangladesh
e-mail: cedas@juniv.edu

© Springer Nature Singapore Pte Ltd. 2019
S. K. Bhatia et al. (eds.), *Advances in Computer Communication and Computational Sciences*, Advances in Intelligent Systems and Computing 924,
https://doi.org/10.1007/978-981-13-6861-5_24

unstructured data. The exactitude and efficiency of target data extraction depend on reasoning and analysis of unstructured data. Here, Universal Networking Language (UNL) has been applied to the proposed method to bring out the desired output. In this method, exceptionally large data sets that are unstructured have been categorized in prespecified relation with the help of UNL, and on these relations, every word of a sentence has been compared in binary relation. Finally, the proposed method extracts information from these binary relations.

Keywords Unstructured data · Natural language processing · Universal Networking Language · Target data

1 Introduction

The rise of information and technology has established a new fact—'information is money.' Internet is considered as the largest source of information. It is estimated that if 20% of the data available to enterprises, computer and the Internet are structured data, then rest of the data, that is 80%, are unstructured. So, the importance of extracting target data from unstructured data does not need any explanation.

Now, it has become a major issue to find very precise, specific and direct information from unstructured data. Search engines work well to find any information. A relevant case is Google, which is probably the most popular and the best search engine to provide information. However, should a search is conducted using Google, the results are listed according to the relevance of keywords in the original query, and the results are published in pages and chronologically appear before the user. Hence, users will have to read the results and, in some cases, rewrite their original query, fine-tune the results and match whichever is best aligned to their inquiry and therefore needs a human intervention that is both time-consuming and prone to mistakes. Sometimes, users fail to get the desired output and get confused in the gargantuan sets of search result. So, in this era, where time and information are considered 'money,' it is very common to look for a useful and smaller set of result. The smaller outcome we can obtain the better for us—it saves time as well as resource.

So, it has become very important to find a more precise, specific and useful way to extract target data from the vast set of unstructured data. But most of the algorithm to extract data performs very well either on numerical data or on structured data. But for extracting target data from natural language, they can solve some specific and limited problems and often lack perfection.

Our work on this topic, on the context of UNL, to derive the meaningful and reasonably succinct answer to a question from data that are not necessarily very organized in nature, will significantly reduce complexity in search engines. Furthermore, we are focusing on Bengali language, spoken by more than 300 million people worldwide, that facilitates the utilization of huge volume of Bengali information in fields such as the literature, where we have secured a couple of noble prizes, science and research, to non-native Bengalese and paves the way for further development of our native language. To obtain data from unstructured data, UNL has been applied as a medium to get UNL expression. The proposed model has extensively relied on UNL expressions. This paper is focused on introducing the advance system of finding information from unstructured data that most correctly matches with on either side of question keywords and the meaning of returned answer. The rest of this paper is organized as follows. Section 2 describes related work. Section 3 shows the fundamental working procedures of Universal Networking Language. Section 4 describes the working procedure of the proposed model. Section 5 demonstrates the experimental procedure. Section 6 shows result analysis, and Sect. 7 describes the conclusion and future work.

2 Literature Review

In the recent past, many researchers investigated different approaches to extract data from unstructured or semi-structured data.

In [1] Semantic Engine using Brain-Like Approach (SEBLA), the prospect to address all the key complexities of next-generation Internet-based data analysis that includes intelligent search, question answering and summarization from natural-language understanding has been explored; there are, however, a significant number of limitation in SEBLA. Jing Jiang researched on information extraction from text showing named entity recognition and relation extraction [2]. Kai Barks Chat proposed an ontology-based knowledge modeling for semantic data extraction [3].

Another approach [4] is proposed for extracting semantic relations from natural language text that uses dependency grammar patterns (DGPs). Raymond J. Mooney and Razvan Bunescu researched on mining knowledge from text using IE where systems can be used to directly extricate abstract knowledge from a text corpus or to extract concrete data from a set of documents which can then be further analyzed with keywords: unstructured data, natural language processing, Universal Networking Language and target data. Traditional data-mining techniques discover more general patterns [5].

Daniel Duma proposed the architecture for generating natural language from linked data that automatically learns sentence templates and statistical document planning from parallel RDF data sets and text [6]. Emdad Khan showed MLANLP for generalization capability of natural language computing.

3 The Universal Networking Language

UNL is a computer language or a form of human interactive language that permits computer to express and exchange any kind of information. UNL can represent semantic data [7] obtained from natural language texts. It also facilitates symbolizing sentences without any ambiguity in logical expressions. The purpose of UNL is to

Table 1 UNL relations with examples

UNL relation	Description	Constituent elements	Examples
Agt	States a thing that starts an action	Agent, intransitive verb like 'work' and 'go'	Karim slept agt(work, Karim)
And	States a partner to have conjunctive relation	Conjunction	Reading and writing and (read(agt > person), write(agt > person))
Aoj	States a thing that is in a state or has an attribute	Verbs like 'adore,' 'hate,' 'love,' 'love,' and 'involve'	Karim believes in Mony, Aoj (believes, Karim) Karim knows Mony, Aoj (knows, Karim)

```
{unl}
agt(eat(icl>consume>do,agt>living_thing,obj>concrete_thing,
ins>thing). @entry.@present,i(icl>person))
obj(eat(icl>consume>do,agt>living_thing,
obj>concrete_thing,ins>thing).@entry.@present,rice(icl>grain>thing))
{/unl}
```

Fig. 1 UNL expression example

build a platform that provides information; the result, consequently, should be derived from natural language with texts as well as numbers—preferably from unstructured data sets. In turns, UNL has the capability to modify data and present them in concise and meaningful sentences. Thus, a reputation of UNL has come forth as a reliable translation tool, since it enables translation without ambiguity [8].

In UNL, sentence information is showed as a hyper-graph [9] where the concepts are represented as nodes and relations as arcs. There will be always binary relation in the hyper-graph. There are 46 predefined relation levels in UNL [10]. Table 1 shows some UNL relations with examples.

With these relation levels, every word of natural language can be represented in UNL expression. In our proposed model, we take these UNL expressions as input.

UNL Expression: The 46 predefined binary relations are the building blocks of UNL expression. This has two universal words (UWs) that are UW1 and UW2. An UNL expression can be recognized by these following tags: unl Starting of UNL the expression/unl End of the UNL expression. The general descriptive structure of the UNL expression is given below: <relation> [:scope-ID>](<from-node>, <to-node>) [10].

For example, if we take a sentence ' "আমি ভাত খাই" ' which translates to 'I eat rice.' In UNL module it will be represented Fig. 1.

Here, there are two UNL expressions. In the first expression, relation between two UWs (UW1-verb, UW2-subject).

The second UNL expression relation between two UWs (UW1-verb, UW2-object).

These relations are constructed based on the input that is given to UNL [10].

4 Proposed Model

Our proposed model correctly obtains the UNL expression from the input by properly utilizing UNL relation; the relations are used from the given 46 use-cases and widely used by UNL-related work and afterward apply our algorithm, that is constructed precisely for data extraction, on the UNL expressions. The length of the sentence is irrelevant because each word will be paired in a binary relationship and our method is to apply the fixed and predefined relation level already established by the UNL. UNL will break the sentence into UNL expressions. In each UNL expression, it will contain binary relation according to the given sentence [11]. And these UNL expressions are the input of our proposed model.

When we get the input, it will clearly have relation levels. So, when user will look for the target data, according to the input it will go to that particular relation level. After that, by extracting UW the user will get the desired output. Figure 2 shows the workflow of the proposed model.

5 Experimental Procedure

Algorithm to extract target data:

A.1:
For the relation 'agt', it defines a thing which initiates an action. Agent is defined as the relation between:

  UW1—do and
  UW2—a thing

When we are looking for that person who has done this, it will go to the 'agt' relation, and then if we extract the UW2, we will get the person or things' name [12].

Example: 'John breaks' using 'agt' relation

$$\text{agt}(\text{break}(\text{icl} > \text{do}), \text{John}(\text{icl} > \text{person})) \tag{1}$$

Equation (1) shows who break by extracting UW2, so that it will give the desire output that the person name is 'John.'

Example: 'computer translates' using 'agt' relation

$$\text{agt}(\text{translate}(\text{icl} > \text{do}), \text{computer}(\text{icl} > \text{machine})) \tag{2}$$

Fig. 2 Workflow of the
proposed model

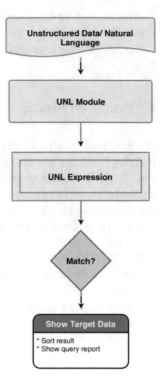

Equation (2) shows which translates (it could be anything) by extracting UW2 so that it will give the expected output 'Computer.' Figure 2 shows the workflow of the proposed model for relation 'agt'.

A.2:
For the relation 'frm', it defines an origin of a thing. From is defined as the relation between:

UW1—a thing and
UW2—an origin of the thing

So, if we are looking for the person who came from somewhere it will go to the 'frm' relation, and by extracting the UW1 we will get the person or thing. And when we want to find from where the thing came, it will extract the UW2.

Example: '… man from Japan' using 'frm' relation

$$frm(man(icl > person), Japan(icl > country)) \tag{3}$$

From Eq. (3) by extracting UW2, we can easily find the desired output from where the man came.

A.3:

For the relation 'ins', it defines the instrument to carry out an event. Instrument is defined as a relation between:

UW1—an event and
UW2—a concrete thing

So, if we are looking for an object or a concrete thing that was used to do an event, it will go for the 'ins' relation. And after that by extracting UW2, the target data will be got.

Example: '… eat … with knife' using 'ins' relation

$$ins(eat(icl > do), knife(icl > thing)) \tag{4}$$

From Eq. (4), if we look forward to extract the data that which instrument was used for eating by extracting UW2, it will give the target data.

A.4:

For the relation 'cag', it defines a thing not in focus which initiates an implicit event which is done in parallel. Co-agent is defined as the relation between:

UW1—an action and
UW2—a thing

So, when we are looking for a thing whose actions are not focused but it accompanies UW1, then it will go to the 'cag' relation, and then by extracting UW2 we are expecting to get the things' name [13].

Example: '… walk with John' using 'cag' relation

$$cag(walk(icl > do), John(icl > person)) \tag{5}$$

Equation (5) described the person who walks with UW1. So when we want to find the name who walks with UW1 by extracting the word UW2, we will get the person or things' name.

Example: '… lives with aunt' using 'cag' relation

$$cag(live(icl > do), aunt(icl > person)) \tag{6}$$

Equation (6) described with whom UW1 lives. So when we want to find the name who lives with UW1 by extracting UW2, we will get the person or things' name.

A.5:

For the relation "plf" states where an event begins or a state becomes true. Place-from or initial place is defined as the relation between:

UW1—an event or state and
UW2—a place or thing defining a place.

Table 2 Result obtained by implementing proposed model

Keyword	Length of text	UNL relation	Matching found	Residual correct
Who	More than 500 words	Mod	19 times	1 time
Where	More than 250 words	Plc, plf, plt	11 times	1 time
From-to	More than 250 words	Fmt	8 times	1 time
With	More than 250 words	Pnt	9 times	1 time
And	More than 700 words	Conj	20 times	1 time

So when it will find an input that shows the initial place of verbs (go, travel, walk, come, etc.) or if we are looking for the place from where the event began, it will go for the 'plf' relation. And after that by extracting UW2, the target data will be got.

Example: '… call from Japan using 'plf' relation

$$plf(call(icl > do), Japan(icl > place)) \tag{7}$$

From Eq. (7), we can see if we extract the UW2, it gives the target data of initial place.

6   Result Analysis

In this paper, we focused on extracting only the target data from unstructured data. For this we used Bengali sentences as the input of the UNL module. After that, when the UNL module gives the UNL expression, we carry that as the input, apply UNL expressions on the sentences again and map to the words that match most accurately to our searching keywords. We have explained 5 relations from the 46 relations of UNL. For each of these 5 relations, we have given input of more than 200 sentences of natural language. So, we have tested more than 1000 sentences. After testing these huge amounts of data set, we came to the conclusion that if we want to extract any data from a data set, it is essential to extract the UW2.

Empirical evidences from the experiments show encouraging signs of improvement and prospect of wider impact on information extraction regardless of sources. To summarize our experiment, we have put together a table with relevant and most interesting data obtained during testing that is shown in Table 2.

Let us consider a full question: ' "পাখিটি খাঁচা থেকে উড়ে যায়" means 'the bird flies from the nest.' The equivalent UNL expression of above sentence is in Fig. 1.

Here, we have found two UNL relations 'obj' and 'plf'. 'obj' relation is used to make relation with word 'bird,' and 'plf' relation is used to make relation with 'nest.' Using the properties of UNL relation, we can make a question using 'plf' relation. 'plf' relation helps us to develop the final answer— পাখিটি কোথা থেকে উড়ে গেল? (Where did the bird fly from?).

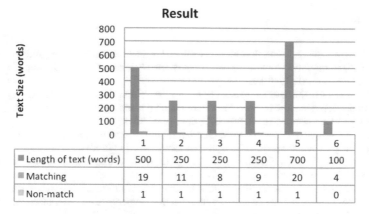

Fig. 3 Results of the proposed method

Let us consider another simple example. Someone named Ratan resides in Dhaka. So, the Bengali pronunciation will be ' "রাতন ঢাকা থাকে।"' and English translation should be 'Ratan lives in Dhaka.' Here, we have found two UNL relations 'aoj' and 'plc'. 'aoj' relation is used to make relation with words 'lives' and 'Ratan,' and 'plc' relation is used to make relation with 'where' and 'Dhaka.' Using the properties of UNL relation, we can make a question using 'plc' relation. 'plc' relation helps us develop a correct question. Table 2 shows the UNL relations that are used to generate question.

Figure 3 and Table 2 show the clear sign of improvement in comparison with extraction methods such as conditional random field (CRF) technique by Peng et al. [14] and features rules learning (FRL) technique by Chen et al. [15], and it is by far producing a lot of consistent outcome, although there are still rooms for improvement in cases where UNL relation is very difficult to establish.

7 Conclusions

It is a fundamental step toward removing linguistic barrier to work on the major issues to obtain information that are requested by the user(s) from unstructured data or natural language. Our proposed model can extract the target information from unstructured data, although from single UNL expression. In our proposed method, we have analyzed the binary relation between the words of a sentence. For this, we used UNL as a medium. After getting the UNL expression, we described what relation should be chosen according to the input from the user. Once it has obtained the relation by extracting the described UW, it will get the desired or target information [16]. In the future, we would like to work for multiple translation layers, where a sentence from Bengali can be translated to more than one language using one single UNL expression.

References

1. Khan, E.: Machine learning algorithms for natural language semantics and cognitive computing. In: 2016 International Conference on Computational Science and Computational Intelligence (CSCI), pp. 1146–1151. IEEE (2016)
2. Jiang, J.: Information extraction from text. In: Mining Text Data, pp. 11–41. Springer (2012)
3. Barkschat, K.: Semantic information extraction on domain specific data sheets. In: European Semantic Web Conference, pp. 864–873. Springer (2014)
4. Akbik, A., Bro, J.: Wanderlust: extracting semantic relations from natural language text using dependency grammar patterns. In: www workshop, vol. 48 (2009)
5. Mooney, R.J., Bunescu, R.: Mining knowledge from text using information extraction. ACM SIGKDD Explor. Newsl. **7**(1), 3–10 (2005)
6. Duma, D., Klein, E.: Generating natural language from linked data: unsupervised template extraction. In: Proceedings of the 10th International Conference on Computational Semantics (IWCS 2013)–Long Papers, pp. 83–94 (2013)
7. Heath, T., Bizer, C.: Linked data: evolving the web into a global data space. Synth. Lect. Semant. Web: Theory Technol. **1**(1), 1–136 (2011)
8. Saha, A.K., Mridha, M., Rafiq, J.I., Das, J.K.: Data extraction from natural language using universal networking language. In: International Conference on Current Trends in Computer, Electrical, Electronics and Communication (ICCTCEEC), 8–9 Sept 2017
9. Saha, A.K., Akhtar, S., Mridha, M.F., Das, J.K.: Attribute analysis for Bangla words for universal networking language (UNL). Editorial Preface **4**(1) (2013)
10. Uchida, H., Zhu, M., Della Senta, T.: UNL: A Gift for a Millennium. The United Nations University (2000)
11. Klein, D., Manning, C.D.: Accurate unlexicalized parsing. In: Proceedings of the 41st Annual Meeting on Association for Computational Linguistics, vol. 1, pp. 423–430. Association for Computational Linguistics (2003)
12. Gagnon, M., Da Sylva, L.: Text compression by syntactic pruning. In: Conference of the Canadian Society for Computational Studies of Intelligence, pp. 312–323. Springer (2006)
13. Cohn, T.A., Lapata, M.: Sentence compression as tree transduction. J. Artif. Intell. Res. **34**, 637–674 (2009)
14. Peng, F., McCallum, A.: Accurate information extraction from research papers using conditional random fields. Retrieved on 13 Apr 2013
15. Chen, J., Chen, H.: A structured information extraction algorithm for scientific papers based on feature rules learning. JSW **8**(1), 55–62 (2013)
16. Filippova, K., Strube, M.: Dependency tree based sentence compression. In: Proceedings of the Fifth International Natural Language Generation Conference, pp. 25–32. Association for Computational Linguistics (2008)

On the Knowledge-Based Dynamic Fuzzy Sets

Rolly Intan, Siana Halim and Lily Puspa Dewi

Abstract In 1965, Zadeh [1] introduced fuzzy set as a generalization of crisp set by considering membership degrees of elements. That membership degree is represented gradually as a real number in an interval [0, 1]. Wang et al. [2] generalized the concept of fuzzy set, called *dynamic fuzzy sets* (DFS). In the DFS, each membership degree of an element is given by a membership function dealing with time variable. Thus, the membership degree of an element in a given dynamic fuzzy set might dynamically change according to the time's variable. In 2002, Intan and Mukaidono [3] proposed an extended concept of fuzzy set, called *knowledge-based fuzzy sets* (KFS). In the KFS, the membership degree of an element given a fuzzy set is subjectively determined by a single knowledge. The membership degree of an element with respect to a given fuzzy set may be different provided by different knowledge of persons. This paper combines both concepts, DFS and KFS, called *knowledge-based dynamic fuzzy sets* (KDFS), by realizing that membership function of a given fuzzy set provided by a certain knowledge may be dynamically changed over time. Three kinds of summary fuzzy sets are proposed and discussed. Some basic operations of KDFS are defined. Their properties are verified and examined.

Keywords Fuzzy sets · Dynamic fuzzy sets · Knowledge-based fuzzy sets · Knowledge-based dynamic fuzzy set

1 Introduction

Classical sets, known as a crisp set, are used to represent collections of objects in which every object as a member of a given set is characterized by a membership degree. Since every element is only recognized either as a member or non-member, membership degree of an element is then expressed by a binary number in {0, 1},

R. Intan (✉) · S. Halim · L. P. Dewi
Petra Christian University, Surabaya 60236, Indonesia
e-mail: rintan@petra.ac.id

© Springer Nature Singapore Pte Ltd. 2019
S. K. Bhatia et al. (eds.), *Advances in Computer Communication and Computational Sciences*, Advances in Intelligent Systems and Computing 924,
https://doi.org/10.1007/978-981-13-6861-5_25

where membership degree of 0 means the element is a non-member, and membership degree of 1 means the element is a member.

In the development of sets theory, several generalized concepts of crisp sets were proposed to be applicable in simulating a complex real-world problem. *Rough sets* theory was introduced by Pawlak in 1982 [4] with a practical goal of representing indiscernibility of objects or elements dealing with an information system, a table of objects characterized by a set of attributes. Rough sets are a generalization of crisp sets by introducing a formulation of sets with imprecise boundaries. A rough set is expressed conceptually as an approximation of a given crisp set into two subsets of approximation, called lower and upper approximations. *Multisets*, also known as *bags*, are also considered as another generalization of crisp sets by allowing multiple occurrences for each of its elements in a multiset as proposed by Blizard in 1991 [5]. The term *multiset* was first time introduced by de Bruijn in 1970 [6]. However, the concept of multisets has been used since many centuries ago, long before Nicolaas Govert de Bruijn introduced the word multiset. The *fuzzy set* was introduced by Zadeh in 1965 [1, 7]. By contrast to crisp sets, membership degrees of elements in a fuzzy set are represented gradually by a real number in an interval (0, 1). So, the membership degree of an element in a given fuzzy set could be gradually started from 0 (non-member) to 1 (member) [8]. Thus, a fuzzy set is also determined as a generalization of a crisp set, i.e., a crisp set is only a special case of fuzzy set.

Membership degree of a given fuzzy set is determined and calculated by a membership function without any consideration to the time variable. In other words, ones when a membership degree of an element is determined by a membership function, dealing with a change of time, it will stay unchangeably. However, it is obviously known that everything in the world is always changing in connection with the change of time. Membership degree of an element given a fuzzy set is possibly changeable with respect to the time variable. Therefore, Wang et al. [2] proposed a concept of set, called *dynamic fuzzy sets* (DFS). The DFS is an extended concept of fuzzy sets, i.e., each membership degree of an element in DFS is given by a membership function with time's variable. Here, the membership degree of an element in a given dynamic fuzzy set might dynamically change according to time's variable. The DFS may be considered as fuzzy multisets dealing with time variable. Practically, the concept of dynamic fuzzy sets can be used to represent or to simulate the change of anything related to the change of time as usually happened in the real-world application. The concept of DFS [2], and discussed by Cai et al. [9] as part of his proposed concept, called shadow fuzzy sets, is different from the concept of DFS discussed by Zhang [10]. In this case, the concept of DFS discussed by Zhang [10] was an extension of intuitionistic fuzzy sets proposed by Atanassov [11].

Considering the membership function of a given fuzzy set is subjectively determined by certain knowledge, Intan and Mukaidono [2, 12, 13] proposed and discussed a generalized concept of fuzzy sets, called *knowledge-based fuzzy sets* (KFS). Here, fuzziness may be regarded as deterministic uncertainty. It means that even in uncertain (unclear) situation or definition of an object, a subject (person) through his/ her knowledge may be subjectively able to determine the object. Thus, a given fuzzy set may have n different membership functions related to n different knowledge.

Similarly, the knowledge-based fuzzy sets may be also regarded as another example of fuzzy multisets dealing with knowledge.

It has possibly happened that a membership function of a fuzzy set given by certain knowledge is changeable over time. Thus, it is necessary to propose a new concept, called *knowledge-based dynamic fuzzy sets* (KDFS) as a hybrid concept of DFS and KFS. Here, the KDFS may also be considered as a concept of two-dimensional fuzzy multisets. Three kinds of summary fuzzy sets are proposed and discussed. They are the knowledge-based summary fuzzy sets, the time-based summary fuzzy sets, and the general summary fuzzy sets. Some basic operations of KDFS such as equality, contentment, union, intersection, and complement are defined. Their properties are verified and examined.

2 Dynamic Fuzzy Sets and Knowledge-Based Fuzzy Sets

Dynamic fuzzy sets and knowledge-based fuzzy sets may be considered as more practical example of fuzzy multisets as proposed by Yager [14] and Miyamoto [15, 16]. Thus, they may be also regarded as the generalization of fuzzy sets in more practical use. The concepts of dynamic fuzzy sets and knowledge-based fuzzy sets are briefly discussed as follows.

2.1 Dynamic Fuzzy Sets

Considering that membership function of a given fuzzy set is possibly changeable dealing with time variable, Wang et al. [2] introduced an extended concept of fuzzy sets, called dynamic fuzzy sets as the following definition.

Definition 1 Let U be a universal set of elements and $T \subseteq R^+$ be a set of time, where $R^+ = [0, \infty)$. Then, a *dynamic fuzzy set* A on U is defined and characterized by the following membership function.

$$A : T \times U \to [0, 1] \tag{1}$$

$A(t, u) \in [0, 1]$ is the membership degree of element u to dynamic fuzzy set A at the time $t \in T$. Obviously, $A(t, u) = 1$ means u is a full member of A at the time $t \in T$. On the other hand, $A(t, u) = 0$ means u is not a member of A. Thus, the membership degree of an element u may vary depending on the time $t \in T$. $A(t) \in \mathcal{F}(U)$ is regarded as a dynamic fuzzy set A at the time $t \in T$ which is a similar concept to the fuzzy set defined by Zadeh in 1965 [1], where $\mathcal{F}(U)$ is a fuzzy power set of U. Set of dynamic fuzzy sets A on U, denoted by $\mathcal{T}(A)$ is given by: $\mathcal{T}(A) = \{A(t) | t \in T\}$.

2.2 Knowledge-based Fuzzy Sets

As discussed by Intan and Mukaidono [3, 12, 13], the uncertainty might be categorized into two types, deterministic uncertainty and non-deterministic uncertainty. Fuzziness may be regarded as a deterministic uncertainty. It means that even in uncertain (unclear) situation or definition of an object, a subject (person) through his/ her knowledge is able to determine the object subjectively. Similarly, as what happened in fuzziness, someone may subjectively determine the membership function of a given fuzzy set using his/ her knowledge. Different knowledge may have different membership function of a given fuzzy set. Thus, n-knowledge may have n different membership function of a given fuzzy set. Here, knowledge plays significant roles in determining the membership function of a given fuzzy set. A concept of knowledge-based fuzzy set is defined as follows.

Definition 2 Let U be a universal set of elements and K be a set of knowledge. Then, a *knowledge-based fuzzy set* A on U based on the knowledge $k \in K$, denoted by A_k is defined and characterized by the following membership function.

$$A_k : U \to [0, 1] \tag{2}$$

Similarly, $A_k(u) \in [0, 1]$ is the membership degree of element $u \in U$ to fuzzy set A based on knowledge $k \in K$. Obviously, $A_k(u) = 1$ means u is a full member of A based on k. On the other hand, $A_k(u) = 0$ means $u \in U$ is not a member of A_k. Thus, the membership degree of an element u to A may vary depending on knowledge $k \in K$. $A_k \in \mathcal{F}(U)$ is regarded as a knowledge-based fuzzy set of A based on knowledge k which is also a similar concept to the fuzzy set defined by Zadeh in 1965 [1, 7], where $\mathcal{F}(U)$ is a fuzzy power set of U. Set of knowledge-based fuzzy sets A on U, denoted by $\mathcal{K}(A)$ is given by $\mathcal{K}(A) = \{A_k | k \in K\}$.

3 Knowledge-Based Dynamic Fuzzy Sets

As the primary objective of this paper, we introduce a new concept, called *knowledge-based dynamic fuzzy sets* as a hybrid concept between dynamic fuzzy sets and knowledge-based fuzzy sets. Practically in the real-world application, even a certain membership function of fuzzy set A has already given by a certain knowledge k, the membership function is possibly changeable over time. The following definition starts the concept of knowledge-based dynamic fuzzy sets.

Definition 3 Let U be a universal set of elements, K be a set of knowledge and $T \subseteq R^+$ be a set of time, where $R^+ = [0, \infty)$. Then, a *knowledge-based dynamic fuzzy set* A on U based on the knowledge $k \in K$, denoted by δ_k^A, is defined and characterized by the following membership function.

Table 1 Relation among
DFS, KFS, and KDFS

	$A(t_1)$	$\cdots$	$A(t_m)$
A_{k_1}	$\delta_{k_1}^A(t_1)$	$\cdots$	$\delta_{k_1}^A(t_m)$
$\vdots$	$\vdots$	$\ddots$	$\vdots$
A_{k_n}	$\delta_{k_n}^A(t_1)$	$\cdots$	$\delta_{k_n}^A(t_m)$

$$\delta_k^A : T \times U \to [0, 1] \tag{3}$$

Related to (3), $\delta_k^A(t, u) \in [0, 1]$ is the membership degree of element $u \in U$ to fuzzy set A based on knowledge $k \in K$ at the time $t \in T$. Similarly, $\delta_k^A(t, u) = 1$ means u is a full member of A based on k at the time t. On the other hand, $\delta_k^A(t, u) = 0$ means $u \in U$ is not a member of $\delta_k^A(t)$. Thus, the membership degree of an element u to A may also vary depending on both the knowledge k and the time t. Here, $\delta_k^A(t) \in \mathcal{F}(U)$ is regarded as a knowledge-based dynamic fuzzy set of A which is based on knowledge k at the time t. $\delta_k^A(t)$ is also a similar concept to the fuzzy set defined by Zadeh [1, 6], where $\mathcal{F}(U)$ is a fuzzy power set of U. Set of knowledge-based dynamic fuzzy sets A on U, denoted by $\mathcal{D}(A)$ is given by $\mathcal{D}(A) = \{\delta_k^A(t) | k \in K, t \in T\}$.

3.1 Summary Fuzzy Sets

The relation among dynamic fuzzy sets (DFS), knowledge-based fuzzy sets (KFS), and knowledge-based dynamic fuzzy sets (KDFS) is represented in Table 1. Let A be a fuzzy set on U, K be a set of knowledge, and T be a set of times, where $K = \{k_1, k_2, \ldots, k_n\}$ and $T = \{t_1, t_2, \ldots, t_m\}$.

For every $t_j \in T$ and $k_i \in K$, it can be further interpreted that both A_{k_i} and $A(t_j)$ are able to be provided by two aggregation functions over $\delta_{k_i}^A(t_j)$ as follows.

$$\forall u \in U, A_{k_i}(u) = \Upsilon\big(\delta_{k_i}^A(t_1, u), \ldots, \delta_{k_i}^A(t_m, u)\big), \text{ where } \Upsilon : [0, 1]^m \to [0, 1] \tag{4}$$

$$\forall u \in U, A(t_j, u) = \Theta\big(\delta_{k_1}^A(t_j, u), \ldots, \delta_{k_n}^A(t_j, u)\big), \text{ where } \Theta : [0, 1]^n \to [0, 1] \tag{5}$$

Here, depending on the context of applications, Υ and Φ may use any existed functions such as maximum, minimum, and average by taking the maximum and the minimum functions as the maximum and the minimum values of both functions, respectively. Equation (5) is exactly similar to the knowledge-based summary fuzzy set as discussed by Intan and Mukaidono [3, 12, 13]. Practically, the knowledge-based summary fuzzy set of A as formulated in (5) means an agreement given by a group of persons represented by a set of knowledge to describe A at the time t_j. Similarly, Eq. (4) may be regarded to provide the time-based summary fuzzy set. For it is sometimes happened in reality, subjective opinion of someone to a given fuzzy set A may be changeable according to the changing of times, and the objective of (4) is to summarize the multiple opinions of a certain knowledge k_i to the fuzzy set A. Therefore, related to the reason behind calculating both summary fuzzy sets as given in (4) and (5), it may be more applicable to use the weighted average as the aggregation function as shown in the following equations.

$$\Upsilon\left(\delta_{k_i}^A(t_1, u), \ldots, \delta_{k_i}^A(t_m, u)\right) = \frac{\sum_{j=1}^m w_j \cdot \delta_{k_i}^A(t_j, u)}{\sum_{j=1}^m w_j}, \text{ where } w_j \in R^+, R^+ = [0, \infty)$$

(6)

$$\Theta\left(\delta_{k_1}^A(t_j, u), \ldots, \delta_{k_n}^A(t_j, u)\right) = \frac{\sum_{i=1}^n w_i \cdot \delta_{k_i}^A(t_j, u)}{\sum_{i=1}^n w_i}, \text{ where } w_i \in R^+, R^+ = [0, \infty)$$

(7)

Related to the summary fuzzy sets, w_i and w_j are the weights to express the prominence of an opinion. For instance, in the case of the knowledge-based summary fuzzy sets, more prominent knowledge k_i is considered in determining the summary fuzzy set, a larger w_i is given to k_i. In the case of the time-based summary fuzzy sets, a larger weight may be given generally to the more recent opinion, since a more current opinion represent a more real-time situation. So, in the case of constructing the time-based summary fuzzy sets, the relation between time and weight should satisfy $t_j > t_p \Rightarrow w_j \geq w_p, \forall t_j, t_p \in T$.

It is also necessary to summarize all interpretation/ opinion based on the knowledge as well as the times into only one summary fuzzy set, called *general summary fuzzy set*. Here, the general summary fuzzy sets may be assumed as an agreement, to sum up all opinions given by multiple pieces of knowledge over time. Given A be a fuzzy set on U. Let $K = \{k_1, \ldots, k_n\}$ and $T = \{t_1, \ldots, t_m\}$. Using the concept of weighted average, we introduce three different equations to provide the general summary fuzzy set as follows.

- Aggregated from the Knowledge-based Summary Fuzzy Sets

$$A^{G_1}(u) = \Psi\left(A_{k_1}(u), \ldots, A_{k_n}(u)\right) = \frac{\sum_{i=1}^n w_i \cdot A_{k_i}(u)}{\sum_{i=1}^n w_i},$$

(8)

where $w_i \in R^+, R^+ = [0, \infty)$.
- Aggregated from the time-based summary fuzzy sets

$$A^{G_2}(u) = \Gamma(A(t_1, u), \ldots, A(t_m, u)) = \frac{\sum_{j=1}^m w_j \cdot A(t_j, u)}{\sum_{j=1}^m w_j},$$

(9)

where $w_i \in R^+, R^+ = [0, \infty)$
- Aggregated directly from the knowledge-based dynamic fuzzy sets

$$A^{G_3}(u) = \Omega \begin{pmatrix} \delta_{k_1}^A(t_1, u) & \cdots & \delta_{k_1}^A(t_m, u) \\ \vdots & \ddots & \vdots \\ \delta_{k_n}^A(t_1, u) & \cdots & \delta_{k_n}^A(t_m, u) \end{pmatrix} = \frac{\sum_{j=1}^n \sum_{i=1}^m w_{ij} \cdot \delta_{k_i}^A(t_j, u)}{\sum_{j=1}^n \sum_{i=1}^m w_{ij}}, \quad (10)$$

where $w_{ij} \in R^+$, $R^+ = [0, \infty)$.

These three different equations may produce different results that it depends on the context of application to decide which one is better to use.

3.2   Basic Operations and Properties

Some basic operations of the knowledge-based dynamic fuzzy sets may be discussed and defined as follows.

Definition 4 Let U be a universal set of elements, K be a set of knowledge, and $T \subseteq R^+$ be a set of time, where $R^+ = [0, \infty)$. A and B are two fuzzy sets on U. Some basic operations and properties of *equality, containment, union, intersection,* and *complementation* are given by the following equations.

Equality

1. $\delta_k^A(t) = \delta_k^B(t) \Leftrightarrow \delta_k^A(t, u) = \delta_k^B(t, u), \forall u \in U$,
2. $A_k = B_k \Leftrightarrow A_k(u) = B_k(u), \forall u \in U$,
3. $A_k \equiv B_k \Leftrightarrow \delta_k^A(t, u) = \delta_k^B(t, u), \forall u \in U, \forall t \in T$,
4. Similarly, $A_k \equiv B_k \Leftrightarrow \delta_k^A(t) = \delta_k^B(t), \forall t \in T$,
5. $A(t) = B(t) \Leftrightarrow A(t, u) = B(t, u), \forall u \in U$,
6. $A(t) \equiv B(t) \Leftrightarrow \delta_k^A(t, u) = \delta_k^B(t, u), \forall u \in U, \forall k \in K$,
7. Similarly, $A(t) \equiv B(t) \Leftrightarrow \delta_k^A(t) = \delta_k^B(t), \forall k \in K$,
8. $A = B \Leftrightarrow \delta_k^A(t, u) = \delta_k^B(t, u), \forall u \in U, \forall k \in K, \forall t \in T$,
9. $A \cong B \Leftrightarrow \delta_k^A(t_{j_1}, u) = \delta_k^B(t_{j_2}, u), \forall u \in U, \forall k \in K, \forall t_{j_1}, t_{j_2} \in T$,
10. $A \triangleq B \Leftrightarrow \delta_{k_{i_1}}^A(t, u) = \delta_{k_{i_2}}^B(t, u), \forall u \in U, \forall k_{i_1}, k_{i_2} \in K, \forall t \in T$,
11. $A \equiv B \Leftrightarrow \delta_{k_{i_1}}^A(t_{j_1}, u) = \delta_{k_{i_2}}^B(t_{j_2}, u), \forall u \in U, \forall k_{i_1}, k_{i_2} \in K, \forall t_{j_1}, t_{j_2} \in T$,
12. $k_{i_1} = k_{i_2} \Leftrightarrow \delta_{k_{i_1}}^A(t, u) = \delta_{k_{i_2}}^A(t, u), \forall u \in U, \forall t \in T, A \in \mathcal{F}(U)$,
    where $\mathcal{F}(U)$ is fuzzy power set on U.

Containment

13. $\delta_k^A(t) \subseteq \delta_k^B(t) \Leftrightarrow \delta_k^A(t, u) \leq \delta_k^B(t, u), \forall u \in U$,
14. $A_k \subseteq B_k \Leftrightarrow A_k(u) \leq B_k(u), \forall u \in U$,
15. $A_k \underline{\subseteq} B_k \Leftrightarrow \delta_k^A(t, u) \leq \delta_k^B(t, u), \forall u \in U, \forall t \in T$,
16. Similarly, $A_k \underline{\subseteq} B_k \Leftrightarrow \delta_k^A(t) \subseteq \delta_k^B(t), \forall t \in T$,
17. $A(t) \subseteq B(t) \Leftrightarrow A(t, u) \leq B(t, u), \forall u \in U$,
18. $A(t) \underline{\subseteq} B(t) \Leftrightarrow \delta_k^A(t, u) \leq \delta_k^B(t, u), \forall u \in U, \forall k \in K$,
19. Similarly, $A(t) \underline{\subseteq} B(t) \Leftrightarrow \delta_k^A(t) \leq \delta_k^B(t), \forall k \in K$,

20. $A \subseteq B \Leftrightarrow \delta_k^A(t, u) \leq \delta_k^B(t, u), \forall u \in U, \forall k \in K, \forall t \in T,$

21. $A \sqsubseteq B \Leftrightarrow \delta_k^A(t_{j_1}, u) \leq \delta_k^B(t_{j_2}, u), \forall u \in U, \forall k \in K, \forall t_{j_1}, t_{j_2} \in T,$

22. $A \preccurlyeq B \Leftrightarrow \delta_{k_{i_1}}^A(t, u) \leq \delta_{k_{i_2}}^B(t, u), \forall u \in U, \forall k_{i_1}, k_{i_2} \in K, \forall t \in T,$

23. $A \subseteqq B \Leftrightarrow \delta_{k_{i_1}}^A(t_{j_1}, u) \leq \delta_{k_{i_2}}^B(t_{j_2}, u), \forall u \in U, \forall k_{i_1}, k_{i_2} \in K, \forall t_{j_1}, t_{j_2} \in T,$

24. $k_{i_1} \unlhd k_{i_2} \Leftrightarrow \delta_{k_{i_1}}^A(t, u) \leq \delta_{k_{i_2}}^A(t, u), \forall u \in U, \forall t \in T, A \in \mathcal{F}(U),$

    where $\mathcal{F}(U)$ is fuzzy power set on U.

Union

25. $\delta_k^{A \cup B}(t, u) = \max\left(\delta_k^A(t, u), \delta_k^B(t, u)\right), \forall u \in U,$

26. $\left(\delta_{k_{i_1}}^A(t_{j_1}) \cup \delta_{k_{i_2}}^B(t_{j_2})\right)(u) = \max\left(\delta_{k_{i_1}}^A(t_{j_1}, u), \delta_{k_{i_2}}^B(t_{j_2}, u)\right), \forall u \in U, t_{j_1}, t_{j_2} \in T,$
    $k_{i_1}, k_{i_2} \in K,$

27. $(A_k \cup B_k)(u) = (A \cup B)_k(u) = \max(A_k(u), B_k(u)), \forall u \in U,$

28. $\left(A_{k_{i_1}} \cup B_{k_{i_2}}\right)(u) = \max\left(A_{k_{i_1}}(u), B_{k_{i_2}}(u)\right), \forall u \in U, k_{i_1}, k_{i_2} \in K,$

29. $(A(t) \cup B(t))(u) = (A \cup B)(t, u) = \max(A(t, u), B(t, u)), \forall u \in U,$

30. $\left(A(t_{j_1}) \cup B(t_{j_2})\right)(u) = \max\left(A(t_{j_1}, u), B(t_{j_2}, u)\right), \forall u \in U, t_{j_1}, t_{j_2} \in T,$

31. $(A \cup B)(u) = \max\left(A^{G_p}(u), B^{G_p}(u)\right), \forall u \in U, p = \{1, 2, 3\},$
    where A^{G_p} and B^{G_p} are general fuzzy sets as given in (8), (9), and (10).

Intersection

32. $\delta_k^{A \cap B}(t, u) = \min\left(\delta_k^A(t, u), \delta_k^B(t, u)\right), \forall u \in U,$

33. $\left(\delta_{k_{i_1}}^A(t_{j_1}) \cap \delta_{k_{i_2}}^B(t_{j_2})\right)(u) = \min\left(\delta_{k_{i_1}}^A(t_{j_1}, u), \delta_{k_{i_2}}^B(t_{j_2}, u)\right), \forall u \in U, t_{j_1}, t_{j_2} \in T,$
    $k_{i_1}, k_{i_2} \in K,$

34. $(A_k \cap B_k)(u) = (A \cap B)_k(u) = \min(A_k(u), B_k(u)), \forall u \in U,$

35. $\left(A_{k_{i_1}} \cap B_{k_{i_2}}\right)(u) = \min\left(A_{k_{i_1}}(u), B_{k_{i_2}}(u)\right), \forall u \in U, k_{i_1}, k_{i_2} \in K,$

36. $(A(t) \cap B(t))(u) = (A \cap B)(t, u) = \min(A(t, u), B(t, u)), \forall u \in U,$

37. $\left(A(t_{j_1}) \cap B(t_{j_2})\right)(u) = \min\left(A(t_{j_1}, u), B(t_{j_2}, u)\right), \forall u \in U, t_{j_1}, t_{j_2} \in T,$

38. $(A \cap B)(u) = \min\left(A^{G_p}(u), B^{G_p}(u)\right), \forall u \in U, p \in \{1, 2, 3\},$
    where A^{G_p} and B^{G_p} are general fuzzy sets as given in (8), (9), and (10).

Complementation

39. $\delta_{k_i}^{\neg A}(t_j, u) = 1 - \delta_{k_i}^A(t_j, u),$

40. $\delta_{\neg k_i}^A(t_j, u) = \begin{cases} \delta_{k_r}^A(t_j, u), r \neq i, |K| = 2, \\ \Theta(\alpha_{k_1}, \ldots, \alpha_{k_{i-1}}, \alpha_{k_{i+1}}, \ldots, \alpha_{k_n}), |K| > 2, \alpha_{k_p} = \delta_{k_p}^A(t_j, u), \end{cases}$

41. $\delta_{k_i}^A(\neg t_j, u) = \begin{cases} \delta_{k_i}^A(t_r, u), r \neq i, |T| = 2, \\ \Upsilon(\beta_{t_1}, \ldots, \beta_{t_{j-1}}, \beta_{t_{j+1}}, \ldots, \beta_{t_m}), |T| > 2, \beta_{t_p} = \delta_{k_i}^A(t_p, u), \end{cases}$

42. $\neg A_{k_i}(u) = 1 - A_{k_i}(u),$

43. $A_{\neg k_i}(u) = \begin{cases} A_{k_r}(u), r \neq i, |K| = 2, \\ \Psi(A_{k_1}(u), \cdots, A_{k_{i-1}}(u), A_{k_{i+1}}(u), \cdots, A_{k_n}(u)), |K| > 2, \end{cases}$

44. $\neg A(t_j, u) = 1 - A(t_j, u),$

45. $A(\neg t_j, u) = \begin{cases} A(t_r, u), r \neq j, |K| = 2, \\ \Gamma(A(t_1, u), \ldots, A(t_{j-1}, u), A(t_{j+1}, u), \ldots, A(t_m, u)), |K| > 2, \end{cases}$

46. $\neg A(u) = 1 - A^{G_p}$, $p \in \{1, 2, 3\}$, where A^{G_p} and B^{G_p} are general fuzzy sets as given in (8), (9), and (10).

There are some properties as a consequence of the basic operations as given in Definition 4 as follows.

- From *Equality* 2, 3, and 4: $A_k \equiv B_k \Rightarrow A_k = B_k, \forall k \in K$.
- From *Equality* 5, 6, and 7: $A(t) \equiv B(t) \Rightarrow A(t) = B(t), \forall t \in T$.
- From *Equality* 8–11: $(A \equiv B) \Rightarrow \left\{ (A \cong B), \left(A \triangleq B \right) \right\} \Rightarrow (A = B)$.
- From *Containment* 14, 15, and 16: $A_k \underline{\in} B_k \Rightarrow A_k \subseteq B_k, \forall k \in K$.
- From *Containment* 17, 18, and 19: $A(t) \underline{\in} B(t) \Rightarrow A(t) \subseteq B(t), \forall t \in T$.
- From *Containment* 20–23: $A \underline{\in} B) \Rightarrow \{(A \underline{\sqsubseteq} B), (A \preccurlyeq B)\} \Rightarrow (A \subseteq B)$.

4 Conclusion

In the real-world application, it is possible that a membership function of a fuzzy set subjectively given by a certain knowledge is changeable over time. Therefore, this paper introduced a generalized concept of fuzzy sets, called knowledge-based dynamic fuzzy sets. The concept is a hybrid concept of dynamic fuzzy sets proposed by Wang et al. [2] and knowledge-based fuzzy sets introduced and discussed by Intan and Mukaidono [3, 10, 11]. The concept of knowledge-based dynamic fuzzy sets is regarded as a practical example of two-dimensional fuzzy multisets dealing with time and knowledge variables. Three kinds of summary fuzzy sets, namely the knowledge-based summary fuzzy sets, the time-based summary fuzzy sets, and the general summary fuzzy sets, are proposed and discussed using aggregation functions. Some basic operations such as equality, containment, union, intersection, and complementation are defined. Their properties are verified and examined.

References

1. Zadeh, L.A.: Fuzzy Sets. Inf. Control **8**, 338–353 (1965)
2. Wang, G.Y., Ou, J.P., Wang, P.Z.: Dynamic fuzzy sets. Fuzzy Syst. Math **2**(1), 1–8 (1988)
3. Intan, R., Mukaidono, M.: On knowledge-based fuzzy sets. International Journal of Fuzzy Systems 4(2), Chinese Fuzzy Systems Association T (CFSAT), pp. 655–664, (2002)
4. Pawlak, Z.: Rough Sets. Int. J. Comput. Inf. Sci. **11**, 341–356 (1982)
5. Blizard, Wayne D.: The development of Multiset Theory. Modern Logic **1**(4), 319–352 (1991)
6. de Bruijn, N. G.: A Generalization of Burnside's lemma, Technical Report, Notitie 55, Technological University Eindhoven (1970)
7. Zadeh, L.A.: Fuzzy sets and systems. Int. J. Gen Syst **17**, 129–138 (1990)
8. Klir, G.J., Yuan, B.: Fuzzy Sets and Fuzzy Logic: Theory and Applications. Prentice Hall, New Jersey (1995)
9. Cai, M., Li, Q., Lang, G.: Shadowed sets of dynamic fuzzy sets. Granular Comput. **2**, 85–94 (2017)

10. Zhang, Z.: A novel dynamic fuzzy sets method applied to practical teaching assessment on statistical software. IERI Procedia **2**, 303–310 (2012)
11. Atanassov, K.T.: Intuitionistic fuzzy sets. Fuzzy Sets Syst. **20**, 87–96 (1986)
12. Intan, R., Mukaidono, M., Emoto, M.: Knowledge- based representation of fuzzy sets. In: Proceeding of FUZZ-IEEE'02, pp. 590–595 (2002)
13. Intan, R., Mukaidono, M.: Approximate reasoning in knowledge-based fuzzy sets. In: Proceeding of NAFIPS-FLINT 2002, pp. 439–444, IEEE Publisher, (2002)
14. Yager, R.R.: On the Theory of Bags. Int. J. Gen Syst. **13**, 23–37 (1986)
15. Miyamoto, S.: Fuzzy multisets and their generalizations. Multiset Proc. **2235**, 225–235 (2001)
16. Miyamoto, S.: Fuzzy Sets in Information Retrieval and Cluster Analysis. Kluwer Academic Publishers (1990)

Analysis and Discussion of Radar Construction Problems with Greedy Algorithm

Wenrong Jiang

Abstract Greedy Algorithm always makes the best choice at moment. In other words, without considering the overall optimality, what it has done is a partial optimal in a situation. The Greedy Algorithm is not the best solution for all problems. For the issues involved in this paper, the advantage of greedy algorithm is more correct and the choice of greedy strategy is only relevant to the current state.

Keywords Greedy algorithm · Greedy strategy · Radar construction

1 Introduction

Research Background [1]: In order to solve practical problems, the algorithms develop rapidly, the algorithms such as Dynamic Programming, Greedy Algorithm, Exhaustive Method, and Partitioning Algorithm. For the complicated problems, we want a simple solution, and greedy algorithm was born for this. It provides a simple and intuitive optimal solution to solve complicated problems. The choices it makes are always the best answer at moment, which simplifies a complicated problem to an easier problem. For the special problems such as The Shortest Path Problem, The Huffman Coding Problem, and The Minimum Spanning Tree Problem, it can even make the overall optimal solution. This trait makes it popular in project.

Problem Description: The topic chosen in this article is POJ1328 title, which is a relatively classic subject in many competition topics. If you do not master the problem solving idea, you may feel that it is difficult to start. However, once you have mastered the rules, such topics can be described as "hands in hand, no expense. effort." The original text is described as follows: "suppose the coasting is an infinite straight line. The land lies on one side of the coast, and the ocean is on the other side. Each island is a point on the coast. Any radar installation installed on the coast can cover only D distance, so if the distance between them is at most d, the islands

W. Jiang (✉)
The School of Computer and Information,
Shanghai Polytechnic University, Shanghai 201209, China
e-mail: wrjiang@sspu.edu.cn

© Springer Nature Singapore Pte Ltd. 2019 303
S. K. Bhatia et al. (eds.), *Advances in Computer Communication and Computational Sciences*, Advances in Intelligent Systems and Computing 924, https://doi.org/10.1007/978-981-13-6861-5_26

in the sea can be covered by radius installation. We use the Descartes coordinate system to define the coast and define the sliding axis as the X-axis. The beach is above the X-axis, and the ground is below. Considering the location of each island in the sea, and considering the radar installation coverage distance, your task is to write a program to find the smallest number of radar devices covering all islands. Note that the location of the island is represented by its X-Y coordinates. In short, the title requires that the top of the x-axis of the map is Shanghai, the bottom is land, and there are n islands in the sea with coordinates $(isl[i].x, isl[i].y)$. There is a radar that detects circles in the radius of D. Ask at least how many radar the coastline can contain. Pay attention to radar is built on the coastline, that is, on the X-axis."

Significance: The greedy algorithm is a simpler and more rapid design technique for some optimal problems. The core of this improved hierarchical processing method is to select a metric based on the topic. The design of the algorithm with greedy algorithm is characterized by step by step, without taking into account the overall situation, it saves a lot of time spent looking for the best solution to the exhaustive, and makes "greedy" choices from top to iterative. Each choice will resolve the overall problem into smaller ones, and ultimately an overall optimal solution to the problem. The algorithm it gives is generally more intuitive, simple, and efficient than the dynamic programming algorithm.

2 Domestic and Foreign Research

According to the data that can be retrieved on the Internet, greedy algorithms are very mature and commonly used algorithms. They have been widely used in various types of algorithms. To a certain extent, the difficulty of the problem is greatly simplified by the optimal solution obtained. Its characteristic is that as the algorithm progresses, two other sets will be accumulated: one contains candidate objects that have been considered and selected, and the other contains candidates that have been considered but are discarded. There is a function to check if a set of candidate objects provides a solution to the problem. This function does not consider whether the solution at this time is optimal. There is also a function to check if a set of candidate objects is feasible, i.e., whether it is possible to add more candidate objects to the set to obtain a solution. As with the previous function, the optimality of the solution is not considered at this time. The selection function can indicate which remaining candidate is the most promising solution to the problem. Finally, the objective function gives the value of the solution [2].

In order to solve the problem, we need to find a set of candidate objects that make up the solution. It can optimize the objective function and the greedy algorithm can be performed step by step. Initially, the set of candidate objects selected by the algorithm is empty. In each of the following steps, based on the selection function, the algorithm selects the object that has the most promising solution from the remaining

Table 1 Machine scheduling problem

Task	A	B	C	D	E	F	G
Start time	0	3	4	9	7	1	6
Finish time	2	7	7	11	10	5	8

candidate objects. If adding this object to the collection is not feasible, then the object is discarded and is no longer considered; otherwise, it is added to the collection. Expand the collection each time and check if the collection constitutes a solution. If the greedy algorithm works correctly, the first solution found is usually optimal.

Shortcomings: Of course, all algorithms are not perfect. Greedy algorithms also have their own shortcomings.

A foreign friend called "Michael Scofield L Q" asked questions. There is such a question: [machine scheduling] Existing n pieces of tasks and unlimited number of machines, tasks can be processed on the machine. The start time of each task is si, and the completion time is fi, si < fi. [si, fi] The time range for processing task i. The two tasks i, j refer to the overlapping of the time ranges of the two tasks, not to the coincidence of the start or end of i, j. For example, the interval [1, 4] overlaps with the interval [2, 4] and does not overlap with the interval [4, 7]. A feasible assignment of tasks means that no two overlapping tasks are assigned to the same machine in the assignment. Therefore, each machine can only process at most one task at any time in a feasible allocation. The optimal allocation refers to the least feasible distribution scheme of the machine used [3].

Suppose there are $n = 7$ tasks, labeled a through g. Their start and finish times are shown in Table 1. If task a is assigned to machine $M1$, task b is assigned to machine $M2$,…, Task g is assigned to machine $M7$. This allocation is a feasible allocation, and a total of seven machines are used. But it is not an optimal allocation because there are other allocation schemes that can make fewer machines available. For example, tasks a, b, and d can be assigned to the same machine, and the number of machines is reduced to five [4].

A greedy method of obtaining optimal allocation is to allocate tasks step by step. One task is assigned to each step and assigned in the non-decreasing order of the task's starting time. If at least one task has been assigned to a machine, the machine is said to be old; if the machine is not old, it is new. When selecting a machine, the following greedy criteria are used: According to the starting time of the task to be distributed, if an old machine is available at this time, the task is distributed to the old machine. Otherwise, assign the task to a new machine. According to the data in the example, the greedy algorithm is divided into $n = 7$ steps, and the task allocation order is a, f, b, c, g, e, d. The first step is that there is no old machine, so assign a to a new machine (such as $M1$). This machine is busy from 0 to 2. In the second step, consider task f. Since the old machine is still busy when f starts, assign f to a new machine (set to $M2$). The third step considers task b. Since the old machine $M1$ is already idle at Sb = 3, b is assigned to $M1$ for execution. The next available time for

$M1$ becomes fb $= 7$, and the available time for $M2$ becomes ff $= 5$. The fourth step, consider task c. Since there is no old machine available at Sc $= 4$, c is assigned to a new machine ($M3$), and the next time this machine is available is fc $= 7$. The fifth step considers task i, which is assigned to machine $M2$, the sixth step assigns task e to machine $M1$, and finally in the seventh step, task 2 is assigned to machine $M3$. (Note: Task d can also be assigned to machine $M2$) [5].

The above-mentioned greedy algorithm can lead to the proof of optimal machine allocation as an exercise. A greedy algorithm with complexity O $(nlogn)$ can be implemented as follows: First, each task is arranged in ascending order of Si using a sorting algorithm (such as heap sort) with a complexity of O $(nlogn)$ and then used a minimum heap for the time available for the old machine.

The netizen noticed that the greedy algorithm does not guarantee optimality in the problem of machine scheduling [6]. However, it is an intuitive tendency and the result is always very close to the optimal value. The rules that it uses are the rules that human machines are expected to use in the actual environment. The algorithm does not guarantee optimal results, but usually the result is almost the same as the optimal solution. This algorithm is also called the heuristic method. Therefore, the LPT method is a heuristic machine scheduling method. The LPT heuristic method has bounded performance. A heuristic method with limited performance is called an approximate algorithm.

In some applications, the result of greedy algorithm is always the optimal solution. However, for some other applications, the generated algorithm is only a heuristic method and may or may not be an approximate algorithm.

Correctness: How can we find the optimal solution from many feasible solutions? In fact, most of them are regular. In the sample, there is a clear rule of greed. But when you get the best solution for [font lang $=$ EN-US] $N = 5$, you only need to find the closest two of the 5 used boards, and then stick a few cowsheds in the middle. Make two wooden planks one piece. The resulting $N = 4$ solution must be optimal, because this board is the least wasted. Similarly, other greedy questions will also have such a nature. Because of its greediness, it can be faster than other algorithms.

To prove the correctness of the greedy nature is the real challenge of the greedy algorithm, because not every local optimal solution will have a connection with the global optimal solution, and the solution generated by greedy is not the optimal solution. In this way, the proof of greedy nature becomes the key to the correct greedy algorithm. A greedy nature that you come up with may be wrong, even though it is feasible in most of the data, you must take into account all possible special situations and prove that your greedy nature is still correct in these special circumstances. This tempered nature can constitute a correct greed.

Summary: If there is a greedy nature, then it must be adopted! Because it is easy to write, easy to debug, extremely fast, and space saving. Almost to say it is the best of all algorithms. However, it should be noted that do not fall into the muddy nature of demonstrating the inaccuracy of greedy nature, because the scope of application of greedy algorithms is not large, and some are extremely difficult to prove. If you are not sure, it is best not to risk because there are other algorithms that will better than it.

3 Solve the Problem

3.1 *Ideas*

In general, the idea of using a greedy algorithm is as follows:

The basic idea of the greedy algorithm is to proceed step by step from an initial solution to the problem. According to an optimization measure, each step must ensure that the local optimal solution can be obtained. Only one data are considered in each step, and his selection should satisfy the conditions for local optimization. If the next data and part of the optimal solution are no longer feasible solutions, the data are not added to the partial solution until all the data have been enumerated or the algorithm can no longer be added.

The process of solving the problem is generally as follows:

(1) Establish a mathematical model to describe the problem;
(2) divide the problem of solution into several subproblems;
(3) For each subproblem, obtain the local optimal solution to the subproblem.
(4) The solution of the suboptimal solution to the local optimal solution is a solution to the original solution problem.

In this question, we need to build a radar in order to determine whether to cover the island, determine whether the radar is built or not, and ensure that the number of radars established is the minimum and satisfies the requirements of the questions. Use greedy algorithms for analysis, build radars from left to right, cover as many islands as possible. Take the island as the center of the circle and d as the radius. If the drawn circle has no intersection with the x-axis, it cannot be achieved. If there is an intersection point, the coordinates of the intersection point between the i-th island and the x-axis are calculated and stored in the structure arrays rad[i].sta and rad[i].end. Sort the array of structures, sort them in ascending order by rad[i].end, and then look for the radar from left to right. For rad[i].end is the current rightmost left coordinate, for rad[j].sta < ran[i].end for the next island, there is no need to create a new radar and the next island needs to be updated. Otherwise, new radars need to be established.

3.2 *Method*

To solve this problem, the common solution is to determine the y-coordinate of a radar placement point. The more positive it is, the better it can be determined so that all radars are placed on the x-axis. Then consider the island A on the left, use a radar to cover it, and make the following islands as far as possible in the range of the radar. Then, the radar should be placed farther to the left under the condition of covering A. That is, A is optimal at the boundary of the radar scan. We can use this to find the coordinates of the radar. Then, we can determine which islands are behind the newly placed radar range following A, and if they are removed from the queue, if Do not use this island again to perform the same operation as A (that is, find the location of the next radar).

Using a greedy algorithm, we can think that since I'm trying to cover the island with a circle (radar range, radius r), why not draw a circle (indicated as a circle O) for the radius of the island centered on r, then As long as the radar is in this circle then the island can be covered. From the previous analysis, we can see that the radar must be placed on the x-axis, so the radar must be placed on the intersection between the circle O and the x-axis, as shown in the schematic diagram of the solution method.

So we can record all of the island's corresponding intervals, and then sort the interval from the left to the right, and then from the first interval, if the second interval intersects, we update the intersection and from the interval 1, 2 is removed from the queue. If the third interval intersects with this intersection, the intersection is updated and the interval 3 is removed until the intersection is not satisfied. Then, continue to simulate this process on the line, each simulation using this process ANS $+ +$ (i.e., the interval selection problem).

The last screenshot is a screenshot of this question.

3.3 Charts and Tables

See Figs. 1, 2, 3.

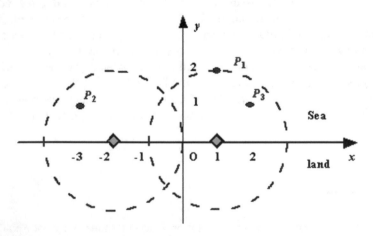

Fig. 1 The illustration is given by the question

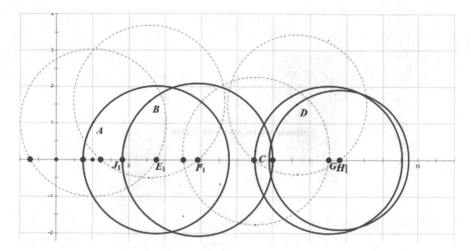

Fig. 2 Illustration shows a schematic diagram

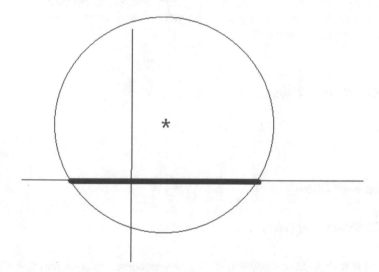

Fig. 3 Illustration shows a schematic diagram of the problem solving method

4 Analysis of Research Results

4.1 Running Time

use ordinary solution to solve this problem program running time is 120 ms;

using greedy algorithm solution to solve this problem program running time is 63 ms;

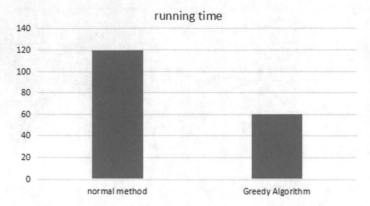

Fig. 4 comparison of the running time of the two methods

Obviously, using a greedy algorithm greatly simplifies the complexity of the program, simplifies the difficulty of the problem, greatly shortens the running time, and further solves the problem.

4.2 Charts and Tables

See Fig. 4.

5 The Key Issue

5.1 Problem Description

When we first thought about this question, we encountered a problem that was not easily noticeable. For a radar placement y-coordinate, it is definitely better to go up, so it can be concluded that all radars are placed on the x-axis and then consider the left-most island A. Use a radar to cover it and make it afterward. The island will be covered as much as possible, so the radar will try to move to the right. Later, this practice was found to be wrong. When placing the first radar, while satisfying Coverage A, it is not better to place it more toward the right, because when the radar moves to the right, it will not cover part of the left side. At this point, point B will need to add another radar to cover, so this idea is wrong!

5.2 Solution

Take the island as its center and r as a radius to draw a circle (marked as circle O) so that the island can be covered as long as the radar is in this circle. From the previous analysis, we can see that the radar must be placed on the x-axis, so the radar must be placed on the intersection of the circle O and the x-axis. In this way, the problem has turned into a problem of interval selection.

6 Summary

If an algorithm is flawed or the algorithm is not properly selected, executing this algorithm will not solve the problem. The same task, different algorithms to perform, may use different time, space, or efficiency to complete. The degree of merit of an algorithm can be measured by the time complexity and the space complexity. The algorithm can be described by many different methods such as natural language, pseudo-code, and flowcharts.

Many specific algorithms can be used to implement operating systems, database management systems, language compilation systems, and a variety of computer application systems.

In this problem, the use of common methods not only takes time and effort, but there are omissions, and the answers obtained are not necessarily the best answers. The use of greedy algorithms can greatly guarantee the optimality of the answer. Since the beginning, the core of the greedy algorithm is to find the optimal solution at a certain moment, which simplifies the answer and further finds the optimal solution.

Of course, not all algorithm problems can use greedy algorithms and can only be used in problems that do not address the full state. In the same way, choosing a greedy strategy is also one of the elements of the solution. In short, proper use of greedy algorithms will greatly help us solve practical problems.

Acknowledgements This work is supported by the Key Disciplines of Computer Science and Technology of Shanghai Polytechnic University under Grant No. XXKZD1604 [7].

References

1. Jiang, W., Chen, J., Tao W.: The development of automotive interior sales website. Commun Comput. Inf. Sci. **268**(12), 342–348 (2011). EI:20122015019844
2. Zhang, D.: Algorithm design and analysis [M], p. 108. National Defense Industry Press, Beijing (2009)
3. AC_hell.POJ-1328Intervalgreedy, geometry[EB/OL], 26 Apr 2016. https://blog.csdn.net/AC_hell/article/details/51250550
4. lbcab. Introduction and use of greedy algorithms [EB/OL], 8 May 2016. https://blog.csdn.net/lbcab/article/details/51347458

5. Evil Cattle. Theory of greedy algorithm details [EB/OL], 27 Nov 2017. https://www.2cto.com/kf/201711/700882.html
6. George, B., Valeva, A.: A database security course on a shoestring. In: ACM SIGCSE Bulletin, Proceedings of the 37th SIGCSE Technical Symposium on Computer Science Education SIGCSE'06, **38**(1): 7–11, (2006)
7. Jiang, W., He, S.: The research of small and medium-sized wedding photography enterprise's product types and consumer participation behavior. In: Proceedings of the 2016 International Conference on Social Science, Humanities and Modern Education(SSHME 2016). Atlantis Press, Guangzhou, China, 30 May 2016, 103-107.WOS:000390400800020

Multilevel ML Assistance to High Efficiency Video Coding

Chhaya Shishir Pawar and SudhirKumar D. Sawarkar

Abstract High efficiency video coding (HEVC) standard also known as H.265 is an emerging international standard for video coding developed by joint collaborative team on video coding which is completed in January 2013. It provides around twice as much compression as H.264 for same the visual quality of the video. This bit rate reduction up to 50% comes with the increased complexity in encoder. The complexity is introduced mainly due to the additional angular modes for intraprediction, introduction of larger coding units and their flexible partitioning structure. HEVC encoders are expected to be much complex than H.264/AVC and will be an area of active research in future. This paper proposes a machine learning-based multilevel assistance to intra- and interprediction stage of video coding so as to reduce the computational complexity and avoid the huge wastage of computational efforts. Also brings the context awareness to improve the overall performance of the system.

Keywords HEVC · Encoder complexity · Intraprediction · Interprediction · Video coding

1 Introduction

Digital video finds its application in digital TV, Internet video streaming, entertainment, education, medicine, video delivery on mobile devices, video telephony and video conferencing, video games, video mail, social media, and so more.

These applications became feasible because of advances in computing power, communication technology, and most importantly because of the efficient video compression technologies. Raw video is huge in size for transmission and storage. Hence, the field is rapidly growing and every improvement in the video compression stan-

C. S. Pawar (✉) · S. D. Sawarkar
Datta Meghe College of Engineering, Navi Mumbai, Maharashtra, India
e-mail: Csp.cm.dmce@gmail.com

S. D. Sawarkar
e-mail: Sudhir_sawarkar@yahoo.com

© Springer Nature Singapore Pte Ltd. 2019   313
S. K. Bhatia et al. (eds.), *Advances in Computer Communication
and Computational Sciences*, Advances in Intelligent Systems and Computing 924,
https://doi.org/10.1007/978-981-13-6861-5_27

dard brings a promising future for video compression. The IP video traffic will have 82% share of the entire consumer video traffic by 2021 which was 73% in 2016 [1].

High efficiency video coding is the next in the family of video compression standards. ITU-T's video coding experts group and the ISO/IEC moving picture experts group (MPEG) are involved in the development of HEVC which was completed in January 2013. It achieves almost 50% bit rate reduction than its predecessor. We propose a method to minimize the time complexity of the futuristic video encoding with the help of machine learning.

We propose a multilevel machine learning assistance along with online training is used here to improve the overall performance of the system. The features used here for the learning and the predicting stage are carefully selected to be low complexity coding metrics so as to avoid the overhead caused by this approach.

2 Computational Complexity in HEVC

HEVC intends to cut down the required bit rate to half without affecting the visual quality at the same time remaining network friendly. Such reduction in bit rate comes with the cost of increased encoder complexity. Introduction of flexible partitioning of the coding units in the form of quadtree structures, as many as 35 intraprediction modes, introduction of more sophisticated deblocking filters and interpolation, several features to support efficient parallel processing are the areas features where complexity is involved.

HEVC supports quadtree partitioning of blocks using different types of blocks like coding tree blocks (CTB), coding units (CU), transform units (TU), prediction units (PU). Figure 1 shows the quadtree-based block partitioning which is the distinguishing feature of HEVC which brings most of the coding efficiency of HEVC relative to earlier standards. CTB is the largest coding block and is the root of the coding tree. Every block is further divided into four equal size blocks every time till the maximum depth is reached. This structuring has some disadvantages like unnecessary partitioning which does not contribute much in the rate distortion cost or redundant motion information getting transmitted and so on. If all such CUs exhaustively undergo RDO process, it will involve huge computational complexity. Hence, different approaches are proposed to take fast decision on CU partitioning depth, i.e., by checking whether we can terminate the splitting early.

During the intraprediction phase, greater flexibility was introduced due to increased number of angular intraprediction modes are made available, which was required by increased resolution videos and quadtree partitioning structures. Figure 2 shows the angular intraprediction modes of HEVC. In H.264, there were only 9 modes available, whereas now total of 35 including DC and Planar intraprediction modes are available to choose from. Now the design decision was to select the optimal mode without letting the CTU undergo testing through all the 35 modes along with its sub CUs, which is most computationally inefficient.

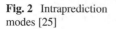

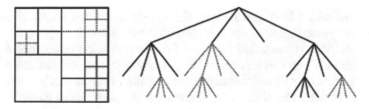

Fig. 1 CTB quadtree partitioning [25]

Fig. 2 Intraprediction
modes [25]

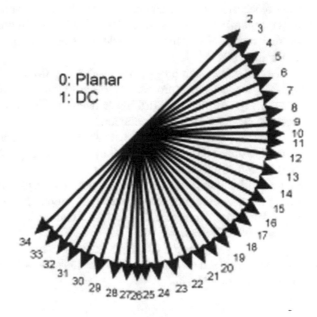

3 Review of Literature

We surveyed through the various research work for the efforts put in order to early
terminate the decision of Coding Unit Splitting, the measures taken in order to reduce
the intraprediction and interprediction complexity. Also reviews the work in the
application of machine learning along with their success rate which provided us with
the research direction and helped deciding which technique is more promising.

3.1 Block Structures

Lei et al. [2] have used CU texture analysis to decide till which depth to carry out the
Coding Unit Partitioning. A down sampling 2:1 filter is used and texture complexity
in terms of its variance is calculated. If texture is flat, it is not further subdivided into

subunits otherwise CU is split further. The algorithm reduced 40.9% of encoding time suffering a negligible loss of average PSNR and bitrate.

Lin et al. [3] have presented the idea of depth correlation of collocated coding tree blocks in the previous and current frame. Using this correlation and using the depth difference, a lot of unnecessary decomposition of blocks can be avoided. The method can provide 16% time saving and maintaining same video quality.

Helle et al. [4] present a method of block merging so that the redundant motion parameters which are being passed due to quadtree partitioning can be minimized. Their block merging core algorithm using the spatial and temporal correlations predicts the merging of blocks and uses the single set of motion parameters for contiguous region of motion compensated blocks. This technique increased average BD rate saving from 6% to 8%.

Lee et al. [5] in their proposed algorithm use SMD, CUSE, and ECUT algorithms together in the overall process. These are based on the RD cost the thresholds. Accordingly, optimal CU partition is decided. The proposed method promised as much as 69% of complexity reduction.

Ahn et al. [6] claimed that the use of spatiotemporal characteristics of CU blocks is effective on the CU partitioning. They used SAO edge parameters as spatial parameters to predict the texture complexity and as motion vector, TU sizes, CBF data to get the motion complexity. Because of use of such parameters which are calculated during the encoding process, no overhead computation is necessary. The presented method provides average of 49.6 and 42.7% of saving on time with minimal loss of 1.4 and 1.0% of bitrate loss.

Park et al. [7] present a CU depth decision method which compares the RD cost and determine the depth using compared information. This achieved an average time saving factor of 44.84%.

3.2 Intrapicture Prediction

Zhang et al. [8], in order to accelerate the process of intraprediction presented a macroscheme for taking the CU depth decision faster and microlevel scheme to select the most effective mode. In the microscheme, the progressive rough mode search based on Hadamard cost is used to select the probable modes along with the early RDOQ skip for avoiding not likely modes to effective modes set. At the macrolevel, they terminate the CU split process if the RD cost of sub-CUs after split is larger than the RD cost of CU without split. The method achieves 2.5 X speed up with just 1% BD rate increase in HEVC test conditions.

Yongfei [9] has proposed an optimal selection of intraprediction mode based on RQT-based coding. They calculate SATD value for all the 35 prediction modes of intraprediction. Then, a subset is formed consisting of the most probable modes and the modes which have smaller SATD value. Then, the RD cost of all the prediction modes in the candidate subset is calculated and the one with the minimum cost is selected as the best prediction mode to encode. It provides 21.29% time saving in intracoding and 15% encoding time is saved in inter coding.

Khan et al. [10] perform the intraangular mode selection by firstly selecting the seed pixel having the measure effect on deciding the intramode decision. These pixels are selected from blocks which have highest impact on the running difference among neighbours. Seed pixels from the current PU are mapped to the seed mode which acts as candidate mode for intraprediction. They proposed to reduce the complexity by estimating the potential intramode by content and priority driven gradient selection and from the earlier computations results. This algorithm offers 44% more time saving as compared to other intramode decision methods.

Kamisli [11] has taken a different approach and exploited a correlation ignored by the standard intraprediction coding algorithms. He modelled the image pixels as Markov processes which uses fewer neighbouring pixels and enables computationally efficient intraprediction algorithm in terms of memory and computational requirements. The much difference between the encoding and decoding times of linear predictions and Markov process are similar in spite of the significant difference between weights and other parameters.

3.3   Inter Picture Prediction

Lin et al. [12] presented a priority-based scheme for spatial motion candidate derived from Motion vector from a different list or picture so as to increase the chances of availability of spatial motion candidates. Here the temporal motion candidate is derived from the previously coded picture in a reference picture list. It is selected based on the predefined order like if the Motion vector that crosses the current picture is selected first and the Motion vector with the same reference list is given a priority over the other reference list. Also, in order to achieve better motion vector prediction, they have presented a surrounding based candidate list where they include temporal candidate from below right corner which is different than HEVC standard which choose from top left corner. Above combination of techniques 3.1% bitrate saving.

Pan [13] proposes an early terminations of TZ search which is adopted in HEVC. By making use of the best point selection in large and small CUs, they proposed to use diamond and hexagonal search to early terminate TZ search. Diamond and hexagonal search find whether the Median predictor is the best initial point. So these help the TZ search decide the best possible initial search point. They evaluated the performance using hit rate and reject rate indicating the complexity and accuracy of search. They claimed 38.96% encoding time saving with acceptable loss in RD performance.

3.4 Machine Learning-Based Approaches

Grellert et al. [14] have made use of Support Vector Machines to propose a fast coding unit partition decision for HEVC. The classifiers were trained and used to decide whether to continue or to early terminate the search for the best CU partitioning structure. Various metrics were studied and selected carefully to decide the set of training features. With this approach, a complexity reduction of 48% is achieved with 0.48%Bjontegaard Delta bitrate (BD-BR) loss.

Zhu et al. [15] presented a fast HEVC algorithm based on binary and multiclass support vector machine. The CU depth decision and PU selection process in HEVC are modelled as hierarchical binary and multiclass classification structures. These processes are optimized by binary and multiclass SVM, without exhaustive Rd cost calculations. This method achieved around 67% time saving on average.

Lin et al. [16] use simple Sobel edge detection to find the strength of the edge in a coding tree block. It helps to limit the maximum depth of the tree which reduces large amount of encoding time. The method gives average time savings of 13.4% with the increase of BD rate by only 0.02%.

Li et al. [17] use the proposed method that separates the intercoding phase into learning stage and predicting stage. In learning stage PU difference, length of motion vector prediction is learned and best searching range is found. In predicting stage, KNN algorithm is used to predict from learned data and 91% complexity is reduced as compared to full search range.

Wu et al. [18], in this work, a Bayesian model based on features is proposed for fast transform unit depth decision. Better results are obtained with the depths of upper, left, and co-located transform unit.

Shen et al. [19] proposed the algorithm computational-friendly features to make coding unit depth decisions. Bayesian rule is used to minimize the RD computations. It saves 41.4% encoding time and minimal loss of 1.88% on RD performance.

Correa et al. [20], in this work, three groups of decision trees are implemented. From the training set, the encoding variable values are extracted to build the decision trees for the early termination of the RD computations by using extracted intermediate data, like encoding variables from a training video sequence. When the three are implemented together, an average time saving of up to 65% is achieved, with a small BD rate increase of 1.36%.

Correa et al. [21] proposes the use of decision trees which decide the early termination of intermode decision. The work predicts the intermode with an accuracy of up to 99% with minimal computational overhead. The method also achieves an overall complexity reduction of 49% with very little increase in BD rate.

Ruiz-Coll et al. [22] used a machine learning classifier to determine the coding unit size for intraprediction phase. The method achieved 30% complexity reduction for high-resolution videos.

Mu et al. [23] propose a support vector machine-based solution which uses the RD cost, MSE, and Number of encoded bits as metrics for speeding up the decision of coding unit depth. It achieves 59% of time saving.

Sun et al. [24] propose a early CU depth decision method which explores the utilization rate of CU at all depths and the neighbouring and collocated CU depth relation. It attains 45% of encoding time improvement with minimal loss of RD performance.

4 Findings of the Literature Review

It is found from the literature survey that larger coding units and additional intraprediction modes has led to increased computational complexity. Recently people have started using machine learning to alleviate the complexity burden on the encoder. It is found that more encoding time saving can be done with machine learning methods.

MPEG is working on requirements for a new video codec which is expected to be completed by 2020. This future generation video codec will try to achieve further bitrate reduction while maintaining the same perceptual quality. But while accommodating the bitrate reduction requirements lot of computational complexity at the encoder is introduced. Thus, reducing the computational complexity while not compromising on quality of video is the key motivation behind this work.

The gap we found is that more attention is always given to offline training of the learning model but being video the work can be enhanced with the use of online training with the help of selected frame that will make the system more context aware. The approach will be helpful for real-time processing and also to account for scene change. Different predictive analytics techniques are useful for different processing for different scenarios. By optimizing the parameters of various learning mechanisms, we can further reduce the computational complexity. The best suitable parameters from visual perception point of view should be mainly considered as ultimate metric for any video coding algorithm is user's viewing experience.

5 Proposed Methodology

The flexible partitioning structure of the coding units increases the coding efficiency of the video compression but it also increases encoding complexity significantly. The frames either undergo intraprediction or interprediction depending upon the type of frame. If the frame is I frame, it has to undergo the process of Intra prediction to identify the spatial redundancy existing in the source frame which can be exploited to improve compression. In order to select the best candidate mode of intraprediction, all the coding units right from size 64×64 till 8×8 and all their possible combinations has to undergo rate distortion optimization process. Every CTU will have as many as 85 number of calculations if exhaustive search for CU partitioning structure is made. Similarly, in interprediction, all the possible combinations of coding units have to undergo RDO optimization process in order to select the optimal partitioning structure of every coding tree unit. Both these processes incur a lot of burden on the entire video compression process.

At the same time with every single frame, a lot of data can be generated from every coding units, their subpartitions, their encoding parameters values, and which amongst all is selected or not selected as best mode or best partitioning structure. Looking at this enormous amount of data, we came up with an idea of using predictive analytics techniques of machine learning to study this huge amount of data offline and analyse this data in order to derive a pattern. Also in future use this learning to help predicting the intraprediction mode for I frame or optimal partitioning structure for all kinds the frames. It is evident from the past that machine learning techniques have been successful in deriving patterns given a huge collection of data sets of appropriate type.

Another concern while using machine learning techniques is the selection of features. Since the problem is of reducing the complexity, so the special care has to be taken while identifying the features. This additional calculation of features should not put any significant overhead on the overall encoding process. Hence, the features selected here are very basic and low complexity features so that it will cost nominal efforts to calculate still provide accurate predictions.

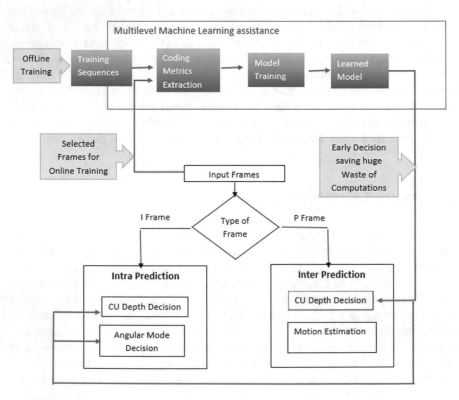

Fig. 3 Proposed framework

Fig. 3 shows the framework proposed to further reduce the complexity of the encoder with the help of machine learning techniques:

- Multilevel ML assistance to intra- and interprediction modes which is essential in the view of even larger coding units in futuristic standards
- Online training of the model along with the offline makes it more context aware
- Use of low complexity coding metrics

This proposed methodology instead of all the coding units undergoing the RDO optimization process for intramode decision or to find the optimal CU splitting we make use of machine learning. In training phase, the various low complexity features like MSE, variance, correlation, standard deviation, etc. are calculated and machine learning algorithms which can be used for predictive analytics like simple decision tree, fine K nearest neighbour, linear support vector machine, ensemble bagged trees, ensemble boosted trees will be implemented for speeding up the coding unit splitting decision.

The algorithm selection will be done on the basis of type of application or the type of characteristics expected in the given context. The compressed video will be assessed with the help peak signal-to-noise ratio, structural similarity index, loss of bitrate and most importantly encoding time reduction.

6 Results and Discussions

When using training the predictive analysis algorithm, it is of utmost Importance to provide the right data set. We have worked on a data set which consists of all the videos of high, medium and low activity. The data set contains varying amount of detail in different videos, different amount of constant background. We have considered raw videos of YUV format of different resolutions.

Table 1 shows the accuracy of various machine learning techniques used. Table 2 shows the speed up we achieved for CU depth decision during Inter prediction process in with the use of machine learning technique over the video encoder without machine learning. It shows the speed up achieved in the futuristic video encoding standard HEVC which currently has rose to 9% share from previous year's 3%. It is still in its evolving stage.

Table 1 Accuracy of various ML methods used

Method	Accuracy
Simple tree	95.8%
Fine KNN	95.6%
SVM linear	94.5%
Ensemble boosted trees	95.1%
Ensemble bagged trees	95.4%

Table 2 % savings in encoding time over video encoding without machine learning

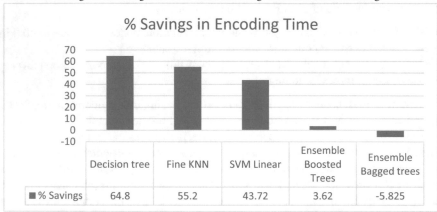

■ % Savings	Decision tree	Fine KNN	SVM Linear	Ensemble Boosted Trees	Ensemble Bagged trees
% Savings	64.8	55.2	43.72	3.62	-5.825

As it can be seen from the above results that machine learning techniques show promising results to be used in video compression domain. Decision tree learning shows around 64–66% of time saving as compared. They are always being easy to implement, interpret. Fine K nearest neighbour saves around 55–57% of encoding time. Support vector machine gives 43–49% of time saving but takes a little longer to train during the training phase. Ensemble methods turn out to be accurate but does not offer much of intended time saving.

7   Conclusion

Being the next generation video codec, high efficiency video coding promises to fulfil the needs of future video coding but at the cost of increased computational complexity. To make it more acceptable and popular, these shortcomings need to be addressed. Looking at the demands of future video applications, new video encoding standards bound to advance. Larger coding units (up to 128×128) with even complicated partitioning structure like quadtree plus binary tree partitioning structure may arise in versatile video coding in future. A machine learning-based multilevel assistance reduces the computational complexity. It avoids the huge wastage of computations in intra- as well as interpredictions by reducing the number of computations during angular mode decision and coding unit depth decision. Also, along with offline training to a learning model, we provide few selected frames of the video and enable online training. This brings more of context awareness which will be a significant requirement of future video coding standards to improve the overall performance of the video compression method. The method achieves a significant reduction in encoding time at the cost of minimal loss of bitrate and PSNR.

References

1. Cisco Visual Networking Index: Forecast and Methodology, CISCO White Paper, pp. 2016–2021, 6 June 2017
2. Lei, H., Yang, Z.: Fast intra prediction mode decision for high efficiency video coding. In: 2nd International Symposium on Computer, Communication, Control and Automation (3CA 2013). Atlantis Press, pp. 34–37
3. Lin, Y.C., Lai, J.C.: A fast depth-correlation algorithm for intra coding in HEVC. Int. J. Sci. Eng. **4**(2), 269–272 (2014)
4. Helle, P., Oudin, S., Bross, B., Marpe, D., Bici, M.O., Ugur, K., Jung, J., Clare, G., Wiegand, T.: Block merging for Quadtree-based partitioning in HEVC. IEEE Trans. Circuits Syst. Video Technol. **22**(12), 1720–1731 (2012)
5. Lee, J., Kim, S., Lim, S., Lee, S.: A fast CU size decision algorithm for HEVC. IEEE Trans. Circuits Syst. Video Technol. https://doi.org/10.1109/tcsvt.2014.2339612 (2013)
6. Ahn, S., Lee, B., Kim, M.: A novel fast CU encoding scheme based on spatio-temporal encoding parameters for HEVC inter coding. IEEE Trans. Circuits Syst. Video Technol. https://doi.org/10.1109/tcsvt.2014.2360031,2014
7. Park, C., Hong, G. S., Kim, B. G.: novel intermode prediction algorithm for high efficiency video coding encoder, vol. 2014, p. 8, Article ID 196035. Hindawi Publishing Corporation Advances in Multimedia. http://dx.doi.org/10.1155/2014/196035
8. Zhang, H., Ma, Z.: Fast intra mode decision for High Efficiency Video Coding (HEVC). IEEE Trans. Circuits Syst. Video Technol. **24**(4), 660–668 (2014)
9. Yongfei, Z., Zhe, L., Mingfei, Z., Bo, L.: Fast residual quad-tree coding for the emerging high efficiency video coding standard. Commun. Software, China Commun. pp. 155–166 (2014)
10. Khan, M. U. K., Shafique, M., Henkel, J.: Fast hierarchical intra angular mode selection for high efficiency video coding. IEEE Int. Conf. Image Proc. ICIP 2014, pp. 3681–3685 (2014)
11. Kamisli, F.: Intra prediction based on markóv process modelling of images. IEEE Trans. Image Proc. **22**(10), 3916–3925 (2013)
12. Lin, J.L., Chen, Y.W., Huang, Y.W., Lei, S.M.: Motion vector coding in the HEVC standard. IEEE J. Sel. Top. Signal Proc. **7**(6), 957–968 (2013)
13. Pan, Z., Zhang, Y., Kwong, S., Xu, L., Wang, X., Xu, L.: Early termination for Tzsearch In Hevc Motion Estimation. In: IEEE International Conference on Acoustic Speech and Signal Processing, ICASSP 2013, pp. 1389–1393. [14] Lee Prangnell, Technical report on Improving Linear Transformation & Quantization in HEVC, HEVC Research, University of Warwick (2014)
14. Grellert, M., Zatt, B., Bampi, S., da Silva Cruz, L. A.: Fast coding unit partition decision for HEVC using support vector machines. IEEE Trans. Circuits Syst. Video Technol. (2018)
15. Zhu, L., Zhang, Y., Wang, R., Kwong, S., Peng, Z.: Binary and multi-class learning based low complexity optimization for HEVC encoding. IEEE Trans. Broadcasting (2017)
16. Lin, Y.C., Lai, J.C.: Feature-based fast coding unit partition algorithm for high efficiency video coding. J. Appl. Res. Technol. **13**, 205–219 (2015)
17. Li, Y., Liu, Y., Yang, H., Yang, D.: An adaptive search range method for HEVC with K nearest neighbour algorithm. IEEE VCIP (2015)
18. Wu, X., Wang, H., Wei, Z.: Bayesian rule based fast TU depth decision algorithm for high efficiency video coding. VCIP 2016, Chengdu, China, 27–30 Nov 2016
19. Shen, X., Yu, L., Che, J.: Fast coding unit size selection for HEVC based on Bayesian Decision rule. In: Picture Coding Symposium 2012, Poland (2012)
20. Correa, G., Assuncao, P.A., Agostini, L.V., da Silva Cruz, L.A.: Fast HEVC encoding decisions using data mining. IEEE Trans. Circuits Syst. Video Technol. **25**(4), 660 (2015)
21. Correa, G., Assuncao, P., Agostini, L., da Silva Cruz, L. A.:Four-step algorithm for early termination in HEVC inter-frame prediction based on decision tree. IEEE Visual Commun. Image Proc. (2014)
22. Ruiz-Coll, D., Adzic, V., Fernández-Escribano, G., Kalva, H., Martínezand, J. L., Cuenca, P.: Fast partitioning algorithm for HEVC intra frame coding using machine learning. ICIP (2014)

23. Mu, F., Song, L., Yang, X., Lu, Z.: Fast coding unit depth decision for HEVC. IEEE Int. Conf. Multimedia Expo Workshops (2014)
24. Leng, J., Sun, L., Ikenaga, T.: Content based hierarchical fast coding unit decision algorithm for HEVC. In: 2011 International Conference on Multimedia and Signal Processing
25. Sullivan, G.J., Ohm, J.R., Wiegand, T., Han, W.J.: Overview of the High Efficiency Video Coding (HEVC) standard. IEEE Trans. Circuits Syst. Video Technol. 22(12), 1649–1668 (2012)

Recognizing Hand-Woven Fabric Pattern Designs Based on Deep Learning

Wichai Puarungroj and Narong Boonsirisumpun

Abstract Hand-woven fabric pattern designs commonly represent the tradition and culture of local communities. Pattern recognition methods can help classify these pattern designs without having to find an expert. The research aims at recognizing woven fabric pattern designs of traditional fabrics called Phasin in Loei province, Thailand based on deep learning methods. In recognizing pattern design, three deep learning models were experimented: Inception-v3, Inception-v4, and MobileNets. The research collected images of real silk fabrics containing 10 pattern designs. For each pattern, there were 180 images segmented and there were 1,800 images in total in the dataset. The data were trained and tested based on 10-fold cross-validation approach. The results of the test show that MobileNets outperforms Inception-v3 and Inception-v4. The test accuracy rates of pattern design recognition for MobileNets, Inception-v3, and Inception-v4 are 94.19, 92.08, and 91.81%, respectively. The future work will be done by carrying out more data training to increase performance, obtaining more pattern designs, and developing an application based on a high-performance model.

Keywords Pattern recognition · Deep learning · Convolutional neural network · Weaving pattern · Loei Phasin · Fabric pattern design

1 Introduction

Traditional woven fabric is gaining higher attraction in Thailand. This is due to the current trend in promoting indigenous fabric conservation, which is similar to the trends in other countries. Furthermore, the governmental agencies also announce

W. Puarungroj (✉) · N. Boonsirisumpun
Computer Science Department, Faculty of Science and Technology, Loei Rajabhat University, Loei 42000, Thailand
e-mail: wichai@lru.ac.th

N. Boonsirisumpun
e-mail: narong.boo@lru.ac.th

© Springer Nature Singapore Pte Ltd. 2019
S. K. Bhatia et al. (eds.), *Advances in Computer Communication and Computational Sciences*, Advances in Intelligent Systems and Computing 924, https://doi.org/10.1007/978-981-13-6861-5_28

325

their encouragement and support the use of local fabrics in daily life. As a result, the hand-woven fabrics and hand-made products using local materials have been highly accepted as essential products. In addition, tourism is another factor that helps promote the textile business. Most tourists prefer woven fabrics that are made up of unique local pattern designs, which present valuable cultures of the place they are being made. Tourism has changed the sluggish hand-woven business into a prosperous, Kimono business in Japan, for example, has been flourishing because of its increasing popularity among tourists [1].

Each piece of fabrics in Loei province in Thailand can contain more than one pattern designs [2]. To recognize and name each design, knowledgeable weavers or village scholars are required to read and interpret. However, it is not possible to find those scholars each time when needed by tourists or other people. Therefore, the use of pattern recognition methods to help automate classification and recognition of pattern designs is required. Image processing and machine learning methods have been employed in previous research in order to recognize fabric pattern designs [3]. Scale-Invariant Feature Transform (SIFT), bags of features, nearest neighbor, and support vector machine (SVM) methods were used to analyze and classify Batik, which is a fabric type commonly produced in Indonesia [4, 5]. This research show some satisfactory results in classifying Batik images with normal and rotated images. Similarly, another work by Nurhaida et al. [6] who applied a SIFT approach to recognize Batik pattern suggest that their improved SIFT approach by combining the Hough transform voting process performed better than the original one. Machine learning methods have also been applied to recognize fabric weaving patterns. Pan et al. [7] constructed weaving fabric pattern database based on Fuzzy C-Mean (FCM) clustering, which keeps standard patterns for comparison. Similarly, Jing et al. [8] classified weaving patterns of fabric by using probabilistic neural network (PNN), which yields a good result from the experiment. More recent recognition and classification methods based on deep learning have been applied to fabric weaving patterns. Boonsirisumpun and Puarungroj [9] conducted an experiment of fabric weaving pattern recognitions based on models of deep neural network, which achieved a good performance.

In this research, the pattern designs of local Thai fabric called Phasin in Loei province were explored. Phasin is a type of fabric normally worn by women in Thailand and neighbor countries. Each piece of local Phasins can contain more than one pattern design, and each design is commonly repetitive. The research attempted to conduct an experiment on classifying and recognizing pattern designs found in Loei's Phasin based on three deep learning models: Inception-v3, Inception-v4, and MobileNets. The performance of these three models was compared in order to find the best model.

2 Related Work

2.1 Pattern Designs of Local Hand-Woven Fabrics

One of the local fabrics in Thailand, worn as a part of traditional clothes, is called Phasin. It has been found in every part of Thailand, especially in the North Eastern. Phasin can be called Sarong due to their similarity in function. However, the appearance of Phasin and Sarong can be different. Women in the North Eastern part of Thailand normally wear two types of Phasins: one for their daily life and another for their special occasions. The first one comes in plain dark colors such as indigo or black with less or no pattern designs, while another is commonly colorful with one or more pattern designs and always worn in some special occasions or religious days in Buddhism. Figure 1 shows Phasin with different pattern designs.

Traditional Phasin in Loei province uses a weaving process called "mud-mee," which defines patterns and their colors. The appearance of the pattern designs is represented in a local style called "mee-kun." Since the term "kun" means separate or has something in between, traditional Loei Phasins add vertical straight lines to divide each pattern design. For example, Phasin in Fig. 1 contains 3 different pattern designs called MhakJub, Phasat, and Naknoi. Each design is separated by some vertical straight lines, which represent a classic identity found in Phasin in Loei province. However, there is no rule for determining how many lines to be used but generally created by the weavers or designers of each piece of Phasin.

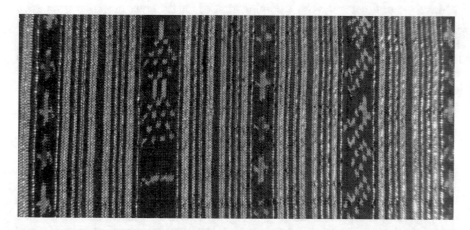

Fig. 1 Example of a piece of Phasin that contains more than one pattern design

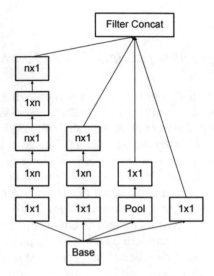

Fig. 2 Example of factorization process in Inception_v3 [11]

2.2 GoogLeNet (Inception-v3)

GoogLeNet or Inception network is the name of a Convolutional Neural Network (CNN) model created by Szegedy et al. in 2014 [10]. The first version consisted of 22 layers neural network with a combination of the different proposed layer such as convolution, max_pooling, fully connected, softmax, and a special layer called inception module.

Later versions were followed almost every year with the improvement in term of accuracy and model architecture. The most often used version now is version 3 or Inception v3 which developed in 2016 [11]. This model changed the convolution node and made them smaller by using the factorization process (See Fig. 2).

2.3 GoogLeNet (Inception-v4)

Inception_v4 was the new design of Inception_v3, by adding more uniform architectures and more inception modules (See Fig. 3). This design accelerated the training module and the performance of the network [12]. The inception_v4 outperformed inception_v3 in terms of accuracy but made the model size larger about double of v3.

Fig. 3 The architecture of Inception_v4 [12]

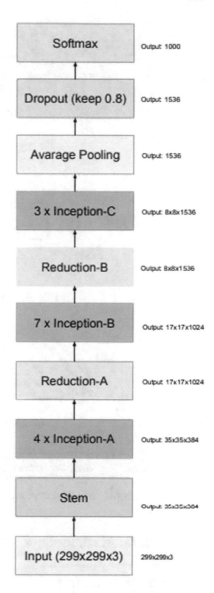

2.4 MobileNets

MobileNets is another model of CNN which has been proposed to decrease the size of previous CNN model making it available to use in mobile platform [13]. Its idea is to replace the standard convolutional filters with two layers (Depthwise and Pointwise convolution) that build a smaller separable filter (See Fig. 4).

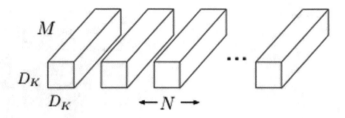

M

D_K

D_K ← N →

(a) Standard Convolution Filters

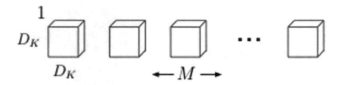

1

D_K

D_K ← M →

(b) Depthwise Convolutional Filters

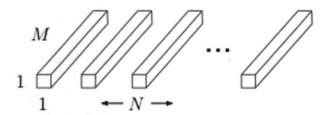

M

1

1   ← N →

(c) 1×1 Convolutional Filters called Pointwise Convolution in the context of Depthwise Separable Convolution

Fig. 4 The depthwise separable convolution filter of MobileNets [14]

3   Research Methodology

3.1   Research Design

In this research, the main research process comprised 4 key steps: (1) Phasin image collection, (2) segmentation of pattern designs images, (3) training based on deep learning models, and (4) testing of trained models. Figure 5 illustrates the key steps carried out in this research in order to recognize and classify pattern designs appeared on Phasin.

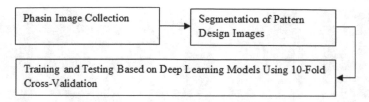

Fig. 5 Research design

3.1.1 Phasin Images Collection

In the first step, the authors surveyed authentic Phasin at various groups of weavers located in different villages in Loei province, Thailand. Images of pattern designs were collected by photo taking. Apart from the images, the name and story behind each pattern were also collected. From this step, there were 10 authentic pattern designs collected on Phasins, which were selected for experimentation.

3.1.2 Segmentation of Pattern Design Images

Phasin images were segmented based on their pattern designs. Since each piece of Phasins contains more than one designs as illustrated in Fig. 1, every pattern appeared on Phasin was segmented. In this process, there were 180 images segmented for each pattern design. In total of 10 pattern designs, there were $180 \times 10 = 1,800$ images in the dataset. These 1,800 images were then trained and tested using 10-fold cross-validation approach in the next step. Figure 6 illustrates examples images in the dataset, which were shown as follows: (1) Mhak Jub, (2) Naknoi, (3) Khor, (4) Aue, (5) Khapia (6) Toom, (7) Nak Khor (8) Phasat (9) Mhak Jub Khoo, and (10) Baipai.

3.1.3 Training and Testing Based on Deep Learning Models

In this step, the training and testing processes were carried out. Segmented training images (180 images for each pattern design) were used to train with three deep learning models called Inception-v3, Inception-v4, and MobileNets. The training and testing were based on 10-fold cross-validation approach where ninefold of training data were trained and the remaining was for testing. To cover all data, 10-fold cross-validation approach allows each fabric pattern design to be tested repeatedly for 10 rounds. The accuracy of this process was calculated. After that, the performance of the three models was evaluated and compared. The highest score was used to determine the pattern design prediction. In this step, the number of correct and incorrect prediction was counted in order to calculate prediction accuracy.

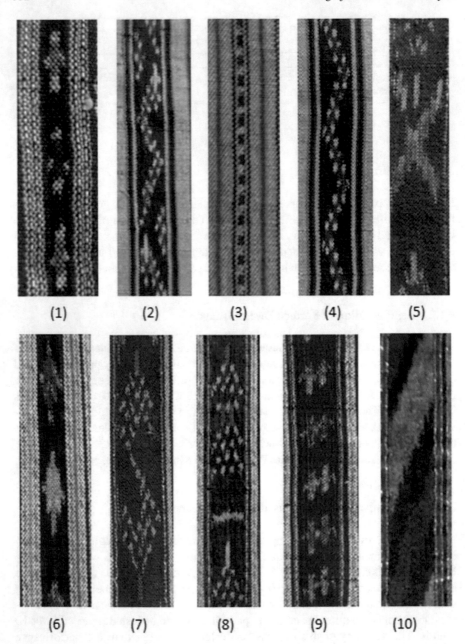

(1)   (2)   (3)   (4)   (5)

(6)   (7)   (8)   (9)   (10)

Fig. 6 Ten authentic pattern design images

4 Results

From the experiment using 10-fold cross-validation approach, train accuracy based on Inception-v3, Inception-v4, and MobileNets is illustrated in Table 1.

From Table 1, the train accuracy of pattern design recognition is reported by each folder. The model that shows powerful prediction is MobileNets with the highest accuracy rate of 95.24%. In descending order, MobileNets, Inception-v3, and Inception-v4 models perform at the rate of 95.24, 93.20, and 93.14%, respectively. Inception-v4, an improved model of Inception-v3, produces the lowest performance. To investigate more on prediction accuracy, Table 2 shows the test accuracy rates of the three models.

In Table 2, the test accuracy results from 10-fold cross-validation experiment have been reported. The results of test accuracy correspond with the results of train accuracy shown in Table 1 where the accuracy rates of MobileNets, Inception-v3, and Inception-v4 are 94.19, 92.08, and 91.81%, respectively. Furthermore, MobileNets shows a surprising result since it can beat other models in both training and testing steps even though its size is the smallest of the three. Interestingly, the experiment yields an unexpected result that Inception-v4, an improved version of Inception-v3, performs worse than its previous version. When viewing by each pattern, the test results of the three models are shown in Table 3.

Table 3 shows the test accuracy by each pattern design. While most of the pattern designs yield high test accuracy, Mhak Jub produces the lowest accuracy rate. From the test, Inception-v3, v4, and MobileNets models can recognize Mhak Jub with the rate of 72.50, 74.72, and 81.39%, respectively. Inception-v3 and Inception-v4 are the models that perform lower than 80% of prediction accuracy.

Table 1 Train accuracy

Folder	Accuracy		
	Inception-v3 (%)	Inception-v4 (%)	MobileNets (%)
Fold0	93.3	94.0	95.3
Fold1	96.3	93.8	96.9
Fold2	92.3	92.3	95.8
Fold3	91.5	92.3	94.4
Fold4	92.3	95.1	98.6
Fold5	92.7	93.3	94.7
Fold6	94.9	93.0	94.3
Fold7	92.3	90.6	94.0
Fold8	94.5	93.8	95.2
Fold9	91.9	93.2	93.2
Average	**93.20**	**93.14**	**95.24**

Table 2 Test accuracy

Folder	Accuracy		
	Inception-v3 (%)	Inception-v4 (%)	MobileNets (%)
Fold0	91.11	90.00	93.33
Fold1	93.33	91.94	95.00
Fold2	95.28	91.67	95.56
Fold3	92.22	91.94	94.17
Fold4	91.94	90.00	93.33
Fold5	92.78	93.06	92.78
Fold6	91.11	93.61	95.28
Fold7	88.06	89.72	93.06
Fold8	91.94	92.78	94.44
Fold9	93.06	93.33	95.00
Average	**92.08**	**91.81**	**94.19**

Table 3 10-Fold average of test accuracy on each pattern

Pattern designs	Accuracy		
	Inception-v3	Inception-v4	MobileNets
Mhak Jub	26.10 (72.50%)	26.90 (74.72%)	29.30 (81.39%)
Naknoi	28.70 (79.72%)	29.50 (81.94%)	30.70 (85.28%)
Khor	31.80 (88.33%)	31.20 (86.67%)	33.50 (93.06%)
Aue	32.20 (89.44%)	31.90 (88.61%)	32.90 (91.38%)
Khapia	36.00 (100%)	36.00 (100%)	36.00 (100%)
Toom	33.10 (91.94%)	31.80 (88.33%)	32.70 (90.83%)
Nak Khor	36.00 (100%)	36.00 (100%)	36.00 (100%)
Phasat	35.60 (98.89%)	35.40 (98.33%)	36.00 (100%)
Mhak Jub Khoo	36.00 (100%)	36.00 (100%)	36.00 (100%)
Baipai	36.00 (100%)	35.80 (99.44%)	36.00 (100%)
Average	33.15 (92.08%)	33.05 (91.81%)	33.91 (94.19%)

5 Conclusion

This research attempted to conduct an experiment on fabric pattern design recognition by using the image dataset collected from the real local hand-woven fabric called Phasin in Loei province, Thailand. Phasin is normally worn by women in Thailand. The pattern designs on Phasin commonly represent the local culture of the village or the region that Phasin has been designed and woven. The experiment was conducted with 10 pattern designs that were found on authentic Phasin in Loei. In total, there were 1,800 images of pattern designs in the dataset. For each pattern, the data were divided into 10-fold where 9-fold were used for training and the remaining was for testing. The training and testing were carried out with three deep learning models called Inception-v3, Inception-v4, and MobileNets. The results of the test show that all models perform well with most of the patterns with more than 90% accuracy rate. However, MobileNets outperforms others even though it has the smallest size. Comparing the 10 pattern designs, Mhak Jub achieves the lowest accuracy rate. Inception-v3, v4, and MobileNets models recognize Mhak Jub with the rate of 72.50, 74.72, and 81.39%, respectively. This is may be caused by the similarity of Mhak Jub to other patterns such as Toom. The test results also show that Inception-v3 performs better than Inception-v4. In the future research, there is more work to be done to improve the recognition accuracy of hand-woven fabric pattern designs. First of all, the research will carry out more training, especially for the pattern design with low accuracy. Secondly, the research will test the dataset with other deep learning models. Furthermore, the deep learning models can be improved to meet the specification of the fabric pattern design images. Thirdly, the research will acquire more pattern designs from weaver groups in order to make the models possible for recognizing most of the pattern design in the province. Finally, the research will develop a mobile app based on high-performance models to make it possible for actual use.

References

1. Russell, L.: Assessing Japan's inbound tourism: a SWOT analysis. Hannan Thesis. Social Sci. Edition **53**(1), 21–50 (2017)
2. Wicaksono, A.Y., Suciati, N., Fatichah, C., Uchimura, K., Koutaki, G.: Modified convolutional neural network architecture for batik motif image classification. IPTEK J. Sci. **2**(2), 26–30 (2017)
3. Zhang, R., Xin, B.: A review of woven fabric pattern recognition based on image processing technology. Res. J. Text. Apparel **20**(1), 37–47 (2016)
4. Azhar, R., Tuwohingide, D., Kamudi, D., Suciati, N.: Batik image classification using SIFT feature extraction, bag of features and support vector machine. Proc. Comput. Sci. **72**, 24–30 (2015)
5. Loke, K. S.: An approach to textile recognition. In: Yin, P. (ed.), Pattern Recognition, pp. 439–459. InTechOpen (2009)
6. Nurhaida, I., Noviyanto, A., Manurung, R., Arymurthy, A.M.: Automatic Indonesian's Batik pattern recognition using SIFT approach. Proc. Comput. Sci. **59**, 567–576 (2015)
7. Pan, R., Gao, W., Liu, J., Wang, H.: Automatic recognition of woven fabric patterns based on pattern database. Fibers Polymers **11**(2), 303–308 (2010)
8. Jing, J., Xu, M., Li, P., Li, Q., Liu, S.: Automatic classification of woven fabric structure based on texture feature and PNN. Fibers Polymers **15**(5), 1092–1098 (2014)
9. Boonsirisumpun, N., Puarungroj, W.: Loei fabric weaving pattern recognition using deep neural network. In: 15th International Joint Conference Computer Science and Software Engineering, IEEE, Thailand (2018)
10. Szegedy, C., Liu, W., Jia, Y., Sermanet, P., Reed, S., Anguelov, D., Erhan, D., Vanhoucke, V., Rabinovich, A.: Going deeper with convolutions. arXiv preprint arXiv:1409.4842, 7 (2014)
11. Szegedy, C., Vanhoucke, V., Ioffe, S., Shlens, J., Wojna, Z.: Rethinking the inception architecture for computer vision. In Proceedings of the IEEE Conference on Computer Vision and Pattern Recognition, pp. 2818–2826 (2016)
12. Szegedy, C., Ioffe, S., Vanhoucke, V., Alemi, A. A.: February. Inception-v4, inception-resnet and the impact of residual connections on learning. In AAAI, 4, 12 (2017)

13. Lee, E., Ji, Y.: Design of fabric pattern recognition search system. Adv. Sci. Technol. Lett. **129**, 133–138 (2016)
14. Howard, A. G., Menglong, Z., Bo, C., Dmitry, K., Weijun, W., Tobias, W., Marco, A., Hartwig, A.: MobileNets: Efficient convolutional neural networks for mobile vision applications. arXiv preprint arXiv:1704.04861 (2017)

Learning Curve as a Knowledge-Based Dynamic Fuzzy Set: A Markov Process Model

Siana Halim, Rolly Intan and Lily Puspa Dewi

Abstract In the fuzzy set theory introduced by Zadeh [15], membership degree of a fuzzy set is determined by a static membership function, i.e., it does not change over time. To improve this condition, then Wang introduced the dynamic fuzzy logic. In this concept, the membership degree of a fuzzy set is changing over time. Intan and Mukaido [5] introduced the knowledge based fuzzy set, by means that the membership degree of a set is dependent on the knowledge of a person. Since the knowledge of a person is not static, the knowledge-based fuzzy set can be measured dynamically over time, so that we have the knowledge-based dynamic fuzzy set. In this paper, we approximate the learning process as a knowledge-based dynamic fuzzy set. We consider that the process of learning is dependent on the knowledge of a person from time to time so that we can model the learning process is a Markov process of dynamic knowledge. Additionally, using the triangular fuzzy number, we follow Yabuuchi et al. (2014), for modeling the time difference in the dynamic knowledge fuzzy set as an autoregressive model of order one.

Keywords Learning curve · Fuzzy dynamic · Knowledge-based fuzzy set · Markov process · Autoregressive

1 Introduction

When life is started, then the learning process has begun. The life is not only the life of a human being, but also, for example, the life of machines, organization, enterprises, and society. In the learning process, either people or machines need to do the same task repeatedly in a series of trials and errors. By doing many exercises, knowledge will be learned over time. The learning curve shows the rate of, e.g., a person, a machine, enterprise's, progress in gaining experience or a new skill. It shows the changes over time or processing cost used in gaining experiences.

S. Halim (✉) · R. Intan · L. P. Dewi
Petra Christian University, Surabaya, Indonesia
e-mail: halim@petra.ac.id

© Springer Nature Singapore Pte Ltd. 2019
S. K. Bhatia et al. (eds.), *Advances in Computer Communication
and Computational Sciences*, Advances in Intelligent Systems and Computing 924,
https://doi.org/10.1007/978-981-13-6861-5_29

So far, there exist several mathematical models for representing the learning curve as the function of experience. The common function that is used to figure the general form of all learning curves is the *S*-curve or the sigmoid function. This curve accumulates the small steps at first slowly, and then, gradually it moves faster, as the learning activity mature and reaches its limit. Kucharavy and De Guio [8] resumed the application of the S-shaped curve, or the sigmoid curve in several types of learning, mainly in the inventive problem solving, innovation, and technology forecasts. Jónás [6] explained the growth of the learning process in manufacturing and service management. He also extended the concept of reliability, derived from fuzzy theory.

Some learning processes can also represent by exponential functions, especially if the proficiency can increase without limit. Ellis et al. [3] depicted the student's emerging understanding as exponential growth. In this paper, Ellis et al. exhibited the student's thinking and learning over time to a set of tasks and activities given by the instructors. The exponential function represents most of the students' learning growth since we all hope that their proficiency is increasing without limit. Another way to present the learning phenomena is by using the power law function. In this point of view, learning is considered as repetitive tasks, so that the more repetitive has done, the smaller the time needed to learn that task. Tanaka et al. [13] used the fuzzy regression model for analyzing the learning process. Since the learning process can also be seen as the process of a person or a machine gain knowledge, in this paper, we proposed a dynamic knowledge-based fuzzy set, for representing the learning curve.

2   Knowledge-Based Dynamic Fuzzy Set

In general, at least there are two types of uncertainty, that is, the fuzziness and the randomness (Intan and Mukaido [5]). The randomness is usually measured using the probabilistic measure, while the fuzziness is measured using membership degrees of elements. In 1965, Zadeh [15, 16] introduced the fuzzy set. In this set, the membership degrees of elements are represented by a real number in an interval [0, 1]. The membership degrees of a set are not always the same along the time. People can change their mind, so that they may have different measurement to the same object from time to time. By these characteristics, then the membership degrees of an object are dynamic.

Researchers in the fuzzy logic see the dynamic problems from two sides. Li [11]; Silva and Rosano [12]; Leonid [10] extended the dynamic terminology in the logic process sense, while the dynamic in the fuzzy set is represented as the changes of the membership degree as time series (Song and Chissom, 1993; Lee and Chou [9]; Cheng and Chen [2]). Intan and Mukaido [5] stated that the membership degrees of elements in the fuzzy set are dependent on the knowledge of a person. Knowledge is essential for defining a fuzzy set.

Table 1 Knowledge-based fuzzy set of A

$K(t, A)$	$u_1 \ldots u_n$
$k_1(t, A)$	$k_1(t, A)(u_1) \ldots k_1(t, A)(u_n)$
$\vdots$	$\ddots$
$k_m(t, A)$	$k_m(t, A)(u_1) \ldots k_m(t, A)(u_n)$

Intan et al. (2018) combined the knowledge-based fuzzy set and dynamic fuzzy set to be a knowledge-based dynamic fuzzy set.

Definition 1: let $U = \{u_1, \ldots, u_n\}$ be a set of elements and $K = \{k_1, \ldots, k_m\}$ be a set of knowledge. A fuzzy set A in U based on a knowledge $k_i \in K$ at time $t \in T$, where T is the set of time is notated as $k_i(t, A))$. To simplify the notation, the $k_i(t, A)$ is also defined as the membership function which map U to a closed interval $[0, 1]$. This mapping can be written as the membership function of a fuzzy set as follow:

$$k_i(t, A) : U \rightarrow [0, 1] \tag{1}$$

A fuzzy set A on U can be written as a fuzzy information table (see Table 1), where $K(t, \Lambda) = \{k_1(t, \Lambda), \ldots, k_m(t, A)\}$ is a knowledge-based dynamic fuzzy time set of A.

where $k_i(t, A)(u_j) \in [0, 1]$ is the membership degree of element u_j in the fuzzy set A which is measured using knowledge k_i at time t.

3 Learning Curve as a Knowledge-Based Dynamic Fuzzy Set

Supposing knowledge of learners on measuring an object A at time t is independent to each other, then we can state that $k_l(t, A)$ is independent to $k_m(t, A)$, $l \neq m$. However, knowledge of a learner on measuring an object A at time t_i is dependent on the basic knowledge that he or she has, let say $k_l(t_0, A)$. Additionally, it is also dependent on the knowledge that he or she gains during the process of learning $k_l(t_{i-j}, A)$, $j = 1, \ldots, i$. For simplicity, we will write this learning process as $k(t_i, A)$, $i = 1, \ldots, T$. Therefore, in this paper, we propose to define the learning curve as a knowledge-based dynamic fuzzy set. The curve can be represented as the dynamic process of knowledge of a person in measuring an object, especially if the object is not in a crisp set.

4 Knowledge-Based Dynamic Fuzzy Set–Learning Curve as a Markov Process

Since the process of learning is dependent on the knowledge of a person from time to time, we then can assume that the learning process is a Markov process of dynamic knowledge, that is:

$$k(t_i|t_{i-1}, \ldots, t_0; A) = k(t_i|t_{i-1}; A) \qquad (2)$$

Equation (2) can be interpreted as the membership degree given by a person when he or she measures an object A at time t_i, given that the membership degree based on his or her knowledge on the past is only dependent on the membership degree his or knowledge at a time. In the Markov model Eq. (2), we also can say that the person's knowledge on measuring of an object is changing step by step.

The Markov process in this model can be written as the autoregressive model of order one (Kallenberg [7]. Since the learning process is not a stationary process, but it has the inclination, and settle at a particular time, then the series of membership degree of knowledge should be adjusted by taking a difference at order one, then we have

$$w(t_i; A) = \Delta k(t_i; A) = k(t_i; A) - k(t_{i-1}; A) \qquad (3)$$

Now, $w(t_i; A)$ can be modeled as the autoregressive of order one, that is

$$w(t_i; A) = \mu + \phi w(t_{i-1}; A) + \varepsilon \qquad (4)$$

There are several ways to model the fuzzy autoregressive models, and Yabuuchi et al. [14] used the triangular fuzzy number for constructing the fuzzy autoregressive models. Other models of fuzzy autoregressive were developed by Fukuda and Sunahara [4]. In this study, we follow Yabuuchi et al. [14] and defined $w(t_i; A)$ as a triangular fuzzy number with $w(t_i; A) = (\alpha_t, \beta_t, \delta_t), \alpha_t \le \beta_t \le \delta_t$.

Following Yabuuchi et al. (2014), the fuzzy autoregressive of order one can be defined as follows:

$$\tilde{w}(t_i; A) = \mu + \phi w(t_{i-1}; A) + \varepsilon \qquad (5)$$

$$w(t_i; A) \subseteq \tilde{w}(t_i; A), \tilde{w}(t_i; A) = \left(\tilde{\alpha}_t, \tilde{\beta}_t, \tilde{\delta}_t \right) \qquad (6)$$

Defined the triangular fuzzy number

$$u = \left(u_\alpha, u_\beta, u_\delta \right) \qquad (7)$$

To minimize the ambiguity of the model, then:

$$\text{Minimize} \sum_{t=2}^{T} \left(\tilde{\delta}_t - \tilde{\alpha}_t \right) \tag{8}$$

$$\text{s.t.} \quad \alpha_t \geq \tilde{\alpha}_t, \delta_t \leq \tilde{\delta}_t, t = 2, \ldots, T$$

$$u_\alpha \leq u_\delta$$

and the parameter, ϕ, can be estimated via Yule–Walker estimate (Brockwell and Davis [1].

5 Conclusion

In this paper, we proposed the knowledge-based dynamic fuzzy set to represent the learning process. The proposed learning curve can be applied to measure the learning process for recognizing objects when a person is not certain about his or her judgment, e.g., "about 10 m," "about 25 years," "about 36 kg," and so on, or if the objects are not in the crisp set, e.g., "smooth," "big," "large," and so on. Additionally, we also model the learning curve as a Markov process and applied the fuzzy autoregressive model of Yabuuchi et al. [14] for modeling the function of a learning curve. In future research, we will consider the dependency on the knowledge of other people in the learning process. We also will relax the assumption in the Markov process so that the dependencies in the autoregressive are not only in order one.

References

1. Brockwell, P. J., Davis, R. A.: Time series: theory and methods, pp. 238–272. Springer, New York (1991)
2. Cheng, C.H., Chen, C.H.: Fuzzy time series model based on weighted association rule for financial market forecasting. Expert Systems. Wiley online Library (2018)
3. Ellis, A.B., Ozgur, Z., Kulow, M., Dogan, M.F., Amidon, J.: An exponential growth learning trajectory: student's emerging understanding of exponential growth through covariation. Math. Thinking Learning **18**(3), 151–181 (2016)
4. Fukuda, T., Sunahara, Y.: Parameter identification of fuzzy autoregressive models. IFAC Proc., **26**(2) Part 5, 689–692, (1993)
5. Intan, R., Mukaidono, M.: On knowledge-based fuzzy sets. Int. J. Fuzzy Syst. **4**(2), 655–664 (2002)
6. Jónás, T.: Sigmoid functions in reliability based management. Soc. Manag. Sci. **15**(2), 67–72 (2007)
7. Kallenberg, O.: Foundation of Modern Probability. Springer, New-York (2002)
8. Kucharavy, D., De Guio, R.: Application of S-shaped Curve. Proc. Eng. **9**, 559–572 (2011)
9. Lee, H.S., Chou, M.T.: Fuzzy forecasting based on fuzzy time series. Int. J. Comput. Math. **81**(7), 781–789 (2004)
10. Leonid, P.: Dynamic fuzzy logic, denotational mathematics, and cognition modeling. J. Adv. Math. Appl. **1**(1), 27–41 (2012)
11. Li, F.: Dynamic fuzzy logic and its applications. Nova-Science Publisher, New York (2008)

12. Silva, J. L. P., Rosano, F. L.: Dynamic fuzzy logic. In: Mexican International Conference on Artificial Intelligence (Micai), Lectures Notes in Computer Science, 1793, 661–670, (2000)
13. Tanaka, H., Shimomura, T., Watada, J.: Identification of learning curve based on fuzzy regression model. Japn. J. Ergon. **22**(5), 253–258 (1986)
14. Yabuuchi, Y., Watada, J., Toyoura, Y.: Fuzzy AR model of stock price. Scientiae Mathematicae Japonicae **10**, 485–492 (2004)
15. Zadeh, L.A.: Fuzzy sets. Inf. Control **8**, 338–353 (1965)
16. Zadeh, L.A.: Fuzzy sets and systems. Int. J. Gen Syst. **17**, 129–138 (1990)

Human Body Shape Clustering for Apparel Industry Using PCA-Based Probabilistic Neural Network

YingMei Xing, ZhuJun Wang, JianPing Wang, Yan Kan, Na Zhang
and XuMan Shi

Abstract Aiming to cluster human body shapes much faster, accurately and intelligently for apparel industry, a new clustering approach was presented based on PCA_PNN model in this study, which referred to a kind of probabilistic neural network combining with principal components analysis. The specific implementation process could include the following steps. Firstly, the human body data were acquired by 3D anthropometric technique. Secondly, after being preprocessed, the human anthropometric data acquired were analyzed by PCA, in order to reduce the data volume. Sequentially, a PNN model for clustering human body shapes was established and evaluated, the input layer of which was composed of the factor scores resulting from PCA and the output layer of which was the body shape category. The experiment results showed that PCA_PNN model presented was a good nonlinear approximately network which was able to cluster male human body shape accurately and intelligently.

Keywords Human body shape clustering · Principal component analysis ·
Artificial neural networks · Probabilistic neural network

Y. Xing · Z. Wang · Y. Kan · N. Zhang · X. Shi
School of Textile and Garment, Anhui Polytechnic University, Wuhu 241000, China

Z. Wang · J. Wang (✉)
College of Fashion and Design, Donghua University, Shanghai 200051, China
e-mail: wangjp@dhu.edu.cn

Y. Xing · Z. Wang
Anhui Province College Key Laboratory of Textile Fabrics, Anhui Polytechnic University,
Wuhu 241000, China

Anhui Engineering and Technology Research Center of Textile, Anhui Polytechnic University,
Wuhu 241000, China
e-mail: hqxiaopan@126.com

© Springer Nature Singapore Pte Ltd. 2019
S. K. Bhatia et al. (eds.), *Advances in Computer Communication
and Computational Sciences*, Advances in Intelligent Systems and Computing 924,
https://doi.org/10.1007/978-981-13-6861-5_30

1 Introduction

In apparel industry, human body shape clustering is a prerequisite for designing garments with appropriate fit. Once the body shapes have been analyzed and understood, it could be easy to enable the production of correctly sized clothing for the potential population under consideration.

In the past few decades, various techniques have been employed for the distribution of the human body shapes into subdivisions, including bivariate analysis and multivariate analysis technique. Bivariate analysis is a statistical method of determining the strength of relationship between two variables, which was popular in the early stage of ready-to-wear apparel industry [1, 2]. Later, multivariate analysis based on principal component analysis (PCA) was introduced into the local or whole body shape clustering [3–8]. However, the traditional and common methods mentioned above, neither bivariate analysis nor multivariate analysis, could represent body shape precisely. Recently, more and more sophisticated and advanced techniques, such as decision tree, fuzzy classification, artificial neural networks (ANNs), have been introduced into classifying population [9–12].

In the field of artificial intelligence and machine learning, artificial neural networks (ANNs) are presumably the best known biologically inspired method, which are inspired by animals' central nervous systems (in particular the brain) that have cognitive abilities. Their goals are to solve hard problems in real-world like pattern recognition, classification, and generalization by constructing simplified computational models [13]. ANNs have the capabilities of properly classifying a highly nonlinear relationship, and once being trained, they can classify new data much faster than it would be possible, by solving the model analytically. Nowadays, the application of ANNs has made progress in textile and apparel areas, such as fiber classification, yarn, fabric, nonwoven and cloth defect detection and categorization, yarn and fabric properties prediction and modeling, process behavior prediction, color coordinates conversion, color separation and categorization, and color matching recipe prediction [14–17]. However, little attention has been devoted to the application of ANNs in garment pattern design, especially in human body shape clustering. In this study, the male human body shapes were taken for instance, and a new clustering approach was proposed by modeling PCA_PNN, which referred to a kind of probabilistic neural network combining with principal component analysis.

2 Research Scheme

2.1 Basic Theory of Probabilistic Neural Network

Probabilistic neural network (PNN) is a kind of ANNs based on the well-known Bayesian classification [18, 19]. PNN has similarity to back-propagation artificial neural network in the way they forwarded, while having dissimilarity in learning

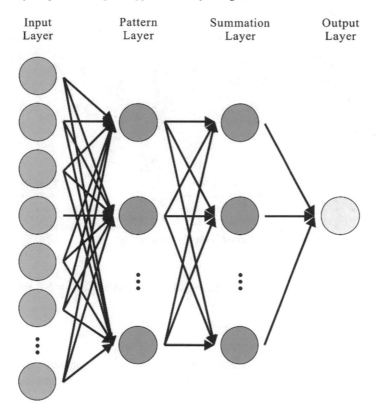

Fig. 1 Architecture of probabilistic neural network (PNN)

procedure. As shown in Fig. 1, the architecture of PNN composes of an input layer, pattern layer, summation layer, and the output layer. The pattern layer, also called hidden layer, has similarity to radial basis function artificial neural network, while the summation layer has similarity to a competitive network. The numbers of neurons in pattern layer are the same as input sample numbers and the numbers of neurons in summation layer are equal to those of target class. Due to its fast training capability and accuracy in classification, PNN was applied as a classifier for human body types in this study.

2.2 General Schemes

The research scheme in this study was described in Fig. 2. Initially, the original human body data were collected by 3D anthropometric technique. Then, after being preprocessed, the data were analyzed by PCA. Afterward, a PNN model for clustering body shapes was established and evaluated. The clustering model framework proposed was illustrated in Fig. 2.

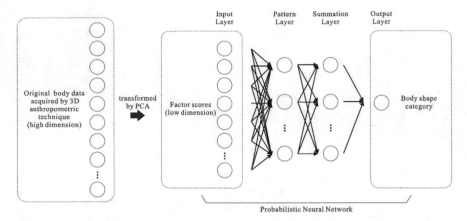

Fig. 2 PCA_PNN model for male human body shape clustering

3 Principal Components Analysis on Human Body Shapes

3.1 Data Acquisition

The anthropometric data were acquired by 3D human body scanner. After that, 27 measurements were selected for further analysis elaborately, as shown in Fig. 3 and Table 1.

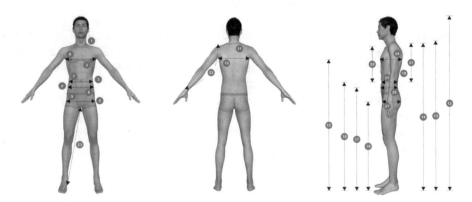

Fig. 3 Illustration of 27 anthropometric measurements for further analysis

Table 1 Twenty-seven anthropometric measurements for further analysis

No	Measurements	No	Measurements	No	Measurements
1	Neck circumference	10	Back breadth	19	Center back length
2	Bust circumference	11	Shoulder breadth	20	Front waist length
3	Waist circumference	12	Statue	21	Back waist length
4	Abdominal circumference	13	Cervical height	22	Arm length
5	Hip circumference	14	Shoulder height	23	Inside leg length
6	Chest breadth	15	Bust height	24	Chest depth
7	Waist breadth	16	Waist height	25	Waist depth
8	Abdominal breadth	17	Abdominal height	26	Abdominal depth
9	Hip breadth	18	Hip height	27	Hip depth

Table 2 Results of KMO and Bartlett's test

Kaiser–Meyer–Olkin measure of sampling adequacy	Bartlett's test of sphericity		
	Approx. chi-square	Df	Sig.
0.921	6557.585	351	0.000

3.2 Data Preprocessing

For principal components analysis, KMO is supposed to be greater than 0.5 and Bartlett's test less than 0.05. As shown in Table 2, the KMO value was up to 0.921, which was greater than 0.9. Bartlett's test of sphericity was highly significant since the observed significance level was $0.000 (p < 0.005)$, which implied that it was likely to factor well for PCA.

3.3 Data Analysis

A new data set which is transformed from the original human body data set by PCA was constructed. The new data set has much smaller dimensionality than the original data set while retaining most of the useful information of human body shape clustering.

As shown in Table 3, the first component had a variance value of 41.2% followed by others having a value of 26.7, 5.3, 4.0, and 3.1%, respectively. The first five rotation sums of squared loading showed a cumulative percentage of 80.3%. Therefore, the factor scores of the first five components were selected for the input neurons of input layer.

Table 3 Total variance explained

Component	Eigenvalues	% of variance	Cumulative %	Component	Eigenvalues	% of variance	Cumulative %
1	11.120	41.184	41.184	15	0.203	0.752	95.612
2	7.215	26.724	67.908	16	0.175	0.649	96.261
3	1.436	5.317	73.225	17	0.152	0.561	96.822
4	1.092	4.043	77.267	18	0.140	0.519	97.341
5	0.824	3.051	80.318	19	0.127	0.471	97.812
6	0.745	2.758	83.077	20	0.123	0.454	98.266
7	0.672	2.489	85.566	21	0.117	0.432	98.698
8	0.487	1.805	87.371	22	0.105	0.390	99.088
9	0.463	1.716	89.087	23	0.085	0.314	99.402
10	0.411	1.523	90.609	24	0.051	0.188	99.590
11	0.321	1.187	91.796	25	0.047	0.176	99.766
12	0.300	1.112	92.908	26	0.040	0.148	99.914
13	0.276	1.023	93.932	27	0.023	0.086	100.000
14	0.251	0.928	94.860				

4 Establishment of PNN Model for Human Body Shape Clustering

In this study, the inputs of the proposed model were the factor scores, and the outputs of the proposed model were a code representing the body type. The concept and data involved in the construction process of the model were formalized as follows:

Let p be a number of input samples.
Let n be a number of neurons in input layer.
Let $D = (d_1, d_2, ..., d_n)$ be a set of input vector in input layer.
Let $C = (c_1, c_2, ..., c_n)$ be a set of center vector formed by input vector.
Let E be a matrix of Euclidean distance between the samples d_i and c_j.
Let $G(\cdot)$ be activation function in pattern layer.
Let P be the initial probability matrix in pattern layer.
Let k be a number of samples for every category in summation layer.
Let S be the sum of initial probability matrix in summation layer.
Let prob be the probability for the ith sample belonging to the jth category.
Let OP be the decision layer output.

4.1 Establishment of PNN Model

The clustering model was established using PNN, through six steps as follows:

Step 1: Distribute the element of D to the input layer:

$$D = \begin{bmatrix} d_{11} & d_{12} & \cdots & d_{1n} \\ d_{21} & d_{22} & \cdots & d_{2n} \\ \cdots & \cdots & \cdots & \cdots \\ d_{p1} & d_{p2} & \cdots & d_{pn} \end{bmatrix} = \begin{bmatrix} d_1 \\ d_2 \\ \cdots \\ d_p \end{bmatrix} \tag{1}$$

The elements of the vectors D are directly passed to the input layer of PNN. And then, the input neurons transmit the input to the neurons in pattern layer, without performing any calculation.

Step 2: Calculate the Euclidean distance matrix E in the pattern layer;

The Euclidean metrics between the vectors D and C are computed in pattern layer. For a given input vector D and a center vector C, the all neurons Euclidean distance matrix E in pattern layer is formed:

$$
E =
\begin{bmatrix}
\sqrt{\sum\limits_{k=1}^{n} |d_{1k} - c_{1k}|^2} & \sqrt{\sum\limits_{k=1}^{n} |d_{1k} - c_{2k}|^2} & \cdots & \sqrt{\sum\limits_{k=1}^{n} |d_{1k} - c_{mk}|^2} \\
\sqrt{\sum\limits_{k=1}^{n} |d_{2k} - c_{1k}|^2} & \sqrt{\sum\limits_{k=1}^{n} |d_{2k} - c_{2k}|^2} & \cdots & \sqrt{\sum\limits_{k=1}^{n} |d_{2k} - c_{mk}|^2} \\
\cdots & \cdots & \cdots & \cdots \\
\sqrt{\sum\limits_{k=1}^{n} |d_{pk} - c_{1k}|^2} & \sqrt{\sum\limits_{k=1}^{n} |d_{pk} - c_{2k}|^2} & \cdots & \sqrt{\sum\limits_{k=1}^{n} |d_{pk} - c_{mk}|^2}
\end{bmatrix}
$$

$$
=
\begin{bmatrix}
E_{11} & E_{12} & \cdots & E_{1m} \\
E_{21} & E_{22} & \cdots & E_{2m} \\
\cdots & \cdots & \cdots & \cdots \\
E_{p1} & E_{p2} & \cdots & E_{pm}
\end{bmatrix}
\tag{2}
$$

Step 3: Calculate the initial probability matrix P in the pattern layer.

The neuron output in pattern layer is activated by a radial Gauss function G:

$$
G(D, c_i) = \exp\left[-\frac{(D - c_i)^{\mathrm{T}}(D - c_i)}{2\sigma^2}\right] = \exp\left[-\frac{E_{pm}}{2\sigma^2}\right], \, i = 1, 2, \ldots, p
\tag{3}
$$

where σ, also known as spread, is the smoothing parameter, and distance between vectors D and c_i is calculated by an Euclidean metrics.

For a given input vector D, the all neurons initial probability matrix P in pattern layer is formed:

$$
P =
\begin{bmatrix}
G(d_{1k}, c_{1k}) & G(d_{1k}, c_{2k}) & \cdots & G(d_{1k}, c_{mk}) \\
G(d_{2k}, c_{1k}) & G(d_{2k}, c_{2k}) & \cdots & G(d_{2k}, c_{mk}) \\
\cdots & \cdots & \cdots & \cdots \\
G(d_{pk}, c_{1k}) & G(d_{pk}, c_{2k}) & \cdots & G(d_{pk}, c_{mk})
\end{bmatrix}
=
\begin{bmatrix}
e^{-\frac{E_{11}}{2\sigma^2}} & e^{-\frac{E_{12}}{2\sigma^2}} & \cdots & e^{-\frac{E_{1m}}{2\sigma^2}} \\
e^{-\frac{E_{21}}{2\sigma^2}} & e^{-\frac{E_{22}}{2\sigma^2}} & \cdots & e^{-\frac{E_{2m}}{2\sigma^2}} \\
\cdots & \cdots & \cdots & \cdots \\
e^{-\frac{E_{p1}}{2\sigma^2}} & e^{-\frac{E_{p2}}{2\sigma^2}} & \cdots & e^{-\frac{E_{pm}}{2\sigma^2}}
\end{bmatrix}
$$

$$
=
\begin{bmatrix}
P_{11} & P_{12} & \cdots & P_{1m} \\
P_{21} & P_{22} & \cdots & P_{2m} \\
\cdots & \cdots & \cdots & \cdots \\
P_{p1} & P_{p2} & \cdots & P_{pm}
\end{bmatrix}
\tag{4}
$$

Step 4: Calculate the sum of probability for each sample belonging to every category in the summation layer.

After receiving the input from pattern layer, the sum of probability for each sample belonging to every category is calculated in the summation layer:

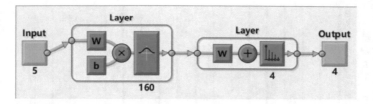

Fig. 4 Four layers' probabilistic neural network

$$S = \begin{bmatrix} \sum_{l=1}^{k} P_{1l} & \sum_{l=k+1}^{2k} P_{1l} & \cdots & \sum_{l=m-k+1}^{m} P_n \\ \sum_{l=1}^{k} P_{2l} & \sum_{l=1}^{2k} P_{2l} & \cdots & \sum_{l=m-k+1}^{m} P_{2l} \\ \cdots & \cdots & \cdots & \cdots \\ \sum_{l=1}^{k} P_{pl} & \sum_{l=k+1}^{2k} P_{pl} & \cdots & \sum_{l=m-k+1}^{m} P_{pl} \end{bmatrix} = \begin{bmatrix} S_{11} & S_{12} & \cdots & S_{1c} \\ S_{21} & S_{22} & \cdots & S_{2c} \\ \cdots & \cdots & \cdots & \cdots \\ S_{p1} & S_{p2} & \cdots & S_{pc} \end{bmatrix} \quad (5)$$

Step 5: Calculate the probability for the ith sample belonging to the jth category.

$$\text{prob}_{ij} = \frac{S_{ij}}{\sum_{l=1}^{c} S_{il}}, i = 1, 2, \ldots, p, j = 1, 2, \ldots, c \quad (6)$$

Step 6: Decision the category of human body type.

$$OP = \arg \max(\text{prob}_{ij}) \quad (7)$$

The output layer is decision layer. According to Bayes classification rule, select the category having the maximum posterior probability as the decision layer output. The PNN model for body shape clustering established was shown in Fig. 4.

4.2 Evaluation

According to the factor scores, the 200 male human anthropometric samples were clustered into four categories by K-means quick cluster method. The clustering results were acted as the expired outputs of the PNN model. The original data set was divided into training sample set and testing sample set. Training sample set consisted of 160 typical samples, in which the number of every category was equal to each other, and testing sample set was composed of the rest 40 samples. The data shown in Table 4 were a part of the training data.

Table 4 Training data set

S. No.	Input learning data					Output learning data
	Factor score 1	Factor score 2	Factor score 3	Factor score 4	Factor score 5	Expired body type code
1	0.29964	0.96205	1.48537	−0.60995	0.28543	1
2	1.26958	−0.24709	1.26733	1.75589	−0.28033	1
3	0.87812	0.26449	1.13550	0.75497	−2.12238	1
4	0.20755	−0.30421	1.55828	0.78658	−2.19149	1
5	1.68994	0.10,282	−0.91959	0.23125	−0.31279	1
6	−0.49350	−0.06882	0.31341	−1.76846	−2.07282	2
7	−0.97509	−0.81889	1.63873	−1.14640	−0.15945	2
8	−1.52392	−1.06525	−0.52289	−1.18106	−1.45423	2
9	−2.19047	−1.52425	−0.18924	−0.69254	0.17615	2
10	1.90883	0.28620	1.48539	−3.08198	0.86457	2
11	−0.57079	0.05746	0.07907	1.76875	−0.23282	3
12	0.97109	−0.88314	0.56237	−0.31169	2.49982	3
13	0.22238	−0.86039	−0.49884	1.76771	0.03011	3
14	−0.17221	−1.46596	1.47230	1.22956	−−0.16668	3
15	0.14522	−1.40408	0.26648	0.02496	1.44204	3
16	0.78727	1.05583	−1.34630	−0.95530	−0.50346	4
17	−0.42185	−0.31614	−0.27431	−0.65660	1.14308	4
18	1.80786	−0.44112	−1.54343	−1.02538	0.91832	4
19	−0.29692	1.80868	−1.23316	−0.05541	0.23603	4
20	1.00379	0.13771	−1.40064	−0.82143	−0.81599	4

Table 5 Accuracy of testing

	1	2	3	4	5	6	7	8	9	10	Average
Accuracy (%)	100	95	95	100	100	95	100	100	100	100	98.5

The well-trained model was tested by samples selected from the testing data set randomly for ten times. For each time, 20 samples were chosen from the testing data set which included 40 samples in all. The testing results were shown in Table 5. The last column showed the average accuracy of testing for ten times, which indicated that the PNN model performed well on human body clustering.

5 Conclusion

In order to meet the needs of enterprises and consumers in the massive customization era, a new approach of clustering the human body shape for apparel industry by PCA_PNN was examined in this paper. Simulation results show that PCA_PNN is a good nonlinear approximately network and clusters male human body shape precisely. PNN combining with principal components analysis can reduce learning time and complexity. With the approaching of digital customization era in apparel industry, it is feasible to adopt PCA_PNN for human body shape cluster. Through being trained by suitable training samples and spread, the accuracy and clustering speed of human body shape clustering are ensured.

Acknowledgements The authors wish to acknowledge the financial support by the Key Research Project Program of Anhui Province College in Humanities and Social Sciences (No. SK2016A0116 and No. SK2017A0119), the Open Project Program of Anhui Province College of Anhui Province College Key Laboratory of Textile Fabrics, Anhui Engineering and Technology Research Center of Textile (No. 2018AKLTF15), the General Research Project Program of Anhui Provincial higher Education Promotion Program in Humanities and Social Sciences (No. TSSK2016B20), Project of National Undergraduate Training Programs for Innovation and Entrepreneurship (No. 201610363035), and Anhui Provincial Undergraduate Training Programs for Innovation and Entrepreneurship (No. 201610363180 and No. 201710363673).

References

1. Staples, M.L., DeLury, D.B.: A system for the sizing of women's garments. Text. Res. J. (1949)
2. Zakaria, N., Gupta, D.: 1—Apparel sizing: existing sizing systems and the development of new sizing systems. In: Gupta, D., Zakaria, N., (eds.) Anthropometry, Apparel Sizing and Design, pp. 3–33. Woodhead Publishing (2014)
3. Song, H.K., Ashdown, S.P.: Categorization of lower body shapes for adult females based on multiple view analysis. Text. Res. J. **81**(9), 914–931 (2011)
4. Park, W., Park, S.: Body shape analyses of large persons in South Korea. Ergonomics **56**(4), 692–706 (2013)

5. Zhuang, Z., Shu, C., Xi, P., et al.: Head-and-face shape variations of U.S. civilian workers. Appl. Ergon. **44**(5): 775–84 (2013)
6. Lee, Y.C., Wang, M.J.: Taiwanese adult foot shape classification using 3D scanning data. Ergonomics **58**(3), 513–523 (2015)
7. Wu, G., Liu, S., Wu, X., et al.: Research on lower body shape of late pregnant women in Shanghai area of China. Int. J. Ind. Ergon. **46**, 69–75 (2015)
8. Jee, S.-C., Yun, M.H.: An anthropometric survey of Korean hand and hand shape types. Int. J. Ind. Ergon. **53**, 10–18 (2016)
9. Hsu, C.-H., Lin, H.-F., Wang, M.-J.: Developing female size charts for facilitating garment production by using data mining. J. Chin. Inst. Ind. Eng. **24**(3), 245–251 (2007)
10. Lin, H.F., Hsu, C.H., Wang, M.J.J., et al.: An application of data mining technique in developing sizing system for army soldiers in Taiwan. WSEAS Trans. Comput. **7**(4), 245 (2008)
11. Bagherzadeh, M., Latifi, M., Faramarzi, A.R.: Employing a three- stage data mining procedure to develop sizing system. World Appl. Sci. J. **8**(8), 923 (2010)
12. Liu, K., Wang, J., Tao, X., et al.: Fuzzy classification of young women's lower body based on anthropometric measurement. Int. J. Ind. Ergon. **55**, 60–68 (2016)
13. Daniel, G.: Principles of artificial neural networks. World Sci. (i) (2007)
14. Bahlmann, C., Heidemann, G., Ritter, H.: Artificial neural networks for automated quality control of textile seams. Pattern Recogn. **32**(6), 1049–1060 (1999)
15. Balci, O., Oğulata, S.N., Şahin, C., et al.: An artificial neural network approach to prediction of the colorimetric values of the stripped cotton fabrics[J]. Fibers. Polym. **9**(5), 604–614 (2008)
16. Fuerstner, I.: Advanced knowledge application in practice (2010)
17. Suzuki, K.: Artificial neural networks—industrial and control engineering applications. (2011)
18. Specht, D.F.: Probabilistic neural networks. Neural Netw. **3**(1), 109–118 (1990)
19. Goh, A.T.: Probabilistic neural network for evaluating seismic liquefaction potential. Can. Geotech. J. **39**(1), 219–232 (2002)

Predicting the Outcome of H-1B Visa Eligibility

Prateek and Shweta Karun

Abstract Lately, there has been a lot of debates and discussions going on about the H-1B visa procedures in the United States. Every year, millions of students and professionals from all across the globe migrate to the USA for higher education or better job opportunities. However, the recent changes and restrictions in the H-1B visa procedure are making it tough for the international applicants to make a firm decision of wishing to work in the United States with this sense of uncertainty. While there are too many papers and journals on the study of the immigration process at the United States or the H-1B visas, there is almost no research carried on this sector in the field of computer science. The paper aims to address this issue of H-1B visa eligibility outcome using different classification models and optimize them for better results.

Keywords Machine learning · Prediction algorithm · Feature selection · Classification · Immigration · H-1B visa

1 Introduction

H-1B visas are work permit that makes it possible for a person to work in the United States under an employer. In order to apply for an H-1B, one must be offered a job by an employer who is also ready to file a visa petition with the US Immigration Office. It is an employment-based, immigrant visa for temporary professionals. The successful grant of an H-1B visa gives an individual an access to work in the USA for three years and it can further be extended to three more years. The US immigration department uses a lottery system to grant H-1B visas to a fraction of applications which is basically a computer-generated random selection process of granting visas.

Prateek (✉) · S. Karun
QR No. 1012, SECTOR 4/c, Bokaro Steel City, Jharkhand, India
e-mail: prateek.ps0794@gmail.com

S. Karun
e-mail: shwetakarun@gmail.com

© Springer Nature Singapore Pte Ltd. 2019
S. K. Bhatia et al. (eds.), *Advances in Computer Communication and Computational Sciences*, Advances in Intelligent Systems and Computing 924,
https://doi.org/10.1007/978-981-13-6861-5_31

With the changes in the laws and procedures, it has become even more difficult to get an H-1B visa. While the final decision will always lie in the hands of immigration department, machine learning can prove to be a helpful tool in deciding the applicants for visa grant based on the past data records of the individual and the visa approvals. This will not only automate the process but will also make it more transparent and fair.

Hence, in this paper, efforts have been made to use classification algorithms to predict the outcome of the H-1B visa eligibility after selecting the most important and relevant features. The H-1B visa petition dataset has about three million petitions for H-1B visas spanning from 2011 to 2016. This dataset has eight attributes, namely, case number, case status, employer name, worksite coordinates, job title, prevailing wage, occupation code, and year filed. The different classification algorithms were used to predict and analyze the dataset and reach to a conclusion using Python. The rest of the paper is organized into different sections. Section 2 talks about the literature review conducted over past research. Section 3 deals with the methodology used to select features and the prediction using the classification algorithms. Section 4 describes the implementation of these algorithms in Python and Sect. 5 has the results obtained. Finally, in Sect. 6, the work is summarized.

2 Literature Review

In [1], payroll data from one of the big four companies, Deloitte, and data from the Labor Conditions Application filling are used to study the difference in payment of accountants on H-1B visa to their peers. Ordinary least square (OLS) and median regressions are used for new hires to understand wage differences at different levels. Using both these methods, they find out that the average starting wages for new hires is not very high and not driver by outliers. Vegesana in [2] addresses the challenges to predict the results of past visa applications using machine learning. She trains various supervised models and evaluates the results using statistical validation methods. Gaussian Naïve Bayes, logistic regression, and random forests are the few algorithms used to predict the case status for each petition based on multiple other factors. Finally, the results are compared to find the most appropriate model for prediction. In this paper [3], Usdansky and Thomas have focused on the legislative and social–political roots to understand the policies of US immigration. Their study is conducted for both the high- and low-skilled workers and they conclude that both these groups have greatly contributed to the evolution of the immigration system. A study of how science, technology, engineering, and mathematics (STEM) non-American workers facilitated by the H-1B program impact the American STEM workers, both college and non-college educated are carried out by Peri, Kevin, and Chad in [4]. They use empirical analysis to study non-native STEM workers in across US cities to find out their effect on native wages and employment. They conclude that STEM workers eventually stimulate productivity and economic growth, especially for college-educated workers. Here, in [5], Peri, Kevin, and chad study the co-relation between the pre-existing non-native workers and the succeeding inflow of immigrants to find out the variation in STEM workers from abroad as a supply-driven increase in STEM workers across all metro cities using empirical analysis. They also talk about the increased house rents that diminished part of their wage profits. In [6], the focus is on the information technology professionals and their visa policies and compensations. It talks about how hiring high-skilled, non-native Americans can prove to be advantageous for the company but can have a negative impact on the wages and employment for the native IT workers. It also studies how supply shocks arising from annual caps on H-1B visas lead to fluctuations in salary for people on work visas. Zavodny in [7] again stress on the implications and effects of the H-1B visa in the IT industry employees. The data from the Department of Labor and current population survey are used for analysis using OLS regression techniques and the instrumental variable (IV) regression technique to determine a relationship between number of H-1Bs and the wages and unemployment in the IT sector. The paper [8] by Lan focuses on the effect of visa status for an entry into a post-doctoral position. The eligibility role for this action is being used as the instrumental variable for visa status. Two-stage least square estimator is used to find the relationship between the permanent or temporary visa holders to the willingness of taking post-doctoral positions. The paper concludes that permanent visa status significantly reduces post-doctoral positions. Kato and Chad in [9] talk about how reducing the number of H-1B visa grants is leading to reduced attraction for US colleges for foreign students. To study the implications of H-1B on international applications, SAT scores are used and simple difference-in-difference model is used for estimation. The results show that there is a decrease in the number of SAT scores sent by the international applicants. Luthra brings in a new perspective in [10] by arguing that the attraction for H-1B is much more beyond the wages and unemployment and that the skilled flow of immigrants in the United States is because of the broad spectrum of available advantages like flexible labor, latest skills, etc. The author studied the market position of H-1Bs using three indicators and used about six different models for analysis to conclude that workers on H-1B are not cheap labors, rather they are utilized as flexible labors. In [11], Gunel and Onur use various prediction algorithms like NB, logit, support vector machine (SVM), and neural networks (NN) to predict the outcome of H-1B visa applications. NN outperformed all other models with about 82% test accuracy and the authors hope to improve this further by increasing the depth of the NN classification model used. In paper [12], the authors perform exploratory data-driven analysis followed by leveraging the features to train and compare the performance of different classifiers, with an aim to lay a foundation for exploring H-1B visas through the lens of computing.

3 Methodology

The methodology used for this research has been shown in Fig. 1. Initially, a detailed analysis of the dataset has been performed. A number of applications are counted based on the different attributes mentioned. Histograms are used to segregate the data into different buckets for better understanding of the division of data based on a given attribute. This also helps in outlier detection. Once the analysis is complete and a clear understanding of the dataset is obtained, data preprocessing is carried out. This is a very important step to remove unwanted data or data with no relevance to the research and also to eliminate missing values from the dataset. Data are then normalized to bring it to a common scale. This is to reduce or even eliminate data redundancy. Next comes the feature extraction, where only those features which are important to relevant to the study are chosen. There are again different methods to do the same. This is crucial as not every attribute has the same contribution toward the results and if not eliminated, the prediction results are not accurate. This completes the preprocessing stage after which comes the application of prediction algorithms. Different classification algorithms used here, logit, RF, and decision tree (DT), have different prediction accuracies with and without feature selection. F-score (3) is calculated to find the accuracy of the test. Precision (1) and recall (2) are calculated to find the result relevancy and to know how many truly relevant results are returned. However, in the results section, the models with feature selection are more optimized and have better accuracy.

Fig. 1 Methodology used for prediction

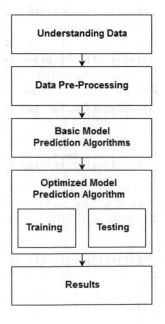

$$\text{Precision} = \text{True Positive}/(\text{True Positive} + \text{False Positive}) \qquad (1)$$

$$\text{Recall} = \text{True Positive}/(\text{True Positive} + \text{False Negative}) \qquad (2)$$

$$\text{F Score} = 2 * ((\text{Precision} * \text{Recall})/(\text{Precision} + \text{Recall})) \qquad (3)$$

4 Experimentation

The H-1B visa petitions 2011–2016 dataset has eight attributes. The dataset is ana-lyzed in depth by finding out the count, mean, standard deviation, and the twenty-fifth and seventy-fifth quartile for each attribute. Histogram analysis is made to find the application counts based on every attribute. For data preprocessing, the irrelevant data are removed. So, rejected, invalidated and pending applications are removed. Outliers found from the histogram analysis are eliminated too. Next, feature selection is done and six most important features are selected based on the maximum node weight. The six selected features are case status, year, worksite, job title, prevailing wages, and employer name. For the classification algorithms, naïve prediction is used on the base model without splitting the data into testing and training or per-forming feature selection. Then logit, RF, and DT algorithms are then applied on the data after cleaning and feature selection forming an optimized model. The data are divided into testing (20%) and training (80%) sets. Cross validation is used to train the data and perform the testing on the test data for assessing how the results of a statistical analysis on the training data will generalize for the test set. The precision, F-score, and recall values are calculated and plotted for the logit, DT, and RF. These three classification models were chosen keeping in mind their robustness and the ease with which the results can be interpreted for further knowledge and research. The scores of the basic and the optimized models are compared.

5 Results and Discussion

Table 1 summarizes the results obtained after applying the classification algorithms. Figures 2, 3, 4, 5, and 6 show how the six most relevant features are distributed across the number of applications. Figures 7 and 8 Show the plot for precision score and F-scores, respectively, for the different classification models used.

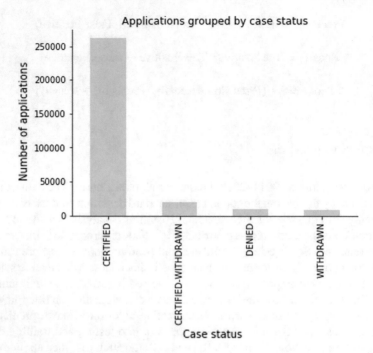

Fig. 2 Number of applications versus case status

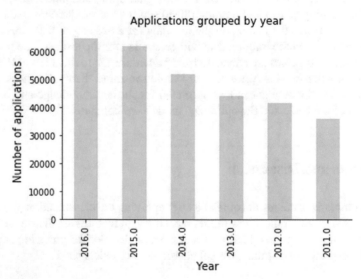

Fig. 3 Number of applications versus year

Table 1 Results of classification algorithms used

	Optimized model			Basic model
	Logit	RF	DT	
Precision	0.86	0.8	0.8	0.76
F-score	0.78	0.81	0.81	0.78

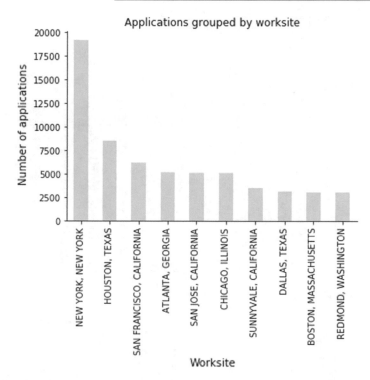

Fig. 4 Number of applications versus worksite

6 Conclusion

This paper has aimed to integrate machine learning concepts with the H-1B visa applications. With no significant work in this field, this paper could be used as a basis for further research and come up with more meaningful insights. Based on this research, the optimized models with feature selection have performed well keeping into consideration the different factors that impact the prediction majorly. For future work, there is tremendous scope. Different classification models could be used, other than the ones used here, and a detailed classification could be performed including more attributes to extract meaningful insights that could aid in granting of H-1B visas.

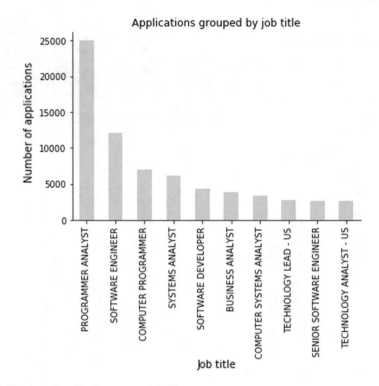

Fig. 5 Number of applications versus job title

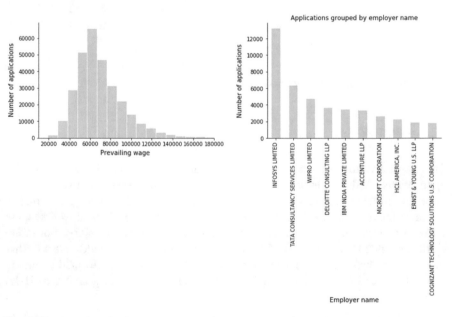

Fig. 6 Number of applications versus prevailing wage and versus employer name

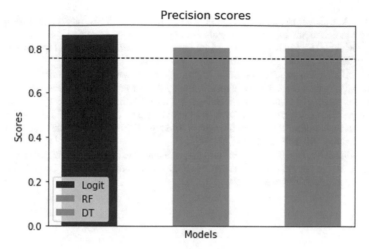

Fig. 7 Precision scores

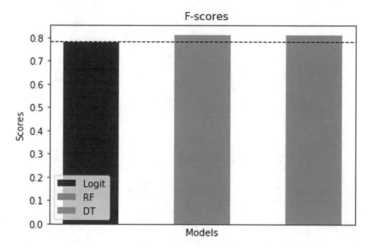

Fig. 8 F-scores

References

1. Bourveau, T., Stice, D., Stice, H., Roger, M.W.: H-1b Visas and wages in accounting: evidence from Deloitte's payroll (2018)
2. Vegesana, S.: Predictive analytics for classification of immigration visa applications: a discriminative machine learning approach (2018)
3. Usdansky, M.L., Espenshade, T.J.: The H-1B visa debate in historical perspective: the evolution of us policy toward foreign-born workers (2000)
4. Peri, Giovanni, Shih, Kevin, Sparber, Chad: STEM workers, H-1B visas, and productivity in US cities. J. Labor Econ. **33**(S1), S225–S255 (2015)
5. Peri, G., Shih, K.Y., Sparber, C.: Foreign STEM workers and native wages and employment in US cities. No. w20093. Nat. Bureau Econ. Res. (2014)

6. Mithas, S., Lucas Jr.H.C.: Are foreign IT workers cheaper? US visa policies and compensation of information technology professionals. Manag. Sci. **56**(5): 745–765 (2010)
7. Zavodny, M.: The H-1B program and its effects on information technology workers. Econ. Rev. Federal Reserve Bank Atlanta **88**(3), 33–44 (2003)
8. Lan, Xiaohuan: Permanent visas and temporary jobs: evidence from postdoctoral participation of foreign Ph. D. s in the United States. J. Policy Anal. Manag. **31**(3), 623–640 (2012)
9. Kato, Takao, Sparber, Chad: Quotas and quality: the effect of H-1B visa restrictions on the pool of prospective undergraduate students from abroad. Rev. Econ. Stat. **95**(1), 109–126 (2013)
10. Luthra, Renee Reichl: Temporary immigrants in a high-skilled labour market: a study of H-1Bs. J. Ethnic Migr. Stud. **35**(2), 227–250 (2009)
11. Gunel, Beliz, Mutlu, Onur Cezmi: Predicting the outcome of H-1B visa applications
12. Hooshmand, H., et al.: An exploration of H-1B visa applications in the United States. arXiv preprint. arXiv:1806.03368 (2018)

Dynamic Neural Network Model
of Speech Perception

Marius Crisan

Abstract Research results in neurobiology showed that the spatial organization of the somatosensory cortex, with linear or planar topology, seems to be the underlying support for the internal representation of the environment. This paper examines the feasibility of constructing self-organizing feature maps (SOFMs) suitable to model speech perception. The objective was to construct a class of dynamic SOFMs that can extract the time–amplitude and time–frequency features of the phonemes that appear in the formation of words. Two approaches are presented. One is based on constructing time-based embedding maps. The second method involved the construction of a dynamic SOFM having the Gabor transform as a transfer function. The time–frequency features of the speech sounds are revealed in the second approach. The results may be useful in applications of speech recognition.

Keywords Self-organizing maps · Speech processing · Dynamic neural networks · Semantic modeling · Time series modeling

1 Introduction

The human cognition is based on language, and our civilization developed through communication. There is no cognition without the function of words, and in the basic forms, the meaning of words can be conveyed using sounds. The role of speech is to convey meaning. This is very effective not only for human brains. Even animals understand and obey commands, and people talk to them. The key element in understanding the speech process is to explain how the vocal and verbal constituents combine to form and convey meaning. The construction of a model able to account for cognition in both speech production and understanding is a difficult task due to the nonlinear character of the phenomenon.

M. Crisan (✉)
Department of Computer and Information Technology, Polytechnic University of Timisoara, Blvd. V. Parvan 2, 300223 Timisoara, Romania
e-mail: marius.crisan@cs.upt.ro

© Springer Nature Singapore Pte Ltd. 2019
S. K. Bhatia et al. (eds.), *Advances in Computer Communication and Computational Sciences*, Advances in Intelligent Systems and Computing 924,
https://doi.org/10.1007/978-981-13-6861-5_32

A recent trend in modeling natural phenomena is to use the dynamic approach. Speech may be also suitable for this approach. Devising models based on dynamic neural networks is a promising direction, supported also by the research results in neurobiology. There are consistent evidences that proved the existence of numerous distinct areas in the cortex that manifests a spatial topographic representation of sensory information. The spatial organization of the cortex, with linear or planar topology, appears to be the way of building a kind of internal representation of the environment. In principle, there are two approaches of dynamic modeling. In one approach, the interest is to put in evidence the neural patterns as they appear as the response of sensory input. The other modeling approach is concerned to account for the internal activity of the neural structure [1]. Substantial experimental evidence has been provided by the event-related brain potentials and event-related fields that were proved to be correlated with the cognitive process of language apprehension. These experimental results suggest a reasonable interpretation from the point of view of dynamical system theory [2]. Also, experiments involving semantic processing of images and words put in evidence the modularity of the processing mechanisms in the brain [3].

The present paper explores the possibility of developing a model based on identifying the dynamic role of speech components (phonemes) in language processing. The speech constituents can be modeled by attractors that are regions of localized structures in phase space. These attractors participate in a superposition process in the formation of patterns in the topology-preserving adaptive neural structure. The paper is organized as follows. The second section explores the possibility of identifying the features of phonemes in embedding space. An embedding self-organizing feature map (SOFM) was constructed and tested experimentally. In the third section, the possibility of extracting the features of phonemes, in the time–frequency domain, was experimented by constructing a SOFM based on the Gabor transform. The conclusions are drawn in the last section.

2   Feature Extraction Using Embedding

We start this section with some insights regarding the approach of modeling the reality of our perceptions (speech being included) from the dynamical systems perspective. This approach is related to the deeper concepts of wholeness and non-locality developed from the traditional interpretations of quantum phenomena. At the quantum level, the matter seems to possess a mind-like quality. The movements and interaction of particles are explained as being a whole by the form of the wave equation. In the classical view, at the contrary, the objects are considered to have a distinct and independent existence and are only related one another by a sort of mechanical interaction. The brain activity, as was suggested by the findings in neuroscience, seems to be more akin to the concepts of wholeness and non-locality. Regarding the speech process, we can observe that when phonemes are uttered, producing a sequence of sounds, the differentiation is only apparent. Ultimately, the perception of the sound sequence appears as a whole that is revealed by the cognitive process of the word meaning. If there are variations of regional accents and pronunciation, or other different vocal parameters, the cognition of the whole meaning is not altered.

A convenient mathematical object for modeling the brain as the seat of linguistic apprehension is the manifold. This has the property to impart a unity to the components of speech, which would have remained a mere collection of distinct sounds. In a concise definition, any object that can be "charted" is a manifold. This means that each point of an n-dimensional manifold has a neighborhood that is homeomorphic to the Euclidean space of dimension n. Usually, a dynamical system can be described by its phase space. The evolution of a variable x can be represented by the equation $x_{n+1} = F(x_n)$, F being a continuous function. According to this equation, iterated maps can be generated as the function can be put in the form of a difference equation $(x_{n+1} - x_n = F(x_n) - x_n)$. In practical applications, the mathematical description of the system and its complexity, or the phase space, is not known. Consequently, a practical method to reconstruct the phase space is to use embedding, which is a map from an n-manifold to $2n + 1$-dimensional Euclidean space. According to Whitney's theorem, each state can be identified uniquely by a vector of $2n + 1$ measurements. Instead of $2n + 1$ generic signals, Takens demonstrated that the time-delayed versions $[s(t), s(t - d), s(t - 2d), ..., s(t - 2nd)]$ of only one available signal can be used to embed the n-dimensional manifold and recreate the same properties as the full original system [4]. In our case, the voice signal $s(t)$, which is a time series, was obtained from s_k samples ($k = 1, 2, 3, ...$) at regular time intervals $t_k = kr$, where $r = 1/[\text{sampling rate}]$. For a given delay d and an embedding dimension m, a new series can be constructed $[s_k, s_{k+d}, s_{k+2d}, ..., s_{k+(m-1)d}]$. Of practical interest is embedding into three-dimensional space (at dimension $m = 3$, $\{s_k, s_{k+d}, s_{k+2d}\}$) and two-dimensional space (at $m = 2$, $\{s_k, s_{k+d}\}$) and obtain projections or "charts" of the manifold. These projections should capture most of the properties of the dynamic system and provide meaningful information about the amplitude and the behavior of the signal in time.

It is a known fact that at the inner ear and subcortical levels in the midbrain, the auditory system performs frequency, timing, and spatial positioning analyses [5]. The information from this level is next processed in the cortex. It appears that even if the acoustic signal has only one physical dimension, this one-dimensional function of time has coded a greater number of higher-dimensional features that are revealed ultimately by the process of cognition in the cortex. Considering the promising research results of using embedding for analyzing sound waveforms [6], we expected that embedding will prove suitable for capturing the short-time perceptual feature of the series of phonemes found in voicing and speech. Therefore, it became a valid assumption that by using the embedding technique and self-organizing maps, we could construct a plausible model.

The original self-organizing map (SOM) model is static and is not suitable for time series. Several temporal SOM approaches were developed, where the time variable was considered by adding the context from the previous time state [7]. However, using directly temporal SOMs for feature extraction of speech is not straightforward. The issue is to find such a topology for the lattice which will allow the formation of continuous trajectories when the sequences of the speech feature vectors are projected to the map. Such continuous trajectories can account for the dynamic evolution of the speech process in the manifold [8].

A self-organizing neural network was constructed to perform a mapping $F: X^n \rightarrow Y^2$. The network was capable to perform feature extraction by preserving the topological relations present in the original multidimensional space. The input information or the training pattern was the vocal signal. The synaptic weights in the SOFM were initialized to random values so that $0 < w < 1$. Each node in the SOFM lattice has two weights, one for each element of the embedding vector: $x = \{s_k, s_{k+d}\}$. Given the input x we defined a function that gives a map from x to $w(i, j)$. The firing neuron is obtained by the condition that $\|x - w\|$ is minimum: $w(i, j) = \text{argmin} \|x_k - w_k\|$, $k = 1, ..., N$. In contrast with the static SOM, the input vector is not chosen at random from the set of training data, but it is presented to the lattice as a time-related sequence. The learning rule is similar to classical competitive learning:

$$w_i(t + 1) = w_i(t) + \varepsilon(t)h_w(t)(s(t) - w(t)), \tag{1}$$

where $\varepsilon(t) = \varepsilon_0 \exp(-t/\lambda)$ is the positive-valued learning function, and $h_w(t) = \exp[-D/(2\sigma^2(t))]$ is the neighborhood function computed for the winning node. According to the learning algorithm, a unit's weights are less affected by distant neurons as time progresses. The value of λ depends on σ_0, and the number of iterations, n: $\lambda = n/\log\sigma_0$. $D = \Sigma(x_k - w_k)^2$ is the Euclidian distance squared.

Due to the sequential nature of the speech process, the current values of the signal are determined in much proportion by the past values. Therefore, using embedding maps is a good solution for discerning the significant patterns of autocorrelation in data. By using such an embedding SOFM, we ensure that nearby points in the weight space will correspond with similar points in the input or feature space. In other words, this arrangement facilitates feature extraction and identification of the salient aspects of time behavior and the amplitude of the sound signal and can be the first stage in modeling the higher compounds of speech. Both planar and linear topologies are of interest and have been explored in the experiments.

The embedding SOFM was tested using three audio signals (Fig. 1a), the vowel /a/, and the words /acute/ and /cute/, sampled at 44.1 kHz (24 bit). Our interest was to prove if the specific patterns of the component phonemes of different words are preserved in the corresponding embedding maps. For getting the resulted structure of autocorrelation more evident, we followed the rule of thumb suggested in [6] and selected the lag at least one-fourth of the period ($d = 108$ samples). Figure 1b shows the two-dimensional embedding plots. As is typical, negative autocorrelation yields the patterns aligned around a line with a slope of -1, and positive autocorrelation tilted with a slope of $+1$ correspondingly. The signal samples average being around zero, the high-amplitude values appear near the periphery of the plot, while the low-amplitude values cluster around the center. The points for vowel /a/ are distributed more uniformly than in the plots for the other two words, which contain consonants. As expected, the vowel sound revealed a clear harmonic spectrum, determined mostly by the first two formants (from IPA $F_1 = 850$ Hz, $F_2 = 1610$ Hz), while the corresponding pitch varies according to the individual. The presence of consonants can be observed in the embedding plots for /acute/ and /cute/. Their effect is to cluster the plot dots around the center due to the higher spectral components at lower amplitudes. This is more visible for the embedding plot of /acute/ than for /cute/because in/acute/ the dominant dynamics is given by the first vowel which is followed by the consonant, which shrinks the plot dots around the center. The vowel also has higher amplitudes which determine the elongation of the embedding plot in the beginning. The effect of the first consonant in /cute/ is not so prominent because the dominant contribution in the word's dynamics is due to the long u vowel pronounced /juː/. The presence of /a/ yields to a positive autocorrelation that can be observed on the slope for /a/ and /acute/, while the effect of /juː/ is a negative autocorrelation. It is remarkable to notice that in /acute/, a very clear distinction between positive and negative autocorrelation can be identified. We can explain this by the presence of the two dominant vowels, /a/ and /u/(/juː/) separated by the consonants.

In Fig. 1c, the embedding maps for a SOFM lattice of 900 neurons are depicted. On the first glance, the SOFMs can reveal the features observed in two-dimensional embedding plots. However, following the details, we can see that the map for /acute/ has a dominant negative autocorrelation in contrast with the corresponding embedding plot. This can be accounted for considering the nature of the neural learning process. Because the long vowel /u/ appears in sequence after /a/, it has a major influence upon the resulted weights.

In Fig. 1d, the linear topology is modeled. The neighborhood is one-dimensional and is defined along a line. If the points are connected in sequence, a trajectory can be generated. This maps out the dynamics of the signal in the embedded phase space as the hearing process takes place. We can observe, for instance, that the trajectory for the vowel /a/ has the tendency to occupy the entire phase space as was expected for the harmonic formants of the vowel. In /acute/ case, the starting point of the trajectory is at some distance from the center due to the higher amplitudes of the vowel /a/, and then, it concentrates in the middle. The word itself was pronounced stressing on the first /a/ as can be seen in the waveforms. Also, in /cute/ the dominant influence of the long vowel /u/ is clearly observed. Also, it appears that a similar trajectory pattern of /cute/ is present on a smaller scale in the middle region of /acute/.

A challenging demand from the SOFMs is the capability to preserve the patterns of the phonetic constituents in words. A simple test was performed trying to extract or reconstruct the dynamic pattern out of the related words. For instance, from the words /acute/ and /cute/ we tried to reconstruct the pattern for the vowel /a/ by a linear combination of corresponding SOFM weights. The pattern in Fig. 2a was obtained by subtracting the weights of /cute/ from /acute/. It is encouraging to notice that the general tendency of the pattern line is to occupy the space similar to the original shape of /a/. Of course, due to the higher harmonic components brought by the consonants, pure harmonic behavior of the vowel /a/ was altered.

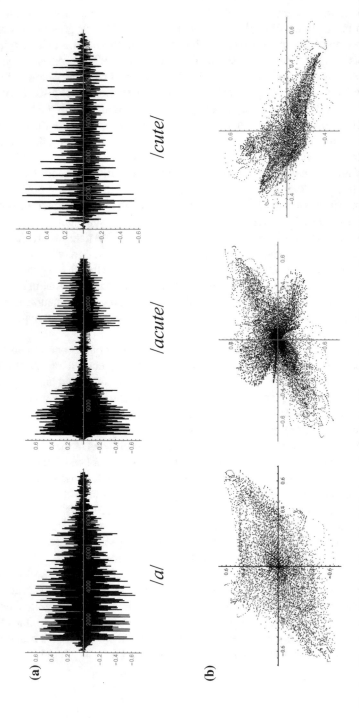

Fig. 1 **a** Waveforms of /a/, /acute/, and /cute/; **b** corresponding two-dimensional embedding plots; **c** two-dimensional SOFMs; and **d** one-dimensional trajectories

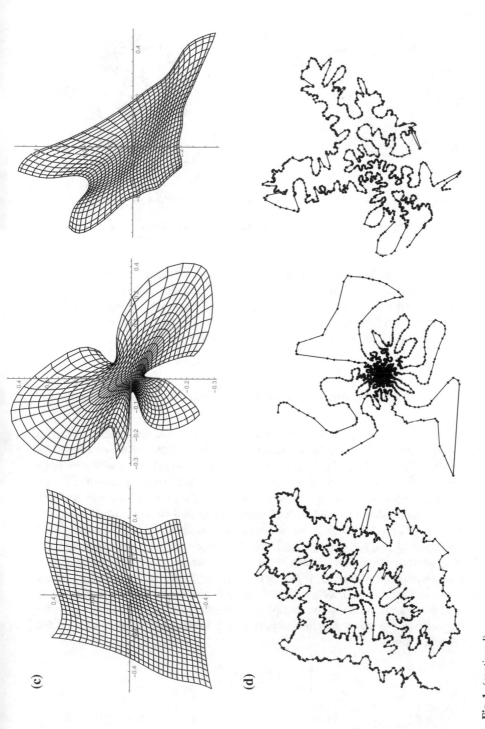

Fig. 1 (continued)

(a)   **(b)**

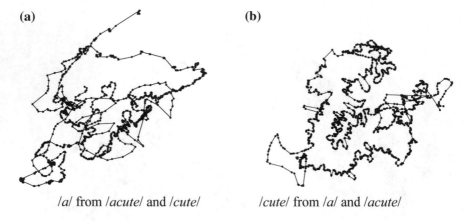

/a/ from /acute/ and /cute/   /cute/ from /a/ and /acute/

Fig. 2 **a** Reconstruction pattern for /a/. **b** Reconstruction pattern for /cute/

A pattern resembling more with the original was obtained by subtracting the weights of /a/ from /acute/ for reconstructing /cute/ (Fig. 2b). The harmonic components of the vowel could be removed easier from the map. Also, an important fact is the nature of neural learning, as noticed above, which has the tendency to emphasize the contribution of the last phonemes in the sound sequence.

3 Feature Extraction Using Dynamic SOFMs

The embedding naturally focuses on time-domain aspects of a signal, and only certain spectral patterns can be readily discerned. In this section, we present another approach to construct dynamic SOFMs. Dynamic neural networks were studied with promising results [9, 10]. One key idea in constructing a dynamic neural network is to modify the classic learning rules, following adaptation as in biological systems. As such, weights become transfer functions, and multiplication is replaced with convolution and the node operator with a nonlinear function. As mentioned above, a manifold is composed of coordinate charts, and the link that holds them together is the transition functions. Gabor transform was found suitable to play such a role. In speech processing, Gabor transforms are used with great efficiency because they facilitate the dynamic spectral analysis. In other words, they show the time–frequency evolution of a signal.

A Gabor transform of discrete samples s_k can be constructed considering a series of steps. First, a window function w is selected. In our experiments, we used the Blackman window:

$$w_b(t) = \begin{cases} 0.5 + 0.5\cos\left(\frac{2\pi t}{L}\right), & |t| \leq \frac{L}{2} \\ 0, & |t| > \frac{L}{2} \end{cases} \tag{1}$$

where L is the length of the window [11]. In the next stage, the discrete samples are multiplied by a sequence of shifted window functions $\{s_k w_b(t_k - d_m)\}$ for $m = (0, M)$ producing several sub-signals that are localized in time. This succession of shifted window functions $\{w_b(t_k - d_m)\}$ provides the partitioning of time. The value of M results from the minimum number of windows needed to cover the time interval $[0, T]$. Finally, the Gabor transform is obtained by applying DFT (discrete Fourier transform) to these sub-signals over the length of their supports:

$$\mathcal{F}\{s_k w_b(t_k - \tau_m)\}, \text{ for } m = \{0, M\} \tag{2}$$

DFTs make the frequency analysis of the signal relative to the above-partitioned sub-signals.

The neural network was constructed as follows. In the first layer, the input vector is the time signal s_k sampled at t_k time intervals. The output of the first layer results as the values of the localized sub-signals. In the second layer, spectral values are computed. The values of the sub-signals are multiplied by $\exp(-jkn2\pi/N)$, the value of weights, using unbiased neurons with linear transfer functions. Next, all the products are summed up. For each required frequency value of the entire spectrum, one such neuron was needed [12]. The DFTs values were then applied to a SOFM of a similar structure as for embedding maps.

The results of the experiments are shown in Fig. 3. We exemplified the process of the dynamic SOFM formation for the same words as in the previous case. In this case, the SOFM provides a time–frequency representation of the speech signal. In Fig. 3a, the time–frequency behavior of the words is shown. We can observe the localized frequencies as they appear in time. Similar patterns can be identified by comparing /a/ and /acute/. Also, the somehow regular frequency strips found in /cute/ can be

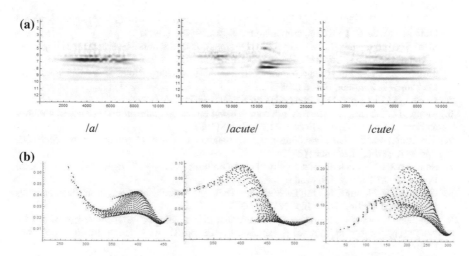

Fig. 3 a Time–frequency dynamics based on Gabor transform. b Two-dimensional patterns of SOFMs

identified in the right part of /acute/. In Fig. 3b, the SOFM patterns are depicted. The heights of SOFMs shapes correspond to the component frequencies of those words. At a higher level, the cognitive structure of the perception is determined by the resultant effect of all the speech components in the temporal sequence.

4 Conclusions

In this work, we investigated the possibility of modeling the cognition process of speech by constructing SOFMs of the component phonemes. Two approaches have been studied and experimented. The first approach involved the method of embedding. Because this method is purely time-based, it proved capable of capturing the short-time perceptual feature of the time series of phonemes found in voicing and speech. The second approach was based on the Gabor transform. As expected, this dynamic neural network has the advantage of extracting features in the time–frequency domain.

In conclusion, the experimental results for both approaches of SOFMs proved that the proposed configurations can be applied in modeling the cognitive process of speech at the word level. The results may be useful in applications of speech recognition.

References

1. Vogels, T.P., Rajan, K., Abbott, L.F.: Neural networks dynamics. Annu. Rev. Neurosci. **28**(7), 357–376 (2005)
2. beim Graben, P., Gerth, S., Vasishth, S.: Towards dynamical system models of language-related brain potentials. Cognitive Neurodynamics, **2**(3): 229–255 (2008)
3. West, W.C., Holcomb, P.J.: Event-related potentials during discourse-level semantic integration of complex pictures. Cognitive Brain Res. **13**(3), 363–375 (2002)
4. Takens, F.: Detecting strange attractors in turbulence. Lecture Notes in Mathematics 898, pp. 366–381, Springer, Berlin (1981)
5. Handel, S.: Listening. MIT Press, Cambridge, Massachusetts (1989)
6. Monro, G., Pressing, J.: Sound visualization using embedding: the art and science of auditory autocorrelation. Comput. Music J. **22**(2), 20–34 (1998)
7. Van Hulle, M.M.: Self-organizing maps, in handbook of natural computing, pp. 585–622. Springer, Berlin, Heidelberg (2012)
8. Somervuo, P.: Speech dimensionality analysis on hypercubical self-organizing maps. Neural Proc. Lett. **17**(2), 125–136 (2003)
9. Gupta, M.M, Homma, N., Jin, L.: Static and Dynamic Neural Networks: From Fundamentals to Advanced Theory. Wiley Inc., NY, (2003)
10. Homma, T., Atlas, L., Marks, R.J.: An artificial neural network for spatio-temporal bipolar patterns: application to phoneme classification. NIPS, pp. 31–40. American Institute of Physics (1987)

11. Walker, J.S.: A primer on wavelets and their scientific applications, 2nd edn. Chapman and Hall/CRC (2008)
12. Velik, R.: Discrete fourier transform computation using neural networks. In: International Conference on Computational Intelligence and Security, CIS 2008, 13–17 Dec 2008, Suzhou, China, Vol. 1, pp. 120–123 (2008)

Leakage Detection of Water-Induced Pipelines Using Hybrid Features and Support Vector Machines

Thang Bui Quy and Jong-Myon Kim

Abstract Pipelines are significant parts of water distribution systems for life and manufacture. Any leakage occurring on those can result in a waste of resource and finance; consequently, detecting early faults for them become necessary. Nowadays, there are many approaches to deal with this problem; however, their results still have limitations. This paper proposes a pattern recognition method that first extracts time-domain and frequency-domain features from vibration signals to represent each fault distinctly, and these features are then utilized with a classifier, i.e. support vector machine (SVM), to classify fault types. To verify the proposed model, the experiments are carried out on different samples of fault in various operating conditions such as pressure, flow rate and temperature. Experimental results show that the proposed technique achieves a high classification accuracy for different leakage sizes, which can be applied in real-world pipeline applications.

Keywords Fault diagnosis · Feature extraction · Machine learning · Pipeline leakage classification · Vibration diagnostic

1 Introduction

Non-destructive testing (NDT) approaches attract many researchers because these can be implemented even when the tested system is operating. One of them is a vibration-based method. The technique analyzes signals sensed by accelerometers mounted on tested pipelines to discover oscillation change between two cases in which one is normal and the other involves defects [1].

Despite its advantage, the vibration-based method faces many challenges. First, pipeline vibration is a complicated phenomenon. The possible causes for the phe-

T. B. Quy · J.-M. Kim (✉)
School of Computer Engineering, University of Ulsan, Ulsan 44610, Republic of Korea
e-mail: jongmyon.kim@gmail.com

T. B. Quy
e-mail: bqthangcndt@gmail.com

© Springer Nature Singapore Pte Ltd. 2019
S. K. Bhatia et al. (eds.), *Advances in Computer Communication and Computational Sciences*, Advances in Intelligent Systems and Computing 924,
https://doi.org/10.1007/978-981-13-6861-5_33

nomenon are from mechanical and hydraulic origins [2, 3]. Mechanical vibration is due to vibration induced from a piece of equipment such as a pump, a compressor, or any other components. A hydraulic-induced vibration is caused by continuous pressure pulses from water flow inside pipelines. The two reasons are puzzling to model and estimate responses [4]. Besides, random factors also make the pipes vibrate. For example, operators walking nearby can cause the vibration. Therefore, finding an irregular point existing on a pipeline from its vibration is not straightforward. So, pipeline leakage detection techniques can be broadly divided into model-based and data-driven-based [5, 6]. Model-based approaches are straightforward and simple if mathematical models of pipelines are available in prior. However, modern-day pipeline applications are highly complicated and involved with various external parameters that restrict to model pipelines. On the other hand, the data-driven method works with data which are collected during operation as health conditions (such as vibration) from these machines.

Therefore, this paper proposes a data-driven technique for pipeline leakage detection that extracts hybrid features about leakage, and these feature vectors are further applied with a classifier. To extract pipeline fault features, we employ time-domain and frequency-domain signal analysis techniques to obtain as much features so that no vital information about leakage is missed. These time-domain and frequency-domain features are highly effective that carefully analyze actual phenomena in pipeline to extract information about faults [7]. Once we have a robust feature pool, we take advantage of support vector machine (SVM) [8] to distinguish pinhole-leakage with different diameters, such as 0 (normal condition), 0.5, 1, and 2 mm. SVM is a robust classifier technique that discriminates an unknown test sample into one of two classes. SVM is a state-of-the-art classifier that outperforms contemporary classifiers such as k-neatest neighborhood (k-NN), artificial neural network (ANN), and naïve Bayesian [8]. The vibration dataset for training is selected in a range of frequency from 0 to 15 kHz. In addition to pre-processes dataset, this paper exploits a rectangle window to divide the dataset into smaller frames, since original vibration signals are a huge number of samples. Therefore, a rectangle window can satisfy the accuracy requirement for the algorithm in framing. To verify our proposed algorithm, we test on datasets that are recorded from self-designed pipeline application.

The rest of this paper is organized as follows: Sect. 2 provides details of the methodology of the proposed technique including data acquisition, hybrid feature extraction models, and fault classification. Sect. 3 summarizes results, and finally, a conclusion is given in Sect. 4.

2 Proposed Methodology for Pipeline Leakage Detection

Proposed pipeline leakage detection technique has three main steps, namely data acquisition, hybrid feature extraction, and fault classification, as shown in Fig. 1.

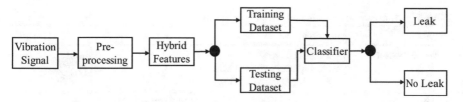

Fig. 1 Generic framework for pipeline leakage detection

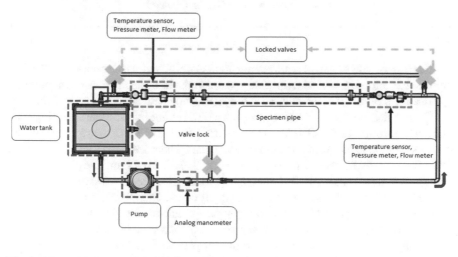

Fig. 2 2D model of our self-designed pipeline system for leakage detection

2.1 Experimental Setup and Vibration Data Acquisition of Pipeline

To simulate different leaks, a self-designed pipeline system is shown in Fig. 2. The red dash rectangle is a tested pipe segment. Water is circulated in the system utilizing a pump. The pump's power is fixed to assure that the experimental condition is the same in all the situations. On the pipe specimen, a pinhole has been drilled throughout the pipe wall to assemble different experimental leakage-support tools. Temperature, pressure, and flow meters are also installed on the two ends of the pipeline to monitor the water flow state inside. There are some external valves in the pipeline system to control water flow mode suitable to the conventional practice. In the paper, they are open entirely to prevent additional vibrations returning from water hammer effect at the valves. Furthermore, the tested segment's two ends are fixed to reduce vibrations induced by other factors like the pump. Before acquiring signals, the systematic noise level is checked to guarantee that measured fluctuations reflect leakage behavior the most integrally.

Table 1 Acquisition equipment and sensor specifications

Device	Features	
Type 3050-A-060	• 2 input channels • DC input range: 51.2 kHz • Sampling rate: 131 kS/s	
622B01 sensor	• Sensitivity: (±5%) 100 mV/g [10.2 mV/(m/s^2)] • Frequency range: 0.2–15,000 Hz • Measurement range: ±50 g (±490 m/s^2)	

The experiment employs equipment (3050-A-060) and two sensors (622B01) to acquire vibration signals from the pipe specimen that are used in real-world pipeline applications. Some of their features and specifications are summarized in Table 1.

The acquisition device interfaces with a PULSE LabShop software running on a computer through LAN-XI Ethernet standard. The vibration data are stored in the computer in the audio format for use directly later. There are three kinds of pinholes with different diameters examined in the work: 2, 1, and 0.5 mm. They demonstrate three different leakage cases: 2, 1, and 0.5-mm leaks. When the valve on the leakage-support tools is closed, it seems that no leakage occurs. Therefore, we have four cases to investigate altogether. The leakage-support tools are displayed in the Figs. 3 and 4.

Furthermore, the effectiveness of our data acquisition system is shown in Fig. 5. According to the result shown in the figure, it is clear that 2-mm leakage signal is more complex than no-leakage signal (normal condition).

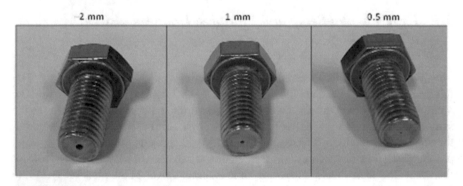

Fig. 3 Different pinholes

Fig. 4 Experimental leakage

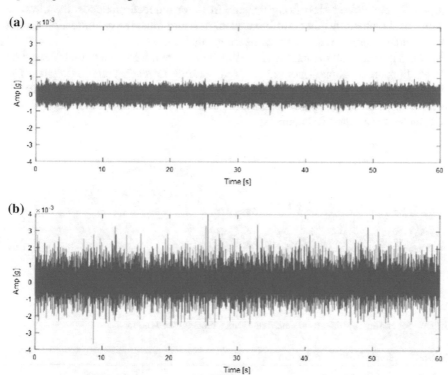

Fig. 5 Examples of recorded vibration signals of **a** 0-mm (normal condition) pinhole and **b** for 2-mm pinhole pipelines

2.2 *Hybrid Feature Model*

As explained in Sect. 1, any abnormal phenomenon occurring in the pipeline can be characterized by extracting statistical proprieties from vibration signals. In addition to pre-processes dataset, this paper exploits a rectangle window to divide the dataset into smaller frames, since original vibration signals are a huge number of samples, as shown in Fig. 6. Therefore, a rectangle window can satisfy the accuracy requirement for the algorithm in framing. This kind of exploration reveals more specific information about leakage. Therefore, we extract the most widely used time- and frequency-domain statistical features, as in [6, 7, 9]. Tables 2 and 3 present types of features which are usually adopted in the algorithm. They are basic features exploited in a lot of other classification applications with a good outcome as diagnosing bearing in [7]. Regarding classifying leakage in water-induced pipelines, the physical nature of the phenomenon is different from bearing; thus, the classification cannot be effectual in comparison with the bearing situation.

This paper uses all the features listed in Tables 2 and 3 to train a model based on SVM. In such a manner, there are eleven features for each vibration signal. The measurement has two sensor channels, one is toward upstream and the other is toward downstream compared with the leak. The algorithm applies twenty-two sets of features for all the two channels.

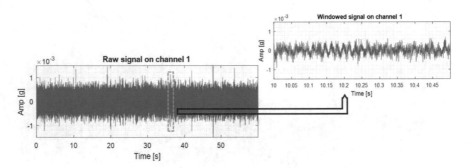

Fig. 6 Seperating raw data from data frames via a rectangular window

Table 2 Features in the time domain

Features	Expressions	Features	Expressions
Short time energy	$E_w = \sum_{n=1}^{N_w} x_n^2$	Standard deviation	$\sigma = \sqrt{\frac{1}{N_w} \sum_{n=1}^{N_w} (x_n - \mu)^2}$
Root mean square	$x_{\text{rms}} = \sqrt{\frac{1}{N_w} \sum_{n=1}^{N_w} x_n^2}$	Zero crossing rate	$\text{zcr} = \frac{1}{N_w} \sum_{n=2}^{N_w} |\text{sign}(x_n) - \text{sign}(x_{n-1})|$
Mean	$\mu = \frac{1}{N_w} \sum_{n=1}^{N_w} |x_n|$		

Table 3 Features in the frequency domain

Features	Expressions	Features	Expressions
Spectral peak	$X(f_{\text{peak}}) = \max(X(f))$	Spectral spread	$ss = \sqrt{\dfrac{\sum_{n=1}^{N_f/2} (f_n - sc)^2 |X_n|^2}{\sum_{n=1}^{N_f/2} |X_n|^2}}$
Spectral kurtosis	$\text{Kurt}[x] = \dfrac{E\left[(x-\mu)^4\right]}{\left(E\left[(x-\mu)^2\right]\right)^2}$	Spectral flatness	$sf_b = \dfrac{\sqrt[N_b]{\prod_{k_b} |X_{k_b}|^2}}{\frac{1}{N_b}\sum_{k_b} |X_{k_b}|^2}$
Spectral centroid	$sc = \dfrac{\sum_{n=1}^{N_f} f_n X_n}{\sum_{n=1}^{N_f} X_n}$	Harmonic spectral centroid	$hsc = \dfrac{\sum_{h=1}^{N_b} f_h A_h}{\sum_{h-1}^{N_h} A_h}$

2.3 Fault Classification

To verify the effectiveness of the hybrid feature pool, we further applied those features with the SVM classifier for identifying fault types. SVM is a state-of-the-art classifier that outperforms contemporary classifiers such as k-NN, artificial neural network, and naïve Bayesian [8]. SVM is a binary classifier that discriminates test samples into one of two classes. In the training phase, SVM constructs a hyperplane that separates two classes with maximum margin-width which ensures a high classification accuracy for the testing phase.

3 Result and Discussion

In this section, we test the efficacy of proposed pipeline leakage detection algorithm including hybrid feature model and classification performance using SVM.

The dataset is loaded directly from audio format into a program which is constructed in MATLAB. Their size and sample frequency are shown in Table 4. Figure 7 presents a signal frame of a channel of the dataset in the time domain and frequency domain. It is possible to claim that the range of frequency is from 4 to 6 kHz; the vibration varies along with the leakage gravity by investigating their frequency spectrum. Hence, the problem is promisingly solved by the algorithm in the document.

One of the main contributions of this study is to extract hybrid features. Figure 8 displays some of the extracted features namely, root mean square, zero crossing

Table 4 Training dataset

Signal type	Size (double)	Sampling frequency [Hz]
No leakage	$7{,}864{,}320 \times 2$	131,072
0.5-mm leak	$7{,}864{,}320 \times 2$	131,072
1-mm leak	$7{,}864{,}320 \times 2$	131,072
2-mm leak	$7{,}864{,}320 \times 2$	131,072

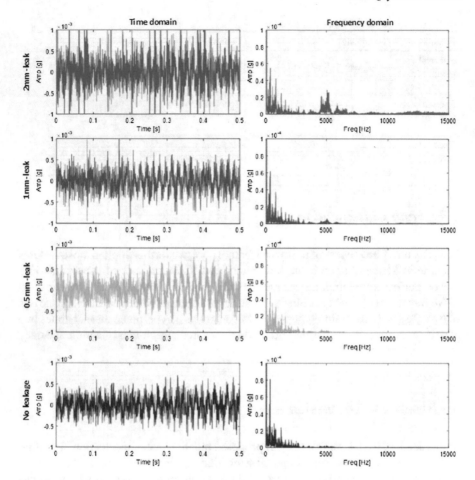

Fig. 7 Vibration signal frame

rate, spectral spread, spectral kurtosis, and spectral peak for different leakage sizes. According to these results, it is evident that the hybrid feature pool accurately represents different leakage sizes.

As our hybrid feature pool is highly discriminative to represent leakage, we apply these feature vectors with SVM to diagnose leakage types. We use a quadratic kernel function to train SVM, which provides a near-optimum result about maximum margin-width. The classification results are calculated as sensitivity (SEN) and overall classification accuracy (OCA) as in [7, 8]. Table 5 provides the result of SEN and OCA for test performance. According to the result, it is clear that the proposed algorithm outperforms about identifying different leakage types with 84.15% average classification accuracy.

In addition, we provide confusion matrix results. The confusion matrix is a robust technique to visualize classification performance by providing actual versus predicted

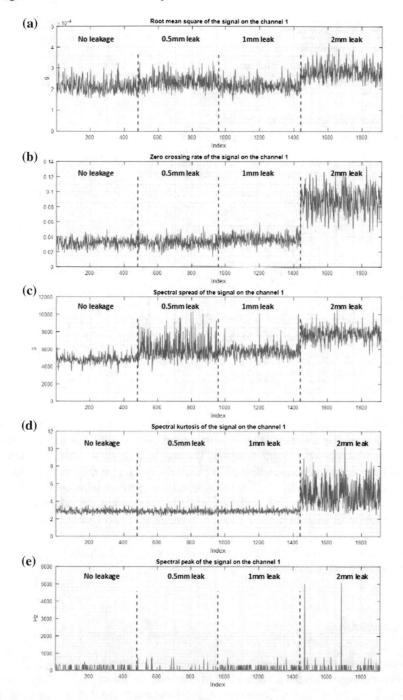

Fig. 8 Extracted features namely, **a** root mean square, **b** zero crossing rate, **c** spectral spread, **d** spectral kurtosis, and **e** spectral peak for different leakage sizes

Table 5 Experimental results

Algorithm	Result (SEN)				OCA
	No leakage	0.5 mm	1 mm	2 mm	
SVM	78.3%	75.4%	82.9%	100%	84.15%

Table 6 Confusion matrix of the trained model

Pipeline leak	Sample	Confusion matrix			
		No leakage	0.5 mm	1 mm	2 mm
No leakage	480 × 16,384 × 2	453 (94.0%)	14 (3.0%)	13 (3.0%)	0
0.5 mm	480 × 16,384 × 2	20 (4%)	366 (76%)	94 (20%)	0
1 mm	480 × 16,384 × 2	18 (4%)	93 (19%)	369 (77.0%)	0
2 mm	480 × 16,384 × 2	0	0	0	480 (100%)

Table 7 Confusion matrix of the test model

Pipeline leak	Sample	Result			
		No leakage	0.5 mm	1 mm	2 mm
No leakage	480 × 16,384 × 2	376 (78.3%)	78 (16.3%)	26 (5.4%)	0
0.5 mm	480 × 16,384 × 2	39 (8.1%)	362 (75.4%)	80 (16.7%)	0
1 mm	480 × 16,384 × 2	15 (3.1%)	67 (14%)	398 (82.9%)	0
2 mm	480 × 16,384 × 2	0	0	0	480 (100%)

deviation. Table 6 presents the confusion matrix result of the trained model. The 2-mm leak case is wholly differentiated, with the accuracy of 100%. The model classifies the others wrongly with several samples. The confusion is reasonable because of the investigated signal restriction. Table 7 presents the result tested by samples received in the same condition as the dataset in training. The accuracy with 2-mm leak is also 100% like the confusion matrix shown in Table 7. However, the others turn out to be a difference shown in Tables 7 and 6, especially the no-leakage case accuracy decreases.

The reason is that 0.5-mm leak and 1-mm leak signals conduct the algorithm to be confusing in classification.

4 Conclusion

This paper presented a pipeline leakage detection algorithm based on hybrid feature models and support vector machine (SVM) to classify different leaks using vibration signals. First, we extracted features from heterogeneous signal processing domains to represent leakage distantly. The algorithm obtains a high classification accuracy to some extent. The experimental result showed that the features extracted from the vibration signals are highly effective for identifying leakage types. The proposed algorithm yielded classification accuracy of 78.3, 75.4, 82.9, and 100% for no leakage, 0.5, 1, and 2 mm, respectively.

Acknowledgements This work was supported by the Korea Institute of Energy Technology Evaluation and Planning (KETEP) and the Ministry of Trade, Industry and Energy (MOTIE) of the Republic of Korea (No. 20172510102130). It was also funded in part by the Basic Science Research Program through the National Research Foundation of Korea (NRF) funded by the Ministry of Education (2016R1D1A3B03931927).

References

1. Mandal, S.K., Chan, F.T.S., Tiwari, M.K.: Leak detection of pipeline: an integrated approach of rough set theory and artificial bee colony trained SVM. Expert Syst. Appl. **39**, 3071–3080 (2012)
2. Xu, M.-R., Xu, S.-P., Guo, H.-Y.: Determination of natural frequencies of fluid-conveying pipes using homotopy perturbation method. Comput. Math Appl. **60**, 520–527 (2010)
3. Dilena, M., Dell'Oste, M.F., Morassi, A.: Detecting cracks in pipes filled with fluid from changes in natural frequencies. Mech. Syst. Signal Process. **25**, 3186–3197 (2011)
4. Zhang, Y., Gorman, D., Reese, J.: Analysis of the vibration of pipes conveying fluid. Proc. Inst. Mech. Eng. Part C J. Mech. Eng. Sci. **213**, 849–859 (1999)
5. Murvay, P.-S., Silea, I.: A survey on gas leak detection and localization techniques. J. Loss Prev. Process Ind. **25**, 966–973 (2012)
6. Blevins, R.D.: Formulas for Dynamics, Acoustics and Vibration. Wiley, Hoboken (2015)
7. Kang, M., Kim, J., Wills, L.M., Kim, J.M.: Time-varying and multi resolution envelope analysis and discriminative feature analysis for bearing fault diagnosis. IEEE Trans. Industr. Electron. **62**, 7749–7761 (2015)
8. Manjurul Islam, M.M., Kim, J.-M.: Reliable multiple combined fault diagnosis of bearings using heterogeneous feature models and multiclass support vector machines. Reliab. Eng. Syst. Saf. **184**, 55–66 (2018)
9. Li, S., Song, Y., Zhou, G.: Leak detection of water distribution pipeline subject to failure of socket joint based on acoustic emission and pattern recognition. Measurement **115**, 39–44 (2018)

Large-Scale Meta-Analysis of Genes Encoding Pattern in Wilson's Disease

Diganta Misra, Anurag Tiwari and Amrita Chaturvedi

Abstract In this paper, we propose an unsupervised learning approach with an objective to understand gene expressions for analysis of Wilson's disease in the liver of Mus musculus organisms. We proceeded to obtain the best parameters for cluster division to correctly classify gene expression sets so as to capture the effect and characteristics of the disease in the genome levels of the organisms in the best possible way. The clustering proved beneficial in capturing the correct genetic analogy of Wilson's disease. Analytical experiments were carried out using various clustering algorithms and were evaluated using performance metrics including silhouette score analysis and Calinski–Harabasz index.

Keywords Wilson's disease · Clustering · Unsupervised learning

1 Introduction

Wilson's disease [1, 2] is a drastic genetic disorder which is a result of copper accumulation in body mostly affecting the brain and liver. It is a result of mutation of the ATP7B protein compound also known as the Wilson disease protein gene and is an autosomal recessive disorder. Symptoms of the disease include vomiting, swelling up of the leg, yellow-tinted skin, and severe itchiness. Arguably not a very common disease, Wilson's disease is still a very harmful one and is capable of catastrophic endings. Proper medication and genetic screening are some precautionary measures taken against the same.

D. Misra (✉)
Kalinga Institute of Industrial Technology, Bhubaneswar 751024, India
e-mail: mishradiganta91@gmail.com

A. Tiwari · A. Chaturvedi
Indian Institute of Technology (BHU), Varanasi 221005, India
e-mail: anuragtiwari.rs.cse17@itbhu.ac.in

A. Chaturvedi
e-mail: amrita.cse@iitbhu.ac.in

© Springer Nature Singapore Pte Ltd. 2019
S. K. Bhatia et al. (eds.), *Advances in Computer Communication and Computational Sciences*, Advances in Intelligent Systems and Computing 924,
https://doi.org/10.1007/978-981-13-6861-5_34

However, for better understanding of how genes are affected because of Wilson's disease [1, 2] and how affected copy of genes are inherited, scientists have put forth various new ways of understanding and interpreting genetic expression of Wilson's disease-affected genomes by incorporating various computerized algorithms mostly in the domain of advanced computing and machine learning. This paper presents a new approach toward understanding Wilson's disease-affected liver model at genetic levels via using an unsupervised machine learning algorithm for clustering the gene sets.

2   Related Research

Genetic expression analysis [3, 4] using advanced machine learning techniques has been in the forefront of interdisciplinary research domains, and scientists are constantly thriving for newer and better solutions taking inspiration from previous works for instances; in [4], the authors have performed an experimental validation of DEGs identified by Cuffdiff2, edgeR, DESeq2, and two-stage Poisson's model which involved mice amygdalae. They also sequenced RNA pools, and then the results were compared. Those results showed the need for the combined use of the sensitive DEG analysis methods, supporting the increase in the number of biological replicate samples. In [5], the authors have presented a genome-wide gene expression along with pathology resources by comparing five transgenic mouse lines with either only plaques or only tangles. Such comparison of the separate effects of these pathologies for the entire life span of the mouse opened up a lot of research aspects. In [6], the authors have compared the classification performance of a number of popularly used machine learning algorithms, such as random forest (RF), support vector machine (SVM), linear discriminant analysis (LDA), and also k-nearest neighbor (kNN). The results showed that LDA gave the best results in terms of average generalization errors and also the stability of error estimates.

In [7], the authors have made use of several supervised machine learning techniques that include neural networks, and then they have built a panel of tissue-specific biomarkers of aging. Their best model achieved an accuracy of 0.80 with respect to the actual bin prediction on the external validation dataset. In [8], the authors presented a genome-wide method for AD candidates in gene predictions. They approached this with the help of machine learning (SVM), which was based upon the integration of gene expression data with the human brain-specific gene network data, in order to discover the wide spectrum of the AD genes in the genome. They achieved an ROC curve of 84.56 and 94%.

For research, we took inspiration from these papers to provide a highly accurate algorithm able to classify gene expression sets to understand Wilson's disease. With previous research work having been concluded using various types of machine learning algorithms like SVM and KNN, we decided to build a heavily evaluated unsupervised approach for our clustering task.

3 Unsupervised Learning

Machine learning has been conventionally defined as the ability of machines to understand and interpret from a constant learning experience without being explicitly programmed. This enables computerized systems to perform human-like tasks that can be perceived in both visual- and audio-based scenarios. Where conventional and traditional algorithms fail, machine learning techniques have established their supremacy in wide variety of tasks ranging from object detection, speech processing, forecasting, and many more.

Machine learning has been divided into two major categories—supervised learning and unsupervised learning. Supervised learning as the name suggests is the branch of machine learning where the ground truth is already available for evaluating the algorithm's efficiency. In a broader perspective, supervised learning is the sector of machine learning where the system is provided with the labels along with the data, thus we already know what should be the correct or desired output. Tasks like regression and classification fall under the category of supervised learning. However, in case of unsupervised learning, an algorithm is given the task to find the underlying pattern in the data or its coherent structure. In simpler terms, there are no labels and the machine is made to predict the labels that mostly involve the procedure of feature learning. Clustering is a common unsupervised learning task, where the optimal number of clusters is unknown and needs to be figured out to best fit the data.

Machine learning was a crucial underlying factor in this whole research for training the algorithm to understand and correctly interpret the underlying features in the genetic data to mark out a clear distinction between healthy and affected samples and then placing them in their corresponding cluster.

4 Clustering Algorithms

In the current scenario, many clustering algorithms have been introduced in machine learning paradigm including K-means clustering [9] and hierarchical clustering [10, 11] which is further subdivided into agglomerative and divisive cluster analysis techniques. Clustering defined in abstract terms is nothing but grouping highly similar data-points or data-points lying within the same category. Clustering helps in profiling unlabeled and unstructured data to better understand the variation in their properties across the various clusters. Clustering algorithms have been incorporated in various use cases in multiple domains from medical sciences to nanotechnology and financial markets. Most notable instance where clustering algorithms are highly useful is customer segmentation where customers are profiled based on various attributes including purchase history, purchase frequency, etc., for better ad-targeting and promotional campaigns.

K-means clustering algorithm [9], proposed by MacQueen in 1967 [9], is one of the most simple and straightforward clustering algorithms where K denotes the number of clusters the data needs to be divided into. The algorithm is an iterative process of defining centroids of the K clusters and assigning data-points in close proximity to the centroid to be a part of that groupage. The same step is repeated by re-assigning K new clusters calculated by the barycenter of the clusters obtained from the previous step. The centroids keep changing their positions in each iterative process until they can move no further. At this point, we define the clustering to have reached its optimal and desired state of value. The algorithm aims to minimize the squared error function which is defined to be as represented in Eq. (1):

$$J = \sum_{j=1}^{k} \sum_{i=1}^{n} \left\| x_i^{(j)} - c_j \right\|^2 \tag{1}$$

In Eq. (1), $\left\| x_i^{(j)} - c_j \right\|^2$ is the distance metric between a data-point represented as $x_i^{(j)}$ and the cluster center represented as c_j. This distance metric is applicable for all the n number of data-points across all the k number of clusters and their respective centroids. For finding the optimal number of clusters in case of K-means clustering, two metrics are used to index and evaluate performance across various number of clusters applied—(1) elbow method and (2) silhouette score analysis. Elbow method counters the fact that K-means is arguably a naïve approach toward clustering unlabeled data. This method incorporates calculating the sum of squared errors (SSE) for various values of k and then plotting a line chart of these values. The plot generally is a vague analogy to the outline of a human arm and suggests the best k value to occur where the plot represents the elbow, hence its name 'elbow method'. The error SSE keeps on decreasing as the value of k is increased. This is because max k value is equal to the number of data-points itself, in which case the SSE will be equal to zero as each data-point will be a cluster itself. So the desired output should be a low value for k which essentially has a low SSE index. One other popular method of assessing the goodness of the clustering fit is the silhouette score analysis method. In this method, an analysis of the Euclidean distance of separation between the obtained clusters is plotted. The plot is an interactive visual way that conveys the measure of how close data-points of one cluster are from data-points of neighboring clusters. On the basis of this closeness, a score is assigned which is also known as silhouette co-efficient in the range of [−1, 1]. The higher the silhouette co-efficient, the better clustered is the data and vice versa. This helps in choosing the right value for number of clusters parameter k.

Hierarchical clustering [10, 11] on the other hand is the process of building a hierarchy of clusters often represented in the form of a dendrogram. This is again divided into two subcategories which are: (1) agglomerative clustering and (2) divisive clustering. Agglomerative clustering is the 'bottom-up' approach as each data-point starts as a cluster themselves and merges with other clusters as one moves up the hierarchy. In contrast, divisive clustering is the 'top-down' approach where all data-points start as members of one single cluster and then break down/split into multiple clusters as one moves down the hierarchy. For evaluating the goodness of the fit, a performance metric called Calinski–Harabasz index is performed on the split. Calinski–Harabasz index or popularly known as CH score incorporates the mathematical formula given in Eq. (2):

$$\frac{SS_B}{SS_W} \times \frac{N - k}{k - 1} \tag{2}$$

In the above-mentioned equation, SS_B represents the overall intercluster variance and SS_W represents the overall intracluster variance; and k is the number of clusters while N is the total number of data-points. According to this metric, maximum k value with highest CH value evaluation is chosen to be the optimal one.

For the research, we applied agglomerative clustering and K-means clustering approach and evaluated their robustness according to the evaluation metrics as discussed earlier. The motivation behind picking these two algorithms particularly is that K-means is a very minimalistic approach and has strong evaluation metrics while agglomerative clustering provides strong visualizations in the form of dendrograms and uses the Euclidean distance between the clusters as an impeding factor. However, our approach of relying on K-means clustering and not only on agglomerative clustering is largely due to the fact that K-means has a lower order of time complexity which is $O(n^2)$ as compared to the time complexity of standard hierarchical clustering which is $O(n^3)$ with a memory complexity of the order of $O(n^2)$.

5   Experimental Setup and Analysis

5.1   Dataset and Preprocessing

The dataset was obtained from public Gene Expression Omnibus (GEO) dataset repository maintained by National Center for Biotechnology Information (NCBI) and is tagged as 'Wilson's disease model: liver' along with its annotation—'Hist1h1c.' The dataset contains gene profiling of Mus musculus organisms and has a unique ID—'34120432'. The data has six GSE5348 expression count samples by array with two genotype sets with normalized expression levels for each gene, namely 'GSM121554,' 'GSM121555,' 'GSM121556,' 'GSM121547,' 'GSM121550,' and 'GSM121552.' Precisely, the dataset contains analyzing factors of livers of copper-transporting ATPase ATP7B animals. The complete dataset is embedded within a .soft file format which upon opening has the annotation file in .annot and .annot.gz formats; and the complete gene profiling file which is converted to a tabular format or a delimiter-separated format like .xlxs or .csv. The main excel sheet contains the complete gene profiling along with their IDs, chromosome tags, platform details, etc. The complete data is encapsulated within 28 columns excluding the serial index column. The data contains 22,691 row entries mapping through the 28 columns each;

however, it contains significant amount of NaN/missing values and thus requires extensive preprocessing (Fig. 1).

Due to the data being in undesired format, R language's `BioConductor` package was used to extract the data from the main SOFT file. `GEOQuery` and `BioBase` were the subsequent modules used in the process of extracting the data from the .gz and .soft files and converting them to suitable analyzable formats. For this research project, the point of focus was the columns containing the six GSE5348 expression gene profiling data. These six columns are all numerical based. Firstly, Python's Pandas package was used to load up the data, and its 'drop' function was used to remove all the unnecessary columns which were not required within the project. Subsequently, the 'dropna' function was used along in axis 0 to remove all the entries containing missing values/NaN values. Although the most generalized way of handling missing values in case of numeric data is to replace it with the mode of the column, however, since the data contained unique gene data, replacing it with the mode would have caused the loss of integrity, uniqueness, and interpretability of the data. After removing all the missing value entries, the data left was in the structure of 6 columns and 11,811 rows containing unique numerical entries.

Once the data had been cleaned, various preprocessing techniques were applied on the data. Firstly, the data was converted into a matrix form using the 'as_matrix' function present within the Pandas package of Python and then was normalized using the L2 normalization techniques as L2 norm is more stable and has built-in feature selection in comparison to L1 norm. The standardized L2 norm is a vector norm defined mathematically for a vector x as represented in Eq. (3):

$$|x| = \sqrt{\sum_{k=1}^{n}|x_k|^2} \tag{3}$$

In the above shown equation, x is a complex vector, while $|x_k|$ is the complex modulus and $|x|$ is the vector norm or L2-normalized value for that vector.

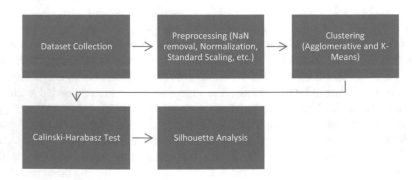

Fig. 1 Flowchart of experimental setup and process

This was achieved using the 'preprocessing.normalize' submodule function of the standard machine learning library in Python called as Scikit-Learn.

5.2 Proposed Algorithmic Architecture and Evaluation Metric Analysis

After obtaining the desired normalized data, agglomerative clustering techniques were applied on the data using the Python's Sklearn and SciPy package. Linkage type was set to be 'Ward' which takes only the Euclidean distance metric for evaluation of distance between clusters, and the number of clusters was set to be 2. The plotting was made both in the form of dendrogram and cluster map using the Seaborn and matplotlib modules in Python which are the standard interactive visualization packages in Python as shown in Figs. 2 and 3, respectively. In Fig. 2, the y-axis represents the cluster index, while the x-axis represents the estimate Euclidean distance between clusters.

Since, the task was unsupervised so to obtain the best number of clusters, Calinski–Harabasz index test was performed which states the higher the score, the better the fit within those number of clusters. The test revealed that 2 is the best dividing cluster fitting the dataset. The line plotting for Calinski–Harabasz index which shows the CH score for number of clusters ranging from 2 to 6 where x-axis represents the

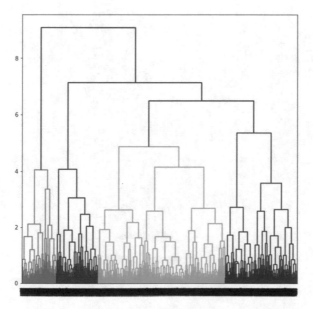

Fig. 2 Dendrogram visualization of agglomerative clustered data

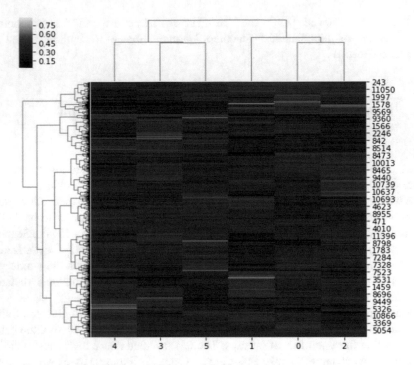

Fig. 3 Cluster map visualization of agglomerative clustered data

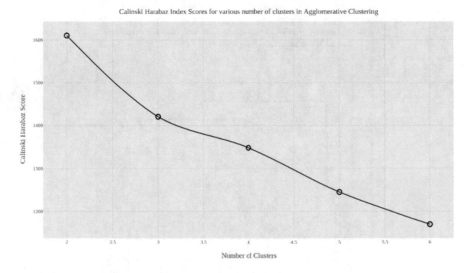

Fig. 4 Calinski–Harabasz score index line plot for various k values

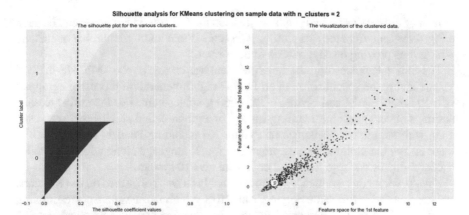

Fig. 5 Silhouette analysis with $k = 2$

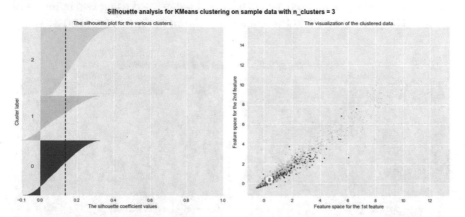

Fig. 6 Silhouette analysis with $k = 3$

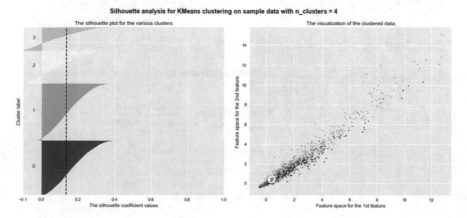

Fig. 7 Silhouette analysis with $k = 4$

number of clusters value and y-axis represents the CH score magnitude can be seen in Fig. 4. As seen from the line plotting, CH score is maximum for number of clusters to be 2, thus proving our hypothesis to be correct.

The complete proceedings of the experiment was carried out on a MSI GP-63 8RE system having the specification of an Intel i7 eighth-generation processor equipped with 16 GB of RAM and Nvidia GTX 1060 graphics card. For the software sim- ulation, standard machine learning library of Python called scikit-learn was used along with subsequent packages of Python 3.6 including Pandas, NumPy, SciPy, Seaborn, and matplotlib. For data preprocessing, R language along with its modules of Bioconductor package was used on a Windows 10 system.

Finally, for further evaluation, silhouette analysis was performed on the K-means clustered dataset with k value ranging from 2 to 5, which also assigns a corresponding co-efficient for various number of clusters. For this, the data was first scaled using the Standard Scaler submodule of Sklearn package in Python. The silhouette score analysis revealed again that the data can be best represented using two numbers of clusters. The plotting obtained from this test is shown in Figs. 5, 6, 7, and 8.

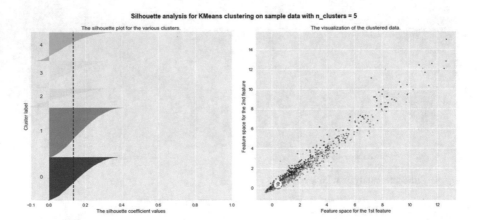

Fig. 8 Silhouette analysis with $k = 5$

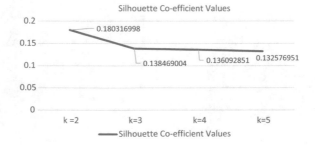

Fig. 9 Silhouette co-efficient for various k values

Figure 9 represents the decreasing trend of the silhouette co-efficient values from 0.18 when the number of clusters equals 2–0.13 when the number of clusters equals 0.13. This confirmed our hypothetical understanding of the data being appropriately clustered and represented correctly when it is clustered into two groups.

6 Conclusion

This research was successful in machine-level understanding of genetic disorder classification, which in this case was Wilson's disease, and opens up new prospects in the domain of genetic expression analysis for understanding genetic disorders by incorporating advanced machine learning automated techniques to amplify and increase the pace of working toward exploring new medical cure for similar genetic disorders and synthesizing new results to better perceive genetic structure. Future scope of research in this domain includes automated machine-level interpretation of biomarkers for genetic ambiguity and disorders and implementation of the same at a lower memory and time-consuming algorithmic models for fast and accurate classification.

References

1. Compston, A.: Progressive lenticular degeneration: a familial nervous disease associated with cirrhosis of the liver, by SA Kinnier Wilson, (From the National Hospital, and the Laboratory of the National Hospital, Queen Square, London) Brain 1912: 34; 295–509. Brain 132(8), 1997–2001 (2009)
2. Rodriguez-Castro, K.I., Hevia-Urrutia, F.J., Sturniolo, G.C.: Wilson's disease: a review of what we have learned. World J. Hepatol. 7(29), 2859 (2015)
3. Link, C.D., Taft, A., Kapulkin, V., Duke, K., Kim, S., Fei, Q., Wood, D.E., Sahagan, B.G.: Gene expression analysis in a transgenic *Caenorhabditis elegans* Alzheimer's disease model. Neurobiol. Aging 24(3), 397–413 (2003)
4. Rajkumar, A.P., Qvist, P., Lazarus, R., Lescai, F., Jia, J., Nyegaard, M., Mors, O., Børglum, A.D., Li, Q., Christensen, J.H.: Experimental validation of methods for differential gene expression analysis and sample pooling in RNA-seq. BMC Genom. 16(1), 548 (2015)
5. Matarin, M., Salih, D.A., Yasvoina, M., Cummings, D.M., Guelfi, S., Liu, W., Nahaboo Solim, M.A., et al.: A genome-wide gene-expression analysis and database in transgenic mice during development of amyloid or tau pathology. Cell Rep. 10(4), 633–644 (2015)
6. Khondoker, M., Dobson, R., Skirrow, C., Simmons, A., Stahl, D.: A comparison of machine learning methods for classification using simulation with multiple real data examples from mental health studies. Stat. Methods Med. Res. 25(5), 1804–1823 (2016)
7. Mamoshina, P., Volosnikova, M., Ozerov, I.V., Putin, E., Skibina, E., Cortese, F., Zhavoronkov, A.: Machine learning on human muscle transcriptomic data for biomarker discovery and tissue-specific drug target identification. Front. Genetics 9 (2018)
8. Huang, X., Liu, H., Li, X., Guan, L., Li, J., Tellier, L.C.A.M., Yang, H., Wang, J., Zhang, J.: Revealing Alzheimer's disease genes spectrum in the whole-genome by machine learning. BMC Neurol. 18(1), 5 (2018)

9. MacQueen, J.: Some methods for classification and analysis of multivariate observations. In: Proceedings of the Fifth Berkeley Symposium on Mathematical Statistics and Probability, vol. 1, no. 14, pp. 281–297 (1967)
10. Johnson, S.C.: Hierarchical clustering schemes. Psychometrika **32**(3), 241–254 (1967)
11. D'Andrade, R.G.: U-statistic hierarchical clustering. Psychometrika **43**(1), 59–67 (1978)

A Dynamic Weight Grasshopper Optimization Algorithm with Random Jumping

Ran Zhao, Hong Ni, Hangwei Feng and Xiaoyong Zhu

Abstract Grasshopper optimization algorithm (GOA) is a novel meta-heuristic algorithm for solving single-objective numeric optimization problems. While it has a simple principle and it is easy to implement, grasshopper optimization algorithm performs badly in some aspects. GOA cannot make full utilization of every iteration, and it is not good at getting rid of local optima. To solve these problems and improve the performance of GOA, this paper proposed an improved grasshopper optimization algorithm based on dynamic weight mechanism and random jumping strategy (DJGOA). The dynamic weight mechanism promoted the utilization of the iterations of the algorithm. The random jumping strategy was introduced to help the algorithm jumping out of the local optima. Several experiments relating to 13 benchmark functions and 4 algorithms were conducted to demonstrate the performance of DJGOA. The results of the experiments demonstrated that DJGOA performed better than GOA and the other algorithms.

Keywords Grasshopper optimization algorithm · Dynamic weight · Random jumping · Global optimization problem

1 Introduction

Optimization problem is the procedure to find the global best value and the global best position of a mathematical function [1]. An optimization problem can occur in many scenarios. With the rapid development of the technology of cloud computing

R. Zhao · H. Ni · H. Feng · X. Zhu (✉)
National Network New Media Engineering Research Center, Institute of Acoustics, Chinese Academy of Sciences, Beijing 100190, China
e-mail: zhuxy@dsp.ac.cn

R. Zhao
e-mail: zhaor@dsp.ac.cn

R. Zhao · H. Ni · H. Feng · X. Zhu
University of Chinese Academy of Sciences, Beijing 100049, China

© Springer Nature Singapore Pte Ltd. 2019
S. K. Bhatia et al. (eds.), *Advances in Computer Communication and Computational Sciences*, Advances in Intelligent Systems and Computing 924,
https://doi.org/10.1007/978-981-13-6861-5_35

and edge computing, the problem of task scheduling comes more prominence. The task scheduling problem is a kind of optimization problem from a mathematical point of view. The performance of the scheduling algorithm is very important for a scheduling system because a better algorithm can help the system to optimize the task execution flow and save much execution makespan and budget.

Recently, many researchers have conducted a lot of researches on the optimization problems of task scheduling system. As the studies going on, some meta-heuristic algorithms are introduced to handle the complex problems of the global optimization. Meta-heuristic algorithms have the capability of global search with simple operators and less limiting constraints. While meta-heuristic algorithms have many advantages, they are easy to fall into local optimum because of the precocity, which prevents the searching process to find a better solution. To overcome the precocity and promote the accuracy of meta-heuristic algorithms handling the optimization problems, a dynamic weight grasshopper optimization algorithm with random jumping is proposed in this paper.

The remainder of this paper is organized as follows: A literature survey about recent optimization algorithms is proposed in the second section. The basic grasshopper optimization algorithm is introduced in Sect. 3. The proposed dynamic weight grasshopper optimization algorithm with random jumping is described in detail in Sect. 4. Several experiments of 13 benchmark functions are implemented, and the results are shown in Sect. 5. Finally, some conclusion and discussion are proposed in the last section.

2   Related Work

A lot of algorithms are introduced for the optimization problems. Genetic algorithm (GA) proposed by Goldberg in 1988 is a classic meta-heuristic algorithm which simulates the behavior of natural selection with some natural operators including mutation, crossover, and selection [2–4]. While the performance of GA is remarkable, the search process and the operators of GA are too complicated to implement for some situations. Simulated annealing algorithm (SA) proposed by Kirkpatrick in 1983 is inspired by the annealing process of solid. SA is based on the greedy strategy and introduces the mechanism of random jumping. It is easy to be trapped in the local optima [5–7].

Some of the algorithms are inspired by nature to simulate the behavior of insects, fishes, herds, and birds. And the algorithms make utilization of swarm intelligence and the experience shared among the members of the search cluster to modify the direction of search and find the best or the approximate best solution of the optimization problem. Those algorithms are called meta-heuristic algorithms. Particle swarm algorithm (PSO) proposed by Kennedy in 1995 is a classical meta-heuristic algorithm. The performance of PSO is pretty good, and the algorithm is simple to implement. PSO is just suitable for continuous space and cannot be directly used for discrete space, so in 1997 Kennedy proposed a discrete binary version of PSO. And afterward, many PSO-based algorithms have been developed and widely used for scheduling problems of many areas [8–15]. Ant colony optimization (ACO) algorithm is inspired by the natural searching behavior of ants between the nest of ants and the food sources. ACO utilizes the communication method among the swarm of ants with chemical pheromone [16–20]. Cuckoo search (CS) algorithm proposed by Yang in 2009 is inspired by the breeding behavior of the cuckoos. CS introduces the mechanism of Lévy flight to jump out of the local optima [21].

In recent years, some novel meta-heuristic optimization algorithms simulating nature behaviors are proposed, and there are rare modification researches about them. Whale optimization algorithm (WOA) proposed in 2016 simulates the natural hunting behavior of whales [22]. Dragonfly algorithm (DA) proposed in 2016 gets inspired from the static and dynamic swarming behaviors of dragonfly in nature [23]. In 2017, Shahrzad Saremi and Seyedali Mirjalili proposed a novel meta-heuristic optimization algorithm called grasshopper optimization algorithm (GOA). GOA simulates the migration behavior of the grasshopper by making utilization of the influence inside the swarm and the wind influence outside the swarm to find the target food [1]. This paper proposed a new algorithm based on GOA to solve optimization problems.

3 Grasshopper Optimization Algorithm

Grasshopper optimization algorithm simulates the insect's swarm behavior of grasshoppers [1]. The grasshoppers form a swarm to migrate over long distance for a new habitat in their adulthood, which is abstracted as an optimization problem. In this process, the grasshoppers influence each other inside the swarm. And the wind and the gravity influence them outside the swarm. The target of food makes impact as well.

With the affection of the three factors mentioned above, the migration process is divided into two stages which are exploration and exploitation. In the phase of exploration, the grasshoppers which are represented as search agents in swarm in optimization problems are encouraged to move rapidly and abruptly to find more potential target areas. In the stage of exploitation, the grasshoppers tend to move locally to find better target areas.

3.1 Mathematical Model of GOA

The two tendencies of exploration and exploitation together with the function of locating a food source are achieved by grasshoppers naturally. Seyedali Mirjalili presented the mathematical model of the process of the grasshopper swarm migration [1]. The simulating equation is showed as follows:

$$X_i = S_i + G_i + A_i \tag{1}$$

where X_i is the position of the ith search agent, S_i represents the force of social interaction, G_i represents the influence factor of gravity force on the ith search agent, and A_i represents the influence factor of the wind. S_i is defined as follows:

$$S_i = \sum_{j=1, j \neq i}^{N} s(d_{ij}) \widehat{d_{ij}} \tag{2}$$

where d_{ij} defines the Euclidean distance between the ith and the jth search agents in space, and it is calculated as $d_{ij} = |x_j - x_i|$, $\widehat{d_{ij}}$ is the unit vector between the ith and the jth search agents, defined as $\widehat{d_{ij}} = \frac{x_j - x_i}{d_{ij}}$, and s is a function which shows the social relationship affection in the grasshopper swarm. The s function is defined as follows:

$$s(r) = f e^{\frac{-r}{L}} - e^{-r} \tag{3}$$

where e is the natural logarithm, f represents the concentration of attraction, and the parameter of L shows the attractive length scale.

When it is used to handle the mathematical optimization problem, some change factors should be added into the Eq. 1 to optimize the mathematical module. The parameters of G_i and A_i which represent the outside influence should be replaced by the parameter of food target. Thus, the equation is reformulated as follows:

$$x_i = c \left(\sum_{j=1, j \neq i}^{N} c \frac{u - l}{2} s(|x_j - x_i|) \frac{x_j - x_i}{d_{ij}} \right) + \widehat{T_d} \tag{4}$$

where u and l represent the upper and lower boundaries of the search space, respectively, and $\widehat{T_d}$ is the food target position which represents the best fitness position which all the search grasshoppers can find in all time in the mathematical module. Besides, the parameter c is the changing factor to balance the process of exploitation and exploration which is calculated as follows:

$$c = c_{max} - iter \frac{c_{max} - c_{min}}{\text{MaxIteration}} \tag{5}$$

where c_{max} and c_{min} are the maximum value and the minimum value of c, respectively, $iter$ represents the current iteration, and $MaxIteration$ represents the maximum number of iterations.

Equation 4 should be iterated over to get the optimal solution. The iteration should be stopped when the destination target is reached or the preset maximum number of iterations is reached. After the iterations end, the algorithm can get the target grasshopper position and the target fitness.

3.2 Advantages and Disadvantages of GOA

GOA is a recent meta-heuristic algorithm inspired by grasshopper natural migration behavior. While GOA has a simple theory foundation and it is easy to implement, the performance of GOA is superior. The original GOA varies the comfort zone of the grasshopper, which can make the target converge toward the global optimum solution through iterations. Comparing with some classic algorithms, the rate of convergence of GOA is much faster. GOA can significantly enhance the average fitness of grasshoppers and improve the initial random population of grasshoppers.

While GOA has those advantages, it also has some disadvantages which prevent the algorithm to get better solutions. After some theoretical analysis of GOA formulas and some experiments conducted with MATLAB codes, several disadvantages of GOA were presented.

First, GOA used linear decreasing parameters to make the search process converge which could hardly distinguish between the two phases of the procedure which were the exploitation stage and the exploration stage. Second, the position of the best solution fluctuated sharply during every iteration of the earlier stage of the search process. It seemed that the final best solution was only influenced by the later stage of the search process regardless of the result of the earlier stage. The theory of GOA could not make full utilization of all the search iterations. The best fitness of GOA was not more outstanding when the maximum number of the iterations increased, for example, from 500 to 1500, which was found from the experiment results. At last, the search process was easy to be stuck in local optimum. The advantage that GOA was simple to implement could not contribute to getting rid of the local optimum, which could be a disadvantage of GOA instead.

4 Dynamic Weight Grasshopper Optimization Algorithm with Random Jumping

The mechanisms of dynamic weight and random jumping were introduced to overcome the disadvantages presented above and promote the performance of GOA.

4.1 Dynamic Weight Mechanism

GOA used linear decreasing parameters to restrict the search space and make all the search agents move toward the target position. The linear decreasing parameters could not enhance the impact of the two phases of the search process which were the exploitation stage and the exploration stage. In the exploration stage, GOA could not converge rapidly around the target search area and the search agents just wandered around the whole search space, which could not make a solid basis for the later search phase. And in the exploitation stage, the parameters often made the search agents slip over the local optima position just as if the search agents had over-speed movements.

The linear decreasing parameter mechanism could not make the algorithm fully use the whole iterations. The dynamic weight parameter mechanism was introduced to improve the utilization of the algorithm. The search progress was divided into three phases, which were the earlier stage, the intermediate stage, and the later stage. In the earlier phase, the weight of the target position should be higher to make the search process converge rapidly. In the intermediate stage, the parameter should be steady to make the search agents search in the whole space. In the later stage, the weight of the gravity force within the search agents should be smaller to exploit deeply around the local optima solution position. The new iterating equation is presented as follows:

$$x_i = mc \left(\sum_{j=1, j \neq i}^{N} c \frac{u-l}{2} s\left(\left|x_j - x_i\right|\right) \frac{x_j - x_i}{d_{ij}} \right) + \widehat{T_d} \tag{6}$$

where m is the dynamic weight parameter to adjust the search process. To correspond to the feature of the three phases of the searching process, the parameter m is calculated as follows:

$$m = \begin{cases} 0.5 - \frac{(0.5-0.1)*\text{iter}}{0.2*\text{MaxIteration}}, 0 < \text{iter} \leq 0.2 * \text{MaxIteration} \\ 0.1, 0.2 * \text{MaxIteration} < \text{iter} \leq 0.8 * \text{MaxIteration} \\ 0.05, 0.8 * \text{MaxIteration} < \text{iter} \leq \text{MaxIteration} \end{cases} \tag{7}$$

4.2 Random Jumping Strategy

The original GOA algorithm did not use a jumping mechanism, and all the search agents just moved according to the influence of the social interaction and the target food attraction. The mechanism of the original GOA might lead the algorithm to be stuck into the local optima position.

A random jumping strategy was introduced to help the GOA algorithm to raise the probability of jumping out of the local optima position. The parameter p was set as the jumping threshold. Before an iteration ended, the current best fitness was compared with the best fitness of the last iteration. If the ratio of the current best fitness and the last best fitness was higher than the threshold p, it could be assumed that the algorithm did not find a better solution, and the random jumping mechanism should startup. A new search agent was generated around the best position according to the random initialization rule. The random initialization rule is presented as follows:

$$\text{tempPos} = \text{curPos} * ((0.5 - \text{rand}(0, 1)) * \text{iniRan} + 1) \tag{8}$$

where *tempPos* is the location of the search agent generated newly and *curPos* is the position of the best solution found presently. *iniRan* is the initialization range parameter that manages the boundary of random jumping. If *iniRan* was higher, the global search ability of the algorithm was enhanced just like the simulated annealing algorithm.

4.3 Procedure of DJGOA

This paper proposed a dynamic weight grasshopper optimization algorithm with random jumping (DJGOA). The procedure of the DJGOA was divided into four phases which were the initialization stage, the parameter setting stage, the calculation stage, and the fitness updating stage. In the initialization stage, some factors including the position boundaries were set and the original swarm position was initialized randomly within the boundaries. In the parameter setting stage, the search process went into the iteration loop, and the dynamic weight parameter was set according to the current iteration by Eq. 7. In the calculation stage, the social interaction within the swarm was calculated by Eq. 6. In the fitness updating stage, the fitness of the target solution was updated if the fitness found currently was superior to the best target optimal value of history. And if the ratio of the current fitness and the fitness of the last iteration was higher than the threshold p set before, the random jumping strategy was used to generate a new search agent to attempt to get away from the local optima position by Eq. 8.

5 Experiment Result

5.1 Experimental Setup

To evaluate the performance of the proposed algorithm, a series of experiments were conducted. We compared the presented DJGOA with the original algorithm of GOA, a recent meta-heuristic algorithm of dragonfly algorithm (DA), and a classical heuristic algorithm of particle swarm algorithm (PSO) via 13 benchmark functions used in [1]. The 13 benchmark functions were divided into two types. Functions $f1$–$f7$ are unimodal functions which test the convergence speed and the local search capability of an algorithm. And functions $f8$–$f13$ are multimodal functions which test the global search ability of an algorithm when there are multiple local optima solutions. The details and the expressions of the 13 benchmark functions are the same as $f1$–$f13$ which are the unimodal and the multimodal benchmark functions described in the literature [1].

To solve the test functions, 30 search agents were employed and the maximum number of iterations was set to 500. Every experiment was conducted for 30 times to generate the statistical results for all the four algorithms which were compared. The parameters of GOA were set as [1] refers. The parameters of DA and PSO were set as [23] refers.

5.2 Computational Result

Average value, standard deviation, the best value, and the worst value of the experiment of 30 times are presented in Table 1 to describe the performance of DJGOA, GOA, DA, and PSO.

From Table 1, it was found that the performance of DJGOA was better than the capability of the other three algorithms. Ten results of all the 13 benchmark functions showed that the average value of DJGOA outperforms to others by several orders of magnitude. And for the results of the other three test functions, DJGOA just performed slightly worse than the others. Actually, DJGOA could get the results of the same order of magnitude with GOA, DA, and PSO. And for the two of the three test functions mentioned above, DJGOA could do much better in getting the best value. The results of the experiment showed that DJGOA could behave better in standard deviation and the worst value, which denoted that DJGOA was more stable in searching. And for 12 results of the experiments, DJGOA could get much better results of the best value, which meant that DJGOA had better ability to find the best fitness than the other three algorithms.

For the unimodal test functions, the performance of GOA was pretty good and DJGOA could greatly enhance the performance of GOA. DJGOA improved the search results by several orders of magnitude. And for the multimodal test functions, DJGOA could help the algorithm to improve a lot, especially in finding more accurate target fitness. In conclusion, the jumping strategy could help the algorithm a lot in searching and DJGOA had better ability to search the optimum than DA, GOA, and PSO in both global searching and local searching.

5.3 Convergence Analysis

The results of the best fitness value along the iterations of PSO, DA, GOA, and DJGOA for all the 13 benchmark functions are shown in Fig. 1.

The convergence curves showed that DJGOA could make the search process converge more rapidly than others, and the results also demonstrated that the dynamic weight mechanism could help the original GOA algorithm to make full utilization of every iteration.

Table 1 Comparison among PSO, DA, GOA, and DJGOA

		PSO	DA	GOA	DJGOA
F1	Mean	0.225563	0.080849	1.36E−07	1.66E−46
	Std	0.237994	0.105599	1.79E−07	9.08E−46
	Best	0.013876	0	3.13E−09	5.89E−68
	Worst	1.152773	0.492894	9.43E−07	4.97E−45
F2	Mean	0.073714	0.068551	6.58E−05	1.11E−32
	Std	0.032231	0.058641	3.38E−05	3.28E−32
	Best	0.018729	0	1.18E−05	6.61E−44
	Worst	0.184727	0.291309	0.000178	1.54E−31
F3	Mean	0.966624	0.344025	1.67E−06	6.32E−26
	Std	0.667126	0.751601	2.33E−06	3.46E−25
	Best	0.039482	0.000341	4.45E−08	1.62E−56
	Worst	2.959222	3.991606	1.16E−05	1.90E−24
F4	Mean	0.493622	0.214187	0.0003332	4.19E−19
	Std	0.220222	0.17597	0.0002152	1.72E−18
	Best	0.175668	0	7.97E−05	1.49E−29
	Worst	1.089921	0.681735	0.000866	9.24E−18
F5	Mean	9.918561	131.6453	174.7015	9.74734
	Std	7.755497	308.8793	387.12699	40.348066
	Best	2.309725	0.225168	0.149193	0.221327
	Worst	35.84272	1537.623	1791.135	221.7863

		PSO	DA	GOA	DJGOA
F8	Mean	−1888.9397	−1951.1	−1835.44	−1640.96
	Std	203.54156	206.5246	225.9528	239.274404
	Best	−2177.87	−2394.72	−2296.76	−2296.76
	Worst	−1430.72	−1566.28	−1347.62	−1191.19
F9	Mean	2.909573	5.783889	9.186762	6.106361
	Std	1.478189	4.579477	7.407304	9.987868
	Best	0.519253	1.014268	1.989918	0
	Worst	6.077489	17.91532	32.83339	34.94222
F10	Mean	0.350586	0.514763	0.944551	1.48E−15
	Std	0.263284	0.598968	3.584623	1.35E−15
	Best	0.075876	7.99E−15	9.25E−05	8.88E−16
	Worst	1.218206	1.95112	19.58689	4.44E−15
F11	Mean	0.396806	0.378879	0.225182	0.078961
	Std	0.140117	0.190647	0.137012	0.117111
	Best	0.1692	0	0.056626	0
	Worst	0.695956	0.772933	0.529255	0.45044
F12	Mean	0.004276	0.043013	3.99E−08	8.56E−11
	Std	0.005508	0.105204	3.08E−08	9.30E−11
	Best	0.000142	0.000125	9.10E−09	3.76E−12
	Worst	0.021633	0.528282	1.30E−07	4.53E−10

(continued)

Table 1 (continued)

F13		PSO	DA	GOA	DJGOA
	Mean	0.0248	0.023843	0.0010989	3.22E−10
	Std	0.022508	0.036932	0.0033526	4.13E−10
	Best	0.00329	9.97E−05	6.37E−09	1.91E−11
	Worst	0.116973	0.152731	0.010988	1.66E−09

		PSO	DA	GOA	DJGOA
F6	Mean	0.22456	0.116021	1.23E−07	2.99E−10
	Std	0.221957	0.138145	1.05E−07	3.72E−10
	Best	0.034077	1.13E−05	4.05E−09	1.45E−11
	Worst	0.891234	0.591843	3.99E−07	1.66E−09
F7	Mean	0.001214	0.001828	0.001188	0.007248
	Std	0.00081	0.001052	0.00114	0.005368
	Best	0.000249	0.000283	0.000155	0.000526
	Worst	0.003066	0.0044	0.00652	0.020287

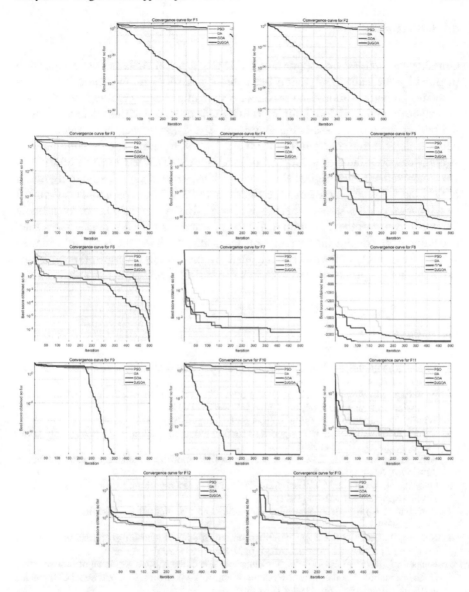

Fig. 1 Comparison of convergence rate of PSO, DA, GOA, and DJGOA for 13 benchmarks

6 Conclusion

Grasshopper optimization algorithm (GOA) is a novel meta-heuristic algorithm inspired by the natural migration behavior of grasshoppers to handle the single-objective numeric optimization problem. A dynamic weight grasshopper optimization algorithm with random jumping (DJGOA) is proposed in this paper to improve the performance of the searching algorithm. The dynamic weight mechanism can help the algorithm to make full utilization of every iteration. The random jumping strategy can contribute to getting rid of local optima. The performance of DJGOA was compared with the original algorithm of GOA, a recent meta-heuristic algorithm of DA, and a classical meta-heuristic algorithm of PSO. The experiment results of 13 benchmark functions showed that the proposed DJGOA could significantly promote the performance of the algorithm in aspects of accuracy, stability, convergence rate, and the ability of finding the best solution. Our future work will focus on further developing the capability of the algorithm to handle more complex optimization problems and applying the proposed algorithm to solve practical scheduling problems.

References

1. Saremi, S., Mirjalili, S., Lewis, A.: Grasshopper optimisation algorithm. Adv. Eng. Softw. **105**, 30–47 (2017)
2. Fonseca, C.M., Fleming, P.J.: Genetic algorithms for multiobjective optimization: formulation discussion and generalization. In: Icga, pp. 416–423 (1993)
3. Whitley, D.: A genetic algorithm tutorial. Stat. Comput. **4**(2), 65–85 (1994)
4. Tanese, R.: Distributed genetic algorithms for function optimization. Ph.D. thesis, Ann Arbor, MI, USA, AAI9001722 (1989)
5. Kirkpatrick, S., Gelatt, C.D., Vecchi, M.P.: Optimization by simulated annealing. Science **220**(4598), 671–680 (1983)
6. Guo, H., Zuckermann, M.J., Harris, R., Grant, M.: A fast algorithm for simulated annealing. Physica Scripta **1991**, 40–44 (1991)
7. Rutenbar, R.A.: Simulated annealing algorithms: an overview. IEEE Circuits Devices **5**(1), 19–26 (1989)
8. Kennedy, J., Eberhart, R.C.: Particle swarm optimization. In: Proceedings of IEEE International Conference on Neural Networks, vol. 4, pp. 1942–1948 (1995)
9. Kennedy, J., Eberhart, R.C.: A discrete binary version of the particle swarm algorithm. In: 1997 IEEE International Conference on Systems, Man, and Cybernetics. Computational Cybernetics and Simulation, vol. 5, pp. 4104–4108 (1997)
10. Shi, Y., Eberhart, R.C.: A modified particle swarm optimizer. In 1998 IEEE International Conference on Evolutionary Computation Proceedings. IEEE World Congress on Computational Intelligence, pp. 69–73 (1998)
11. Eberhart, R.C., Shi, Y.: Particle swarm optimization: developments, applications and resources. In Proceedings of the 2001 Congress on Evolutionary Computation, vol. 1, no. 1, pp. 81–86 (2001)
12. Eberhart, R.C., Shi, Y.: Comparing inertia weights and constriction factors in particle swarm optimization. In: Proceedings of the 2000 Congress on Evolutionary Computation, vol. 1, pp. 84–88 (2000)
13. Liao, C.-J., Tseng, C.-T., Luarn, P.: A discrete version of particle swarm optimization for flowshop scheduling problems. Comput. Oper. Res. **34**, 3099–3111 (2007)

14. Gomathi, B., Krishnasamy, K.: Task scheduling algorithm based on hybrid particle swarm optimization in cloud computing environment. Int. J. Adv. Comput. Sci. Appl. **55**(1), 12–16 (2013)
15. Qiao, N., You, J., Sheng, Y., et al.: An efficient algorithm of discrete particle swarm optimization for multi-objective task assignment. IEICE Trans. Inform. Syst. 2968–2977 (2016)
16. Dorigo, M., Gambardella, L.M.: Ant colonies for the travelling salesman problem. BioSystems **43**(2), 73–81 (1997)
17. Dorigo, M., Di Caro, G.: Ant colony optimization: a new meta-heuristic. In: Proceedings of the 1999 Congress on Evolutionary Computation-CEC992, vol. 2, pp. 1470–1477 (1999)
18. Dorigo, M., Blum, C.: Ant colony optimization theory: a survey. Theor. Comput. Sci. **344**, 243–278 (2005)
19. Blum, C.: Ant colony optimization: introduction and recent trends. Phys. Life Rev. **2**(4), 353–373 (2005)
20. Hao, H., Jin, Y., Yang, T.: Network measurement node selection algorithm based on parallel ACO algorithm. J. Netw. New Media **7**(01), 7–15 (2018)
21. Yang, X., Deb, S.: Cuckoo search via lévy flights. In: Nature and Biologically Inspired Computing, pp. 210–214 (2009)
22. Mirjalili, S., Lewis, A.: The whale optimization algorithm. Adv. Eng. Softw. **95**, 51–67 (2016)
23. Mirjalili, S.: Dragonfly algorithm: a new meta-heuristic optimization technique for solving single-objective, discrete, and multi-objective problems. Neural Comput. Appl. **27**(4), 1053–1073 (2016)

Simulation Model and Scenario to Increase Corn Farmers' Profitability

Erma Suryani, Rully Agus Hendrawan, Lukman Junaedi and Lily Puspa Dewi

Abstract Corn demand in Indonesia is quite high; this commodity is useful as food, as animal feed ingredients, and as industrial raw materials. The main problem in corn farming is not enough production to meet the demand as staple foods and industry. Therefore, it is necessary to increase the amount of corn production to meet demand as well as to increase farmers' income and profits. Based on this condition, in this research we propose to develop a simulation model and scenario to increase farmers' income through land productivity improvement. As a method use for model development, we utilize system dynamics framework based on consideration that system dynamics is a scientific framework for addressing complex and nonlinear feedback systems. System dynamics can use both qualitative and quantitative techniques such as computer simulations. It also facilitates the adoption of nonlinear mental models so that they can search and describe the feedback process of problem dynamics. In particular, system dynamics has proven useful in overcoming agricultural problems. Simulation results show that increasing farmers' income can be done through increasing land productivity. With the increase in land productivity, corn production will increase, hence the income of corn farmers will also increase. Increased productivity can be done by carrying out structural and non-structural approaches. Structural approach can be carried out through rehabilitation of watersheds and irrigation networks. Meanwhile, non-structural approach can be carried out through the application of new technologies, strict land conversion rules, dynamic planting calendars, dissemination of climate information, and the development of climate field schools.

Keywords Simulation · Model · System dynamics · Corn · Farmers' profitability

E. Suryani (✉) · R. A. Hendrawan
Department of Information Systems, Institut Teknologi Sepuluh Nopember, Surabaya, Indonesia
e-mail: erma.suryani@gmail.com

L. Junaedi
Department of Information Systems, Universitas Narotama, Surabaya, Indonesia

L. P. Dewi
Informatics Department, Petra Christian University, Surabaya 60236, Indonesia

© Springer Nature Singapore Pte Ltd. 2019
S. K. Bhatia et al. (eds.), *Advances in Computer Communication and Computational Sciences*, Advances in Intelligent Systems and Computing 924,
https://doi.org/10.1007/978-981-13-6861-5_36

1 Introduction

For Indonesia, corn is the second food crop after rice. Even in some places, corn is the main staple food as a substitute for rice or as a mixture of rice. Corn demand in Indonesia is currently quite large reaching more than 10 million tons per year [6]. One of the developments of the agricultural sector besides rice is corn. This commodity is an important food ingredient because it is the second source of carbohydrates after rice. Besides that, corn is also used as animal feed ingredients and industrial raw materials [8].

From the market side, the marketing potential of corn continues to experience enhancement. This can be seen from increasingly the development of the livestock industry that will eventually increase the corn demand as a mixture of animal feed.

Currently, the problem of corn farmers is the amount of imported corn which has caused a fall in the price of local corn so that it can cause losses on farmers. The price of imported corn is often cheaper than local corn. The imported corn price is often cheaper than local corn. This is because the demand for animal feed entrepreneurs who lack local corn supply, so the government must import corn. Corn imports should not be done when farmers are harvesting and must also be stopped during post-harvest especially in July–September and January–March.

2 Literature Review

In this section, a literature review will be discussed about the structure of farm costs, prices, and farm income.

2.1 Farming Cost Structure

Farming is the science that studies how to cultivate and coordinate production factors in the form of land, natural resources, and capital to provide benefits as well as possible [14]. Farming cost structure is influenced by two factors: fix cost and variable cost. Farming costs can be calculated using the formula as given in Eq. (1) [9].

$$TC = VC + FC \tag{1}$$

where
  TC = total cost; VC = variable cost; FC = fixed cost.

2.2 Price

Price is generally applied by buyers and sellers who mutually negotiate. The seller asks for a price higher than the production costs that must be paid. Therefore, through this bargain will reach the agreed price.

Prices act as a major determinant of buyer choice, especially in developing countries [12]. Although non-price factors have become increasingly important in buyer behavior over the past few decades, prices are still one of the most important elements that determine market share and probability of a company.

2.3 Farming Income

Farming income is the difference between total revenue and total cost. Farming income can be calculated using the formula as given in Eq. (2) [9].

$$I = TR - TC \tag{2}$$

where
  I = income; TR = total revenue; TC = total cost.

Farmers as producers of agricultural products not only aim to achieve high production results but also aim to obtain high income as well.

3 Model Development

This section demonstrates the model development which consists of causal loop diagram development, stock and flow diagrams, model formulation, as well as model validation.

3.1 Causal Loop Diagram Development

The price of corn is influenced by the total cost of corn production, which consists of fixed costs and variable costs. Fixed costs are the cost of using tools used for corn farming such as land rent and fuel costs. Meanwhile, variable costs include costs for the purchase of seeds, fertilizers, pesticides and labor wages for land processing, planting and refining, maintenance, fertilization, control of pest (Plant Disturbing Organisms), harvesting, threshing, transportation, as well as other agricultural services [11]. Most of the costs of corn farming are spent on labor costs (39.51%), then fertilizer costs (26.01%), fixed costs (24.81%), and finally the cost of seeds (9.67%)

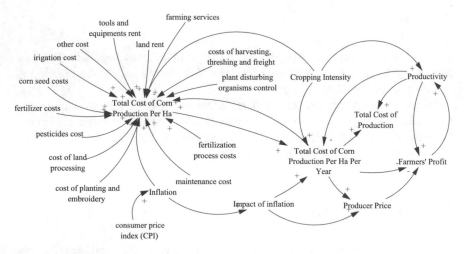

Fig. 1 Causal loop diagram of corn farmers' profitability

[7]. Factors that significantly influence the total cost of corn production is the rate of inflation. Based on the above researches, we develop a causal loop diagram of farmers' profit as shown in Fig. 1. As we can see from Fig. 1, farmers' profit depends on land productivity, total cost of corn farming production, and producer price.

3.2   Stock & Flow Diagrams and Model Formulation

From the causal loop diagram (CLD) that has been developed at the beginning of the model development, this CLD is converted into stock & flow diagrams (SFD). SFD is a technique that represents a precise quantitative specification of all systems components and their interrelation [15]. Stock and flow diagrams of corn farmer profitability can be seen in Fig. 2. As we can see from Fig. 2, farmers' profitability is influenced by land productivity, total cost of corn farming production, and producer price. Cropping intensity and productivity will influence the corn production per year.

The cost of corn seeds is one of the factors that influence the production costs of farming. The use of corn seeds on 1 ha of land is about 20–25 kg. It depends on the type of seed used [10]. The average corn seed in 1 kg contains 2,200 seeds, and the other seeds contain 3,000 seeds [1]. The average seed price that apply in the area is about Rp28,000/kg, so that the average cost of purchasing corn seeds is about Rp700,000/ha [11].

In corn farming, it is necessary to cultivate corn cultivation such as weeding and refining. The process of fertilizing corn plants is conducted by using NPK fertilizer with a uniform composition. Fertilization process then requires third fertilization after corn which is 30–40 days after planting with the composition of urea 200 kg/ha

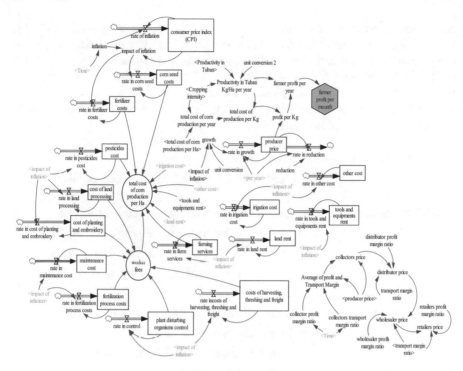

Fig. 2 Stock & flow diagrams of corn farmers' profitability

[4]. Costs incurred for the purchase of fertilizer both urea fertilizer and NPK fertilizer for corn production per hectare amounted to Rp1,213,500/ha/planting season. The cost of corn fertilization has grown around 3.4% per year because of inflation.

Pesticides used by corn farmers to prevent severe diseases and pests vary in type. To prevent disease attacks, fungicides/fungi chemical compounds such as Manzate, Dithane, Antracol, Cobox, and Vitigran Blue are used. For pest control, insecticides/insecticides, which are sprayed liquids such as Diazinon 60 EC, Baycard 500 EC, Hopcin 50 EC, Clitop 50 EC, Mipcin 50 WP, Azodrin 15 WSC, are used. Meanwhile, insecticides in the form of granules include 3G as well as Dharmafur and Curater [5]. The costs incurred by corn farmers to purchase pesticides in 2017 averaged Rp110,300/ha/planting season with a growth rate 3.2% per year because of inflation.

Costs for payment of labor wages include costs for land processing activities, planting and refining, maintenance, fertilization, control of plant disturbing organisms, harvesting, threshing, and transportation, as well as other agricultural services in the amount of Rp3,800,000 in 2014/ha/planting season. With the growth rate of pesticide purchase costs due to inflation of 3.0% per year, the cost of purchasing pesticides in 2017 is Rp4.150.200/ha/planting season. Other major expenses incurred arise from fixed costs such as irrigation costs, agricultural service costs, land rental costs, rental costs of business equipment or other facilities, and other costs which are accumulated on average Rp3,600,000/ha/planting season in 2014. In 2017, the fixed cost reached around Rp3,971,300/ha/planting season as a result of inflation. The total cost of corn production in 2014 is estimated to be an average of Rp9,300,000/ha/planting season, and in 2017, the fixed costs incurred by corn farmers were Rp10,217,500/ha/planting season because of inflation.

Model formulation of several variables that have significant impact on farmers' profitability can be seen in Eqs. (3)–(9).

$$\text{Productivity} = \text{Initial Productivty}$$
$$+ \int (\text{Increase Productivity} - \text{Decrease Productivity}) \quad (3)$$

$$\text{Total cost of corn production per ha} = \text{corn seed costs} + \text{farming services}$$
$$+ \text{fertilizer costs} + \text{irrigation cost}$$
$$+ \text{land rent} + \text{other cost} + \text{pesticides cost}$$
$$+ \text{tools and equipment rent} + \text{worker fees} \quad (4)$$

$$\text{Worker fees} = \text{cost of land processing} + \text{cost of planting and embroidery}$$
$$+ \text{costs of harvesting, threshing and freight}$$
$$+ \text{fertilization process costs} + \text{maintenance cost}$$
$$+ \text{plant disturbing organisms control} \quad (5)$$

$$\text{Costs of harvest, threshing and freight} = \text{Initial Cost}$$
$$+ \int \text{rate in costs of harvest,}$$
$$\text{threshing, \& freight} \quad (6)$$

$$\text{Rate in costs of harvesting, threshing and freight} = \text{costs of harvesting,}$$
$$\text{threshing and freight} * \text{impact of inflation} \quad (7)$$

$$\text{Impact of inflation} = \text{IF THEN ELSE}$$
$$(\text{consumer price index (CPI), inflation , 0})\quad (8)$$

$$\text{Consumer price index (CPI)} = \text{Initial CPI} + \int (\text{rate of inflation}) \quad (9)$$

Simulation result of corn farmers' profitability can be seen in Fig. 3. As we can see from Fig. 3, farmers' profitability is influenced by productivity and profit/kg. Farmers' profitability in 2017 was around Rp26.37 million as the impact of productivity which was around 10,143 kg/ha and profit/kg that was around Rp2600/kg.

Fig. 3 Farmers' profitability

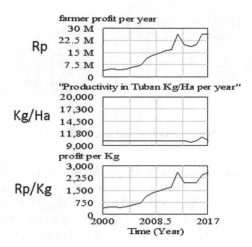

4 Model Validation

Model validation constitutes a very important step in system dynamics framework. According to Barlas [2], a model will be valid if the error rate is ≤5% and the error variance is ≤30%. We validate the model by checking the error rate and error variance. The formulations of error rate and error variance are demonstrated in Eqs. 10 and 11.

$$\text{ErrorRate} = \frac{\lceil \bar{S} - \bar{A} \rceil}{\bar{A}} \tag{10}$$

$$\text{ErrorVariance} = \frac{\lceil S_s - S_a \rceil}{S_a} \tag{11}$$

where

$\bar{S}$ the average rate of simulation
$\bar{A}$ the average rate of data
A data at time t
S simulation result at time t
S_s the standard deviation of simulation
S_a the standard deviation of data

Error rate and error variance of some variables that have significant impact to the corn farmers' profit can be seen in Table 1.

Based on the above calculation, all the error rates are less than 5% and error of variances are less than 30% which means that our model is valid.

Table 1 Error rate and error
variance

Variable	Error rate	Error variance
Consumer price	0.000025	0.0029
Total production cost	0.0467	0.0725
Producer price	0.000052	0.000052

5 Scenario Development

Scenario is a method to predict some possible future outcomes by changing the model
structure and parameters. Scenario development consists of three dimensions—those
are information acquisition, knowledge dissemination, and strategic choices [3]. Several potential alternative scenarios can be obtained from a valid model by changing
the model structure or by changing the model parameters [13]. In this research, we
develop scenario to increase farmers' profitability through increasing land productivity. Increased productivity can be done by carrying out structural and non-structural
approaches as shown in Fig. 4.

As we can see from Fig. 4, corn farmers' profitability which is projected could
be increased to be around Rp38.67 million in 2019 and would become Rp131.4
million in 2030 as the impact of productivity improvement that would be around
13,100 kg/ha in 2019 and would be around 15,133 kg/ha in 2030.

Fig. 4 Scenario result of
corn farmers' profitability

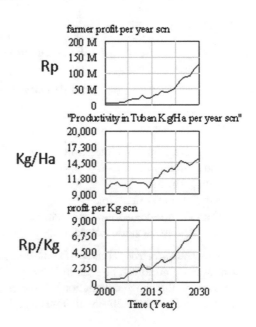

6 Conclusion and Further Research

This paper discusses the corn farmers' profitability as the impact of productivity, total farming production cost, and corn price. Several main factors are impacting the total cost of corn production—those are fixed costs and variable costs. Fixed costs consist of the fee of using tools used for corn farming such as land rent and fuel costs. Meanwhile, variable costs include costs for the purchase of seeds, fertilizers, pesticides and labor wages for land processing, planting and refining, maintenance, fertilization, control of pest (Plant Disturbing Organisms), harvesting, threshing, transportation, as well as other agricultural services. Most of the costs of corn farming are spent on labor cost, fertilizer cost, fixed costs, seed cost, and inflation.

Simulation result shows that farmers' profitability is influenced by productivity and profit/kg. Farmers' profitability in 2017 was around Rp26.37 million as the impact of productivity which was around 10,143 kg/ha and profit/kg that was around Rp2,600/kg. To increase farmers' profitability, we can conduct land productivity improvement. Land productivity improvement can be conducted by carrying out structural and non-structural approaches. The structural approach can be done through rehabilitation of watersheds and irrigation networks. Meanwhile, the non-structural approach can be conducted through the application of new technologies, strict land conversion rules, dynamic planting calendars, dissemination of climate information, and the development of climate field schools. Further research is required to explore other strategies to improve the corn supply chain profitability.

Acknowledgements This research is supported by Directorate of Research and Community Service—Ristekdikti, ITS (Institut Teknologi Sepuluh Nopember) Research Center, Enterprise Systems Laboratory in Information Systems Department, Department of Agriculture in East Java, as well as Faculty of Information and Communication Technology of ITS.

References

1. Asrofi: Corn Seeds Demand Per Ha (Kebutuhan Benih Jagung Per Hektar): Retrieved from KampusTani.Com: https://www.kampustani.com/kebutuhan-benih-jagung-per-hektar/ (2018)
2. Barlas, Y.: Formal aspects of model validity and validation in system dynamics. Syst. Dyn. Rev. **12**, 183–210 (1996)
3. Bouhalleb, A., Smida, A.: Scenario planning: an investigation of the construct and its measurement. J. Forecast. **37**, 419–517 (2018)
4. Bijibersemi: Abundant corn harvest, this is what farmers must do (Panen Jagung Melimpah, Ini Yang Harus Dilakukan Petani): Retrieved from Pioneer: https://www.pioneer.com/web/site/indonesia/Panen-Jagung-Melimpah-Ini-Yang-Harus-Dilakukan-Petani (2018)
5. Ihsan: Corn pests and plant diseases and their control (Hama dan Penyakit Tanaman Jagung dan Pengendaliannya): Retrieved from Petani Hebat: https://www.petanihebat.com/hama-dan-penyakit-tanaman-jagung-dan-pengendaliannya/ (2018)
6. Khalik, R.S.: Diversification of Food Consumption in Indonesia: Between Hope and Reality (Diversifikasi Konsumsi Pangan di Indonesia: Antara Harapan dan Kenyataan). Pusat Analisis Sosial Ekonomi dan Kebijakan Pertanian, Bogor (2010)

7. Nizar, R., Siswati, L., Putri, A.: Analysis of the function of cubic total costs in corn farming in Rumbai Subdistrict, Pekanbaru City (Analisis Fungsi Biaya Total Kubik pada Usahatani Jagung di Kecamatan Rumbai Kota Pekanbaru). Jurnal Agribisni*s* **18**(1) (2016)
8. Purwono, Hartono, R.: Grow Corn Superior (Bertanam Jagung Unggul). Penebar Swadaya, Jakarta (2005)
9. Rahmi, C., Sebayang, T., Iskandarini: Analysis of corn farming and marketing (Case study: Pamah Village, Tanah Pinem District, Dairi Regency) (Analisis Usahatani Dan Pemasaran Jagung (Studi Kasus: Desa Pamah, Kecamatan Tanah Pinem, Kabupaten Dairi)), pp. 1–35 (2012). Retrieved from https://media.neliti.com/media/publications/15050-ID-analisis-usahatani-dan-pemasaran-jagung-studi-kasus-desa-pamah-kecamatan-tanah-p.pdf
10. Sidabutar, P., Yusmini, Y., Yusri, J.: Analysis of corn farming (zea mays) in the village of Dosroha, Simanindo sub-district, Samosir district, North Sumatra province (Analisis Usahatani Jagung (zea mays) di Desa Dosroha Kecamatan Simanindo Kabupaten Samosir provinsi Sumatera Utara). Jurnal Online Mahasiswa Fakultas Pertanian Universitas Riau **1**(1) (2014)
11. Statistics Indonesia: Production value and production costs per planting season per hectare cultivation of paddy rice fields, field rice, corn, and soybeans (Nilai Produksi dan Biaya Produksi per Musim Tanam per Hektar Budidaya Tanaman Padi Sawah, Padi Ladang, Jagung, dan Kedelai) (2015). Retrieved from Tanaman Pangan: https://www.bps.go.id/statictable/2015/09/25/1855/nilai-produksi-dan-biaya-produksi-per-musim-tanam-per-hektar-budidaya-tanaman-padi-sawah-padi-ladang-jagung-dan-kedelai-2014.html
12. Syukur, A.: Analysis of farmers' income in the corn marketing system in Jeneponto (Analisis Pendapatan Petani dalam sistim Pemasaran Jagung di kabupaten Jeneponto). Thesis (2007)
13. Suryani, E., Chou, S.Y., Chen, C.H.: Dynamic simulation model of air cargo demand forecast and terminal capacity planning. Simul. Model. Pract. Theory **28**, 27–41 (2012)
14. Suratiyah: Farming Science (Ilmu Usahatani). Penebar Swadaya, Jakarta, (2006)
15. Transentis Consulting: Step-by-step tutorials, introduction to system dynamics-stock and flow diagrams, articles (2018). Available at https://www.transentis.com/step-by-step-tutorials/introduction-to-system-dynamics/stock-and-flow-diagrams/

Comparison of Bayesian Networks for Diabetes Prediction

Kanogkan Leerojanaprapa and Kittiwat Sirikasemsuk

Abstract A Bayesian network (BN) can be used to predict the prevalence of diabetes from the cause–effect relationship among risk factors. By applying a BN model, we can capture the interdependencies between direct and indirect risks hierarchically. In this study, we propose to investigate and compare the predictive performances of BN models with non-hierarchical (BNNH), and non-hierarchical and reduced variables (BNNHR) structures, hierarchical structure by expert judgment (BNHE), and hierarchical learning structure (BNHL) with type-2 diabetes. ROC curves, AUC, percentage error, and F1 score were applied to compare performances of those classification techniques. The results of the model comparison from both datasets (training and testing) obtained from the Thai National Health Examination Survey IV ensured that BNHE can predict the prevalence of diabetes most effectively with the highest AUC values of 0.7670 and 0.7760 from the training and the testing dataset, respectively.

Keywords Bayesian network · Diabetes · Machine learning

1 Introduction

Diagnostic analysis is a very important application especially for reducing diagnostic errors and serving as a basis for the development of better final strategic plans [1]. It can generate a very high impact, but there are many complex and uncertain events that are relevant. Different modeling techniques for diagnostic decision support were surveyed in PubMed for computer assisted diagnosis [2]. It was found that the Bayesian

K. Leerojanaprapa (✉)
Department of Statistics, Faculty of Science, King Mongkut's Institute of Technology Ladkrabang (KMITL), Bangkok 10520, Thailand
e-mail: kanogkan.le@kmitl.ac.th

K. Sirikasemsuk
Department of Industrial Engineering, Faculty of Engineering, King Mongkut's Institute of Technology Ladkrabang (KMITL), Bangkok 10520, Thailand
e-mail: kittiwat.sirikasemsuk@gmail.com

© Springer Nature Singapore Pte Ltd. 2019
S. K. Bhatia et al. (eds.), *Advances in Computer Communication and Computational Sciences*, Advances in Intelligent Systems and Computing 924,
https://doi.org/10.1007/978-981-13-6861-5_37

network is one of the defined techniques for diagnostic decision making, and it is ground for Bayes' theorem. Some models are employed to support decision making such as regression model and logistic regression model. However, they were used to investigate the correlation between a dependent variable and a set of independent variables such as the model to define all risks factors as being directly related to disease status. As a result, these models do not consider causal relationships [3]. In epidemiological studies, the causes of disease or factors associated with the diseases status may have already been identified by many studies. However, a hierarchical model should be a natural explanation of the causal relationship between relevant risk factors.

The Bayesian network (BN) model can depict the complexity of dependencies between variables as causal networks. Furthermore, the BN shows the dependency hierarchically as networks rather than only showing the non-hierarchical relationship between a dependent variable and a set of independent variables as with the regression model. The hierarchical relationships can be constructed to represent causal relationship with direct and indirect effects by using BN structure. The hierarchical relationships can show root causes so we can truly define a suitable plan to reduce disease prevalence more effectively. It is clear from the literature that different BN modeling structures can provide different predictions. This research aims to investigate different concepts of BN model structuring for diabetes prediction. The remainder of this paper is organized as follows. Section 2 shows relevant factors related to diabetes. The data collection and instrument construction are described in Sect. 3. Results are analyzed and described in Sect. 4. Finally, Sect. 5 provides the interpretation of results.

2 Relevant Factor Identification

2.1 Bayesian Network in Literature

In the literatures, there are 2 different approaches to constructing BN models. One is experts' knowledge; the other is machine learning from datasets. The former can build BN models from combining multiple experts' knowledge and experience. The methods to construct the model are varied and not standard, although some studies proposed protocols such as [4, 5]. The latter technique implements learning techniques from data to construct BN models from proposed algorithms. It is considered to be less time-consuming, but also inefficient [5]. Different proposed algorithms for structure learning and parameter learning for BN models are presented and compared, as shown in the systemic review of applying Naïve Bayesian network for disease prediction [6]. Furthermore, Xie et al. [7] applied K2 algorithm to construct BNs for predicting type 2 diabetes based on electronic health record dataset. The both approaches are compared and combined to improve the quality of the BN model. The comparison of both techniques is proposed in literature, e.g., predicting football results [8]. Furthermore, the combination of these methods is also presented in the prediction of heart failure [9]. It is difficult to conclude the single best model for every dataset. Therefore, this paper aimed to develop and compare a series of techniques to construct BNs using either expert knowledge or other machine learning techniques from the Thai National Health Examination Survey dataset, maintained by the Thai Public Health Survey Institute for Health Systems Research.

2.2 Bayesian Network Conceptual Framework

We imply the key conceptual model behind an approach to model NCD presented by Leerojanaprapa et al. [10]. They defined the causal framework of noncommunicable diseases (NCDs) by three main root sources: biomarkers, behavioral markers, and demographic markers. This section shows the relevant factors which were identified as 7 relevant causal factors to diabetes and shown in Fig. 1.

Behavioral Markers

1. Physical activity was measured using the Global Physical Activity Questionnaire (GPAQ) by considering the level of energy for specific activities [11]. *Low*: any activity less than the medium level. *Medium*: a. Heavy activity $\geq$ 3 days/week and $\geq$ 20 min/day, or b. Moderate activity or walking $\geq$ 5 days at least 30 min/day, or c. Moderate activity or walk $\geq$ 5 days/week and total MET—minutes per week $\geq$ 600. *High*: a. Heavy activity $\geq$ 3 days/week and MET—minutes per week $\geq$ 1,500, or b. Moderate activity $\geq$ 7 days/week and total MET—minutes per week $\geq$ 3,000.
2. Fruit and vegetable consumption concern the following: frequency of consumption, and amount of fruits and vegetables consumed in one week over the past 12 months. Two states are defined as Less and Normal to high. *Less*: < 5 servings/day. *Normal to high*: $\geq$ 5 servings/day.

Biomarker

1. Obesity, specifically abdominal obesity, is defined by 2 states: Normal and Over. *Normal*: body mass index (BMI) < 25 and waist circumference < 90 cm (male)

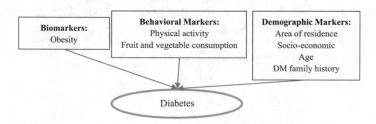

Fig. 1 Variable identification

or < 80 cm (female). *Over*: body mass index (BMI) $\geq$ 25 or waist circumference $\geq$ 90 cm (male) or $\geq$ 80 cm (female).

Demographic Marker

1. The area of residence is classified as rural and urban area.
2. Socio-economic status is determined based on the wealth index according to the DHS methodology by considering household income. The levels of the index were divided into quintile (dividing the population into 5 income groups).
3. Age is categorized into two groups: 15–34 years and 35–59 years.
4. DM Family history shows whether participants have direct relatives (including parents and siblings) with a history of diabetes.

3 Materials and Methods

3.1 Data

Data for this study were collected from a cross-sectional survey in the Thai National Health Examination Survey IV. This study focuses on the Thai population aged 15–59 years. The final collected sample comprised 11,240 individuals. The data were randomly partitioned in 70:30. The major part was for establishing a predictor model as a training dataset, and the remainder was utilized for testing. As a result, the training dataset comprises 8,003 individuals and training dataset is 3,237 individuals.

3.2 Bayesian Network

A Bayesian network is a reasoning process under a set of random variables, X, of a problem domain from a joint probability distribution [12]. The chain rule is applied to represent a set of conditional probability distributions, P, which specifies a multiplicative factorization of the joint probability distribution over X, as shown in Eq. (1). The joint probability becomes larger in size when the domain is more complex. When implementing learning structure or learning parameter from data, the inferential analysis for this research was performed with GeNIe version 2.0, Hugin Lite version 8.2, and Weka 3.6.

$$P(X) = \prod_{v \in V} P\left(X_v | X_{pa(v)}\right) \tag{1}$$

3.3 BN Model Structuring with Experts Versus Structured Learning

Bayesian network modeling is in paradigms which could be conveyed in a graphic form called the mental model from an expert domain. Bayesian networks experts such as Kudikyala et al. [13] used this technique to represent causal relationships between probability nodes. The experts then created a chart using the Pathfinder algorithm. It may be possible to create a united pathfinder network if there are many experts involved. The model can also use machine learning algorithms to determine the structure and estimate the probability of creating a model. The process of constructing a Bayesian model by machine learning is based on [14].

3.4 BN Model Comparison with F1 Score and AUC

A confusion matrix is a two-way table that is often employed to describe the performance of a classification model on a dataset for which the actual values of category are known. The simple confusion matrix is used in this study for binary classification, as shown in Table 1.

F1 is the first selected main criterion for model performance evaluation in this research. The F1 score is defined as the harmonic mean of precision (TP/[TP + FP]) and recall (TP/[TP + FN]) as shown in Eq. (2).

$$F1 = 2\left(\frac{\text{Precision} * \text{Recall}}{\text{Precision} + \text{Recall}}\right) \quad (2)$$

In addition, the second criterion for model performance evaluation is the receiver operating characteristic (ROC) curve. The ROC curve illustrates the relationship between the true-positive rate and the false-positive rate. In addition, the area under the ROC curve (AUC) is used to show the overall performance of the classification model. The AUC value lies between 0 and 1. An AUC value close to 1 denotes a good performance with no error. On the other hand, when the AUC value is close to 0, the model usually provides a wrong prediction. When the AUC value is close to 0.5, the model performs randomly.

Table 1 Confusion matrix for binary classifier

Actual	Predictive	
	Positive	Negative
Positive	True Positive (TP)	False Negative (FN)
Negative	False Positive (FP)	True Negatives (TN)

4 Results

It was found that 5.24% of the respondents from the training dataset and 5.53% of those from the testing dataset suffer from diabetes. These rates are similar, so they confirm the random selection for dividing the datasets into 2 parts.

4.1 BN Model Structures

A comparative study of BN models for predicting diabetes for this study consisted of 4 models based on different methods of BN structuring, as shown in Fig. 2. Model structuring was defined by types and methods, as can be seen in Table 2.

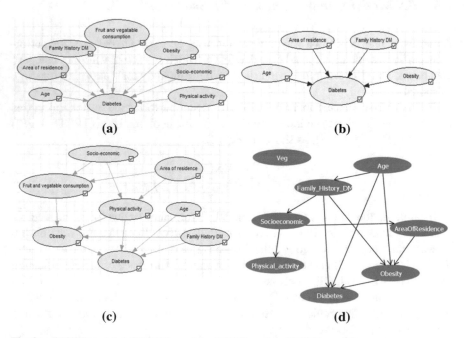

Fig. 2 a BNNH model. **b** BNNHR model. **c** BNHE model. **d** BNHL model

Table 2 Characteristics of focused BN models

Type of structure	Methods	Model
Non-hierarchical	Simple	BNNH
Non-hierarchical	Simple with variable reduced	BNNHR
Hierarchical	Expert	BNHE
Hierarchical	Learning from dataset	BNHL

Diabetes BN Model with Non-hierarchical Structure (BNNH) is a simple model with a Bayesian network non-hierarchical structure linking all 7 defined risk factors to diabetes status, as shown in Fig. 2a.

Diabetes BN Model with Non-hierarchical Structure and Reduced Model (BNNHR) can be improved by reducing those variables from the simple model structure. Four variables (age, area of residence, family history DM, and obesity) from seven variables are significantly dependent on the diabetes status at a significance level of 0.05 (p-value ≤ 0.05) by considering the result of Pearson's chi-square test of independence, as shown in Table 3. On the other hand, p-values from fruit and vegetable consumption, socio-economic status, and physical activity are greater than 0.05, so there is no relationship between these factors and the diabetes status. Therefore, the second model is proposed as the BN model with non-hierarchical structure with reduced model (BNNHR), as presented in Fig. 2b.

Diabetes BN Model with Hierarchical Structure by Expert Judgment (BNHE) is a Bayesian network (BN) that takes into account the hierarchical structures among direct and indirect causes of diabetes and is constructed by expert judgment by [10]. Figure 2c shows the relationship between 7 cause factors and the prevalence of diabetes. The experts participated in the research determined all possible states of particular variables and all relationships of particular variables in the model. The BNHE model was quantified by considering both available information from the National Health Examination Survey data and experts' opinion.

Diabetes BN Model with Hierarchical Learning Structure (BNHL) is derived from machine learning by using a training dataset, with Weka 3.7 software. The Generic Search is the selected Search Algorithm to determine the structure of the model as explained in [15]. The next step is teaching the machine to compute the values of probability tables of each variable and then choosing simple estimator method called the BNHL model in this study, as illustrated in Fig. 2d.

Table 3 Pearson's chi-square and p-value for test of independence

Factors	Pearson chi-square	p-value
Age	22.3382	0.0000*
Area of residence	199.2318	0.0000*
Family history DM	266.1512	0.0000*
Fruit and vegetable consumption	1.5101	0.2191
Obesity	188.1879	0.0000*
Socio-economic	4.8520	0.3028
Physical activity	0.9525	0.3291

*p-value < 0.05

4.2 Performance Comparison

For the evaluation of three BN models, the first performance criterion is defined by comparing F1, as shown in Eq. (2). The 0.1 and 0.2 threshold classification levels are employed to compare F1 scores, since true negatives and false positives are available. It is found that F1 score of the BNHE model is the highest.

We observe from Table 4 that ROC curves and AUC obtained from the test dataset are used for the comparison of performance in predicting the probability of positive diabetes status. Figure 3a, b depicts the ROC curves of the training and testing dataset. BNNHR, BNHE, and BNHL show the areas under the ROC curves with similar patterns on the upper left corner, as presented in Fig. 3. By considering the

Table 4 F1 scores comparison by threshold levels and AUC values by dataset

Model	F1					AUC	
	Threshold					Training dataset	Testing dataset
	0.5	0.4	0.3	0.2	0.1		
BNNH	0.1460	0.1456	0.1941	0.2011	0.1872	0.6281	0.6698
BNNHR	–	–	–	–	0.2157	0.7667	0.7743
BNHE	–	–	–	0.0965	0.2383	0.7670	0.7760
BNHL	–	–	–	–	0.2383	0.7607	0.7749

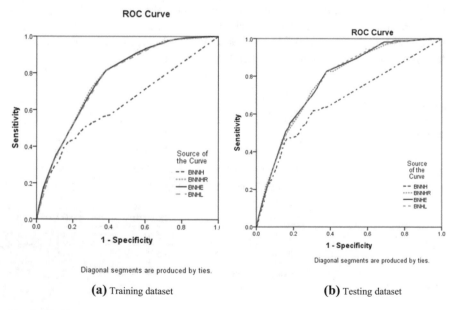

(a) Training dataset (b) Testing dataset

Fig. 3 ROC curve for classification models

area under the ROC curve (AUC), the BNHE model achieves the highest AUC values of 0.7670 and 0.7760 from the training and the testing dataset, respectively.

5 Conclusions and Further Studies

The results of this study have highlighted the comparison between different BN modeling structures applied in the context of diabetes. The Bayesian network is a naturally hierarchical structure which shows the causal relationship between focal variables. The result of investigation shows that BN modeling can be used to identify the prevalence of diabetes either by the hierarchical or non-hierarchical modeling structure. The result emphasizes that the hierarchical structure is able to perform better classification than the two non-hierarchical models. Furthermore, the BNHE model constructed by experts also perform better than the BNHL model constructed by learning from data, as was also concluded by Xiao-xuan et al. [5]. The BN modeling with hierarchical structure is unlike logistic regression [3], since the independent variables for logistic regression are analyzed by providing all variables in the same order. For this reason, it is not possible to identify the root cause of type 2 diabetes. Therefore, the application of BN models will be useful in understanding and identifying the root cause of diabetes in order to determine suitable strategic plans.

We considered two criteria for comparing models: F1 score and AUC, rather than using one criterion since selecting a single overall performance measure poses some difficulties [16]. In this paper, we used the F1 score, since it includes precision and recall in its formula. Therefore, this score takes both false positives and false negatives into account. Furthermore, F1 is usually more useful than accuracy, especially if we agree that false positives and false negatives of diabetes prediction have unequal costs. Otherwise, accuracy works best if false positives and false negatives are defined at the same cost. Further research should aim to improve and update the model by means such as including some other important variables, for example, smoking, alcohol consumption, etc. In addition, implementing different methods of BN learning should be explored further.

Acknowledgements This research is supported by King Mongkut's Institute of Technology Ladkrabang [No. 2559-02-05-050]. We would like to thank the Thai Public Health Survey Institute for Health Systems Research for providing helpful datasets.

References

1. Thammasitboon, S., Cutrer, W.B.: Diagnostic decision-making and strategies to improve diagnosis. Curr. Probl. Pediatr. Adolesc. Health Care **43**(9), 232–241 (2013)
2. Wagholikar, K.B., Sundararajan, V., Deshpande, A.W.: Modeling paradigms for medical diagnostic decision support: a survey and future directions. J. Med. Syst. **36**(5), 3029–3049 (2012)

3. Nguefack-Tsague, G.: Using bayesian networks to model hierarchical relationships in epidemiological studies. Epidemiol. Health **33**, 1–8 (2011)
4. Leerojanaprapa, K., van der Meer, R., Walls, L.: Modeling supply risk using belief networks: a process with application to the distribution of medicine. In: 2013 IEEE International Conference on Industrial Engineering and Engineering Management, pp. 201–205 (2013)
5. Xiao-xuan, H., Hui, W., Shuo, W.: Using expert's knowledge to build bayesian networks. In: 2007 International Conference on Computational Intelligence and Security Workshops (CISW 2007), pp. 220–223 (2007)
6. Atoui, H., Fayn, J., Gueyffier, F., Rubel, P.: Cardiovascular risk stratification in decision support systems: a probabilistic approach. Application to pHealth (2006)
7. Xie, J., Liu, Y., Zeng, X., Zhang, W., Mei, Z.: A Bayesian network model for predicting type 2 diabetes risk based on electronic health records. Mod. Phys. Lett. B **31**, 1740055 (2017)
8. Joseph, A., Fenton, N.E., Neil, M.: Predicting football results using Bayesian nets and other machine learning techniques. Knowl. Based Syst. **19**(7), 544–553 (2006)
9. Julia Flores, M., Nicholson, A.E., Brunskill, A., Korb, K.B., Mascaro, S.: Incorporating expert knowledge when learning Bayesian network structure: a medical case study. Artif. Intell. Med. **53**(3), 181–204 (2011)
10. Leerojanaprapa, K., Atthirawong, W., Aekplakorn, W., Sirikasemsuk, K.: Applying Bayesian network for noncommunicable diseases risk analysis: implementing national health examination survey in Thailand. In: IEEE International Conference on Industrial Engineering and Engineering Management, vol. 2017–December (2018)
11. Aekplakorn, W., Satheannoppakao, W., Putwatana, P., Taneepanichskul, S., Kessomboon, P., Chongsuvivatwong, V., Chariyalertsak, S.: Dietary pattern and metabolic syndrome in thai adults. J. Nutr. Metab. **2015**, 468759 (2015)
12. Jensen, F.V., Nielsen, T.D.: Bayesian Networks and Decision Graphs, 2nd edn. Springer, New York (2007)
13. Kudikyala, U.K., Bugudapu, M., Jakkula, M.: Graphical Structure of Bayesian Networks by Eliciting Mental Models of Experts, pp. 333–341. Springer, Singapore (2018)
14. Duijm, N.J.: Safety-barrier diagrams as a safety management tool. Reliab. Eng. Syst. Saf. **94**(2), 332–341 (2009)
15. Bouckaert, R.R.: Bayesian Network Classification in Weka for Version 3-5-7. University of Waikato (2008)
16. Kurt, I., Ture, M., Kurum, A.T.: Comparing performances of logistic regression, classification and regression tree, and neural networks for predicting coronary artery disease. Expert Syst. Appl. **34**(1), 366–374 (2008)

Part IV
Web and Informatics

A Survey on the Detection of Android Malicious Apps

Sanjay K. Sahay and Ashu Sharma

Abstract Android-based smart devices are exponentially growing, and due to the ubiquity of the Internet, these devices are globally connected to the different devices/networks. Its popularity, attractive features, and mobility make malware creator to put number of malicious apps in the market to disrupt and annoy the victims. Although to identify the malicious apps, time-to-time various techniques are proposed. However, it appears that malware developers are always ahead of the anti-malware group, and the proposed techniques by the anti-malware groups are not sufficient to counter the advanced malicious apps. Therefore, to understand the various techniques proposed/used for the identification of Android malicious apps, in this paper, we present a survey conducted by us on the work done by the researchers in this field.

Keywords Android · Malicious apps · Dangerous permissions · Anti-malware

1 Introduction

In 1992, International Business Machines Corporation came up with a prototype mobile computing device incorporating with the personal digital assistant features (demonstrated it in the Computers Dealer's Exhibition). Later on, Simon Personal Communicator first designed a device that was referred as smart device/phone which receives calls, sends faxes, emails, and more. The smart devices/phone technology continued to advance throughout early 2000, and in 2007 Android-based smart device was unveiled by Google. Since then the popularity/demand for Android-based smart devices is continuously growing. An estimate shows that more than 15 billion smart

S. K. Sahay (✉)
Department of CS and IS, BITS, Pilani, Goa Campus, Sancoale, India
e-mail: ssahay@goa.bits-pilani.ac.in

A. Sharma
C3i, CSE, IIT Kanpur, Kanpur, India
e-mail: ashush@cse.iitk.ac.in

© Springer Nature Singapore Pte Ltd. 2019   437
S. K. Bhatia et al. (eds.), *Advances in Computer Communication
and Computational Sciences*, Advances in Intelligent Systems and Computing 924,
https://doi.org/10.1007/978-981-13-6861-5_38

devices are connected globe-wise and are expected to be reaching 200 billion by the end of the year 2020 [8]. Also, due to the mobility and the attractive features of these devices have drastically changed our day-to-day life. Many of these functionalities are very much similar to our other information technology systems and are capable to remotely access the enterprise's data for the work. In addition, because of the ubiquity of Internet, the user uses these devices for the shopping, financial transactions, share private information/data, etc. [21]. Hence, the security risks of the smart devices are now at never seen before levels, hence an attractive target for online attackers. Also, nowadays online criminals are investing more and more for the sophisticated attacks, e.g., ransomware or to steal the valuable personal data/information from it.

Today in the growing smart devices, Android is the most popular operating system (OS) ($\sim$70% of the market share) [14] and is connected to different devices/networks through the Internet (five out of six mobile phones are Android based [36]). The popularity of Android OS is because of its open source, java-supported free rich software developer's kit, exponential increase in the Android-based apps, and third-party distribution. According to Statista, in addition to the third-party Android apps, at Google Play Store, there are $\sim$2$\times$10^6 apps are available for the users [40], and some of these apps may be malicious [1]. Therefore, the probability of the malware in the Android smart devices is now at never seen before levels. Thus attacks on the Android-based smart devices are increasing exponentially, basically due to the ease of creating the variants of malware [6, 42]. In 2013, there was a 200% increase in the malicious apps, and 3.7 million of variants added in McAfee's database [7]. In 2015, Kaspersky reported that the growth rate of new malware variant is 300% with 0.88 million new variants [20]. The number of malicious installations found in 2015 was around three million, and around seven thousand mobile banking Trojans were also found in the same year [19]. In the third quarter of 2015, Quick Heal Threat Research reported that per day they had received $\sim$4.2 $\times$ 10^5 samples of the Windows and Android platforms [21]. Trend Micro estimated that the number of malicious mobile apps would reach 20 million by the end of 2017 [9]. In this, stepping up its fight against bad apps and malicious developers, Google has removed over 700,000 Android apps (i.e., 70% more apps that violated the Google Play policies in 2017 than the apps they removed in 2016) from Play Store and also took down as many as 100,000 bad developers in 2017 [2].

The recent attacks on Android devices show that the security features in these devices are not good enough to stop the adversary [37]. Therefore, its a need of time to design a robust anti-malware, in particular, to counter zero-day attack [31]. Also, it is an open question, how to detect the variants of malicious Android apps which are concealed in the third-party apps markets [1], and how to find the repackaged apps in the ocean of Android apps. In addition, to avoid the deployed detection methods, malware developer uses various obfuscation techniques [35]. Hence, any gap in the security of the Android smart devices will allow the attacker to access the information stored in it. To defend the attack/threat from the Android malicious apps, a number of static and dynamic methods were proposed [4, 12, 22, 25]. But still, it appears that to defend the malicious apps, the proposed techniques are not good enough in the growing smart devices usage in our day-to-day life [31]. Thus, Android-based smart devices security is one of the important fields to be addressed, and understanding the market share of the Android-based smart devices, in this paper to know the various techniques proposed/used to identify the Android malicious apps, in Sect. 2, to understand security mechanism of the Android devices we discuss in brief that how user data and its resources are secure by providing the features, viz. sandbox, permission, secure inter-component communication, and signing the apps. In Sect. 3, we present the survey conducted by us on the work done by the researchers in this field. Finally, in Sect. 4, we summarize our conclusion.

2 Security Mechanisms

The security of the Android-based devices is mainly focused to protect the user data, its resources, and isolation of the application by providing the features, viz. sandbox, permission, secure inter-component communication, and signing the apps. In this, the **sandbox** in the Android-based system isolates the apps from each other by the user ID (UID) and permissions. After installation, the apps run in its assigned sandbox and can access only its own resources, unless other apps give explicit access permission to this applications. However, if the apps are designed by the same developer, then such apps can share the same UID and can run in the same sandbox to share resources/data between them.

Android apps consist of four components, viz., services, broadcasts, activities, and providers. Similar to inter-process communication in the Linux system, it provides a secure **inter-component communication** by binder mechanism (middleware layer of Android). Inter-component communication is achieved by the intents/message, and these intents are explicitly used for the communication, if it identifies the receiver name, or used for implicit communication that allows the receiver to know that can it access this intent or not.

Application signing creates a certification between developers and their applications to ensure the security of the apps, and before putting it in its sandbox, it makes a relationship between the apps and UID. Without application signing, apps will not run. If two or more apps have the same UID, then all the apps which have the same UID can communicate with each other, share the permissions, and can run in the same sandbox. By the application singing, apps update process can be simplified. The updated new version will have all the permissions that the old version has, and also the certificate does not change so that the package manager can verify the certificate. It also makes sure that without using inter-component communication, apps cannot communicate with another apps. However, if the apps are developed by the same developer, then the developer without changing the application signing can enable the direct communication between the same developer apps.

In these smart devices, there are four levels of **permissions**, viz. signatureOrSystem (granted to the apps that are installed by the root or pre-installed apps), signature (granted within the same sandbox), dangerous (granted by the users), and normal (granted automatically) permissions. In total, there are 235 permissions out of which

163 are hardware accessible and remaining are for user information access [24]. Before installation of the apps, the system asks the users to grant all requested permissions; if users agree and grant the requested permissions to the apps, then in general installation becomes successful else it may get canceled. These permission mechanisms put some restriction when the apps want to access the application programming interface (API) which are sensitive to the OS. The Android apps run in the sandbox, and if it needs to access resources/data outside its sandbox which could potentially affect the user's privacy/data, viz. *short message service*, contacts, camera, location, etc., then the user has to approve/reject the permission.

3   Detection of Android Malicious Apps

To identify the Android malicious apps, static and dynamic analysis are the two basic methods that are used [18]. In both the methods, classifiers are first trained with a dataset for the classification of apps. However, in static analysis, apps are analyzed without executing it to extract some patterns, viz. APIs used, data flow, permissions, control flow, intents, etc. Whereas, in the dynamic analysis the codes are analyzed during the execution of the apps, and monitoring its dynamic behavior (resources usage, tracing system calls, API call, network traffic, etc.) and the response of the system. In this, in 2009, understanding the users do not understand what applications will do with their data/resources, and thus not able to decide which permissions shall be allowed to the application to run with. Fuchs et al. [13] proposed a tool called SCANDROID (suppose to be the first program analysis tool for the Android-based devices) which can extracts security specifications from the applications, and can identify that the data flow of such apps are consistent with the specification or not.

In 2012 Sanz. et al. [27] based on machine learning techniques proposed a method to detect the Android malicious apps by automatically characterize the applications. For the classification, the feature sets used are the printable strings, apps permissions, and the apps permissions extracted from the Android Market. Their experiment with seven different categories (820 samples) and five classifiers shows that among the selected five classifiers, Bayes Tree Augmented Naive Bayes is the best classifier (0.93 area under the curve (AUC)), while random forest (RF) stands second in the investigated classifier (0.9 AUC), and among the analyzed classifier, the worst was Decision Tree with J48 (0.64 AUC). Wu. et al. based on the static feature proposed a technique called *DroidMat*, to detect the Android malicious apps which analyze AndroidManifest.xml and the systems calls. For the experiment, they used 238 malicious and 15000 benign programs and claimed that their approach is very effective (97.87% accuracy), scalable, and efficient [43].

In 2013, Michael et al. proposed a mobile sandbox to automatically analyze the Android apps in two steps. In the first step (static analysis), applications Manifest files are parsed and decompiled; then they find that the applications are using suspicious permissions/intents or not. In the next step, dynamic analysis is performed, where the apps are executed to know all the actions including those originating from the associated API calls. They experimented with ~36,000 apps from the third-party Asian mobile markets and reported that 24% of all analyzed apps use associated calls. [39]. Min Zheng et al. [45] developed a signature-based system called *DroidAnalytics* for collecting the malware automatically, and to generate signatures for the identification of the malicious code segment. They conducted extensive experiments with 150,368 apps and detected 2,494 malicious apps from one hundred two different families, in which three hundred forty-two of them were zero-day malicious samples from the six different families. They claimed that their methods have significant advantages over the traditional MD5 hash-based signature and can be a valuable tool for the classification of Android malicious apps.

In 2014, a detection method was proposed by Quentin et al. which depends on the opcode sequences. They tested libsvm and SVM classifier with the reduced dataset (11,960 malware and 12,905 benign applications) and obtained 0.89% F-measure. However, their approach is not capable to detect completely different malware [16]. Kevin Allix et al. [3] devised classifiers that depend on the features set that are designed from the apps control flow graphs. They analyzed their techniques with ~50,000 Android apps and claimed that their approach outperformed existing machine learning approaches. Also from the analysis, they concluded that for the realistic malware detectors, the tenfold cross-validation approach on the usual dataset is not a reliable indicator of the performance of the classifier.

The smartphone can act as like a zombie device, controlled by the hackers via command and control servers. It has been found that mobile malware is targeting Android devices to get root-level access so that from the remote server they can execute the instructions. Hence, such type of malware will be a big threat to the homeland security. Therefore, Seo et al. [29] discuss the defining characteristics which are inherent in the devices and show the feasible mobile attack scenario against the Homeland security. They analyze various mobile malware samples, viz. monitoring the home and office, banking, flight booking and tracking, from both the unofficial and official market to identify the potential vulnerabilities. Their analysis shows that the two banking apps (Axis and Mellat bank apps) charge SMS for malicious activities and two other banking apps were modified to get permissions without the consent. Finally, they discuss an approach that mitigates the homeland security from the malware threats.

In 2015, Jehyun Lee et al. developed a technique to detect the malicious apps that use automated extraction of the family signature. They claimed that compared to earlier behavior analysis techniques, their proposed family signature matching detection accuracy is high and can detect variants of known malware more accurately and efficiently than the legacy signature matching. Their results were based on the analysis done with the 5846 real Android malicious apps which belong to 48 families collected in April 2014 and achieved 97% accuracy. Smita et al. [23] addressed the problem of system-call injection attack (inject independent and irrelevant system calls when programs are executing) and proposed a solution which is evasion-proof and is not vulnerable to the system-call injection attacks. Their technique characterizes the program semantics by using the property of asymptotic equipartition, which allows to find the information-rich call sequences to detect the malicious apps. According to their analysis, the semantically relevant paths can be applied to know the malicious behavior and also to identify the number of unseen/new malware. They claimed that the proposed solution is robust against the system-call injection attacks and are effective to detect the real malware instances.

In 2016, a host-based Android malicious apps detection system called Multi-Level Anomaly Detector for Android Malware (*MADAM*) has been proposed by Saracino et al. [28]. Their system at the same time can analyze and correlate the features at four levels, viz., application, kernel, user, and package to identify. It stopped the malicious behaviors of one hundred twenty-five known malware families and encompassed most of the malware. They claimed that MADAM could understand the behavior characteristics of almost all the real malicious apps which can be known in the wild, and it can block more than 96% of Android malicious apps. Their analysis on 9,804 clean apps shows low false alarm rate, limited battery consumption (4% energy overhead), and negligible performance overhead. BooJoong et al. [17] proposed an n-opcode-based static analysis for the classification and categorizing the Android malware. Their approach does not utilize the defined features, viz. permissions, API calls, intents, and other application properties, rather it automatically discovers the features that eliminate the need of an expert to find the required promising features. Empirically, they showed that by using the occurrence of n-opcodes, a reasonable classification accuracy can be achieved, and for $n = 3$ and $n = 4$, they have achieved F-measure up to 98% for categorization and classification of the malware.

Based on inter-component communication (ICC)-related features, Ke Xu et al. [44] proposed a method to identify the malicious apps called ICCDetector, which can capture the interaction between the components or cross-application boundaries. They evaluated the performance of their approach with 5264 malware and 12026 benign apps and achieved an accuracy of 97.4%. Also, after manually analyzing, they discovered 43 new malware in the benign data and reduced the number of false positive to seven. Jae-wook Jang et al. [15] proposed a feature-rich hybrid anti-malware system called *Andro-Dumpsys*, which can identify and classify the malware groups of similar behavior. They claimed that *Andro-Dumpsys* could detect the malware and classify the malware families with low false positive (FP) and false negative (FN). It is also scalable and capable to respond zero-day threats. Gerardo Canfora et al. [5] evaluated a couple of techniques for detecting the malicious apps. First one was based on *hidden markov model*, and the second exploits the structural entropy. They claimed that their approach is effective for desktop viruses and can also classify the malicious apps. Experimentally, they achieved a precision of 0.96 to differentiate the malicious apps, and 0.978 to detect the malware family [5]. For the detection of malware in runtime, Sanjeev Das et al. [10] proposed a hardware-enhanced architecture called *GuardOL* by using *field programmable gate arrays* and processor. Their approach after extracting the system calls made the features from the high-level semantics of the malicious behavior. Then the features are used to train the classifier and multilayer perceptron to detect the malicious apps during the execution. The importance of their design was that the approach in the first 30% of the execution detects 46% of the malware, and after 100% of execution 97% of the samples have been identified with 3% FP. [10].

In 2017, Ali Feizollah et al. [12] proposed AndroDialysis to evaluate how effective is the Android intents (implicit and explicit) as a feature to identify the malicious apps. They showed that the intents are semantically rich features compared to well studied other features, viz. permissions, to know the intentions of malware. However, they also concluded that such features are not the ultimate solution, and it should be used together with other known promising features. Their result was based on the analysis of the dataset of 7406 apps (1846 clean and 5560 infected apps). They achieved a 91% detection accuracy by using Android Intent, 83% using Android permission and by combining both the features they obtained the detection rate of 95.5%. They claimed that for the malware detection, Intent is more effective than the permission. Bahman Rashidi et al. [25] proposed an Android resources usage risk assessment called *XDroid*. They claimed that the malware detection accuracy could be increased significantly by using the temporal behavior tracking, and security alerts of the suspicious behaviors can be generated. Also, in real time their model can inform users about the risk level of the apps and can dynamically update the parameters of the model by the user's preferences and from the online algorithm. They conducted the experiment on benchmark *Drebin* malware dataset and demonstrated that their approach could estimate the risk levels of the malware up to 82% accuracy and with the user input it can provide an adaptive risk assessment.

Recently, Ashu et al. [30] with the benchmark *Drebin* dataset and first without making any groups, they examine the five classifiers using the opcode occurrence as a feature for the identification of malicious applications and achieved detection accuracy up to 79.27% by functional tree classifier. They observed that the overall accuracy is mainly affected by the FP. However, highest true positive (99.91%) is obtained by RF and fluctuates least with the number of features compared to the remaining four classifiers. Their analysis shows that overall accuracy mainly depends on the FP of the investigated classifiers. Later on, in 2018, similar to the analysis of Windows desktop malware classification [26, 32, 34], they group-wise analyzed the dataset based on permissions and achieved up to 97.15% overall average accuracy [33]. They observed that the MICROPHONE group detection accuracy is least while CALENDAR group apps are detected with maximum accuracy, and top 80–100 features are good enough to achieve the best accuracy. As far as the true positive (TP) is concerned, RF gives best TP for the CALENDAR group.

4 Summary

The attack/threat by the malware to the Android-based smart devices which are connected to different devices/networks through the Internet is increasing day by day. Consequently, these smart devices are highly vulnerable to the advanced malware, and its effect will be highly destructive if an effective and timely countermeasures are not deployed. Therefore, time to time, various static and dynamic methods that have been proposed by the authors for the identification of the Android malicious apps have been discussed in this paper to counter the advanced malicious apps in the fast-growing Internet and smart devices usage into our daily life. Although in the literature various survey, viz. [11, 38, 41] are available on the detection Android malicious but in this survey, we presented the comparative study of various approaches and the observations done by the various authors understanding that all malware are not of the same type and cannot be detected with one algorithm.

References

1. 9apps: Free android apps download (2016). http://www.9apps.com
2. Ahn, A.: How we fought bad apps and malicious developers in 2017. Technical report, Google Play (2018). https://android-developers.googleblog.com/2018/01/how-we-fought-bad-apps-and-malicious.html
3. Allix, K., Bissyandé, T.F., Jérome, Q., Klein, J., Le Traon, Y., et al.: Large-scale machine learning-based malware detection: confronting the 10-fold cross validation scheme with reality. In: Proceedings of the 4th ACM Conference on Data and Application Security and Privacy, ACM, pp. 163–166 (2014)
4. Arp, D., Spreitzenbarth, M., Hubner, M., Gascon, H., Rieck, K.: Drebin: effective and explainable detection of android malware in your pocket. In: NDSS, pp. 1–15 (2014)
5. Canfora, G., Mercaldo, F., Visaggio, C.A.: An HMM and structural entropy based detector for android malware: an empirical study. Comput. Secur. **61**, 1–18 (2016)
6. Christiaan, B., Douglas, F., Paula, G., Yashashree, G., Francisca, M.: Mcafee threats report. Technical report, McAfee (2012). https://www.mcafee.com/in/resources/reports/rp-quarterly-threat-q1-2012.pdf
7. Christiaan, B., Douglas, F., Paula, G., Yashashree, G., Francisca, M.: Mcafee labs threats report. Technical report, McAfee (2015). https://www.mcafee.com/ca/resources/reports/rp-quarterly-threat-q1-2015.pdf
8. Christiaan, B., Douglas, F., Paula, G., Yashashree, G., Francisca, M.: McAfee labs threats report. Technical report (2017). https://www.mcafee.com/in/resources/reports/rp-threats-predictions-2016.pdf
9. Clay, J.: Trend micro, continued rise in mobile threats for 2016 (2015). http://www.blog.trendmicro.com/continued-rise-in-mobile-threats-for-2016
10. Das, S., Liu, Y., Zhang, W., Chandramohan, M.: Semantics-based online malware detection: towards efficient real-time protection against malware. IEEE Trans. Inf. Forensics Secur. **11**(2), 289–302 (2016)
11. Faruki, P., Bharmal, A., Laxmi, V., Ganmoor, V., Gaur, M.S., Conti, M., Rajarajan, M.: Android security: a survey of issues, malware penetration, and defenses. IEEE Commun. Surv. Tutor. **17**(2), 998–1022 (2015)
12. Feizollah, A., Anuar, N.B., Salleh, R., Suarez-Tangil, G., Furnell, S.: Androdialysis: analysis of android intent effectiveness in malware detection. Comput. Secur. **65**, 121–134 (2017)
13. Fuchs, A.P., Chaudhuri, A., Foster, J.S.: Scandroid: automated security certification of android. Technical report, University of Maryland Department of Computer Science (2009)
14. Gandhewar, N., Sheikh, R.: Google android: an emerging software platform for mobile devices. Int. J. Comput. Sci. Eng. **1**(1), 12–17 (2010)
15. Jang, J.W., Kang, H., Woo, J., Mohaisen, A., Kim, H.K.: Andro-dumpsys: anti-malware system based on the similarity of malware creator and malware centric information. Comput. Secur. **58**, 125–138 (2016)
16. Jerome, Q., Allix, K., State, R., Engel, T.: Using opcode-sequences to detect malicious android applications. In: 2014 IEEE International Conference on Communications (ICC), IEEE, pp. 914–919 (2014)

17. Kang, B., Yerima, S.Y., Mclaughlin, K., Sezer, S.: N-opcode analysis for android malware classification and categorization. In: 2016 International Conference On Cyber Security And Protection Of Digital Services (Cyber Security), pp. 1–7 (2016)
18. Kapratwar, A.: Static and dynamic analysis for android malware detection. Master's thesis, San Jose State University (2016)
19. Lab, K.: Red alert: Kaspersky lab reviews the malware situation in Q3. Technical report (2014)
20. Lab, K.: Securelist: mobile malware evolution. Technical report (2015)
21. Lab., Q.H.: Threat report 3rd quarter, 2015. Quick heal lab (2015). http://www.quickheal.co.in/resources/threat-reports
22. Narayanan, A., Yang, L., Chen, L., Jinliang, L.: Adaptive and scalable android malware detection through online learning. In: 2016 International Joint Conference on Neural Networks (IJCNN), IEEE, pp. 2484–2491 (2016)
23. Naval, S., Laxmi, V., Rajarajan, M., Gaur, M.S., Conti, M.: Employing program semantics for malware detection. IEEE Trans. Inf. Forensics Secur. **10**(12), 2591–2604 (2015)
24. Olmstead, K., Atkinson, M.: Apps permissions in the Google Play store. Technical report, Pew Research Center (2016)
25. Rashidi, B., Fung, C., Bertino, E.: Android resource usage risk assessment using hidden Markov model and online learning. Comput. Secur. **65**, 90–107 (2017)
26. Sahay, S.K., Sharma, A.: Grouping the executables to detect malwares with high accuracy. Procedia Comput. Sci. **78**, 667–674 (2016)
27. Sanz, B., Santos, I., Laorden, C., Ugarte-Pedrero, X., Bringas, P.G.: On the automatic categorisation of android applications. In: 2012 IEEE Consumer Communications and Networking Conference (CCNC), IEEE, pp. 149–153 (2012)
28. Saracino, A., Sgandurra, D., Dini, G., Martinelli, F.: Madam: effective and efficient behavior-based android malware detection and prevention. IEEE Trans. Dependable Secur. Comput. **99**, 1–1 (2017)
29. Seo, S.H., Gupta, A., Sallam, A.M., Bertino, E., Yim, K.: Detecting mobile malware threats to homeland security through static analysis. J. Netw. Comput. Appl. **38**, 43–53 (2014)
30. Sharma, A., Sahay, K.S.: An investigation of the classifiers to detect android malicious apps. In: Proceedings of ICICT 2016 Information and Communication Technology, vol. 625, pp. 207–217. Springer, Berlin (2017)
31. Sharma, A., Sahay, S.K.: Evolution and detection of polymorphic and metamorphic malwares: a survey. Int. J. Comput. Appl. **90**(2), 7–11 (2014)
32. Sharma, A., Sahay, S.K.: An effective approach for classification of advanced malware with high accuracy. Int. J. Secur. Appl. **10**(4), 249–266 (2016)
33. Sharma, A., Sahay, S.K.: Group-wise classification approach to improve android malicious apps detection accuracy. Int. J. Netw. Secur. (2018) (In Press)
34. Sharma, A., Sahay, S.K., Kumar, A.: Improving the detection accuracy of unknown malware by partitioning the executables in groups. In: Advanced Computing and Communication Technologies, pp. 421–431. Springer, Berlin (2016)
35. Shaun, A., Tareq, A., Peter, C., Mayee, C., Jon, D.: Internet security threat report. Technical report, Symantec (2014). https://www.symantec.com/content/dam/symantec/docs/reports/istr-22-2017-en.pdf
36. Shaun, A., Tareq, A., Peter, C., Mayee, C., Jon, D.: Internet security threat report 2016. Technical report, Symantec Corporation (2016)
37. Shaun, A., Tareq, A., Peter, C., Mayee, C., Jon, D.: Internet security threat report 2017. Technical report, Symentec (2017)
38. Souri, A., Hosseini, R.: A state-of-the-art survey of malware detection approaches using data mining techniques. Hum.-Centric Comput. Inf. Sci. **8**(3), 22 (2018)
39. Spreitzenbarth, M., Freiling, F., Echtler, F., Schreck, T., Hoffmann, J.: Mobile-sandbox: having a deeper look into android applications. In: Proceedings of the 28th Annual ACM Symposium on Applied Computing, ACM, pp. 1808–1815 (2013)
40. Statista: number of available applications in the Google Play store from December 2009 to February 2016 (2016). https://developer.android.com/guide/topics/security/permissions.html

41. Tam, K., Feizollah, A., Anuar, N.B., Salleh, R., Cavallaro, L.: The evolution of android malware and android analysis techniques. ACM Comput. Surv. **49**, 1–41 (2017)
42. Vidas, T., Christin, N., Cranor, L.: Curbing android permission creep. In: Proceedings of the Web, vol. 2, pp. 91–96 (2011)
43. Wu, D.J., Mao, C.H., Wei, T.E., Lee, H.M., Wu, K.P.: Droidmat: android malware detection through manifest and API calls tracing. In: 2012 Seventh Asia Joint Conference on Information Security (Asia JCIS), IEEE, pp. 62–69 (2012)
44. Xu, K., Li, Y., Deng, R.H.: ICCdetector: ICC-based malware detection on android. IEEE Trans. Inf. Forensics Secur. **11**(6), 1252–1264 (2016)
45. Zheng, M., Sun, M., Lui, J.C.: Droid analytics: a signature based analytic system to collect, extract, analyze and associate android malware. In: 2013 12th IEEE International Conference on Trust, Security and Privacy in Computing and Communications (TrustCom), IEEE, pp. 163–171 (2013)

A Survey on Visualization Techniques Used for Big Data Analytics

Sumit Hirve and C. H. Pradeep Reddy

Abstract According to the recent developments and efficient technology trends, Big Data has become a vital asset for all industries and organizations in modern times. Big Data is trending due to few of the main reasons such as cloud migration initiated by companies, aggregation of digital unstructured and machine data, strong administration of data security permission, and many others. As we all know, analytics is the process of drawing conclusions and finding insights from a big pile. Big Data analytics is defined as the process of querying, simplifying, obtaining insights from the huge set of data integrated in the file systems of Big Data framework. The insights obtained as the outcome of analytics should reach the end users operating on the application platform in the form of visual representation techniques such as reports, line graphs, and bar charts for better understanding and exploration about the data. We propose a case study consisting of comparison of all the existing data visualization tools and techniques available and suitable with Big Data. The paper outlines all the advantages and disadvantages of Data Visualization tools and recommends to use the one which outclasses the comparison test. The later part of paper explains about the methodology proposed using Big Data Hadoop and CDH's Apache Impala, Hue, and Hive. The dataset chosen is imputed and fed to the Cloudera Engine for query processing and report generation. Further, the generated 2D output is given as input to Unity 3D engine for generating QR codes and 3D visualization using Marker-based technique of augmented reality.

Keywords Big data · Visualization · Apache hadoop · Apache hive · Apache sqoop · Data analytics · Google charts

S. Hirve (✉) · C. H. Pradeep Reddy
VIT AP Amaravati, Amaravati, India
e-mail: sumitarun.h@vitap.ac.in

C. H. Pradeep Reddy
e-mail: pradeep.ch@vitap.ac.in

© Springer Nature Singapore Pte Ltd. 2019
S. K. Bhatia et al. (eds.), *Advances in Computer Communication and Computational Sciences*, Advances in Intelligent Systems and Computing 924,
https://doi.org/10.1007/978-981-13-6861-5_39

447

1 Introduction

In the field of computer science, the recent trends and developments in areas such as data science, artificial intelligence, and machine learning have achieved enormous appreciation and have set a benchmark for the current developments. Visualization is a research area that allows and helps user in better understanding and exploration of data or records. Many specific institutions, I.T firms, schools and colleges, medical and healthcare load and store a tremendous amount of digital data. This data may be dirty, noisy, and can get pretty hard on the applications for outputting a smooth representation to the end users. The term 'visualization' is referred to filter out the necessary fields of data, chop them into a category, and display it to the end users in the form of reports, charts, tree diagrams, or any diagrammatical representation which can help user to quickly understand the data and derive conclusions from the data. Being a powerful and productive scientific term, visualization has led to a rapid increase in growth of innovations and heavily data-equipped projects. Visualization can be categorized into analytical visualization and information visualization (InfoViz). InfoViz is similar to data visualization, except for the fact when data are not directly processed and conveyed as information. Ranging from schools to sports, InfoViz is used for visual analysis of student data, medical data, business records, and all other public and private repositories [1]. D. A. Keim et al. in their research article mentioned the use of visual analytics for information age-solving problems [2]. People who leave their traces on the social media help other investigators to track them via social media data. Siming Chen et al. mentioned the process of analytics of geo-sampled social media data along with the movement patterns and visual analytics of sampled trajectories [3]. The authors have designed an uncertainty-aware microblog retrieval toolkit and a composite visualization to understand the retrieved data. This toolkit is helpful in estimating the uncertainty which arises by the analysis algorithm and also for quickly retrieving striking posts, users of social media, and hashtags used in social media [4]. Talking about urbanization's progress, it has impacted the modern civilization in a very positive way but has also endured many challenges in people's life. Yu Zheng et al. in their research article reviewed and published in ACM Transactions on Intelligent Systems and Technology, explained the general framework of Urban Computing consisting of four layers and principles of Data Analytics [5]. The concept of 'urban computing' is described as a process of acquiring and integrating the data from different sources such as sensors, embedded devices, man-made machines, and use it to tackle the issues such as traffic, air pollution, and water problems. To improve the environment, urban computing works on connecting the robust and pervasive technologies. The frequent challenges and data types used in urban visual analytics were described by Yixian Zheng et al. in their IEEE overview Transaction on Big Data entitled 'Visual Analytics in Urban Computing: An Overview'. The challenges of Visual Analytics for Urban Computing such as heterogeneity, hug volumes of data were explained with the existing work on visual analytics and visualization tools [6]. With the boom in data science fields, Big Data analytics and data mining, the various visualization techniques are

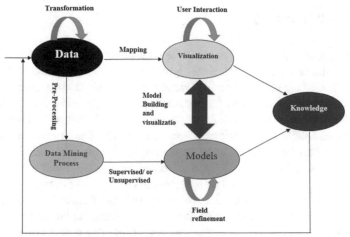

Fig. 1 Process of visual analytics

used in the above-mentioned aspects. Big Data is being researched and implemented to overcome the urbanization hurdles such as traffic pollution, weather forecast, air pollution, water depletion, and others. The massive data are gathered from sources, analytics is conducted on the data, and results are derived. The authors in their survey paper stated that results are the filtered part of data obtained from the entire piece of data collected. The results can also be defined as the insights or conclusions which are displayed to the users in form of diagrammatical representation using visualization tools and techniques [7]. There is a plethora of open source tools which are mostly used to visualize the analytical data and seek the user's attention. The process of providing a transparent layer over data and processing of information is called as visual analytics. Being a multi-disciplinary approach, visual analytics combines several research areas such as visualization, HCI, data analysis, and management. The concept proposes a combination of visual and automatic method with a robust bonding of HCI to obtain knowledge form data. The entire process is described in Fig. 1 for better understanding of visual analytics.

Daniel Keim et al. in their book 'Solving problem with Visual Analytics' explained the building blocks and applications of visual analytics in depth and focused on the infrastructure, perception, cognitive aspects, and challenges possessed by visual analytics [8]. All the previous work done related to visualization tools and techniques are discussed in details in the literature survey. The comparison of existing tools with their drawbacks and advantages, existing systems, and suggested hypothesis are explained in the latter half of the paper.

2   Literature Survey

Since the centralized servers proved to be a bottleneck to store and process huge volume of data, Google introduced an algorithm called MapReduce: The paradigm which divides the big task into small parts and assign them to many computer nodes. The end result is a combination of integration of small segments separated from a big task. As MapReduce was favorable for analytics and visualization, there has been a lot of improvement and development in fields of Big Data analytics and visualization in the past few years. Fernanda B. Viegas et al. in their IEEE transaction: 'Participatory visualization with Wordle' stated the use of 'Wordle', a web-based tool which was developed for visualizing formats, texts, and different layouts of the text. The authors have mentioned a few new versions of cloud tags and propose a concept of Analytical and emotional use [9]. J. Feinberg explained about 'Wordle' in detail through an online web source in 2008 [10]. The authors of the blog [11] define a unique visualization software 'Google Charts' as an easy and interactive tool for all browsers. The software can read data from Excel, MySQL databases, Google spreadsheets, CSV files, and transform an efficient version of visualization. The source has mentioned disadvantages stated as customizability is problem if JavaScript programmer then preferable, does not give sophisticated statistical processing, requires network connection. Journalism++ Cologne, 'Data wrapper', 2016 [12] explains all the characteristics and principles of the tool 'Data wrapper'. The link states the applications of the tools and its essentials such as zero coding, easily upload of data, and publishing charts and map. The website has mentioned custom layouts to integrate your visualizations perfectly, and access to local area maps is also available. The description given in the page [13] states that tableau is a visualization tool which is mostly used in corporate industries and is scalable with Big Data for producing charts, graphs, and maps. A desktop application version of Tableau is available for visual analytics of datasets residing in small workstations or PC's. Things have become so convenient that if a particular user does not want to install the software on machine, then the server solution itself lets you to visualize reports online and on smart handhold gadgets. Some advantages of Tableau consist of a good option for large-scale business and produce multiple PDF files from the report. Considering other technical aspects, Tableau is not a good tool for creating list reports and is also expensive as it needs a desktop and server license separately. After the huge success of Tableau in fields of visual analytics, came the e-charts of Baidu. The literature survey of the same states that Baidu's echarts are compatible with all types of browsers such as chrome, safari, Firefox, IE6, and above [14]. The variations supported for visualization by Baidu include line graphs, pie charts, scatter plots, maps, and others. The added advantages and features consist of drag-recalculate, which allows extraction and integration from multiple charts. Magic and legend switches are meant to switch from one visualization technique to another. The other embedded features into e-charts are area zoom, chart linkage, scale roaming, glow effect, big data mode, data streaming, auxiliary line, multi-series stacking, sub-region mode, mark point and mark line, high customizability, event interaction, and timeline. From

[15] it is clearly described that IBM will close its new tool 'Many eyes' collaborative data visualization service very soon. As compared to Baidu's echarts, many eyes do not have features such as flexibility, switching of charts, and variation for different user insight usage. No less than other its advantages include application prototyping, dashboards, infographics, and automated reports. The above described tools were and are used for visualizing sorted data into graphical representation techniques which are considered to be reports or analytical results. Previous work and research papers by research legends prove that there are various applications and approaches to data visualization. Adding more to the visual analytics, the authors in [16] and [17] mentioned the use of visual analytics principles for anomaly detection in GPS data and visual causality analyst system. The authors in their survey paper 'An approach toward data visualization based on AR principles' describes all the visualization techniques, applications, and highlights the points where Big Data meets Augmented Reality for Data Visualization [18]. Following the conceptual hypothesis of AR with Big Data, Ekaterina Olshannikova et al. elaborated about the challenges and research agenda of implementing the principles of Augmented Reality for Big Data Visualization [19]. More about Big data analytics was conveyed through a survey written by Tsai et al. [20]. Dilpreet Singh and Chandan K Reddy focused on the various platforms suggested for Big Data analytics and explained about the roles and platforms in detail [21].

3 Methodology

The below given methodology describes the overall procedure and system workflow. The design comprises of Cloudera Engine (CDH), a Big Data framework: Hadoop and its relative plug-ins dedicated to Apache. The data sources obtained from public and private repositories may vary in file format extensions such as .csv, .arff, and .txt. The tabular data can be queried using Apache Impala or Hive with the help of Apache Hue's manual input procedure (Fig. 2).

There are two methods of transferring data into Cloudera's base. The first and traditional methods comprise of manual input procedure from Hue and second involves of using Apache Hive's command line interface. Cloudera Engine (CDH) is an open source distribution by Cloudera, which was built to meet enterprise demands. The box engine framework was built with intention of integrating all Apache frameworks for ease of processing data. Apache Hive can be defined as a data warehouse software which is located on top of Hadoop distributed structure and provide querying and analysis of datasets residing in the same. It also provides basic MySQL functionality and compatibility for querying and analysis purposes. Whereas Apache Impala is a new SQL engine used by Cloudera to transfer data from relational database system for query processing and analysis. It is also the first open source SQL engine invented for Hadoop by Apache. Residing in one of the clusters of CDH, this SQL engine can be used precisely for obtaining faster query searching and results.

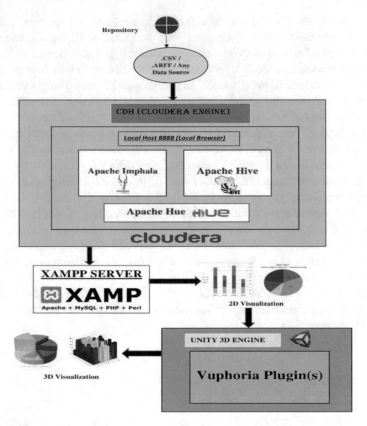

Fig. 2 Methodology of the conceptual hypothesis

4 Bag of Data

In modern applications, trending features require a huge number of input from users end. The different types of data provided to the application demand several analysis and visualization methods. In this section, we categorize the frequently used data such as market and E-commerce data, population and environmental data, respectively. Table 1 explains the data attributes and types of data in each category.

5 Visualization Based on Locations

Due to growing urbanization and geographical locations, there has been a tremendous amount of increase in environment and location sectors. As specified in [22], the locations in the subject of interest are categorized into co-ordinate, division, and linear

Table 1 Datasets and their data types

Date category	Data set	Data property	Type-N	Type-C	Type-T
E-commerce and retail data	Instacart online grocery	Order number	✓		
		Order name			✓
		Weight	✓		
		Feedback		✓	
	Bigmart	Item type		✓	
		Item outlet sale	✓		
		Item outlet location			✓
		Discount	✓		
	Grocery store US	Product name			✓
		Product discount	✓		
		Loyalty points	✓		
		Fat content		✓	
Population data	Educational attainment for females over 15 years	Country		✓	
		Aadhaar number	✓		
		Sex		✓	
		Region code			✓
	UN world bank dataset	No. of children	✓		
		Region		✓	
		Family name			✓
		Immigrants data		✓	
	China world bank data	City		✓	
		Gender		✓	
		Address			✓
		Account	✓		
Environmental data	Energy utilization	Time	✓		
		User ID	✓		
		Consumption		✓	
		Appliance			✓
	Khadakwasla water conservation	Data			✓
		Level	✓		
		Storage (BCM)	✓		
		Purity		✓	
	Weather forecast	Rainfall (mm)	✓		
		Temperatures (Degree/Celsius)	✓		
		Day			✓
		Sunlight		✓	

Note that N, C, and T stand for Numerical, Categorical, and Textual, respectively

Fig. 3 Dot distribution
technique

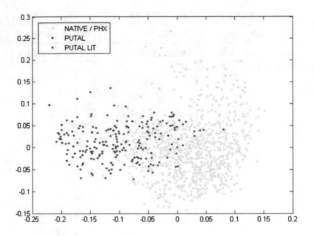

referencing. Following are the visualization techniques described for the above-mentioned referencing categories.

- Visualization based on Points

This visualization technique is the simplest type of object mapping technique in which one single point on a spatial plane (ex—map, area chart) represents a unique object. The objects represented by points can be relative and interlinked with each other. The wild card of point-based method is that it enables users to identify the major element in the bag full of elements on a plane. N. Ferreira et al. in their research script mentioned the use of point-based visualization [22] to identify the regular pattern or anomalies. The points in the graph represented the drops off and pick-up points for taxi trips in Manhattan. Graduated circle and dot distribution are the types of point-based visualization which are explained with the figure examples (Fig. 3).

- Visualization based on Regions

Region-based visualization is the technique which is used when there is massive information and filtered aggregated information is supposed to be shown. H. Liu et al. in their conference of visual analytics science and technology elaborated on the concepts of visual analytics for route diversity [23]. Whereas the region-based visualization techniques were used for composite density maps for multivariate trajectories [24]. Also, a review 'A Survey on Traffic data Visualization' described the implementation of region based method [25]. For showing trends and statistics for geographical areas, cities, and codes, Choropleth graphs is one of the best methods of region-based visualization advised by data scientists and experts. Region-based visualization is also favorable for visualizing temporal data patterns and was concluded by Bach et al. [26] (Fig. 4).

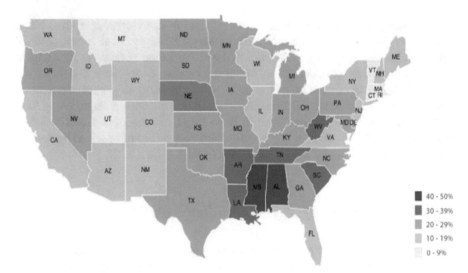

Fig. 4 Example of Choropleth graph

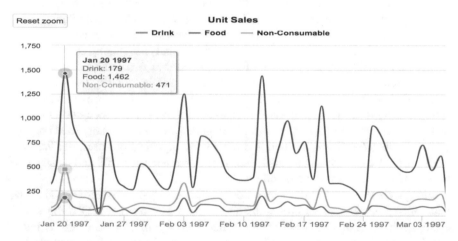

Fig. 5 Line-based visualization

- Visualization based on Lines

Line-based visualization is the common type of technique which is used to specify locations based on networks in context with linear referencing. This technique is mostly used to display the progression (Slopes and Dropdowns) in terms of percentage, number count, and other numeric terms. Trajectory conventions are made possible using the line visualization method. H. Zou et al. in their edition explained the concept of edge bundling in information visualization using the principles of line visualization [27]. The authors in the paper have presented a scalable multi-featured architecture for high volume maps to enable multi-purpose exploration methods using various customizable fields [28]. Apart from the listed techniques, time-based visualization is a less focused visualization concept. W. Aigner et al. highlighted the time visualization technique by describing the visualization of time-oriented data in their research article [29]. As we know the source code of software keeps on fluctuating in the cycle of runtime of the code. S. L. Voinea et al. proposed a version centric layout of line representations and represented the code changes using line-based visualization [28]. Below given is an example of line-based visualization in which you can see the rise and dropdowns of food consumed represented by lines of different color ranging in terms of increasing months (Fig. 5).

6 Comparative Study of Visualization Tools

The visualization tools have been compared and evaluated on the basis of their features, support, characteristics, and facilities provided. Several datasets with size varying densities were tested on all these tools, and Google Charts and Apache Zeppelin were considered on priorities as they outclassed the comparison test. Tableau: As discussed in the literature survey, Tableau is a visualization tool which is used in many corporate industries and areas of engineering. Its main purpose is used to create charts, pie diagrams, and maps for graphical analysis. Talking about expense, Tableau is available for $200 per year for a single user access. Fusion Charts XL are the new e-charts developed using javascripts for web and mobile by Infosoft Global private limited offer a great feature of compatibility with internet browsers (Mozilla Firefox, Google Chrome, and Safari). Also allows the soft mapped code to equip techniques such as charts, line graphs, pie diagrams, and others. Many social organizations such as Facebook, Twitter, and other huge organizations use Data Wrapper tools for report generation and analytics. It has an interactive GUI and is very easy to use and requires almost zero coding.

Google charts is an API made by Google which is very easy to learn and visualize data. Works on all modern browsers and can read data from MySQL, Big Data's HDFS, and excel spreadsheet. The only disadvantage of Google charts is that it requires a network connection to display the visualized content. This visualization technique would be an ideal technique to use in the proposed concept as the API is suitable to integrate with an integrated development environment (IDE) such as eclipse. The data generated in SQL can be imported to eclipse using plug-ins and .jar files connectivity methods. The data values can be embedded in the API code and displayed on the web interface in 2D form. Ying Zhu in the year 2012 highlighted the concepts of Google maps and tools used for visualization. The proposed theory proved to be pretty impressive for students pursuing studies of relevance [30] (Table 2).

Many small organizations, beginners, and students pursuing machine learning and Big Data interests prefer using Apache Zeppelin as a web-based notebook. It is a multi-purpose web notebook which supports several backend languages such as MySQL, Hive, Postgresql, Spark, Python, and others. Other operations which Zep-

Table 2 Comparative study of visualization tools

Features	Tableau	Fusion charts XL	Data wrapper	Google charts	Apache Zeppelin
Multiple chart accessibility	✓	✓	✓	✓	✓
Remote access to server	✓	✗	✗	✓	✓
One click export	✓	✓	✓	✗	✓
Zero coding	✓	✗	✗	✓	✗
Visualization types support	✓	✓	✓	✓	✓
Compatibility with MySQL	✗	✓	✗	✓	✓
Compatibility with IDE's	✓	✓	✗	✓	✓
Embed in API	✗	✓	✗	✓	✗
Web based notebook	✗	✗	✗	✗	✓
Loading time less than 5 s	✓	✗	✓	✓	✗
Interactive GUI	✓	✗	✗	✗	✓
Other plug-in support	✗	✗	✗	✓	✓
Backend technologies support	✗	✓	✗	✗	✓

pelin emphasizes on are data discovery, data ingestion, data visualization, analytics, and collaboration. It has an interactive GUI and provides ease of visualization and requires almost zero coding. Moreover, Zeppelin supports more than 40 interpreters such as Alluxio, HDFS, Hbase, PSQL, Spark, and others.

7   Conclusion

A large number of digital data are produced in industries and sectors of computer science such as military, telecommunication, aviation, health care and diagnostics, and other related areas. Apache Hadoop is a Big Data framework which was invented to process and feature a massive amount of data (in Gb's and Tb's). The analytics or results filtered from the target data are meant to display to the end users in forms of reports, charts, and scatter plots for a better understanding and exploration of the data. The above discussed visualization tools such as Wordle, Data Wrapper, Baidu e-charts, and Google Charts had one mutual purpose of delivering analysis and queried data results to the viewers or end uses. According to the results of comparative study, Google chart outclassed other tools in terms of reliability, ease in visualization and other aspects. In the proposed case study, it is suggested to use Google Charts API or Apache Zeppelin parallel to latest version of Hadoop, Apache Hive, Apache Sqoop, and other Apache Plug-ins. Due to its high compatibility and IDE integration feature, Google charts have become most favorable for domestic and commercial use.

8 Future Scope

The proposed concept can be extended to a specific level using principles and software tools of augmented reality (AR). The code of visualized content in 2D form can be given to the Unity AR Engine by using a PHP host file. The host file will act as an input data to the engine, which will process the data and convert it into a mobile application. The statistics generated will be converted into small QR codes and end users will be able to view it using any smart device. End users will be able to install the mobile application and point it toward the barcodes adhered on the walls of any room. The primary purpose is to deploy the application in conference and meeting rooms where high statistics are displayed to board members for knowledge imposition.

Acknowledgements This research was extensively supported by the university VIT-AP, Amaravati, Andhra Pradesh, India. We are thankful to Mr. Ajinkya Kunjir who provided additional knowledge and information that greatly assisted the research.

References

1. Liu, S., Cui, W., Wu, Y., Liu, M.: A survey on information visualization: recent advances and challenges. Visual Comput. **30**(12), 1373–1393 (2014)
2. Keim, D.A., Kohlhammer, J., Ellis, G., Mansmann, F.: Mastering the Information Age-Solving Problems with Visual Analytics. Florian Mansmann, Konstanz, Germany (2010)
3. Chen, S., Yuan, X., Wang, Z., Guo, C., Liang, J., Wang, Z., Zhang, X., Zhang, J.: Interactive visual discovering of movement patterns from sparsely sampled geo-tagged social media data. IEEE Trans. Vis. Comput. Graph. **22**(1), 270–279
4. Liu, M., Liu, S., Zhu, X., Liao, Q., Wei, F., Pan, S.: An uncertainty-aware approach for exploratory microblog retrieval. IEEE Trans. Vis. Comput. Graph. **22**(1), 250–259 (2016)
5. Zheng, Y., Capra, L., Wolfson, O., Yang, H.: Urban computing: concepts, methodologies, and applications. ACM Trans. Intell. Syst. Technol. **5**(38) (2014)
6. Zheng, Y., Wu, W., Chen, Y., Qu, H., Ni, L.M.: Visual analytics in urban computing: an overview. IEEE Trans. Big Data **2**(3), 276–296 (2016)
7. Quinan, P., Meyer, M.: Visually comparing weather features in forecasts. IEEE Trans. Vis. Comput. Graph. **22**(1), 389–398 (2016)

8. Keim, D.: Information visualization and visual data mining. IEEE Trans. Vis. Comput. Graph. **8**(1), 1–8 (2002)
9. Viegas, F.B., Wattenberg, M., Feinberg, J.: Participatory visualization with wordle. IEEE Trans. Vis. Comput. Graph. **15**(6), 1137–1144 (2009)
10. Feinberg, J.: Wordle. [Online]. Available: http://www.wordle.net/ (2008)
11. Google: Google Charts. [Online]. Available: https://developers.google.com/chart/ (2016)
12. Journalism++ Cologne: Data wrapper. Available: https://datawrapper.de/
13. Tableau Software: Tableau. [Online]. Available: http://www.tableau.com/ (2016)
14. Baidu: Baidu Echarts. [Online]. Available: http://echarts.baidu.com.cn/ (2016)
15. IBM: Many Eyes. [Online]. Available https://manyeyes.alphaworks.ibm.com/manyeyes/ (2015)
16. Liao, Z., Yu, Y., Chen, B.: Anomaly detection in GPS data based on visual analytics. In: Proceedings of IEEE Symposium on Visual Analytics Science and Technology, pp. 51–58 (2010)
17. Wang, J., Mueller, K.: The visual causality analyst: an inter-active interface for causal reasoning. IEEE Trans. Vis. Comput. Graph. **22**(1), 230–239 (2016)
18. Hirve, S., Kunjir, A., Shaikh, B., Shah, K.: An approach towards data visualization based on AR principles. In: ICBDACI, IEEE (2017)
19. Olshannikova, E., Ometov, A., Koucheryavy, Y., Olsson, T.: Visualizing big data with augmented and virtual reality: challenges and research agenda. J. Big Data **2**(1), 22 (2015)
20. Tsai, C.W., Lai, C.-F., Chao, H.-C., Vasilokos, A.V.: Big data analytics: a survey. J. Big Data **2**(1), 21 (2015)
21. Singh, D., Reddy, C.K.: A survey on platforms for big data analytics. J. Big Data **2**(1), 8 (2015)
22. Ferreira, N., Poco, J., Vo, H.T., Freire, J., Silva, C.T.: Visual exploration of big spatio-temporal urban data: a study of New York city taxi trips. IEEE Trans. Vis. Comput. Graph. **19**(12), 2149–2158 (2013)
23. Liu, H., Gao, Y., Lu, L., Liu, S., Qu, H., Ni, L.M.: Visual analysis of route diversity. In: Proceedings of the IEEE Conference on Visual Analytics Science and Technology, pp. 171–180 (2011)
24. Scheepens, R., Willems, N., Van de Wetering, H., Andrienko, G., Andrienko, N., Van Wijk, J.J.: Composite density maps for multivariate trajectories. IEEE Trans. Vis. Comput. Graph. **17**(12), 2518–2527 (2011)
25. Chen, W., Guo, F., Wang, F.-Y.: A survey of traffic data visualization. IEEE Trans. Intell. Transport. Syst. **16**(6), 2970–2984 (2015)
26. Bach, B., Shi, C., Heulot, N., Madhyastha, T., Grabowski, T., Dragicevic, P.: Time curves: folding time to visualize patterns of temporal evolution in data. IEEE Trans. Vis. Comput. Graph. **22**(1), 559–568 (2016)
27. Zhou, H., Xu, P., Yuan, X., Qu, H.: Edge bundling in information visualization. Tsinghua Sci. Technol. **18**(2), 145–156 (2013)
28. Voinea, S.L., Telea, A., van Wijk, J.J.: A line based visualization of code evolution. In: 11th Annual Conference of the Advanced School for Computing and Imaging (ASCI 2005)
29. Aigner, W., Miksch, S., Schumann, H., Tominski, C.: Visualization of Time-Oriented Data. Springer Science & Business Media, Berlin, Germany (2011)
30. Zhu, Y.: Introducing Google chart tools and Google maps API in data visualization courses. IEEE Comput. Graph. Appl. **32**(6), 6–9 (2012)

User Behaviour-Based Mobile Authentication System

Adnan Bin Amanat Ali, Vasaki Ponnusamy and Anbuselvan Sangodiah

Abstract Android is one of the most popular operating systems being used in smartphones and is facing security issues. Many authentication techniques are being used, and most of them are based on username, password or PIN. These techniques are considered weak because of several drawbacks such as passwords and PIN can be guessed, forgotten or stolen. When the mobile phone is stolen, misplaced or in the possession of the third party, an unauthorized user can get access to the applications and features of the mobile phone. Furthermore, after a one-time login, no further credentials are required. Therefore, a better security authentication system is needed to overcome this problem. This study proposes an identity management framework that can secure mobile application's data from an unauthorized user. The proposed framework consists of various features that are extracted from phone swiping behaviour. The framework is comprised of enrolment and authentication phases. In the enrolment phase, the system learns the user behaviour, and in the authentication phase, it is able to accept or reject the current user based on his behaviour.

Keywords Continuous authentication · Behavioural · Touch screen · Sensors · Security

---

The original version of the chapter was revised: Author's belated correction has been incorporated. The correction to this chapter is available at https://doi.org/10.1007/978-981-13-6861-5_64

A. B. A. Ali (✉) · V. Ponnusamy · A. Sangodiah
Faculty of Information and Communication Technology,
Universiti Tunku Abdul Rahman, 31900 Kampar, Malaysia
e-mail: adnanbinamanat@1utar.my

V. Ponnusamy
e-mail: vasaki@utar.edu.my

A. Sangodiah
e-mail: anbuselvan@utar.edu.my

© Springer Nature Singapore Pte Ltd. 2019
S. K. Bhatia et al. (eds.), *Advances in Computer Communication
and Computational Sciences*, Advances in Intelligent Systems and Computing 924,
https://doi.org/10.1007/978-981-13-6861-5_40

461

1 Introduction

Most authentication methods are based on the single-entry point of an application such as PIN, password or some physical biometrics, and once the user successfully entered the application, there is no further check on the user authentication. Furthermore, the above authentication methods have several drawbacks—PIN or passwords can be stolen, and frequently changing the PIN or passwords encourages the user to keep the simple PIN or passwords because complicated passwords are difficult to remember. Physical biometrics is impractical in some cases like fingerprint scanning and iris scanning are device-specific approaches. Most of the applications require only single-time authentication, for example Gmail.

Therefore, there is a need for continuous authentication system based on behavioural biometrics that keeps on checking the user authenticity continuously. Behavioural biometrics is the study of human activities. It consists of keystroke dynamics, touch dynamics, gait, signature, etc. Behaviour biometrics authentication is a process of measuring human interaction behaviour and extracting behavioural features, then evaluating the authenticated or unauthenticated user.

From the mobile touch screen, we can find data that is unique in every individual. The data is related to touch actions of a user, and it consists of the touch area, touch size, touch position, touch duration. In this paper, we develop an application to extract all the data related to swipe movement and further use it in identifying a user for authentication.

This research aims at designing a new framework for identity management using human behaviour patterns' classification. Although psychological and physiological identities such as biometric techniques have been in use, they are still subjected to environmental conditions and human physiological factors. This proposed research differs from biometric techniques by identifying distinctive behaviour patterns on the mobile phone to uniquely identify the individuals. The framework consists of various user behaviours related to swipe that will be used for classifying unique individuals. By proposing this identity management framework, mobile device users can secure their applications from unauthorized users when the device is stolen or misplaced.

2 Related Work

Keystroke dynamics have been extensively studied in distinguishing users by the way they type their personal identification number (PIN)-based passwords [1]. Another study was done on PIN-based password while holding the phone normally, on the table and walking. Different features i.e. key hold time, interval time by two nearby keys, 4 features of movement along x-, y- and z-coordinates using the accelerometer, 4 features using the gyroscope, tab size, and touch coordinates were collected. Best EER achieved by using the mobile phone in the hand, on the table and walk postures were 7.35, 10.81, 6.93%, respectively. So best EER achieved is for the walk posture

[2]. Research done on the analysis of keystroke dynamics for identifying users as they type on a mobile phone can be found in [3–8], whereas [7] has performed behaviour classification on mobile phones by identifying behaviour patterns when the user types the password. This work is limited to authorize the user by analysing the way the password is entered. Keystroke analysis is effective for a physical keyboard, but nowadays most of the smartphones contain virtual keyboard that has interaction only with the touch screen interface.

Four features from the tapping data on the touch screen are used to verify the authenticating user whether the person who enters the password is the authorized user or someone who happened to know the password. The tapping behaviour can be classified into the pressure on the touch screen, the size, the time and the acceleration using machine learning. This work focuses only on the patterns of the keypad by for a fixed length passcode entered by various users. It does not look into keypad use on other areas of the phone, for example dialling a number, tapping on the application or sliding the phone screen. This system does not work effectively when the user changes the passcode frequently. Whenever the user decides to change passcode, the system needs to be retrained to build a new set of tapping features. Moreover, the system can be bypassed by having an imposter mimicking the tapping behaviour of the legitimate user. However, not all the features can be exactly observed and reproduced by an imposter [7].

Using password-based authentication, a different research was done while moving by car or any other transportation. Features used in this research were hold time, press-press time, release-release time, release-press time, flick distance, flick angle, flick speed, pressure, size, event counter, position, acceleration, angular velocity and orientation. By using a different combination of these features, good EER was achieved [9]. But this work is also limited to password only.

Behaviour analysis based on graphical passwords was also conducted by [8, 10, 11] as they argue that graphical-based password is much easier to remember than text-based passwords. The graphical-based password is suitable when pressure data is used for classification and various usability tests reveal that graphical-based passwords produce better results. However graphical-based passwords require more computing power for classification and also require more storage compared to the text-based passwords. Some research works in [3, 4, 12] argue that user behaviour based on motion sensor on the phone is sufficient to identify the individual user. Later on, more dimensions of data from accelerometer, gyroscope and orientation sensors were used for better classification. [13] Conducted a similar research as [7] that enhances password patterns on smartphones. The application was developed on all data available on the touch screen such as pressure, size, X- and Y-coordinates and time.

A multi-touch screen sensor data classification based on iPad was conducted by [14]. The classification was based on palm movement which may not be very suitable for mobile phones as mobile phone screen is too small for palm movements. A slightly different approach is used in [15] by performing progressive authentication on the need basis. Instead of authenticating by using a password or token-based identification, the system decides when to authenticate and for which applications to authenticate. Non-continuous use of mobile phone switches the phone to silent mode which requires constant authentication. Authentication is not required when the user is in physical contact with the phone, whereas authentication is required when the phone loses physical contact with the user. If the phone can infer the authenticated uses through surrounding sensors of the phone, the users are allowed to use the applications without further validation. When the phone has a low confidence value of the physical contact of the user and with the phone, the system would require further validation to use high content applications. The goal of the research is to avoid authentication for low-value contents but requires authentication for high-value contents such as financial data. This is done by associating a confidence value for each application from low confidence to high confidence depending on the applications. The system uses multiple authentication signals and multiple authentication levels. Although the system sounds sophisticated, it requires heavy processing as it works on multiple sensor data and has to continuously monitor the physical contact and presence of the user using various sensors.

A very much closer work similar to what has been proposed is conducted in [16]. The author uses keystroke authentication which is a behavioural biometric authentication mechanism. Similar keystroke pattern has been used in [17], looking into keystroke duration as an important parameter for authentication decision. A threshold value for the keystroke duration hinders session hijacking, and this method shows 97.014% correct user authentication. The work of [18] differs from others by including user profile data as an important criterion for user authentication. User profile, such as name, ID, hobby and service details such as service name, provider name, is used for the authentication decision.

Fusion of four models 'type', 'tap', 'swipe' and 'pinch' have been used in [19]. For type, five letters were typed and from the five letters, 38 features that contain five hold times features, four down–down times features, five pressures features, five finger areas, five accelerations, five magnetometers, and five features of avg. values of hold, down–down, up–down, pressure and finger area were extracted. For tap, the author used five features: hold time, pressure, touch area, magnetometer and acceleration. Swipe is further subdivided into four features, swipe up, swipe down, swipe right and swipe left. For each swipe, nine features—time, pressure start, pressure end, finger area starts, finger area end, distance, velocity, acceleration and magnetometer—are used. For pinch, 6 features—time, length, avg finger area, avg pressure, avg delta, and avg scale—are used. FAR 0.26 and FRR 0.26 have been achieved, also with minimum resources usage, i.e. avg CPU usage increases from 13.55 to 13.81%, avg memory usage from 70.5 to 78.9 MB, app + data from 11.96 to 13.07 MB and cache used most from 4.58 to 24.59 MB. This research was limited to a number of letters, and there is a need to improve memory usage.

Behavioural authentication is done in [17] based on username, password and keystroke duration. Even if an unauthorized user knows username and password of the authorized user, he will not be able to access the application because of his different keystroke duration. This method works only to access the application and does not work if an unauthorized user gains access to the application.

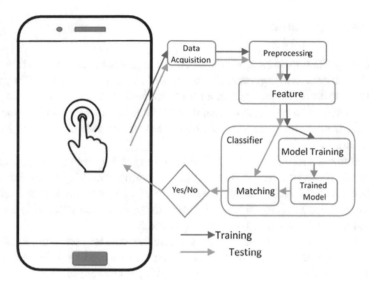

Fig. 1 Methodology

The proposed framework checks the user authenticity continuously even after gaining access to the application. Once the user opens the application, his/her touch behaviour is captured and sent to the classifier for continuous authentication. In this way, the application will be disabled if it is being accessed by an unauthorized user.

3 Proposed Methodology

In the proposed methodology, once the user starts using the mobile application, his/her user behaviour data related to swipe will be observed and recorded. This data will go through a preprocessing stage where outliers from the data will be removed for better classification results. After data preprocessing, feature extraction will be done from the raw data. In training mode, the set of features will be sent to the classifier where a model will be created from the trained data. This step will be done once at the time of first use of the application. And for the subsequent use of the application, this created model will be used for testing the new data received on the real-time usage. In the testing phase, the newly received data will be matched with the trained model and according to the matching result the user is set to be authorized or an imposter. Figure 1 is the proposed framework of the methodology.

4   Data Acquisition

After running the application, user's touch dynamics data related to swipe will be acquired. This data will be generated after multiple inputs over a specific time period. The collected raw data may contain a set of repeated input samples. The basic purpose of this step is to collect raw data, directly acquired from the mobile device, to preprocess. Data acquisition step will be carried out in both enrolment phase and authentication phase.

A customized android application is required to get the user behaviour data using a smartphone. For this purpose, an android application is developed and installed in the mobile devices. Figure 2 shows the interface of this application. This interface is just for data collection purposes, and later on, this application will run as a separate service on the background of the main application.

Using this application, a user can do swipe right, swipe left, scroll up, scroll down; upon each swipe, the number of swipes will be incremented by 1. For example, in Fig. 2, there are six swipes right, three swipes left, seven scrolls up and four scrolls down. Data samples were collected from ten users using android device Samsung Galaxy J7 Pro. Participants were chosen randomly, and they were all postgraduate students from different departments. All the participants were asked to do around 400 swipe movements in all directions in one interval. In order to ensure the purity of data collection, three different time intervals were selected in a day, i.e. morning, afternoon and evening, and in each interval, the user was not bound to keep the mobile in his hand, i.e. the user can place and pick mobile multiple times. The purpose of taking the data in different times is to capture data in different handgrips, in different moods (relaxed, tired, active) and in different positions (free posture sit, stand, lying).

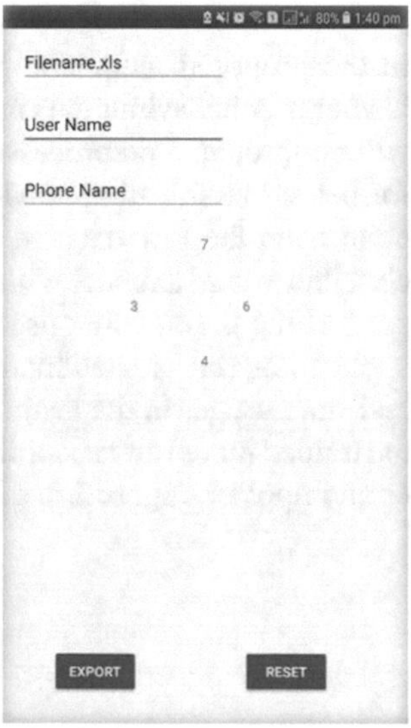

Fig. 2 Interface of the application

For each interval, data is collected in a separate file using different user names and file names, and saved in the mobile device. The total moves are 24,000 (400 × 3 × 2 × 10).

4.1 Preprocessing and Feature Extraction

Swipe feature is most commonly used gesture feature in smartphones. Usually, with swipe function, the user scrolls the screen using up–down or left–right. With swipe function, many behavioural biometrics features can be obtained, e.g. swipe length, swipe absolute length, swipe starting position, swipe ending position, the time taken for one swipe.

During experiments, the values of some instances were equal to zero and very small swipe length that is probably was a touch event; all these types of values were discarded because it can affect the data quality, accuracy and performance of the authentication model. The preprocessing will be carried out in both enrolment phase and authentication phase. Some feature values are taken as it is from raw data, and some are calculated using some mathematical formulas. Following is the detail of these features.

Touch Size

Touch size represents the area that is being touched by the user. The touch size of an adult male usually has a larger touch size than a child or an adult female. According to android documentation, the value of size is measured in pixels and it ranges from 0 to 1. To get the size value, the function getsize() is used. This value differs in different users according to the fingertip size of the user. It is difficult for an imposter to mimic as an authorized user. The touch area is the scaled value of the approximate size of the screen being pressed when touched with the finger. The actual value in pixels corresponding to the finger touch is normalized with a device-specific range of values and scaled to a value between 0 and 1. Usually, the touch size value is used as a feature without further manipulation. Three area features—average area, area at the start of the swipe and area at the end of the swipe—are used. The area values at the start and end of the swipe are different. Usually, during swipe movement, the user touches on more area at the start and release with low touch area that is why start area is higher than end area.

Touch Pressure

Touch pressure is the value that measures the force exerted on the screen. It is directly related to the touch area; the more the area is covered, more the pressure is exerted on the touch screen. The minimum value for the touch pressure is 0, and the normal value of press is 1. The values higher than 1 depend on the calibration of the input device. To get the pressure value, the function MotionEvent.getpressure() is used. Touch pressure value depends on the force of finger muscle of the user that is distinctive in every user. So, it is also very hard to mimic for an imposter. Touch pressure value is also used as it is without further manipulation as a feature. Three pressure features—average pressure, the pressure at the start of the swipe and pressure at the end of the swipe—are used. The pressure values at the start and end of the swipe are different. Usually, during swipe movement, the user presses hard at the start and release with low pressure, that is why start pressure is higher than end pressure.

Touch Position
Touch position is a location on the screen where the user touches the screen, and it is measured in two coordinates, i.e. x and y. It is the screen position according to the user's palm and finger size and dimensions of the phone. So every person has different screen touch areas according to their hand geometry. Due to this discriminative feature, users can be differentiated for identification. Four features—start x, start y, end x and end y—were extracted for the classification.

The Length and Absolute Length
Due to the different hand geometry of each user, the trajectory of swipe and its absolute length will be different. So two features, i.e. length and absolute length, are calculated. The length is calculated using the android built-in function. Absolute length is the straight line from the swipe starting position to end position. The absolute length is calculated from the start and end points of swipe. Equation 1 is used to find the absolute length.

$$\text{Length} = \sqrt{(x_2 - x_1)^2 + (y_2 - y_1)^2} \tag{1}$$

Direction
Swipe direction is the angle that is calculated from the horizontal line of the starting position of the swipe. Due to different user behaviour, the value of this direction is different. This is used using Eq. 2.

$$\text{Angle} = \tan^{-1}\left(\frac{y_2 - y_1}{x_2 - x_1}\right) \tag{2}$$

The Rectangular Area Covered Under Swipe
The area covered in swipe movement is a rectangular area shown in Fig. 3. This is also different for different users. This can be calculated using Eq. 3.

$$\text{SwipeArea} = (x_2 - x_1) \times (y_2 - y_1) \tag{3}$$

Swipe Duration
Swipe duration is the time taken to complete a swipe action. Swipe duration is calculated by taking the difference of swipe start and end time. To get the time value, the function gettime() is used. Time is calculated in milliseconds.

Fig. 3 Swipe features

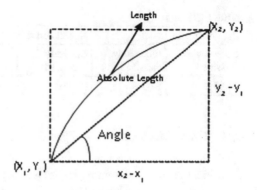

Average Speed

Average swipe speed is calculated by adding the speed instances and then dividing the number of instances. To get the speed value of an instance, the function getspeed() is used.

4.2 Template Generation and Storage

After getting the values of these features, a profile is built for each user. Extracted features are concatenated into a sequence of n-dimensional feature vectors where n is the number of feature element selected and is stored in a database for model training. A template is a compact form of transformation representation of user's touch dynamic characteristics. This representation will be helpful in the authentication process to identify a user. In the enrolment phase, these templates are stored in the database and in the authentication phase, these templates are sent to classifiers.

4.3 Training and Testing

Training is usually done once before using the application. In the training phase, multiple inputs are fed into the system. Training data consists of a combination of legitimate user data and imposter data. The system is trained with the values of both legitimate user and imposter and then tested by providing some test data of both legitimate and imposter data.

Table 1 Results

Random forest		Neural network		K-nearest neighbour	
Correctly classified	Incorrectly classified	Correctly classified	Incorrectly classified	Correctly classified	Incorrectly classified
94.97%	5.03%	93.13%	6.87%	91.98%	8.02%

4.4 Classification

Classification is a data mining technique that assigns items in a collection to target categories or classes. The goal of classification is to accurately predict the target class for each case in the data. After the enrolment, the system will store and compare the template. After the template storage, data classification is carried out to compare the feature data with the data stored in the database. For classification purpose, different classifiers are used like support vector machine (SVM), random forest (RF), neural network (NN), decision tree (DT), k-nearest neighbour (kNN) and distance measure (DM) by different authors [19–23]. Classifier selection depends on the type and size of the data set. All of the above classifiers are tested, and RF found the best fit for the data set. The classifiers with default parameters are used. Following are the results of three closely fitted classifiers as shown in Table 1.

4.5 Decision Making

The result of the classification is termed as decision score [7]. Decision-making is a process of finalizing the user's originality. According to the decision score, provided by the classifier, the decision is taken with respect to the threshold value.

5 Limitations

The work is done only for the swipe movements because for tap function; only a few features are available, i.e. touch pressure, size and position, that are not sufficient in distinguishing the users. The values of touch pressure and size are between 0 and 1. It will be very hard for the classifier to distinguish between different users only on the basis of these values. For tap, the position of a button is the same for all the users, so position values will be the same for all the users. Hence, the chances of mimicking an authorized user will be high for tap function.

6 Conclusion

This research proposed a continuous user authentication system for smartphone applications based on user's interacting behaviour. Using this authentication system, we are able to distinguish an authenticated user from an unauthenticated user based on the swipe movements. We showed the performance of each individual classifier, and among them, the random forest classifier gave the best results. Using this classifier, the system achieved an accuracy of 94.97%. The results show that every user can be authenticated based on his unique interactive behaviour.

Acknowledgements This research is funded by Universiti Tunku Abdul Rahman (UTAR) under the UTAR Research Fund (UTARRF): 6200/V02.
Informed consent was obtained from all individual participants included in the study.

References

1. Killourhy, K.S., Maxion, R.A.: Comparing anomaly-detection algorithms for keystroke dynamics. In: Proceedings of International Conference on Dependable Systems and Networks, pp. 125–134 (2009)
2. Roh, J.H., Lee, S.H., Kim, S.: Keystroke dynamics for authentication in smartphone. In: 2016 International Conference on Information and Communication Technology Convergence, ICTC 2016, pp. 1155–1159 (2016)
3. Clarke, N., Karatzouni, S., Furnell, S.: Flexible and transparent user authentication for mobile devices. IFIP Adv. Inf. Commun. Technol. **297**, 1–12 (2009)
4. Clarke, N.L., Furnell, S.M.: Authenticating mobile phone users using keystroke analysis. Int. J. Inf. Secur. **6**(1), 1–14 (2007)
5. Nauman, M., Ali, T.: TOKEN: trustable keystroke-based authentication for web-based applications on smartphones. Commun. Comput. Inf. Sci. **76**, 286–297 (2010)
6. Zahid, S., Shahzad, M., Khayam, S.A.: Keystroke-based user identification on smart phones. In: International Workshop on Recent Advances in Intrusion Detection, pp. 224–243 (2009)
7. Zheng, N., Bai, K., Huang, H., Wang, H.: You are how you touch: user verification on smartphones via tapping behaviors. In: Proceedings—International Conference on Network Protocols, ICNP 2014, pp. 221–232 (2014)
8. Biddle, R., Chiasson, S., Van Oorschot, P.C.: Graphical passwords: learning from the first twelve years. Security **V**, 1–43 (2009)
9. Takahashi, H., Ogura, K., Bista, B.B., Takata, T.: A user authentication scheme using keystrokes for smartphones while moving. In: International Symposium on Information Theory and Its Applications, no. C, pp. 310–314 (2016)
10. Chang, T.Y., Tsai, C.J., Lin, J.H.: A graphical-based password keystroke dynamic authentication system for touch screen handheld mobile devices. J. Syst. Softw. **85**(5), 1157–1165 (2012)
11. Jermyn, I., Mayer, A., Monrose, F., Reiter, M.K., Rubin, A.D.: The design and analysis of graphical passwords. In: Proceedings of 8th USENIX Security Symposium, vol. 8, p. 1 (1999)
12. Bergadano, F., Gunetti, D., Picardi, C.: User authentication through keystroke dynamics. ACM Trans. Inf. Syst. Secur. **5**(4), 367–397 (2002)
13. De Luca, A., Hang, A., Brudy, F., Lindner, C., Hussmann, H.: Touch me once and i know it's you! In: Proceedings of 2012 ACM Annual Conference on Human Factors in Computing Systems, CHI '12, p. 987 (2012)

14. Sae-Bae, N., Ahmed, K., Isbister, K., Memon, N.: Biometric-rich gestures: a novel approach to authentication on multi-touch devices. In: SIGCHI Conference on Human Factors in Computing Systems, p. 977 (2012)
15. Riva, O., Qin, C., Strauss, K.: Progressive authentication: deciding when to authenticate on mobile phones. In: Proceedings of the 21 st USENIX Conference on Security Symposium, pp. 1–16 (2011)
16. Teh, P.S., Teoh, A.B.J., Yue, S.: A survey of keystroke dynamics biometrics. Sci. World J. **2013** (2013)
17. Babaeizadeh, M., Bakhtiari, M., Maarof, M.A.: Keystroke dynamic authentication in mobile cloud computing. Int. J. Comput. Appl. **90**(1), 975–8887 (2014)
18. Jeong, H., Choi, E.: User authentication using profiling in mobile cloud computing. AASRI Procedia **2**, 262–267 (2012)
19. Putri, A.N., Asnar, Y.D.W., Akbar, S.: A continuous fusion authentication for Android based on keystroke dynamics and touch gesture. In: Proceedings of 2016 International Conference on Data and Software Engineering, ICoDSE 2016 (2017)
20. Abdulhakim Alariki, A., Abdul Manaf, A., Mojtaba Mousavi, S.: Features extraction scheme for behavioral biometric authentication in touchscreen mobile devices. Int. J. Appl. Eng. Res. **11**(18), 973–4562 (2016)
21. Teh, P.S., Zhang, N., Teoh, A.B.J., Chen, K.: A survey on touch dynamics authentication in mobile devices. Comput. Secur. **59**, 210–235 (2016)
22. Saini, B.S., Kaur, N., Bhatia, K.S.: Keystroke dynamics for mobile phones: a survey. Indian J. Sci. Technol. **9**(6), 1–8 (2016)
23. Bhardwaj, I., Londhe, N.D., Kopparapu, S.K.: Study of imposter attacks on novel fingerprint dynamics based verification system. IEEE Access **5**, 595–606

Cyber Social Media Analytics and Issues: A Pragmatic Approach for Twitter Sentiment Analysis

Sanur Sharma and Anurag Jain

Abstract Social media analytics is one of the most important forms of data analytics in current scenario of cyber world. To leverage the use of social media analytics for branding and image building and its negative impact on misuse of such data for scamming and defamation activities, it is necessary to understand the content posted on social media and adopt comprehensive analytical techniques to predict and evaluate relevant information. This paper discusses social media analytics at data level, its challenges and issues that arise in analysis of social media data at large. It also presents a study on analysis of twitter data, where tweets are collected in real time and machine learning and lexicon-based techniques are applied to understand the sentiments and emotions behind the text that is posted on social media. The results efficiently classify the social media users opinions into positive, negative and neutral to better understand their perspectives.

Keywords Social media analytics · Sentiment analysis · Twitter analysis · Machine learning techniques

1 Introduction

Social media presents a platform to users to connect, interact and communicate with others to share their views openly. It has become a business tool that is used for gaining insights regarding businesses, campaigns and polls. Social media types include various social networking sites Facebook, twitter, Instagram, etc. Social media analytics is concerned with development of framework and tools to collect, visualise and analyse the humungous amount of social media data in order to mine useful information and patterns to predict user behaviour. Social media covers various fields

S. Sharma (✉) · A. Jain
Guru Gobind Singh Indraprastha University, Delhi, India
e-mail: sanursharma@gmail.com

A. Jain
e-mail: anurag@ipu.ac.in

© Springer Nature Singapore Pte Ltd. 2019
S. K. Bhatia et al. (eds.), *Advances in Computer Communication and Computational Sciences*, Advances in Intelligent Systems and Computing 924,
https://doi.org/10.1007/978-981-13-6861-5_41

or subjects like social media marketing, social media reviews, election campaigning, macroeconomics, disaster relief and stock market prediction [1, 3]. Social media analytics includes opinion mining to present sentiments of social media users, evolution of opinion over time, identify opinion leaders and their influence on other social media users. Sentiment analysis has been very efficiently used in various fields like CRM, recommendation system, spam detection, business intelligence, detection of campaign promoters, opinion leaders and influencers and government intelligence as well [9]. The research challenges in social media includes the text cleaning and preprocessing issues (incorrect data, missing and noisy data) and security-related issues (public availability of data which makes social media users vulnerable to attacks like data dwindling, creation of fake profiles, negative campaigning, promoting distrust, etc.). According to Cisco, the most common malware attacks in 2015 were related to Facebook scams and the social media-related cyber-crimes had quadrupled over the past five years. One in eight enterprises has suffered social media-related security attacks and cyber-crimes [5].

This paper presents the social media analytics, issues and challenges that arise in social media, various techniques that are used for analytics like sentiment analysis. Sentiment analysis has been used very extensively on social media to find out the peoples emotion about a particular subject, product or event etc. This work discusses sentiment analysis on twitter data by extracting tweets in real time and applying machine learning techniques for prediction and identification of sentiments and the polarity of the text. This paper is organised as follows: Sect. 2 discusses the cyber social media analytics, challenges and issues. Section 3 discusses the various techniques used for social media analytics and sentiment analysis. Section 4 presents an experimental study on twitter sentiment analysis where tweets are collected in real time and machine learning techniques are applied to evaluate the sentiment polarity of tweets. Finally, Sect. 5 presents conclusions and future work of this study.

2   Cyber Social Media Analytics: Challenges and Issues

Cyber social media deals with the use of social media with the help of internetworking and its analysis to characterise and predict various user behaviours, patterns, events, etc. The abundant amount of social media data present these days raises wider security issues and challenges that present some important factors to be considered in detail.

2.1   Social Media Analytics

The research in social media analytics majorly revolve around content analysis, network or group analysis and prediction of events. The prediction results are obtained by evaluating the content and network on various metrics or predictors.

Social Network Analysis: Social network analysis basically revolves around formation of groups oftenly known as community formation. It comprises of user interactions within a group and understands and analyses the user behaviour and detects influential users present in the network. The various predictors for social network analysis are centrality metrics (like degree betweenness and closeness), density measure, network centralisation and temporal metrics [18]. These measures help in understanding the characteristics of complex networks and in identification of influential users and leaders that propagate or disseminate information in the network.

Message or Content Analysis: Message or content analysis deals at the data level where the content posted on social media is studied and analysed to identify trending topics and discussions, users opinions and sentiments about the network they are in (like comments on particular topic of discussion on social media, product reviews of some website etc.). These kinds of analysis helps in promoting social media events, campaigns and marketing strategies by predicting important features and results like users inclination towards certain product, person or event and how it can affect it in a positive or negative way. The predictors used for content analysis are sentiment analysis metrics like sentiment difference, sentiment index and sentiment polarity. It also covers time series analysis which dynamically tracks the speed and process of message generation by computing post generation rate [20].

2.2 Challenges in Social Media Analytics

Social media is a time-sensitive network where detection of communities, social media monitoring, reputation management, branding and innovation are major tasks as well as challenges in it. Analysis of social media has its own issues and challenges due to its uncertainty and time sensitivity. Following are the various challenges that arise in social media analytics:

- Social media contains humongous amount of unstructured data, which is difficult to collect and process. Due to its diverse nature, it is difficult to crawl from multiple sources. This makes the processing and analysis quiet challenging.
- Due to its unstructured form and free flow of text, social media data are noisy, inconsistent, and contain missing data and misrepresentation of facts. This further makes the identification of patterns, trends and events difficult.
- Social media data are temporal or dynamic in nature, which makes its visualisation and real-time presentation difficult.
- The language styles used on social media are diverse and lack proper contextual representation. The relevance of social media topics have been rapidly diverging. All this leads to the problems of data aggregation and linking of factual information.
- Social media analytics at data level is concerned with the identification of opinions, sentiments and emotion which is a very complex task. It uses computational, machine learning and NLP techniques to efficiently automate and classify the

data. This complete process is complex and accuracy in prediction of events is always under scanner.

Some other issues in content analysis include rumour identification, identification of implicit sentiments like sarcasm and irony, opinion leader identification and entity identification.

2.3 Cyber Security Issues in Social Media

Social media has become a reconnaissance tool for hackers and cyber criminals to social engineer their victims. Cyber criminals have used facebook updates posted by third parties containing malwares which offer free merchandise to everyone participating and forwarding it to others as well. Twitter has been subject to scams like free vouchers and LinkedIn has been attacked by the ZBot malware [2]. The trust issues and privacy behaviour of social media users are under scanner, as public display of data within social media applications lead to misuse and unwary behaviour. Some important security issues are:

- Selling of users' data to third parties for profit making or for performing analysis which in turn leads to revelation of sensitive information of users.
- Increased level of anonymity and masquerading in social media is another threat to benign and naive users.
- Revelation of too much information on social media makes the user more vulnerable to attacks and specifically children.
- Creation of fake profiles by imposters to trap users and tarnish their image.
- There are issues like sock puppetting, data provenance and trust that have more recently increased.
- A high occurrence of deception and spam is another security concern.

3 Techniques Used for Social Media Analytics: Sentiment Analysis

The various techniques used for social media analytics and sentiment analysis are discussed below:

3.1 Lexicon Based

Lexicon-based techniques are based on the orientation of the words and phrases based on the context and polarity. It identifies the semantic orientation of the text based on some dictionary or corpus. The basic approach of lexicon-based techniques is to first preprocess the textual data, initialise the overall polarity and then check is the word token is present in the dictionary or not and assign the score according to the threshold value [17].

3.2 Machine Learning

Machine learning techniques rely on supervised and unsupervised learning for classification of data. The classification is done by learning from past experience or from analytical observations. For sentiment analysis, the basic approach that is followed is the creation of feature vector by performing POS tagging, identifying adjectives and evaluating measures like tf, idf in order to evaluate the polarity of the text. Following are some of the machine learning techniques that are used for sentiment analysis:

Logistic Regression: Logistic Regression is a discriminative classifier that makes use of logistic function that classifies an observation in to one of the predefined classes. The generalised linear model is represented as follows:

$$F(x) = e^x/1 + e^x \tag{1}$$

$$A_i/B_i = b_i \tilde{\ } binary(p(b_i)) \tag{2}$$

$$p(b_i) = P(A_i = 1/B_i = b_i) = \frac{e^{b^t_i \beta}}{1 + e^{b^t_i \beta}} \tag{3}$$

For evaluating the likelyhood between positive and negative labels for sentiment classification, the case of binary-dependent variable for regression analysis is considered to give better results than Naive Bayes classifier [8]. In case of binomial response, generalised linear model is represented as follows:

$$A_i \tilde{\ } Binomial(x_i, P(b_i)) \tag{4}$$

$$P(A_i = a_i/B_i = b_i) = \binom{x_i}{a_i} p(b_i)^{a_i} (1 - p(b_i))^{x_i - a_i} \tag{5}$$

$$\left\{ \frac{A_i}{x_i} \right\} \tag{6}$$

is a generalised linear model

Naive Bayes: Naive Bayes classifier is based on the Bayes theorem that computes the posterior probabilities of a class. For sentiment analysis, it maps the feature set to particular sentiment class labels. It evaluates the probability of the text to belong to each label based on the training and testing data [11, 16]. The classification is performed as follows:

$$P(l/f) = \frac{P(l) * \prod^{m}_{i=1} P(f_i/l)}{P(f)} \qquad (7)$$

SVM: SVM is a classification technique that uses linear separators to identify classes. SVM is used for identifying sentiment polarity by constructing hyperplanes for classification. SVM is used to identify user credibility and subjectivity of microblog opinions [12]. SVM is used on emoticon data set for feature presence by [4]. For twitter sentiment analysis, SVM is very effectively used to extract and analyse the tweets [15, 19]. It further minimises the misclassification error for prediction tasks as defined by [8]

$$y = [1, if x_i\phi - a > 0; -1, otherwise] \qquad (8)$$

$$argmin_\phi \sum{}^{n}_{i} \delta(y_i(x_i\phi - a) <= 0) \qquad (9)$$

Max Entropy: Maximum entropy is used for encoding of vectors to label features and find distribution over classes [13]. The weights of the features are calculated by using the encoded vectors. Reference [4] used Stanford classifier for max entropy classification. For calculating the weights, they used Gradient ascent and L2 regularisation. For tweet classification using max entropy, following formula is used:

$$P(class/Tweet, wt) = \frac{e^{\sum wt, f(class, Tweet)}}{\sum_c lasse^{\sum_w t.f(class, Tweet)}} \qquad (10)$$

Decision Tree: Tree classifier uses hierarchical classification by dividing the data based on all the possible conditions on the attributes of the data set. It works recursively until the condition reaches the leaf node to finally classify the data. There are various decision tree algorithms like ID3, C4.5, C5, etc. that have been used extensively for text classification [7, 10].

3.3 Hybrid

Hybrid techniques include combination of machine learning techniques and lexicon-based techniques and these techniques. Such techniques are formal concept analysis (FCA) and techniques based on the fuzzy theory like fuzzy formal concept analysis (FFCA) [14].

4 Experimental Study

This section presents a study on twitter sentiment analysis, where sentiments can be evaluated from the tweets which present useful indicators for prediction. Sentiments are characterised as positive, negative and neutral. Consider an example of twitter sentiment analysis where the tweets are extracted, collected and preprocessed in real-time and machine learning techniques are applied for identifying the sentiment polarity of the text.

4.1 Data Set

Sentiment 140 data set is used which contains 1.6 million tweets extracted via twitter api [4]. This set is used for training the model which contains 800000 positive tweets and 800000 negative tweets. For testing purposes, the data are manually collected that contains 177 negative tweets, 181 positive tweets and 139 neutral tweets. There are six fields in the data set. First is the target which defines the polarity of the text (0-negative, 2-neutral and 4-positive). Second is the ids of the tweets, third is the date of the tweet, fourth defines the query, fifth is the user of the tweet and sixth is the actual text of the tweet. Lexicon Dictionary: An opinion lexicon of 6789 words is used, out of which 4783 words are negative and 2006 words are positive [6].

4.2 Data Pre-processing

The preprocessing step includes tokenisation, vectorisation, parts of speech (POS) tagging, removal of stop words, etc. Tokenisation means assigning a value to each word, it also includes tokenisation of abbreviations and emoticons. POS tagging is used to define words and phrases with multiple meanings and removal of stop words that do not contribute to the opinion making like and, the, etc.

4.3 Features

There are various features that have been used for the classification of text like unigrams, bigrams and specific features for sentiment analysis like POS features, lexicon-based features which include a list of positive and negative words. These features also extract the context from phrases. Next important feature that has been used is term presence and frequency that creates document term matrix based on the vocabulary-based vectorisation. TF-IDF feature is used for text preprocessing which evaluates the importance of word in the corpus. TF refers to term frequency which is the ratio of number of times a word occurs in a text to the total number of words in the text. IDF refers to inverse document frequency is the ratio of log probability of the total number of documents to the number of documents with the particular word in it. TF-IDF is evaluated as:

$$Tf - IDF = TF(w) * IDF(w) \tag{11}$$

where,

TF(w) = number of times word w appears/total no of words,

IDF(w) = log(Total no of doc/ no. of docs with word w).

4.4 Results

The aim of this work is twofold in terms of sentiment analysis in social media. First the social media that is covered is the twitter data, where the tweets are gathered in real time for over a period of time and are processed and stored in a file. A dynamic visualisation and analysis of tweet data are presented. Second is the application of a machine leaning approach to evaluate the sentiments of the gathered data and the effectiveness of the features are shown in terms of the accuracy and ROC of the model. For the first set of experiment, we collected the tweets form the twitter api and the search term we used was on the recent Facebook data theft by Cambridge Analytica. The word cloud of the collected data at different times is shown in Figs. 1 and 2. Figure 3 presents the word frequencies of the 10 most recent words from the wordlist of the extracted tweets of Cambridge Analytica Corpus. Next tweet sentiments based on the lexicon-based approach where the list of positive and negative words are used to categorise the sentiment polarity of the tweets. The tweets are collected over a period of time and are dynamically shown (see Fig. 4). In this set-up, another approach

Fig. 1 Word cloud of 100 frequent words

Fig. 2 Word cloud of 200 frequent words

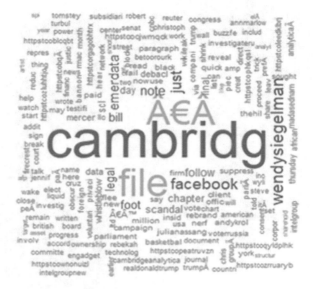

Fig. 3 Word frequencies of top 10 words in 'Cambridge Analytica' tweets

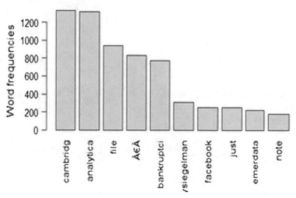

that has been used is logistic regression which is a machine learning technique that classifies the sentiment polarity. The model is trained on the sentiment 140 data set discussed above which uses 1.6 million tweets for training the model.

After the model is trained, the collected tweet corpora is fed into the model to evaluate the tweet sentiment rate of the data. Figure 5 presents the tweet sentiment rate which classifies the collected tweet sentiments into positive, negative and neutral with their tweet sentiment values ranging between 0 and 1, where 0 represents negative sentiment rate and 1 represents positive sentiment rate. The accuracy of the model is 87.28% and ROC curve is shown in Fig. 6.

In general machine learning techniques like logistic regression, Naive Bayes, SVM give good prediction results for classification of sentiments. But in case where evaluation of sentiment classification is based only on positive and negative sentiments, logistic regression gives better results than Naive Bayes and other linear classifiers.

Fig. 4 Lexicon-based
sentiment classification of
real-time tweets on
'Cambridge Analytica'
tweets

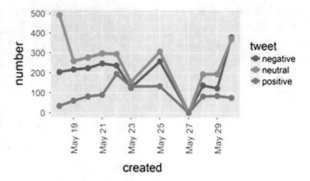

Fig. 5 Tweet sentiment Rate
of 'Cambridge Analytica'
data set

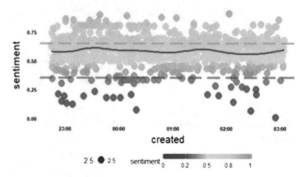

Fig. 6 ROC curve depicting
the accuracy of the
regression model

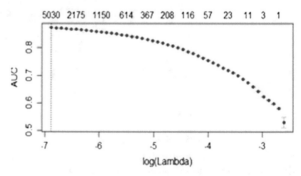

5 Conclusions and Future Work

This study discusses the social media analytics techniques, challenges, issues and
applications in various social media domain. Twitter sentiment analysis is presented
by crawling real-time twitter data and applying machine learning techniques to efficiently classify the data. In the experimental set-up, lexicon-based and logistic regression techniques are implemented based on lexical dictionaries and feature extraction.
The results classify the tweet sentiments efficiently into positive and negative with
an accuracy of 87.2%. This shows that twitter sentiment analysis can be beneficial

in extracting opinions of various users like customers, promoters, opinion leaders, etc which further enables to understand their perspectives towards critical success in various application domains.

In future, the research can be further extended to evaluate various machine learning techniques on twitter data and their performance evaluation on various parameters. The research will further explore the scope of social media (which includes social networking sites like Facebook, twitter, product review sites like epinions, photo and video sharing sites like Instagram, etc.), its security and challenges that occur in analysing and vulnerability assessment of social media data. Detection of deceptive social media messages and posts based on content analysis will also be conducted.

Acknowledgements This publication is an outcome of the R & D work undertaken project under the Visvesvaraya Ph.D. Scheme of Ministry of Electronics & Information Technology, Government of India, being implemented by Digital India Corporation and with the cooperation of GGSIP University.

References

1. Beigi, G., Hu, X., Maciejewski, R., Liu, H.: An overview of sentiment analysis in social media and its applications in disaster relief. In: Sentiment Analysis and Ontology Engineering, pp. 313–340. Springer, Berlin (2016)
2. Cooper, C.: Social media is a cyber security risk for business. https://www.csoonline.com/article/3198715/data-breach/social-media-is-a-cybersecurity-risk-for-business.html (2017). Accessed 20 April 2018
3. Fan, W., Gordon, M.D.: The power of social media analytics. Commun. ACM **57**(6), 74–81 (2014)
4. Go, A., Bhayani, R., Huang, L.: Twitter sentiment classification using distant supervision. CS224N Project Report, Stanford **1**(12) (2009)
5. Hayes, N.: Why social media sites are the new cyber weapons of choice. https://www.darkreading.com/attacks-breaches/why-social-media-sites-are-the-new-cyber-weapons-of-choice/a/d-id/1326802? (2016). Accessed 10 April 2018
6. Hu, M., Liu, B.: Mining and summarizing customer reviews. In: Proceedings of the Tenth ACM SIGKDD International Conference on Knowledge Discovery and Data Mining. pp. 168–177. ACM (2004)
7. Hu, Y., Li, W.: Document sentiment classification by exploring description model of topical terms. Comput. Speech Lang. **25**(2), 386–403 (2011)
8. Huang, C.L., Hu, Y.C., Lin, C.H.: Twitter sentiment analysis. https://cseweb.ucsd.edu/classes/wi17/cse258-a/reports/a080.pdf (2017). Accessed 20 April 2018
9. Jindal, N., Liu, B.: Opinion spam and analysis. In: Proceedings of the 2008 International Conference on Web Search and Data Mining. pp. 219–230. ACM (2008)
10. Kim, J.W., Lee, B.H., Shaw, M.J., Chang, H.L., Nelson, M.: Application of decision-tree induction techniques to personalized advertisements on internet storefronts. Int. J. Electron. Commer. **5**(3), 45–62 (2001)
11. Kim, S.B., Rim, H.C., Yook, D., Lim, H.S.: Effective methods for improving naive bayes text classifiers. In: Pacific Rim International Conference on Artificial Intelligence. pp. 414–423. Springer, Berlin (2002)
12. Li, Y.M., Li, T.Y.: Deriving market intelligence from microblogs. Decis. Support Syst. **55**(1), 206–217 (2013)

13. Lu, Y., Hu, X., Wang, F., Kumar, S., Liu, H., Maciejewski, R.: Visualizing social media sentiment in disaster scenarios. In: Proceedings of the 24th International Conference on World Wide Web. pp. 1211–1215. ACM (2015)
14. Medhat, W., Hassan, A., Korashy, H.: Sentiment analysis algorithms and applications: a survey. Ain Shams Eng. J. **5**(4), 1093–1113 (2014)
15. Pak, A., Paroubek, P.: Twitter as a corpus for sentiment analysis and opinion mining. In: LREc. vol. 10 (2010)
16. Rout, J.K., Singh, S., Jena, S.K., Bakshi, S.: Deceptive review detection using labeled and unlabeled data. Multimed. Tools Appl. **76**(3), 3187–3211 (2017)
17. Sarlan, A., Nadam, C., Basri, S.: Twitter sentiment analysis. In: 2014 International Conference on Information Technology and Multimedia (ICIMU), pp. 212–216. IEEE (2014)
18. Sharma, S., Jain, A.: Dynamic social network analysis and performance evaluation. Int. J. Intell. Eng. Inform. (2018)
19. Sharma, S.: Application of support vector machines for damage detection in structures. Ph.D. thesis, Worcester Polytechnic Institute (2008)
20. Yu, S., Kak, S.C.: A survey of prediction using social media. CoRR arXiv:abs/1203.1647 (2012). http://dblp.uni-trier.de/db/journals/corr/corr1203.htmlabs-1203-1647

Internet of Things for Epidemic Detection: A Critical Review

S. A. D. S. Kaushalya, K. A. D. T. Kulawansa and M. F. M. Firdhous

Abstract An epidemic is a spread of an infectious disease in a community or a region affecting a large number of people. Prompt detection of epidemic outbreaks can help reduce the impact saving many lives and enable controlling future outbreaks. Hence detecting and identifying the cause of the outbreak quickly and accurately plays a vital role in controlling the consequences of the outbreak. In recent times, computer-based models and simulations have increasingly been used for detecting and characterizing disease outbreaks in a community. Several models and systems have been proposed for detecting a possible disease outbreak from data collected from individuals seeking health care at individual clinics before the problem becomes serious. In this paper, the authors present a critical review of Internet of Things-based epidemic detection models presented in the literature with special emphasis on the strengths and weaknesses of each model proposed.

Keywords Internet of things · Epidemic outbreaks · Epidemic detection · Automated surveillance

1 Introduction

The world has been witnessing the emergence of new diseases as well as reemergence of previously known diseases in new environments recently [1]. Emergence of such threats leads to loss of human life and limbs as well as requires many resources to

S. A. D. S. Kaushalya
Faculty of Information Technology, University of Moratuwa, Moratuwa, Sri Lanka
e-mail: sandamini.kaushalya@gmail.com

K. A. D. T. Kulawansa
Department of Computational Mathematics, University of Moratuwa, Moratuwa, Sri Lanka
e-mail: dilinik@uom.lk

M. F. M. Firdhous (✉)
Department of Information Technology, University of Moratuwa, Moratuwa, Sri Lanka
e-mail: firdhous@uom.lk

© Springer Nature Singapore Pte Ltd. 2019
S. K. Bhatia et al. (eds.), *Advances in Computer Communication and Computational Sciences*, Advances in Intelligent Systems and Computing 924,
https://doi.org/10.1007/978-981-13-6861-5_42

485

control the further spread of them. Spread of such infectious diseases affecting large number people in a short time in an area is commonly known as an epidemic [2]. Early detection of epidemic outbreaks can help control the effects of such epidemics in a cost-effective manner with the least amount of losses.

The increase in population density, increased social contacts and interactions between people resulting from rapid urbanization has resulted in controlling epidemics more difficult [3]. Conventional epidemic control techniques generally consist of both offline control and model-based approaches that isolate and treat people to limit the spreading of an epidemic. These traditional approaches suffer from many shortcomings including lack of data for proper analysis and detection and simplified social behavior models with many assumptions that lead to improper predictions. In order to overcome the limitations of the traditional manual techniques, researchers have proposed to use advanced computing technologies to detect the epidemic outbreaks at the earliest possible time enabling to reduce the consequences [1].

Automated epidemic detection systems continuously collect data on many aspects of human behavior and the surroundings and then analyze them using mathematical or statistical models with minimum human intervention. These models can handle large amount of data of different types increasing the detection probability of epidemic outbreaks and spreads [4]. In recent times, the availability of powerful computing facilities including high end hardware, simulation software, and data collection methods have contributed to increased use of computer-based simulations for studying different aspects of epidemic outbreaks [5]. In this paper, the authors carry out a critical review of the work that has proposed to use Internet of Things (IoT) for detecting epidemics outbreaks. The paper takes a critical look at each proposal with special emphasis on the issue they are focusing on along with their strengths and weaknesses.

This paper consists of four main sections: Sect. 1 provides an overview of the paper with a brief analysis of the problem area to be explored. Sections 2 provides a general description of IoT and how they can used in healthcare applications along with a discussion on similar work carried out by other researchers. Section 3 starts with a brief discussion on epidemics and their characteristics and then presents the critical analysis of the IoT-based epidemic detection schemes followed by a summary of each work along with their strengths and weaknesses. Lastly, Sect. 4 completes the paper with a discussion on the results of the analysis.

2   Internet of Things and Its Application in Health care

Internet of Things is the interconnection of heterogeneous set of generic objects embedded with intelligence and communicates with each other and other objects using wireless communication techniques with Internet protocols [6]. The complete IoT architecture is a system made up of both physical, virtual, and hybrid objects along with communication links and protocols, network devices, servers for storing analyzing, and presenting data and applications. The physical objects are generally made of sensors, actuators, RFID tags, and smart phones and while the virtual objects

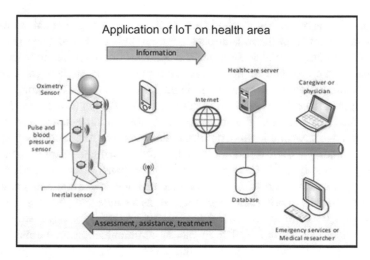

Fig. 1 Application of Internet of Things in health care

consist of computer applications, application middleware, multimedia contents, and web services [7].

With the rapid increase in elderly population all over the world, the healthcare system has been put under heavy pressure to take care of them demanding many resources including hospital beds, healthcare professionals, drugs, and rehabilitation systems [8]. This demands a solution that could minimize the burden on healthcare systems while continuing with the provision of better health care to patients. With the emergence of IoT and the recognition of its capabilities, it has been widely viewed as a potential solution to reduce the burdens on healthcare systems [6, 9]. Hence, much effort has been devoted to crating new IoT-enabled systems and applications focusing on solving the problems associated with the healthcare industry. A considerable number of research efforts have been dedicated for the supervision of patients with special requirements, such as diabetes or Parkinson's disease, aiding rehabilitation through continuous observations of progress, and emergency health care [9].

Figure 1 presents a generic architecture of an IoT-enabled healthcare system. The system identifies the main components of such a system starting from the sensors attached to the body of a patient to the communication links for transferring data and the computers that store, analyze, and transform the data into information and knowledge. The information thus created by the data processing applications in the computers is transferred back to the relevant parties through the communication medium as shown in the back arrow. An epidemic monitoring system needs not only to monitor patients, but also the environment where the patient currently is as the environmental factors plays a major role in transforming an infectious disease into an epidemic [10].

Several researchers have presented surveys and reviews of specific areas and technologies related to IoT in health care in the literature. This paragraph takes a brief look at some of these works and seeks to show the work presented in this paper differentiates from all of them. Islam et al., have in [11] presented a comprehensive survey of commonly available solutions and potential future applications along with issues that need immediate attention in IoT in health care. Each item has been given an individual and comprehensive attention but in a disconnected fashion. Dimitrov has in [12] presented a review of Medical Internet of Things (mIoT) and Big Data in health care. He discusses in detail about data storage, mining, and analysis with special reference to wearables and mobile applications. The main shortcoming of this work is the treating of these technologies independently without integrating them into a single system. Poon et al., have presented a comprehensive review of sensor types used in body sensor networks along with a focus on inter-device communication [13]. This paper also fails to bring the topics together as a single and complete system. In [14], Yin et al., have presented review with specific concentration on sensing and data management in healthcare IoT. The impact of communication on the performance has not been given sufficient attention in this paper. In [15], Ahmed et al., have taken an in depth look at IoT-based health monitoring systems with special emphasis on wireless standards and technologies for remote health monitoring and security issues and challenges. This work also concentrates on a few specific aspects of IoT and hence lacks completeness in terms of a discussion on a fully fledged IoT-based healthcare system. Alqahtani has broadly discussed the application of IoT in different areas including industry, business, and health care and different technologies that make IoT applications possible [16]. But this work lacks focus as it tries to focus on many areas within a short space. This is a good generic paper on possible applications and enabling technologies on IoT, but nothing much on the application of IoT in health care. The work presented by Baker et al., in [9] has identified all key components of an end-to-end IoT-based healthcare system. They also propose a generalized model for an IoT-based healthcare system with remote monitoring. This model can be extended to meet any requirements of emerging IoT-powered healthcare systems. The shortcoming of this work is that the entire review has been limited to building the model rather than providing a comprehensive discussion on the strengths and weaknesses of using IoT in healthcare applications. From the above discussion, it can be seen that all the researchers have concentrated reviewing the work carried out on individual disease monitoring. The authors could not find any literature in any database reviewing the work carried out on the IoT applications in detecting epidemic outbreaks. In this sense, the critical review presented in this paper is unique and could serve as a starting point for researchers working in this challenging area.

3   Internet of Things in Epidemic Detection

This section starts with a brief discussion on what an epidemic is and then presents the critical review of the IoT assisted epidemic detection and management systems proposed in the literature. The subsection on epidemics will discuss the difference between the normal diseases, infectious diseases, epidemics, and pandemics in terms of their features and how to identify them. Then, the next subsection is essentially the critical review of published work on IoT-assisted epidemic detection systems.

3.1 Epidemics

During their life time, human beings suffer from many illnesses including mild ones to very serious ones. These illnesses can be broadly divided into two groups as communicable diseases and non-communicable diseases [17]. A non-communicable disease is a medical condition that is not caused by infectious agents such as viruses or bacteria [18]. On the other hand, communicable diseases are caused by microorganisms such as viruses, bacteria, fungi or parasites that can spread, directly or indirectly from one person to another [19]. Infectious diseases spread from one individual to another through the direct transfer of disease causing microorganisms. This happens when an infected person touches, kisses, coughs, or sneezes at a healthy person. When an infectious (communicable) disease spreads rapidly affecting a large number of people in a given region within a short period of time, it is known as an epidemic [20]. A pandemic is an epidemic that happens at global proportions affecting large number of people across a large geographic area [21]. This generally occurs when a previously unknown emerges among people causing serious illnesses easily spreading among the human population.

Ebola, Influenza A (H1N1), Zika virus outbreak, Chikungunya virus, Dengue fever, Cholera, and Yellow fever are some of the epidemics that killed many people in the past. The main reason for their rapid spread followed by large casualties was the inability to predict, detect, and control them in time. The traditional offline disease detection and control methods such as model-based approaches quarantine and immunize people to contain the spreading of epidemics [22]. These traditional approaches suffer from many shortcomings including lack of timely data and inaccurate predictions resulting from impractical assumptions included in the models about the dynamics of social contact networks. Thus, it has been recognized the need for continuous healthcare monitoring for improving health conditions and prevent epidemic outbreaks. Many researchers have identified the IoT-based technology would be able to overcome issues with traditional methods in continuous monitoring through the different types of devices and systems that are capable of predicting and diagnosing epidemics along with raising public awareness on them. The following subsection takes a critical look at some of the major researches conducted in this area with special emphasis on their strengths, weaknesses, challenges faced, and real-world applications used for detecting disease.

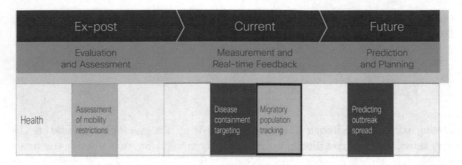

Fig. 2 Phases of epidemic detection

3.2 Internet of Things Powered Epidemic Detection

During the last few decades, the drastic reduction in cost and size of electronic devices along with the increase in performance has enabled the production of many wearable low power computing devices helping the creation of many powerful healthcare applications [23]. Currently there are many systems available in the market for different purposes including personal health care, activity monitoring, and fitness. Researchers are continuing to propose new clinical applications using these technologies for remote health monitoring including functionalities for long-term status recording, and medical access to physiological information of patients. Leveraging the capabilities of IoT-based technologies (applications) for the purpose of detecting epidemic outbreaks along with emerging trends of IoT-based technologies in epidemic detection are discussed in the following paragraphs. Figure 2 shows the different phases of epidemic detection that must be strictly followed in identifying an epidemic outbreak.

Masuda and Holme have in [24] proposed an infectious disease prediction and controlling system using temporal networks. They suggest that temporal data have the ability to understand the dynamics of social groups with respect to time and epidemic spreading and can help to find better ways to analyze, predict, and control epidemics. The proposed method requires large amount of temporal data without any missing values. Missing values and temporary outliers may affect the accuracy of the results. Cheedella has reviewed the applicability of IoT in healthcare system [25]. In his study, he has found that IoT communication acts as an important enabler for worldwide distribution of reputed healthcare applications. This study also focuses on how a patient's data can be securely transferred without being modified, tampered with or lost on the way. The importance of ensuring the reliability of the information and networks has been given special attention in the study. The requirement of a reliable transmission system increases the overhead of data transmission and cost of implementation.

Sood and Mahajan have in [26], proposed a portable IoT-based system for detecting and managing the spread of Chikungunya virus. The proposed system has been tuned only for detecting a single virus and hence detecting a new threat may not be possible. Sandhu et al., have in [27] proposed a monitoring and controlling Influenza A (H1N1) using social network analysis. The proposed mechanism is very effective in collecting and analyzing data using IoT-enabled sensors and cloud computing, but the accuracy and the reliability of the predictions needs further validation. Zhao et al., have in [28] investigated the medical applications of IoT. They have critically analyzed the possibilities of and issues related to providing better healthcare services for each person all over the world. They have identified that the newly established research directions in medical Internet of Things has the ability to meet this demand. They also identify the need for the formulation of uniform standards, protocols along with establishing secured data transferring method is a prerequisite for proper operation of such systems.

Turcu and Turcu in [29] have identified IoT as the key enabler in reducing errors in identifying and treating infectious diseases. The patient records presently stored in different geographical locations can be effectively brought together and accessed using advanced technologies. In this direction, they propose an IoT-based solution coupled with RFID for proper identification. The main downside of the proposal includes the lack of definition on security problems, nonstandardization of communication protocols, and arising of scalability issues as the network grows in size. Sareen et al., have in [30] proposed a secure IoT-based cloud framework for controlling Zika virus outbreak. This framework works effectively for detecting Zika virus but lack generalizability to detect and future outbreaks of unknown origins.

Mathew et al., have in [2] proposed a smart disease monitoring system based on IoT that provides an efficient monitoring of diseases identifying and tracking their trends. However, it suffers from several shortcomings including inability to integrate with other systems, high implementation cost, and inadequate feedback mechanism. It also requires improving the effective use of tools for timely data collection and analysis for accurate epidemic prediction.

Huang et al., have investigated a method for identifying the transmission of flu by social sensors in China [31]. They have come up with a strategy for detecting the contagion transmission based on social networks. The main drawback of the proposal is the high resource requirement for processing data and the prediction of transmission. In [32], Wilson and Brownstein have proposed a method for the early identification of epidemic outbreaks with the use of the Internet. This method provides an efficient epidemic intelligence collection scheme for offering effective solutions but the requirement for uninterrupted Internet access along with the limited evidence of detecting threats compared to traditional methods can be considered as the weakness of this proposal.

Table 1 given provides the summary of the comparative study carried out on a set of similar research publications reported in literature and discussed above.

Table 1 Summary of the comparison of relevant researches

Work cited	Benefits	Drawbacks
[2]	Provides an efficient disease surveillance by identifying the trends	Cannot be integrated with other systems and provides insufficient feedback. Needs further improvement for effective use of tools for timely data collection and analysis for accurate epidemic prediction
[11]	State-of-the-art network architecture for IoT in health care with safety solutions/privacy attributes and security/policies and regulations	Lacks ability monitor continuously. Also requires low power protocols and data protection issues needs to be addressed separately
[24]	Temporal data capturing in social networks using timescales to detect and contain the spreading of an epidemic	Requires large amounts of data. Missing values and outliers affect the accuracy of predictions
[25]	Continuous healthcare monitoring for improving health conditions	Lacks a proper system architecture for obtaining efficient outputs. Data transmission impairments may affect the accuracy of results
[26]	Diagnosis of CHV in accordance with patient's health symptoms and the surrounding conditions	Designed only for detecting CHV virus. Detecting a new threat may not be possible
[27]	Predicts H1N1 infectors earlier and provide others with prevention strategies that help reduce infection rate	Accuracy and reliability of the prediction have not been validated
[28]	Has the ability to provide advanced services for healthcare management in an effective manner using IoT	Uniform standards and protocols along with establishing secure data transferring method is a prerequisite for proper operation
[29]	Radio frequency-based identification for reducing medical errors along with multi-agent and IoT technologies at an affordable cost. Improves safety and optimizes the healthcare process	Suffers from security risks and privacy issues. Different types of communication standards may affect the implementation. Scalability becomes an issue at large-scale deployments
[30]	Prevents and controls the spread of Zika virus effectively	Lacks generalizability to detect future outbreaks of unknown disease types
[31]	A strategy for detecting contagion transmission based on social networks	Data processing and prediction of transmission are very costly in terms of resources required
[32]	Efficient collection of epidemic intelligence for offering effective solutions	Requires uninterrupted Internet access. Limited evidence of detecting threats compared to traditional methods

(continued)

Table 1 (continued)

Work cited	Benefits	Drawbacks
[33]	An IoT application to reach excellence in health care at very affordable costs	Utilization of similar functionalities with regards to identifiable wireless embedded systems in health care
[34]	IoT communication framework as the leading enabler in distributed worldwide applications of health care has been proposed	Unavoidable performance degradations due to low standard constrained device optimizations. Hence, solutions offered may not be feasible enough
[35]	Investigated using eight sets of empirical human contact data sets	When the temporal information is neglected, over-estimation takes place

4 Conclusions

The world has been suffering from epidemic outbreaks from time immemorial. In recent times, despite the advances made in many technical fronts, including detection and treatment of infectious and contagious diseases, new epidemics are emerging threatening the whole population. Once an epidemic outbreak happens, the consequences are very disastrous with many lives lost. Hence, early detection and containment of epidemic is very vital for the human well-being. Computer technologies have been continuously explored for effective detection of epidemics by many researchers. This paper presented a critical review of the work carried out by many researchers on IoT-enabled epidemic detection systems. The paper critically analyzed the basic principle on which the proposed systems have been based on along their strengths and weaknesses. Finally, the paper presented the summary of all the observation in an easily readable and comprehensible tabular format.

References

1. Jafarpour, N., Izadi, M., Precup, D., Buckeridge, D.L.: Quantifying the determinants of outbreak detection performance through simulation and machine learning. J. Biomed. Inform. **53**, 183–187 (2015)
2. Mathew, A., Amreen, S.A.F., Pooja, H.N., Verma, A.: Smart disease surveillance based on Internet of Things (IoT). Int. J. Adv. Res. Comput. Commun. Eng. **4**(5), 180–183 (2015)
3. Zhang, Z., Wang, H., Wang, C., Fang, H.: Cluster-based epidemic control through smartphone-based body area networks. IEEE Trans. Parallel Distrib. Syst. **26**(3), 681–690 (2015)
4. Lessler, J., Cummings, D.A.T.: Mechanistic models of infectious disease and their impact on public health. Am. J. Epidemiol. **183**(5), 415–422 (2016)
5. Orbann, C., Sattenspiel, L., Miller, E., Dimka, J.: Defining epidemics in computer simulation models: how do definitions influence conclusions? Epidemics **19**, 24–32 (2017)
6. Firdhous, M.F.M., Sudantha, B.H., Karunaratne, P.M.: IoT enabled proactive indoor air quality monitoring system for sustainable health management. In: 2nd IEEE International Conference on Computing and Communication Technologies, Chennai, India, pp. 216–221 (2017)
7. Elkhodr, M., Shahrestani, S., Cheung, H.: A middleware for the Internet of Things. Int. J. Comput. Netw. Commun. (IJCNC) **8**(2), 159–178 (2016)

8. Smits, C.H.M., van den Beld, H.K., Aartsen, M.J., Schroots, J.J.: Aging in the Netherlands: state of the art and science. The Gerontologist **54**(3), 335–343 (2014)
9. Baker, S.B., Xiang, W., Atkinson, I.: Internet of Things for smart healthcare: technologies, challenges, and opportunities. IEEE Access **5**, 26521–26544 (2017)
10. Nii-Trebi, N.I.: Emerging and neglected infectious diseases: insights, advances, and challenges. Biomed. Res. Int. **2017**(5245021), 1–15 (2017)
11. Islam, S.M.R., Kwak, D., Kabir, M.H., Hossain, M., Kwak, K.S.: The Internet of Things for health care: a comprehensive survey. IEEE Access **3**, 678–708 (2015)
12. Dimitrov, D.V.: Medical Internet of Things and Big Data in healthcare. Healthc. Inf. Res. **22**(3), 156–163 (2016)
13. Poon, C.C.Y., Lo, B.P.L., Yuce, M.R., Alomainy, A., Hao, Y.: Body sensor networks: in the era of big data and beyond. IEEE Rev. Biomed. Eng. **8**, 4–16 (2015)
14. Yin, Y., Zeng, Y., Chen, X., Fan, Y.: The Internet of Things in healthcare: an overview. J. Ind. Inf. Integr. **1**, 1–13 (2016)
15. Ahmed, M.U., Bjorkman, M., Causevic, A.D., Fotouhi, H., Linden, M.: An overview on the Internet of Things for health monitoring systems. In: 2nd EAI International Conference on IoT Technologies for HealthCare, Rome, Italy, pp. 429–436 (2015)
16. Alqahtani, F.H.: The application of the Internet of Things in healthcare. Int. J. Comput. Appl. **180**(18) (2018)
17. Abebe, S.M., Andargie, G., Shimeka, A., Alemu, K., Kebede, Y., Wubeshet, M., Tariku, A., Gebeyehu, A., Bayisa, M., Yitayal, M., Awoke, T., Azmeraw, T., Birku, M.: The prevalence of non-communicable diseases in northwest Ethiopia: survey of dabat health and demographic surveillance system. BMJ Open **7**(10), 1–9 (2017)
18. Hyder, A.A., Wosu, A.C., Gibson, D.G., Labrique, A.B., Ali, J., Pariyo, G.W.: Noncommunicable disease risk factors and mobile phones: a proposed research agenda. J. Med. Internet Res. **19**(5), 1–16 (2017)
19. Ditmar, M.F.: Pediatric secrets, Chap. In: Infectious Diseases, 5 edn., pp. 354–422. Elsevier, Milton, ON Canada (2011)
20. Dimitri, N.: The economics of epidemic diseases. PLoS ONE **10**(9), 1–8 (2015)
21. Morens, D.M., Folkers, G.K., Fauci, A.S.: What is a pandemic? J. Infect. Dis. **200**(7), 1018–1021 (2009)
22. Zhang, Z., Wang, H., Lin, X., Fang, H., Xuan, D.: Effective epidemic control and source tracing through mobile social sensing over WBANs. In: IEEE INFOCOM, Turin, Italy, pp. 300–304 (2013)
23. Azzawi, M.A., Hassan, R., Bakar, K.A.A.: A review on Internet of Things (IoT) in healthcare. Int. J. Appl. Eng. Res. **11**(20), 10216–10221 (2016)
24. Masuda, N., Holme, P.: Predicting and controlling infectious disease epidemics using temporal networks. F1000 Prime Rep. **5**(6), 1–12 (2013)
25. Cheedella, P.: Internet of Things in Healthcare System, M.Tech. thesis, Division of Software Systems, VIT University, India (2016)
26. Sood, S.K., Mahajan, I.: Wearable IoT sensor based healthcare system for identifying and controlling Chikungunya virus. Comput. Ind. **91**, 33–44 (2017)
27. Sandhu, R., Gill, H.K., Sood, S.K.: Smart monitoring and controlling of pandemic influenza A (H1N1) using social network analysis and cloud computing. J. Comput. Sci. **12**, 11–22 (2016)
28. Zhao, W., Wang, C., Nakahira, Y.: Medical application on Internet of Things. In: IET International Conference on Communication Technology and Application (ICCTA 2011), Beijing, China, pp. 660–665 (2011)
29. Turcu, C.E., Turcu, C.O.: Internet of Things as key enabler for sustainable healthcare delivery. Proc. Soc. Behav. Sci. **73**(27), 251–256 (2013)
30. Sareen, S., Sood, S.K., Gupta, S.K.: Secure Internet of Things-based cloud framework to control Zika virus outbreak. Int. J. Technol. Assess. Health Care **33**(1), 11–18 (2017)
31. Huang, J., Zhao, H., Zhang, J.: Detecting flu transmission by social sensor in China. In: 2013 IEEE International Conference on Internet of Things, Beijing, China, pp. 1242–1247 (2013)

32. Wilson, K., Brownstein, J.S.: Early detection of disease outbreaks using the Internet. Can. Med. Assoc. J. **180**(8), 829–831 (2009)
33. Kulkarni, A., Sathe, S.: Healthcare applications of the Internet of Things: a review. Int. J. Comput. Sci. Inf. Technol. **5**(5), 6229–6232 (2014)
34. Bui, N., Zorzi, M.: Health care applications: a solution based on the Internet of Things. In: 4th International Symposium on Applied Sciences in Biomedical and Communication Technologies, Barcelona, Spain, pp. 131:1–131:5 (2011)
35. Holme, P.: Temporal network structures controlling disease spreading. Phys. Rev. E **94**(2), 1–8 (2016)

Semantic Recognition of Web Structure to Retrieve Relevant Documents from Google by Formulating Index Term

Jinat Ara and Hanif Bhuiyan

Abstract Nowadays among various search mechanism, Google is one of the best information retrieval mechanisms for retrieving useful result as per user query. Generally, the complete searching process of Google is completed by crawler and indexing process. Actually, Google performs the documents indexing process considering various features of web documents (title, meta tags, keywords, etc.) which helps to fetches the complementary result by exactly matching the given query with the index term with user interest. Though appropriate indexing process is much difficult but it essential as extracting relevant document completely or partially depend on how much relevant the index term with the document is. However, sometimes, this indexing process is influenced by assorted number of feature of web documents which produced variegated result those are either completely or partially irrelevant to the search and seems unexpected. To reduce this problem, we analyze web documents considering its unstructured (web content and link features) data in terms of efficiency, quality, and relevancy with the user search query and present a keyword-based approach to formulate appropriate index term through semantic analysis using NLP concept. This approach helps to understand the current web structure (effectiveness, quality, and relevancy) and mitigate the current inconsistency problem by generating appropriate, efficient, and relevant index term which might improve the search quality; as satisfied search result is the major concern of Google search engine. The analysis helps to generate appropriate Google web documents index term which might useful to retrieve appropriate and relevant web documents more systematically than other existing approaches (lexicon and web structure based approaches). The experimental result demonstrates that, the proposed approach is an effective and efficient methodology to predict about the competence of web documents and finding appropriate and relevant web documents.

J. Ara
Jahangirnagar University, Dhaka, Bangladesh
e-mail: aracse2014@gmail.com

H. Bhuiyan (✉)
University of Asia Pacific, Dhaka, Bangladesh
e-mail: hanif_tushar.cse@uap-bd.edu

© Springer Nature Singapore Pte Ltd. 2019　　　　　　　　　　　　　　　497
S. K. Bhatia et al. (eds.), *Advances in Computer Communication
and Computational Sciences*, Advances in Intelligent Systems and Computing 924,
https://doi.org/10.1007/978-981-13-6861-5_43

Keywords Google · Search engine · Information retrieval · Page ranking algorithm · Index term and feature

1 Introduction

With rapid growth of Internet, the demand of web is growing dramatically [1]. Among various interactivity of modern web, search engine is a magnificent invention to access the web resources according to user's interest. Search engine has an incredible capability to fetch all the information from the World Wide Web (WWW) within a second. For this outstanding competence of search engine, information searching on the web becomes a common activity for most of the people [2]. For finding relevant and useful information, people are using various search engines like Yahoo, MSN, Bing, and so forth. Among the entire search engine, Google is the most popular to user which known as Google search. Google has billions of web pages with billions of topic which are the primary focus and core element since 1998 [3, 4]. It is handling more than three billion searches each day which rewarded it as most used and most preferable search engine over the web [5].

Since 1998 to till, Google have almost 4.5+ billion active users (December 31, 2017). For some prominent categories such as Keywords/query-based search, image based search, voice search, etc., now Google becomes top search engine throughout the world. Google search engine completes the searching process with the help of crawling and indexing [6]. Process of crawling is completed by crawler and acts as an identifier which identifies and stored document in search index. Generally, indexing is a collection of billions of web pages which helps to find relevant records and also improve the speed of search [7, 8]. Surprisingly, number of web documents on Google index is increasing dramatically which approximately crossed Trillions of web documents (Mar 1, 2013; 30 trillion unique web documents). Every web documents in Google index have various feature such as title, content, tags, meta tags, dates, links, etc. are referred as useful criteria for web page indexing. These all are indexing through different attributes like Documents type (text base, encoded base), Documents format (HTML/XML), Web page Title, Link, Metadata, Document term, Language, etc.

Although Google's search engine has an outstanding capability to search and fetch the content through query, some unexpected problem occurs during searching time. As an example, we submit a set of queries like Algorithm Evaluation, in return search engine extract several pages as result which containing various web documents. As though most of the extracted documents are relevant, few of them are totally irrelevant to the search with junk results. When a query q placed on search box, search engine parse the query q and find similarity between query and document text [similarity:$(S(q{:}dj))$] with every index term to find the most relevant document to the search, some snapshot of Google search through query shows in Fig. 1 where few of extracted web documents are not relevant with the search query those we are looking for. So still now, search engine has some relevancy shortcoming.

Fig. 1 Snapshots of Google search through query

In addition, every web documents are indexing through various criteria and notion, but there are quite lacks of document relevancy concern for information retrieval and during searching every word might not be used. As the number of web documents with relevant information on the web is growing rapidly, it becomes a significant area not only for the researchers but also for every user on the web. Many researchers, information analyzer, search engine optimizers enhance various methods [9–11] to improve existing searching process to extract more relevant content from the web. But still there are some unimproved sectors (like indexing) because only query matching result does not return satisfied result until indexing is appropriate. Moreover, most of the index term used by Google might be irrelevant and inappropriate with the document. We believe that, every content carry some significant words those are completely unique and assist to categorize the individual content. Our goal is to solve the search inconsistency problem by generating an automated process to analyze the web content and generate relevant key terms of the document for indexing.

2 Related Work

From mid 90s, invention of first search engine yet efficiency, quality, and relevancy are the principal criteria for any search engine [1]. Though, with the time, different search engines are trying to place as top search engine, but still now Google search engine is the most popular to user and also first priority for majority of researchers [12]. So, especially for Google search engine, several researches have been taken for increasing searching relevancy considering various aspects of criteria. Among them Google index term is one of the major criteria to make more convenient result. These index terms are act like a reflection (document topic, meaning, valuable words, etc.) of the whole content of the document under an index. Moreover, more relevant and appropriate index term could increase the possibility of appearing more relevant documents. Based on Google feature, Babu et al. and Michael et al. evaluates search engine using relevant information such as users search history and geographic locations [3, 4]. For extracting more relevant information through search, Holzle et al. proposed an approach using clustering technique [5], Talib et al. work on knowledge base technique [7] and Vaid et al. perform the analysis through neighboring technique [8]. These three techniques predict that, statistical technique can fetch more relevant documents through search. Furthermore, Page et al. analysis using the additional information such as hypertext to produce better search results [1]. Chari et al. perform their analysis using meta data or information such as title and keywords for improving the search relevancy [2]. Though these all approaches showed impressive work especially for relevant information [3, 4] and for additional information [1, 2] but they only focus on lexicon data and web structure which may lead to irrelevant result. On the contrary, we analyze web documents considering its unstructured features and use documents text to formulate proper index term as valuable words are insights on the documents text instead of its others feature. This approach might be helpful for Google user to retrieve relevant documents through search more statistically than before.

3 Research Methodology

This section describes the analysis process in detail in order to solve the searching problem and improves the search quality. The work has been completed into three steps shown in Fig. 2: first classification process performed on web documents (On-page). Next, evaluate the web documents through content-based feature (through title for measuring the efficiency) and link-based feature (through page rank for measuring the quality) for finding useful feature as index feature. Finally, keyword-based approach has been adopted which semantically generates key index terms through NLP concept in order to reduce the inconsistency problem and find useful content through search.

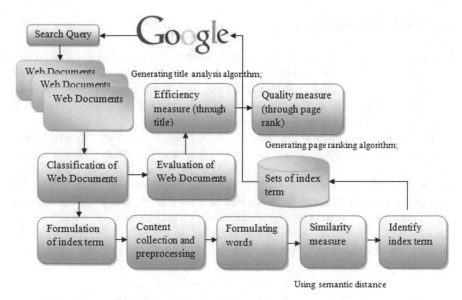

Fig. 2 Work flow of optimizing Google search

3.1 Classification of Web Documents

Nowadays, information retrieval among billions of web pages is the major challenge for any search engine [13, 14]. As web documents are uncontrolled nature, classification is essential. World Wide Web (WWW) has been created a platform [5] where web documents are classified into various categories like plaintext, knowledge based, neighboring selection, etc. Among them, "On-page" categorization perform a vital role for web page research. On-page feature splits into HTML tags (title, heading, metadata, and main text) and URL. This paper focus on On-page feature (HTML tags and URLs) where web page content and link feature are the prime concern. In order to extract this information, focus crawler is implemented which extract information using web API through HTTP Get method.

3.2 Evaluation of Web Documents

After classifying the web documents, evaluation process performed on extracted web documents through its Title and Page ranking (backlinks), in order to find the most useful feature as index term. Though several numbers of features exist in web documents, these two features have interactive capability to represent the web documents more effectively and efficiently than others.

3.2.1 Efficiency Measure (Title)

Generally, almost every search engines updates their index frequently. Most of the time not only Google but also other search engines prefer to select documents title as index term. Moreover, title is an effective feature which explains the full content in short with user interest. Google search engine uses various semantic methods to analyze these effective feature.

Though semantic method has been improved constructively but still now human interests are not completely understandable by the computer. Among several incompetence, major one is assigning operator ("!", ":", "?") as they have the ability to change the full content meaning. As an example, searching query like "Algorithm development and evaluation," in return Fig. 3a Shows that in title with the searching query, assign operator has existing which totally change the meaning of the search. For another search query "Algorithm Evaluation" in return Fig. 3b, shows that the title of the web document "Algorithm runtime prediction: Methods and evaluation" where search query become a set [(algorithm, evaluation) or (evaluation, algorithm)] but not maintain an order [order should like: algorithm evaluation].

From our observation, it concluded that, still now, these kinds of uncertainties have been existed which impact on search quality. So title should be relevant with the content as well as search query which helps to improve the search result and find relevant content more efficiently. In this paper, we proposed an algorithm (Fig. 4) to analyze web page title for understanding how effective the title is as index term. The full process described through algorithm in below.

Here, (in Fig. 4) in line 1, check the searching query in the title, if exist (2) then check the assign operator in line 3. In line 4, if exist any operator, return false, either true (5). In line 6, check the query sequence, either the searching query maintains order return true (7) either false (8).

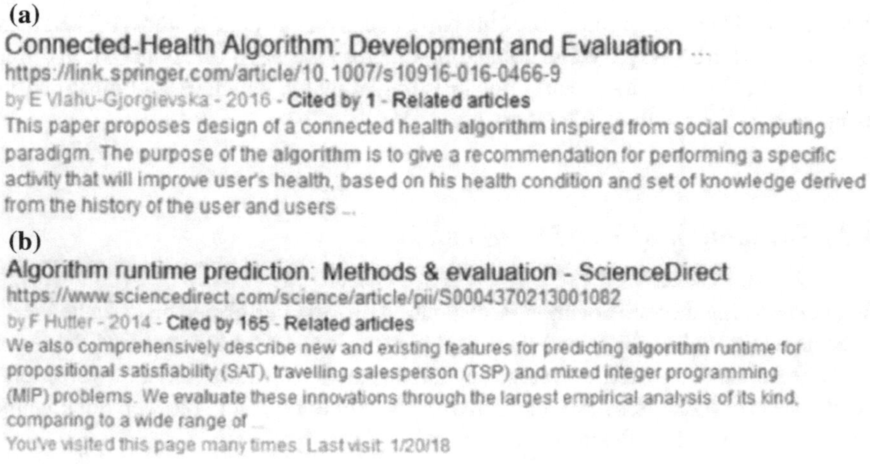

(a)

Connected-Health Algorithm: Development and Evaluation ...
https://link.springer.com/article/10.1007/s10916-016-0466-9
by E Vlahu-Gjorgievska - 2016 - Cited by 1 - Related articles
This paper proposes design of a connected health algorithm inspired from social computing
paradigm. The purpose of the algorithm is to give a recommendation for performing a specific
activity that will improve user's health, based on his health condition and set of knowledge derived
from the history of the user and users ...

(b)

Algorithm runtime prediction: Methods & evaluation - ScienceDirect
https://www.sciencedirect.com/science/article/pii/S0004370213001082
by F Hutter - 2014 - Cited by 165 - Related articles
We also comprehensively describe new and existing features for predicting algorithm runtime for
propositional satisfiability (SAT), travelling salesperson (TSP) and mixed integer programming
(MIP) problems. We evaluate these innovations through the largest empirical analysis of its kind,
comparing to a wide range of ...
You've visited this page many times. Last visit: 1/20/18

Fig. 3 Some incompetence of Google search

Algorithm: Web Page Title Analysis
Input: web page title, searching query
Output: efficient/inefficient
1. existence of search query [sq]
2.   if sq exist in title, then check
3. existence of assign operator [; : ? !] within sq
4.   if exist at least one, return false;
5. else, return true;
6. check the sequence of sq
7.   if maintain order, return true;
8. else, return false;

Fig. 4 Title analysis algorithm

3.2.2 Quality Measure (Back Links)

In order to evaluate the quality of web documents page ranking [15] is an effective way. A good number of web page researches have been performed on page ranking by applying various link analysis algorithms because link is the vital resource to represent the importance of web documents. Link analysis is actually the analysis of hyperlinks [16], which helps to develop the web search. Several existing web search engine also used various link analysis algorithm such as simple iterative algorithm using link graph [6], weighted page rank algorithm based on popular web pages [17], HITS algorithm focusing on the useful query and relevant pages link for predicting the web page quality. Among them, Google also evaluates the web pages depend on page ranking considering various types of links such as Text link, Image link, Mixed link, Duplicate links, etc. Though a large number of researches have been conducted on various link analysis algorithm, most of them are focusing on link structure and semantic text pattern.

Here, we proposed a page ranking algorithm which perform through analyzing the ratio of the backlinks (external and internal) of a single document through normalizing technique because the backlinks (referred as reference) of a single web document helps to appropriate approximation of document quality. By extracting all back links including internal link [snapshoot Fig. 5 (upper)] and external link [snapshoot Fig. 5 (lower)] from HTML pages, we normalize the ratio of internal and external links in order to find the weight of the web documents.

Page Ranking Algorithm:

$$R(A1) = (1 - d) + d[N(IL)/N(EL)]$$
$$R(A2) = (1 - d) + d[N(IL)/N(EL)]$$

$$R(A3) = (1 - d) + d[N(IL)/N(EL)] \ldots \tag{1}$$

$$PR(A) = R(A1) + R(A2) + \cdots + R(An) \tag{2}$$

```
<ul class="mgn-links">
    <li>
        <a href="https://moz.com/about/contact">Contact</a>
    </li>
    <li class="mgn-hide-small">
        <a href="https://moz.com/products">Products</a>
    </li>
    <li class="mgn-hide-small">
        <a href="https://moz.com/products/api">API</a>
    </li>
    <li>
        <a href="https://moz.com/terms-privacy">Terms &amp; Privacy</a>
    </li>
    <li>
```
```
</li>
<li class="mgn-line-block mgn-hide-large-down"></li>
<li class="mgn-pro mgn-products mgn-blue mgn-selectable ">
    <a href="https://moz.com/products/pro" title="Moz Pro">Moz Pro</a>
</li>
<li class="mgn-local mgn-products mgn-teal mgn-selectable ">
    <a href="https://moz.com/local/overview" title="Moz Local">Moz Local</a>
</li>
<li class="mgn-tools mgn-selectable mgn-products">
    <a href="https://moz.com/tools" title="Free Tools">Free SEO Tools</a>
</li>
```

Fig. 5 Example of back links on HTML page format

Here, numbers of web pages are A, B, C to N, those have several web documents $A = A1, A2$, to An; $B = B1, B2$, to Bn, to $N = N1, N2$, to Nn. Compute the ranking point of each web document through Eq. 1 where "d" referred as constant, called damping factor which should be between 0 and 1, because low damping factor estimate the more external link than internal links and higher damping factor estimate the reverse. Here, set the damping factor $d = 0.75$.

Equation 2 performs summation of all ranking point and estimate the page rank. After performing the observation it noticed that, those documents have less ranking point they have more effectiveness than higher, here consider effective documents within page rank between 0.25 and 3.5; which estimate that below 0.25 and up 3.5 represent the poor quality of web documents. So, page ranking has the potentiality to predict about web documents in terms of quality and relevancy.

3.3 Formulation of Index Term

After completing the evaluation process through web content feature, it seems that, large number of web documents have been existed with inefficient title and low page rank. In spite of this, these documents are surfaces on search result which called junk result and it happens for some inappropriate index term those are totally irrelevant with the content meaning. So this section is concluded a keyword-based approach which formulate the appropriate and relevant index term which might reduce junk result during search.

3.3.1 Content Collection and Preprocessing

Though title and page rank are major feature for web content to appear on search, majority of web documents appears on search for index term. Though formulation of proper index term is quite difficult, it is essential. So, here we proposed an automated system to generate appropriate index term through web document text. Text of the document is the most valuable resource which impact on web page usefulness and relevancy. After extracting web document text, filtering is performed, as document text consists of variegated information. All punctuation's are removed, eliminate all dates and numbers from the text, and convert the content into their raw format.

3.3.2 Formulating Words

In order to retrieve the most predominant words form the document, all the text of the document are converted into their singular form. MySQL stop word removal applied for removing all the stop words from the text. After removing all stop word, make the first dataset of words. To make second dataset, consider the term frequency (TF) of the word within text. Term frequency is a determination which represents the words which carrying significant meaning of the document. Here, considered the word those TF has $>= 5$. Those words occurred at least five times, and they select as the words for second dataset.

3.3.3 Similarity Measure

After generating two datasets, find the semantic similarity between two datasets words. In this section, WordNet semantic similarity (WS4j) for java is applied for finding the similarity between two sets of words. Semantic similarity is a technique, which perform a set of documents in order to find the distance between two sets of words.

3.3.4 Identify Index Term

After finding similarity between two sets of words, it assumes that, there are lots of words which semantic score is null and some of them achieve high semantic score. Our considered semantic score is Sc $>= 0.5$, from the datasets consider the words as index terms those have selected semantic score because containing < 0.5 scorer word has less relatedness with the meaning and those have $>= 0.5$, they have higher relatedness with the meaning. After formulating the words, rank all the words

Fig. 6 Summery of index
term identifying algorithm

Algorithm: Formulation of Index Term
Input: web document text
Output: index term

1. make the first set of words w1
        by converting singular form and
                removing stop words
2. make the second set of words w2
        using Term frequency [TF] of words
                consider the word, those TF>=5
3. measure similarity between two sets of words
        by applying WordNet semantic similarity
4. identify index term
        if semantic similarity>=0.5
                select the corresponding word of w1 as index term
5. else return null
6. make the set of index term

according to their searching volume (in Sect. 4) which ranked by the Google. The process of formulating index term concludes through algorithm which is given below (Fig. 6).

In line 1 made the first dataset by converting text into their singular form. Second dataset made through words term frequency; selected score is >= 5, in line 2. Apply WordNet semantic similarity for java in line 3. Those words from second dataset have semantic score >= 0.5, select as index term in line 4. Finally make a set of index term with selected words in line 5.

4   Experimental Result

This section explains the experiment result of proposed approach. Experiment performed in order to improve the performance of the search as search quality is the major concern of search engine. Here we demonstrated our approaches based on web content and link-based feature instead of benchmark approaches where most of the existing approaches works on lexicon and web structure-based approach [18]. To evaluate the proposed approach, experiment performs on 700 web documents those are selected randomly from the web. First evaluate web pages considering content-based feature (title, text) and link-based feature (backlinks) to understand the efficiency, quality, and relevancy of web documents. Second present the overall statistical result of the current web documents. At last, evaluate the keyword-based approach considering the generated keywords and previous keywords according to their volume and volunteer judgments in order to understand the effectiveness of two systems keywords and also for selecting the most appropriate index term.

After measuring web documents efficiency (Sect. 3) and quality (Sect. 3), to evaluate the document relevancy, we perform manual inspection on document text with the user search query. Table 1 shows that, according to title analysis, 75% web

Table 1 Accuracy of web documents regarding efficiency, quality, and relevancy

No. of web pages	Useful web pages considering title analysis	Useful web pages considering page ranking	Relevant web documents
700	75%	60.5%	65.7%

No of web documents	Average Indexing			Bad Indexing			Total no (%)
	Title	Link	Document text	Title	Link	Document text	15%
700	√	√	×	×	×	×	
	√	×	√	Good Indexing			Total no (%)
	×	√	√	Title	Link	Documenttext	
Total no (%)	55%			√	√	√	30%

Fig. 7 Statistical result of current web documents

document has efficient title which makes the search more efficient, regarding to page ranking, 60.5% web document has useful ranking which depicts about their quality and 65.7% web documents (according to manual inspection) has better relevancy to fetch relevant content on search result. From the analysis, it assumes that, web content and link feature are effective to predict about web documents effectiveness, quality, and relevancy which impact on search result. In addition, a large number of web documents with inappropriate features are dominated on the web those are totally irrelevant to the content. Moreover, our proposed approach has capability to analyze and classify the web documents according to their quality and efficiency and also for understanding the current web documents structure. From the observation, it speculated that, by considering only distinct feature (content/link), it reveals the preliminary estimation about web structure but it cannot reflect the current standard statistics of the web. In contrast, considering three feature, the result reflects the insights of modern web framework (indexing) more effectively. Figure 7 shows the exact statistics of current web page indexing. Among 700 web documents, percentage of average indexing, bad indexing, and good indexing are 55, 15, and 30%, respectively, which seems that huge number of web documents have been existed with bad indexing which impact on search result and responsible for junk result also.

By considering these three features (title/page ranking/content text), result represents the standard scenario of current web documents indexing as these three features have intellectual ability to represent the whole web document more effectively.

For justifying the keywords effectiveness, extract the previous keywords (those are cited on web document as keyword) and formulating keywords [through semantic similarity (Sect. 3)] of web documents and perform the evaluation process into two steps. First evaluation performs through manual inspection on both keywords; generated keywords and previous keywords for measuring the effectiveness

Table 2 Accuracy of relevant keywords for finding relevant index term

Total no. web page	700
Most searching words according to volume	
Proposed keywords	Previous keywords
82%	72%
Volunteer result based on keywords relevancy with content	
Proposed keywords	Previous keywords
88%	77%

of two systems keywords. Here we rank all the keywords (new and previous) according to their search volume. Several keywords research tool have been existed those rank the keywords according to their search volume which are free for all. Among them, Google Trend is one of the biggest repositories which classify the keywords according to their volume, year base, and searching region. Further evaluation performed by hiring volunteer to analyze the keywords relevancy with the document text for both system keywords. Table 2 shows that, for first evaluation considering keywords searching volume (through Google Trend), proposed keywords or generated keyword gained 82% accuracy, in contrast, previous keywords gained 72% accuracy which depict that generated keywords have better efficiency than previous to fetching relevant document. Regarding to keywords relevancy, volunteer suggests that 88% proposed keywords has better effectiveness to reflect the text meaning where previous keywords has only 77% accuracy.

5 Conclusion

Extracting relevant information from innumerable web documents is one of the challenging approaches. For extracting appropriate information from these innumerable web documents, necessity of a powerful mechanism is indispensable. Google search engine conducts this function fluently. But sometimes some error prone result appears because of inappropriate indexing process and features selection which might not satisfactory. This paper demonstrates an automated process for finding relevant web documents through generating appropriate, effective, and efficient index term using NLP concept. This approach evaluates the efficiency, quality, and relevancy of web documents considering two features (content and link). Here we perform the analysis process on 700 web documents. This large-scale study revealed the importance of these features to understand the current web structure as well as to generate appropriate index term. The major goal of this proposed approach is to reduce the searching problem by formulating appropriate index term using NLP concept and semantic distance. The experimental results show that content- and link-based features are more effective than others. Moreover, generated index term has good accuracy than previous index term which indicate that new generated index term has potentiality to improve the search quality than before index term.

References

1. Brin, S., Page, L.: The anatomy of a large scale hypertextual web search engine. Comput. Netw. ISDN Syst. **30**(1–7), 107–117 (1998)
2. Madhavi, A., Chari, K.H.: Architecture based study of search engines and meta search for information retrieval. Int. J. Eng. Res. Technol. (IJERT) (2013)
3. Speretta, M., Gauch, S.: Personalized search based on user search histories. In: Proceedings of the 2005 IEEE/WIC/ACM International Conference on Web Intelligence, pp. 622–628. IEEE Computer Society (2005)
4. Lieberman, M. D., Samet, H., Sankaranarayanan, J., Sperling, J.: STEWARD: architecture of a spatio-textual search engine. In: Proceedings of the 15th Annual ACM International Symposium on Advances in Geographic Information Systems, p. 25. ACM (2007)
5. Barroso, L.A., Dean, J., Holzle, U.: Web search for a planet: the Google cluster architecture. IEEE Micro. **2**, 22–28 (2003)
6. Horowitz, D., Kamvar, S.D.: The anatomy of a large scale social search engine. In: Proceedings of the 19th International Conference on World Wide Web, pp. 431–440. ACM (2010)
7. Ilyas, Q.M., Kai, Y.Z., Talib, M.A.: A conceptual architecture for semantic search engine. In: 9th IEEE International Multi Topic Conference Pakistan (2004)
8. Jones, C.B., Abdelmoty, A.I., Finch, D., Fu, G., Vaid, S.: The spirit spatial search engine: architecture, ontologies and spatial indexing. In: International Conference on Geographic Information Science. Springer, Berlin, Heidelberg, pp. 125–139 (2004)
9. AlShourbaji, I., Al-Janabi, S., Patel, A.: Document selection in a distributed search engine architecture. arXiv preprint (2016). arXiv:1603.09434
10. Verma, D., Kochar, B.: Multi agent architecture for search engine. Int. J. Adv. Comput. Sci. Appl. **7**(3), 224–229 (2016)
11. Bhuiyan, H., Ara, J., Bardhan, R., Islam, M.R.: Retrieving youtube video by sentiment analysis on user comment. In: 2017 IEEE International Conference on Signal and Image Processing Applications (ICSIPA), pp. 474–478. IEEE (2017)
12. Brophy, J., Bawden, D.: Is google enough? Comparison of an internet search engine with academic library resources. In: Aslib Proceedings, vol. 57, no. 6, pp. 498–512. Emerald Group Publishing Limited (2005)
13. Joachims, T.: Optimizing search engines using click through data. In: Proceedings of the Eighth ACM SIGKDD International Conference on Knowledge Discovery and Data Mining, pp. 133–142. ACM (2002)
14. Trotman, A., Degenhardt, J., Kallumadi, S.: The architecture of eBay search. In: Proceedings of the SIGIR 2017 Workshop on eCommerce (ECOM 17) (2017)
15. Tyagi, N., Sharma, S.: Weighted page rank algorithm based on number of visits of links of web page. Int. J. Soft Comput. Eng. (IJSCE) (2012). ISSN 2231-2307
16. Kleinberg, J.M.: Authoritative sources in a hyperlinked environment. J. ACM (JACM) **46**(5), 604–632 (1999)
17. Chau, M., Chen, H.: A machine learning approach to web page filtering using content and structure analysis. Decis. Support Syst. **44**(2), 482–494 (2008)
18. Bhuiyan, H., Oh, K. J., Hong, M.D., Jo, G.S.: An unsupervised approach for identifying the infobox template of wikipedia article. In: 2015 IEEE 18th International Conference on Computational Science and Engineering (CSE), pp. 334–338. IEEE (2015)

An Effective and Cost-Based Framework for a Qualitative Hybrid Data Deduplication

Charles R. Haruna, MengShu Hou, Moses J. Eghan, Michael Y. Kpiebaareh and Lawrence Tandoh

Abstract In real world, entities may occur several times in a database. These duplicates may have varying keys and/or include errors that make deduplication a difficult task. Deduplication cannot be solved accurately using either machine-based or crowdsourcing techniques only. Crowdsourcing were used to resolve the shortcomings of machine-based approaches. Compared to machines, the crowd provided relatively accurate results, but with a slow execution time and very expensive too. A hybrid technique for data deduplication using a Euclidean distance and a chromatic correlation clustering algorithm was presented. The technique aimed at: reducing the crowdsourcing cost, reducing the time the crowd use in deduplication and finally providing higher accuracy in data deduplication. In the experiments, the proposed algorithm was compared with some existing techniques and outperformed some, offering an utmost deduplication accuracy efficiency and also incurring low crowdsourcing cost.

Keywords Qualitative hybrid data deduplication · Edge-Pivot clustering · Entity resolution · Crowdsourcing

C. R. Haruna (✉) · M. Hou · M. Y. Kpiebaareh · L. Tandoh
University of Electronic Science and Technology of China, Chengdu, China
e-mail: charuna@ucc.edu.gh

M. Hou
e-mail: mshou@uestc.edu.cn

M. Y. Kpiebaareh
e-mail: michael.kpiebaareh@sipingsoft.com

L. Tandoh
e-mail: 3065550735@qq.com

C. R. Haruna · M. J. Eghan
University of Cape Coast, Cape Coast, Ghana
e-mail: meghan@ucc.edu.gh

© Springer Nature Singapore Pte Ltd. 2019
S. K. Bhatia et al. (eds.), *Advances in Computer Communication and Computational Sciences*, Advances in Intelligent Systems and Computing 924,
https://doi.org/10.1007/978-981-13-6861-5_44

1 Introduction

Identical records are clustered into numerous groups and this tends to be a problem in data deduplication due to the non-existence of common identifiers, thus, accurate data deduplication is a challenge. For a record pair, a machine-based method uses metrics to calculate similarities between the pair. Examples are Cosine, Jaccard similarity metrics, and Euclidean distance. Record pairs with similarity values greater or equal to a given threshold are considered to be same. Machine-based methods often face challenges when records with great similarities but having different entities are being processed. Humans, when solving complicated tasks are better than machines, and thus are used in improving the data deduplication accuracy drawbacks which the machine-based methods offer [1]. In data deduplication, crowdsourcing are used on a dataset to identify possible identical pairs of records. In the addition, the crowd examines each pair, whether they are duplicate records. However, due to large dataset being examined by humans, this approach incurs huge overheads. With available crowdsourcing platforms, Human–machine-based (hybrid) techniques [2–4] have been proposed by researchers.

2 Related Work

Works that fuse machine-based and crowdsourcing approaches to create hybrid deduplication algorithms have been proposed. There is a survey [5] on deduplication in entity conciliation. Recently, proposed works under several crowdsourcing platforms in academia dn industry have been captured in a survey [6]. Reference [7] presented CrowdDB by extending Sequential Query Language to CrowdSQL which enables crowd-based operators. Deco [8] is a database system which answers queries about relational data by employing a crowdsourcing platform. Crowd-clustering [9] presented a technique that was used in identifying records belonging to the same classification. A data deduplication method presented in [2] provided a higher deduplication accuracy but with a huge crowdsourcing cost. In [3, 10, 11], the authors developed techniques TransM, TransNode, and GCER, respectively, that significantly reduced crowdsourcing overheads but had poor deduplication accuracy. In ACD [4], a hybrid deduplication method offering better accuracy and having reasonable crowd overheads was proposed. A progressive entity relational approach, ProgressER [12], was presented to produce results with the highest quality and accuracy. Reference [1] used human power to generate for datasets, classifiers. In [2], human power was used as well but the algorithm, when used on datasets with different entities, always had problems executing.

3   Problem Definition

Let the set of records and a function, respectively, be $R = (r_1, r_2, \ldots, r_n)$ and g, such that they map r_i to the records they represent. In most cases, it is often difficult to gain g; therefore, a similarity value function $f : R \times R \longrightarrow [0, 1]$ is assumed to exists, so that $f(r_i, r_j)$ is the probability that r_i and r_j are the same entity. The objective of deduplication is to divide a set of records R, into clusters $C = \{C1, C2, \ldots, Ck\}$ and pass the clusters to the crowd to query. Therefore, for any selected edge $(r_i, r_j) \epsilon R$, if $g(r_i) = g(r_j)$, then r_i and r_j are in one cluster and are assumed to be the records as well. Otherwise, the records are dissimilar and are thus not put in the same cluster. A machine data deduplication method metric cost $\Lambda(R)$ [4, 13] is adopted. The metric cost $\Lambda(R)$ states;

$$\Lambda(R) = \sum_{r_i, r_j \epsilon R. i < j} x_{i,j} \cdot \left(1 - f\left(r_i, r_j\right)\right) + \sum_{r_i, r_j \epsilon R. i < j} \left(1 - x_{i,j}\right) \cdot \left(1 - f\left(r_i, r_j\right)\right)$$

(1)

If a record pair r_i and r_j are considered to a cluster, then $r_{i,j} = 1$, if not $r_{i,j} = 0$. $\Lambda(R)$ imposes a penalty of $1 - f(r_i, r_j)$ on a pair in the same cluster.

For example, in Table 1, we have a set of records $R = \{m, n, o, p, q, r, s, t\}$, with similarity values. The pairs with less than similarity scores threshold 0.4 are unused. Triangles $\langle m, n, r \rangle$ and $\langle p, q, t \rangle$ can be validated that the metric cost $\Lambda(R)$ is minimized. References [4, 14] proved that it is NP-hard to minimize the metric cost $\Lambda(R)$.

A similarity-based function [15] to calculate the similarity values and reduce the number of pairs the crowd has to evaluate was employed. In this approach, a similarity function and a threshold t are thus required. After all similarity values of all the pairs are computed, record pairs with similarity values greater than or equal to the threshold t are assumed to be of the same entity. In this work, the Euclidean distance, D_E, is used, defined [16] as

Euclidean distance similarity function is defined as

$$D_E\left(\overrightarrow{t_a}, \overrightarrow{t_b}\right) = \left(\sum_{t=1}^{m} \left|w_{t,a} - w_{t,b}\right|^2\right)^{1/2}$$

(2)

Table 1 Similarity values of some pairs of record

Record pairs	Similarity value	Record pairs	Similarity value
(m, n)	0.80	(p, q)	0.55
(n, r)	0.75	(q, t)	0.55
(o, t)	0.65	(a, g)	0.45
(o, s)	0.60	$\ldots$	< 0.4

4 Contributions of the Work

The proposed deduplication technique consists of:

1. Firstly, to obtain the candidate set S made up of the pairs $(r_1, r_2) \geq$ threshold, Euclidean distance is used as the machine-based approach, on the set R.
2. Furthermore, a proposed algorithm clusters the set S into disjoint sets $C = \{C1, C2, \ldots, Ck\}$. The disjoint sets are issued to the crowd to examine for similarities and issue their results.
3. Finally, in the experiments on dedeuplication accuracy, efficiency, and crowd overheads, we compared our model to some work.

4.1 Lone Cluster Algorithm

Reference [4] defined crowd confidence $f_c(r_1, r_2)$ as the proportion of humans who examined a pair of record (r_1, r_2) and concluded they are the same. $R \times R \rightarrow [0, 1]$ is set as the crowd-based similarity score, such that $f_c(r_1, r_2)$ implies that $r_1 = r_2$. In the first step, that is using the machines, if a pair of record was eliminated, then $f_c(r_1, r_2) = 0$ is set. In clustering using chromatic correlation clustering, triangular clusters are generated in a categorical manner [14]. Reference [17] proposed and described a score $f_t(r_1, r_2, r_3)$ where $((r_1, r_2), (r_1, r_3), r_2, r_3)) \epsilon E$ and forms a triangular cluster. $f_t(r_1, r_2, r_3)$ is the fraction of $f_c(r_1, r_2)$, $f_c(r_1, r_3)$ and $f_c(r_2, r_3)$ over the number of edges in the triangular cluster (3). Therefore, a triangular cluster is formed when $f_t(r_1, r_2, r_3)$ is greater than the threshold set.

a. If the pivot selected in Fig. 1a was edge (A,B) and choosing objects $x \epsilon R'$ so that triangle $\langle A, B, C \rangle$ forms a cluster, Fig. 1b. Though $f_c(A, C) = 0.4$ is less than the threshold 0.5, calculating the triangular similarity score $f_t(A, B, C)$ gives 0.53 which is greater than the triangular similarity score threshold 0.5. Though (A,C) has a crowd's confidence of 0.4 which is less than the threshold, a triangular cluster is formed due to their triangular similarity score being greater than the threshold. It is inferred that the records maybe the same record.

b. Whereas in choosing an edge (D, F) and all other objects $x \epsilon R'$ from Fig. 1a, to form a triangle $\langle D, E, F \rangle$, it is impossible to generate a cluster like in Fig. 1c. This is because the triangular similarity score $f_t(D, E, F)$ gives 0.46, lesser than the triangular similarity score threshold 0.5. Thus, the three records are not of the same entity.

$$\Lambda'(R) = \sum_{r_i, r_j \epsilon R. i < j} x_{i,j} \cdot (1 - f_c(r_i, r_j)) + \sum_{r_i, r_j \epsilon R. i < j} (1 - x_{i,j}) \cdot (1 - f_c(r_i, r_j)) \quad (3)$$

From 3, if a record pair r_i and r_j are grouped as a cluster, then $x_{i,j} = 1$, $x_{i,j} = 0$, if otherwise. In [14], minimizing the metric cost $\Lambda'(R)$ in Eq. 3 is NP-hard problem. If clusters in Fig. 1b, c were generated from Fig. 1a, there will be a cost of 6, amounting

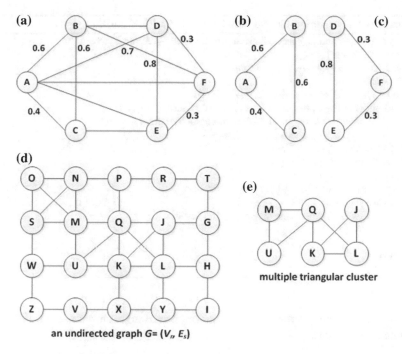

Fig. 1 **a–c** and **d–e** Represent an undirected graph with crowd's confidence values an example of a multiple clustering, respectively

to the total number of edges deleted during the generation process. We use a pair of objects (r_i and r_j) instead of a single object as a pivot under human–machine-based technique. the categorical way of clustering in [14] is non-trivial because the construction of clusters are done individually and the i−th($i > 1$) cluster is dependent on the first $i − 1$ clusters. It is therefore time-consuming and expensive constructing the clusters individually. To prevent these drawbacks, construction of multiple clusters will be presented.

A MCEP pseudo code is the clustering generation phase algorithm adopted from [14]. During each iteration, an edge rather than a single object is chosen as a pivot. The inputs are $G = (V_r$ and $E_s)$ and S where R is a record of a vertex V_r and an edge $(r_i, r_j) \epsilon E_s$ in S. The set of individual clusters $C = \{C1, C2, \ldots, Ck\}$ is the output. R' which is at first equal to R is initialized. R' stores the vertices yet to be put in clusters. For each iteration, an edge (r_i, r_j) from S is randomly chosen as pivot. With edge (r_i, r_j) as pivot, all other edges $r_k \epsilon R'$ for which a triangle $\langle r_i, r_j, r_k \rangle$ is formed, irrespective of the records being the same or not, that is $(r_i, r_j), (r_i, r_k)$ and (r_j, r_k) are selected and put in *Ctemp* and later posted to the crowd to infer the similarities. The crowd's confidence $f_c(r_i, r_j)$ for all the edges are calculated in lines 9–11. Further, the triangular similarity score $f_t(r_i, r_j, r_k)$ is also calculated. If the result $f_t(r_i, r_j, r_k) \geq 0.5$, the three records are grouped as a cluster and these records and edges are eliminated for G. Otherwise, a new edge is selected as pivot.

Algorithm 1 A Machine-Crowd-Edge-Pivot(MCEP) Algorithm

Require: $G = (V_r$ and $E_s)$, an undirected graph, where $V_r \epsilon R$ and $E_s \epsilon S$.
Ensure: C cluster set.
1: **Initialize** two empty sets, C and *Ctemp*.
2: $R' \leftarrow R; i \leftarrow 1$.
3: **While** $(r_i, r_j) \epsilon R'$ exists such that $(r_i, r_j) \epsilon E$, do.
4: **Randomly** choose a pivot edge $(r_i, r_j) \epsilon R'$, such that $(r_i, r_j) \epsilon E$.
5: **Select** all other records $r_k \epsilon R'$ for which there is a triangle $\langle r_i, r_j, r_k \rangle$.
6: **Insert** (r_i, r_j), (r_i, r_k) and (r_j, r_k) to *Ctemp*
7: **Post** to the crowd, the set *Ctemp*
8: **For** each pivot edge in *Ctemp* do.
9: **Calculate** $f_c(r_i, r_j)$.
10: **If** the crowd decides that $f_t(r_i, r_j, r_k) \geq 0.5$ then
11: **Put** *Ctemp* into C.
12: **Delete** from G all other remaining edges and vertices.
13: $C(r_k) \leftarrow i$.
14: $i \leftarrow i + 1$.
15: **return** C.
end

Finally, the remaining vertices in R' are grouped as a single cluster and the algorithm terminates by returning C.

Computational Complexity The computational complexities of MCEP are selecting the edge pivot and the clusters generations. A time of $O(m \log n)$ is needed when choosing a pivot randomly, where $n = |V|$ and $m = |E|$. In MCEP, to identify the pivot, we use random priorities with queues to select an edge. Edges are eliminated from the queue whether chosen as pivot or not. Furthermore, MCEP iterating accesses all of the pivot's (r_i, r_j) neighbors. The neighbors are, thus, not used again in subsequent iterations because after a current cluster has been formed, they are cut off. In each iteration to form a cluster, MCEP visits each edge once using a time of $O(m)$. The complexity of MCEP is $O(m \log n)$.

4.2 Multiple Clustering

In each iteration in the MCEP algorithm, a lone triangular cluster is given to the crowd to cross-question and publish their answers. The number of iterations of MCEP uniquely determines the execution time. To decrease the time significantly, during each iteration, the number of iterations must be reduced by concurrently generating multiple triangular clusters with only one pivot.

Using MCEP, all edges in Fig. 1d have the same chance of being a pivot. Once an edge is selected, there is a possibility that others may be eliminated in the process, thus not being examined by the crowd. If MCEP chooses (M, Q) as the pivot, $\langle M, Q, U \rangle$ forms a triangle and posted to the crowdsourcing platform. Depending on the answers from the humans, a lone triangular cluster may be formed. So if $\langle M, Q, U \rangle$ is removed form G, edges connected to the three records may be broken form the graph. This is a problem in using MCEP to build lone clusters. Next, we present the algorithm for the generation of multiple clusters by just selecting only one pivot in each iteration. $d(r_i)$ is the total edges incident to vertex $r_i \epsilon R$. Let $\Delta(r_i) = \max d(r_i)$ and $\lambda(r_i)$ = arg max $d(r_i)$. When the algorithm is executing, in each iteration, the vertex r_i with the topmost number of edges and a probability which is directly proportional $\Delta(r_i)$ is chosen as first vertex. Then, a neighboring vertex of r_i with the next highest number of edges with a probability proportional to $d(r_j, \lambda(r_i))$ is selected as the second vertex r_j. The two vertices (r_i, r_j) form an edge are used as the pivot. With (r_i, r_j), the algorithm generates clusters by attaching all other vertices r_k, in the process forming a triangular cluster $\langle r_i, r_j, r_k \rangle$. It continues to attach r_k vertices to the cluster formed so far as the new vertices form triangular clusters $\langle A, B, C \rangle$ as well. A is r_i or r_j and C, any vertex in the current triangular cluster created. From Fig. 1d, Q and M have five edges each making each vertex a potential first vertex to be chosen. Our first vertex $r_i = Q$ is chosen and from the neighboring nodes, a vertex r_j with probability proportional to $d(r_j, \lambda(r_i))$ is selected as the second. Here, K is chosen as the second vertex, thus our pivot in the iteration is (K,Q). A provisional triangular cluster $\langle K, L, Q \rangle$ is generated. U is chosen and forms a triangular cluster $\langle M, Q, U \rangle$, since Q belongs to the first provisional cluster. Next, because of K, J is selected and forms $\langle J, K, L \rangle$. Figure 1e shows the cluster formed from the first iteration of the algorithm. This provisional cluster is sent to the crowd and comparisons are done for duplicate records. The crowd computes crowds confidence $f_c(r_1, r_2)$ and the triangular similarity scores $f_t(p, q, r)$. A triangle with a less $f_t(p, q, r)$ than the threshold is removed and attached to $G = (V_r, E_s)$. The algorithm terminates when there are no more triangular clusters to be formed and added to multiple clusters. The unexploited vertices and edges form a lone cluster, then is passed to the crowdsourcing platform for comparisons to be made for matching records.

Computational Complexity The time to execute and the generation of the multiple clusters are the complexity of the algorithm. Just like in MCEP, when choosing the pivots, priority queues are implemented having priorities of $\Delta x rnd$, with a random number, rnd. To compute Δ for the vertices, it needs a time of $O(n+m)$. The priority queue also needs a time of $O(n \log n)$, because a vertex, at an iteration, is added to the queue or removed from once. For a first vertex u already selected, all of its neighbors are visited once to select a second vertex v, and this takes $O(m)$ time. It takes a time of $O(m)$ and $O(1)$ to generate the multiple clusters and to retrieve each edge, respectively. Therefore, our multiple cluster generation algorithm has a complexity of $O(n(\log n) + m)$.

5   Experiments

Three benchmark datasets: Paper [18], Product [19], and Restaurant [20] were employed. To calculate for the machine-based similarity values, we used the Euclidean distance for each pairs of records. We set the threshold of machine-based similarity value at $t = 0.5$. Motivated by ACD [4], we contacted the authors for their answers obtained from the crowd to reuse in our experiments.

5.1 Analysis with Some Existing Methods

Our algorithm was compared with some of the existing methods based on:

Analysis of Accuracy of Deduplication:

In Fig. 2, under both 3-worker($3w$) and 5-worker($5w$) settings on the three datasets, ACD had the highest, our algorithm, the second highest, thus performed better than TransNode and TransM. On all three datasets, between TransNode and TransM, the former had the least accuracy followed by the latter. From our test on paper, our algorithm had equal accuracy with regard to ACD, with only marginal differences. Finally, on Product and Restaurant, all methods being compared, to some degree, performed equally.

Analysis of Overheads of Crowdsourcing:

As shown (Fig. 2), though our algorithm had almost the same accuracy as ACD, it incurred a little cost in the crowdsourcing overheads than ACD as shown in Fig. 3. Under $3w$ and $5w$ settings, using paper and Product datasets, TransNode and TransM incurred the least cost compared to our algorithm and ACD. On the other hand, on Restaurant dataset and under both $3w$ and $5w$ settings, they incurred the highest cost with regard to our algorithm and ACD.

Analysis of Efficiency of Crowdsourcing:

Comparing all the methods under both settings and on all datasets as Fig. 4 represents, the efficiency of our algorithm is better than TransM. ACD 's efficiency outperformed our method.

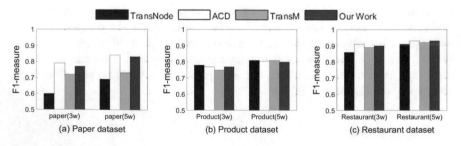

Fig. 2 Deduplication accuracy analysis

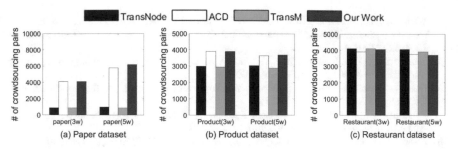

Fig. 3 Crowdsourcing overheads analysis

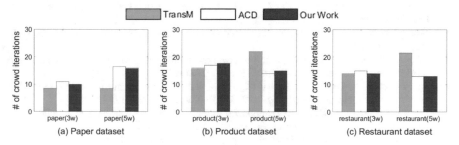

Fig. 4 Crowdsourcing efficiency analysis

6 Conclusion

In this work, a machine-based function, the Euclidean distance and a clustering algorithm, were used to propose a qualitative human–machine (hybrid) data deduplication technique. In the experiments, we compared our algorithm to a few existing ones, with our algorithm having a high accuracy of deduplication incurring very low crowdsourcing overheads. The deduplication efficiency of our algorithm, comparatively, is almost the same as the other techniques. Under qualitative human–machine deduplication, our algorithm can be considered a cost-effective technique to use. In future, we will focus on improving the clustering by refining the clusters for the crowd to examine the deleted edges while executing the multiple clustering algorithm.

References

1. Arasu, A., Gotz, M., Kaushik, R.: On active learning of record matching packages. In: SIGMOD Conference, pp. 783–794 (2010)
2. Wang, J., Kraska, T., Franklin, M.J., Feng, J.: Crowder: crowdsourcing entity resolution. Proc. VLDB Endow. **5**(11), 1483–1494 (2012)
3. Wang, J., Li, G., Kraska, T., Franklin, M.J., Feng, J.: Leveraging transitive relations for crowdsourced joins. In: Proceedings of the 2013 International Conference on Management of Data, ACM, pp. 229–240 (2013)

4. Wang, S., Xiao, X., Lee, C.H.: Crowd-based deduplication: an adaptive approach. In: Proceedings of the 2015 ACM SIGMOD International Conference on Management of Data, ACM, pp. 1263–1277 (2015)
5. Elmagarmid, A.K., Ipeirotis, P.G., Verykios, V.S.: Duplicate record detection: a survey. IEEE Trans. Knowl. Data Eng. **19**(1), 1–16 (2007)
6. Doan, A., Ramakrishnan, R., Halevy, A.Y.: Crowdsourcing systems on the world-wide web. Commun. ACM **54**(4), 86–96 (2011)
7. Franklin, M.J., Kossmann, D., Kraska, T., Ramesh, S., Xin, R.: Crowddb: answering queries with crowdsourcing. In: SIGMOD, pp. 61–72 (2011)
8. Parameswaran, A., Park, H., Garcia-Molina, H., Polyzotis, N., Widom, J.: Deco: declarative crowdsourcing. Technical report, Stanford University. http://ilpubs.stanford.edu:8090/1015/
9. Gomes, R., Welinder, P., Krause, A., Perona, P.: Crowdclustering. In: NIPS, pp. 558–566 (2011)
10. Whang, S.E., Lofgren, P., Garcia-Molina, H.: Question selection for crowd entity resolution. Proc. VLDB Endow. **6**(6), 349–360 (2013)
11. Vesdapunt, N., Bellare, K., Dalvi, N.: Crowdsourcing algorithms for entity resolution. Proc. VLDB Endow. **7**(12) (2014)
12. Altowim, Y., Kalashnikov, D.V., Mehrotra, S.: ProgressER: adaptive progressive approach to relational entity resolution. ACM Trans. Knowl. Discov. Data (TKDD) **12**(3), 33 (2018)
13. Hassanzadeh, O., Chiang, F., Lee, H.C., Miller, R.J.: Framework for evaluating clustering algorithms in duplicate detection. Proc. VLDB Endow. **2**(1), 1282–1293 (2009)
14. Bonchi, F., Gionis, A., Gullo, F., Tsourakakis, C.E., Ukkonen, A.: Chromatic correlation clustering. ACM Trans. Knowl. Discov. Data (TKDD) **9**(4), 34 (2015)
15. Bayardo, R.J., Ma, Y., Srikant, R.: Scaling up all pairs similarity search. In: WWW, pp. 131–140 (2007)
16. Huang, A.: Similarity measures for text document clustering. In: Proceedings of the Sixth New Zealand Computer Science Research Student Conference (NZCSRSC 2008), Christchurch, New Zealand, pp. 49–56 (2008)
17. Haruna, C.R., Hou, M., Eghan, M.J., Kpiebaareh, M.Y., Tandoh, L.: A hybrid data deduplication approach in entity resolution using chromatic correlation clustering. In: International Conference on Frontiers in Cyber Security, pp. 153–167. Springer, Singapore (2018)
18. http://www.cs.umass.edu/~mccallum/data/cora-refs.tar.gz
19. http://www.cs.utexas.edu/users/ml/riddle/data/restaurant.tar.gz
20. http://dbs.uni-leipzig.de/Abt-Buy.zip

Ontology-Based Natural Language Processing for Thai Herbs and Thai Traditional Medicine Recommendation System Supporting Health Care and Treatments (THMRS)

Akkasit Sittisaman and Naruepon Panawong

Abstract Herbs become increasingly popular in health care and treatments these days, because they are safe under careful usages. Although Thai herbs information can be retrieved from many websites, it takes long searching times for discovering the information that matches with users' requirements. Therefore, this research aimed to develop a Thai herbs and Thai traditional medicines recommendation system for health care and treatments using ontology-based natural language processing (THMRS). In addition, users can use the proposed recommendation system via web or Windows application. The proposed THMRS is composed of ontology-based databases called Thaiherb, searching process using natural language combined with word tokenization and spell checking by ISG (Index of Similarity Group) Algorithm, translation from natural language to SPARQL commands is designed for semantic search, and the decision making whether information found will be delivered to users is done by co-occurrence density analysis. The search result was displayed in a single website or a window which is fast and user-friendly. The experiments show that the average *F*-Measure is 95.07%, the average precision is 93.76%, and the average recall is 96.42% which indicate that THMRS is very high efficiency.

Keywords ISG · Natural language · Ontology · SPARQL · Thai herbs

1 Introduction and Related Work

Thai herbs have been incorporated into Thai ways of life for a very long time. They are natural products which Thai people use to treat symptoms and diseases. Parts of herbs are ingredients of herbal medicine, e.g., their flowers, leaves, fruits, bark, and

A. Sittisaman · N. Panawong (✉)
Department of Applied Science, Faculty of Science and Technology, Nakhon Sawan Rajabhat University, Nakhon Sawan, Thailand
e-mail: jnaruepon.p@gmail.com

A. Sittisaman
e-mail: akkasit@gmail.com

© Springer Nature Singapore Pte Ltd. 2019
S. K. Bhatia et al. (eds.), *Advances in Computer Communication and Computational Sciences*, Advances in Intelligent Systems and Computing 924, https://doi.org/10.1007/978-981-13-6861-5_45

roots. In addition, herbs are food ingredients and can do health promotion. Every regions of Thailand are abundant of various types of herbs. Each type of Thai herbs has different medicinal properties; therefore, users need to study and understand how to appropriately select and use them. Thai herbs are gaining in popularity. They also provide various options for health care, health improvement, preventing diseases, and treatment. Furthermore, the optimal dosages of herbal medicines have less side effects than conventional medicines, for instance, if we use Paracetamol, a medication used for pain and fever relief, for more than 5 consecutive days, it may damage the liver. If we use chronic gout medication, it may cause kidney problem. However, most people lack Thai herb knowledge. On the other hand, Thai herb information can be found from websites or documentations but each information source does not provide the complete set of herb knowledge. Many sources are not connected together. Hence, users usually take a long time to retrieve a complete set of Thai herb information. Sometimes, users may make typographic mistakes while searching for herb information; consequently, they cannot get the right information. Sometimes, users need to use Thai herbal medicines as the alternative medication, but they are inconvenient to compound the herbal medicine themselves. Hence, users will search for the ready-made medicines from the marketplaces. For example, a headache relief herbal medication is Coolcap (wormwood herbal medicine) by C.A.P.P group (Thailand) ltd., Ouay Un company Ltd.'s garlic capsules for reducing lipid level in the blood. Java tea by Abhaibhubejhr can treat suppression of urine, diuretics, nourish the kidney, and reduce uric which is the cause of gout.

In 2013, Sivilai and Namahoot [1] proposed the development of question answering food recommendation system for patients using web application. The system was developed by PHP with RAP API. Users can ask questions by natural language, for example, "ช่วยแนะนำอาหารสำหรับคนป่วยโรคความดันโลหิตสูงและไม่กินเนื้อวัวหน่อยครับ(What kind of foods that are good for high blood pressure and do not eat beef patients?)" Then, the question will be transformed to SPARQL in order to search from food ontology database. In addition, an expert system was applied to deduction mechanism that can improve food recommendation system efficiency. Song et al. [2] proposed "Natural Language Question Answering and Analytics for Diverse and Interlinked Datasets." Its concept is that users ask questions using natural language, for example, "How many companies develop drugs for pain?" This system will analyze the question and automatically create a set of SQL and SPARQL commands for searching the database that are associated with ontology. This research is a good prototype. Kongthon et al. [3] proposed a semantic-based question answering system for Thailand tourism information such as place, accommodation, restaurant, and attraction. The system will accept a natural language question, for instance, "รบกวนแนะนำที่พักที่เกาะกูดหน่อยครับไปกัน10 คน(Could you suggest accommodation at Koh Kood?), going with 10 people", and then Thai word segmentation system will analyze and construct the segmented word patterns. These patterns are further converted to SPARQL queries for searching from the tour ontology. This research shows that natural language processing can be a good practice for question answering system. While Kato et al. [4] proposed Thai herbal recommendation based on ontology and reasoning. Instead of entering question, a user has to select both type of gender and disease. Then, the proposed system will perform searching using SPARQL. The experimental results show that the system can derive the related information to the query. In addition, Tungkwampian et al. [5] developed the Thai herbal database using ontology which consisted of 323 concepts, i.e., herbal materials (plants, animals, and minerals), formulation, taste, clinical warning, and use method and health problem. This system can search for medicinal herbs for disease treatments. The experiments by ontology experts and traditional doctors showed that the answers are relevant with each other. However, the system delivered some irrelevant information to the question. On the other hand, the accuracy of the system is not tested.

Therefore, this research proposed a Thai herbs and Thai traditional medicine rec-ommendation system supporting health care and treatments (THMRS). The proposed system adopts natural language processing for analyzing questions from users, ISG algorithm for spell checking, word co-occurrence density for selecting recommen-dations for users. Moreover, the proposed system suggests the ready-made herbal medications provided by drug companies as well. On the other hand, ontology is utilized for storing Thai herbal database. Thai herbal medications and words will indicate the medicinal properties. Hence, THMRS facilitates patients for retrieving Thai herbs and herbal medication information within one website or one window. The remainder of this paper is organized as follows. Section 2 describes methodology. The system architecture is proposed in Sect. 3. Section 4 presents the experimental setup and result found. Finally, Sect. 5 states the conclusion of our work and future research.

2 Methodology

A framework of THMRS is composed of 5 components, i.e., Thaiherb Ontology, Word Ontology, ISG Algorithm, Question and Answer Processing and Word Co-occurrence Density. The details of each component are described in the following sections.

2.1 Thaiherb Ontology

The authors design Thaiherb Ontology and Thai herbs database storing using Protégé 5.0. There are 6 classes specifically Thaiherbs, HerbalMedicines, MedicineGroups, Symptom, Disease, and MedicinalPart (Fig. 1). They are described as follows:

1. Thaiherbs: The basic information about Thai herb use methods, i.e., Thai herbal names, scientific name, family, common names, general characteristics, proper-ties, use methods, notes/warning for medicinal uses, compounding medicines, medicine group, medicinal parts, symptom treatment parts, and diseases that should not use herbs.

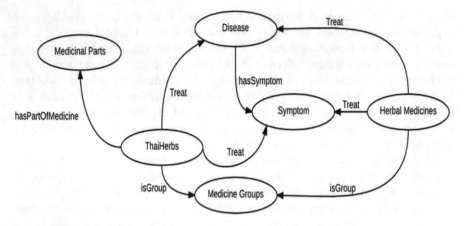

Fig. 1 Thaiherb ontology

2. HerbalMedicines: Details about herbal medication, i.e., medicine name, properties, manufacturers/distributors, adverse drug reaction, warning/caution, toxic,
   notes/suggestions, medicine groups, symptom treatments, and diseases that
   should not use herbs.
3. MedicineGroups: Medicines are divided into 12 groups which are der-
   matologic agents, laxatives, expectorants, analgesics, antidiarrheals, haem-
   agogue medicines, heart/blood/strengthening/nourishing nutrition, healing
   wound/inflammatory/ulcer treatment, lice killers/scabies treatments/ sea animal
   antidotes, diuretic/gallstone dissolution agents, and miscellaneous drugs, e.g.,
   Insomnia treatments, hypolipidaemic agents, vermifuges, travel sickness treatment [6].
4. Symptom: Collection of symptoms that can be treated by herbs and herbal
   medicines.
5. Disease: Collection of diseases which will be used for recommendation or caution
   of using Thai herbs or herbal medicine. For example, this herbal medicine is not
   recommended for diabetes or high blood pressure.
6. MedicinalPart: Collection of medicinal parts of herbs, for example, flowers, fruits
   and roots, etc.

2.2 Word Ontology

The word collection is applied to comparison and tokenizing processes. The words
are collected from the analysis of herb properties and herbal medicine using for
treatment and health care. Word characteristics indicate the property of the considered
medicine, including cure, treat, relief, exclusion, nourish, prevent, symptom, relax,

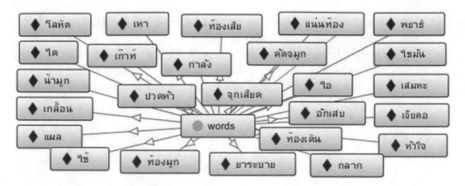

Fig. 2 Examples of word ontology

heal. The authors collect total 150 words ontology for question analysis. Examples of word ontology are shown in Fig. 2.

2.3 ISG Algorithm

ISG (Index of Similarity Group) is the Name Matching Algorithm which measure or compare similarities between names by scaling from 0 to 1 [7]. If the similarity value is 1, it means that they are the same names. The higher the similarity value, the higher similarities between names. However, if the similarity is 0, it means that the comparing names are completely different. This algorithm also was applied to a web-based Thai name checking system which utilized Thai astrology naming and clustering techniques [8]. In addition, ISG algorithm was applied to provinces' name verification and spell checking of the ontology-based tourism information system in order to discover the wrong spelling websites [9]. The ISG index calculation is as follows:

$$ISG = \frac{I}{I + D} \tag{1}$$

I is the number of similar characters.
D is the number of different characters between names.

The proposed THMRS system uses ISG Algorithm for finding similarity between words entered by users. This stage is called "spell checking" which help improving the information found from the search. The implementation of ISG is as follows.

1. Calculate name lengths of 2 received message. The longer name is the principle name.
2. Initialize a character pointer, similarity value and difference to be 0.

3. Compare the character pointer with the longest name length. If the pointer value is higher than the length, go to step 8.
4. If any characters in the considered two names are the same, increase similarity value at that position.
5. If any characters in the considered two names are different, increase difference value at that position.
6. Increase pointer value.
7. Go back to stage 3.
8. Determine ISG from (1).

2.4 Question and Answer Processing

Natural language processing (NLP) concerns interactions between computers and human languages. It enables computers to understand human (natural) languages. The process starts with word tokenizing, question analyzing, and translation for providing the answers to users whether by speaking or text. Differences between computer language and natural language are computer language that has limited boundary, limited words, exact syntaxes, and clear translation, or it is formal language. While the natural language is very large bounded, it is really hard to define the exact formats. The natural language can be categorized into units, i.e., characters, words, and sentences. Therefore, the NLP needs to do structure analyzing, meaning analyzing, and interpretation [10, 11].

The THMRS system facilitates users by entering the natural language questions. Then, the Thai herbal ontology has searched for required information using the follow steps.

1. A User enter natural language question, for example, " ช่วยแนะนำสมุนไพรที่ช่วยบรรเทาอาการ ไอ เป็นไข้และช่วยเกี่ยวกับอาการท้องผูกหน่อยครับ(Could you suggest herbs for cough, fever relief and help easing constipation, please?)"
2. Bring the question from step 1 to check spelling for Thai language using ISG algorithm (Eq. 1). The corrected question sentence is " ช่วยแนะนำสมุนไพรที่ช่วยบรรเทาอาการ ไอ เป็นไข้และช่วยเกี่ยวกับอาการท้องผูกหน่อยครับ"
3. Word Segmentation will be applied to the question sentence from step 2 using a Java Library. The example question from step 1 was segmented as ช่วย|แนะนำ|สมุนไพร|ที่|ช่วย|บรรเทา|อาการ|ไอ||เป็น|ไข้|และ|ช่วย|เกี่ยว|กับ|อาการ|ท้องผูก|หน่อย|ครับ
4. Classify the type of question using question analysis which is POS-Base bigram [12] incorporated with word ontology. The question may be classified as question for information search or recommendation. This step obtained three words, from example, for search, i.e., cough, fever, and constipation.
5. Translate natural language to SPARQL: This process translates word found in word ontology, from question analysis step, to SPARQL command, and then send the command to search herbal ontology. For example,

```
SELECT ?name ?sciname ?othername ?properties
?remark ?hasTreatWHERE {?a myont:hasName ?name;
myont:hasSciName ?sciname; myont:hasOtherName
?othername; myont:hasProperties ?properties;
myont:hasRemark ?remark;
myont:isTreat ?Treat . ?Treatmyont:hasTreat
?hasTreat
. FILTER (langMatches(lang(?name), 'th') &&
(CONTAINS(str(?hasTreat),'ไอ')||CONTAINS(str(?hasTreat),
'ไข้')||CONTAINS(str(?hasTreat), 'ท้องผูก')))}
```

6. Determine the word co-occurrence density between the results from step 5 and the question. Then, the word co-occurrence density is the criteria for selecting which information should be the answer. The details of the word co-occurrence density are described below.

2.5 Word Co-occurrence Density

The criteria of information retrieving are the relationship between search words and search results. The relationship is determined by the word co-occurrence density. If any information or document has higher density, it will be selected as an answer for the query [13, 14]. The word co-occurrence density can be calculated by Eq. 2. An example of the calculation is shown in Table 1.

$$R = \frac{\text{Number of keywords appeared in contents}}{\text{Total number of keywords}} \times 100 \qquad (2)$$

Table 1 An example of the word co-occurrence density calculation

Thai herb name	Word co-occurrence density
Question-> ช่วยแนะนำสมุนไพรที่ช่วยบรรเทาอาการไอ เป็นไข้และช่วยเกี่ยวกับอาการท้องผูกหน่อยครับ Could you suggest herbs for cough, fever relief and help easing constipation, please? **Search word->** ไอไข้ท้องผูก Cough fever constipation	
ชะมวง ราก ใบ ผล: ขับเสมหะ แก้ไอ ยาระบาย แก้ท้องผูกแก้ไข้ Cowa root leaf fruit: expectorate cough relief laxative constipation fever relief	$= \frac{3}{3} \times 100 = 100\%$
ชะพลู ใบ ราก:ขับเสมหะ แก้ไอ Wildbetal leafbush leaf root: expectorate cough relief	$= \frac{1}{3} \times 100 = 33.33\%$
มะเขือขื่น ราก: ขับเสมหะ แก้ไข้ Cockroach berry root: expect rate fever relief	$= \frac{1}{3} \times 100 = 33.33\%$
ขี้เหล็ก ใบ ดอก:ยาระบาย แก้ท้องผูกทำให้นอนหลับ Siamese cassia leaf flower: flaxative constipation sleep aids	$= \frac{1}{3} \times 100 = 33.33\%$

3   System Architecture

Figure 3 demonstrates the THMRS system architecture. The principles of THMRS system are as follows.

1. A user inputs natural language question to the system, such as herb names or symptoms via web or Windows applications. For instance, user input " ช่วยแนะนำ สมุนไพรที่ช่วยลดไขมันในเลือดและช่วยเกี่ยวกับอาการท้องผูกหน่อยครับ(Please suggest herbs that help reduce blood lipid)" or " มะแว้งมีสรรพคุณอย่างไรบ้างครับ(What is Ma-wang properties)."
2. The system check spelling of the question from step 1. If it is wrong, the system automatically corrects the typographical mistakes.
3. Word tokenizing using word ontology.
4. The question is analyzed and transformed to SPARQL in order to search Thaiherb Ontology.
5. Determine the relationship between search words and search results using word-occurrence density. The information with highest co-occurrence density will be selected as an answer. The advantage of word co-occurrence density is the information provide for users is met their needs.

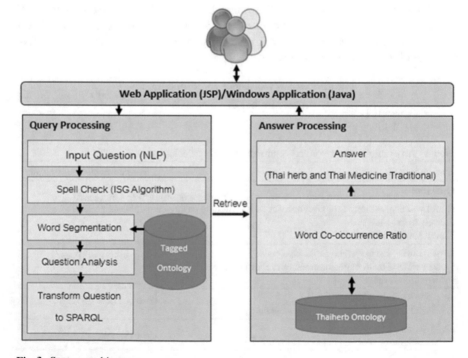

Fig. 3 System architecture

6. Displaying answer which is the basic information related to Thai herbs and medicines, e.g., names, properties, usages, and compounding method.

4 Testing and Results

This section shows experiments and results of THMRS. THMRS was implemented by web application using JSP and Windows application using Java. Users can ask for suggestions about Thai herbs and medicines by natural language questions. Hence, THMRS processes and display answers. There are three types of query specifically Thai herb data, herbal medicine, and suggestion about Thai herb and medicine usages.

4.1 Query for Thai Herbs Information

A user input a question "มะแว้งมีสรรพคุณอย่างไร (What are Ma-Wang properties?)," and then, the answer was displayed in Fig. 4.

THMRS showed information about Ma-Wang which consisted of properties, pictures, alias names, and compounding method for the user. This information can be used for the disease treatment.

4.2 Query for Ready-Made Herbal Medicines

A user asked the following question "ยาสมุนไพรช่วยลดคอเลสเตอรอลมีอะไรคะ (What are herbal medicine for lowering cholesterol?)" Then the answer was displayed in Fig. 5.

THMRS displayed ready-made Thai herbal medicines list in Fig. 5. The information consisted of ready-made herbal medicine names, properties, manufacturers or distributers, and registration no.

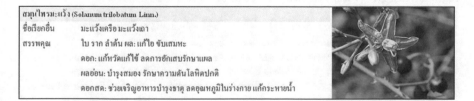

Fig. 4 Displayed answers for the question "What are Ma-Wang properties?"

ยาสมุนไพรช่วยลดคอเลสเตอรอล	
ยาแคปซูลระย่อมลมสูง	ช่วยให้นอนหลับสบาย ลดความดันโลหิตสูง/ร้านยาไทยโพธิ์เงิน - อภัยภูเบศร โอสถ
สมุนไพรกระเทียมแคปซูล	ลดคอเลสเตอรอล ช่วยย่อยอาหาร แก้ท้องอืดแน่น ลดไขมันในเลือด/ชุมชนปฐมอ
อ้วยอัน การ์ลิค กระเทียมสกัด	ลดระดับคอเลสเตอรอลในเลือด ละลายลิ่มเลือด ป้องกันการอุดตันของหลอดเลือด

Fig. 5 Displayed answer for the question "What are herbal medicine for lowering cholesterol?"

4.3   Query for Thai Herb and Medicine Usage Recommendations

A user's query was " ช่วยแนะนำสมุนไพรที่ช่วยบรรเทาอาการไอ เป็นไข้และช่วยเกี่ยวกับอาการท้องผูกหน่อยครับ (Could you suggest herbs for cough, fever relief and help easing constipation, please?)". THMRS displayed the request information as shown in Figs. 6 and 7.

Figure 6 shows the display of Thai herb and herbal medicine recommendations. The user obtained the request information from THMRS which is Thai herb names, parts with treatment properties. For example,

คำถาม : ช่วยแนะนำสมุนไพรที่ช่วยบรรเทาอาการไอ เป็นไข้และช่วยเกี่ยวกับอาการท้อ งผูกหน่อยครับ	
คำสำคัญ : ไอ ไข้ ท้องผูก	
แนะนำสมุนไพรบรรเทาอาการไอ ไข้ ท้องผูก	
ชะมวง	ราก ใบ ผล: ขับเสมหะ แก้ไอ ยาระบาย แก้ท้องผูก แก้ไข้
ชะพลู	ใบ ราก: ขับเสมหะ แก้ไอ
มะขามป้อม	ผล: แก้ไอ ขับเสมหะ
มะเขือขื่น	ราก: ขับเสมหะ แก้ไข้
แพงพวยฝรั่ง	ใบ: บำรุงหัวใจ รักษาโรคเบาหวาน แก้ท้องผูก
ขี้เหล็ก	ใบ ดอก: ยาระบาย แก้ท้องผูก ทำให้นอนหลับ

Fig. 6 An example for Thai herbs and herbal medicine recommendations

แนะนำยาสมุนไพรบรรเทาอาการไอ ไข้ ท้องผูก		
ยาแก้ไอมะขามป้อม	บรรเทาอาการไอ ขับเสมหะ และช่วยให้ชุ่มคอ	สมุนไพรอภัยภูเบศร
ยาน้ำแก้ไอมะขามป้อม ตราไอยรา	บรรเทาอาการไอ ระคายคอ คันคอ ช่วยขับเสมหะและทำให้ชุ่มคอ	บริษัท ที.แมน ฟาร์มา จำกัด
ยาน้ำแก้ไอมะแว้ง	บรรเทาอาการไอ ขับเสมหะ ทำให้ชุ่มคอ	บริษัท อ้วยอัน โอสถ จำกัด
คลูแคป	บรรเทาไข้ ขับเหงื่อ ลดอาการอักเสบ เจ็บคอ ถ่ายล้าง ระบายพิษ	บริษัท จี.เอ.พี.พี กรุ๊ป (ประเทศไทย) จำกัด
แคปซูลอระเพ็ด	แก้ไข้ แก้ร้อนใน	สมุนไพรอภัยภูเบศร
แคปซูลฟ้าทะลายโจร	แก้ไข้ ร้อนใน บรรเทาอาการเจ็บคอ ท้องเสีย	สมุนไพรอภัยภูเบศร
ยาแคปซูลผสมมะขามแขก	บรรเทาอาการท้องผูก	สมุนไพรอภัยภูเบศร

Fig. 7 A display example for herbal medicine recommendations

1. Cha-Muang roots, leaves, and fruits properties are expectorant, cough reliever, relaxative, fever reducer.
2. Cha-Ploo leaves and roots properties are expectorant, cough reliever.
3. Kee-Lekleaves and flowers properties are relaxative, sleeping aid.

Figure 7 shows the display of herbal medicine recommendations which are consisted of names, properties, and manufacturers or distributers. For example,

1. Compound Makham Pom Cough Mixture coughs relief, expectorant, freshening throat Abhaibhubejhr brand.
2. Coolcap fever relief, sweating, inflammatory, toxin drainage C.A.P.P. group (Thailand) Co, Ltd.
3. Compound Makham Kak relaxative agent Abhaibhubejhr brand.

4.4 The Efficiency of Answering Process

The efficiency of answering process was measured by precision, recall, and F-measure. This experiment performs with 200 questions from users. Those experimental questions are classified into three types, i.e., query for Thai herbs information, query for ready-made herbal medicines, query for Thai herb, and medicine usage recommendations. F-Measure calculation formula is in Eq. 3 and shown in Table 2.

$$F\text{-Measure} = \frac{2 * \text{Precision} * \text{Recall}}{\text{Precision} + \text{Recall}} \tag{3}$$

Precision	is True Positive/(True Positive + False Positive).
Recall	is True Positive/(True Positive + False Negative).
True Positive	is the right answer and was displayed.
False Positive	is the wrong answer and was displayed.
False Negative	is the right answer, but was not displayed.

Table 2 shows that the average F-Measure is 95.07%, the average precision is 93.76%, and the average of recall is 96.42%. However, the ambiguity of questions has made THMRS delivered wrong answers. For example, " ทำอย่างไรดีเหมือนจะปวดหัวแต่ปวดท้องมาก(What should I do with my severe stomach-aches

Table 2 The efficiency of answer processing

Question types	P (%)	R (%)	F (%)
Thai herbs information	93.88	97.87	95.83
Ready-made herbal medicines	95.92	97.92	96.91
Thai herb and medicine recommendations	91.49	93.48	92.47
Average	93.76	96.42	95.07

but I rather have a headache too)", "กินยาอะไรดีที่ช่วยให้ไม่ปวดท้อง" Therefore, if the system is incorporated with sentence checking and correcting process, the precision should be higher. Thus, the system efficiency is improved.

5   Conclusion and Further Work

This research proposed Ontology-Based Natural Language Processing for Thai Herbs and Thai Traditional Medicine Recommendation System Supporting Health care and Treatments. The applications of the proposed system are web application and Windows application. The natural language processing is applied to queries from user, spell checking by ISG algorithm and question analyzing and constructing SPARQL commands for search Thai herb ontology. Thai herbs word ontology database consists of basic information about Thai herbs, herbal medicine, syndromes, and words for tokenizing. Then, the answers retrieving from the database is based on word co-occurrence density which can match user's requirements. The experiments were performed using 200 queries which can be categorized into three types, i.e., query for Thai herbs information, ready-made herbal medicines, and Thai herb and medicine usage recommendations. The results shown that the average F-Measure is 95.07%, the average precision is 93.76%, and the average recall is 96.42. Therefore, THMRS is high efficiency.

However, if the questions are vague, the proposed system cannot deliver the answer that meets user's requirements. Therefore, the question analysis process should be modified. In addition, the application of the system should be extend to android application.

References

1. Sivilai, S., Namahoot, C.S.: Development of question answering for patient health food recommendation system. In: Proceedings of the 5th National Conference on Information Technology, Bangkok, Thailand, pp. 167–172 (2013) (in Thai)
2. Song, D., Schilder, F., Smiley, C.: Natural language question answering and analytics for diverse and interlinked datasets. In: Proceedings of the North American Chapter of the Association for Computational Linguistics—Human Language Technologies, Denver, Colorado, USA, pp. 101–105 (2015)
3. Kongthon, A., Kongyoung, S., Haruechaiyasak, C., Palingoon, P.: A semantic based question answering system for Thailand tourism information. In: Proceedings of the KRAQ11 Workshop: Knowledge and Reasoning for Answering Questions, Chiangmai, Thailand, pp. 38–42 (2011)
4. Kato, T., Kato, Y., Bjorn, K., Maneerat, N., Varakulsiripunth, R.: Providing thaiherbal recommendation based on ontology and reasoning. In: Proceedings of the 1st International Symposium on Technology for Sustainability, Bangkok, Thailand, pp. 509–512 (2011)
5. Tungkwampian, W., Theerarungchaisri, A., Buranarach, M.: Development of Thai herbal medicine knowledge base using ontology technique. Thai J. Pharm. Sci. **39**(3), 102–109 (2015)

6. MIS Editorial Board: Stop Sick With Herb, Bangkok. MIS Publishing, Thailand (2014) (in Thai)
7. Snae, C.: A comparison and analysis of name matching algorithms. Int. J. Appl. Sci. Eng. Technol. **4**(1), 252–257 (2007)
8. Namahoot, K., Snae, C.: Web-based Thai name checking system using Thai astrology naming and clustering techniques. NU Sci. J., **4**(1), 89–104 (2007) (in Thai)
9. Panawong N., Namahoot, C.S.: Performance analysis of an ontology-based tourism information system with ISG algorithm and name variation matching. NU Sci. J. **9**(2), 47–64 (2013) (in Thai)
10. Booncharoen, S.: Artificial Intelligence. Top Publishing, Bangkok, Thailand (2012). (in Thai)
11. Kumpaor, N., Ketcham, M.: System of book recommendation in library by using natural language processing. Pathumwan Acad. J. **5**(13), 45–60 (2015) (in Thai)
12. Poria, S., Agarwal, B., Gelbukh, A., Hussain, A., Howard, N.: Dependency-based semantic parsing for concept-level text analysis. In: Gelbukh, A. (ed.) Computational Linguistics and Intelligent Text Processing. CICLing 2014. Lecture Notes in Computer Science, Springer, Berlin, Heidelberg. vol. 8403, pp. 113–127 (2014)
13. Shao, M., Qin, L.: Text similarity computing based on LDA topic model and word co-occurrence. In: Proceedings of the 2nd International Conference on Software Engineering, Knowledge Engineering and Information Engineering (SEKEIE 2014), pp. 199–203. Singapore (2014)
14. Panawong, N., Sittisaman, A.: A responsive web search system using word co-occurrence density and ontology: Independent Study Project Search. Adv. Intell. Syst. Comput. **760**, 185–196 (2018)

Tourism Web Filtering and Analysis Using Naïve Bay with Boundary Values and Text Mining

Naruepon Panawong and Akkasit Sittisaman

Abstract This paper proposes a technique to filter tourism websites and datamining to analyze keywords to improve searching on tourism websites as Search Engine Optimization (SEO). This work focuses on tourist sites or attractions, hotel or accommodation, and restaurant in the tourist provinces as a core keywords. Content in websites is retrieved from Google with queries of 11 Thai famous tourism provinces. From all retrieved results, filtering using Naïve Bayes algorithm with Boundary Values is performed to detect only relevant content as 6,171 filtered websites (66.55%) from 9,273 retrieved websites. From keyword analysis method, we compared three methods including (1) keywords from Apriori algorithm, (2) keywords from frequent terms within websites, and (3) keywords by frequency of terms from the ontology. The experiment results are conclusive that keywords from frequent terms within websites performed best. The keywords are usable to customize the websites to improve search ranking.

Keywords Apriori algorithm · Web filtering · Naïve Bayes · SEO · Text mining

1 Introduction and Related Work

In the era of Internet, the exponential growth in the quantity and complexity of information sources on the internet is inevitable. It leads to information overload and becomes increasingly difficult to find relevant information with simple queries with a large amount of possible matching. Most of the provided information is not organized and classified, and it results in irrelevant retrieved information. To make the retrieved results more relevant to the query, much effort has been spent to deal

N. Panawong (✉) · A. Sittisaman
Department of Applied Science, Faculty of Science and Technology,
Nakhon Sawan Rajabhat University, Nakhon Sawan, Thailand
e-mail: jnaruepon.p@gmail.com

A. Sittisaman
e-mail: akkasit@gmail.com

© Springer Nature Singapore Pte Ltd. 2019
S. K. Bhatia et al. (eds.), *Advances in Computer Communication
and Computational Sciences*, Advances in Intelligent Systems and Computing 924,
https://doi.org/10.1007/978-981-13-6861-5_46

effectively with this complexity. Among those researches, filtering a retrieved result [1] is one of effective method to reduce irrelevant results. Furthermore, this can be performed without interfering in a development of the search engines which services are already available online.

The query results with search engine can be more than a thousand, while from the study [2], users tentatively see through the only top 10–20 ranks at best. Hence, it is essential for web content developers to aim for top ranks in searching result [3], and they apply search engine optimization (SEO). One of the methods to increase a rank of a website is to apply 'Tags' helping for matching to users' query; however, the difficult part is to generate an effective keyword as Tags. Effective keywords are normally required times and efforts to test out in a real situation [4, 5]. Although the general term is a keyword, it can regardless be a word, a phrase or a sentence that can potentially represent content and help in classification [6, 7]. In Thailand, tourism is one of the most anticipated focus since it is a major market contributor. From reports [8], estimates of tourism receipts directly contributing to the Thai GDP of 12 trillion baht range from 9% (one trillion baht) (2013) to 17.7% (2.53 trillion baht) in 2016. Thus, the need of support in tourism has been urged in several domains including information technology. Previously, the work of Panawong et al. [9] proposed to use a keyword to classify 475 tourism websites using Naïve Bayes algorithm. The results were promising, but it is yet unpractical with the small selected testing set. Namahoot et al. [10] published their work on using frequency of terms in a classification of tourism website. They also report that the content within the body part of the website is the most informative content and should be focused. Moreover, Tosqui-Lucks and Silva [11] presented the apply of ontology to represent important terms/concepts in English related to tourism and used them to instantiate tourism sites. Their work shows the potentials of conceptual terms to represent expressed terms in the tourism/travelling web content.

From the above-mentioned work projects and issues, we aim to reduce the irrelevant retrieved results of Thai Tourism website by using filtering with Naïve Bayes with a Boundary Values algorithm. Additionally, we attempt to find a keyword(s) to help improving in search ranking by exploiting term frequency with text-mining from the given web content. We will compare the results of techniques including Apriori algorithm and frequency based statistical approach. We also consider using ontological terms from existing Thai Tourism Ontology as a predefined keyword list.

2 Methodology

This research consists of 3 components, i.e., Filtering Websites Using Naïve Bayes Algorithm with Boundary Values, Text Mining Technique and Apriori Algorithm. The details of each component is described the following sections.

2.1 Filtering Websites Using Naïve Bayes Algorithm with Boundary Values

Naïve Bayes algorithm applies probabilistic method for categorization. It is simple in implementation and is reported to perform well in several tasks such as spam detection and classification [12]. Naïve Bayes algorithm with Boundary Values (NBB) is an enchanted version of common Naïve Bayes to increase proficiency in classification as given in Eqs. 1 and 2 for filtering websites [13].

$$C_{\text{map}} = \arg\max_{c \in C} \left(p(c) \sum_{k=1}^{n} \log(P(t_k|c)) \right) \tag{1}$$

$C_{i\text{Max}} <= \text{threshold} >= C_{i\text{Min}}$,

$C_{i\text{Max}} = \max(C_{\text{map}})$ and $C_{i\text{Min}} = \min(C_{\text{map}})$, with $1 <= i => 3$ (2)

Let C_{map} is a probability result of web classification, and is calculated by multiplying $P(c)$ and $\underline{P}(t_k|c)$. C refers to class for classification (c_1, c_2, c_3). While $P(c)$ is a probability of class, $P(t_k|c)$ is a probability of a term given in a class. In this work, there are three categories including attractions, accommodations and restaurants. Last, n is a number of terms used in classification of a targeted website.

This work exploits Naïve Bayes algorithm with Boundary Values to filter only the websites relevant to tourism domain. The step-wise method is as follows.

1. Reading website's content.
2. Applying Eq. 1 to calculate probability of the website.
3. Using results from the previous step and filtering websites within boundary and treating them as tourism domain website by Eq. 2.
4. Repeating step 1 to all websites.
5. Classifying the websites to defined category (attractions, accommodations and restaurants).

The results of this method will be used for keywords analysis in the later processes.

2.2 Tourism Keyword Analysis Using Text-Mining Technique

Text-mining is a learning method from training text data to find relation with in data using statistical approach. It can be used in many tasks such as Web document-based text clustering, Information Retrieval, Topic tracking, Summarization and concept linkage [14]. There are several research projects using datamining such as patent analyzing [15] and sentimental analysis from online hotel review [16]. In this work, we apply text-mining to find keywords in website content for tags to improve ranks of the website in searching. The methods are as follows.

1. Calculating for term frequency and sorting from high to low, respectively.
2. Selecting the top 2 frequent terms from each website as keyword set.
3. From keyword set, counting frequency based on existing from all websites.
4. Evaluating the keyword set using F-measure.
5. Repeating step 2 by adding more top-2 terms (next frequent rank) to keywords and repeating step 3–4 until no term left.
6. Choosing the best keyword set with highest F-measure result.

2.3 Apriori Algorithm

The Apriori Algorithm is an influential algorithm for mining frequent itemsets for boolean association rules. Apriori uses a bottom up approach, where frequent subsets are extended one item at a time. Working step of Apriori algorithm is summarized as follows [17].

1. Reading through content of each website and counting term frequency
2. Examining term frequency if it is greater than or equal to threshold value or not. If positive, this indicates that the terms are collocated (L_1). Else, the terms will be discarded. In this work, threshold is set as 0.3.
3. Making a group using the obtained L_1 where the size is $2(C_2)$. C_2 is created by considering L_1 that consists of two terms.
4. Running through web content again to count frequency of C_2. If the terms are greater than or equal to threshold value, making them L_2. Else, they are discarded.
5. Repeating step 3 and 4 until C_k cannot be created from L_{k-1} while k stands for the size of terms.

3 Development of Tourism Web Filtering and Analysis

From Fig. 1, the processes can be explained in step-wise as follows.

1. A website search result by province name and a combination of province name and concepts of attractions, accommodations and restaurants (in Thai) are gathered. The examples of combining queries are: Chiangmai attractions (ท่องเที่ยวเชียงใหม่), Chiangmai Hotel (ที่พักเชียงใหม่), Loei restaurant (ร้านอาหารจังหวัดเลย), Phuket attractions (ท่องเที่ยวภูเก็ต). The retrieved websites are extracted to gather the content given within Body tag and processed to remove HTML tag; hence, only text content will be remained. Thai Word segmentation is applied, and stop-words (Thai function word) are removed for storing into a database.
2. The result from step 1 is processed with NBB (Sect. 2.1) to filter only relevant contents as targeted documents. The irrelevant contents are discarded.
3. The result from step 2 is separately text-mined with two methods; (1) Apriori algorithm (Sect. 2.3) and (2) frequency-based text-mining (Sect. 2.2). The

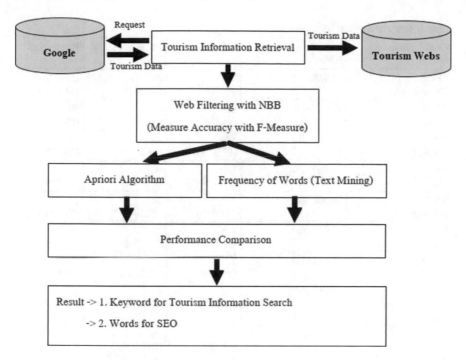

Fig. 1 Overview of tourism web filtering and analysis

frequency-based text-mining uses the list of keywords from all concepts in Tourism Ontology [13] and within websites.

4. The comparison of two methods is performed based on the result of *F*-measure.
5. The better result in comparison will be available as web tag for Search Engine Optimization (SEO) to improve search ranking.

4 Experiments and Results

4.1 Results of Filtering Websites Using Naïve Bayes with Boundary Values

From the scope, we gathered 9,273 websites using Google search engine (accessed 31 March 2018) with the queries mentioned in Sect. 3. Table 1 shows the query results using province name (in Thai) with Google recommended categories while Table 2 shows the retrieved results using combination query.

Then, we applied our method "Filtering Websites Using Naïve Bayes with Boundary Values" and obtained the filtering results as given in Tables 3 and 4 based on data from Tables 1 and 2, respectively.

Table 1 Retrieved results of province name

Query (province name)	Records found (Google)	Attractions	Accommodation	Restaurant
พิษณุโลก (Phitsanulok)	5,760,000	90 (95.74%)	21 (22.34%)	42 (44.68%)
เชียงใหม่ (Chiangmai)	22,400,000	119 (95.20%)	40 (32.00%)	70 (56.00%)
ภูเก็ต (Phuket)	7,880,000	120 (93.02%)	49 (37.98%)	71 (55.04%)
กระบี่ (Krabi)	11,400,000	100 (94.34%)	40 (37.74%)	58 (54.72%)
เชียงราย (Chiangrai)	13,300,000	113 (95.76%)	38 (32.20%)	59 (50.00%)
ชลบุรี (Chonburi)	15,800,000	111 (94.87%)	28 (23.93%)	46 (39.32%)
แม่ฮ่องสอน (Maehongson)	7,110,000	121 (90.30%)	42 (31.34%)	73 (54.48%)
ประจวบคีรีขันธ์ (Prachuapkhiri khan)	7,560,000	122 (96.06%)	43 (33.86%)	53 (41.73%)
สุราษฎร์ธานี (Suratthani)	9,450,000	116 (97.48%)	20 (16.81%)	47 (39.49%)
ระยอง (Rayong)	13,600,000	117 (89.31%)	42 (32.06%)	67 (51.14%)
เลย (Loei)	91,300,000	95 (91.35%)	20 (19.23%)	55 (52.89%)

Table 2 Retrieved results of a combination of defined concepts and province name (in Thai)

Category + province name	Attractions	Accommodation	Restaurant
สถานที่ท่องเที่ยว (Attractions)	1,609 (76.24%)	858 (40.65%)	1,031 (48.85%)
ที่พัก (Accommodation)	1,285 (85.90%)	1,368 (91.44%)	842 (56.28%)
ร้านอาหาร (Restaurant)	1,085 (86.11%)	578 (45.87%)	1,080 (85.71%)

From the results, we found that there are total of 6,171 retrieved websites past the filtering as targeted websites. From observation, the rest of the retrieved contains very few terms in common while some websites present the tourism using images; hence, they cannot be handled with the proposed method and considered as out of scope.

Table 3 Filtering results of websites from Table 1

Query	Google (web sites)	Filtered by NBB	% Filtered by NBB	F-measure (%)		
				P	R	F
Phitsanulok	213	94	44.13	100.00	69.12	81.74
Chiangmai	224	125	55.80	100.00	78.62	88.03
Phuket	219	129	58.90	100.00	81.13	89.58
Krabi	213	106	49.77	100.00	74.13	85.14
Chiangrai	236	118	50.00	100.00	72.39	83.99
Chonburi	244	117	47.95	100.00	83.57	91.05
Maehongson	225	134	59.56	100.00	83.75	91.16
Prachuapkhiri khan	227	127	55.95	100.00	77.91	87.59
Suratthani	222	119	53.60	100.00	77.27	87.18
Rayong	247	131	53.04	100.00	74.86	85.62
Loei	203	104	51.23	100.00	75.91	86.31
Average			52.72	100.00	77.15	87.04

Table 4 Filtering results of websites from Table 2

Category + province name	Google (web sites)	Filtered by NBB	% filtered by NBB	F-measure (%)		
				P	R	F
Attractions	2,989	2,111	70.63	100.00	92.60	96.12
Accommodation	1,874	1,260	67.24	100.00	97.57	98.77
Restaurant	1,937	1,496	77.23	100.00	96.46	98.19
Average			71.70	100.00	95.54	97.69

4.2 Results of Term Frequency

In this part, we attempt to test the keywords assignment using our proposed text-mining. The data to be tested are content of 4,937 websites in which can be separated based on three concepts as follows. (1) Attractions: 2,237 websites (2) Accommodation: 1,400 websites and (3) Restaurant: 1,300 websites.

These websites were pre-processed with word segmentation and stop-word removal. They were used as training data. The frequency of the terms in the training data is given in Table 5.

Since this work uses two different set of keywords as terms in websites and terms given in the tourism ontology, we show the frequency of terms from both sources separately in Figs. 2 and 3 respectively.

Based on method given in Sect. 2.2, we attempted to create a list of significant keywords and gained the result of keywords based on frequency. The example of

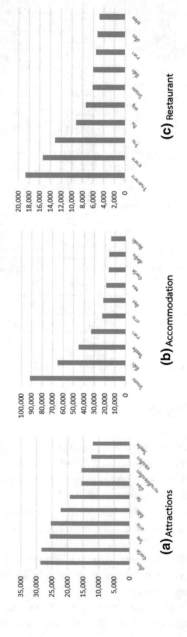

Fig. 2 Some top frequent terms given in web content

Table 5 Example of frequency of terms in the training data (in Thai)

Web sites	Word frequency							
	โรงแรม Hotel	ที่พัก Accommodation	ท่องเที่ยว Tourism	เที่ยว Travel	ร้านอาหาร Restaurant	อาหาร Food	:	ทัวร์ Tour
Web#1	8	14	34	44	18	98	:	0
Web#2	7	22	17	31	4	4	:	20
Web#3	43	14	10	42	0	0	:	1
Web#4	17	132	9	39	1	2	:	13
Web#5	9	23	22	8	113	87	:	5
Web#6	21	16	32	35	88	111	:	3
:	:	:	:	:	:	:	:	:
Web#4937	11	20	25	36	7	18	:	3

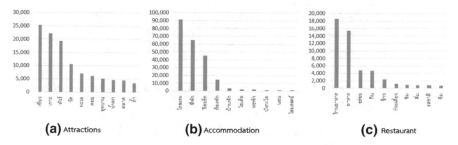

Fig. 3 Some top frequent terms given in web content based on terms given in tourism ontology

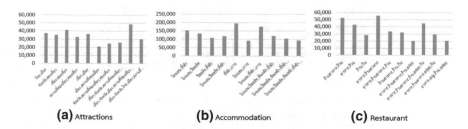

Fig. 4 Some top keywords generated by frequency of terms in training data

gained keywords based on frequency within web content is demonstrated in Fig. 4 and keyword results based on the ontology terms is given in Fig. 5.

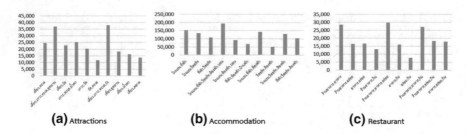

(a) Attractions   **(b)** Accommodation   **(c)** Restaurant

Fig. 5 Some top keywords generated by frequency of terms from the ontology in training data

Table 6 *F*-measure results of keywords from Apriori algorithm

Category	Keywords	F-measure (%)		
		P	R	F
Attractions	จังหวัด,เกาะ,ท่องเที่ยว	87.18	81.86	84.44
Accommodation	โรงแรม,ที่พัก,รีสอร์ท	95.54	88.71	92.00
Restaurant	ร้านอาหาร,ร้าน,อาหาร,กิน	99.97	89.44	94.41

Table 7 Example of the *F*-measure results of keywords from frequent terms within websites

Keywords	F-measure (%)		
	P	R	F
Attractions			
1. เที่ยว,ไทย,เกาะ,วัด,เมือง,ท่องเที่ยว	99.93	91.70	95.55
2. เที่ยว,ไทย,เกาะ,วัด,เมือง	99.93	91.69	95.54
3. เที่ยว,จังหวัด,ไทย,เกาะ,วัด,เมือง	99.93	91.69	95.54
4. เที่ยว,จังหวัด,ไทย,ที่พัก,เมือง	99.93	91.58	95.47
5. เที่ยว,จังหวัด,ไทย,เกาะ,เมือง,ท่องเที่ยว	99.93	91.45	95.41
Accommodation			
1. โรงแรม,ห้องพัก,ที่พัก,เกาะ,เมือง,บ้าน	99.97	89.99	94.72
2. โรงแรม,ห้องพัก,ที่พัก,จอง,เมือง,บ้าน	99.97	89.92	94.68
3. โรงแรม,รีสอร์ท,ห้องพัก,ที่พัก,เมือง,บ้าน	99.97	89.80	94.61
4. โรงแรม,รีสอร์ท,ที่พัก,เมือง,วิว,บ้าน	99.97	89.80	94.61
5. โรงแรม,ห้องพัก,ที่พัก,เมือง,บ้าน	99.97	89.65	94.53
Restaurant			
1. อาหาร,ร้านอาหาร,ร้าน,อร่อย,กิน	99.97	89.70	94.56
2. อาหาร,ร้าน,อร่อย,กิน	99.97	89.67	94.54
3. อาหาร,ร้าน,กิน	99.97	89.52	94.46
4. อาหาร,ร้านอาหาร,ร้าน,กิน	99.97	89.44	94.41
5. อาหาร,ร้าน,หมู	99.94	89.44	94.40

4.3 Results of Keyword Efficacy

In this experiment, 6,171 websites were tested with the generated keywords using F-measure given in (2). Precision (P) and Recall (R) in this work were calculated in the following specification [18].

P is a total number of websites that the keywords can retrieve/all websites

R is a total number of websites that the keywords cannot retrieve/all websites

The testing is separated to three groups as (1) keywords from Apriori algorithm, (2) keywords from frequent terms within websites, and (3) keywords by frequency of terms from the ontology. Hence, we obtained three results given in Tables 6, 7 and 8, respectively.

Table 8 Example of the F-measure results of keywords from frequent terms based on terms given in the tourism ontology

Keywords	F-measure (%)		
	P	R	F
Attractions			
1. เที่ยว,ทะเล,เกาะ,วัง	59.53	55.89	57.66
2. เที่ยว,เกาะ,วัง	57.77	54.25	55.95
3. เที่ยว,ทะเล,วัง	52.96	49.73	51.29
4. เที่ยว,หนัง	52.37	49.18	50.72
5. เที่ยว,ทะเล,เกาะ,อุทยาน	44.50	41.78	43.10
Accommodation			
1. ที่พัก,โรงแรม,รีสอร์ท,ห้องพัก,นอน	95.54	88.71	92.00
2. โรงแรม,รีสอร์ท,ห้องพัก,นอน	82.00	76.14	78.96
3. ที่พัก,รีสอร์ท,ห้องพัก,นอน	80.85	75.07	77.85
4. ที่พัก,โรงแรม,ห้องพัก,นอน	77.69	72.14	74.81
5. ที่พัก,โรงแรม,รีสอร์ท,ห้องพัก	61.08	56.71	58.81
Restaurant			
1. ร้านอาหาร,อาหาร,อร่อย	57.42	53.00	55.12
2. อาหาร,ย่าง,ข้าว	39.50	36.46	37.92
3. อาหาร,กิน,ชิม	34.92	32.23	33.52
4. อาหาร,กิน,ย่าง,อร่อย	34.08	31.46	32.72
5. อาหาร,ร้านอาหาร,กิน,ย่าง,อร่อย	32.83	30.31	31.52

5   Conclusion and Further Work

In this paper, we propose a technique to filter tourism websites and datamining to analyze keywords to use for assisting searching on tourism websites as Search Engine Optimization (SEO). The focused concepts in this work are about attractions or tourist sites, hotel or accommodation, and restaurant in the tourist provinces. Since this work is about content in websites, we use retrieval results from Google with queries of 11 Thai provinces. From all retrieved results, filtering using Naïve Bayes algorithm (NBB) is performed to detect only relevant content as 6,171 filtered websites (66.55%) from 9,273 retrieved websites.

From keyword analysis method, we compared three methods including (1) keywords from Apriori algorithm, (2) keywords from frequent terms within websites, and (3) keywords by frequency of terms from the ontology. The experiment results can be concluded that keywords from frequent terms within websites performed best. The keywords are expectedly usable to customize the websites to improve search ranking.

In the future, we plan to include semantic network to improve our filtering and keyword analyzing method. The semantic network of terms involves in expanding terms into local dialects such as Northern Thai dialect and Northeastern Thai dialect.

References

1. Cao, T., Nguyen, Q.: Semantic approach to travel information search and Itinerary recommendation. Int. J. Web Inf. Syst. **8**(3), 256–277 (2012)
2. The Electronic Transactions Development Agency (Public Organization).: Thailand Internet User Profile 2017. Retrieved 27 March 2018 from https://www.aripfan.com/thailand-internet-user-profile-2017 (2017) (in Thai)
3. Lu, P., Cong, X.: The research on webpage ranking algorithm based on topic-expert documents. Adv. Intell. Syst. Comput. **361**, 195–204 (2015)
4. Deka, R.: Increasing website visibility using search engine optimization. Int. J. Eng. Technol. Sci. Res. **1**(1), 18–22 (2014)
5. Vignesh, J., Deepa, V.: Search engine optimization to increase website visibility. Int. J. Sci. Res. **3**(2), 425–430 (2014)
6. Jain, A., Sharma, S.: An efficient keyword driven test automation framework for web application. Int. J. Eng. Sci. Adv. Technol. **2**(3), 600–604 (2012)
7. Ozbal, G., Pighin, D., Strapparava, C.: Automation and evaluation of the keyword method for second language learning. In: Proceedings of the 52nd Annual Meeting of the Association for Computational Linguistics (Volume 2: Short Papers), Baltimore, USA, pp. 352–357 (2014)
8. Theparat, C.: Tourism to continue growth spurt in 2017 (2017). Retrieved 27 Mar 2018 from https://www.bangkokpost.com/business/tourism-and-transport/1199925/tourism-to-continue-growth-spurt-in-2017
9. Panawong, N., Namahoot, C.S., Brueckner, M.: Classification of tourism web with modified Naïve Bayes algorithm. Adv. Mater. Res. **931–932**, 1360–1364 (2014)
10. Namahoot, C.S., Lobo, D., Kabbua, S.: Enhancement of a text clustering technique for the classification of Thai tourism websites. In: Proceedings of Computer Science and Engineering Conference (ICSEC), IEEE, Khon Kaen, Thailand, pp. 203–208 (2014)

11. Tosqui-Lucks, P., Silva, B.C.D.D.: Structuring an ontology of the basic vocabulary of tourism. Int. J. Inf. Educ. Technol. **2**(4), 221–334 (2012)
12. Khan, A., Baharudin, B., Lee, L.H., Khan, K.: A review of machine learning algorithms for text-documents classification. J. Adv. Inf. Technol. **1**(1), 4–20 (2010)
13. Namahoot, C.S., Brueckner, M., Panawong, N.: Context-aware tourism recommender system using temporal ontology and naïve bayes. Adv. Intell. Syst. Comput. **361**, 183–194 (2015)
14. Hashimi, H., Hafez, A., Mathkour, H.: Selection criteria for text mining approaches. Comput. Hum. Behav. **51**, 729–733 (2015)
15. Noh, H., Jo, Y., Lee, S.: Keyword selection and processing strategy for applying text mining to patent analysis. Expert Syst. Appl. **42**, 4348–4360 (2015)
16. Berezina, K., Bilgihan, A., Cobanoglu, C., Okumus, F.: Understanding satisfied and dissatisfied hotel customers: text mining of online hotel reviews. J. Hospitality Market. Manag. **25**(1), 1–24 (2015)
17. Wasilewska, A.: APRIORI algorithm (2016). Retrieved 2 Apr 2018 from http://www.cs.sunysb.edu/~cse634/lecture_notes/07apriori.pdf
18. Lemnitzer, L., Monachesi, P.: Extraction and evaluation of keywords from learning objects—a multilingual approach. In: Proceedings of the Language Resources and Evaluation Conference (LREC 2008), Morocco, pp. 1–8 (2008)

One Novel Word Segmentation Method Based on N-Shortest Path in Vietnamese

Xiaohua Ke, Haijiao Luo, JiHua Chen, Ruibin Huang and Jinwen Lai

Abstract Automatic word segmentation of Vietnamese is the primary step in Vietnamese text information processing, which would be an important support for cross-language information processing tasks in China and Vietnam. Since the Vietnamese language is an isolating language with tones, each syllable can not only form a word individually, but also create a new word by combining with left and/or right syllables. Therefore, automatic word segmentation of Vietnamese cannot be simply based on spaces. This paper takes automatic word segmentation of the Vietnamese language as the research object. First, it makes a rough segmentation of Vietnamese sentences with the N-shortest path model. Then, syllables in each sentence are abstracted into a directed acyclic graph. Finally, the Vietnamese word segmentation is obtained by calculating the shortest path with the help of the BEMS marking system. The results show that the proposed algorithm achieves a satisfactory performance in Vietnamese word segmentation.

Keywords N-shortest path model · Automatic segmentation · Vietnamese text information processing

1 Introduction

With the establishment of China-ASEAN Free Trade Area and the implementation of the "Belt and Road" initiative, China and Vietnam have witnessed closer economic and trade relations and more frequent cultural and academic exchanges. The success-

X. Ke (✉) · H. Luo · R. Huang · J. Lai
School of Information Science and Technology, Guangdong University of Foreign Studies, Guangzhou, Guangdong 510006, People's Republic of China
e-mail: carrieke@gdufs.edu.cn

J. Chen
Faculty of Asian Language and Cultures, Guangdong University of Foreign Studies, Guangzhou, Guangdong 510420, People's Republic of China
e-mail: chenjihua@gdufs.edu.cn

© Springer Nature Singapore Pte Ltd. 2019
S. K. Bhatia et al. (eds.), *Advances in Computer Communication and Computational Sciences*, Advances in Intelligent Systems and Computing 924,
https://doi.org/10.1007/978-981-13-6861-5_47

ful 2007 China-Vietnam Economic and Trade Forum indicates future increase in the scale and level of trade investment between China and Vietnam. Against this back-drop, Vietnamese-speaking talents are in greater demand in very social sector, so are the tasks to process Vietnamese texts. Word segmentation of a text is the first step of information process. Since Vietnamese is an isolating language with tones, each syllable in Vietnamese could be an independent word or in combination with syllables left and/or right to become a new word. Thus, it is impossible to segment Vietnamese sentences simply based on blank spaces. This paper studies word segmentation of Vietnamese texts. The aim of this research was to contribute to information process of Vietnamese texts by coming up with a proved new automatic word segmentation algorithm in line with the linguistic features of Vietnamese based on past investigations into automatic word segmentation algorithms for the Vietnamese language and on the ample study results of automatic segmentation of Chinese words.

2 Status Quo of Vietnamese Automatic Word Segmentation Study

In NLP tasks, automatic word segmentation, tagging, and disambiguation are the most fundamental informatization steps which would lay a solid foundation for subsequent processing tasks. At present, in NLP research and application, the Chinese and English text information processing technologies are relatively mature. Therefore, researches on other languages often refer to the results of pertaining studies on Chinese and English. As Vietnamese is a minority language, it faces a status quo of fewer research teams and results, scarce corpus source, heavy proofreading workload, and increasing demand of application. Thus, research on Vietnamese is a promising field deserving attention and scientific research from the perspective of natural language processing.

Modern Vietnamese in use is written in Latin characters to present pronunciation plus additional diacritics to mark tones [1], with each syllable as a writing unit. Many syllables are words by themselves, which are the most common text forms in Vietnamese. In addition, there are phrases and polysyllable words as the former mainly composed of disyllable words and the latter mostly proper nouns. In other words, in terms of word structure, each syllable can either be an independent word, or combine with left and/or right syllables to form a new word. Vietnamese word structure shares some similarity with that of Chinese. Table 1 shows the contrast of three groups of Vietnamese–Chinese single words, disyllable words, and polysyllable phrases (proper nouns).

Written in Latin alphabets, the Vietnamese language, like English, separate syllables and phrases with blank spaces. However, there are spaces within the Vietnamese disyllable words and polysyllable words, that is, separating the multiple syllables of the same word by spaces, instead of connecting the syllables of that word with underscores "_". Obviously, Vietnamese words cannot be segmented simply according to

Table 1 Comparing three groups of Vietnamese-Chinese simple words, disyllable words, and polysyllable phrases

Single words	Disyllable words	Polysyllable phrases
哭 (Cry) Khóc	中国 (China) Trung Quốc	数据结构 (Data Structure) Cấu trúc dữ liệu
笑 (Smile) Cười	玩 (Play) ăn chơi	一心一意 (Undivided Attention) Một lòng một dạ
说 (Speak) Nói	货币 (Currency) Tiền tệ	市场经济 (Market Economy) Nền kinh tế thị trường

Table 2 Comparing the use of blanks spaces in Vietnam, Chinese, and English texts

Comparing the use of blanks spaces		
资本主义	Chủ nghĩa tư bản	Capitalism
古生物学家	Nhà cổ sinh vật học	Paleontologist
经济危机	Khủng hoảng kinh tế	Economic crisis

blank spaces. Table 2 shows three words in Vietnamese, Chinese, and English, and in each text, the blank spaces serve distinctive functions.

Therefore, the difficulty in the automatic word segmentation of Vietnamese lies in the determination of word boundaries. Existing researches involve methods of both rules and statistics. In 2008, Prof. Lê Hồng Phương of Hanoi National University developed vnToolkit 3.0, a Vietnamese text processing toolkit [2]. The Vitk is a probability-based word segmentation software developed on the basis of the maximum entropy principle and the conditional random field theory [3]. The corpora for testing the tagging algorithm in this paper were preprocessed by Vitk word segmentation tool in the toolkit. In recent years, there have also been many researches on Vietnamese text information processing. Nguyen et al. [4] explore Vietnamese word segmentation through "the detection and correction of the problematic annotations in the VTB corpus, and then by combining and splitting the inconsistent annotations." They construct linguistic rules to carry out the POS tagging in Vietnamese. Yu Zhengtao and Gao Shengxiang of Kunming University of Science and Technology have achieved impressive results in the study of Chinese and Vietnamese languages. They have published several papers in various academic journals, delving into not only the fields of machine translation and corpus construction, but also word annotation and word alignment. They have proposed in these papers Vietnamese word segmentation method based on CRFs, ambiguity models [5], the Chinese-Vietnamese bilingual word alignment method based on deep neural network DNN_HMM [6], and Chinese-Vietnamese word similarity computation based on Wikipedia [7]. All these have made great contributions to the study of Vietnamese word segmentation and laid a technical foundation for the work of bilingual parallel corpus and automatic translation.

Apart from the studies mentioned, there are many researches on Chinese–Vietnamese information retrieval, manifesting the significant role the research and application of Vietnamese text information processing plays in not only teaching and scientific research, but also in various specialized fields, especially in areas like social network and public opinion calculation, decision analysis. In early researches, Luo [8] conducted investigations into web news extraction, topic sentence extraction, and cross-language text similarity, and put forward the WEB news extraction method based on the density of text blocks. In recent years, some researchers have proposed the Chinese–Vietnamese bilingual event correlation analysis based on bilingual topic and factor graph model [9]. Taking bilingual events as the graph factor, they propose a local intimacy propagation algorithm based on the factor graph. Some researchers also use sentences as nodes. They constructed an undirected graph model based on the nodes to study the abstract sentence extracted from Chinese–Vietnamese news [10]. These works have laid a solid foundation for future Chinese–Vietnamese bilingual information technology research and application.

3 N-Shortest Path Algorithm for Vietnamese Word Segmentation

The significance of word segmentation task in this paper mainly lies in allowing computer to process corpora based on the linguistic features of Vietnamese with higher efficiency and accuracy, including tasks of sentence alignment, word segmentation, and annotation. Since Vietnamese is an isolating language with tones [11], each syllable morpheme alone can be a word, and the combination with neighboring morphemes can also form new words. Therefore, the segmentation of Vietnamese sentences, like in Chinese, cannot be simply based on blank spaces, making it imperative to figure out a method to single out independent words in Vietnamese natural language processing. In the study of Chinese word segmentation, the N-shortest path word segmentation method has a longer application history and can achieve a good result with lower computing costs. However, as Vietnamese is written in alphabets, there are various expressions such as word confusion, phonetic transcription, initial letters, numbers, etc. As a result, it is impossible to copy the Chinese word segmentation mechanism. A targeted summarization and categorization of the various situations would be necessary to develop segmentation rules and algorithms able to resolve ambiguity and suited to Vietnamese NLP tasks.

3.1 Build N-Path Map Model

In the task of word segmentation, the N-shortest path model is mostly used in situations such as rough word segmentation or preprocessing [12]. The advantage lies in

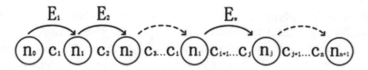

Fig. 1 The N-shortest path map model

its low calculation cost. In the process of constructing an N-path map model, the syllable of each Vietnamese sentence is abstracted into a directed acyclic graph (Fig. 1), in which the space between each syllable is recorded as a node. If a syllable between two nodes can constitute a word, then a reachable path is recorded between them. With the help of the annotated corpus, a training model is constructed to obtain the conditional distribution between syllables. Based on this distribution, weights are assigned to all reachable paths so as to get the shortest path in the map. The general steps are as follows:

(a) Extracting syllable morphemes in Vietnamese sentences, and supposing that there is a sentence consisting of n syllable morphemes.

$$S = (c_1, c_2, c_3, \ldots, c_n)$$

(b) Extracting all possible morphemes sets which can constitute words from the syllable morpheme set S through maximized matching in forward and reverse directions based on the Vietnamese corpus.

(c) Organizing the sets in the form of a directed acyclic graph G with $n + 1$ nodes, naming each node as $(n_1, n_2, \ldots, n_n, n_{n+1})$. If $(c_i, \ldots, c_j)$ appears in the corpus, add a path in graph G to connect $[n_{i-1}, n_j]$ and set its weight as E_w.

(d) Using as prior knowledge, the conditional distribution F of Vietnamese words appearing in sentences by training annotated corpora.

$$F(\text{Word}_2) = \{\text{Word}_2 | \text{Word}_1\}$$
$$E_w = F(c_i \ldots c_j)$$

(see Fig. 1).

(e) Calculating its shortest path and obtain segmentation results.

From the above process, the statistic-based N-shortest path model can receive more accurate word segmentation results than the dictionary matching method in forward and reverse directions in that the path weight, or the conditional distribution probability of each word, is derived from the statistical data of a large number of corpora. This segmentation model can well resolve the ambiguity in the division of Vietnamese phrases. For example, nghiên cứu sinh mệnh (research life) can be divided into nghiên cứu (research) sinh mệnh (life) and nghiên cứu sinh (postgraduate) mệnh (life). Such division ambiguity is more common in Sino-Vietnamese words which are Vietnamese words introduced from Chinese. Dictionary-based matching methods fail to distinguish this type of ambiguity effectively. On the contrary, with reference to the word conditional distribution in the corpus, the N-shortest path model can determine whether it is a proper division according to the conditional probability of that word. Therefore, the N-shortest model is able to eliminate ambiguity to some extent.

3.2 BEMS Marking System

The word segmentation experiment in this paper is based on the Vietnamese phonetic unit, and the accuracy of the calculating algorithm is judged by whether the algorithm correctly distinguishes the position of syllables in words. This paper employed the BEMS marking system where syllables and punctuation marks in a Vietnamese sentence can be divided into B (Begin), E (End), M (Middle), and S (Single). For example, Tư tưởng Chủ nghĩa xã hội đặc sắce Trung Quốc (Socialism with Chinese characteristics). The results of its word segmentation is Tư tưởng (idea) Chủ nghĩa xã hội (socialism) đặc sắc (characteristic) Trung Quốc (China). (period); then, the sentence's BEMS should be marked as Tư(B) tưởng(E) Chủ(B) nghĩa(M) xã(M) hội(E)đặc(B) sắc(E) Trung(B) Quốc(E). (S)

It is obvious that the BEMS marks contain the syllable attribute and position information in each Vietnamese word and can accurately represent the word segmentation information of the syllable. At the same time, the system can conveniently count and compare the results of word segmentation.

3.3 Data Collection and Experiment

This paper has implemented the algorithm of the above N-shortest path segmentation. The training corpus is composed of 8840 sentences from Vietnamese news texts, novel texts, as well as legal documents, and 2000 sentences of varying lengths are selected as test sets. Tasks such as de-duplicating, cleaning, regularization, and recoding have been completed during the implementation; in terms of the basic dictionary, a total of 41,084 Vietnamese expressions have been collected and organized in the form of Tire trees, which has greatly accelerated dictionary retrieval.

In experiments, this paper has selected 1000 Vietnamese sentences from the annotated corpus for test. The sentences used for the experiment were collected from the Vietnamese news network, novel network, and written texts with a total of 25,033 syllables (Table 3).

Table 3 The corpus description of the experiment

Source	Number of sentences	Average number of syllables
Việt Nam	168	23
Báo Nhân Dân	279	21
Jade Dynasty in Vietnamese	321	28
The Constitution of Vietnam	148	33
Subtitles of Vietnam Films	84	15

4 Experimental Results

4.1 Evaluation Standard

This paper evaluates the word segmentation algorithm by comparing the algorithm for segmentation with the BEMS marks judged by humans. The evaluation standard uses three common indicators: precision, recall rate, and $F1$-score. When comparing with the results of manual segmentation which is regarded as the correct criterion, this paper calculated the BEMS results of algorithm segmentation according to the following three formulas. The $F1$-score is the harmonic mean of the precision and recall rate, which could reflect the consistency between the algorithm segmentation and manual segmentation

$$\text{Precision} = \frac{\text{the number of hand marks in algorithm}}{\text{the number of marks in alogrithm}}$$

$$\text{Recall} = \frac{\text{the number of hand marks in algorithm}}{\text{the number of hand marks}}$$

$$F1 \text{ Score} = \frac{2 \times \text{precision} \times \text{recall}}{\text{precision} + \text{recall}}$$

4.2 Experimental Results

The N-shortest path segmentation model has a good effect on Vietnamese word segmentation tasks. When the $F1$-score is 0.89, it means that the results of the N-shortest path segmentation algorithm are basically consistent with the results of manual segmentation. At the same time, the $F1$-score was good in identifying morphemes of the B (Begin), E (End), and S (Single) or punctuation. The algorithm is excellent in terms of the precision of the B (Begin) and E (End) as well as the recall rate of the S (Single). However, the result is poor in the recognition of polysyllable words, that is, in the segmentation of words with more than two syllables (Fig. 2).

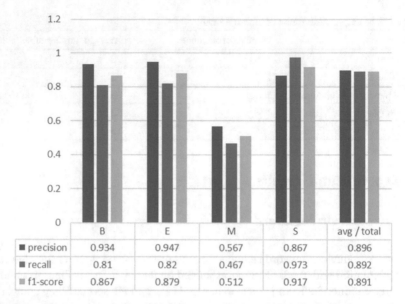

	B	E	M	S	avg / total
■ precision	0.934	0.947	0.567	0.867	0.896
■ recall	0.81	0.82	0.467	0.973	0.892
■ f1-score	0.867	0.879	0.512	0.917	0.891

Fig. 2 The experimental result of the *N*-shortest path segmentation algorithm

From the above experimental results, the scores of the M (Middle) are very low in terms of the three indicators partly because that there are too few polysyllable samples in the training corpus. In Vietnamese, syllable combinations are flexible as such linguistic phenomena as polysyllable idioms, proverbs and idioms are common in Vietnamese. Therefore, in collecting the basic dictionaries, more attention should be paid to polysyllable words. At the same time, the sentence of polysyllable words should also be added to enrich the testing sample before and after training words.

5 Conclusion

To automatically segment words in Vietnamese, this paper proposes an automatic word segmentation method based on the *N*-shortest path. The method uses the *N*-shortest path model to make rough segmentation of Vietnamese sentences. It is on this basis that each syllable is abstracted into a directed acyclic graph, and the shortest path is calculated by combining Vietnamese language features and BEMS marking system so as to realize the automatic word segmentation in Vietnamese. The results show that the method proposed in this paper is more efficient than previous researches and can effectively promote Vietnamese text information processing.

References

1. Võ, T., Xuân, V.: A brief exposition of the influence of Chinese on Vietnamese. J. Jinan Univ. **5**, 56–57 (2001)
2. Ngo, Q.H., Dien, D., Winiwajrter, W.: Automatic searching for english-Vietnamese documents on the internet. In: 24th International Conference on Computational Linguistics, pp. 211 (2012)
3. Do, T.N.D., Le, V.B., Bigi, B., et al.: Mining a comparable text corpus fora Vietnamese-French statistical machine translation system. In: Proceedings of the Fourth Workshop on Statistical Machine Translation. Association for Computational Linguistics, pp. 165–172 (2009)
4. Nguyen, Q.T., Nguyen, N.L.T., Usuke Miyao, Y.: Comparing different criteria for Vietnamese word segmentation. In: Proceedings of 3rd Workshop on South and Southeast Asian Natural Language Processing (SANLP), pp 53–68 (2012)
5. Xiong, M., Li, Y., Guo, J., Mao, C., Yu, Z.: Vietnamese word segmentation with conditional random fields and ambiguity model. J. Data Acquisition Process. 636–642 (2017)
6. Mo, Y., Guo, J., Mao, C., Yu, Z., Niu, Y.: A bilingual word alignment method of Vietnamese-Chinese based on deep neutral network. J. Shandong Univ. (Nat. Sci.) 51(1), 78–82 (2016)
7. Yang, Q., Yu, Z., Hong, D., Gao, S., Tang, Z.: Chinese-Vietnamese word similarity computation based on Wikipedia. J. Nanjing Univ. Sci. Technol. 40(4), 462–466 (2016)
8. Luo, L.: Research on the construction of web-based comparable corpora of Chinese and Vietnamese. Kunming University of Science and Technology Master's Thesis, P17–20 (2015)
9. Tang, M., Zhu M., Yu, Z., Tang P., Gao, S.: Chinese-Vietnamese Bilingual Event Correlation Analysis Based on Bilingual Topic and Factor Graph. ACL 2017 (2017)
10. Generation of Summarization for Chinese-Vietnamese Bilingual News Event Differences. ACL2017 (Note: Oral presentation at conference)
11. Yuan, L., Yangxiu, Z.: Modern Vietnamese grammar. World Publishing Guangdong Corporation, Guangzhou (2012)
12. Huaping, Z., Qun, L.: Model of Chinese words rough segmentation based on N-shortest-paths method. J. Chin. Inf. Process. **5**, 1–7 (2002)

The Analysis of Student Performance Using Data Mining

Leo Willyanto Santoso and Yulia

Abstract This paper presents the study of data mining in the education industry to model the performance for students enrolled in university. Two algorithms of data mining were used. Firstly, a descriptive task based on the K-means algorithm was utilized to select several student clusters. Secondly, a classification task supported two classification techniques, known as decision tree and Naïve Bayes, to predict the dropout because of poor performance in a student's first four semesters. The student academic data collected during the admission process of those students were used to train and test the models, which were assessed using a cross-validation technique. Experimental results show that the prediction of drop out student is improved, and student performance is monitored when the data from the previous academic enrollment are added.

Keywords Data mining · Education · Drop out · Student performance

1 Introduction

Data mining represents a significant computational advance in obtaining information from hidden relationships between variables. This discipline aims to extract useful knowledge from a high volume of data in which initially this knowledge is unknown, but when applying mining techniques, these relationships are discovered. The application of the technologies and tools of data mining in various educational contexts is known as educational data mining (EDM) or data mining in education [1].

The contributions of data mining in education have been used to increase understanding of the educational process, with the main objective of providing teachers and researchers with recommendations for the improvement of the teaching-learning process. By implementing data mining applications in education, teachers and administrators could organize educational resources in a more efficient way.

L. W. Santoso (✉) · Yulia
Petra Christian University, Surabaya, Indonesia
e-mail: leow@petra.ac.id

© Springer Nature Singapore Pte Ltd. 2019
S. K. Bhatia et al. (eds.), *Advances in Computer Communication and Computational Sciences*, Advances in Intelligent Systems and Computing 924,
https://doi.org/10.1007/978-981-13-6861-5_48

The objective of the EDM is to apply data mining to traditional teaching systems—in particular to learning content management systems and intelligent web-based education systems. Each of these systems has different data sources for knowledge discovery. After the pre-processing of the data in each of these systems, the different techniques of data mining are applied: statistics and visualization, grouping and classification, association rules and data mining.

The amount of academic information stored in the databases of educational institutions is very useful in the teaching and learning process; that is why nowadays there has been significant research interest in the analysis of the academic information. This research focuses to apply data mining techniques to the academic records of the students that entered the academic periods between July 2010 and June 2014 through the construction of a mining model of descriptive data, which allows to create the different profiles of the admitted students with socioeconomic information. For the development of the research, the CRISP-DM methodology was used to structure the lifecycle of a data mining project in six phases, described in four levels, which interact with each other during the development of the research [2].

This paper is organized as follows: Sect. 2 contains the background of the research and a review of the state of the art of data mining and the use of its techniques in the educational industry sector. In Sect. 3, the understanding of the data is made, in order to perform a preliminary exploration of the data. The preparation of these covers all the activities necessary for the construction of the final dataset, the selection of tables, records, and attributes. Section 4 focuses on the design and evaluation of a descriptive and classification model. Finally, in Sect. 5, the conclusions and future work are presented.

2 Literature Review

Data mining is widely used in many interdisciplinary fields [3], including in the education sector. There have been many researches in data mining for education. Araque et al. [4] conducted a study on the factors that affect the university dropout by developing a prediction model. This model could measure the risk of abandonment of a student with socioeconomic information and academic records, through the technique of decision tree and logistic regression, to quantify students at high risk of dropping out.

Kotsiantis et al. [5] present the study of a learning algorithm for the prediction of student desertion—i.e., when a student abandons studies. The background of their research is the large number of students who do not complete the course in universities that offer distance education. A large number of testing were carried out with the academic data, the algorithms of decision tree, neural network, Naive Bayes, logistic regression, and support vector machines were compared to know the performance of the proposed system. The analysis of the results showed that the Naive Bayes algorithm is the most appropriate to predict the performance of students in a distance education system.

Kuna et al. [6], in their work "The discovery of knowledge obtained through the process of Induction of decision trees," used decision trees to model classifications of the data. One of the main results obtained was the characterization of students at high risk of abandoning their university studies.

Kovacic [7] studied socioeconomic variables such as age, gender, ethnicity, disability, employment status, and the distance study program. The objective of the research was to identify students at high risk of dropping out of school. Data mining techniques, decision trees, and logistic regression were used in this research.

Yadav et al. [8] presented a data mining project to generate predictive models and identify students at high risk of dropping out taking into account student records at the first enrollment. The quality of the prediction models was examined with the algorithms ID3, C4.5, and ADT of the decision tree techniques. ADT machine learning algorithms can learn from predictive models with student data from previous years. With the same technique, Quadril and Kalyankar [9] presented the study of data mining to construct and evaluate a predictive model to determine the probability of desertion of a particular student; they used the decision tree technique to classify the students with the application of the algorithm C4.5.

Zhang and Oussena [10] proposed the construction of a mining course management system based on data mining. Once the data were processed in the system, the authors identified the characteristics of students who did not succeed in the semester. In this research, support vector machine, Naive Bayes and decision tree were used. The highest precision in the classification was presented with the Naive Bayes algorithm, while the decision tree obtained one of the lowest values.

The evaluation of the important attributes that may affect student performance could improve the quality of the higher education system [11–14]. Radaideh et al. [15] presented a classification model by implementing ID3 and C4.5 algorithms of the decision tree techniques and the Naive Bayes. The classification for the three algorithms is not very high, to generate a high-quality classification model, it is necessary to add enough attributes. In the same study, Yudkselturk et al. [16] examined the prediction of dropout in online academic programs, in order to classify students who dropped out, three mining techniques were applied: decision tree, Naive Bayes, and neural network. These algorithms were trained and evaluated using a cross-validation technique. On the other hand, Pal [17] presented a data mining application to generate predictive models taking into account the records of the students of the first period. The decision tree is used for validation and training to find the best classifier to predict students who dropped out.

Bhise et al. [18] studied the evaluation factors of students to improve performance, using grouping technique through the analysis of the K-Means algorithm, to characterize the student population. Moreover, Erdogan and Timor [19] presented the relationship of university students between the entrance examinations and the results of success. The study was carried out using algorithm techniques of group analysis and K-Means. Bhardwaj and Bhardwaj [20] presented the application of data mining in the environment of engineering education, the relationship between the university, and the results obtained by students, through the analysis of K-algorithm techniques.

3 Data Analysis and Modeling

This chapter focuses on the understanding of the data where visualization techniques are applied, such as histograms, in order to perform a preliminary exploration of the records and verify the quality of the data. Once the analysis is done, we proceed with the data preparation phase, which includes the tasks of selecting the data to which the modeling techniques will be applied for their respective analysis.

The first task is collecting the initial data. The objective of this task is to obtain the data sources of the academic information system of the University. The first set of data grouped the socioeconomic information and the result of the admission tests (Language, English, Mathematics, and Logic). The second set of data is made up of the academic and grading history obtained by the students: the academic year and period of the student's admission; the program in which he/she is enrolled; the student's academic situation (academic blocking due to low academic performance and no academic blocking); and number of academic credits registered, approved, lost, canceled, and failed. The generated queries were made through the PostgreSQL database management system. A process of concatenation of the two datasets was performed, obtaining a flat file with 55 attributes and 1665 records of students admitted and enrolled in the systems and electronics engineering programs.

The next task is data exploration. Exploratory analysis is a task that allows detailed analysis of some variables and identifying characteristics; for this, some of the visualization tools such as tables and graphs were used, with the purpose of describing the data mining objectives of the comprehension phase.

The task of checking the quality of the data specifies a revision of the same as the lost or those that have missing values committed by coding errors. In this section, the quality of the data corresponding to the socioeconomic information of the admitted student is verified.

The next task is data selection. In this task, the process of selecting the relevant data for the development of the data mining objectives is carried out. A first pre-processing, for the final selection of the data, is the selection of attributes. It was obtained that there are 55 attributes or variables that contain values that may or may not contribute to the study; this is based on the exploration initial of the data and in the description of the fields defined in the variable dictionary. In the dataset selected for the modeling, no errors were found in the fields; differences in the selected records, the errors that were presented in some cases were missing, due to the fact that the processing was not adequate at the time of the typing such as email, residence address, telephone number, date of birth, type of blood, and ethnicity that are attributes considered not relevant to the case under study.

To develop the model, the application RapidMiner was used for automatic learning for analysis and data mining; this program allows the development of data analysis processes through the linking of operators through a graphic environment. For the implementation of the algorithm, the K-Means operator of the grouping and segmentation library using the Euclidean distance was used to evaluate the quality of the groups found. The algorithm is responsible for both numerical and categorical values.

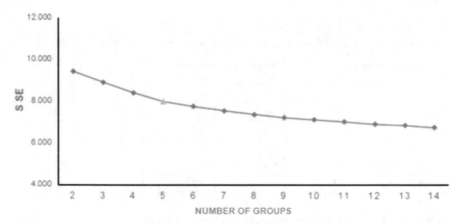

Fig. 1 Selection of the Group number (K) for students admitted

Table 1 Distribution of the registration number in the application of the K-means algorithm

	Group 0	Group 1	Group 2	Group 3	Group 4
The number of record	317	130	418	389	409
Percentage (%)	19	8	25	23	25

However, additional pre-processing was performed to normalize all the numerical attributes between 0 and 1 with the normalize operator. All attributes must have the same scale for a fair comparison between them.

A grouping model was applied to the dataset for the characterization of the admitted students, create the different profiles of the students in the different groups found, and determine what other factors define the separation of groups produced by the K-Means algorithm.

Repeated interaction was performed to determine the value of K or the number of groups. The value of K varied from 2 to 14. The results were evaluated based on the quadratic error of each iteration; for the selection of the group number, the elbow method was used.

Figure 1 shows the iterations performed to find the value of k in the first dataset of the admitted students, the k with value of 5 was selected, where the SSE is equal to 7954. The K-Means algorithm produced a model with five groups, from the description of these groups is expected to characterize the profiles of admitted students. Table 1 shows the distribution of the number of records and the percentage of each of the resulting groups. Group 2 and 4 group the largest number of records, on the contrary, the lowest percentage of records are in group 1.

The model was necessary to "de-normalize" it, to put each one of the values of the variables in their original ranges. The analysis of the model was made with the socioeconomic information and the results of the admission tests; then, an analysis was made about the academic situation of the students who in each group had academic block with four enrollments.

Table 2 Distribution of the number of students with four enrollments

No. of groups	Enrollment 1	Enrollment 2	Enrollment 3	Enrollment 4	Total
Group 0	32	13	11	10	66
Group 1	127	63	47	29	266
Group 2	178	58	72	23	331
Group 3	160	90	48	41	339
Group 4	66	32	45	52	195
Number of records	563	256	223	155	1197
Percentage (%)	47	21	19	13	100

Table 3 Distribution of the number of students with four enrollments

No. of groups	Enrollment 1		Enrollment 2		Enrollment 3		Enrollment 4	
	No block	Block acad.	No block	Block acad.	No block	Block acad.	No block	Block acad.
Group 0	23	26	18	2	14	3	15	0
Group 1	17	30	13	11	15	2	9	2
Group 2	43	11	15	3	22	0	7	0
Group 3	22	26	21	6	14	0	11	1
Group 4	16	17	12	5	23	0	26	1

Table 2 shows the distribution of the number of records in each of the groups in the first four semesters or academic enrollments. Groups 2 and 3 are characterized by grouping the largest number of records with 28% in each group. In group 0, on the other hand, there is the smallest number of records with 6%. 47% of the registers are students of the first semester, 21% present second enrollment, 19% with third enrollment, and 13% of the remaining registers have four academic enrollments.

Table 3 shows in each group the academic status of the students in the first four enrollments. Group 1 is characterized by grouping the highest percentage of students with academic block, in contrast to group 2 where you can see the lowest percentage of students with academic block. The group 0 is characterized by good performance in the admission tests, grouped 26% of students with blocking in the first enrollment, 2% in the second and 3% in the third enrollment. Group 1 groups the students with the lowest performance of the admission tests similar to group 2. 30% of the students present blocking in the first enrollment, 11% in the second, and 2% are in the third and fourth enrollment.

Group 2 is characterized by grouping the students with the lowest performance of the admission tests and the least number of students with blocks. About 11% of the students have a block in the first enrollment and 3% in the second enrollment. Group 3 is characterized by grouping the students with good performance in the admission tests similar to group 0.26% of students present blockage in the first enrollment, 6% with two and 1% with four enrollments. Finally, group 4 is characterized by grouping the smallest number of students with blocks. About 17% of students with a registration have a block, 5% correspond to students with two enrollments, and 1% with four enrollments.

4 Result and Analysis

In this section, two models of data mining to analyze the academic and non-academic data of the students are presented. The models used two classification techniques, decision tree, and Naïve Bayes, in order to predict the loss of academic status due to low academic performance in their study. The historical academic records and the data collected during the admission process were used to train the models, which were evaluated using cross-validation.

Table 4 presents the total number of registrations or students with first enrollment or enrollment and number of students with academic block due to underperformance.

Table 5 shows the number of students with academic block in each period or enrollment. The largest number of students with academic block is presented in the first enrollment. The second, third, and fourth enrollment shows a decrease in the number of students with blocks. In the 2010–01 entry period, the highest number of students with academic blocks was presented in each academic enrollment.

The classification model proposed in this research uses the socioeconomic information. The classification model uses two widely used techniques, decision trees and a Bayesian classifier. The reason for selecting these algorithms is their great simplicity and interpretability.

The decision tree is the first technique used to classify the data; this algorithm generates a recursive decision tree when considering the criterion of the highest proportion of information gain—that is, it chooses the attribute that best classifies the data. It is a technique where an instance is classified following the path of conditions, from the root to a leaf, which will correspond to a labeled class. A decision tree can easily be converted into a set of classification rules. The most representative algorithm is C4.5, which handles both categorical and continuous attributes. It generates a decision tree recursively when considering the criterion of the highest proportion of information gain. The root node will be the attribute whose gain is maximum. Algorithm C4.5 uses pessimistic pruning to eliminate unnecessary branches in the decision tree and to improve classification accuracy.

The second technique to be considered for the construction of the model is a Bayesian classifier. It is one of the most effective classification models. Bayesian classifiers are based on Bayesian networks; these are models probabilistic graphs that allow modeling in a simple and precise way the underlying probability distribution to a dataset. Bayesian networks are graphic representations of dependency and independence relationships between the variables present in the dataset that facilitate the understanding and interpretability of the model. Numerous algorithms have been proposed to estimate these probabilities. Naive Bayes is one of the practical learning

Table 4 Registration number and academic blocks per academic period

Academic condition	Academic period									
	2010–1	2010–2	2011–1	2011–2	2012–1	2012–2	2013–1	2013–2	2014–1	2014–2
No block	115	145	137	119	100	146	131	151	110	166
Acad. block	70	32	49	30	31	22	25	35	27	23
Total record	185	177	186	149	131	168	156	186	137	189

Table 5 Academic block by entry period or first enrollment

Income period	Academic block									
	2010-1	2010-2	2011-1	2011-2	2012-1	2012-2	2013-1	2013-2	2014-1	2014-2
2010-1	**40**	15	7	5	0	1	2	0	0	0
2010-2		**28**	0	0	3	0	0	0	0	0
2011-1			**39**	6	2	2	0	0	0	0
2011-2				**22**	8	0	0	0	0	0
2012-1					**20**	11	0	0	0	0
2012-2						**14**	8	0	0	0
2013-1							**15**	10	0	0
2013-2								**28**	7	0
2014-1									**26**	1
2014-2										**23**

Table 6 Academic situation in the first four enrollments

Academic situation	Enrollment 1	Enrollment 2	Enrollment 3	Enrollment 4
No block	309	190	214	145
Acad. block	255	66	9	10
Total record	564	256	223	155

Table 7 Test, training and validation dataset

Number of enrollment	Total record	Training and validation data 80%		Test data 20%	
		No block	Acad. block	No block	Acad. block
Enrollment 1	564	247	204	62	51
Enrollment 2	256	152	53	38	13
Enrollment 3	223	171	7	43	2
Enrollment 4	155	116	8	29	2

algorithms most used for its simplicity, resistance to noise, short time for processing, and high predictive power.

Different models were trained and tested to predict if a student will be blocked in a particular enrollment. The first model analyzed the loss of academic status based on socioeconomic information and the results of the tests collected during the admission process. The second model was analyzed with the initial information of the enrollment process and the academic records of the first four registrations. Table 6 describes the number of registrations in the first four enrollments with academic status (No Block and Academic Block).

For the design of the model, the RapidMiner application was used; this is a program for automatic learning and data mining process, through a modular concept, which allows the design of learning models using chain operators for various problems. For the validation of the classification model, stratified sampling technique was used. The operator to partition the dataset called split data; this operator creates partition to the dataset in subsets according to the defined size and the selected technique. For the implementation of the decision tree algorithm, the decision tree operator and the Bayesian algorithm Naive Bayes were used. Table 7 shows the number of records in the first four enrollment, and 80% of the records were taken as training set and 10-fold cross-validation and 20% of the sample was used as a test set.

To estimate the performance of the model, the X-Validation operator was used. This operator allows to define the process of cross-validation with 10-fold on the input dataset to evaluate the learning algorithm. The performance of the model was measured with the operator performance binomial classification. This operator presents the performance results of the algorithm in terms of accuracy, precision, recall, error, and ROC curve. To analyze the errors generated from a classification model, the confusion matrix is used. It is a visualization tool that is used in supervised learning.

Table 8 Prediction model of the loss of academic condition with the training and validation dataset

Prediction	Decision tree			Naïve Bayes		
	Enrollment 2	Enrollment 3	Enrollment 4	Enrollment 2	Enrollment 3	Enrollment 4
Measure-*F*	0	0	30.77%	0	53.41%	38.51%
Precision	0	0	33.33%	0.00%	56.27%	44.17%
Exhaustive	0.00%	0.00%	28.57%	0.00%	51.45%	36.00%
Accuracy	54.76%	74.14%	94.93%	92.76%	59.43%	69.81%
Error	45.24%	25.86%	5.07%	7.24%	40.57%	30.19%
Curve (AUC)	0.5	0.5	0	0	0.608	0.63
Kappa	0.0	0.0	0.282	−0.015	0.177	0.19
Specificity	100%	100%	97.61%	99.23%	66.02%	81.65%
Sensitivity	0.00%	0.00%	28.57%	0.00%	51.45%	36.00%
False positive	0%	0%	2%	1%	19%	14%
False negative	45%	26%	3%	6%	22%	17%

Each column of the matrix represents the number of predictions of each class, while each row represents the instances in the real class.

The following measurements are calculated during the experiment: accuracy, classification error, exhaustiveness (recall), precision, f_measure, specificity, sensitivity, false-negative rate, false-positive rate, and area under the curve (AUC).

In this stage, different models were trained and tested to classify students with academic block in the first four academic enrollments, using the socioeconomic information. For the configuration of the experiments, we used cross-validation with 10-fold to train the models and the evaluation of the model we used the test dataset. The performance of the model was evaluated with 80% of the training and validation data, and 20% of the sample was used as a test set. In the decision tree technique with training and validation data, the tree depth was varied from 1 to 20; the lowest classification error was found in depth 3, where the error begins to show some stability in each of the four academic periods. Finally, the training and validation models were evaluated with the test dataset.

Table 8 presents the results of the pre-condition model of the loss of the academic condition with training and validation data, comparing the different classification techniques in terms of the different performance parameters.

Analyzing the results of the training and validation dataset with the admission information of the admission process, it is observed how the Bayesian classifier presents the best accuracy of academic block records that were correctly classified. In the third, enrollment increased by 7% with respect to the decision tree. Similarly, after by reviewing the area under the curve (AUC), the decision tree in the

Table 9 Prediction model of the loss of the academic condition using the training and validation data

Prediction	Decision tree			Naïve Bayes		
	Enrollment 2	Enrollment 3	Enrollment 4	Enrollment 2	Enrollment 3	Enrollment 4
Measure-F	74.42%	0	11.11%	71.00%	41.67%	40.00%
Precision	66.40%	0.00	10.00%	60.75%	29.41%	33.33%
Exhaustive	86.67%	0.00%	12.50%	87.00%	71.43%	50.00%
Accuracy	84.45%	94.31%	87.05%	81.02%	92.06%	90.19%
Error	15.55%	5.69%	12.95%	18.98%	25.86%	9.81%
Curve (AUC)	0.851	0	0	0.912	0	0
Kappa	0.637	−0.224	0.042	0.578	0.295	0.350
Specificity	83.58%	98.20%	92.12%	78.90%	92.93%	92.94%
Sensitivity	86.67%	0.00%	12.50%	87.00%	71.43%	50.00%
False positive	13%	2%	7%	16%	7%	6%
False negative	1%	4%	6%	3%	1%	3%

first and second enrollment shows a poor performance below 0.5. The Naive Bayes algorithm presents the highest percentage of cases with no academic blockade that were classified incorrectly with academic block. The decision tree presents the highest proportion of class with academic block that were classified incorrectly with no academic block.

Table 9 presents the results of the model of the pre-condition of the loss of the academic condition with the admission information of the admission process and the academic record of the previous semester with the data of training and validation, the different classification techniques are compared in terms of different performance parameters.

Analyzing the results of the training and validation dataset, we observe how the decision tree increased its level of accuracy in the second and fourth enrollment. The Bayesian classifier increased the accuracy of records with academic blocks that were correctly classified. Similarly, by reviewing the area under the curve (AUC), both algorithms in the second enrollment have a good performance above 0.7.

Table 10 presents the results of the model of pre-condition of the loss of the academic condition with the admission information of the admission process and the academic record of the previous semester with the test data; the different classification techniques are compared in terms of performance parameters.

Analyzing the results of the test dataset, we observe how the decision tree presents the highest number of predictions with academic blocking that were correctly classified in the second enrollment. Likewise, by reviewing the area under the curve

Table 10 Prediction model of the loss of academic condition using the test data

Prediction	Decision tree			Naïve Bayes		
	Enrollment 2	Enrollment 3	Enrollment 4	Enrollment 2	Enrollment 3	Enrollment 4
Measure-F	74.29%	0	0	70.27%	0%	0%
Precision	59.09%	0.00%	0.00%	54.17%	0.00%	0.00%
Exhaustive	100%	0.00%	0.00%	100%	0.00%	0.00%
Accuracy	17.65%	4.44%	6.45%	21.57%	6.67%	9.68%
Error	82.35%	95.56%	93.55%	78.43%	93.33%	90.32%
Curve (AUC)	0.882	0.500	0.534	0.913	0.907	0.828
Kappa	0.622	0.00	0.000	0.556	−0.031	−0.045
Specificity	76.32%	100%	100%	71.05%	97.67%	96.55%
Sensitivity	100%	0.00%	0.00%	100%	0.00%	0.00%
False positive	18%	0%	0%	22%	2%	6%
False negative	0%	4%	6%	0%	4%	3%

(AUC), the Naive Bayes algorithm presents a good performance with an area greater than 0.9 in comparison with the algorithm of the decision tree.

5 Conclusion

In recent years, there has been great interest in data analysis in educational institutions, in which high volumes of data are generated, given the new techniques and tools that allow an understanding of the data. For this research, a set of data was compiled from the database of the "X" University with socioeconomic information and the academic record of the previous enrollment, for the training and validation of the descriptive and predictive models.

The objective of the application of the K-Means algorithm of the descriptive model was to analyze the student population of the university to identify similar characteristics among the groups. It was interesting to establish that some initial socioeconomic characteristics allowed to define some profiles or groups. In the evaluation of the model, it was observed that the student's socioeconomic information affects the results of their academic performance, showing that the groups with the highest academic performance in the knowledge test results were found in the schools with low socioeconomic status.

The classification model presented in this paper analyzed the socioeconomic information and the academic record of the student's previous enrollment. The decision tree algorithm with the test data presented a better performance with the addition of the academic record of the previous semester compared to the Naive Bayes algorithm. The analysis of the data could show that there are different types of performance according to the student's socioeconomic profile and academic record, demonstrating that it is feasible to make predictions and that this research can be a very useful tool for decision making.

This research can be used for decision making, by the permanency and graduation program of the University and can be used as a starting point for future data mining research in education. Another important recommendation is that to improve the performance of the model, other sources of data should be integrated, such as the information of the student who is registered as a senior in high school in the senior high school, before entering the university.

References

1. Romero, C., Ventura, S.: Educational data mining: a survey from 1995 to 2005. Expert Syst. Appl. **33**(1), 135–146 (2007)
2. Chapman, P.: CRISP-DM 1.0: Step-by-Step Data Mining Guide. SPSS, New York (2000)
3. Khamis, A., Xu, Y., Mohamed, A.: Comparative study in determining features extraction for islanding detection using data mining techniques: correlation and coefficient analysis. Int. J. Electr. Comput. Eng. (IJECE) **7**(3), 1112–1134 (2017)
4. Arague, F., Roldan, C., Salguero, A.: Factors influencing university drop-out rates. Comput. Educ. **53**(3), 563–574 (2009)
5. Kotsiantis, S., Pierrakeas, C., Pintelas, P.: Preventing student dropout in distance learning systems using machine learning techniques. In: Proceedings of the 7th International Conference Knowledge-Based Intelligent Information and Engineering System (KES), Oxford, pp. 267–274 (2003)
6. Kuna, H., Garcia-Martinez, R., Villatoro, F.: Pattern discovery in university students desertion based on data mining. In: Proceedings of the IV Meeting on Dynamics of Social and Economic Systems, Buenos Aires, pp. 275–285 (2009)
7. Kovacic, Z.: Predicting student success by mining enrolment data. Res. Higher Educ. J. **15**, 1–20 (2012)
8. Yadav, S., Bharadwaj, B., Pal, S.: Mining educational data to predict student's retention: a comparative study. Int. J. Comput. Sci. Inf. Secur. (IJCSIS) **10**(2), 113–117 (2012)
9. Quadril, M., Kalyankar, N.: Drop out feature of student data for academic performance using decision tree techniques. Global J. Comput. Sci. Technol. **10**(2), 2–5 (2010)
10. Zhang, Y., Oussena, S., Clark, T., Kim, H.: Use data mining to improve student retention in higher education—a case study. In: Proceedings of the 12th International Conference on Enterprise Information Systems, pp. 190–197 (2010)
11. Santoso, L., Yulia.: Predicting student performance using data mining. In: Proceedings of the 5th International Conference on Communication and Computer Engineering (ICOCOE) (2018)
12. Rao, M., Gurram, D., Vadde, S., Tallam, S., Chand, N., Kiran, L.: A predictive model for mining opinions of an educational database using neural networks. Int. J. Electr. Comput. Eng. (IJECE) **5**(5), 1158–1163 (2015)
13. Santoso, L., Yulia.: Data warehouse with big data technology for higher education, Proc. Comput. Sci. **124**(1), 93–99 (2017)
14. Santoso, L., Yulia.: Analysis of the impact of information technology investments—a survey of Indonesian universities. ARPN JEAS **9**(12), 2404–2410 (2014)

15. Al-Radaideh, Q., Al-Shawakfa, E., Al-Najjar, M.: Mining student data using decision trees. In: Proceedings of the 2006 International Conference on Information Technology (ACIT), pp. 1–5 (2006)
16. Yudkselturk, E., Ozekes, S., Turel, Y.: Predicting dropout student: an application of data mining methods in an online education program. Eur. J. Open Distance E-Learning **17**(1), 118–133 (2014)
17. Pal, S.: Mining educational data to reduce dropout rates of engineering students. Int. J. Inf. Eng. Electron. Bus. **4**(2), 1–7 (2012)
18. Bhise, R., Thorat, S., Superkar, A.: Importance of data mining in higher education system. IOSR J. Humanit. Soc. Sci. **6**(6), 18–21 (2013)
19. Erdogan, S., Timor, M.: A data mining application in a student database. J. Aeronaut. Space Technol. **2**(2), 53–57 (2005)
20. Bhardwaj, A., Bhardwadj, A.: Modified K-means clustering algorithm for data mining in education domain. Int. J. Adv. Res. Comput. Sci. Softw. Eng. **3**(11), 1283–1286 (2013)

Efficient CP-ABE Scheme for IoT CCN Based on ROBDD

Eric Affum, Xiasong Zhang, Xiaofen Wang and John Bosco Ansuura

Abstract A content centric network (CCN) is a promising network proposed for fifth generation (5G) network paradigm. This will improve the efficient distribution of the future Internet of Things (IoT) media content and allow nodes communication based on content names. End users can obtain content from any intermediate caches which possess difficulties in securing data cached. Therefore, the ability for self-contained protection is paramount. A ciphertext-policy attribute base encryption (CP-ABE) scheme had been identified as a preferable solution. To achieve an efficient performance of CP-ABE scheme, we exploited access tree representation of access structure which defines access policy and modified it into a unique CP-ABE scheme based on the reduced ordered binary decision diagram (ROBDD), and proposed an efficient non-monotonic ROBDD CP-ABE scheme for IoT CCN. The size of the key is not affected by the number of users. The ROBDD legitimate path is used to determine the size of the ciphertext instead of the number of the nodes. This scheme has a high decryption stage and offers resistance to collision attack. The efficiency of the proposed scheme is based on the efficiency of the ROBDD. In accordance with performance analysis, ROBDD CP-ABE scheme achieves a high efficiency as compared to the existing CP-ABE schemes.

Keywords Content centric network · Internet of things · Ciphertext policy · Reduce ordered binary decision diagram · Access policy

E. Affum (✉) · X. Zhang · X. Wang · J. B. Ansuura
School of Computer Science and Technology, University of Electronic Science
and Technology of China, Chengdu 611731, China
e-mail: affrico23@yahoo.com

X. Zhang
e-mail: johnsonzxs@uestc.edu.cn

X. Wang
e-mail: xfwang@uestc.edu.cn

J. B. Ansuura
e-mail: jansuura@gmail.com

© Springer Nature Singapore Pte Ltd. 2019
S. K. Bhatia et al. (eds.), *Advances in Computer Communication
and Computational Sciences*, Advances in Intelligent Systems and Computing 924,
https://doi.org/10.1007/978-981-13-6861-5_49

1 Introduction

This immense user generated traffic along with the amount of current and projected machine-to-machine traffics are driving researches into domain such as 5G technologies, aimed at providing a complete wireless communication network with almost no limitations, with a higher bandwidth and ultra-low latency for mobile applications [1]. According to [2], fifty billion devices will be connected to the Internet through IoT with a tremendous increase in data size generated by these devices. This implies that the majority of internet traffics will be multimedia and will be originated from IoT devices [3], which will impose limitations on having an efficient data flow needed by IoT applications [4]. This rapid growth driven by the use of P2P software such as BitTorrent, Ares, and many others allow every Internet user to become a data server. With this phenomenon, the Internet users are interested in the data content (video, music, movies) but indifferent about the source of content as long as they are sure about the content. These concerns had been a strong push to restructure the IoT networks where each data item is being named and routing is performed by using the name of the data [2]. There are several proposals such as CCN/NDN [5], DONA [6], PSIRP [6], PURSUIT [7], and NetInf [8], aimed to attain the above objectives. In CCN, the packets are contents and interests. The sender does not directly send contents to the receivers. The advertisement of messages is published by senders to all networks for content to be shared without knowing intended users of the content. A receiver declares interest for data without knowing the publisher of the content. The CCN establish a delivery path between publisher and receiver for delivery of contents based on the receiver's subscription. A CCN is a unique paradigm regarding its naming, routing, caching and security. One of the main aspects of CCN is the security aspect, which changes to securing the content instead of the path. This new security model adding to the current attacks may affect CCN. There are three main entities of CCN scenario in Fig. 1: content server (CS), content centric network (CCN) and end users (EU). Suppose an end user EU A first access a content, it is being accessed from the CS. The content is then cached in each of the intermediate CCN node along the path, indicated in (red) between EU A and the content server. When EU B want to access the same content, it is accessed from the cached in the CCN 2. Similarly, end user EU C will find the content cached in CCN 1. This gives efficient content delivery. However, in the consideration of a complicated setting such as content centric IoT environment where users access cached content from nodes situated in diverse geographical domains, an effective and flexible management of content is highly challenging. The security in CCN will be a basic part of the design but not an extra feature, which requires content to be thoroughly self-dependent, not to rely on any third party for security but to be self-secured.

The most intuitive way to achieve this objective is to encrypt contents before publishing, which could be accessed by only legitimate clients. Most authors have proposed many schemes to achieve authentication and confidentiality based on encryption for ICN architectures [9]. The traditional public key encryption schemes adopted by some of these works are for one-to-one communication. This may prohibit efficient

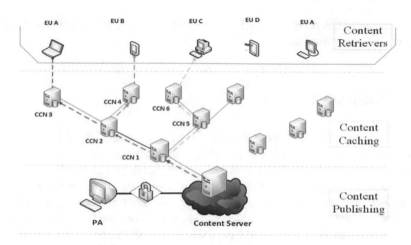

Fig. 1 IoT CCN distribution scenario

caching and effective management since any cached content needs to be encrypted and retrieved with different pairs of keys for different users and do not support different subscriptions. This limitation has created a way for the invention of Attribute base encryption. Sahai and Water [10] proposed an ABE scheme for the first time extended from IBE with two main branches: CP and KP ABE schemes. As in [11], researchers proposed security for sensor and IoT. Authors of this work [12] investigated into authentication, authorization, and access control scheme of IoT. However, these works are not suitable for IoT based on CCN since they are design for a single cache server and support end-to-end communications. Considering IoT scenarios based on CCN environment, CP-ABE schemes maintain advantages for efficient and secure shared contents. ABE was first introduced into ICN architecture by Papanis et al. [13] and Rao et al. [14]. With this concept, integrating CP-ABE into IoT based on CCN. IoT content is encrypted with policy to achieve a self-contained data protection as well as enforcing access control to make content accessible by only authorized users.

Various researchers have proposed security scheme for IoT based on ICN. To enhance on the efficiency of CP-ABE scheme, and for that matter, an efficient IoT CCN secure data retrieval, the access structure used for access policy needs to be optimized. There are several access structures such as LSSS matrix [15], distributed matrix [16], AND gate [17], threshold gates [18], and many others which are used to represent access policy. Viera et al. [19] proposed a protocol to provide a Security Coverage Network designed for Smart Grid on ICN platform. The Key management in this agreement requires more inquiries and protocols need a revise with respect to scalability. In addition, an ordered binary decision diagram (OBDD) was used to describe the access policies in CP-ABE for IoT. The system completely utilizes the authoritative descriptive control and OBDD high computational efficiency and improves performance and efficiency [20]. To ensure a significant improvement of the efficiency of CP-ABE, this work will employ a reduced ordered binary decision diagram for access structure representation.

1.1 Contribution

We constructed a modified access structure for CP-ABE using ROBDD and then presented a ROBDD CP-ABE scheme for efficient IoT CCN content dissemination. The access structure based on ROBDD is non-monotonic access structure. An ordered binary decision diagram was further reduced to optimize the access policy. It supports every Boolean operations and can support multiple users of the same attributes of subscriptions in strategy. The proposed scheme has a more compact access structure and has less nodes and paths. Therefore, the key generation, encryption, and decryption process are much more efficient. The integration of ROBDD CP-ABE scheme with IoT CCN will provide self-secured and efficient content demand system and only legitimate users can access contents effectively.

1.2 Organization

The remaining of the paper is organized as follows: The preliminaries of the research are discussed in Sect. 2. The demonstration of access structure used in the proposed scheme and our main construction is conducted in Sect. 3. The security and performance analysis is evaluated in Sect. 4. Finally, this paper concluded in Sect. 5.

2 Preliminaries

2.1 Bilinear Maps

Let G_0 and G_1 be two multiplicative cyclic group of prime number order p. Let g be the generator of G_0 and e be a bilinear map, $e: G_0 \times G_0 \to G_1$. The bilinear map has the following properties.

1. *Bilinearity*: for all $u, v \in G_0$ and $a, b \in Z_P$ we have $e(u^a, v^b) = e(u, v)^{ab}$
2. *Non-degeneracy*: $e(g, g) \neq 1$.

For G is a bilinear if the group operation in G_0 and the bilinear map: $e: G_0 \times G_0 \to G_1$ are both efficiently computable.

2.2 Access Structure

Let A be collection of an access structure for non-empty subset of S, thus $A \in 2^s\{\Phi\}$. The subsets are known as authorized sets. Any set which does not belong to A is termed as unauthorized sets.

2.3 Ciphertext-Policy Attribute Base Encryption

The CP-ABE basic algorithm includes these four fundamental operations.

Setup (λ): The authority executes this process by taking security parameter λ as input and output the public key *PK* and master key *MK*.

Encrypt (*PK, MK, A*): The data owner executes this algorithm. It takes public parameter *PK*, plaintext M and access policy A, to output Ciphertext *CT*

Keygen (*MK, S*): The authority executes this algorithm. It takes masters key *MK* and attribute set S as input and outputs the secret key *SK*.

Decrypt (*PK, CT, SP*): This takes in public parameter *PK*, ciphertext *CT*, and a secret key *SK* as input and output the message M for the user

2.4 Decision Bilinear Diffie-Hellman

Let e denote a bilinear map e: $G_0 \times G_0 \rightarrow G_1$ with g as a generator and a, b, c, z as random numbers chosen from Z.

The assumption of DBDH says no probabilistic polynomial time adversary can distinguish tuple $(g^a, g^b, g^c, e(g, g)^{abc})$ from tuple $(g^a, g^b, g^c, e(g, g)^c)$.

2.5 CPA Security Model

Init: The adversary chooses an access structure A and sends to the Challenger

Setup: The setup algorithm runs by the challenger to generate freshly *PK* and *MK* and then sends *PK* to the adversary.

*Phase*1: The adversary makes secret request to *Keygen* algorithm using the attribute set S, with a condition that $S \notin A$.

Challenge : Two messages, M_0 and M_1, of equal length send to the challenger by the adversary. Upon receiving M_0 and M_1, the challenger randomly chooses $b \in (0, 1)$ and encrypts M_b using A to obtain the ciphertext *Ct*. The challenger then sends *Ct* to the adversary.

*Phase*2: This *Phase* is the same as *Phase*1 with the same restriction.

Guess: A guess $b^1 \in (0, 1)$ is output by the adversary. The advantages in the above security game is define as: $\text{Adv} = \left| \Pr[b = b^1] - \frac{1}{2} \right|$.

2.6 Reduced Ordered Decision Diagram (ROBDD)

A *binary decision diagram* (*BDD*) [21] is a rooted directed acyclic graph with the following properties: There are one or two terminal nodes of out-degree zero and are labeled as $0(False)$ or $1(True)$. A set of variable nodes u of out-degree two. The two outgoing edges are given by two functions $low(u)$ and $high(u)$. A variable $var(u)$ is associated with each variable node.

A BDD is *ordered binary decision diagram* (*OBDD*) if on all paths through the graph, the variables respect a given linear order $x_1 < x_2 < \cdots < x_n$.

An OBDD is ROBDD if:

No two distinct nodes u and v have the same variable name and *low* and *high* successor, i.e., if $var(u) = var(v)$, $low(u) = low(v)$ and $high(u) = high(v)$, implies $u = v$ and No variable node u has identical low and high successor, i.e., $low(u)$ and $high(u)$.

ROBDD is based on a fixed ordering of variables and have the additional property of being reduced. This means it is irredundant, unique, and recover the important canonicity property. Thus, for a fixed variable ordering, each Boolean function has a canonical (unique) representation as an ROBDD. This means we can compare Boolean functions by constructing their ROBDDs and checking if they are equal same successors.

3 CP-ABE Scheme Based on Robdd

3.1 Access Structure Based on ROBDD

A non-monotonic special access structure is proposed. This is more efficient and flexible, and it is based on ROBDD. This can support dissimilar attributes in a repeated way.

3.2 Boolean Function of Access Policy

Let $(0 \leq i \leq n - 1)$ represent the whole attributes in access policy numbered in a predefined sequence. These represent as $w_i (0 \leq i \leq n - 1)$ where the total number of the attributes is n. We can then represent access policy by Boolean function, $f(w_0, w_1, \ldots, w_{n-1})$. The Boolean function can be converted to logical operations which entails AND, OR, and NOT. However, the transformation is more complicated for threshold gate in such Boolean operation. The legitimate users who access and decrypt messages in certain security systems are those who are able to accomplish specified threshold operations. Such operation is termed as a threshold gate $T(t, n)$. Let N represent an attribute set where n is the number of attributes in N. The steps below are used to construct a Boolean function of given threshold.

First Step: All subsets in N are chosen. This consists of t different attributes. Compute the total number of the subsets $C(n, t)$, separately denoted as $Com_1, Com_2, \ldots, Com_{n,t}$ using the formula of permutation and combination.

Second Step: Conduct a set level conjugate operation for each subset with size t. These formulas are individually denoted as $Com_1, Com_2, \ldots, Com_{n,t}$.

Third Step: Perform a disjunctive operation on $C(n, t)$, conjugate formulas to obtain Boolean function $T(t, n)$. Finally, Boolean function is obtain as $f(t, n) = V_{i=1}^{C(n,t)}$.

3.3 ROBDD Access Structure

To obtain a Boolean construction for access policy, ROBDD was used. It has a compact size, and it can represent a Boolean function in an efficient way. With a given Boolean function, a tight upper bound for worst case size of ROBDD is $((2^n/n))(2 + \mathcal{E})$ where n is variables of the function [22]. In practice, it is very lesser than worst case and it almost proportional to the number of the variables. The construction of ROBDD for Boolean function is performed by applying recursive process based on Shannon's expression theorem: $f_1(w_0 w_1 \ldots w_n) = w_{i=1} \cdot f|w_{i=1} + w' \cdot f|w_{i=0} (0 \leq i \leq n - 1)$. The variable ordering chosen for Boolean function may have impact on the number of nodes contained in ROBDD. Therefore, the order or variable must be stated before designing to ensure a specific ROBDD. Let τ represent access structure of ROBDD. The Node denotes the nodes of ROBDD. The Table T denotes the dictionary which stores previous constructs. The final ROBDD expression after construction is given as $\{Node_{id}^i \, id\}$ where $i \in I$, the attribute set of all the attributes in the access structure and $id \in ID$ is the set of the ID numbers of the non-terminal nodes of the ROBDD. The tuple $Node_{id}^i$ is $<i, id, h, l>$ where h denotes the high ID branch and l low ID branch. To ensure the relationship between parents and children nodes, h and l parameters are used. The leaf nodes 0 and 1 will be deleted in ROBDD structure to minimize storage cost since they have a constant meaning, and i, h, and l domain of these nodes have no meaning.

3.4 Satisfying a ROBDD

Let us assume ROBDD is an access structure with root node R and R_w represent non-terminal nodes. S denotes attribute set. A comparison is made starting from R.

If attribute of R_w satisfied the attributes in S, the comparison is sent to branch node 1, else it will be sent to branch node (0). The procedure will be constructed recursively till it gets to any of the terminal nodes. The set S of attribute satisfied ROBDD is denotes as S = ROBDD if it finally gets to terminal node 1. Otherwise, S does not satisfy ROBDD, denoted as $S|$ = ROBDD.

3.5 Obtaining ROBDD Access Structure from Access Policy

Converting Access Policy into an Access Structure Based on ROBDD: Assuming $\{w_1, w_2, w_3\}$ is attribute set of the access policy A, and threshold gate $T(2, 3)$. Users who can complete decryption successfully are those who own attribute w_0 or any of the two attributes. The corresponding Boolean function based on the method used to obtain the Boolean function above is obtained by $f_1(w_0, w_1, w_2, w_3) = w_0 + w_1 w_2 + w_1 w_3 + w_2 w_3$. Considering variable ordering to be π: $w_0 < w_1 < w_2 < w_3$, the ROBDD representation of Boolean function is obtained by the of recursive process based on Shannon's expansion theorem $f_1(w_0 w_1 \ldots w_n) = w_i \cdot f|w_{i=1} + w_i' \cdot f|w_{i=0}(0 \leq i \leq n - 1)$. All nodes are numbered from top to bottom and from left to right. The expression of ROBDD is finally calculated as ROBDD = $\text{Node}_{id}^i|id \in ID, i \in I\}$. Finally, the expression for access structure is obtained as: $\{\text{Node}_2^0, \text{Node}_3^1, \text{Node}_4^2, \text{Node}_5^3\}$.

Valid Path and Invalid Path: The path from the root node to the terminal node (1) represents a variable assignment for which the Boolean function is true, known as the valid path P. Invalid path P' is from the root node to the leaf node (0) shown in (Sect. 3.6).

3.6 The Main Construction of CP-ABE Based on ROBDD

The symbols $(i, \bar{i})$ represent the positive and negative attributes in our construction. Let n be elements in the attribute set N that has id numbers $\{0, 1, 2, \ldots, n - 1\}$. The following basic algorithms below are used to construct CP-ABE based on the ROBDD.

Setup: The Key Generation Authority will execute the algorithm in Fig. 2 to obtain ROBDD correspondent of Boolean function (Fig. 3) and complete the subsequence operations.

Let G_0 be a bilinear group with of prime order p and let g be a generator with a bilinear map $e : G_0 \times G_0 \rightarrow G_1$. Randomly choose $\alpha, t_i, t_i' \in Z_p$, where $(t_i \in, T_i t_i' \in T_i')$ match with $(i, \bar{i})$ and define $Y = e(g, g)^\alpha$.

Where $i \in N$, set $T_i := g^{t_i}$ and $T_i' := g^{t_i}$. The public key and the master secret key are generated as $PK = e, g, Y, (T_i, T_i')$ and MSK = $\alpha, (t_i, t_i')$.

Fig. 2 Obtaining ROBDD
from Boolean function

```
Input: A Boolean function f(w) and maximum index of valuable n − 1
Output: ROBDD representing f(w)
1)   BUILD_ROBDD [T1, T2](h)
2)   function BUILD_ROBDD (char h, int i){
3)      initialize (T1); initialize (T2) to be empty set
4)      insert new node u (id, l.low, h.high)
5)      if (i > n − 1)
6)      {if (h is false)} return 0
7)      else return 1
8)      else u1 = BUILD_ROBDD (f[0\w_i], i + 1)
9)      else u2 = BUILD_ROBDD (f[1\w_i], i + 1)
10)     return ROBDD (id, u1, u2)
11)  }
12)  end BUILD ROBDD (h, i)
13)  ROBDD[T1, T2](id, l.low, h.high){
14)     if (l.low = h.high)
15)     return l.low
16)     else if create tuple (T2, id, l.low, h.high) exist
17)     return (T2, id, l.low, h.high)
18)     else if u= tuple (T1, id, l.low, h.high)
19)     store (T2, id, l.low, h.high, u)
20)     return u
21)  }
22) }
```

Encryption (PK, M, ROBDD): This algorithm takes inputs of public key PK, message to be encrypted M, and the access structure ROBDD. The owner of message or an authority who represent the data owner executes this algorithm. To encrypt message M, under $\text{ROBDD} = \{\text{Node}_{id}^i | (id \in ID, i \in I)\}$, let the number of valid path to be (1) which is P, expressed as $R = (r_0, r_1, \ldots, r_{P-1})$. Randomly choose β, and compute $C_0 = g^\beta$ and $C_1 = M \cdot Y^\beta$, the related ciphertext element related to the path rP is set as $g^{(\sum_{ip_L} \cdot \beta)}$, which is defined as $(C_{rp} = \prod_i P_i)^\beta$, where T_i corresponds to P_i. The information of the attributes of the elements bounded together in rP is $\sum i \in I$. Finally, the ciphertext is set as:

$$C_t = \{\text{ROBDD}, C_0, C_1, \forall rp \in R : C_{rP}) \tag{1}$$

KeyGen (S, MSK): The authority executes the algorithm in Fig. 2 and it takes an input of the set of attributes S and the MSK. This will output the SK based on the attributes submitted by the users. For $i \in S$, except a default value $\bar{i}$. Define p_i as $i \in S \wedge \underline{i}$, while $t_{\underline{i}} = p_i$ else set $t_{\underline{i}} = p_i'$. Randomly select $r \in Z_p$ and compute $D_0 = g^{\alpha+r}$ and $D_1 = g^{\sum_{i \in I} p_i}$. The private key is finally return as:

$$SK = \left\{ D_0 = g^{\alpha+r}, D_1 = g^{(-r)\backslash(\sum_{i \in I} p_i)} \right\} \tag{2}$$

Decrypt(C_t, SK): The user of the data executes the decryption algorithm. It consists of a recursion algorithm. It takes input of ciphertext C_t and secrete key SK.

Fig. 3 ROBDD representation of $f(w_0, w_1, w_2, w_3)$ and the relationship between P_i, P_i' and i

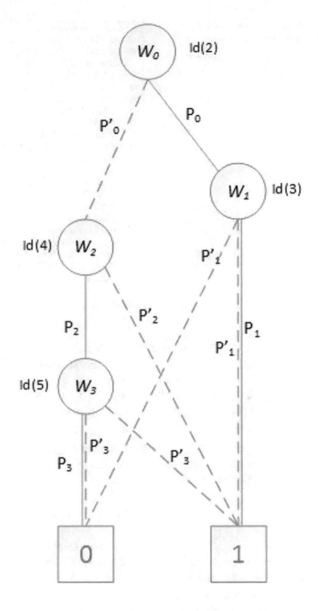

Let $C_t = \{\text{ROBDD}, C_0, C_1 \forall \in R : C_{rp}\}$, the following recursive algorithm will be executed

Step1: The algorithm search for the root node and defines the current node.
Step 2: The information about current node is extracted. For $i \in I$; if $i \in S = i$, go to the next stage; else, if $i \in S \wedge i = \bar{i} \vee i \notin S$; go to *Step 3*: Based on *high*, seek the current node's high branch node (*high bn*). If a *high bn* is the leaf node (0), the algorithm is aborted $\perp$; else if *Step 3*: Based on *high*, seek the current node's *high*

bn. If a *high bn* is a leaf node (0), the algorithm is aborted ⊥; else if the *high bn* is a leaf node (1); go to *step 5*: else if a *high bn* is a non-leaf node; go to *Step 2*.

Step 3: Based on *high*, seek the current node's *high bn*. If a *high bn* is the leaf node (0), the algorithm is aborted ⊥; else if a is a leaf node (1); *high bn* go to *step 5*: else if a *high bn* is a non-leaf node; go to *Step 2*.

Step 4: According to *low* Search the current node's *high bn*. If the *low bn* is leaf node (0), the algorithm is terminated ⊥. Else if the *low bn* is the leaf node (1); go to *Stage 5*; else if *low bn* is a non-leaf node, consider it as current node and go to *Step2*.

Step 5: The algorithm stores the path R_P from the root to the terminal node (1) and compute;

$$e(C_0, D_0) \cdot e(C_{rp}, D_1) = e(g, g)^{\beta(\alpha+b)} \cdot e(g, g)^{\beta.b}$$
$$= e(g, g)^{\beta.\alpha} = Y^\beta \tag{3}$$

The message is computed as follows:

$$M = C_1 \backslash Y^\beta = C_1 \backslash e(g, g)^{\beta \cdot \alpha} \tag{4}$$

4 Analysis and Evaluation

4.1 Security Proof

This scheme can withstand collusion attack since random number b was used in the generation of the keys. The users attributes $i \in I$ have a link to the private key. The rest of the security proof will be done by the reduction Chosen Ciphertext Attack (CPA) security to the Decisional Diffie–Hellman assumption *(DBDH)* security assumption.

Theorem 1 An algorithm can be built to solve the Decisional Diffie–Hellman assumption problem with non-negligible advantage if there is a probabilistic polynomial time that can break the CP-ABE scheme with non-negligible advantage.

Proof Assuming an adversary can break the CP-ABE scheme with advantage ε. A simulator can be built to solve the *DBHD* problem with $\varepsilon \backslash 2$.

The challenger *chl.* selects elements which consist of G_0 and G_1 with a bilinear map $e = G_0 \times G_0 \to G_1$, randomly selects s, b, c, z, Z; flip a coin to select a random $v = \{0, 1\}$ and a generator $g \in G_0$. If $v = 0$, $Z = e(g, g)^{abc}$, else $Z = e(g, g)^z$. The *chl* then select the tuple $(g, A, B, C, Z) = (g, g^a, g^b, g^c, g^z, Z)$ and then send (g, A, B, C, Z) to the simulator *sim* and the *sim* will play the role of the challenge upcoming process.

Init.: The *adv.* gives access tree and the identity $ROBDD = \{Node_{id}^i \backslash id\}$, to the simulator.

Setup: The *sim* defines $Y = e(A, B) = e(g, g)^{ab}$ and chooses $(t_i, t_i') \in Z$, for all $i \in Z_p$ while t_i has inverse of t_i.

Phase 1: The *adv.* request for the secret key by sending a set of attribute S that he wants to challenge with a restriction that S cannot satisfy any valid path of *ROBDD*. For any valid path, attribute $A_j \in I$ which satisfy these two cases; $A_j \in S \land A_j = A_{\bar{j}}$ or $A_j \notin S \land A_j = A_j$. Supposing the *sim* choose $A_j \notin S \land A_j = A_j$, the components associated with each attribute are allocated as follows; for $A_j \notin S \land A_j = A_j$; including several cases; for $i \neq A_j$; $i \in S \land \underline{i} = i, t_{\underline{i}} = t_i$; $i \in S \land \underline{i} = \bar{i}, t_{\underline{i}} = b \cdot t_i$; $i \notin S \land \underline{i} = \bar{i}, t_{\underline{i}} = t_i'$; $i \notin S \land \underline{i} = i, t_{\underline{i}} = b \cdot t_i'$. The *sim* then defines the components of the secret key as: $(D_0, D_1) = \left(g^{ab}, g^{\left((-r) \setminus \sum_{i \in I} t_i' \right)} \right)$.

Challenge: *Adv.* send two plaintext M_0 and M_1 with the same length to *sim*. *Sim* select $b \in (0, 1)$ and set $C_0 = M_b \cdot Z$, and encrypt C_1 under the challengers information submitted. Then, the computed ciphertext $C_t = \{\text{ROBDD}, C_0, C_1, \forall rt \in R : C_{rt} = g^{\sum_{i \in I_L} c}\}$ by the simulator to the *Adv.*

Phase 2: *Adv.* repeats the process in Phase 1.

Guess: *Adv.* output a guess $b = (0, 1)$. If $b' = b$, *sim* output $v' = 0$, indicating $Z = (g, g)^{abc}$, otherwise, $v' = 1$, indicating $Z = (g, g)^z$. The ciphertext C_t is valid, If $Z = (g, g)^{abc}$ and therefore, the advantage of the *Adv.* winning this game is ε. $Pb' = b|v = 0 = \frac{1}{2} + \mathcal{E}$. The ciphertext is entirely random if $Z = (g, g)^z$ and the *Adv.* cannot have access to information about M_b. Hence, the advantage of the Adversary wining this game is $1/2$. That is, $P[b' \neq b|v = 1]\frac{1}{2}$. The above analysis indicates that, the simulator will solve *DBDH* with the advantages of $\left(\frac{1}{2} \right) P\left[v' = v|v = 0 \right] + \left(\frac{1}{2} \right) P[v' = 1] - \frac{1}{2} = \varepsilon/2$. Therefore, if the adversary has the advantage of winning the challenge game, with the help of the *adv's* advantages, the simulator will use the advantage of $\varepsilon/2$ to solve DBDH assumption.

4.2 Analysis of Performance Evaluation

The propose scheme supports several attributes occuring in the same policy and can handles negative and positive attributes in defining of the policy without the increase of system cost. In addition, this scheme uses any Boolean operation to describe a free-form access policy. In the consideration of the analysis of our scheme and most popular used access structures used in access policies, our scheme gives a better performance. Table 1 shows the analysis of the access structures of a function $f_2(w_0, w_0, w_0) = w_0 w_1 + w_0' w_2 + w_1' w_2'$ in Fig. 4. In threshold gate, as indicated, more variables and values are needed to express Boolean functions, which makes it relatively simple. For instance, in disjunction description of $f_2(w_0, w_1, w_2)$; two different values $(w_i w_1')$ where $i = (0, 1, 2)$ of each variables are set in a multiple portion of a given disjunction. Multiple number of nodes and variables must be used in the construction of the structure, and two values are not represented by a single variable of threshold gate. Threshold gates do not support NOT operation since the

Table 1 Comparing the capacity analysis access structures

Scheme	Access structure	Decryption				Variables supported	Number of nodes
		AND	OR	NOT	Threshold		
[17]	AND gate	Yes	No	No	No	–	–
[18]	Threshold gate	Yes	Yes	Yes	No	6	10
[14]	Threshold gate	Yes	Yes	Yes	No	6	10
[20]	OBDD	Yes	Yes	Yes	No	3	6
Our	ROBDD	Yes	Yes	Yes	No	3	3

Table 2 Comparing the efficiency analysis of CP-ABE scheme

Scheme	KeyGen	Encryption	Decryption		Ciphertext	Secret key size
	EG	$EG\mu$	$EG\mu$	P_e		
[17]	$2n + 1$	$n + 1$	$0(n)$	$0(n)$	$(n + 1)D_G + D_{G\mu}$	$(2n + 1) \cdot D_{G0}$
[18]	$2\sigma + 2$	$2w + 1$	$0(\varphi)$	$0(\varphi)$	$(2w + 1)D_G + D_{G\mu}$	$(2\sigma + 1) \cdot D_{G0}$
[14]	$\sigma + 3$	$2w + 1$	$0(\varphi)$	$0(\varphi)$	$(2w + 1)D_G + D_{G\mu}$	$(\sigma + 1) \cdot D_{G0}$
[16]	$\sigma + 1$	$w + 1$	$0(\varphi)$	$0(\varphi)$	$(w + 1)D_G + D_{G\mu}$	$(\sigma + 1) \cdot D_{G0}$
Our	2	$P + 1$	$0(1)$	$0(\varphi)$	$(P + 1)D_G + D_{G\mu}$	$2 \cdot DG0$

w_i and w'_1 are independent variables, and there is a break of relationship in terms of reciprocal. Therefore, the AND gate may not be considered in this scenario since it is limited in the description of function $f_2(w_0, w_1, w_2)$. Our work is similar to the proposal in [22]; however, the ordered binary decision diagram was further reduced which is more compact and represent a Boolean function in more efficient way was used in the construction of the access policy. The nodes used in ROBDD are significantly lesser than the nodes used in OBDD, and the terminal nodes with fixed meanings were discarded to reduce the storage capacity. Therefore, half of the nodes used in OBDD will be required for ROBDD access policy representation. Table 2 represents the performance comparisons of our work and other relevant works. The notations used in this work and their meaning are as follows. w indicates the attribute numbers of the structure, n represents the numbers of global attributes required to make a successful decryption E_G and $E_{G\beta}$ are the exponential in G_0 and G_1, respectively. P_e is the number of bilinear pairings computation. D_G and $D_{G\beta}$ denote each elements size in the group G_0 and G_1, respectively. P indicates the number of path contains in the reduced ordered binary decision diagram. φ represents the least number which was used in a successful decryption σ is attributes used to generate end user's private key.

From Table 2, our scheme performs better with respect to key generation, decryption, and many aspects. The key generation algorithm requires two bilinear pairing computation, and there are only two exponentials in G_0 for key generation algorithm. In addition, computation time for key generation and decryption is $0(1)$. There is a significant improvement in the efficiency of encryption and ciphertext algorithm of

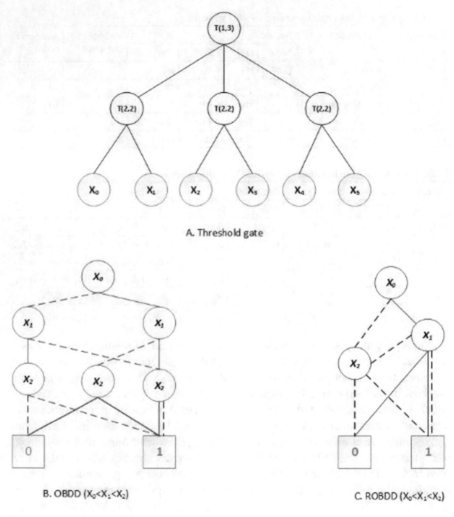

Fig. 4 Representation of $f_2(x_0, x_1, x_2)$

our scheme since the path P is small, and encryption and decryption is based on the number P in ROBDD instead of the number of attributes.

5 Conclusion

To improve on the efficiency of IoT data sharing on CCN while ensuring data security and denying unauthorized data accessibility when data are cached in an ICN approach, we proposed a CP-ABE based on ROBDD, which is more efficient with significant improvement of capacity in the access policy expression. This scheme is based on DBDH assumption and can resist collision attack. The keys generated are of a constant size, and the decryption is very efficient and is not affected by the increased number of attributes. This scheme also supports multiple occurrences and all Boolean operations in the expressing of the access policy. The analysis of performance indicates that ROBDD can perform much better than most of the existing schemes proposed for the IoT CCN. For future work, we will investigate on how to improve the efficiency of the ROBDD CP-ABE scheme and further study how to improve the encryption process and to reduce the size of the ciphertext.

Acknowledgements This work is supported by the National Natural Science Foundation of China under Grants 61502086, the foundation from the State Key Laboratory of Integrated Services Networks, Xidian University (No. ISN18-09).

References

1. Wang, X., Chen, M., Taleb, T., Ksentini, A., Leung, V.: Cache in the air: exploiting content caching and delivery techniques for 5 g systems. IEEE Commun. Mag. **52**(2), 131139 (2014)
2. Jacobson, V., Smetters, D.K., Thornton, J.D., Plass, M.F., Briggs, N.H., Braynard, R.L.: Networking named content. In: International Conference on Emerging Networking Experiments and Technologies, p. 112. ACM (2009)
3. Cisco.: Cisco visual networking index forecast (2019). http://www.cisco.com/c/en/us/solutions/serviceprovider/visualnetworkingindex-vni/vni-forccast.html (2016)
4. Yin, H., Jiang, Y., Lin, C., Luo, Y., Liu, Y.: Big data: transforming the design philosophy of future internet. IEEE Netw. **28**(4), 1419 (2014)
5. Koponen, T., Chawla, M., Chun, B., Ermolinskiy, A., Kim, K., Shenker, S., Stoica, I.: A data-oriented (and beyond) network architecture. ACM SIGCOMM Comput. Commun. Rev. **37**(4), 181192 (2007)
6. Tarkoma, S., Ain, M., Visala, K.: The publish/subscribe internet routing paradigm (psirp): designing the future internet architecture. In: Towards the Future Internet, p. 102 (2009)
7. Fotiou, N., Nikander, P., Trossen, D., Polyzos, G.C.: Developing information networking further: from PSIRP to PURSUIT. In: ICST Conference on Broadband Communications, Networks, and Systems, p. 113 (2010)
8. Dannewitz, C.: NetInf: an information-centric design for the future Internet. In: 3rd GI/ITG KuVS Workshop on The Future Internet (2009)
9. Kuriharay, J., Uzun, E., Wood, C.: An encryption-based access control framework for content-centric networking. In: IFIP Networking Conference, pp. 19 (2015)
10. Sahai, A., Waters, B.: Fuzzy identity-based encryption. In: Proceedings of EUROCRYPT, Aarhus, Denmark, pp. 457–473 (2005)
11. Granjal, J., Monteiro, E., Silva, J.S.: Security for the internet of things: a survey of existing protocols and open research issues. IEEE Commun. Surveys Tuts. **17**(3), 1294–1312 (2015)
12. Calhoun, P., Loughney, J., Guttman, E., Zorn, G., Arkko, J.: Diameter base protocol, IETF, Fremont, CA, USA, RFC 3588 (2003)
13. Papanis, J.P., Papapanagiotou, S.I., Mousas, A.S., Lioudakis, G.V., Kaklamani, D.I., Venieris, I.S.: On the use of attribute-based encryption for multimedia content protection over information-centric networks. Trans. Emerg. Telecommun. Technol. **25**(4), 422–435 (2014)
14. Rao, Y.S., Dutta, R.: Dynamic ciphertext-policy attribute-based encryption for expressive access policy. In: Proceedings of ICDCIT, Bhubaneswar, India, pp. 275–286 (2014)

15. Waters, B.: Ciphertext-policy attribute-based encryption: an expressive, efficient, and provably secure realization. In: Public Key Cryptography PKC, Berlin, Germany, p. 5370 (2011)
16. Balu, A., Kuppusamy, K.: An expressive and provably secure ciphertext-policy attribute-based encryption. Inf. Sci. **326**(4), 354–362 (2014)
17. Ling, C. Newport, C.: Provably secure ciphertext policy ABE. In: Proceedings of ACM CCS, New York, NY, USA, pp. 456–465 (2007)
18. Bethencourt, J., Sahai, A., Waters, B.: Ciphertext-policy attributebased encryption. In Proceedings of IEEE SP, Oakland, CA, USA, May, pp. 321–334 (2007)
19. Vieira, B., Poll, E.: A security protocol for information-centric networking in smart grids. In: Proceedings of SEGS, Berlin, Germany, p. 110 (2013)
20. Li, L., Gu, T., Chang, L., Xu, Z., Lui, Y., Qian, J.: A ciphertext-policy attribute-base encryption based on an ordered binary decision diagram. In: IEEE Access (2017)
21. Andersen, H.R.: Binary decision diagrams. Lecture Notes for Advanced Algorithms. Available at http://www.itu.dk/people/hra/notesindex.html (1997)
22. Sharma, R.K., Sahu, H.K., Pillai, N.R., Gupta, I.: BDD based cryptanalysis of stream cipher: a practical approach. IET Information Security (2016)

Authority-Based Ranking in Propaganda Summary Generation Considering Values

Kun Lang, Wei Han, Tong Li and Zhi Cai

Abstract The era of big data provides a favorable condition for the study of news report and public opinions. By using keyword search, language processing, data mining and information techniques, websites, blogs, forums can be further analyzed for public opinion monitor and precisely guidance. Thus, in this paper, we propose a novel ranking methodology for object or propaganda summary generation, i.e., ValueRank, which incorporates tuples values in authority flow and facilitating the approximation of importance for arbitrary databases. According to ObjectRank and other PageRank fashioned techniques that consider solely authority flow through relationships and therefore their effectiveness is limited on bibliographic databases. Experimental evaluation on DBLP and Northwind databases verified the effectiveness and efficiency of our proposed approach.

Keywords ValueRank · Propaganda summary · Keyword search

1 Introduction

Information extraction is an important technique in relational textual representing and mining, e.g., news abstract, propaganda, etc. Our previous work [1–3] provided a novel approach for size-l object summaries generation by using pre-computation of rankings of each tuples in the dataset, i.e., a PageRank-style approach. Namely every database has different and specific semantics that infer different properties. Unlike the web pages, all edges are hyperlinks [4], in a database data–graph different relationships and values of attributes represent different properties. ObjectRank [5] appropriately extends and modifies the PageRank algorithm to perform keyword

K. Lang · W. Han
China Three Gorges International Corporation, Xicheng District, Beijing, China

T. Li · Z. Cai (✉)
College of Computer Science, Beijing University of Technology, Chaoyang District, Beijing, China
e-mail: caiz@bjut.edu.cn

© Springer Nature Singapore Pte Ltd. 2019
S. K. Bhatia et al. (eds.), *Advances in Computer Communication and Computational Sciences*, Advances in Intelligent Systems and Computing 924,
https://doi.org/10.1007/978-981-13-6861-5_50

search in databases for which there is a natural flow of authority between their tuples.

However, even ObjectRank disregards values of attributes of tuples that affect global importance of tuples. For example, a customer with many orders of high total value should have more importance than another customer with the same amount of orders but less total value.

In view of this limitation, we propose ValueRank algorithm that also considers such values contributing to object or propaganda summary generations. Similarly to ObjectRank approach, we use the schema and specify the ways in which authority flows across the nodes of the data graph considering also tuples values. The rest of this paper is organized as Sect. 2 reviews our preliminary and related work, Sect. 3 introduces ValueRank methodology, Sect. 4 provides an experimental evaluation while Sect. 5 concludes the paper.

2 Related and Preliminary Work

In this section, we firstly present preliminary work and their fundamental semantics that we build on, i.e., authority-based rankings. Then, we present and compare other related work in relational keyword search techniques.

The majority of authority-based techniques are influenced by PageRank which is considered to be the most effective ranking technique on the web. Other similar techniques include work by Kleinberg's HITS [6], Doyle and Snell [7], TupleRank [8], XRank [9] etc. The adaptation of PageRank on databases can be effective for ranking tuples (or nodes) of relational databases where their corresponding relationship edges are associated with authority flow semantics, e.g., bibliographic databases such as DBLP. However, as we explain below the direct application of PageRank is not suitable for databases [5].

PageRank in which all web pages are set as tuples while their corresponding hyperlinks are assumed as edges among such connected graph. In the graph, let G (V, E) representing a graph that including a set of nodes or tuples $V = (v_1, ..., v_n)$, and a set of edges E and it set r denoting the vectors with the PageRank r_i of each node v_i, then PageRank vector r is calculated as:

$$r = d \cdot Ar + (1 - d) \cdot \frac{e}{|V|} \tag{1}$$

where d is a factor (generally set to 0.85 from web pages calculations), A is an $n * n$ matrix with $A_{ij} = 1/\text{Outdeg}(v_j)$, if there is an outgoing edge from $v_j \rightarrow v_i$ (OutDeg(v_i) is the outgoing degree of v_j) and 0 otherwise and $e = [1, ..., 1]^T$.

ObjectRank ObjectRank [5] as an interesting technique extension of PageRank on relational databases, it provides the definition of *Authority Transfer Rates* between the tuples of each relation of the database (also called *Authority Transfer Schema Graph*). This is because of the observation that mapping solely a relational database

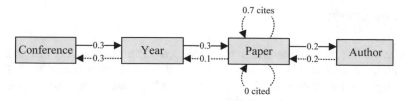

Fig. 1 *Authority Transfer Schema Graph* of DBLP database

to a data graph (as in the case of the web pages) is not accurate. Authority Transfer Schema Graph is required to control the flow of authority in neighboring tuples, while in this paper, we also conducted our proposed approach with such authority transfer schema.

Different from PageRank, in ObjectRank, there are two more definitions, i.e., backward reflection edge and transfer rate. For each edge (like in PageRank) in the structured database graph, it is determined as a backward edge between such two relations (or tables) that demonstrating the reflections back, e.g., the dash line in Fig. 1. While regarding to the transfer rate, it is defined as $\alpha(e^f)$, where ranges from 0 to 1, each transfer rate can be calculated considering its corresponding tuples that out going edges or degrees, e.g., the number that on the arrows.

Meanwhile, in the case of ObjectRank, the base set S presenting a set of tuples or nodes that including the keywords that user interested in, while r denotes the vector with ObjectRank r_i of each node v_i, thus r can be calculated:

$$r = d \cdot Ar + (1 - d) \cdot \frac{s}{|S|} \qquad (2)$$

where $A_{ij} = \alpha(e)$ if there is an edge from v_i to v_j in the relational graph and 0 otherwise, d the same as that from PageRank generally set to 0.85 value, it controls the Base Set importance and s is a $n * n$ matrix, while by setting S to be the set of bookmarks of a user. In [1], it was used to perform topic-specific PageRank on the Web.

Relational Keyword Search techniques facilitate the discovery of joining tuples (i.e., Minimal Total Join Networks of Tuples [10]) that collectively contain all query keywords and are associated through their keys [11, 8], for this purpose the concept of Candidate Networks is introduced, e.g., DISCOVER [10, 12], and Relational Stream Keyword Search [13–15], BANKS [16, 17]. Meanwhile, in our previous work [1, 2], an proposed size-*l* Object Summary does not only differentiate semantically from traditional relational keyword search, but also technically. An size-*l* Object Summary is a tree structured set of tuples while keywords are included in the root node, as it is also a standalone meaningful semantic snippet, e.g., the techniques in [6, 13].

3 ValueRank

In relational databases, if two relations are connected by primary keys or foreign keys, it means that there should be one or more specific semantics between them. Thus, such PageRank-style approaches, e.g., ObjectRank [5] are considered to be effective to rank their corresponding tuples according to different number of connections and relations. However, in contrast for business or trading databases (e.g., Northwind or TPC-H), although such PageRank-style studies can provide indications of importance, they completely ignoring the values of their tuples, which may greatly affect the rankings. As one of the most important motivations, in such databases, we propose and investigate an efficient solution, i.e., ValueRank that can address the importance of values in business, economic databases, or other databases that including tuples that can be valued.

As an extension of RageRank styled approach, by modifying the Base Set S, we want to update the nature of 0 (no connection) or 1 (has connection) to a [0, 1] range that considering tuples attributes and normalized values, which have significant influence on other tuples authority. Take the Northwind database as an example in Fig. 2, in relation `OrderDetails` (value of UnitPrice * Quantity) and `Products` (value of UnitPrice * Quantity), all tuples in them can be a function of these values.

More precisely, we set s_i value of each relation describes the importance of their normalized corresponding tuples. Thus, the s_i of a single tuple v_i in base set S are defined as the following formula:

$$s_i = \alpha \cdot f(v_i) \tag{3}$$

where α ranges from 0 to 1, it is a tuning threshold and $f(v_i)$ is a normalization function of their corresponding values. Thus, we guarantee the base set s_i produces values in the range of (0, 1]. Comparing with the PageRank styled approach, e.g.,

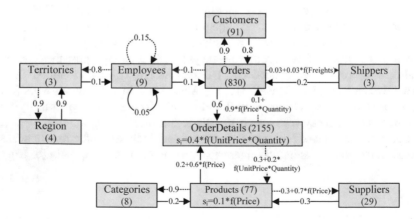

Fig. 2 The value-based authority transfer schema graph G^A for the Northwind database

Fig. 3 Summary of
ValueRank scores

Northwind database			
R_i	*Minimum*	*Median*	Maximum
Employees	*0.002*	**0.193**	0.384
Territories	*0.003*	**0.003**	0.079
Region	*0.017*	**0.025**	0.723
Orders	*0.138*	**0.152**	0.736
Customer	*0.021*	**0.180**	0.691
Shipper	*0.199*	**0.214**	0.272
OrderDetails	*0.007*	**0.025**	0.242
Product	*0.147*	**0.230**	1.000
Supplier	*0.003*	**0.017**	0.043
Categories	*0.039*	**0.083**	0.112

ObjectRank, the base set is 0 or 1, our proposed method can provide a more specific base set considering values.

Moreover, we further combine the idea on authority transfer rates, because we want such effect further influence the ranking results. Thus, we updated the $\alpha(e)$ in Eq. (1) and (2) with the following formula:

$$\alpha(e) = \beta + \gamma \cdot f(v_i \rightarrow v_j) \qquad (4)$$

where β γ are denoting the constants and we restricted such $\beta + \gamma \leq 1$, note that when γ *is* set to 0 means ignoring the affect on such corresponding values while f $(v_i \rightarrow v_j)$ is a normalization function in the range [0, 1] similar to $f(.)$ in Eq. (3) of the values of v_i and v_j.

3.1 Inter-relational Tuple Ranking

The transfer rates should be selected accordingly in order to give us the desired inter-relational ranking. For example, we would like to rank the most important employee higher than the most important customer (or vice versa). The following table summarizes the ValueRank scores per relation for Nortwind produced by the G^A of Fig. 3 and $d = 0.85$. Such statistical metrics are a good indication of these ranking issues. Although sometimes intra-relational ranking is very good, bad inter-relational ranking yields to bad size-l OSs. Earlier work, i.e., ObjectRank, did not investigate inter-relation ranking of tuples in depth.

3.2 ValueRank Versus ObjectRank

The results of Fig. 4 show the results generated by our proposed ValueRank against ObjectRank scores for the selected tuples of the Microsoft Northwind dataset.

Tuple ID	ValueRank	ObjectRank	Total Orders	{UnitPrice*Quantity, Freight Price$^{\dagger\,\dagger}$}
Customer SAVEA	0.654	**0.702**	**31**	115,673.4
Customer QUICK	**0.691**	0.616	28	**117,483.4**
....				
Shipper 1	0.198	0.359	249	16,185.3†
Shipper 2	**0.272**	**0.470**	**326**	**28244.8†**
....				
Product 38	**1.000**	0.486	24	**149,984.2**
Product 59	0.495	**1.000**	**54**	76,296.0
....				
Employee 4	**0.384**	**0.375**	**156**	**250,187.4**
Employee 3	0.352	0.304	127	213,051.3
....				
Supplier 18	**0.043**	0.086	2	**281.5††**
Supplier 7	0.023	**0.132**	**5**	177.8††
....				

Fig. 4 Samples of OR and VR scores

In Fig. 4 (the maximum number in each line is set to bold), the ObjectRank results were produced by the corresponding ObjectRank version of this G^A in Fig. 2, in which authority transfer rate $\alpha(e)$ and base set S were set to the same as PageRank. Obviously, our proposed ValueRank approach provides a group of comparative ranking results than that from ObjectRank calculations, and we also summary as: ObjectRank scores generally higher if such relations or tuples have a greater number of correlations, while ValueRank results performance better by adjusting such ranking results considering their corresponding values. For instance, Customer SAVEA has 31 Orders with total UnitPrice * Quantity = 115,673.30 whereas Customer QUICK has 28 Orders and 117,483.39. The former has bigger ObjectRank, whereas the latter bigger ValueRank. Similar results are observed for Products and Suppliers. Whereas Employee 4 and Shipper 2 are first for both ValueRank and ObjectRank because they have more Orders and bigger UnitPrice * Quantity or Freight values, respectively.

4 Experimental Evaluation

Firstly, the effectiveness of ValueRank and our previously proposed size-l OSs [1, 2] is thoroughly investigated against human evaluators. For more complete evaluation of effectiveness, correctness, and efficiency issues discussed in this paper, a variety of databases has been used, i.e., DBLP, TPC-H (scale factor 1) and Northwind. For instance, the DBLP database was very interesting for evaluating effectiveness on bibliographic data, whereas the Northwind database for evaluating general databases

Table 1 System parameters (ranges and default values)

Parameter	Range
G^A	G^{A1}, G^{A2}, G^{A3}
$d\ (d_1, d_2, d_3)$	0.85, 0.10, 0.99

with values (e.g., trading, etc.). In addition, the TPC-H and DBLP databases, because of their scale, were very useful for evaluating the correctness and efficiency of the two greedy algorithms. The Northwind database because of its rich schema (comprising of more relations and more attributes), it was very useful for presenting and also evaluating the techniques with more understandable examples. The cardinalities of the three databases are 2,959,511, 8,661,245, and 3,300 tuples, whereas their sizes are 319.4 MB, 1.1 GB, and 3.7 MB, respectively.

We measure the impact of (1) damping factor d and (2) *Authority Transfer Rates* (i.e., by using several G^As). We imitate and extend the setting parameters used in the evaluation of ObjectRanks [5]. More precisely in [5], the impact of d is investigated with values 0.85, 0.10, and 0.99 (with 0.85 as default), whereas *Authority Transfer Rates* are investigated with only one G^A (i.e., the G^A of Fig. 1). We investigated the same values for d and three different G^As per database. More precisely, for the bibliography DBLP dataset (collected data until the year of 2014), we use the same default G^A as in [5] (i.e. the G^A of Fig. 1, also denoted as G^{A1}). The G^{A2} had common transfer rates (i.e., $a(e)= 0.3$) for all edges while G^{A3} had common forward (i.e., $a(e)= 0.3$) and common backward (i.e., $a(e)= 0.2$) transfer rates. For both G^{A2} and G^{A3}, transfer rates had only one exception, i.e., (similarly with G^{A1}) the citing-Paper edge was set to 0. The omission of this exception would result to very bad results as Papers would have received Importance by citing other Papers, and this would not give useful comparison. For the Northwind database, the default G^{A1} is the G^A of Fig. 5; the corresponding G^{A2} had common β and γ set to 0.3 for all edges (i.e., $\alpha(e)= 0.3 + 0.3 \cdot f(.)$). Whereas G^{A3} had all α (consequently $s_i= 0$) and γ set to 0, therefore producing ObjectRank values (Table 1).

4.1 Effectiveness of ValueRank

In order to evaluate the quality of ValueRank, we conducted a similar evaluation survey with ObjectRank [5]. Namely five professors and researchers from our University, who had no involvement in this project and knew nothing about our ranking algorithms, participated in this survey. Each participant was asked to compare and rank lists of 10 randomly selected tuples assigning a score of 1 to 10.

Meanwhile, a group of semantic details and statistics were also given. The provision of such details was necessary as to provide a better concept of the dataset to the evaluators. The survey investigated the impact of d and G^A on both intra-relational and inter-relational tuple ranking on the Northwind database (Table 2).

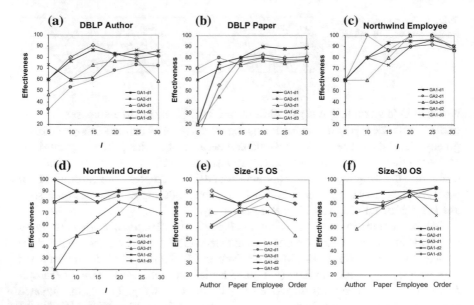

Fig. 5 Effectiveness of size-l OSs with ValueRank

Table 2 Evaluation of ValueRank effectiveness

	d_1-G^{A1}	d_2-G^{A1}	d_3-G^{A1}	d_1-G^{A2}	d_1-G^{A3}
Intra-relational	7.8	8.1	7.8	7.8	3
Inter-relational	7	6.3	7	5	3

The survey revealed that the damping factor had a moderate impact on the intra-relational ranking on the particular G^{A1}. Since d_1 and d_3 values produced almost similar intra-relational ranking, whereas d_2 had influenced only Product and OrderDetails intra-relational tuples ranking and more precisely favored them according to their s_i values, e.g., Price. This is because the Base Set includes tuples only from two relations (i.e., Product and OrderDetails) and also their s_i values are small to influence general ranking, i.e., as values of α were 0.4 and 0.1, respectively.

The impact of d_2 was viewed more positively than the other ds by evaluators because it ranked higher Products with greater Price when they had similar Unit-Price * Quantity values. d_1-G^{A1} and d_1-G^{A2} produced similar intra-relational ranking and evaluators were equally satisfied by both. The good intra-relational ranking produced by G^{A2} (which is a simplification of G^{A1}) infers that as long as we know the parameters that influence importance (e.g., price, freight), it is not very difficult to choose a G^A that will produce effective intra-relational ranking. ObjectRank ranking (produced by Northwind G^{A3}) did not satisfy the evaluators, since ObjectRank had a big correlation with connectivity of tuples rather than their values.

4.2 *Effectiveness of Size-1 OS*

Due to the nature of the contents of our databases, we conducted two different evaluation methodologies for DBLP and Northwind datasets.

Evaluation for the DBLP Database

Because the database included real content about real people (i.e., data subjects) with their existing papers, in this work we asked the Authors themselves to size-l their own Author and Paper OSs for l values 5, 10, 15, 20, 25, 30 (i.e., to propose their own size-5, ..., size-30 OSs). The results of Fig. 5 represent the average of the percentage of the correct tuples that our system found (using Importance scores produced by the settings of Table 1) in comparison with the evaluators' size-l OSs. The rationale of this evaluation methodology is that the DSs themselves have in fact the best knowledge of their work and therefore can be the best judges of their work. More precisely, we asked seven different database researchers who had entries of their papers in our database to participate (their names are not disclosed because of anonymity requirements). We firstly familiarized evaluators with the concepts of ValueRank-based Object Summary in general and size-l OSs.

The various values of d and G^As had only inter-relational rather than intra-relational impact on ObjectRank ranking. Regarding d, it was expected since no Base Set was used, whereas regarding G^As since the variations of G^As were in fact minor.

In general, d_1-G^{A1} and d_3-G^{A1} are safe options for size-l OS generation for DBLP Author and Paper (this is because these two settings have very similar ObjectRank scores). As they always dominate on large and more meaningful ls, this is depicted by Fig. 5e, f which summarize the average effectiveness per relation for size-15 and size-30 OS.

User's evaluation revealed that the inter-relational properties affected the approximation quality of selected size-l OSs. For example, on DBLP.Author OSs, where evaluators firstly selected important DBLP.Papers tuples (i.e., to include in size-5 OSs) and then additional tuples such as Co-authors, Year, Conferences/Journals (i.e., to include in larger size-ls, e.g. size-10 etc.), was favored by the setting d_2-G^{A1} (although in general this setting was not very effective). From the results in (a) of Fig. 5, thus settings achieves more than 70% (in comparison with 60% of the default setting) for size-5.

Evaluation for the Northwind Database

For the evaluation of the Northwind database, we presented a small amount of complete OSs together with some explanatory descriptive information for each tuple to the evaluators (similarly with the evaluation of ValueRank effectiveness) and asked evaluators to select top-l themselves (similarly with the evaluation on the DBLP dataset). The evaluators were academic staffs from universities from our town (some of them also participated in the previous evaluations as well). More precisely, we presented to them 16 object summaries and we asked evaluators to select size-l the results with l equals 5, 10, 15, 20, 25, 30.

In summary, d_1-G^{A1} and d_3-G^{A1} are the safest options as they produce good size-l OSson both Employee and Order (this is again for the same reasons, that the minor difference of d does not have a significant impact on ValueRank scores). d_1-G^{A2} also returns good results but not as good as d_1-G^{A1} and d_3-G^{A1}. As expected G^{A3}, which is the ObjectRank version of the G^{A1}, did not produce satisfactory size-l OSs. For instance, G^{A3} size-15 OSs are very bad for both OSs (Fig. 5e). Also, the d_2-G^{A1} is not very effective on either relation, and this is due to the inter-relational ranking.

Obviously, the results of effectiveness of ValueRank scores (Table 2) have some similarity with the results of effectiveness of size-l OSs. For instance, the good ValueRank scores produced by settings d_1-G^{A1} and d_3-G^{A1} resulted to good size-l OS while poor ValueRank scores produced by ObjectRank resulted poor size-l OSs. An interesting observation is that the effectiveness of size-l OSs produced by the setting d_2-G^{A1} is not very good. Although d_2-G^{A1} gave the best intra-relational ranking (i.e., 0.81), its reduced inter-relational ranking eventually results to poorer size-l OSs.

5 Conclusion and Future Work

This paper addresses the challenges of the effective ranking of a database tuples considering also their values in the relational generation of size-l OSs for textual abstract or propaganda. For this purpose, as one of the extensions of ObjectRank, our proposed method ValueRank is introduced. ValueRank is the first attempt in authority-based ranking that also considers values in order to rank tuples. An experimental evaluation was conducted on databases (i.e., DBLP, Northwind) that verifies the effectiveness of our technique.

Acknowledgements This work was partially supported by the Beijing Natural Science Foundation under Grant number 4172004, Beijing Municipal Science and Technology Project under Grant number Z161100001116072, and Beijing Municipal Education Commission Science and Technology Program under grant number KM201610005022.

References

1. Fakas, G., Cai, Z., Mamoulis, N.: Diverse and proportional size-l object summaries for keyword search. SIGMOD conference, 363–375 (2015)
2. Fakas, G., Cai, Z., Mamoulis, N.: Diverse and proportional size-l object summaries using pairwise relevance. VLDB J. **25**(6), 791–816 (2016)
3. Fakas, G.: Automated generation of object summaries from relational databases: a novel keyword searching paradigm. In: ICDE Workshop on Ranking in Databases (DBRank'08), 564–567 (2008)
4. Brin, S., Page, L.: The anatomy of a large-scale hypertextual web search engine. WWW Conference, 107–117 (1998)

 5. Balmin, A., Hristidis, V., Papakonstantinou, Y.: Objectrank: authority-based keyword search in databases. VLDB, 564–575 (2004)
 6. Koutrika, G., Simitsis, A., Ioannidis, Y.: Précis: the essence of a query answer. ICDE, 69 (2006)
 7. Doyle, P.G., Snell, J.L.: Random Walks and Electric Networks. Mathematical Association of America, Washington, D. C. (1984)
 8. Huang, A., Xue, Q., Yang, J.: TupleRank and implicit relationship discovery in relational databases. WAIM, 445–457 (2003)
 9. Guo, L., Shao, F., Botev, C., Shanmugasundaram, J.: XRANK: ranked keyword search over XML documents. SIGMOD, 16–27 (2003)
10. Hristidis, V., Papakonstantinou, Y.: DISCOVER: keyword search in relational databases. VLDB, 670–681 (2002)
11. Fagin, R., Lotem, A., Naor, M.: Optimal aggregation algorithms for middleware. PODS, 614–656 (2001)
12. Hristidis, V., Gravano, L., Papakonstantinou, Y.: Efficient IR-style keyword search over relational databases. VLDB, 850–861 (2003)
13. Simitsis, A., Koutrika, G., Ioannidis, Y.: Précis: from unstructured keywords as queries to structured databases as answers. VLDB J. **17**(1), 117–149 (2008)
14. Liu, F., Yu, C., Meng, W., Chowdhury, A.: Effective keyword search in relational databases. SIGMOD, 563–574 (2006)
15. Markowetz, A., Yang, Y., Papadias, D.: Keyword search on relational data streams. SIGMOD, 605–616 (2007)
16. Aditya, B., Bhalotia, G., Chakrabarti, S., Hulgeri, A., Nakhe, C., Sudarshan, P.S.: BANKS: browsing and keyword searching in relational databases. VLDB, 1083–1086 (2002)
17. Bhalotia, G., Hulgeri, A., Nakhe, C., Chakrabarti, S., Sudarshan, S.: Keyword searching and browsing in databases using BANKS. VLDB, 431–440 (2002)

A Concept to Improve Care for People with Dementia

Mary Sio Lai Karppinen, Jori Karppinen and Raija Halonen

Abstract Dementia is one of the severe causes of mortality in elder groups of human population, and the number of dementia patients is expected to increase all over the world. Globally, the cost to take care of patients with the illness is high. Moreover, the lives for dementia patients are commonly impacted by a variety of challenges, for example, in terms of communication, emotional behaviour, confusion and wandering. Therefore, the future aim is to develop a serious games package which can help to slow down the progression of dementia, better life for the patients and their loved one and reduce the use of medications and the cost for the government sectors. In this study, a conceptual model was formed to present the problems the dementia patients and their caregivers are facing, and non-pharmacological approaches are suggested to slow down the progression of dementia-related impairments and behavioural symptoms by using well-structured serious games with personalised data.

Keywords Dementia · Alzheimer's disease · ICT technologies · Elderly health care · Gamification · Serious games

1 Introduction

The purpose of the current study was to twofold: first, find knowledge about the problems that people suffering from dementia and their caregivers are facing, and second, see if there are means to slow down the progression of dementia-related symptoms with the help of non-pharmacological approach, in this case with the help of ICT-enabled games.

The number of dementia cases is increasing year by year all over the world. There are approximately 50 million people worldwide diagnosed with dementia [1], and

M. S. L. Karppinen · J. Karppinen · R. Halonen (✉)
Faculty of ITEE, M3S, University of Oulu, Oulu, Finland
e-mail: raija.halonen@oulu.fi

M. S. L. Karppinen
e-mail: ssiolai@gmail.com

© Springer Nature Singapore Pte Ltd. 2019
S. K. Bhatia et al. (eds.), *Advances in Computer Communication
and Computational Sciences*, Advances in Intelligent Systems and Computing 924,
https://doi.org/10.1007/978-981-13-6861-5_51

Alzheimer's disease is the most common cause of dementia [2]. There are many direct and indirect costs for the society in taking care of this group of patients, and this has become a huge burden for every country in the world [3].

At the time of the study, no cure for Alzheimer's disease is reported. Since 2003, not even a new treatment option is published in the clinical market [4]. However, there are already interventions for people suffering from dementia focusing on different technologies. On the other hand, the role of social health and social participation as objectives is not established so far. There are studies that propose positive findings related to computer-based cognitive interventions carried out with people with dementia, and earlier knowledge proposes that people suffering from dementia can benefit from ICT-based applications that support the demented people to facilitate and create social networks [5, 6].

In general, the behavioural and psychological symptoms of dementia should be managed with non-pharmacologic care if possible, and the management of those symptoms form an important part of the care plan. In this sense, the practices and treatment should be evaluated frequently [7]. A recent study [8] of serious games in health care showed that the role of serious games is increasing especially to influence education, cognitive rehabilitation, psychology and physical rehabilitation.

The current study wanted to find out how the issues related to ageing people with dementia and other memory-related diseases are discussed in earlier research, especially from the viewpoint of non-medical approach, considering also their caretakers. To answer the research question, the earlier knowledge was approached by a mapping study. In addition, a pre-study was carried out to give pre-understanding of the research context and elderly suffering from memory issues.

The current study proposes that carefully designed serious games can both support communication skills and reduce emotional problems of the dementia patients. With these improvements, dementia patients are to experience easier, healthier and better lives with their families.

2   Research Approach

A mapping study is about doing classification and thematic analysis of the literature on the chosen topic. In mapping studies, research questions are rather generic than specific. The research question typically asks "how" and "what type". The actual research process is defined by the research area. In the beginning, the scope can be wide and then sharpen when the process progresses. In mapping studies, the researcher can choose to focus, e.g., on targeted sets of articles or to one or two digital libraries [9]. A systematic mapping study aims at finding research directions, while the systematic literature study seeks for ways to combine evidence [10]. Papers with empirical research included into the study can be classified into two categories: validation research and evaluation research, and papers with non-empirical research can be classified into four categories: solution proposals, philosophical papers, opinion papers and experience [11].

A prospective descriptive design was conducted to analyse the use of ICT design and gamification with personalised data to slow down the progression of dementia-related impairments and to control behavioural symptoms.

In this study, the findings from studies of researches by other scholars and authors were concluded and combined with the result from the primary research. The Google search engine and Google Scholar search engine were used to look for studies within the same or related context. Articles and papers were found with the keywords "dementia", "cost for dementia", "caregivers", "behaviour of dementia", "symptoms for dementia", "stages of dementia", "gamification", "gamification health", "gamification dementia" and "personalised information".

These references were selected based on the relevance of the article for the study and the number of citations. The final number of analysed articles was 338 in total. The articles were read through, and based on their focus or main content, 118 articles were included in the analysis.

3 The Study/Literature Review

According to a 2015 report by the World Health Organization (WHO), 47.5 million people by an estimate were living with dementia. Moreover, since then, the number of people with illness has only been increasing. In the report, WHO is projecting the diagnosed cases to triple by the year 2050. The cost on dementia patients in the United States was around 818 billion U.S. dollars in the year 2015, and by a forecast, the total cost on this group of people will be around 1 trillion U.S. dollars in 2018 [12, 13]. This sum equals to 1.09% of the global GDP, which is comparative to the entire annual GDP of countries like Indonesia, the Netherlands or Turkey. Nevertheless, this cost is also larger than the market value of the world's largest companies (2017), such as Apple Inc. (752 billion U.S. dollars), Microsoft (507.5 billion U.S. dollars), Amazon.com (427 billion U.S. dollars) and Facebook (407.3 billion U.S. dollars) [14]. Furthermore, the estimated cost on dementia patients will be around 2 trillion by the year 2030, which corresponds to the current total market value of all these companies together [15]. Studies also show the average cost for taking care the group of patients is around 81% more expensive than patients without dementia, for example, while comparing to the costs on heart disease or cancer patients [16, 17]. Finally, by these figures on diagnosed cases and their tendency of growth, many studies acknowledge dementia as a globally substantial issue [18, 19].

The percentage of people with dementia in Finland is higher than that in the European Union (EU) on average. In 2013, this proportion was 1.7%, representing 92,000 people of the entire Finnish population [20]. Furthermore, annually roughly 13,000 new diagnoses are being made, while a significant proportion of patients die. In addition to personal tragedies caused by disease, from economic and social perspectives, a person with dementia produces 5,088 euros higher healthcare costs per year than an average citizen [21].

3.1 Family Caregivers' View in Taking Care of Demented Family Members

The cost of family caregivers to patients at different stages of dementia is also a significant factor as these groups of people are constantly under stress, for example, on psychological, practical and financial levels, to mention a few. Moreover, informal caregivers from family can easily develop a problem with depression by taking care of dementia patients [22–24]. As the family caregivers realise they might need to spend a long time to take care their loved one whose situation with health is getting worse day by day, this might increase their worry and uncertainty for the future. In addition, they may need to alter their plans for everyday life, which may not be able to work out the same as before. Caregivers might also reduce their participation in different kinds of social activities, only to avoid their loved ones behaving inappropriately due to the cause of dementia symptoms. For some people, such arrangements may not be that challenging; however, for others, it can cause a lot of stress and harm the caregiver's health [23, 25]. Moreover, taking care of a patient with memory disorders takes a lot of time. Therefore, many caregivers need to reduce their working hours or even resign from their job to be able to offer enough support to their loved ones [26–28]. The costs are emphasised on mild and moderate patients with dementia who are mainly taken care of by their spouses or kids (in-law) at home [29]. These indirect costs for taking care of dementia can be even doubled to direct costs for a society [23, 29–31]. Therefore, this is clearly increasing the burden of both the communities and the entire world. Although the cost of in-home care is high, there are many positive reasons supporting it. As long as a patient's partner or close relatives are able to look after one, the society should still provide its maximum support for an informal caregiver to look after their patient at a regular home rather than a nursery one, especially at the early stage of the illness. Patients with dementia are more willing to rely on and to be taken care of by the people they find familiar and trustable [32]. Family caregivers can give more mental support to the patients and—if they are receiving enough information, support and professional instructions from the society—it is possible for the patient to release their stress by taking care the loved ones and even improve the overall relationships within the family [25, 31].

3.2 Difficulties of People Having Symptoms of Dementia

The fact sheet published by WHO [33] describes dementia as commonly a longer-term and gradually decreasing ability to think and remember, having an impact on a person's daily functioning [33]. Thus, in addition to issues with memory loss being individual to every patient, it also impacts on the surrounding society they live in [19, 22, 34]. As the disease evolves, a varying set of difficulties are either directly or indirectly faced by both patients and their loved ones. These are commonly experienced in the form of symptoms such as—but are not limited to—communication [35–38], emotional [39, 40], behavioural [41–43], wandering [44–48] and comprehension [36, 49–51] problems.

Communication is the most common challenge people diagnosed with dementia are facing each day. Generally, patients with dementia start to lose the ability in finding words [52–55]. Moreover, they may also not be able to fully understand and interpret surrounding conversations and words in a usual manner. Dementia is causing it difficult for some patients to catch part of the instruction or information in the whole discussion and therefore might understand it incorrect [56]. As the illness progresses, patients with dementia might no longer communicate with other people as fluently as usual. Due to this, it is possible for patients to lose their confidence, increase depression and feel more moody [52, 57–59]. Furthermore, losing the ability to communicate can path dementia patients way to emotional problems, having to experience anxiety, frustration, anger and depression [45, 60, 61]. These symptoms may also exist when the patient starts to reply more to others to complete their daily tasks, unable to maintain their interest and participating in different activities, brain disorder and presence of physical symptoms such as pain, loss of appetite and lack of energy to deal with the dementia disease [62, 63].

Furthermore, the behaviour of people diagnosed with dementia also changes as the illness progresses [61]. The dementia patients might start doing the same things repeatedly, as well as time to time behaving restless, sleepy, wandering, screaming and being physically aggressive [63–66]. It is common for dementia patients to wander. Nevertheless, this behaviour is very exhaustive to the caretakers as patients might put themselves into danger, such as getting into an accident, lost and confused. This inappropriate behaviour might interfere notably with the daily activities of the patients and their caretakers [45].

Finally, brain degeneration effects on the patient's ability to comprehend. Impairment of comprehension is a factor very necessary to be considered while observing the performance of patients. Based on the findings, processing words is a significant problem for most of these people with dementia [67–69]. Complex, long, passive sentences or ones with new words are always difficult for dementia patients to understand [68, 70].

In all, the difficulties bring caretakers a burden to apply more physical efforts while looking after their patients. In addition, the risk of their mental stress is increasing.

3.3 Different Diagnosis in Early, Middle and Later Stages of Dementia

Dementia progress is often divided into three different stages, which are early dementia, middle dementia and late dementia [71–73]. In the early stage of dementia, symptoms might not be very visible. Patients might only make mistakes—as an example—with days, names of people or items, daily tasks, metaphysical thinking or having difficulties in drawing simple pictures. Moreover, the patients may not sleep restlessly and have as stable emotions as normally. At their early stage of dementia, people are able to take care of themselves independently, only not as smoothly as before with their normal health. People around the patient may not immediately recognise such behaviour as signs for the patient get diagnosed with dementia [74–76].

Typically, patients will move to the middle stage of their disease. At this stage, symptoms are getting more obvious, and dementia patients might start to lose a direction in areas familiar to them. Patients commonly experience difficulties in understanding instructions and information, problems in concentration and emotions, for example, becoming more moody or violent [2, 77, 78]. The patients may also behave inappropriately in public areas, experiencing frustration [79].

Finally, at the late stage of dementia, individuals with the illness may not be able to take care themselves while performing daily tasks, such as eating, bathing and dressing up. At this terminal phase, some may no longer be able to read, write, talk, walk or recognise people they used to know. Normal conversations with the patients are almost impossible. Their personality might totally change and become more emotional. Typically to these patients, crying and screaming might happen. Furthermore, the patients might not have much energy to participate in their usual hobbies [80, 81]. As the level and nature of the symptoms at different stages of dementia patients are various, it is also assumed the impact of serious games varies from one patient to another.

3.4   Using Personalised Information to Decrease the Problems

Although the symptoms of dementia reduce a person's ability to maintain their daily routines, the studies found that communication, emotion, behaviour and comprehension problems can be improved if subjects of discussions are related to their personal preferences. Patients are more interested to speak about topics which they are familiar with. These topics are usually related to one's own life stories or memories, which are having a positive and significant role or impact on the person directly [82–85]. Topics which are related to the patient's history are always easier to chat and share with the patient [86], and better communication will increase the patient's self-confidence [87].

Moreover, this interest can be encouraged with pictures or video recognisable by the patients. Patients with dementia can remember better and have a more clear picture of their past than what is happening around them recently [85]. Creating an environment with pictures, video or music, which match with the individual's interest or background, can help in stabilising the patient's emotion and behaviour [87]. Older, personal memories relevant to the individual always have a higher chance to be remembered by patients diagnosed with dementia [88].

Furthermore, discussions on special events belonging to the dementia patient's own culture can also stimulate one's memory and improve performance on communication [89]. As dementia individuals can remember things in the past better, the best contextual to support this group of patients are the items or pictures that they are familiar with. The suggestion of writing a life story books, memory books, family albums and memory boxes have been widely introduced in many researches, associations and communities [90, 91]. People from different nations and environment will develop different habits, interests and values to their lifestyle and history [92]. Understanding the culture, historical background and interests of a dementia patient can let them feel more willing and comfortable to maintain their interest in communication. This can stimulate the patient's verbal skills and exploration. Finally, due to the brain degeneration issue, language comprehension ability of people diagnosed with dementia will decline; however, with contextual support, the patient can also have a better understanding of conversations and instructions [68].

3.5   Gamification Impact in Health Care

Gamification is about applying game design elements in non-game contexts [93]. Generally, gamification is a process for applying game design elements in motivating users to participate and achieve a positive result in their performance [94, 95]. It can also change the behaviour of a person through "playing" with the "games" [94, 96–98]. A survey reported that 65% of the cases under study showed gamification having a positive impact on encouraging end-user participation, and 32% of the cases claimed that gamification elements have a capability to change behaviour [99].

Thanks to the development of modern technological solutions, such as laptops, tablets and smartphones, gamification has been widely used in many different areas [100–103]. For example, in the marketing industry, the role of gamification is to modify customer's behaviour, increase customer engagement and awareness and maintain customers' loyalty [104–106]. Moreover, gamification is used for educational purpose. It can motivate students to learn and achieve better results in their studies by receiving more accurate feedback [103, 107, 108]. Studies show that through "gaming", users can be educated to become more efficient and productive. Furthermore, it will also motivate them to practice learning and apply the skills in their studies or work [106, 109]. Many researches and studies indicate that positive elements—such as passion in learning and engagement [110], which comes along with gamification—are very valuable and beneficial in improving a service and the health and well-being of patients [111]. Gamification can help in educating the public to understand and promote different health-related information [98, 112]. More importantly, gamified systems can also be applied to healthcare-designed training by motivating the patients to exercise more, as a result of better improvement in their health [18, 113–115].

Related to gamification in supporting patients with dementia, there are suggestions that non-pharmacological therapies might offer appealing alternatives for the

Table 1 Concepts and papers

Communication	34; 35; 36; 37; 51; 52; 53; 54
Emotion	38; 39; 44; 51; 56; 57; 58; 59; 60; 61; 62
Behaviour	40; 41; 42; 43; 44; 45; 46; 47; 60; 62; 63; 64; 65
Comprehension	35; 48; 49; 50; 66; 67; 68; 69
Personalised information	67; 81; 82; 83; 84; 85; 86; 87; 89; 90; 91
Gamification	92; 93; 94; 95; 96; 97; 98; 99; 100; 101; 102; 103; 104; 105; 106; 107; 108; 109; 110; 111; 112; 113; 114

treatment of cognitive symptoms. In addition, serious games can positively address people's condition with dementia and improvements can be found particularly from emotional, behavioural and social aspects [114]. Gamified training can increase the enjoyment for patients to continue their participation in brain training exercises, and as a result, both "self-confidence" and "self-rated memory" are advanced [115]. Combined with the importance of personalised information in improving the communication of dementia patients [86], serious game systems with personalised content have the potential to improve the power of gamification and enhance the life quality of dementia patients.

Table 1 summarises the grounds of this study by identifying references reviewed for each topic of this paper.

4 Findings

In this study, the findings are presented as a conceptual framework (Fig. 1).

The literature studies suggested that problems related to communication [36, 38], emotional [40], behavioural [42], wandering [45, 47] and comprehension [49, 51], are the most common symptoms for individuals in early, middle and later stages of dementia [71]. On the other hand, personal content can motivate dementia patient's interest in communication [86], and serious games may have a positive impact on patient's emotion, behaviour and sociability ability [114]. Based on the findings from referenced literature studies and the approach introduced in the conceptual framework of this paper, serious games tailored with personalised content is expected to influence on the patient's daily routines and potentially the quality of life.

Future development and studies will evaluate if the proposed conceptual framework shows potential in managing behavioural symptoms at different stages of dementia, while applied to well-designed serious games with personalised content.

5 Discussion and Conclusions

The purpose of the paper was to identify potential means to slow down the progression of dementia-related symptoms with the help of a non-pharmacological approach. As a part of the conceptual framework, a technological approach to the non-pharmacological method was introduced.

Dementia is one of the severe causes of mortality in elder groups of the human population. The studies estimate that the number of dementia patients will increase all over the world [2]. Moreover, the cost for the world to take care of patients with dementia is high and further growing [12]. In order to reduce the countries' burden for taking care of the illness and improve the quality of their lives, solutions of helping this group of people are needed. The current study proposed a concept (see Fig. 1) that opens new possibilities to support people with lower costs.

Senior citizens are willing to use new technologies if they are well informed about the provision of usable technologies and have gathered positive experiences in using them [116]. The advantages of applying gamification on patients in healthcare sectors have already been identified and reported in many researches [18, 111, 113]. On the other hand, based on the literature studies, the chances for people to get diagnosed with dementia will start to increase rapidly from 65 years on [19, 117–119]. This is the age group at risk and in target for researches. Therefore, it is presumable that after experiencing the initial phase of dementia, some of their skills for using information systems are still present. Furthermore, personal content can motivate dementia patient's interest in communication [86], and serious games have a positive

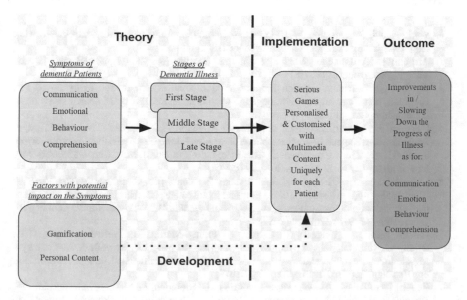

Fig. 1 Proposed conceptual framework

effect on patient's emotion, behaviour and sociability [114]. However, the power of personal content with serious gaming has not been researched and tested. Therefore, a conceptual framework (Fig. 1) was constructed. The conceptual framework will guide analysing if serious games with content personalised for dementia patients diagnosed into one of the three different phases of dementia (first stage, middle stage and late stage) can bring new ways to meet the patients and to support their lives. The current study proposes that the conceptual framework will offer new knowledge to be applied when seeking for reliable and realistic data on the impact of serious gaming with personalised content on dementia and explore its possibilities to be used as a non-pharmacological method of treatment.

In the future, further interviews, prototyping and targeted field trials on audiences possessing a risk for dementia are needed. Academic collaborators from the field of research and development (R&D) are introduced with the concept and invited to work together on the second phase of the project, development of a prototype for the target group. Future studies are aiming at developing a real-life prototype consisting of an interactive, gamified Web/mobile-based application for storing and consuming personal, familiar memories presented in form of images, videos, sounds (voice), graphics and colours. The findings and feedback are concluded with the resulting prototypes of a serious game and its impact on dementia patients at different stages of their illness. More studies are needed to conclude the value of gamified, interactive and personal Web/mobile solutions for the lives of dementia patients.

References

1. World Health Organization, WHO launches Global Dementia Observatory. http://www.who.int/mental_health/neurology/dementia/en/ (2018)
2. Alzheimer's Association, What Is Alzheimer's? https://www.alz.org/alzheimers_disease_what_is_alzheimers.asp (2018)
3. Boseley, S.: Dementia research funding to more than double to £66 m by 2015. The Guardian. London. https://www.theguardian.com/society/2012/mar/26/dementia-research-funding-to-double (2012)
4. Mishra, C.B., Kumari, S., Manral, A., Prakash, A., Saini, V., Lynn, A.M., Tiwari, M.: Design, synthesis, in-silico and biological evaluation of novel donepezil derivatives as multi-target-directed ligands for the treatment of Alzheimer's disease. Eur. J. Med. Chem. **125**, 736–750. https://doi.org/10.1016/j.ejmech.2016.09.057 (2017)
5. Pinto-Bruno, Á.C., García-Casal, J.A., Csipke, E., Jenaro-Río, C., Franco-Martín, M.: ICT-based applications to improve social health and social participation in older adults with dementia. A systematic literature review. Aging Mental Health **21**(1), 58–65. http://dx.doi.org/10.1080/13607863.2016.1262818 (2017)
6. García-Casal, J.A., Loizeau, A., Csipke, E., Franco-Martín, M., Perea-Bartolomé, M.V., Orrell, M.: Computer-based cognitive interventions for people living with dementia: a systematic literature review and meta-analysis. Aging Mental Health **21**(5), 454–467. https://doi.org/10.1080/13607863.2015.1132677 (2017)
7. Robert, P.H., König, A., Amieva, H., Andrieu, S., Bremond, F., Bullock, R., Ceccaldi, M., Dubois, B., Gauthier, S., Kenigsberg, P.A., Nave, S., Orgogozo, J.M., Piano, J., Benoit, M., Touchon, J., Vellas, B., Yesavage, J., Manera, V.: Recommendations for the use of Serious

Games in people with Alzheimer's Disease, related disorders and frailty. Front. Aging Neurosci. **6**, 54 (2014)

8. Korhonen, T., Halonen, R.: Serious games in healthcare: results form a systematic mapping study. In: 30th Bled eConference: Digital Transformation—From Connecting Things to Transforming Our Lives, pp. 349–368 (2017)

9. Kitchenham, B.A., Budgen, D., Brereton, O.P.: Using mapping studies as the basis for further research–a participant-observer case study. Inf. Softw. Technol. **53**(6), 638–651 (2011)

10. Petersen, K., Vakkalanka, S., Kuzniarz, L.: Guidelines for conducting systematic mapping studies in software engineering: an update. Inf. Softw. Technol. **64**, 1–18 (2015)

11. Wieringa, R., Maiden, N., Mead, N., Rolland, C.: Requirements engineering paper classification and evaluation criteria: a proposal and a discussion. Requir. Eng **11**(1), 102–107. https://doi.org/10.1007/s00766-005-0021-6 (2006)

12. Alzheimer's Disease International, Dementia Statistics. https://www.alz.co.uk/research/statistics (2015)

13. Wimo, A., Guerchet, M., Ali, G.C., Wu, Y.T., Prina, A.M., Winblad, B., Jonsson, L., Liu, Z., Prince, M.: The worldwide costs of dementia 2015 and comparisons with 2010. Alzheimers Dement **13**, 1–7 (2017)

14. Statista, Inc.: The 100 Largest Companies in the World by Market Value in 2017 (in billion U.S. dollars). https://www.statista.com/statistics/263264/top-companies-in-the-world-by-market-value/ (2017)

15. Prince, M., Wimo, A., Guerchet, M., Ali, G.C., Wu, Y.-T., Prina, M.: The world Alzheimer report 2015. In: The Global Impact of Dementia: an analysis of prevalence, Incidence, Cost and Trends. Alzheimer's Disease International. https://www.bupa.com/~/media/files/site-specific-files/our%20purpose/healthy%20ageing%20and%20dementia/war%202015/world-alzheimer-report-2015.pdf (2015)

16. Donna, R., Time Inc.: Coping With the Costs of Dementia: The Middle Stage. http://time.com/money/4120981/dementia-costs-middle-stage/ (2015)

17. Gina, K.: Costs for Dementia Care Far Exceeding Other Diseases. The New York Times. https://www.nytimes.com/2015/10/27/health/costs-for-dementia-care-far-exceeding-other-diseases-study-finds.html (2015)

18. McCallum, S.: Gamification and serious games for personalized health. Stud. Health Technol. Inf. **177**, 85–96 (2012)

19. Alzheimer's Association: Alzheimer's disease facts and figures. Alzheimer's Dementia **13**(4), 325–373 (2017)

20. Alzheimer Europe: 2013: The Prevalence of Dementia in Europe (Finland). http://www.alzheimer-europe.org/Policy-in-Practice2/Country-comparisons/2013-The-prevalence-of-dementia-in-Europe/Finland (2014)

21. ScienceNewsline: Alzheimer's Patients' Health Care Costs Higher Already Before Diagnosis: Finnish Study. http://www.sciencenewsline.com/news/2015111617430034.html (2015)

22. Beach, S.R., Schulz, R., Yee, J.L., Jackson, S.: Negative and positive health effects of caring for a disabled spouse: longitudinal findings from the Caregiver health effects study. Psychol. Aging **15**(2), 259–271 (2000)

23. Carroll, L.: Alzheimer's extracts a high price on caregivers, too. Today. http://www.today.com/health/alzheimers-extracts-high-price-caregivers-too-8C11070658 (2013)

24. Alzheimer's Association and National Alliance for Caregiving: Families Care: Alzheimer's Caregiving in the United States 2004. Alzheimer's Association. https://www.alz.org/national/documents/report_familiescare.pdf (2004)

25. Acton, G.J., Wright, K.B.: Self-transcendence and family caregivers of adults with dementia. J. Holist. Nurs. **18**(2), 143–158 (2000)

26. Chambers, M., Ryan, A.A., Connor, S.L.: Exploring the emotional support needs and coping strategies of family carers. J. Psych. Mental Health Nurs. **8**(2), 99–106 (2001)

27. Prince, M.: Care arrangements for people with dementia in developing countries. Int. J. Geriatr. Psych. **19**(2), 170–177 (2004)

28. Papastavrou, E., Kalokerinou, A., Papacostas, S.S., Tsangari, H., Sourtzi, P.: Caring for a relative with dementia: family caregiver burden. J. Adv. Nurs. **58**(5), 446–457 (2007)
29. Schwarzkopf, L., Menn, P., Kunz, S., Holle, R., Lauterberg, J., Marx, P., Mehlig, H., Wunder, S., Leidl, R., Donath, C., Graessel, E.: Costs of care for dementia patients in community setting: an analysis for mild and moderate disease stage. Value Health **14**(6), 827–835 (2011)
30. Beeri, M.S., Werner, P., Davidson, M., Noy, S.: The cost of behavioral and psychological symptoms of dementia (BPSD) in community dwelling Alzheimer's disease patients. Int. J. Geriatr. Psych. **17**(5), 403–408 (2002)
31. Brown, R.M., Brown, S.L.: Informal caregiving: a reappraisal of effects on caregivers. Social Issues Polic. Rev. **8**(1), 74–102 (2014)
32. Carbonneau, H., Caron, C., Desrosier, J.: Development of a conceptual framework of positive aspects of caregiving in dementia. Dementia **9**(3), 327–353 (2010)
33. World Health Organization, WHO. Dementia. https://www.who.int/en/news-room/fact-sheets/detail/dementia (2017)
34. Lamping, D.L., Banerjee, S., Harwood, R., Foley, B., Smith, P., Cook, J.C., Murray, J., Prince, M., Levin, E., Mann, A., Knapp, M.: Measurement of health-related quality of life for people with dementia: development of a new instrument (DEMQOL) and an evaluation of current methodology. Health Technol. Assess. **9**(10), 1–93 (2005)
35. Burgio, L., Allen-Burge, R., Stevens, A., Davis, L., Marson, D.: Caring for Alzheimer's disease patients: issues of verbal communication and social interaction. In Clair, J.M., Allman, R.M. (eds.), The Gerontological Prism: Developing Interdisciplinary Bridges, pp. 231–258 (2000)
36. Altmann, L.J., McClung, J.S.: Effects of semantic impairment on language use in Alzheimer's disease. Semin. Speech Lang. **29**(1), 18–31 (2008)
37. Bayles, K.A., Tomoeda, C.K., Trosset, M.W.: Relation of linguistic communication abilities of Alzheimer's patients to stage of disease. Brain Lang. **42**(4), 454–472 (1992)
38. Honig, L.S., Mayeux, R.: Natural history of Alzheimer's disease. Aging Clin. Exp. Res. **13**(3), 171–182 (2001)
39. Magai, C., Cohen, C., Gomberg, D., Malatesta, C., Culver, C.: Emotional expression during mid- to late-stage dementia. Int. Psychogeriatr. **8**(3), 383–395 (1996)
40. The Alzheimer Society of Ireland: Understanding Late Stage Dementia—Alzheimer Society of Ireland. https://www.alzheimer.ie/Alzheimer/media/SiteMedia/Helpline%20and%20Information%20Resources/Info%20Pack%20PDF%27s/Understanding-late-stage-dementia_Section-A6.pdf
41. Finkel, S.: Behavioral and psychological symptoms of dementia: a current focus for clinicians, researchers, and caregivers. J. Clinic. Psych. **62**(Suppl 21), 3–6 (1998)
42. Cipriani, G., Vedovello, M., Nuti, A., Di Fiorino., M.: Aggressive behavior in patients with dementia: correlates and management. Geriatr. Gerontol. Int. **11**(4), 408–413 (2011)
43. Teri, L., Truax, P., Logsdon, R., Uomoto, J., Zarit, S., Vitaliano, P.P.: Assessment of behavioral problems in dementia: the revised memory and behavior problems checklist. Psychol. Aging **7**, 622–631 (1992)
44. Robinson, L., Hutchings, D., Dickinson, H.O., Corner., L., Beyer., F., Finch., T., Hughes, J., Vanoli, A., Ballard, C., Bond, J.: Effectiveness and acceptability of non-pharmacological interventions to reduce wandering in dementia: a systematic review. Int. J. Geriatr. Psych. **22**, 9–22 (2007)
45. Cipriani, G., Lucetti, C., Nuti, A., Danti, S.: Wandering and dementia. Psychogeriatrics **14**(2), 135–142 (2014)
46. Hope, R.A., Fairburn, C.G.: The nature of wandering in dementia—a community based study. Int. J. Geriatr. Psych. **5**, 239–245 (1990)
47. Yong, T.K., Young, S.Y., Koo, M.S.: Wandering in dementia. Dement Neurocogn. Disord **14**(3), 99–105 (2015)
48. Rosato, D.: Coping with the Costs of Dementia: The Middle Stage. Money. http://time.com/money/4120981/dementia-costs-middle-stage/ (2015)
49. Jefferies, E., Patterson, K., Jones, R.W., Lambon Ralph, M.A.: Comprehension of concrete and abstract words in semantic dementia. Neuropsychology **23**(4), 492–499 (2009)

50. Grossman, M., D'Esposito, M., Hughes, E., Onishi, K., Biassou, N., White-Devine, T., Robinson, K.M.: Language comprehension profiles in Alzheimer's disease. Neurology **47**(1), 183–189 (1996)
51. de Carvalho, I.A., Mansur, L.L.: Validation of ASHA FACS-functional assessment of communication skills for Alzheimer disease population. Alzheimer Dis. Assoc. Disord. **22**(4), 375–381 (2008)
52. Killick, J., Allen, K.: Communication and the care of people with dementia. Open University Press, Buckingham (2001)
53. Blair, M., Marczinski, C.A., Davis-Faroque, N., Kertesz, A.: A longitudinal study of language decline in Alzheimer's disease and frontotemporal dementia. J. Int. **13**, 237–245 (2007)
54. Bohling, H.R.: Communication with Alzheimer's patients: an analysis of caregiver listening patterns. Int. J. Aging Hum. Dev. **33**(4), 249–267 (1991)
55. Dementia Care Central, How to Communicate with Someone with Dementia, https://www.dementiacarecentral.com/about-us/ (2018a)
56. Stokes, G.: Tackling communication challenges in dementia. Nursing Times **109**(8), 14–15 (2013)
57. Thomas, P., Lalloué, F., Preux, P.M., Hazif-Thomas, C., Pariel, S., Inscale, R., Belmin, J., Clément, J.P.: Dementia patients caregivers quality of life: the PIXEL study. Int. J. Geriatr. Psych. **21**(1), 50–56 (2006)
58. Yahya, A., Chandra, M., An, K.S., Garg, J.: Behavioral and psychological symptoms in dementia and caregiver burden. Clinic Med. Res. **4**(2–1), 8–14 (2015)
59. Prado-Jean, A., Couratier, P., Druet-Cabanac, M., Nubukpo, P., Bernard-Bourzeix, L., Thomas, P., Dechamps, N., Videaud, H., Dantoine, T., Clément, J.P.: Specific psychological and behavioral symptoms of depression in patients with dementia. Int. J. Geriatr. Psych. **25**(10), 1065–1072 (2010)
60. Dementia Care Central, Dementia Emotional Problems—Depression. https://www.dementiacarecentral.com/caregiverinfo/coping/emotions/ (2018b)
61. Kim, S.Y.: Behavioral and psychological symptoms of dementia. Korean. Dementia Association **3**(1), 14–17 (2004)
62. Herbert, J., Lucassen, P.J.: Depression as a risk factor for Alzheimer's disease: genes, steroids, cytokines and neurogenesis—What do we need to know? Front. Neuroendocrinol. **41**, 153–171 (2016)
63. Rongve, A., Boeve, B.F., Aarsland, D.: Frequency and correlates of caregiver-reported sleep disturbances in a sample of persons with early dementia. J. Am. Geriatr. Soc. **58**(3), 480–486 (2010)
64. Kim, H.: Behavioral and psychological symptoms of dementia. Ann. Psych. Ment. Health **4**(7), 1086 (2016)
65. Cerejeira, J., Lagarto, L., Mukaetova-Ladinska, E.B.: Behavioural and psychological symptoms of dementia. Front. Neurol. **3**(73), 1–21 (2012)
66. Yesavage, J.A., Friedman, L., Ancoli-Israel, S., Bliwise, D., Singer, C., Vitiello, M.V., Monjan, A.A., Lebowitz, B.: Development of diagnostic criteria for defining sleep disturbance in Alzheimer's disease. J. Geriatr. Psychiatry Neurol. **16**(3), 131–139 (2003)
67. Reilly, J., Martin, N., Grossman, M.: Verbal learning in semantic dementia: Is repetition priming a useful strategy? Aphasiology **19**(3–5), 329–339 (2005)
68. Weirather, R.R.: Communication Strategies to assist comprehension in dementia. Hawaii Med. J. **69**(3), 72–74 (2010)
69. Kempler, D.: Language changes in dementia of the Alzheimer type. In: Lubinski, R. (ed.), Dementia and Communication: Research and Clinical Implications, pp. 98–114 (1995)
70. Ni, W., Constable, R.T., Mencl, W.E., Pugh, K.R., Fulbright, R.K., Shaywitz, S.E., Shaywitz, B.A., Gore, J.C., Shankweiler, D.: An event-related neuroimaging study distinguishing form and content in sentence processing. J. Cogn. Neurosci. **12**(1), 120–133 (2000)
71. Muck-Seler, D., Presecki, P., Mimica, N., Mustapic, M., Pivac, N., Babic, A., Nedic, G., Folnegovic-Smalc, V.: Platelet serotonin concentration and monoamine oxidase type B activity in female patients in early, middle and late phase of Alzheimer's disease. Prog. Neuropsychopharmacol. Biol. Psychiatry **33**(7), 1226–1231 (2009)

72. Dijkstra, K., Bourgeois, M., Petrie, G., Burgio, L., Allen-Burge, R.: My recaller is on vacation: Discourse analysis of nursing-home residents with dementia. Discourse Process. **33**(1), 53–74 (2002)
73. Dijkstra, K., Bourgeois, M., Burgio, L., Allen, R.: Effects of a communication intervention on the discourse of nursing home residents with dementia and their nursing assistants. J. Med. Speech-Lang. Pathol. **10**(2), 143–157 (2002)
74. Luke, A.: Dementia Symptoms: Doing this Could be an Early Warning Sign. https://www.express.co.uk/life-style/health/950356/dementia-symptoms-signs-types-causes (2018)
75. Morris-Underhill, C.: Hantsport women team up to create fidget quilts for people with dementia. http://www.hantsjournal.ca/community/hantsport-women-team-up-to-create-fidget-quilts-for-people-with-dementia-204557/ (2018)
76. Van Gool, W.A., Weinstein, H.C., Scheltens, P., Walstra, G.J.: Effect of hydroxychloroquine on progression of dementia in early Alzheimer's disease: an 18-month randomised, double-blind, placebo-controlled study. Lancet **358**, 455–460 (2001)
77. Shimokawa, A., Yatomi, N., Anamizu, S., Torii, S., Isono, H., Sugai, Y., Kohno, M.: Influence of deteriorating ability of emotional comprehension on interpersonal behavior in Alzheimer-type dementia. Brain Cogn. **47**(3), 423–433 (2001)
78. Warshaw, G.A., Bragg, E.J.: Preparing the health care workforce to care for adults with Alzheimer's disease and related dementias. Health Aff. **33**(4), 633–641 (2014)
79. Alzheimer Society of Canada: Middle Stage—What to Expect. http://alzheimer.ca/en/Home/Living-with-dementia/Caring-for-someone/Middle-stage (2017)
80. Jefferies, E., Bateman, D., Lambon Ralph, M.A.: The role of the temporal lobe semantic system in number knowledge: evidence from late-stage semantic dementia. Neuropsychologia, **43**(6), 887–905 (2005)
81. Kovach, C.R., Magliocco, J.S.: Late-stage dementia and participation in therapeutic activities. Appl. Nurs. Res. **11**(4), 167–173 (1998)
82. McKeown, J., Clarke, A., Ingleton, C., Ryan, T., Repper, J.: The use of life story work with people with dementia to enhance person-centred care. Int. J. Older People Nurs. **5**(2), 148–158 (2010)
83. Westmacott, R., Black, S.E., Freedman, M., Moscovitch, M.: The contribution of autobiographical significance to semantic memory: evidence from Alzheimer's disease, semantic dementia, and amnesia. Neuropsychologia **42**(1), 25–48 (2004)
84. Martin, L.S., Beaman, A.: Communication strategies to promote spiritual well-being among people with dementia. J. Pastoral Care Couns **59**(1–2), 43–55 (2005)
85. Cheston, R.: Stories and metaphors: talking about the past in a psychotherapy group for people with dementia. Ageing Soc. **16**(5), 579–602 (1996)
86. Subramaniam, P., Woods B.: Towards the therapeutic use of information and communication technology in reminiscence work for people with dementia: a systematic review. Int. J. Comput. Healthcare **1**(2), 106–125 (2010)
87. Yasuda, K., Kuwabara, K., Kuwahara, N., Abe, S., Tetsutani, N.: Effectiveness of personalised reminiscence photo videos for individuals with dementia. Neuropsych. Rehabil. **19**(4), 603–619 (2009)
88. Graham, K.S., Hodges, J.R.: Differentiating the roles of the hippocampal complex and the neocortex in long-term memory storage: evidence from the study of semantic dementia and Alzheimer's disease. Neuropsychology **11**(1), 77–89 (1997)
89. Beard, R.L., Knauss, J., Moyer, D.: Managing disability and enjoying life: how we reframe dementia through personal narratives. J. Aging Stud. **23**(4), 227–235 (2009)
90. Ryan, E.B., Schindel Martin, L.: Using narrative arts to foster personhood in dementia. In: Backhaus, P. (ed.) Communication in elderly care, London, England: Continuum Press, pp. 193–217 (2011)
91. UK Dementia Directory: How to Make a Memory Book. https://www.dementia.co.uk/products/how-to-make-a-memory-book
92. Day, K., Cohen, U.: The role of culture in designing environments for people with dementia: a study of Russian Jewish immigrants. Environ. Behav. **52**(3), 361–399 (2000)

93. Deterding, S, Dixon, D, Khaled, R, Nacke, L.: From game design elements to gamefulness: defining gamification. In: Proceedings of the 15th International Academic MindTrek Conference: Envisioning Future Media Environments 28 Sept 2011, Association for Computing Machinery, pp. 9–15 (2011)

94. King, D, Greaves, F, Exeter, C, Darzi, A.: 'Gamification': influencing health behaviours with games. J. R. Soc. Med. **106**(3), 76–78 (2013)

95. Reeves, B., Read, J.L.: Total engagement: using games and virtual worlds to change the way people work and businesses compete. Harvard Business School Press, Boston, MA (2009)

96. Cugelman, B.: Gamification: what it is and why it matters to digital health behavior change developers. JMIR Serious Games, **1**(1) (2013)

97. Hamari, J., Koivisto, J., Sarsa, H.: Does gamification work?—a literature review of empirical studies on gamification. In: 2014 47th Hawaii International Conference on System Sciences (HICSS). IEEE, pp. 3025–3034 (2014)

98. Bamidis, P.D., Gabarron, E., Hors-Fraile, S., Konstantinidis, E., Konstantinidis, S., Rivera, O.: Gamification and behavioral change: techniques for health social media. In: Participatory Health Through Social Media, pp. 112–135 (2016)

99. Seaborn, K., Fels, D.I.: Gamification in theory and action: a survey. Int. J. Hum. Comput. Stud. **74**, 14–31 (2015)

100. Chin, S.: Mobile technology and gamification: the future is now! In: 2014 Fourth International Conference on IEEE. Digital Information and Communication Technology and it's Applications (DICTAP), pp. 138–143 (2014)

101. Bunchball, I.: Gamification 101: an introduction to the use of game dynamics to influence behavior. White Paper (2010)

102. Combéfis, S., Beresnevičius, G., Dagienė, V.: Learning programming through games and contests: overview, characterisation and discussion. Olymp. Inf. **10**(1), 39–60 (2016)

103. Muntean, C.I.: Raising engagement in e-learning through gamification. In: Proceedings of 6th International Conference on Virtual Learning ICVL, vol. 1, pp. 323–329 (2011)

104. Lucassen, G., Jansen, S.: Gamification in Consumer Marketing-Future or Fallacy? Procedia-Soc. Behav. Sci. **148**, 194–202 (2014)

105. Hamari, J., Järvinen, A.: Building customer relationship through game mechanics in social games. In: Cruz-Cunha, M. Carvalho, V., Tavares, P. (eds.), Business, Technological and Social Dimensions of Computer Games: Multidisciplinary Developments, IGI Global, Hershey, PA (2011)

106. Werbach, K., Hunter, D.: For the win: how game thinking can revolutionize your business. Wharton Digital Press (2012)

107. Zhuang, Y., Ma, H., Xie, H., Leung, A.C.M., Hancke, G.P., Wang, F.L.: When innovation meets evolution: an extensive study of emerging e-learning technologies for higher education in Hong Kong. In: International Symposium on Emerging Technologies for Education, pp. 574–584 (Springer 2016)

108. Strmeki, D., Bernik, A., Radoevi, D.: Gamification in e-Learning: introducing gamified design elements into e-learning systems. J. Comput. Sci. **11**(12), 1108–1117 (2015)

109. Topîrceanu, A.: Gamified learning: a role-playing approach to increase student in-class motivation. Procedia Comput. Sci. **112**, 41–50 (2017)

110. Lenihan, D.: Health games: a key component for the evolution of wellness programs. Games Health: Res. Develop. Clinic. Appl. **1**(3), 233–235 (2012)

111. Pereira, P., Duarte, E., Rebelo, F., Noriega, P.: A review of gamification for health-related contexts. In: International Conference of Design, User Experience, and Usability, vol. 8518, pp. 742–753. Springer International Publishing (2014)

112. Menezes, J., Gusmão, C., Machiavelli, J.: A proposal of mobile system to support scenario-based learning for health promotion. Procedia Technol. **9**(3), 1142–1148 (2013)

113. Pannese, L., Wortley, D., Ascolese, A.: Gamified wellbeing for all ages—how technology and gamification can support physical and mental wellbeing in the ageing society. In: XIV Mediterranean Conference on Medical and Biological Engineering and Computing 2016. Springer, Berlin, pp. 1281–1285 (2016)

114. Groot, C., Hooghiemstra, A.M., Raijmakers, P.G., van Berckel, B.N., Scheltens, P., Scherder, E.J., van der Flier, W.M., Ossenkoppele, R.: The effect of physical activity on cognitive function in patients with dementia: a meta-analysis of randomized control trials. Ageing Res. Rev. **25**, 13–23 (2016)

115. Savulich, G., Piercy, T., Fox, C., Suckling, J., Rowe, J.B., O'Brien, J.T., Sahakian, B.J.: Cognitive training using a novel memory game on an iPad in patients with amnestic mild cognitive impairment (aMCI). Int. J. Neuropsychopharmacol. **20**(8), 624–633 (2017)

116. Eisma, R., Dickinson, A., Goodman, J., Syme, A., Tiwari, L., Newell, A.F.: Early user involvement in the development of information technology-related products for older people. Univ. Access Inf. Soc. **3**(2), 131–140 (2004)

117. Canadian Study of Health and Aging Working Group: The incidence of dementia in Canada. Neurology, **55**(1), 66–73 (2000)

118. Lobo, A., Launer, L.J., Fratiglioni, L., Andersen, K., Di Carlo, A., Breteler, M.M., Copeland, J.R., Dartigues, J.F., Jagger, C., Martinez-Lage, J., Soininen, H., Hofman, A.: Prevalence of dementia and major subtypes in Europe: a collaborative study of population-based cohorts. Neurology **54**(11 Suppl 5), S4–S9 (2000)

119. Robert, P., Leone, E., Amieva, H., Renaud, D.: Managing behavioural and psychological symptoms in Alzheimer's disease. In: Alzheimer's Disease. Chapter 9, pp. 71–84 (2017)

Analysis of Centrality Concepts Applied to Real-World Big Graph Data

Soyeon Oh, Kyeongjoo Kim and Minsoo Lee

Abstract Graphs are mathematical models to represent relationships, and graph theories have an important role in recent research in the computer science area. These days, there are many kinds of graph-structured data such as social network service and biological and location data. And the graph data are now easily considered big data. Analyzing such graph data is an important problem. In this paper, we apply four major centralities and PageRank algorithms to real-world undirected graph data and find some empirical relationships and features of the algorithms. The results can be the starting point of many data-driven and theoretical link-based graph studies as well as social network service analysis.

Keywords Graph centrality · PageRank · Graph mining

1 Introduction

Graphs are useful mathematical models consisting a set of nodes and edges to represent relationships [1]. Graph theory is one of the basic research areas in computer science. These days, there are many kinds of graph-structured data such as social network service and biological and sensor network data. The size of these kinds of data is quite huge and is usually considered big data. The forms of the data are various. So analyzing such real-world big graph data is an important problem these days.

There are many methods to analyze graph data. Centrality and PageRank analysis are basic analysis methods for graph data. Centrality analysis is a method to find central or important nodes in the graph. PageRank is an algorithm to rank nodes in the

S. Oh · K. Kim · M. Lee (✉)
Department of Computer Science and Engineering, Ewha Womans University, Seoul, Korea
e-mail: mlee@ewha.ac.kr

S. Oh
e-mail: soyeon.oh@ewhain.net

K. Kim
e-mail: kjkimkr@ewhain.net

© Springer Nature Singapore Pte Ltd. 2019
S. K. Bhatia et al. (eds.), *Advances in Computer Communication and Computational Sciences*, Advances in Intelligent Systems and Computing 924,
https://doi.org/10.1007/978-981-13-6861-5_52

graph, assuming nodes with many in-links are important nodes. Centrality and PageRank are similar in the point of view that both treat nodes with link characteristics.

Many researches about link-based graph analysis are being done these days, for example, studies about the calculation of centralities in the large graph [2, 3] and the studies about the calculation of PageRank [4, 5]. These are studies about calculation methods. We can obtain the values of centralities and PageRank results fast or correctly with such studies. But from another point of view, especially data-driven, the meaning and relationships about centralities and PageRank are not focused on so far. Graph analysis is part of the data analysis where it has both theoretical and empirical or data-driven features.

In this paper, we compare and analyze the four major centralities, which are degree centrality, betweenness centrality, closeness centrality, and eigenvector centrality along with the PageRank algorithm applying them to real undirected social network graph data. We found some empirical or data-driven features and relationships as well as theoretical features and relationships between each centrality and PageRank. The results can be the starting point of many data-driven and theoretical link-based graph studies and also social network service analysis.

2   Related Work

2.1   A YELP Dataset

For the experiments, we used the YELP dataset composed of information about local businesses in 11 cities across 4 countries. It includes 4.1M reviews and 947K tips by 1M users for 144K businesses. There are also 1.1M business attributes such as address, hours, categories, and name. Among the datasets, we used only User.json dataset which has user name, user id, friends list, review counts, type, etc. There are 2,102,231 undirected edges and 298,361 nodes [6]. Figure 1 shows the YELP data graph which has only part of the whole social graph containing 500 edges and 524 vertices. We confirmed that this dataset is an undirected graph. Edges are shown in red color. Therefore, the vertices which are connected with high density have red colored areas between the vertices. The center circle part of Fig. 1 shows a densely connected group of vertices with red colored area between the vertices.

2.2   Data Preprocessing

Before doing the experiment, we had to preprocess the dataset which is too large and wide to get the desired results clearly and quickly. We used Hadoop MapReduce Jobs which were written in JAVA, additional C Programs, and R scripts to preprocess the

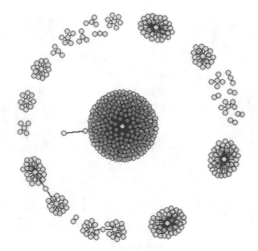

Fig. 1 YELP data graph (500 edge and 524 vertices)

data. We first generated UserId-UserId edge lists and UserId-Name lists. Then, we processed duplicate edges and names. Lastly, we generated undirected graph edge lists and vertex name lists for visualization.

2.3 igraph R Package

There are many packages in R for network analysis. The igraph library provides a set of data types and functions for easy implementation of graph algorithms, fast handling of large graphs with millions of vertices and edges, allowing rapid prototyping via high-level languages like R [7]. Using this igraph library, we calculated and analyzed node centrality and PageRank, and we defined our own plot functions (plotMeWithOthers(), plotMeAndMyFriends(), plotMeAndMyFriendsWithOthers()) for visualizing the result.

3 Empirical Characteristics of Centralities

There are many centralities used to find important nodes [8]. But we focused on the four major centralities which are degree centrality, betweenness centrality, closeness centrality, and eigenvector centrality. We studied the empirical characteristics of the four centralities on undirected graph data, which is real-world social network data.

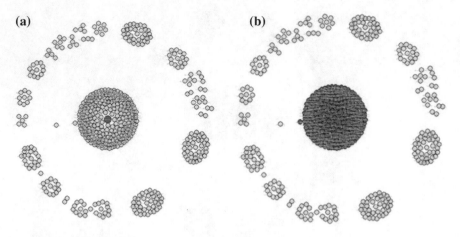

Fig. 2 Node with many edges or with high degree

3.1 Empirical Characteristics of Degree Centrality

Nodes with many edges or with high degree get the high value of degree centrality and nodes with few edges or with low degree get the low value of degree centrality. We calculated the degree centrality value on our YELP dataset and got the distribution, where the values are very heavily weighted for very few people. One person had very high degree centrality values, and about 10 people had notable degree centrality values. Others had very small degree centrality values near to 0. In Fig. 2a, the single vertex that has the most densely connected edges is shown in red. In order to show all of the connections from this vertex, all of the neighboring vertices that are directly connected to this single vertex are shown in red in Fig. 2b.

3.2 Empirical Characteristics of Betweenness Centrality

For some nodes, you can count the number of occurrences where the node is on the shortest path of every pair of nodes. Nodes with high count value get high value of betweenness centrality and nodes with high betweenness centrality value are the bottlenecks or connection points of two networks. We calculated the betweenness centrality value on our YELP dataset and got the distribution where the value is heavily weighted in some people. One person had very high betweenness centrality values and about 10 people had notable betweenness centrality values. Others had very small betweenness centrality values near to 0. Figure 3a shows a node in red color which we shall consider. Figure 3b shows that this node has a degree of 2 and connects two small clusters of vertices, making it the node that needs to be included

(a)   **(b)**

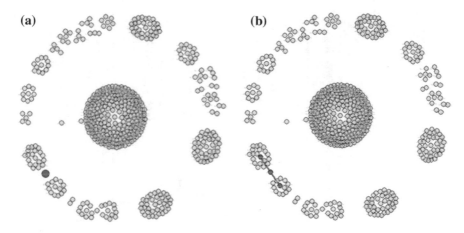

Fig. 3 Node with high betweenness centrality value

in all shortest paths that connect all combinations of node pairs in the two different sides of clusters. Therefore, it has a considerably high betweenness centrality even though it only has a low degree of 2.

3.3 Empirical Characteristics of Closeness Centrality

For some node, the farness can be calculated by the sum of all shortest paths from the node to other nodes. And the closeness centrality value is the inverse of the farness value.

Therefore, nodes with high closeness value can be thought that they are very close to other nodes. And nodes with low closeness value can be thought that they are very far from other nodes. Figure 4a shows a node that is very close to many vertices in the center, and thus the sum of all shortest paths from this node to all other nodes is small. In other words, as the farness value is small, the inverse value which is the closeness becomes large. Figure 4b shows a node which is quite far from the central vertices cluster, and the sum of all shortest paths to all other nodes becomes quite large. This means the farness value is large and in turn the closeness becomes small. We calculated the closeness centrality value on our YELP dataset and got the distribution where the distribution is quite uniform compared to other distribution of centralities. There is one group with high closeness centralities and other groups with low closeness centralities. But the difference of the closeness values between the two groups is not quite large.

We calculated each centrality values on datasets which has different size of nodes. The execution time is shown in Table 1.

Table 1 Runtime of each centrality

Number of edges	Calculation centrality time (degree)	Calculation centrality time (betweenness)	Calculation centrality time (closeness)	Calculation centrality time (eigen vector)	Plotting time	Visibility/intelligibility of result
All (2,102,231)	Under 1 s	Over 10 h	4 h	2 s	Over 10 h	Visibility of visualized graph was unreadable, because of too many nodes and edges
Over 1,000,000						
50,000		46 s	23 s	2 s	55 s	
10,000		2 s	2 s	Under 1 s	17 s	
1,000		Under 1 s	Under 1 s		3 s	
500					Under 1 s	Clear
300						Result is not intelligible, because network size is too small
100						

(a)   (b)

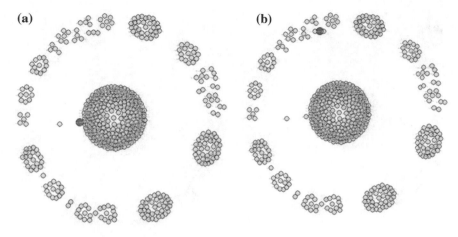

Fig. 4 **a** Nodes with high closeness, **b** nodes with low closeness

Closeness centrality has important characteristics, that is, it is not suitable for big graph data. In the distribution, we got the difference of the value which is not quite big. Because in our dataset, the differences of shortest path distances with the node in the central subgraph and in the edge subgraph are not quite big. In a big graph, this situation is more severe. Therefore, the values of the closeness centrality in big graph are not quite notable. Moreover, we need to calculate the shortest path distance to get closeness centrality values. This takes quite a long time when the size of the data gets larger.

3.4 Empirical Characteristics of Eigenvector Centrality

We can calculate the eigenvector centrality by calculating the eigenvector of the adjacency matrix of the graph. The value of eigenvector centrality is the connection degree with the nodes with high centralities. Therefore, nodes connected to nodes with high centrality or degree have high eigenvector centrality values.

We calculated the eigenvector centrality value on our YELP dataset and got the distribution where one person has the highest eigenvector centrality and some people have notable eigenvector centralities similar to each other. They are people connected with the person with the highest eigenvector centrality value.

And eigenvector centrality can be calculated very fast compared with closeness centrality or betweenness centrality. So it is suitable for big graph data.

Figure 5a shows a node that has high eigenvector centrality. This is due to the fact that the neighboring node shown in Fig. 5b has a high centrality value and thus is also assuming a high centrality.

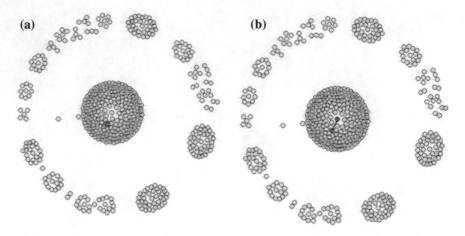

Fig. 5 **a** Nodes with high eigenvector, **b** neighbor node with high centrality

Table 2 Correlation coefficient of each centrality and PageRank

	Degree	Betweenness	Closeness	Eigenvector	PageRank
Degree	–	0.98	0.01	0.76	1.00
Betweenness		–	0.04	0.80	0.97
Closeness			–	0.62	0.00
Eigenvector				–	0.76
PageRank					–

4   Relationship of Centralities and PageRank

Centralities and PageRank both take the link characteristics of the graph data. So we think that there can be some relationships between the centralities and PageRank [9]. We calculated the correlation coefficient of each pair of calculation results. The result is given in Table 2.

5   Conclusion

In this paper, we applied four major centralities and PageRank algorithms to real-world undirected graph data. Through the graph analysis, we could found some empirical relationships and features of the algorithms. In general, the eigenvector centrality has some relation with PageRank because of the algorithmic or mathematical reason. In addition to that, we found that degree centrality has high relation with PageRank because of the undirected graph. Degree centrality has also high relation with betweenness centrality because of the data-driven characteristics. Therefore, degree centrality, betweenness centrality, and PageRank are correlated with eigenvector centrality. Based on the analysis, we can figure out who is an important person in social network services. We expect that the results could contribute to many data-driven and theoretical link-based graph studies and social network service analysis.

Acknowledgements This research was supported by Basic Science Research Program through the National Research Foundation of Korea (NRF) funded by the Ministry of Education (2017R1D1A1B03034691).
This work was supported by the "Convergence Female Talent Education Project for Next Generation Industries" through the MSIP and NRF (2015H1C3A1064579).

References

1. Graph: In Wikipedia, The Free Encyclopedia. Retrieved 16:36, June 6, 2017, from https://en.wikipedia.org/w/index.php?title=Graph&oldid=771368109 (21 Mar 2017)
2. Kang, U., Papadimitriou, S., Sun, J., Tong, H.: Centralities in Large Networks: Algorithms and Observations
3. Bonchi, F., De Francisci Morales, G., Riondato, M.: Centrality Measures on Big Graphs: Exact, Approximated, and Distributed Algorithms
4. Page, L. et al.: The PageRank Citation Ranking: Bringing Order to the Web (1999)
5. Gleich, D., Polito, M.: Approximating personalized pagerank with minimal use of web graph data. Internet Math. **3**(3), 257–294 (2006)
6. YELP Homepage. https://www.yelp.com/dataset_challenge. Last accessed 2017/06/06
7. R igraph Manual Pages. http://igraph.org/r/doc/aaa-igraph-package.html. Last accessed 2017/06/06
8. Centrality: In Wikipedia, The Free Encyclopedia. Retrieved 20:44, June 6, 2017, from https://en.wikipedia.org/w/index.php?title=Centrality&oldid=783572769 (3 June 2017)
9. PageRank: In Wikipedia, The Free Encyclopedia. Retrieved 21:14, June 6, 2017, from https://en.wikipedia.org/w/index.php?title=PageRank&oldid=782965157 (30 May 2017)

How Does Grooming Fit into Social Engineering?

Patricio Zambrano, Jenny Torres and Pamela Flores

Abstract In this research, we propose to formally include the technique of grooming within the processes of social engineering, validating its insertion with the phases of social engineering designed by Kevin Mitnick. Scientific evidence has shown that grooming is the technique applied by child sexual harassers and research in this field has generated significant contributions in this study. Nevertheless, researchers do not interrelate the contributions generated by grooming and social engineering, which allows us to infer that the studies are isolated and incomplete and must be consolidated with complementary scientific evidence.

1 Introduction

Nowadays, social engineering attacks are more common, specific, and sophisticated. Cyber attackers take advantage of social behavior and norms such as reciprocity or social subtleties to gain access to information through human vulnerability. In the same way, they take advantage of the fact that most people never expect to be victims of social engineering, nevertheless, they are harmed by this type of attack, delivering unconsciously sensitive information. Some social engineering attacks have proven to be extremely expensive. In the UK, it is estimated that crimes related to identity theft cost around 1.2 billion pounds in 2009. Phishing losses were around 23.2 million pounds in 2005 [1]. In 2004, the Department of Justice of USA concluded that one in three people could become a victim of social engineering during their lifetime [2].

This is a turning point where new questions are generated. Is only personal data the goal of social engineers? What types of techniques have been studied against

P. Zambrano (✉) · J. Torres · P. Flores
Faculty of System Engineering, Escuela Politcnica Nacional, Quito, Ecuador
e-mail: patricio.zambrano@epn.edu.ec

J. Torres
e-mail: jenny.torres@epn.edu.ec

P. Flores
e-mail: pamela.flores@epn.edu.ec

© Springer Nature Singapore Pte Ltd. 2019
S. K. Bhatia et al. (eds.), *Advances in Computer Communication
and Computational Sciences*, Advances in Intelligent Systems and Computing 924,
https://doi.org/10.1007/978-981-13-6861-5_53

human vulnerability? Are only professional people vulnerable to social engineering techniques? To answer these questions, we evaluated the types of Internet users and found that: in USA, two-thirds of households with children use the Internet. From these, 84% of children and 97% of young people between 12 and 18 are permanently connected due to the massive use of social networks [3]. This alarming data leads to a more detailed analysis of the different attacks to which this group of users would be vulnerable. According to the Child Protection and Online Protection Agency, online child harassment, known as grooming, was the most reported crime in the UK between 2009 and 2010 [4, 5]. Grooming is considered a process of preparation to approach, persuade and compromise the privacy and intimacy of a child or adolescent. After a more detailed analysis in the literature, we determined that grooming is not considered a social engineering technique, despite the efforts made to automatically determine the syntactic and semantic behavioral patterns of pedophiles online, based on the analysis of chats and P2P networks, when applying grooming as a technique of harassment [6, 7]. Krombholz et al. [8] proposes a taxonomy of social engineering that cover users, channels, and techniques. Nevertheless, grooming is not included as a social engineering technique.

In this paper, we propose an extension of this taxonomy to include the grooming as part of the social engineering processes. Our main objective is to expand the field of social engineering based on studies related to grooming [6, 7] thus creating greater knowledge that can be used as a tool to teach and as a framework to develop a holistic protection strategy. The rest of this document has been organized as follows. Section 2 describes the different criteria and concepts of social engineering and grooming. Section 3 highlights the research methodology based on related studies. Section 4 shows the interrelationship between social engineering and grooming. Sections 5 and 6 conclude the study by presenting a discussion of the results obtained and conclusions, respectively.

2 Background

2.1 Social Engineering

The identification and classification of threats are the basis for building defense mechanisms. However, humans are considered the weakest link in information security due to their susceptibility to different manipulation techniques [1]. Concepts associated with "social engineering", define it as a dark art in computer science where, the use of deception is the main tool to induce a person to divulge private information involuntarily and, in this way, attackers gain access to computer systems [2]. The techniques employed by a social engineer include persuasion, coercion, urgency, authority, supplanting or requesting assistance, among others; taking advantage of human weakness [9]. In the case of social networks, it is very common for users to downplay the importance of the security of their information, as they rely

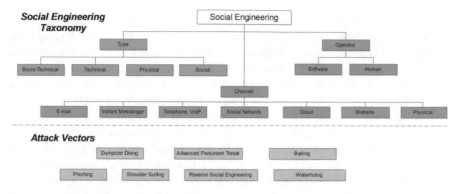

Fig. 1 Krombholz et al. social engineering taxonomy

on the protection mechanisms of the companies that run the social networking sites. Users tend to believe that popular companies like Facebook and Twitter will not allow anyone to exploit their information. However, instead of using technical means to exploit the user, social engineers use deception techniques to convince users to accept an attack. Kevin Mitnick said it is much easier to trick someone to present his credentials than using sophisticated piracy maneuvers [9]. The field of social engineering is still in its early stages of standardization with formal concepts, frameworks, and work templates [10]. The "art" of influencing people to divulge sensitive information is known as social engineering and the process of doing it, is known as social engineering attack. Nowadays, there are definitions of social engineering and several different models of social engineering attacks [1]. Many scientists strive in their studies to associate social engineering with non-technicality, while others include technical attacks that do not imply an important role for traditional social engineering from human to human. Academics, such as Krombholz et al. [8], have proposed a theoretical/technical taxonomic analysis including technicality, as can be seen in Fig. 1.

This study is based on two previous proposals [11, 12]. Our study considers taking this taxonomy as a reference point, for its theoretical and technical support, as well as the four phases proposed by Mitnick [1] to determine if an attack is associated with social engineering: (1) information gathering, (2) development of relationship, (3) exploitation of relationship; and (4) execution to achieve objective.

2.2 Grooming

Grooming has been studied for more than a decade, and these studies have generated significant contributions to society [6, 7, 13]. In the scientific field, grooming has been conceptualized as a procedural technique used by cybernetic attackers, who in some cases are pedophiles or pederasts. On the other hand, it is also considered as an operational concept, through which an aggressor applies affinity search strategies, while acquiring information and sexually desensitizing the victims to develop relationships that result in the satisfaction of the attacker's needs [6]. Through different studies, it is evident that the grooming can be applied for several years, in order to prepare the victims and guarantee cooperation, thus minimizing the risk of exposure by the attacker to the victim. In some cases, it is also considered the preparation of relatives close to the victims to create an atmosphere of acceptance and normalization of the attack. Perverted-justice is an online tool, which aims to eradicate online predators [6]. Since its inception, they planned to publish chats of real predators, thus exposing them. This was the starting point where scientists began to study and analyze the text chains published by this web portal, thus determining psychological and technical behavioral traits when applying grooming as a preparation for victims. One of the biggest challenges evidenced in the studies of chat chains is phonetics and phonology, where the fields of study of morphology, syntax, semantics, pragmatics, and discourse are derived. Within these studies, it has been determined that online pedophiles tend to seduce their victim through attention, affection, kindness, and even gifts. In a survey applied to 437 schoolchildren between 11 and 13 years that use habitually chat rooms. 59% of the participants agreed to have regularly participated in chats with people through the Internet. The 24% of people who chatted online admitted having delivered some type of personal information. These include the house telephone number, the mobile phone number, and the address of the house. The most alarming fact in this study was that 37 children admitted to decide to meet the person they were chatting with MacFarlane and Holmes [14]. In our analysis and technical questioning of the proposals of the grooming stages, see Fig. 2, we observe that the terms are very general or ambiguous. Olson, Welner, O'Connell, evidence and justify different stages of grooming, these being the most cited by related research [4, 5, 15]. Based on that, we use this set of technical criteria to propose four stages that show a level of access to the victims: (A) Selection and beginning of relationship with the victim. In every attack, the perpetrator analyzes and selects his objective, for our case a child or adolescent who provides the ideal conditions (ethnicity, deprivation, lack of attention, etc.) to start a relationship of trust. (B) Analysis of vulnerabilities or deficiencies of the victim. Once the relationship starts, the attacker will proceed to analyze their vulnerabilities and strategically exploit each of them, for our case, the attacker will proceed to fill spaces where the victims feel isolated, lacking attention or suffer from some type of economic need. (C) Handling the victim in sexual aspects and privacy of the relationship. The victim will be involved in the sexual field once the attacker has achieved a space of marked confidence. This process will require time and confidentiality in the exchange of information since the attacker will sometimes share photographs of their own sexual content or that of other victims to desensitize them and to establish confidentiality rules so that the victim does not disclose their conversations. And (D) Keep control over the relationship. In this stage, the attacker has become a very important person for the victim and in his eagerness to preserve the domain of the relationship he will apply again the techniques of vulnerability.

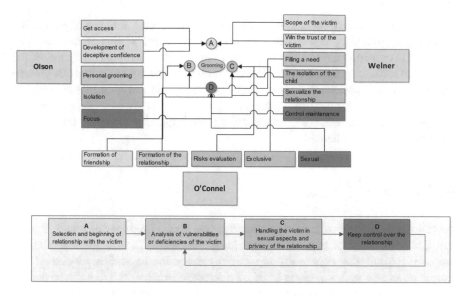

Fig. 2 Grooming stages

3 Methodology

The inductivist approach is the general basis of this research, since it draws general conclusions, starting from particular observational statements. This method is characterized by having four basic stages: observation and recording of all the facts, analysis and classification of the facts, inductive derivation of a generalization from the facts, and contrast [2]. (i) *Observation and registration.* We begin by observing the problem, online pedophilia in contrast to social engineering, and the need to generalize a model that encompasses them by recording all the scientific documentation that deals with these topics. (ii) *Analysis and classification.* To raise this generalization, the study assumed that this could be created using the existing knowledge. The observations were made on studies of attacks of real online pedophiles that use grooming as an attack technique, and the analysis of studies and taxonomies related to social engineering [8]. The study of the literature showed that the topics under discussion are part of present and future research. Various characteristics that describe them have been investigated, analyzed, and published, except for their interrelation. Krombholz et al. [8], Bakhshi et al. [9], Gupta et al. [16]. (iii) *Inductive derivation of a generalization.* To better understand cyber-pedophilia and the use of grooming as an attack technique, we endeavor to demonstrate well-documented and representative studies of grooming and cyber-pedophilia [6, 13, 15]. A particularity of the investigations related to pedophilia is that they analyze the texts of real conversations of pedophiles published in the preverted-justice website. These studies are very diverse and range from neuro-linguistic programming (NLP), textmining, mathematical models to a wide gamma of artificial intelligence techniques to determine a profile or pattern of pedophile behavior. These studies demarcate a considerable analysis of pedophilia and refer to grooming as the technique used to access the victims. Under this criterion, it is considerable to infer that grooming is part of the generalization of the concept of social engineering. And finally (iv) *Testing*. It should be noted that grooming, in the investigations that were part of our study, did not infer that it is part of social engineering. Under this premise, a more detailed study of social engineering was acomplished, considering vectors of attacks, applied techniques, taxonomies, and communication channels.

4 Results

As stated above, for an attack to be classified as a Social Engineering attack, it must fulfill Mitnick's four phases: (1) Information gathering, (2) Development of relationship, (3) Exploitation of relationship, and (4) Execution to achieve objective. The grooming as an attack technique has shown certain characteristics that can be clearly classified within two Mitnick phases: development of relationship; and exploitation of relationship.

4.1 *Information Gathering*

This phase consists of collecting as much information as possible about the target. This information is used in the next phases and is of vital importance for the quality of the result in cases of targeted attacks. In the study about grooming, we consider it relevant to understand the psychological-technical profile of the attacker, since the type of attack and the information gathered depend on it.

Types of attackers In the literature, we can identify different criteria of the attackers since they are treated as child sexual abusers or pedophiles. The latter is sometimes considered a phenomenon under constant study, however, in other cases it is treated as a disorder of sexual preference "pedophile" and the consequence of their behavior is named as adolescent sexual abuse (CSA) [17, 18]. Pedophilia is defined by the diagnostic manual of the World Health Organization (WHO) and the International Classification of Diseases (ICD-10) as: "a disorder of sexual preference". It should be mentioned that ICD-10 indicates that pedophile behaviors are very heterogeneous. These behaviors can be harmless, even those that reach levels of child sexual abuse or pedophilia [19]. However, there is another type of attacker. This one does not have a disorder of sexual preference toward children but sees in child pornography a profitable business type and can use the grooming to obtain child pornographic material, for its subsequent sale and distribution. In the scientific field, the pedophile is considered the only user who uses grooming to persuade his victim and commit an act of rape. Under this consideration, our investigation raises three types of criminals and their interests in the field of pedophilia: the cyber-pedophile, cyber-pederast or child sexual abuse, and the cyber-offender. After the analysis of the potential attackers, it is possible to infer the type of information that these will gather in the process of attack: files with pornographic content, personal information, address of domicile and places of frequent visit, etc., being their potential targets, the personal use and/or commercialization of pornographic material obtained and in the worst scenario, the violation.

4.2 Development of Relationship

In this phase, related researches describe characteristics of development of relationship without linking them to social engineering. It is noted that, to locate children or people with access to a child, pedophiles use social networking sites, blogs, chat rooms for children, IM, email, discussion forums and online children's play websites, to start a relationship. The online games of PC, Xbox, PlayStation, and Wii, are spaces where dangerous relationships are developed between a child and a delinquent [20]. An important aspect of development of relationship is mentioned in Yong et al. [21] where pedophiles seduce their young prey through attention, affection, kindness, and even gifts. Two additional elements to the relationship and trust are: simulation of child behavior (slang), it is generally used by predators to copy their linguistic style and the second aspect is the sexual language, where predators gradually change a normal conversation to a sexual one, starting with more ordinary compliments [18]. The development of the relationship and trust by stages has been evaluated in Cano et al. [15], from the point of view of the development of deceptive trust that consists of building a relationship of trust with the child. This stage allows the predator to build a common ground with the victim.

4.3 Exploitation of Relationship

The exploitation of the relationship is another feature considered by Mitcknic in Gupta et al. [16], given that, it demonstrates the success of this exploitation by achieving the exchange of pornographic content files with underages. Harms mention that an aggressor applies affinity search strategies, while acquiring information and sexually desensitizing its specific victims to develop relationships that lead to the satisfaction of needs [6]. After the previous stage, the predators maintain a fixed discourse, in which they are not willing to depart from the sexual conversation [18]. The elimination of communication records and the transfer of responsibility to the victim are two traits marked in this investigation. Cano et al. [15] represent the attack by cycles and in one of them it, shows the isolation of the victim from his friends and relatives once the relationship has intensified and, consequently the predator seeks to physically approach the minor. In this stage, the predator requests information such as, the child's and father's schedules, and the child's location.

4.4   Execution to Achieve Objective

The first use of the information gathered is the sale and distribution of pornographic material. The profile of the associated attacker, to this process, is the cybercriminal that based on aggression, which in some cases is agreed with family members, obtains audio or visual material. The private use of pornographic material is for the exclusive use of cyber-pedophiles who do not intend to commit rape, however, within the development of relationship, they exchange pornographic material to desensitize their victims and even use aggression to persuade them to keep the relationship secret. The pedophile or pederast, who in some cases is considered a child sexual abuse is the most dangerous because within the fields of exploitation of relationship and the development of relationship is the one that uses more mechanisms to get information of his victim with the aim of sexually abusing her.

5   Discussion

The correlation between the phases of Mitnick and the processes associated with grooming, begin with the definition of the types of attackers and their objectives (information collection), because they will be responsible for collecting the necessary information to proceed with the next phase. In the development of the relationship, Fig. 3, the attackers apply tactics such as persuasion, offline meetings, and alternative communications. In this way, if they achieve their objectives, they move on to the next phase of "exploitation of the relationship", where the attackers achieve the isolation of their victims, sexual desensitization of these, obtain the information they request, etc. Finally, they achieve their objectives, such as the sale and distribution of pornographic material, private use of the material and the execution of pedophilia. The grooming by its nature must be located within the attack vectors of the proposed taxonomy, as shown in Fig. 4, establishing the relationship with the various channels that allow access to its victims: email, instant messaging, social networks, websites, and in some cases physical encounters. The social aspect is established as the type of attack and the human aspect as operator.

The adaptation of grooming to the taxonomy of Krombholz et al. can be used in the scientific field in different ways. It can be used to educate about social engineering based on real experiential knowledge. As the taxonomy covers the actors and activities related to the field of social engineering (victims, attackers, and the protective organization), it offers a holistic and comprehensible vision. It also facilitates a deeper understanding of the grooming process, and perhaps most importantly, offers an easy way to understand how to develop a protection strategy. From the academic point of view, this research presents a start point for security researchers who try to position themselves in various fields related to pedophilia. By complementing the taxonomy, from the computer science point of view, computer applications can be generated to facilitate the early detection of possible attacks on children, and adolescents using computer learning.

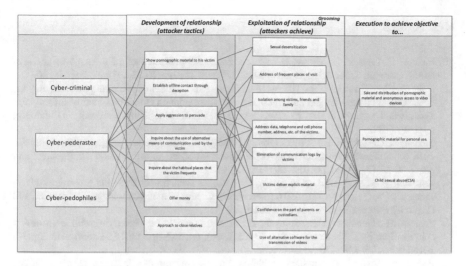

Fig. 3 Summary of Mitnick phases and grooming

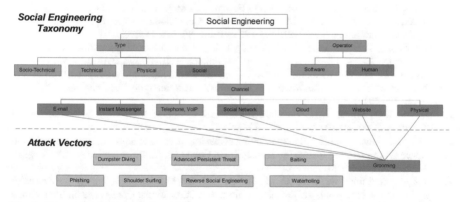

Fig. 4 Proposed taxonomy and grooming

6 Conclusions

Grooming as an access technique has shown in our research that it focuses on a very sensitive group of society, children and adolescents. It is possibly the most dangerous technique within security information since it not only attacks against material assets but also against the emotional and mental stability of people that could bring sequels for a lifetime. Computer science in its continuous contribution to society is establishing new aspects of social engineering, for a better understanding and research of this phenomenon. We established the interrelationship between social engineering and grooming through the four phases of the Mitnick cycle. In our eagerness to position and encourage the study of grooming, it is considered that it should be part of a taxonomy already agreed among scientists, thus generating a significant contribution to the development of previously established knowledge.

References

1. Mouton, F., Leenen, L., Venter, H.S.: Social engineering attack examples, templates and scenarios. Comput. Secur. **59**, 186–209 (2016). http://dx.doi.org/10.1016/j.cose.2016.03.004
2. Nohlberg, M., Kowalski, S.: The cycle of deception: a model of social engineering attacks, defenses and victims. In: Second International Symposium on Human Aspects of Information Security and Assurance (HAISA 2008), Plymouth, UK, 8–9 July 2008. University of Plymouth, pp. 1–11 (2008)
3. Mohammed, S., Apeh, E.: A model for social engineering awareness program for schools. In: 2016 10th International Conference on Software, Knowledge, Information Management and Applications (SKIMA), pp. 392–397. IEEE (2016)
4. Gunawan, F.E., Ashianti, L., Candra, S., Soewito, B.: Detecting online child grooming conversation. In: Proceedings - 11th 2016 International Conference on Knowledge, Information and Creativity Support Systems, KICSS 2016 (2017)
5. Pranoto, H., Gunawan, F.E., Soewito, B.: Logistic models for classifying online grooming conversation. Procedia Comput. Sci. **59**, no. Iccsci, pp. 357–365 (2015). http://dx.doi.org/10.1016/j.procs.2015.07.536
6. Pendar, N.: Toward spotting the pedophile telling victim from predator in text chats. In: ICSC 2007 International Conference on Semantic Computing, no. c, pp. 235–241 (2007)
7. McGhee, I., Bayzick, J., Kontostathis, A., Edwards, L., McBride, A., Jakubowski, E.: Learning to Identify Internet Sexual Predation. Int. J. Electron. Commer. **15**(3), pp. 103–122 (2011). http://www.tandfonline.com/doi/full/10.2753/JEC1086-4415150305
8. Krombholz, K., Hobel, H., Huber, M., Weippl, E.: Advanced social engineering attacks. J. Inf. Secur. Appl. **22**, pp. 113–122 (2015). http://dx.doi.org/10.1016/j.jisa.2014.09.005
9. Bakhshi, T.: Social engineering: revisiting end-user awareness and susceptibility to classic attack vectors. In: 2017 13th International Conference on Emerging Technologies (ICET), pp. 1–6, IEEE (2017)
10. Mitnick, K.D., Simon, W.L., Wozniak, S.: The art of deception: controlling the human element of security. 2002 (2006). ISBN 0-471-23712-4
11. Ivaturi, K., Janczewski, L.: A taxonomy for social engineering attacks. In: International Conference on Information Resources Management. Centre for Information Technology, Organizations, and People (2011)
12. Foozy, F.M., Ahmad, R., Abdollah, M., Yusof, R., Masud, M.: Generic taxonomy of social engineering attack. In: Malaysian Technical Universities International Conference on Engineering and Technology, pp. 527–533 (2011)
13. Hall, R.C., Hall, R.C.: A profile of pedophilia: definition, characteristics of offenders, recidivism, treatment outcomes, and forensic issues. Mayo Clin. Proc. **82**(4), 457–471 (2007)
14. MacFarlane, K., Holmes, V.: Agent-mediated information exchange: child safety online. In: Proceedings - International Conference on Management and Service Science, MASS 2009 (2009)
15. Cano, A., Fernandez, M., Alani, H.: Detecting Child Grooming Behaviour Patterns on Social Media. Lecture Notes in Computer Science (including subseries Lecture Notes in Artificial Intelligence and Lecture Notes in Bioinformatics) vol. 8851, pp. 412–427 (2014)
16. Gupta, S., Singhal, A., Kapoor, A.: A literature survey on social engineering attacks: phishing attack. In: 2016 International Conference on Computing, Communication and Automation (ICCCA), pp. 537–540. IEEE (2016)
17. Bogdanova, D., Rosso, P., Solorio, T.: Modelling Fixated Discourse in Chats with Cyberpedophiles. In: Proceedings of the Workshop on Computational Approaches to Deception Detection, pp. 86–90 (2012). http://www.aclweb.org/anthology/W12-0413
18. Bogdanova, D., Petersburg, S., Rosso, P., Solorio, T.: On the impact of sentiment and emotion based features in detecting online sexual predators. In: Proceedings of the 3rd Workshop on Computational Approaches to Subjectivity and Sentiment Analysis, no. July, pp. 110–118 (2012). http://www.aclweb.org/anthology/W12-3717

19. Vartapetiance, A., Gillam, L.: Our little secret: pinpointing potential predators. Secur. Inform. **3**(1), pp. 1–19 (2014). http://link.springer.com/article/10.1186/s13388-014-0003-7
20. Penna, L., Clark, A., Mohay, G.: A framework for improved adolescent and child safety in MMOs In: Proceedings - 2010 International Conference on Advances in Social Network Analysis and Mining, ASONAM 2010, pp. 33–40 (2010)
21. Yong, S., Lindskog, D., Ruhl, R., Zavarsky, P.: Risk mitigation strategies for mobile Wi-Fi robot toys from online pedophiles. In: Proceedings - 2011 IEEE International Conference on Privacy, Security, Risk and Trust and IEEE International Conference on Social Computing, PASSAT/SocialCom 2011, pp. 1220–1223 (2011)

Part V
Intelligent Image Processing

Towards Improving Performance
of Sigma Filter

Mayank Tiwari, Subir Singh Lamba and Bhupendra Gupta

Abstract Sigma filter is one of the widely used filters for noise removal. This is
popular because of its simple implementation and computationally in-expensiveness
and better filtering accuracy than many other existing filters. In this work, we have
proposed an improved Sigma filter, where the filtering method is separately applied
to high- and low-frequency segments of the image and then restored image is constructed by combining the two filtered images. The basic assumption of this work is
the fact that effect of noise is different on high-frequency (more) and low-frequency
(analogously less) segments of the image and hence more amount of noise can be
removed by separately applying a noise removal filter on high- and low-frequency
segments of the image. Comparative results between the proposed noise removal filter and other noise removal filters show that proposed noise removal filter is capable
of removing noise from images without affecting their visual details.

Keywords Image restoration · Noise detector · Improved sigma filter · Discrete
cosine transform

1 Introduction

'Image noise' is what might as well be called film grain for simple cameras. Then
again, one can consider it closely resembling the subtle foundation hiss you may
get notification from your sound framework at full volume. For digital images
(IMGs), this noise shows up as irregular spots on a generally smooth surface and can

M. Tiwari (✉) · S. S. Lamba · B. Gupta
Department of Mathematics, PDPM Indian Institute of Information Technology,
Design and Manufacturing Jabalpur, Jabalpur 482005, MP, India
e-mail: mayanktiwariggits@gmail.com

S. S. Lamba
e-mail: subirs@gmail.com

B. Gupta
e-mail: gupta.bhupendra@gmail.com

© Springer Nature Singapore Pte Ltd. 2019
S. K. Bhatia et al. (eds.), *Advances in Computer Communication
and Computational Sciences*, Advances in Intelligent Systems and Computing 924,
https://doi.org/10.1007/978-981-13-6861-5_54

essentially debase IMG quality. In spite of the fact that noise frequently degrades an IMG, it is some of the time attractive since it can include an out-dated, grainy look which is reminiscent of early film. Some noise can likewise build the obvious sharpness of an IMG. Noise increments with the sensitivity setting in the camera, length of the exposure, temperature and even changes among various camera models.

The existence of noise degrades the quality of the IMG, and it creates problems for further post-processing IMG-dealing applications. Hence, it is necessary to remove noise from the IMG before sending it for further IMG-dealing applications. To remove noise from given IMG, a number of noise removal filters have been proposed by the various group of researchers. These filters can be broadly divided into two main classes, namely (i) spatial domain filters and (ii) transformed domain filters [6–17]. The spatial domain filters are further divided as (i–a) linear filters such as mean filter (MF), Gaussian filter (GF) and (i–b) nonlinear filters such as median filter (MdF), Sigma filter (SF) [12], bilateral filter (BF) [24], NL mean filter (NLMF) [5], JSTCPW-NL mean filter (JSTCPW-NL-MF) [26], probabilistic NL mean filter (PNLMF) [27] etc. On the other hand, transformed domain filters are notch-filter (NF) and the bandpass filter (BPF) [13].

One of the extensively used nonlinear filter available in the open literature is the SF [12]. This filter is popular because of its simple implementation, computationally in-expensiveness and good filtering accuracy than many other filters. However, the performance of the SF is poor in preserving edge details. This is due to the fact that the SF is not able to work for IMG patches whose variance is close to the variance of additive noise [7]. In order to improve the detail preservation performance of the SF, other approaches have been proposed [3–25].

Here, we have proposed an improved SF. The basic assumption of this work is the fact that effect of noise is different on high-frequency HF (more) and low-frequency LF (analogously less) segments [4, 8] of the IMG and hence more amount of noise can be removed by separately applying the noise removal filter on high and LF segments of the IMG. Under our this observation, we have improved the SF. The improved SF is capable of removing noise from given IMG without extracting its scene details.

The remaining part of this work is organized as follows: Sect. 2 describes few preliminaries concepts related to our work. In Sect. 3, we have shown complete working of the improved SF. In Sect. 4, comparative results of improved SF with other noise removal filters are presented. Section 5 concludes the work. At last, Sect. 1.6 contains acknowledgement by the authors.

2 Few Essential Concepts

This part explains few essential concepts of digital IMG processing which are necessary for better understanding of the improved SF.

2.1 Sigma Filter

In this section, we will define the overall working of the SF; let $x_{p,q}$ be the intensity of pixel (p, q), and $\hat{x}_{p,q}$ be the smoothed pixel (p, q). Also, we assume that the noise is additive with zero mean (ZM) and variance σ. The SF procedure is then characterized as follows:

$$\delta_{k,l} = \begin{cases} 1, & (x_{p,q} + \Delta \leq x_{k,l} \leq x_{p,q} - \Delta); \\ 0, & otherwise; \end{cases} \tag{1}$$

where $\Delta = 2\sigma$. Now, the resultant smoothed pixel is given as:

$$\hat{x}_{p,q} = \frac{\sum_{k=p-n}^{p+n} \sum_{l=q-m}^{q+m} \delta_{k,l} x_{k,l}}{\sum_{k=p-n}^{p+n} \sum_{l=q-m}^{q+m} \delta_{k,l}} \tag{2}$$

2.2 Discrete Cosine Transform (DCT)

The DCT was developed to work as quick as, fast Fourier transform [1]. DCT changes an IMG into basic frequency segments [1, 2, 10, 20] and energy compaction property of DCT makes it a suitable choice over discrete wavelet transform (DWT) [16].

2.3 Noise Estimation Method

For estimation of noise level from a given corrupted IMG, we are utilizing method suggested by [14]. This method chooses low-rank patches without HF from IMG affected by Gaussian noise. This method applies the principle component analysis system to appraise the noise level in view of the information of chosen patches. The eigenvalues of the IMG gradient covariance matrix are utilized as the metric for texture quality and how it changes with various noise levels is investigated. Based on this method, the noise is estimated.

3 Improved Sigma Filter

In our work, we are improving the performance of the SF. This improvement depends on the fact that effect of noise is different on HF (more) and LF (comparatively less) segments [4, 8] of the IMG, and hence more amount of noise can be removed by applying the noise removal filter on high and LF segments of the IMG separately. Under this observation, we have improved the SF. The improved SF is capable of removing noise from given IMG without extracting its scene details.

In the improved version of the SF, we firstly decompose the given IMG into elementary frequency segments (high and low) using DCT. Then, we convert the high and LF segments back to spatial domain using inverse DCT and name it lf_{img} and hf_{img}. Now, we apply the SF on each IMG lf_{img} and hf_{img} separately and combine the results to get the resultant IMG. The overall working of the improved SF can be explained using the following two algorithms.

Algorithm 1 improved SF

Require: in_img;

$\sigma_{est} = noiseestimation(in\_img)$;

% $noiseestimation(\cdot)$ is the algorithm which is used for estimation of noise parameter.

$[lf_{im}, hf_{im}] = high\_low\_dct2(in\_img)$;

% $high\_low\_dct2(\cdot)$ is the algorithm which is used for decomposition of IMG into elementary frequency segments.

$rest\_lf_{im} = sigmafilter(lf_{im}, \sigma_{est}/2)$;

$rest\_hf_{im} = sigmafilter(hf_{im}, \sigma_{est})$;

% $sigmafilter(\cdot)$ is the MALTAB implementation of the SF.

$rest\_lf\_comp = dct2(rest\_lf_{im})$;

$rest\_hf\_comp = dct2(rest\_hf_{im})$;

% $dct2(M)$ returns the DCT of M.

$rest_{im} = idct2(rest\_lf\_comp + rest\_hf\_comp)$;

% $idct2(M)$ returns the inverse DCT of M.

Next, we show algorithm which is used to access LF and HF segments using DCT.

Algorithm 2 $high\_low\_dct2$ algorithm

Require: in_im;

$orig_t = dct2(in\_im)$;

% $dct2(A)$ returns the discrete cosine transform of A.

$row = size(in\_im, 1)$;

% function $size(X)$ for matrix X, returns the number of rows and columns in X as separate output variables.

$cutoff = round(0.5 * row)$;

% function $round(X)$ rounds each element of X to the nearest integer.

$high_t = fliplr(tril(fliplr(orig_t), cutoff))$;

% function $fliplr(X)$ returns X with the order of elements flipped left to right along the second dimension.

% function $tril(X)$ returns the lower triangular part of X.

$low_t = orig_t - high_t$;

$hf_{img} = idct2(high_t)$;

$lf_{img} = idct2(low_t)$;

% $idct2(A)$ returns the inverse discrete cosine transform of A.

lf_{img} and hf_{img} are output of the algorithm;

---

4   Experimental Results

Numerical evaluation of the improved SF and other conventional filters are presented in this section. Here, we have selected four standard greyscale IMGs 'elaine', 'house', 'lena' and 'plane', and we have added ZM Gaussian noise with different values σ to them. In the all experiments, a 7×7 SF is used. Figure 1 shows performance of the improved SF in terms of noise removal from given IMG.

In order to visually show the smoothing impact of the improved SF, the following results are shown in Fig. 2.

1. Figure 2a shows total nine IMGs. Here from top to bottom first IMG is input IMG, second IMG is noisy IMG with added Gaussian noise and rest seven IMGs are results of various noise removal filters.
2. Figure 2b shows intensity profiles of a straight line which is marked in lena's forehead in Fig. 2a.
3. Figure 2c shows difference of intensity profile plots in between input IMG and various restored IMGs which are obtained after applying noise removal filters.
4. Figure 2c also shows mean and variance values of difference of intensity profile plots in between input IMG and various restored IMGs.

Figure 2 clearly shows that restoration result of the improved SF is able to maintain fine details of input IMG in the restored IMG, also as mean and variance results of the improved SF at Fig. 2c has least values than other seven noise removal filters. This shows that the improved SF is able to preserve edge and texture information of input IMG in restored IMG more accurately than other de-nosing filters.

Figure 3 shows comparison of results of mean values of amount of noise left in the IMG by different noise removal filters. For this, the restored IMG is subtracted from the original noise-free IMG and mean value of the subtraction result is shown. It is clear that results obtained by the improved SF have least values as compared to other conventional filters.

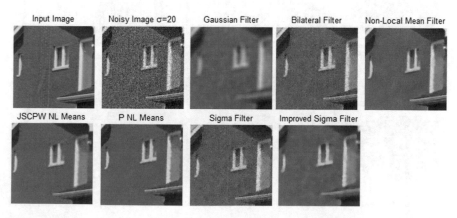

Fig. 1 Comparison of results of improved SF with other conventional filters for noise removal

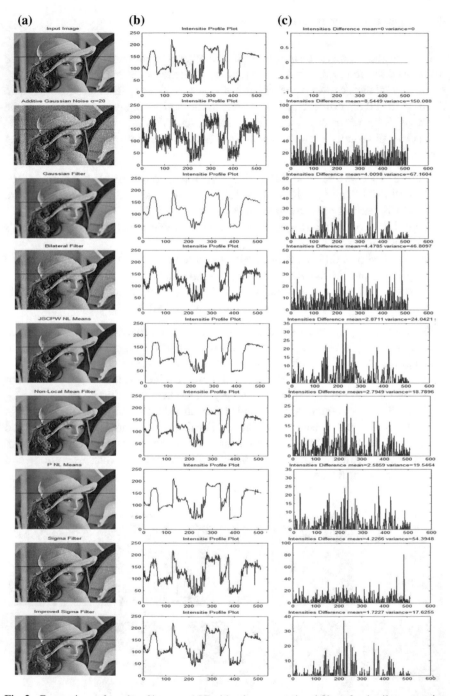

Fig. 2 Comparison of results of improved SF with other conventional filters for detail preservation

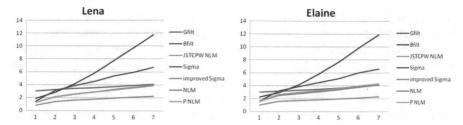

Fig. 3 Comparison of results of mean values of amount of noise left in the IMG by conventional filters. Here, x-axis has noise values as $\sigma = \{10, 15, 20, 25, 30, 35, 40\}$

Table 1 Average PSNR comparison of each algorithm according to the σ of added Gaussian noise

Methods $\Downarrow$ & $\sigma_{est} \Rightarrow$	20	25	28	33	37
GF	29.67	29.53	29.34	29.10	28.90
BF	30.67	27.83	25.44	23.39	21.74
SF	30.79	29.34	28.17	28.01	28.17
NLMF	32.28	32.57	31.27	31.18	30.15
PNLM	32.54	32.59	31.36	31.29	**30.42**
JSCPW-NLMF	32.74	32.66	31.56	31.50	30.16
Proposed	32.98	32.85	31.71	31.73	30.27

The peak signal-to-noise ratio (PSNR which is a widely used matrix to measure signal restoration [19, 21]) is used by us to measure quality of IMG restoration. Average PSNR results (for lena, elaine, house, plane) between the restored IMGs and the input IMGs are shown Table 1. Here, some unknown amount of noise is added in clean IMGs to convert them into noisy IMGs and then the amount of noise added to the IMG is estimated by the noise estimation (σ_{est}) algorithm. The σ_{est} is used by us in different noise removal algorithms as noise parameters.

Table 1 demonstrates the average PSNR examination of every method as per the σ_{est} of included Gaussian noise. Similar results from tests demonstrate that the proposed method accomplishes higher increases, by and large than GF, BF [24], NLML [5], JSTCPW-NLM-F [26], PNLM [27] and the SF [12]. As given by [21] that performance of [5, 26, 27] filters depends on prior estimation of amount of noise added in the IMG. An exact estimation of this value is still a difficult problem.

5 Conclusion

The proposed work presents an improved version of the SF. The improvement is based on our observation that the effect of noise is different on HF (more) and LF (analogously less) segments of the IMG [4, 8] and hence more amount of noise can be removed by applying the noise removal filter on HF and LF segments of the IMG separately. Figures 1, 2 and 3 support our this claim. Moreover, results of Table 1 show that up to $\sigma_{est} = 33$ the proposed filter is able to work better than other conventional filters and for $\sigma_{est} = 37$ it is able to work better than five conventional filters out of six. We have shown that sometimes a very simple idea produces good results. As the improved SF is capable in removing the noise from given image so in future this filter can be used for photo-response-nonuniformity-(PRNU)-based [9, 22, 23] digital image forensics applications. Also as there are chances of further improvement, in future the proposed method can be improved in many ways such as:

1. Instead of DCT, any other method such as fast cosine transform (FCT) [11] or discrete cosine transform pyramid (DCTP) [18] can be used.
2. The noise estimation method [14] can be replaced with any other method, whose noise estimation accuracy is higher than the currently used method.
3. Instead of Sigma filter, other noise removal filters can be used.

Acknowledgements We are very grateful to all the reviewers for giving their precious time in order to review our work. We have found their suggestion very useful in improving the quality of this research article and have humbly incorporated all their suggestions.

References

1. Ahmed, N., Natarajan, T.R., Rao, K.R.: Discrete cosine transform. IEEE Trans. Comput. **23**(1), 90–93 (1974)
2. Bhandari A.K., Kumar A., Singh G.K.,: SVD based poor contrast improvement of blurred multispectral remote sensing satellite images. In: International Conference on Computer and Communication Technology, pp. 156–159 (2012)
3. Bilcu, R.C., Vehvilainen, M.: A modified sigma filter for noise reduction in images. In: Circuits, Systems, Communications and Computers Multiconference 15, pp. 1–6 (2005)
4. Biswas P.K., Video lecture on 'Image Enhancement in Frequency Domain,' https://www.youtube.com/watch?v=Qcl5g1utev4&List=PL1F076D1A98071E24&Index=21, Accessed 28 March 2018
5. Buades, A., Coll, B., Morel, J.M.: A review of image denoising algorithms, with a new one. Multiscale Model. Simul. **4**, 490–530 (2006)
6. Gonzalez, R.C., Woods, R.E.: Digital Image Processing, 3rd edn. Addison Wesley, Reading (2008). ISBN 9780131687288
7. Gu M.R., Kang D.S.: Modified sigma filter using image decomposition. In: International Conference on Multimedia Systems and Signal Processing, pp. 193–198 (2010)
8. Gupta, B., Tiwari, M.: Improving source camera identification performance using DCT based image frequency segments dependent sensor pattern noise extraction method. Digit. Investig. **24**, 121–127 (2018)
9. Gupta B., Tiwari M.: Improving performance of source camera identification by suppressing peaks and eliminating low-frequency defects of reference SPN. In: IEEE Signal Processing Letters, pp. 1340–1343 (2018)
10. Khayam S.A.: The discrete cosine transform (DCT): theory and application, department of electrical and computing engineering (2003)
11. Kok, C.W.: Fast algorithm for computing discrete cosine transform. IEEE Trans. Signal Process. **45**(3), 757–760 (1997)
12. Lee, J.S.: Digital image smoothing and the sigma filter. Comput. Graph. Image Process. **24**, 255–269 (1983)

13. Lim J.S.: Two-Dimensional Signal and Image Processing. Prentice (1990)
14. Liu, X., Tanaka, M., Okutomi, M.: Single-image noise level estimation for blind denoising. IEEE Trans. Image Process. **22**(12), 5226–5237 (2013)
15. Pratt, W.K.: Digital Image Processing. Wiley, New York (1978)
16. Rao K., Yip P.: Discrete Cosine Transform: Algorithms, Advantages, Applications. Academic, Boston (1990) ISBN 0-12-580203-X
17. Sonka M., Hlavac V., Boyle R.: Image Processing, Analysis, and Machine Vision, ITP (1999)
18. Tan, K.H., Ghanbari, M.: Layered image coding using the DCT pyramid. IEEE Trans. Image Process. **4**(4), 512–516 (1995)
19. Tiwari, M., Gupta, B.: Image denoising using spatial gradient based bilateral filter and minimum mean square error filtering. Procedia Comput. Sci. **54**, 638–645 (2015)
20. Tiwari M., Lamba S.S., Gupta B.: An approach for visibility improvement of dark color images using adaptive gamma correction and DCT-SVD. In: Proceedings of SPIE 10011, First International Workshop on Pattern Recognition (2016)
21. Tiwari M., Gupta B.: A consistent approach for image de-noising using spatial gradient based bilateral filter and smooth filtering. In: Proceedings of SPIE 10011, First International Workshop on Pattern Recognition (2016)
22. Tiwari, M., Gupta, B.: Image features dependant correlation-weighting function for efficient PRNU based source camera identification. Forensic Sci. Int. **285**, 111–120 (2018)
23. Tiwari M, Gupta B.: Enhancing source camera identification using weighted nuclear norm minimization de-noising filter. In: Advances in Computer Communication and Computational Sciences, pp. 281–288 (2018)
24. Tomasi C., Manduchi, R.: Bilateral filtering for gray and color images. In: International Conference on Computer Vision, pp. 839–846 (1998)
25. Ville, D.V.D.: Noise reduction by Fuzzy image filtering. IEEE Trans. Fuzzy Syst. **11**(4), 429–436 (2003)
26. Wu, Y., Tracey, B., Natarajan, P., Noonan, J.P.: James-Stein type center pixel weights for non-local means image denoising. IEEE Signal Process. Lett. **20**(4), 411–414 (2013)
27. Wu, Y., Tracey, B., Natarajan, P., Noonan, J.P.: Probabilistic non-local means. IEEE Signal Process. Lett. **20**(8), 763–766 (2013)

Video-Based Facial Expression Recognition Using a Deep Learning Approach

Mahesh Jangid, Pranjul Paharia and Sumit Srivastava

Abstract This research aims at classifying facial expression of humans in a video. Facial expressions were classified into one of the following common facial expression classes that are anger, disgust, fear, happiness, sadness, surprise, and neutral. To accomplish this task, convolutional neural networks were developed, to classify each facial image extracted from a frame into one of the seven classes of facial expressions we have chosen. The model was developed in Keras and tensorflow. Frames were extracted using OpenCV to detect location of facial image from each frame. Face detector was used based on SSD framework (single-shot multi-box detector), using a reduced ResNet-10 model. On all the facial images detected, their expressions were classified using the developed CNN model and based on results of entire images, a table is prepared to classify in which expression has been identified most in the video. Finally, we compare the results of frame extracted at two different rates, i.e., 1 fps and 10 fps.

Keywords Facial expression recognition · Deep learning · Deep convolution neural network · Tensorflow

1 Introduction

The facial expressions (non-verbal communication) are a meaningful way to observe the feelings of any human being without any word or verbal communication. The facial expression recognition also helps to understand the intentions of humans. This can be used at any places such as in education system to get the true response of students [1], in office to observe the emotion of the employees or the customers [2], in public areas such as airport lobby to identify the suspicious traveler [3], and in service industry to get the feedback with any written feedback [4]. There are many other areas, where facial expression recognition (FER) system is also used including virtual

M. Jangid (✉) · P. Paharia · S. Srivastava
Manipal University Jaipur, Jaipur, Rajasthan 303007, India
e-mail: mahesh_seelak@yahoo.co.in

© Springer Nature Singapore Pte Ltd. 2019
S. K. Bhatia et al. (eds.), *Advances in Computer Communication
and Computational Sciences*, Advances in Intelligent Systems and Computing 924,
https://doi.org/10.1007/978-981-13-6861-5_55

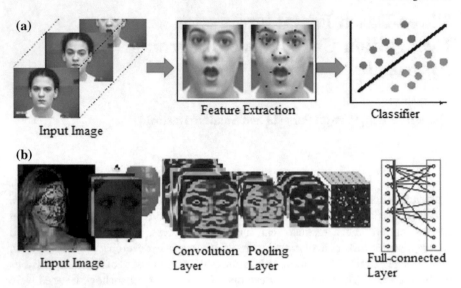

Fig. 1 **a** Phases of the handcrafted feature approach and **b** phases of the deep learning approach used for facial expression recognition

reality [5], driver assistant systems [6], human–computer interaction [7, 8], and online gaming [9]. Owing to the various applications of facial expression recognition system (FER), many research works have been reported in past years. The facial expressions have been classified into six different categories broadly (happiness, disgust, fear, sadness, anger, and surprise) [10]. A standardized set of muscle movements have also been created by [11] and known as facial action units (FAUs).

In the recent years, the deep learning approaches have been applied in many areas of computer vision and other. The many researchers have applied the deep learning approach on facial expression recognition and obtained a significant improvement in the recognition accuracy. The previous works on FER can be divided into two groups such as handcrafted features approach and deep learning approach. The phases of these approaches can be seen in Fig. 1. The face components (eyes, nose, lips, and eyebrows), landmarks, etc., are detected and features are extracted in the handcrafted features approach. However, in deep learning approach, the features are automatically extracted by the deep learning network.

2   Dataset

I have used an FER-2013 dataset which was provided in Kaggle for facial expression recognition challenge. Dataset comprises of total 35,887 grayscale images of 48 × 48 pixels. It contains labeled images from seven different categories, i.e., anger, disgust, fear, happiness, sadness, surprise, and neutral. The dataset has been divided into three different sets with 28,709 examples in training set and 3,589 examples in each validation set and test set. All images in the dataset and already preprocessed and contained images are with different angle, lighting, and objects so no data processing has been required in the dataset. The sets also contain images from all the classes in an almost equal ratio so there was no bias issue with the dataset.

3 Methodology

The complete system architecture is shown in Fig. 2. In the first phase, we trained the deep learning model using available facial expression dataset. In the next phase, we extracted the face image from the frame captured from the video using the pre-trained model face detection model and then our trained facial expression recognition model has been applied to classify the face expression.

3.1 Frame Extraction

To extract frames from a video, we have used CAP_PROP_POS_MSEC method of OpenCV, which sets the current position of the video file in milliseconds and using read() method to capture the frame of the video at that position. We have stored all the frames in every fixed time interval and the extracted frames have been passed to face detection.

3.2 Face Detection

We have used a pre-trained face detector model [12], which was created with SSD framework. This model is a similar network like ResNet-10 and used 1,40,000 images of size 300 × 300 to train the model. This model detects entire faces from the input frame and resizes them to size 48 × 48 pixels.

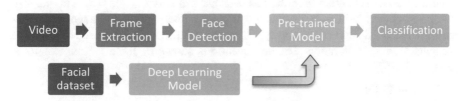

Fig. 2 System architecture to recognize facial expression recognition from the video

Table 1 Architecture of five layers of CNN model and layerwise trainable parameters

Layer (type)	Output shape	Param #
conv2d_1 (Conv2D)	(126, 126, 64)	640
conv2d_2 (Conv2D)	(124, 124, 64)	36,928
conv2d_3 (Conv2D)	(122, 122, 64)	36,928
max_pooling2d_1	(61, 61, 64)	0
dropout_1 (Dropout)	(61, 61, 64)	0
conv2d_4 (Conv2D)	(59, 59, 128)	73,856
conv2d_5 (Conv2D)	(57, 57, 128)	147,584
max_pooling2d_2	(28, 28, 128)	0
dropout_2 (Dropout)	(28, 28, 128)	0
flatten_1 (Flatten)	(100352)	0
dense_1 (Dense)	(512)	51,380,736
dropout_3 (Dropout)	(512)	0
dense_2 (Dense)	(7)	3,591
Total params: 51,680,263		
Trainable params: 51,680,263		
Non-trainable params: 0		

3.3 Expression Recognition Model

Entire extracted face images have given as input to the CNN model for classification of the face image into one of the seven classes. We have trained different CNN models by varying the model architecture and hyper-parameters. Readers can refer previous paper [13] to know more about the CNN model and different optimizers.

5-Layered CNN

We defined a 5-layered CNN model with three convolutional layers of 64 filters followed by a max-pooling layer and a dropout layer with dropout 0.3, that was further followed by two convolutional layers of 128 filters with a max-pooling layer and a dropout layer of 0.4 dropout. At last, a dense layer with 512 neurons was used followed by a dropout layer of 0.5 dropout and a softmax classifier with seven neurons. ReLU activation function has been used in entire layers with a kernel size of 3 × 3 in convolutional layers and 2 × 2 in max-pooling Layer. Categorical cross-entropy has been used as the loss function and Adadelta has been used to optimize weights after each iteration. The architecture is shown in Table 1 Architecture of 5-layered CNN.

7-Layered CNN

The next model was defined a 7-layered CNN model with four convolutional layers of 64 filters followed by a max-pooling layer and a dropout layer with dropout 0.3, that was further followed by three convolutional layers of 128 filters with a

Table 2 Architecture of seven layers of CNN and layerwise trainable parameters

Layer (type)	Output shape	Param #
conv2d_1 (Conv2D)	(126, 126, 64)	640
conv2d_2 (Conv2D)	(124, 124, 64)	36,928
conv2d_3 (Conv2D)	(122, 122, 64)	36,928
conv2d_4 (Conv2D)	(120, 120, 64)	36,928
max_pooling2d_1	(60, 60, 64)	0
dropout_1 (Dropout)	(60, 60, 64)	0
conv2d_5 (Conv2D)	(58, 58, 128)	73,856
conv2d_6 (Conv2D)	(56, 56, 128)	147,584
conv2d_7 (Conv2D)	(54, 54, 128)	147,584
max_pooling2d_2	(27, 27, 128)	0
dropout_2 (Dropout)	(27, 27, 128)	0
flatten_1 (Flatten)	(93312)	0
dense_1 (Dense)	(512)	47,776,256
dense_2 (Dense)	(512)	262,656
dropout_3 (Dropout)	(512)	0
dense_3 (Dense)	(512)	262,656
dense_4 (Dense)	(7)	3,591
Total params: 48,785,607		
Trainable params: 48,785,607		
Non-trainable params: 0		

max-pooling layer and a dropout layer of 0.3 dropout. After this, two dense layers with 512 neurons are used followed by a dropout layer of 0.5 dropout, that was further followed up by a dense layer with 512 neurons and a softmax classifier with 7 neurons. ReLU activation function has been used in all layers with a kernel size of 3×3 in convolutional layers and 2×2 in max-pooling layer. Categorical cross-entropy has been used as the loss function and Adadelta has been used to optimize weights after each iteration. The architecture of 7-layered CNN is shown in Table 2.

4 Experiments and Results

The entire experiments have been performed on the ParamShavak supercomputer system having two multicore CPUs with each CPU consists of 12 cores along with two number of accelerator cards. This system has 64 GB RAM with CentOs 6.5 operating system. The first model of five layers has applied to the dataset. Where five convolutional layers of three with 64 filters and two with 128 filters with kernel size 3 * 3, 2 max-pooling layers with filter size 2 * 2, 3 dropout layers with dropout

Table 3 Modelwise loss and accuracy of training, validation, and testing datasets

Model	Training loss	Training accuracy	Validation loss	Validation accuracy	Testing loss	Testing accuracy
5 Layer-1	0.425	86.22	1.85	62.24	1.806	61.88
5 layer -2	0.0059	99.78	3.101	62.42	3.203	**62.12**
7 Layer	0.143	96.86	1.4131	59.18	1.523	57.61

Fig. 3 Training and validation loss of 5 layer-2 CNN model at each epoch

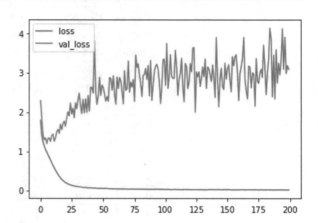

0.4, 0.4, and 0.6, respectively, in each layer, two fully connected layer; one with 512 neurons and the second is the output layer with seven neurons (number of classes). All layers used ReLU activation function and final output layer has used foftmax classifier. The categorical cross-entropy loss function has used as a loss function. This model applied with AdaDelta optimizer and the batch size kept 128. This model has applied with (5 layer-2) and without batch normalization (5 layer-1). The accuracy and loss of training, validation, and testing are reported in Table 3. The accuracy of the training and validation dataset at each epoch up to 200 epochs is shown in Fig. 3. Figure 4 has shown the recorded accuracy of training and validation at each epoch to 200 epochs. Further, a deeper model of seven layers has applied to observe the impact of an additional layer on accuracy. But the obtained accuracy was less than the previous model as shown in Table 3. This built model of five layers with batch normalization has applied on a video to evaluate the model. This model has applied at different fps rate and the distribution of recognized facial expression from the video at different frame rate is shown in Table 4 and Fig. 5.

Fig. 4 Training and validation accuracy of 5 layer-2 CNN model at each epoch

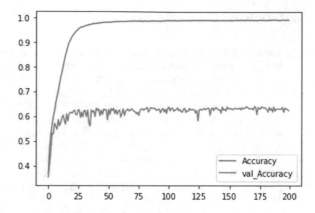

Table 4 Distribution of recognized facial expression from a video at a different frame rate

S. No.	Expression name	Accuracy (10 fps)	Accuracy (1 fps)
1	Angry	2.7	4.47
2	Disgust	0.0	0.0
3	Fear	3.53	2.44
4	Happy	33.36	40.65
5	Sad	44.98	37.8
6	Surprise	0.19	0.41
7	Neutral	15.24	14.23

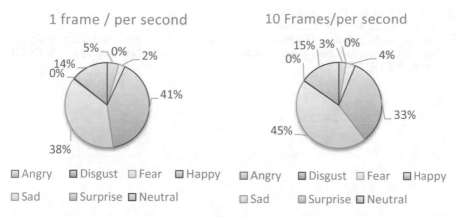

Fig. 5 Facial expression distribution on a video at one frame per second (left side) and 10 frames per second (right side)

References

1. Bartlett, M.S., et al.: Real time face detection and facial expression recognition: development and applications to human computer interaction. In: Conference on Computer Vision and Pattern Recognition Workshop, 2003. CVPRW '03. vol. 5. IEEE (2003)
2. Oliver, N., Pentland, A., Bérard, F.: LAFTER: a real-time face and lips tracker with facial expression recognition. Pattern Recogn. **33**(8), 1369–1382 (2000)
3. Scherer, K.R., Ceschi, G.: Criteria for emotion recognition from verbal and nonverbal expression: studying baggage loss in the airport. Pers. Soc. Psychol. Bull. **26**(3), 327–339 (2000)
4. Wood, A., et al.: Fashioning the face: sensorimotor simulation contributes to facial expression recognition. Trends Cognit. Sci. **20**(3), 227–240 (2016)
5. Chen, C.-H., Lee, I.-J., Lin, L.-Y.: Augmented reality-based self-facial modeling to promote the emotional expression and social skills of adolescents with autism spectrum disorders. Res. Dev. Disabil. **36**, 396–403 (2015)
6. Assari, M.A., Rahmati, M.: Driver drowsiness detection using face expression recognition. In: 2011 IEEE International Conference on Signal and Image Processing Applications (ICSIPA). IEEE (2011)
7. Bartneck, C., Lyons, M.J.: HCI and the Face: towards an art of the soluble.In: International Conference on Human-Computer Interaction. Springer, Berlin, Heidelberg (2007)
8. Dornaika, F., Raducanu, B.: Efficient facial expression recognition for human robot interaction. In: International Work-Conference on Artificial Neural Networks. Springer, Berlin, Heidelberg (2007)
9. Blom, P.M., et al.: Towards personalised gaming via facial expression recognition. In: AIIDE (2014)
10. Ekman, P., Friesen, W.V., Ellsworth, P.: Emotion in the human face: guidelines for research and an integration of findings. Elsevier (2013)
11. Kanade, T., Tian, Y., Cohn, J.F.: Comprehensive database for facial expression analysis. In: fg. IEEE (2000)
12. https://github.com/thegopieffect/computer_vision/blob/master/CAFFE_DNN/res10_300x300_ssd_iter_140000.caffemodel. Access date 12 June 18
13. Jangid, M., Srivastava, S.: Handwritten devanagari character recognition using layer wise training of deep convolutional neural network and adaptive gradient methods. J. Imagin. (2017). ISSN 2313-433X

Zernike Moment-Based Facial Expression Recognition Using Two-Staged Hidden Markov Model

Mayur Rahul, Rati Shukla, Dinesh Kumar Yadav and Vikash Yadav

Abstract Facial expression recognition is an application used for biometric software that can be used to recognize special expressions in a digital image by analysing and comparing different patterns. This software is popularly used for the purpose of security and is commonly used in other applications such as medicines, home security, human–computer interface, credit card verification and surveillance systems. Recognizing faces becomes very difficult when there is a frequent change occurs in facial expressions. In this paper, new improved extension of HMM is used to identify continuous effective facial expressions. Extraction of features is done using Zernike moments. The new improved version of HMM consists of two stages: first stage denotes the expression made by individual features like nose, mouth, etc., and second stage denotes the mixture of individual features like smile, fear, etc. Seven universal facial expressions are recognized, i.e. sadness, surprise, neutral, anger, disgust, fear and joy. Experimental result shows the better performance of the proposed system using JAFFE data sets.

Keywords Zernike moments · HMM · JAFFE · Feature extraction · Classifier · Facial expressions · Pattern recognition · Human–computer interface

M. Rahul (✉)
Department of Computer Applications, UIET, CSJMU, Kanpur, India
e-mail: mayurrahul209@gmail.com

R. Shukla
GIS Cell, MNNIT, Allahabad, India
e-mail: mca.rati@gmail.com

D. K. Yadav
Delhi Technological University, Delhi, India
e-mail: dineshyadavdtu@gmail.com

V. Yadav
Department of Computer Science and Engineering, ABES Engineering College, Ghaziabad, India
e-mail: vikash.yadav@abes.ac.in

© Springer Nature Singapore Pte Ltd. 2019　　　　　　　　　　　661
S. K. Bhatia et al. (eds.), *Advances in Computer Communication and Computational Sciences*, Advances in Intelligent Systems and Computing 924,
https://doi.org/10.1007/978-981-13-6861-5_56

1 Introduction

A good human–computer interaction system is very useful in the present scenario for the facial expression recognition (FER) [1]. The FER is the process by which the facial expression created by the user can be identified by the receiver. Extractions of emotions from the face are the region surrounding the eyes, mouth and nose. These facial regions are used to generate suitable activities. The changes, location and place of the faces are very useful in recognizing facial expressions. In order to recognize all facial expressions, the human face and its movement need to be identified. This can be done by a device which has some sensing capability. Devices can be magnetic field suits, gloves, cameras and computer vision methods. Each sensor has different characteristics, for example they can differ in accuracy, resolution, latency, range and comfort. Each requires different techniques and approach. Our objective is to develop a robust system which is capable of recognizing facial expressions and has good efficiency.

Human faces are non-rigid objects which are differ in size, shape, colour, etc., and can be easily detected by human beings. The recognition of facial expression is the process of identifying facial gestures regardless of their illumination, scales, orientation, poses, etc. Facial expressions' recognition can be easily applicable in human—computer interface, credit card verification, security in offices, criminal identification, surveillance systems, medicines, etc. They can also be used in information retrieval techniques such as query refinement database indexing, content-based image retrieval and similarity-based retrieval. The facial expression can be categorized in two categories: transient and permanent. Eye, eyebrows, cheeks and facial hair are permanent features, and region surrounding the mouth and eye is called transient features. Difference of appearance due to different identity is called face detection, and change of appearance due to different movement is called facial expression recognition.

The HMM is the best method for facial expression recognition. On comparing HMM with human being, it is found to be similar. Hidden state of the HMM is the mental state of human being, and facial expression is the observations. The hidden Markov model used acoustic state model for speech pattern analysis [2]. HMM becomes very powerful in recent years due to its capability of time-related data handling and also capability of quick learning. It is also used to classify feature vectors which are not known to us. Another benefit of HMM is that it requires only few data to train [3, 4]. It is also capable of matching signal rate due to its principle of statistical modelling and pattern matching [5, 6]. The disadvantage of HMM is its discriminative power. Due to this reason, it is less powerful. To overcome this disadvantage, we introduced newly improved HMM as a classifier.

Rest of the paper is organized as follows. Paper has seven sections. Basics of Zernike moments has been discussed in Sect. 2, proposed system have been discussed in Sect. 3, experiments and results are in Sect. 4, and finally conclusion and future works in Sect. 5.

2 Basics of Zernike Moments

A moment describes the pattern of the pixel in the image. It is used in many areas as region-based descriptors for shape. It also shows the compactness, irregularity, area and higher-order pattern descriptors in the form of bits [7]. Moment of an image is calculated by integration of polynomial of region-defined [8, 9]. The region-defined is the area where the image is not invalid. From [10], the generalized moment M_{pq} of any given image

$$M_{pq} = \int \int \text{pol}_{pq}(x, y) f(x, y) dx dy \tag{1}$$

where p and q are greater than zero and $p + q$ is the order of image function $f(x, y)$.
    Where $\text{pol}_{pq}(x, y)$, $i = 1(0)p$, $j = 1(1)q$ are defined on domain D.
    Zernike moments are the functions set and are complex, orthogonal and defined over the disc. They are first used in the analysis of the image [11]. They are based on Zernike polynomials. The meaning of orthogonal in this paper is that there is no overlapping or redundancy of information between two different moments. Therefore, moments can be calculated according to their orders [12, 13]. The special and important feature of ZM is its invariance due to rotation [14, 15, 16, 17]. The ZM is

$$V_{nm}(\rho, \theta) = R_{nm}(\rho) e^{im\theta}, \theta <= 1 \tag{2}$$

where

(m, n) denotes the order of the ZM and phase angle multiplicity.
$\rho = (x^2 + y^2)^{1/2}$ is the image pixel radial vector

$$\theta = \arctan(y/x) \text{is the angle} \tag{3}$$

$$R_{nm}(\rho) = \sum_{a=0}^{(n-|m|)/2} .(-1)^a \frac{(n-a)!}{a!.\left(\frac{n+|m|}{2} - a\right)!.\left(\frac{n-|m|}{2} - a\right)!} \rho^{n-2a} \tag{4}$$

The R_{nm} is the basis Zernike polynomial. The following conditions hold (Fig. 1).

(1) $n \in Z^+$
(2) $n - |m|$ is even
(3) $|m| <= n$
(4) $\int_0^{2\pi} \int_0^1 V_{nm}^*(\rho, \theta) \rho d\rho d\theta = \frac{\pi}{n+1} \delta_{np} \delta_{mq}, \delta_{zv} = \begin{cases} 1 & z = v \\ 0 & \text{otherwise} \end{cases} \tag{5}$

(a)   **(b)**   **(c)**

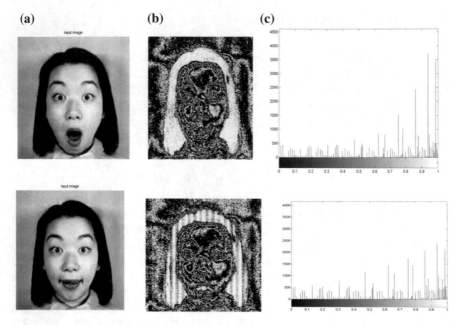

Fig. 1 a Original image from JAFFE [29–31], **b** filtered image using ZM and **c** corresponding histogram

3 Proposed Method

Pattern recognition is very important part of any image analysis system. Most of these systems contain four general steps: image acquisition, pre-processing of the images, feature extraction and classification [18]. Our goal is to improve classification results under various situations. Many approaches have been applied including feature selection, decision theory, learning, etc. [19]. An effective shape descriptor is very important in the description of multimedia content, since shape is a basic property of an object. Contour-based and region-based are the two shape descriptors [20]. Moment invariants are the contour-based shape descriptors developed by Hu (1962) which consist of set of equations [21]. Further, these moment invariants are extended to larger sets [22] and to some other forms [23, 24]. Zion et al. have developed an image processing algorithm, which has been used to discriminate between three fish species in the freshwater fish farms [25]. Shutler et al. developed Zernike velocity moments which not only used as shape descriptor but also used as a motion descriptor [26].

We have followed the method of Oluleye et al. [27]. They used the Zernike moments in their system. The incorporation of ZM with the modified HMM improves the recognition result when compared the results with MCE-HMM [28].

The newly improved version of HMM consists of two stages: first stage denotes the expression made by individual features like nose, mouth, etc., and second stage denotes the mixture of individual features like smile, fear, etc. (see Fig. 2). The proposed framework introduces extraction of features using Zernike moments and classification using modified HMM (see Fig. 3). All experiments are performed in MATLAB2016. In this simulation, observed sequence probability calculation is done by forward method, state sequence is calculated by Viterbi, and parameter estimation is done by Baum–Welch methods. The proposed framework is shown in Fig. 3.

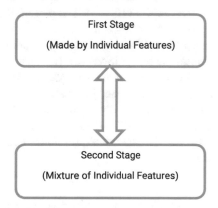

Fig. 2 Layered extension of HMM [34]

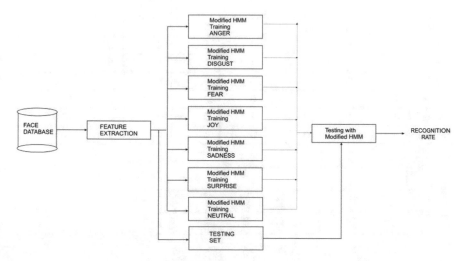

Fig. 3 Block diagram of proposed framework

4 Experiments and Results

JAFFE database has the 140 images from 13 subjects [29–31]. Suppose that if an expression has n images (n folds), then $n - 1$ images are used to train the HMM classifier and 1 image is used for training. Further, final result is the mean of all the results from each fold.

The recognition results using Zernike moments and modified HMM are shown in Table 1 and its graphical representation in Fig. 4.

The receiver operation characteristics is a very popular curve to represent graphically the performance of the classifier (Fig. 5). The plot of TPR and FPR above the 45° line shows the goodness of our classifier.

The comparison between MCE-HMM and proposed work is depicted in Fig. 6. In this figure, the X-axis denotes the recognition rates (in %age) and Y-axis denotes the seven facial expressions. The recognition rates of MCE-HMM are 81.5, 84.6, 78.2, 76.2, 91.5 and 93.5 for anger, disgust, fear, joy, sadness and surprise, respectively. The recognition rates of proposed framework are 70.3, 90.4, 53.8, 100, 88, 91.3 and 100 for anger, disgust, fear, joy, sadness, surprise and neutral, respectively. The MCE-HMM is unable to recognize neutral expressions. This proposed framework introduces the recognition of neutral.

Table 1 Confusion matrix for FER using proposed framework from JAFFE data sets

	Anger	Disgust	Fear	Joy	Sadness	Surprise	Neutral	Recognition rate (%age)
Anger	**11**	0	2	0	0	2	0	73.3
Disgust	0	**19**	0	0	1	1	0	90.4
Fear	2	0	**7**	2	2	0	0	53.8
Joy	0	0	0	**26**	0	0	0	100
Sadness	1	0	0	0	**22**	0	2	88
Surprise	0	0	0	2	0	**21**	0	91.3
Neutral	0	0	0	0	0	0	**17**	100

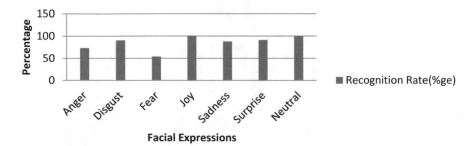

Fig. 4 Graphical representation of Table 1

Fig. 5 ROC curve for the given confusion matrix

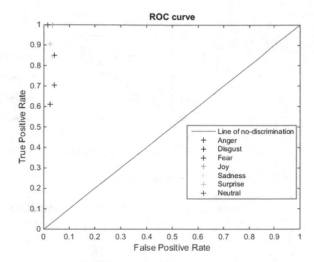

Fig. 6 Comparison between MCE-HMM [28] and proposed framework using JAFFE (Table 1)

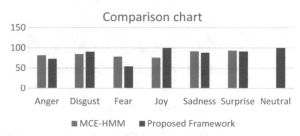

Table 2 Comparison of error rates between MCE-HMM [28] and proposed framework

No. of training samples required		2	4	6	8
Error rates (%)	MCE-HMM [28]	21.3	18.6	16.3	15.8
	Proposed framework	20.2	18.1	15.7	14.9

The error rates for FER are depicted in Table 2. The MCE-HMM denotes the error rates of 21.3, 18.6, 16.3 and 15.8, and proposed framework denotes the error rates of 20.2, 18.1, 15.7 and 14.9 for 2, 4, 6 and 8 images taken as a training sets to train HMM, respectively. The table shows the proposed framework is also able to minimize error rates to some extent.

The time analysis for the facial expression recognition is given in Fig. 7. The x-axis denotes the no. of images taken from the database, and y-axis denotes the time taken in seconds. The ACA-HMM [32] takes 0.24, 0.49, 0.95, 1.48 and 2.12, the PCA-HMM [33] method takes 0.245, 0.52, 0.9, 1.52 and 2.02, the MCE-HMM [28] takes 0.25, 0.53, 1.06, 1.43 and 2.05, and proposed method takes 0.256, 0.57, 1.11, 1.6 and 2.18 for 6, 12, 18, 24 and 30 images in JAFFE database, respectively.

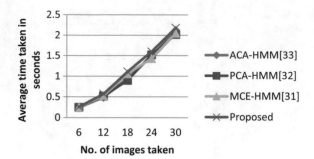

Fig. 7 Time analysis using different methods

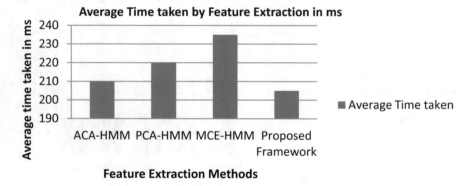

Fig. 8 Time analysis for different feature extraction techniques

The average time analysis for the different feature extraction methods is shown in Fig. 8. The x-axis denotes the feature extraction methods, and y-axis denotes the average time taken in MS. The ACA-HMM takes 210 ms, and PCA-HMM takes 220 ms for feature extraction. The MCE-HMM takes 235 ms for feature extraction. The proposed framework takes 205 ms for feature extraction in JAFFE database. The proposed framework takes less time from existing framework, and it gives excellent results.

5 Conclusion and Future Works

In this paper, we are able to prove that our proposed framework is able to improve recognition results. We tested our framework in terms of recognition rate, error rates, time analysis of various methods and time analysis of various feature extraction techniques and found the overall accuracy of 87.85%.

In the future, this modified HMM will be incorporated with other feature extraction methods such as partition-based, moment invariants-based and geometric moment-based feature extraction methods and compared it with other state-of-the-art methods to improve the accuracy to some extent.

Acknowledgements Japanese Female Facial Expression (JAFFE) is publicly available dataset. It is available free of charge from website http://www.kasrl.org/jaffe.html. The database was planned and assembled by Michael Lyons, Miyuki Kamachi and JiroGyoba.

References

1. Rahul, M., Kohli, N., Agrawal, R.: Facial expression recognition using hidden Markov model: a review. In: National Conference on RAICT, HBTU, Kanpur, March 25–26 (2017). ISBN: 978-93-86256-82-9
2. Katagiri, S., Lee, C.-H.: A new hybrid algorithm for speech recognition based on HMM segmentation and learning vector quantization. IEEE Trans. Speech Audio Process. **1**(4), 421–430 (1993)
3. Tsapatsoulis, N., Leonidou, M., Kollias, S.: Facial expression recognition using HMM with observation dependent transition matrix. In: 1998 IEEE Second Workshop on Multimedia Signal Processing, pp. 89–95, 7–9 Dec 1998
4. Rabiner, L.R.: A tutorial on hidden Markov models and selected applications in speech recognition. Proc. IEEE **77**(2), 257–286 (1989)
5. Atlas. L., Ostendorf, M., Bernard, G.D.: Hidden Markov models for monitoring, machining tool-wear. In: IEEE International Conference on Acoustics, Speech, and Signal Processing, vol. 6, pp. 3887–3890 (2000)
6. Hatzipantelis, E., Murray, A., Penman, J.: Comparing hidden Markov models with artificial neural network architectures for condition monitoring applications. In: Fourth International Conference on Artificial Neural Networks, vol. 6, pp. 369–374 (1995)
7. Nixon, M.S., Aguado, A.S.: Feature Extraction and Image Processing for Computer Vision. Elseviet Ltd. The Boulevard, Langford Lane, Kindlington, Oxford, OX5 1 GB, UK (2012)
8. Flusser, J.: On the independence of rotation moment invariants. Pattern Recogn. **33**, 1405–1410 (2000)
9. Flusser, J., Suk, T., Zitova, B.: Moments and moment invariants in pattern recognition, pp. 1–303. Publication, Wiley (2009)
10. Simon, X.L.: Image analysis by moments. Ph.D. thesis at the department of Electrical and Computer Engineering, The University of Manitoba Winnipeg, Manitoba, Canada (1993)
11. Teague, M.R.: Image analysis via the general theory of moments. J. Optical Soc. Am. **70**, 920930 (1980)
12. Chen, B.J., Shu, H.Z., Zhang, H., Chen, G., Toumoulin, C., Dillenseger, J.L., Luo, L.M.: Quaternion Zernike moments and their invariants for color image analysis and object recognition. Signal Process. **92**(2), 308–318 (2012)
13. Thawar, A., Zyad, S., Lala, K., Sami, B.: Object classification via geometric, Zernike and Legendre moments. J. Theor. Appl. Inf. Technol. **7**(1), 31–37 (2009)
14. Hasan, S.Y.: Study of Zernike moments using analytical Zernike polynomials. Adv. Appl. Sci. Res. **3**(1), 583–590 (2012)
15. Sabhara, R.K., Chin-Poo, L., Kian-Ming, L.: Comparative study of Hu moments and Zernike moments in object recognition. Smart Comput. Rev. **3**(3), 166–175 (2013). 2235 B. J. Math. Comput. Sci. **4**(15), 2217–2236 (2014)
16. Xin, Y., Miroslaw, P., Simon, X.L.: Image reconstruction with polar Zernike moments. ICARPR 2005, LNCS 3687, 394–403 (2005)

17. Zhao, Y., Wang, S., Fend, G., Tang, Z.: A robust image hashing method based on Zernike moments. J. Comput. Inf. Syst. **6**(3), 717–725 (2010)
18. Khotanzad, A., Lu, J.-H.: Classification of invariant image representations using a neural network. IEEE Trans. Acoust., Speech Signal Process. **38**, 1028–1038 (1990)
19. Kim, Y.K., Han, J.H.: Fuzzy K-NN algorithm using modified K-selection. Proceedings of 4th IEEE International Joint Conference on Fuzzy Systems (FUZZ-IEEE/IFES'95), vol. 3, 1673–1680 (1995)
20. Kim, W.-K., Sung, Y.: A region-based shape descriptor using Zernike moments. Signal Process. Image Commun. **16**, 95–102 (2000)
21. Hu. M.: Visual pattern recognition by moments invariants. IRE Trans. Inform. Theory, IT8, 179–187 (1962)
22. Wong, W.H., Siu, W.C.: Improved digital filter structure for fast moment computation. IEEE Proc. Vis. Image Signal Process. **46**, 73–79 (1999)
23. Dudani, S.A., Breeding, K.J., Mcghee, R.B.: Aircraft identification by moment invariants. IEEE Trans. Comput. C-26, 39–46 (1977)
24. Liao, S.X., Pawlak, M.: On the accuracy of Zernike moments for image analysis. IEEE Trans. Pattern Anal. Mach. Intell. 20, 1358–1364 (1998)
25. Zion, B. et al.: Sorting fish by computer vision. Comput. Electron. Agric. 23: 175–187 (1999)
26. Shutler, J.D., Nixon, M.S.: Zernike velocity moments for description and recognition of moving shapes. Proc. BMVC, 705–714 (2001)
27. Babatunde, O., Armstrong, L., Leng, J., Diepeveen, D.: Zernike moments and genetic algorithm: tutorial and application. B. J. Math. Comput. Sci. (2013)
28. Guojiang, W.: Facial expression recognition method based on Zernike moments and MCE based HMM. In: 9th International Symposium on Computational Intelligence and Design (2016)
29. Lyons, M.J., Akamatsu, S., Kamachi, M., Gyoba, J.: Coding facial expressions with gabor wavelets. In: Proceedings, Third IEEE International Conference on Automatic Face and Gesture Recognition, Nara Japan. IEEE Computer Society, pp. 200–205, 14–16 Apr 1998
30. Lyons, M.J., Budynek, J., Akamatsu, S.: Automatic classification of single facial images. IEEE Trans. Pattern Anal. Mach. Intell. **21**(12), 1357–1362 (1999)
31. Dailey, M.N., Joyce, C., Lyons, M.J., Kamachi, M., Ishi, H., Gyoba, J., Cottrell, G.W.: Evidence and a computational explanation of cultural differences in facial expression recognition. Emotion **10**(6), 874–893 (2010)
32. Pardas, M., Bonafonte, A.: Facial animation parameters extraction and expression recognition using Hidden Markov Models. Sig. Process. Image Commun. **17**, 675–688 (2002)
33. Aleksic, P.S., Katsaggelos, A.K.: Automatic facial expression recognition using facial animation parameters and multi stream HMMs. IEEE Trans. Inf. Forens. Secur. **1**(1), 3–11 (2006)
34. Rahul, M., Kohli,N., Agrawal, R.:Facial Expression Recognition using Multi-Stage Hidden Markov Model. J. Theor Appl. Inf. **95**(23) (2017). ISSN: 1992-8645, E-ISSN: 1817-3195

Automated Facial Micro-expression Recognition Using Local Binary Patterns on Three Orthogonal Planes with Boosted Classifiers: A Survey

Kennedy Chengeta

Abstract Micro-expressions are used to understand one's mind and also show hidden intentions through facial subtle changes. SwanCare is using facial recognition technology to identify micro-expressions that indicate the presence of pain in elderly patients with dementia who cannot verbally express the pain. Recognition of such micro-expressions needs trained experts as evidenced by lie detection and polygraph experts. Automatic expression recognition not only removes subjectivity but also saves resources and time. The paper proposes the detection of micro-expressions using the local binary patterns on three orthogonal planes(LBP-TOP) feature extractor due to its capability to detect temporal facial expression features. Other variants of the algorithm, namely LBP-MOP which is a mean of the feature extractions and LBP-SIP were also analyzed. The weighted classifiers proposed included support vector machines, k-nearest neighbor and random forest as a weighted ensemble voting classifier. To evaluate the accuracy and performance, the CASME II and SMIC 3D datasets were used in the study. A higher accuracy was obtained which was an improvement to other modern micro-expression methods. The study also leveraged Gabor wavelet filters, principal components analysis (PCA) and linear discriminant analysis (LDA) for dimension reduction.

1 Introduction

There is limited research done in facial micro-expression recognition compared to macro-expression recognition. Paul Ekman classifies micro-expressions as brief facial expressions that last for fractions of a second driven by involuntary concealed feelings or deliberate emotions [1, 2]. The expressions' durations are short and low in intensity than macro-expressions which makes them difficult to measure [1, 2]. There has been increased usage of video-based micro-expression recognition in psychology, computer science, security, police interrogations, clinical diagnosis,

K. Chengeta (✉)
School of Mathematics, Statistics and Computer Science,
University of KwaZulu-Natal, West View Campus, Durban, South Africa
e-mail: 216073421@ukzn.ac.za

© Springer Nature Singapore Pte Ltd. 2019
S. K. Bhatia et al. (eds.), *Advances in Computer Communication and Computational Sciences*, Advances in Intelligent Systems and Computing 924,
https://doi.org/10.1007/978-981-13-6861-5_57

neuroscience and gaming sectors [1, 2]. Emotions trigger the seven basic expressions which include happiness, sadness, surprise, disgust, anger, fear and contempt. The critical component involves extracting histogram sequences using local binary patterns from TOP [3] and Gabor filters. The main facial units are broken into local regions of interests, namely nose, mouse, forehead and cheekbones. For classification, support vector machine classifiers and k-nearest neighbors weighted in a voting ensemble classifier with different weightings are used. Our experimental results on key databases used in micro-expressions including the CASME II dataset with higher temporal and spatial resolution showed improved results [4, 5]. These were generated in a laboratory-controlled environment with proper lighting.

The study used localized feature extraction methods to identify the hidden feelings based on the micro-expressions. The baseline evaluation involved using a local binary pattern variant called LBP-TOP with boosted and weighted machine learning classifiers, namely support vector machines, k-nearest neighbor, random forest as part of voting classifiers. Video sequences were used, and the best performance for a five-class classification achieved around 72%. The study explored the use of different machine learning classification algorithms, such as SVM, RF and kNN as base classifiers and extreme gradient boosting, random forest and voting as Level 2 meta-classifiers. Voting classification using soft voting was implemented. This involved combining the output of classifiers using voting techniques to build a combined classification which outperformed the base classifiers.

2 Literature Review

People communicate and express themselves using facial expressions. While the macro-expressions are widely applied and researched on, micro-expressions are also a fulcrum for behavioral scientists, pediatricians and psychologists. With technological advancements, there is a wide need to analyze micro-expressions though they are subtle, short in nature and random as well as difficult to detect with the naked eye [1, 2]. As opposed to macro-expressions, micro-expressions last only a quarter of a second to a fifth of a second. The subtleness of facial expression changes is difficult to identify with the naked eye making facial recognition difficult [1, 2]. Computer vision offers a solution by detecting the facial movements in video or image sequences through computer-aided techniques. Significant research on micro-expressions has prioritized classification with single classifiers [6–8]. Less focus has been applied to feature extraction methods based on 2D/3D histograms. Support vector machines and k-nearest neighbor algorithms have been used to classify with success micro-expressions [4, 5].

Macro-expressions are characterized by facial skin muscle movements and deformations. Local binary patterns and HOG feature extraction algorithms have been used with success to extract macro-facial deformations [4, 5]. Of late convolutional neural network (CNN) approaches have also been used to extract spatial features with great success. The fact that these algorithms only consider spatial features means they do not cater for the dynamism of the facial expressions and subtle movements. Accurate representations using video sequences of images have been done with dynamic variants of local binary patterns like LBP-TOP with relative success [4, 5]. Deep learning approaches with dynamic encoding capability in video sequences have also been researched on with success where convolutional neural networks (CNN) encoded dynamics in the sequence for classification of facial expressions [4, 5].

2.1 Micro-expression Recognition

Haggard and Isaacs followed by Ekman popularized micro-expression research [1, 2]. The former pair published their findings on psychotherapeutic interview results as micro-momentary expressions [1, 2]. The latter published the importance of micro-expression on deception and depression recognition based on hospital patient data [1, 2]. The focus was on the subtle facial movements. The expressions were classified into simulated micro-, neutralized and masked micro-expressions. The simulated micro-expressions were not followed by genuine expressions, while the neutralized micro-expressions involved suppressing the emotions and showing a neutral face. Masked micro-expressions involved masking expressions with a falsified expression (Figs. 1, 2, 3 and 4).

Micro-expressions' measurement difficulties are well documented because of their temporal nature and short existence. Optical strain data has been extended to LBP-TOP as weighting functions to be able to identify smaller motion movements [4, 5]. An integral projection providing the benefits of texture and shape was introduced [5]. Extensions of LBP based on the three intersecting lines on a central point also improved micro-expression recognition [5]. Spatial–temporal completed quantized patterns were used to provide more information by extracting joint data that characterizes magnitudes and orientations [4, 5]. To counter low intensity of micro-

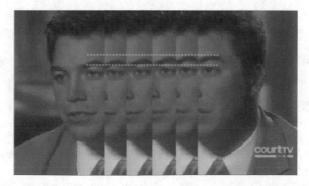

Fig. 1 Eye micro-expressions [4, 5]

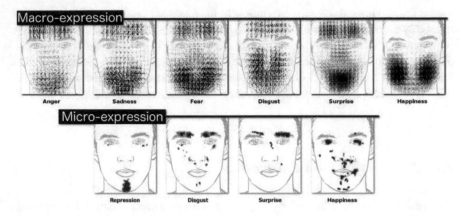

Fig. 2 Macro- and micro-expressions [4, 5]

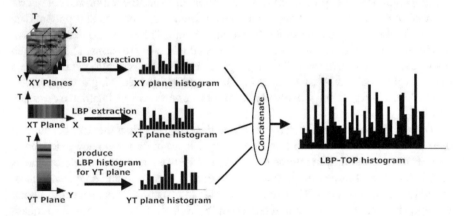

Fig. 3 Feature extraction graphs from LBP from three orthogonal planes represented in the XY, YT and XT orthogonal planes [5, 9]

$$\begin{bmatrix} -3 & -3 & 5 \\ -3 & 0 & 5 \\ -3 & -3 & 5 \end{bmatrix} \begin{bmatrix} -3 & 5 & 5 \\ -3 & 0 & 5 \\ -3 & -3 & -3 \end{bmatrix} \begin{bmatrix} 5 & 5 & 5 \\ -3 & 0 & -3 \\ -3 & -3 & -3 \end{bmatrix} \begin{bmatrix} 5 & 5 & -3 \\ 5 & 0 & -3 \\ -3 & -3 & -3 \end{bmatrix} \begin{bmatrix} 5 & -3 & -3 \\ 5 & 0 & -3 \\ 5 & -3 & -3 \end{bmatrix} \begin{bmatrix} -3 & -3 & -3 \\ 5 & 0 & -3 \\ 5 & 5 & -3 \end{bmatrix} \begin{bmatrix} -3 & -3 & -3 \\ -3 & 0 & -3 \\ 5 & 5 & 5 \end{bmatrix} \begin{bmatrix} -3 & -3 & -3 \\ -3 & 0 & 5 \\ -3 & 5 & 5 \end{bmatrix}$$

Fig. 4 Local directional pattern (LDP) includes edge detection using the Kirsch algorithm [25]

expressions, Li introduced motion or micro-expression magnification and temporal interpolation [9]. He also concluded that longer interpolated sequences did not lead to improved performance since dilution occurred. The directional mean optical flow (MDMO) research showed that magnitude of micro-expressions was more prominent than direction [9].

Previous methods focused on recognizing micro-expression on the whole face. Some argue that micro-expression is prominent in the eye region [1, 4, 5, 9, 10].

Previous researches were in isolating the eye region and using several classifiers trained to recognize micro-expressions using CASME2 dataset [2].

Macro- and micro-expressions The difference between macro- and micro-expression depends essentially on the duration and the intensity of expression. Macro-expressions are voluntary facial expressions and cover a large face area. The underlying facial movements and the induced texture deformations can be clearly discriminated from the noises [1, 5]. The typical duration of macro-expression is between 0.5 and 4 s. Macro-expressions are also characterized by high intensities, in terms of facial muscle movements and texture changes. Motion propagation is prevalent in the facial area in recent studies [5]. On the contrary, micro-expressions are involuntary facial expressions [1, 2, 5, 11]. Often, they convey hidden emotions that determine true human feelings and state of mind. Micro-expressions tend to be subtle manifestations of a concealed emotion under a masked expression. Micro-expressions are characterized by rapid facial movements and cover restricted fragments of the facial area [1, 5]. The typical duration of micro-expressions is between 65 and 500 ms. Micro-expressions are also characterized by low intensities in terms of facial muscle movement and texture changes [1, 3, 5].

Micro-expression dataset data creation is difficult to create and generate. This is because macro-expressions can be recorded in normal environments, whereas micro-expressions can only be generated in laboratory environments [1, 5, 9, 10]. Only high speed cameras can be used with an assumption that less noise is generated . It is also difficult to distinguish true hidden and fake hidden emotion-driven micro-expression movements [1, 5]. Different cultures of participants also introduce complexity to measure the subtle facial movements. Micro-expression spotting also makes it difficult to distinguish natural facial expression movements like blinking, eye gazing, reaction to nearby noise, visual effects or audio effects and group micro-expression reactions [1, 5].

2.2 Micro-expression Recognition Using 3D-HOG and Deep Learning

3D histograms of oriented gradient (3D-HOG) algorithms have been used with success to extract micro-expressions [12, 13]. The facial region was split into 12 regions of interest on posed micro-expression data captured with a high-speed camera [12, 13]. Thirteen expressions were identified, and their duration was measured. K-means unsupervised machine classifiers were used. Frame-by-frame classification was used to detect AUs in 8 video regions [12]. Different regions had different contributions to the overall accuracy. Offsets and apex were then measured with AU timing characteristics included to distinguish posed and spontaneous micro-expressions [12]. The regions' contribution was also identified as not isolated but to be overlapping [12, 13]. Thus, fuzzy micro-expression (fuzzy histogram of optical flow orientation (FHOFO)) classification of the regions on CASME II was executed with better accuracy. SMIC, CASME and CASME II datasets were used in the evaluation [13].

Deep learning approaches have been used with success in micro-expression recognition using spatiotemporal 3D video sequence convolutional neural networks [14]. The core feature of convolutional neural networks is the network architecture used to represent input data for instance LeNet,Googlenet and AlexNet. These approaches also focused on static image classification and object detection [14]. Since these approaches rely on huge datasets, it is a challenge in micro-expression classification since there are limited datasets that recognize subtle movements. Recent findings also argue that not all video sequences are needed to recognize micro-expressions [15, 16]. Successful researches done included where only the apex and onset video frames were chosen and used to recognize the micro-expressions [14]. The earlier frame had the highest intensity of expressions of all the frames, and the onset frame identified had a neutral expression.

3 Local Binary Patterns in Micro-expressions

Local binary patterns are based on facial images being split into local subregions. The challenges of local binary patterns include facial occlusion and rigidness [17–19]. Local binary patterns are invariant to gray-level images which also reduce illumination. Localized feature vectors derived are then used to form the histograms which are used by machine learning classifiers or deep learning methods. The local features are position-dependent [11, 17–19]. For local binary patterns, the facial region is divided into small blocks like mouth, eyes, ears, nose and forehead [6, 11]. The basic local binary pattern non-center pixels have a threshold on the central pixel value which takes binary values 0 or a 1 [17–19]. Uniform binary patterns are characterized by a uniformity measure corresponding to the bitwise transition changes. The local binary pattern has 256 texture patterns. The local binary LBP V, N operator is represented mathematically as follows:

$$LBP_{(V,N)}(m_y, n_y) = \sum_{K-1}^{K=0} q(p_y - p_c)2^K. \tag{1}$$

The neighborhood is depicted as an m-bit binary string leading to V unique values for the local binary pattern code. The gray level is represented by 2V-bin distinct codes. Different LBP variants used in micro-expression recognition include ternary local binary pattern and the central symmetric local binary patterns [6, 11]. Over-complete local binary pattern (OCLBP) is another key variant that includes overlapping into adjacent image blocks. The rotation-invariant LBP is designed to remove the effect of rotation by shifting the binary structure [6, 19]. Other variants include the monogenic and central symmetric (MCS-LBP). LBP basic algorithm is limited in micro-expression recognition due to low-intensity motions. The LBP-TOP is widely used in micro-expression identification [4, 10, 20].

3.1 Local Gabor Binary Patterns on Three Orthogonal Planes

3D dynamic texture recognition which concatenates three histograms from LBP on three orthogonal planes was proposed [11]. The three orthogonal planes, namely XT, XY and YT have been widely used [4, 10, 20]. LBP-TOP extracts features from the local neighborhoods over the three planes. The spatial–temporal data is classified as volume sets in the (X, Y and T) domain space where X and Y are the spatial coordinates and T represents the time index in the temporal domain [4, 10, 21]. The feature descriptor is represented in three-dimensional space X, Y and T and computes the local binary patterns of given center pixels by finding a threshold of surrounding pixels [4, 10, 20]. The orthogonal planes include the XY (similar to regular LBP static image), XT and YT planes [4, 22]. The YT plane indicates the features of motion in the temporal space domain [8, 11].

The spatial plane XY is similar to the regular LBP in static image analysis. The vertical spatiotemporal YT plane and horizontal XT plane are the other two planes in the three-dimensional space. The resulting descriptor enables encoding of spatiotemporal information in video images. The performance and accuracy of the latter were also comparable to the LBP-TOP. The LBP STCLQP or spatiotemporal completed local quantized pattern (STCLQP) was also used to consider the pixel sign, orientation and size or magnitude [11]. The LGBP-TOP algorithm added Gabor filtering to improve accuracy. With the added filtering algorithm, rotational misalignment of consecutive facial images was mitigated [4, 22, 23]. To avoid LBP-TOP statistical instability, a re-parameterization technique was proposed [4, 10, 20].

$$k = (H_L, X_Y, H_L, XT, H_L, Y_T, H_C, X_Y, HC, XT, HC, YT) \tag{2}$$

(LBP/C)T OP feature is denoted in vector form where Hv, m (v = LBP or C, and m = XY, XT, YT where m = XY, XT, YT) are the six LBP subhistograms in the three orthogonal planes [4, 10, 20]. The algorithm describes video sequence changes in both spatial and temporal domains and hence captures structural information of the former domain and longitudinal data of the latter [4, 5, 23]. LBP histogram features encode spatial data in the XY plane, and the histograms from the YT plane and XT plane include the temporal and spatial data. With facial actions causing local and expression changes over time, the dynamic descriptors have an edge in facial expression analysis over the static descriptors [4, 5, 23].

$$H_{i,j} = \sum x, y, t f_j(x, y, t) = i \tag{3}$$

The contrasts Cm are computed by incorporating three subhistograms as Hx, y (x = C and y = XY, XT, YT). Image device quality of the facial expression videos also impacts frame rates and spatial resolution quality as well [4, 5, 23].

The local binary pattern six interception points (LBP-SIP) considered six unique points along the decussating or converging lines of the three orthogonal planes to derive the binary pattern histograms where there is sparse data using compact features in high-dimensional feature spaces [4, 22, 23]. The LBP-SIP is represented by $AB, DF, EG = L_A \cap L_B \cap L_C$ where AB, DF and EG are intersection points. Six unique neighbor points carry sufficient information to describe the spatiotemporal textures centered upon point C [4, 22, 23]. LBP-three mean orthogonal planes is another variant to have been successfully used by concatenating mean images from image stacks derived along the three orthogonal planes [4, 22, 23]. It also preserves essential image patterns and reduces redundancy which affects encoded features [4].

4 Facial Micro-expression Recognition Implementation

The micro-expression implementation involved analyzing video streams to track facial feature points over time. The feature vectors were calculated, and emotions were detected from the trained models. Training and classification of the models was done using k-nearest neighbor, random forest, support vector machines and ensemble voting machine learning classifiers. The section describes the approach, databases selected and then classification algorithms chosen and implemented. The study's objective was to recognize facial expressions from video sequences. The approach involved locating and tracking the faces and expressions during video segmentation and sequential modeling phase. The video sequence detection involved landmark detection and tracking, which define the facial shapes [21]. Viola Jones OpenCV detection was used. The features were extracted using various 3D video feature extraction variants of the LBP-TOP algorithm. Gabor filters [4, 22–25] were applied during preprocessing. Geometric features were normalized to reduce illumination changes. The study analyzed a sequence of frames that change from one form to another to detect faces from a live video based on the CASME II and SMIC datasets [4, 22, 23].

4.1 Preprocessing and Classification

Preprocessing was achieved using Gabor filters [8] which act like a Gaussian kernel which is modulated by a sinusoidal plane wave as shown in the equation below [26] where K and Q are derived normalizing factors. The Gabor filters extract expression-invariant features.

Algorithm 1: Local Gabor Binary Patterns on TOP [4, 26]

Data: Copy and preprocess video image CASME II and SMIC datasets
Result: Facial expression classification results for the CASME II and SMIC image datasets
while *There are unprocessed images inside the CASME II and SMIC datasets* **do**
  1. divide the data into training and test sets;
  2. identify region of interest, namely eye, mouth, forehead 3. for each image region of interest ROI inside the given micro expression datasets;
  4. extract with Viola Jones algorithm and preprocess the image using Principal Component Analysis;
  5. extract LBP-TOP, LBP-XY, LBP-XT, LBP-YT, LBP-MOP, LBP-SIP features
  7. apply Gabor Filters to get the LGBP-TOP and LGBP-MOP features
  8. apply the classification on each with SVM, kNN, random forest and ensemble weighted voting classifier;
  End While'
end

$$G_c[x, y] = Kq^{-\frac{(x^2+y^2)}{2\sigma^2}} \cos(2\pi f(x\cos\theta + y\sin\theta)); \tag{4}$$

$$G_s[x, y] = Qq^{-\frac{(x^2+y^2)}{2\sigma^2}} \sin(2\pi f(x\cos\theta + y\sin\theta)); \tag{5}$$

Two commonly used validation techniques in computer vision are n-fold cross-validation and leave-one-subject-out (LOSO). The study used the latter due to its widespread success. Filtering face images was achieved using Kirsch compass masks. The eight directional Kirsch masks are used for the filtering of edges from face by convolving the 3-by-3 neighborhoods of the image with the Kirsch masks. These eight directional masks result in eight responses for each pixel. For local directional patterns or LDP, a key edge detection local feature extractor, the images were divided into LDPx histograms, retrieved and then combined into one descriptor [4, 15, 17, 24, 27].

$$LDP_x(\sigma) = \sum_{K}^{r=0} \sum_{L}^{r=0} f(LDP_q(o, u), \sigma). \tag{6}$$

4.2 Micro-expression Classification

Base and meta-Classifiers [8] used in the study included k-nearest neighbor, random forest, neural networks, support vector machines and voting classifier [19, 24, 25, 28]. The study used the base classifiers on each region of interest. For SMIC, ME samples were classified into three categories, namely positive, negative and surprise. For CASME II, ME samples were classified into five categories which include happiness, disgust, surprise, repression and others.

Voting Meta-classifiers The research applied voting classifier with different weights of the respective classifiers, namely random forest, k-nearest neighbor and support vector machines with different ratios and weights. The EnsembleVoteClassifier, a meta-classifier, was used to combine weak machine learning classifiers through a majority or weighted voting pattern [27, 28]. The hard voting model involves predicting the final class as the modal predicted model and in soft voting the class labels are average class probabilities based on a weighting mechanism [25, 28].

$$wM = \hat{y} = \arg\max_i \sum_{j=1}^{m} w_j \chi_A\big(T_j(\mathbf{x}) = i\big), \tag{7}$$

where χ_A is the characteristic function $[T_j(\mathbf{x}) = i \; \in A]$ and A is the set of unique class labels for the weighted majority classifier wM.

$$sV = \hat{y} = \arg\max_i \sum_{j=1}^{m} w_j p_{ij}, \tag{8}$$

where w_j is the weight assigned to the jth classifier for the soft voting meta-classifier sV. Different modal ratios found included 3:5:2, 5:1:4 and 4:3:3 for support vector machines, random forest and kNN for LBGP-TOP, LBGP-MOP and LBGP-SIP to the voting classifier, respectively.

4.3 Macro- and Micro-expression Datasets

Datasets of facial micro-expressions used were CASME II and SMIC. The Chinese Academy of Sciences Micro-Expression (CASME) database has 195 spontaneous facial micro-expression recorded from dual 60 fps cameras [8]. There were 26 Asian subjects with average age of 22 years. The micro-expressions had a duration less than 500 ms and coded with onset, apex and offset frames as well tagging with action units and onset duration also less than 250 ms. The CASME II dataset had five classes, namely happiness, disgust, surprise, repression and others. Tense and fear were grouped as other micro-expressions. One category or Class A was recorded with BenQ M31 camera with 60 fps and 1280 by 720 pixel resolution in natural light [8]. These were recorded by a 200 fps camera in a controlled laboratory. The recordings were then saved in JPEG format without any inter-frame compression [8]. The SMIC dataset has HS dataset which was recorded with a 100 fps high-speed camera and 164 video sequences belonging to three classes (positive, negative and surprise) from 16 participants [8]. It also has the VIS dataset recorded from a 25 fps standard camera and NIR dataset recorded with an infrared 25 fps camera [8]. The last two datasets recorded 77 video sequences. The datasets included micro-expression clips of videos from onset to offset with each sequence segmented by the original videos and labeled either surprise, negative and positive expressions [8].

Table 1 Video sequence classification on CASME II dataset with 593 video image sequences—6 regions ROI weighted

Eye(s) region	R	XY	XT	YT	Classifier	Weight	Acc (CASME II)	Acc (SMIC)	Length (CASME II)	(SMIC)
LGBP-TOP	8	3	3	3	SVM	1	0.657	0.656	0.234	0.28
LGBP-TOP	8	3	3	3	kNN	1	0.655	0.658	0.234	0.28
LGBP-TOP	8	3	3	3	Random forest	1	0.679	0.662	0.234	0.28
LGBP-TOP	8	3	3	3	Fuzzy neural classifier	1	0.685	0.689	0.234	0.28
LGBP-TOP	8	3	3	3	Voting classifier (SVM, RF, kNN)	3;5;2	0.698	0.699	0.234	0.28
LGBP-TOP	8	3	3	3	ROI (SVM-Eye, RF-Mouth, kNN-Other)	1	0.697	0.701	0.234	0.28
LGBP-MOP	8	3	3	3	kNN	1	0.671	0.663	0.234	0.28
LGBP-MOP	8	3	3	3	Random forest	1	0.683	0.685	0.234	0.28
LGBP-MOP	8	3	3	3	Voting classifier	5;1;4	0.698	0.704	0.234	0.28
LGBP-MOP	8	3	3	3	SVM	1	0.687	0.689	0.234	0.28
LGBP-MOP	8	3	3	3	ROI (SVM-Eye, RF-Mouth, kNN-Other)	1	0.713	0.713	0.235	0.28
LGBP-SIP	8	3	3	3	Random forest	3;5;2	0.679	0.674	0.234	0.28
LGBP-SIP	8	3	3	3	Fuzzy neural classifier	6;1;3	0.663	0.665	0.234	0.28
LGBP-SIP	8	3	3	3	Voting classifier (SVM, RF, kNN)	4;3;3	0.696	0.703	0.234	0.28
LGBP-SIP	8	3	3	3	SVM	1	0.674	0.682	0.234	0.28
LGBP-SIP	8	3	3	3	ROI (SVM-Eye, RF-Mouth, kNN-Other)	1	0.698	0.703	0.234	0.28
3DHOG	8	3	3	3	Voting classifier(SVM, RF, kNN)	1	0.623	0.647	0.234	0.28

Table 2 Video sequence classification on facial expressions database SMIC dataset—eye and mouth regions

Algorithm	R	XY	XT	YT	Classifier	Weight	Acc (CASME II)		Acc (SMIC)		Acc (SMIC)			CASME II~ (s)	SMIC~ (s)
							Eye	Mouth	Eye	Mouth	HS	VIS	NIR		
LGBP-TOP	8	3	3	3	SVM	1	0.647	0.624	0.637	0.634	1	1	1	0.234	0.28
LGBP-TOP	8	3	3	3	kNN	1	0.645	0.627	0.639	0.638	1	1	1	0.234	0.28
LGBP-TOP	8	3	3	3	Random forest	1	0.646	0.613	0.638	0.631	1	1	1	0.234	0.28
LGBP-TOP	8	3	3	3	Fuzzy neural classifier	1	0.655	0.615	0.621	0.639	1	1	1	0.234	0.28
LGBP-TOP	8	3	3	3	Voting classifier (SVM, RF, kNN)	3;5;2	0.659	0.636	0.644	0.643	1	1	1	0.234	0.28
LGBP-MOP	8	3	3	3	kNN	1	0.651	0.621	0.630	0.637	1	1	1	0.234	0.28
LGBP-MOP	8	3	3	3	Random forest	1	0.653	0.611	0.636	0.633	1	1	1	0.234	0.28
LGBP-MOP	8	3	3	3	Voting classifier (SVM, RF, kNN)	5;1;4	0.655	0.632	0.645	0.649	1	1	1	0.234	0.28
LGBP-MOP	8	3	3	3	SVM	1	0.645	0.623	0.633	0.635	1	1	1	0.234	0.28
LGBP-SIP	8	3	3	3	Random forest	1	0.648	0.621	0.634	0.632	1	1	1	0.234	0.28
LGBP-SIP	8	3	3	3	Fuzzy neural classifier	1	0.654	0.610	0.636	0.634	1	1	1	0.234	0.28
LGBP-SIP	8	3	3	3	Voting classifier (SVM, RF, kNN)	4;3;3	0.656	0.637	0.633	0.642	1	1	1	0.234	0.28
LGBP-SIP	8	3	3	3	SVM	1	0.645	0.614	0.637	0.632	1	1	1	0.234	0.28
3DHOG	8	3	3	3	Voting classifier (SVM, RF, kNN)	4;3;3	0.623	0.622	0.645	0.627	1	1	1	0.234	0.28

Table 3 LBP-TOP, SMIC dataset facial micro-recognition dataset of 400 webcam videos image sequences—full face and mouth region alone

CASME II	Support	Confusion matrix					
Happy	95	[91,	2,	2,	1,	0	0]
Disgust	69	[1,	64,	0,	2,	1,	0]
Surprise	57	[1,	1,	54,	1,	1,	0]
Repression	65	[1,	1,	1,	62,	0,	0]
Other	71	[1,	1,	0,	0,	69,	0]
Avg/total	400						

SMIC full face	Support	Confusion matrix		
Negative	69	[64,	0,	0]
Surprise	57	[1,	54,	0]
Positive	56	[2,	3	2]
Avg/total	400			

CASME II	Support	Confusion matrix					
Happy	95	[91,	2,	2,	1,	0	0]
Disgust	69	[69	45	1	1	1	1]
Surprise	57	[57	1	43	1	1	1]
Repression	65	[65	1	1	21	1	1]
Other	78	[78	1	1	1	22	1]
Avg/total	322						

SMIC mouth	Support	Confusion matrix			
Negative	69	[45	1	1	1]
Surprise	57	[57	23	2	1]
Positive	67	[67	1	22	1]
Avg/total	322				

Table 4 CASME II dataset classifier

CASME II dataset classifier	Ratio	LBP-TOP	LBP-MOP	LBP-SIP
SVM linear kernel	1	90.23	90.13	90.11
SVM RBF kernel	1	90.11	90.43	90.13
Multiperceptron classifier	1	1	1	1
k-nearest neighbor (radius = 1)	1	1	1	1
k-nearest neighbor (neighbor = 1)	1	1	1	1
Random forest	1	1	1	1
Voting classifier (SVM, kNN, RF)	0.4, 0.5, 0.1	1	1	1

5 Experimental Results on CASME II and SMIC 3D Datasets

Recognition experiments were executed for the LBP-XY plane, LBP-XT plane as well as the LBP-YT planes. The combined recognition rate for the LBP-TOP was also calculated. The four scenarios used an ensemble classifier of support vector machines, k-nearest neighbor as well as random forest classifiers with different weights. The CASME II dataset, combined LGBP-TOP with Gabor filtering and an ensemble of voting classifier combination of support vector machines, k-nearest neighbor and random forest achieved a higher accuracy of 70.19% from a sequence of 593 video sequences. For the SMIC database with 400 video image sequences, the corresponding accuracy was 70.5%. The combined feature extractor LBP-TOP achieved higher classification rates compared to specific dimensions LBP-XT, LBP-XT and LBP-YT accuracy rates. Better variation was experienced for the LBP-XT-based plane. The second experiments evaluated the efficiency of using Gabor filters to enable multi-orientation fusion. The combined classifier with Gabor filters and LBP-TOP feature extractor showed greater accuracy to the normal LBP-TOP algorithm. The other LBP-TOP variants like SIP and MOP achieved greater accuracy, but the LGBP-TOP with parameters of 8, 3 on each dimension achieved better accuracy to all the LGBP-TOP variants. The confusion matrix obtained from the video datasets for showed an overall success of 69.51 and 69.2% when Gabor filtered on the CASME II and SMIC database, respectively (Tables 1, 2, 3 and 4).

6 Conclusion and Future Work

The feature extraction methods based on LBP-TOP variants applied to major facial components were used to analyze facial expressions in video datasets and showed marked improved method compared to traditional methods used before. For each facial component angle, namely XY, XT and YT-3D dimension, a different variant of the classification algorithm was used. The classification rate was a weighted ensemble classifier composed of a support vector machine, k-nearest neighbor classifier and a random forest classifier. The contribution of each algorithm to the ensemble classifier had k-nearest neighbor as the majority contribution in the XY axis. For YT domain, the random forest dominated the ensemble classification algorithm. The Gabor filters improved the accuracy, and the LBP-TOP variants also showed great accuracy.

Future improvements include establishing to what degree is it possible to distinguish genuine from fake facial expressions. There is also a need to judge one's authenticity of facial expression. Factoring of interpersonal and intrapersonal differences as well as group effect on facial expression is also crucial. The research also recommends analyzing expressions in an African context where there are different cultures with each culture having different ways of expressing themselves. Some of the key cultures suggested include the Zulu culture in South Africa, Swahili culture in Eastern Kenya and Tanzania, as well as Shona culture in Zimbabwe.

References

1. Ekman, P., Friesen, W.V.: The repertoire of nonverbal behavior: categories, origins, usage, and coding. Semiotica 1, 49–98 (1969)
2. Haggard, E.A., Isaacs, K.S.: Micro-momentary facial expressions as indicators of ego mechanisms in psychotherapy. In: Gottschalk, L.A., Auerbach, A.H. (eds.) Methods of Research in Psychotherapy, pp. 154–165. Appleton-Century-Crofts, New York (1966)
3. Kellokumpu, V., Zhao, G., Pietikäinen M.: Recognition of human actions using texture descriptors. Mach. Vis. Appl. 22(5), 767–780 (2011)
4. Wang, Y., See, J., Phan, R.C.-W., Oh, Y.-H.: LBP with six intersection points: reducing redundant information in LBP-TOP for micro-expression recognition. In: Computer Vision–ACCV 2014, Singapore, pp. 525–537 (2014)
5. Wang, Y., Yu, H., Stevens, B., Liu, H.: Dynamic facial expression recognition using local patch and LBP-TOP. In: 2015 8th International Conference on Human System Interaction (2015)
6. Pavithra, P., Ganesh, A.: Detection of human facial behavioral expression using image processing
7. Acevedo, D., Negri, P., Buemi, M.E., Gómez Fernández, F., Mejail, M.: A citation k-NN approach for facial expression recognition (2018)
8. Hao, Z.: Micro-expression recognition based on 2D Gabor filter and sparse representation. J. Phys. Conf. Ser. 787(1), 012013 (2017). https://doi.org/10.1088/1742-6596/787/1/012013
9. Liu, Y., Zhang, J., Yan, W., Wang, S., Zhao, G., Fu, X.: A main directional mean optical flow feature for spontaneous micro-expression recognition. IEEE Trans. Affect. Comput. 7(4), 299–310 (2016). https://doi.org/10.1109/TAFFC.2015.2485205
10. Huang, X., Zhao, G., Pietikainen, M., Zheng, W.: Dynamic facial expression recognition using boosted component-based spatiotemporal features and multiclassifier fusion. In: Advanced Concepts for Intelligent Vision Systems, pp. 312–322. Springer, Berlin (2010)
11. Pietikinen, M., Hadid, A., Zhao, G., Ahonen, T.: Computer Vision Using Local Binary Patterns (2011)
12. Dalal, N., Triggs, B.: Histograms of oriented gradients for human detection. In: Proceedings of the IEEE Computer Society Conference on Computer Vision and Pattern Recognition (CVPR'05), Washington, DC, USA, pp. 886–893 (2005)
13. Happy, S.L.: Fuzzy histogram of optical flow orientations for micro-expression recognition. Department of Electrical Engineering, Indian Institute of Technology Kharagpur. https://doi.org/10.1109/TAFFC.2017.2723386

14. Li, X., Yu, J., Zhan, S.: Spontaneous facial micro-expression detection based on deep learning. In: IEEE 13th International Conference on Signal Processing (ICSP), Chengdu, pp. 1130–1134 (2016)
15. Sanin, A., Sanderson, C., Harandi, M.T., Lovell, B.C.: Spatio-temporal covariance descriptors for action and gesture recognition. IN: IEEE Workshop on Applications of Computer Vision (2013)
16. Jain, S., Hu, C., Aggarwal, J.K.: Facial expression recognition with temporal modeling of shapes. In: Proceedings of the IEEE International Conference on Computer Vision Workshops (ICCV Workshops), Barcelona, pp. 1642–1649 (2011)
17. Aung, M.S., Kaltwang, S., Romera-Paredes, B., Martinez, B., Singh, A., Cella, M., Valstar, M., Meng, H., Kemp, A., Shafizadeh, M., et al.: The automatic detection of chronic pain-related expression: requirements, challenges and a multimodal dataset. Trans. Affect. Comput. (2015)
18. Lemaire, P., Ben Amor, B., Ardabilian, M., Chen, L., Daoudi, M.: Fully automatic 3D facial expression recognition using a region-based approach. In: Proceedings of the 2011 Joint ACM Workshop on Human Gesture and Behavior Understanding, J-HGBU '11, New York, USA, pp. 53–58. ACM (2011)
19. Padgett, C., Cottrell, G.W.: Representing face images for emotion classification. In: Advances in Neural Information Processing Systems, pp. 894–900 (1997)
20. Jiang, B., Valstar, M., Martinez, B., Pantic, M.: A dynamic appearance descriptor approach to facial actions temporal modelling. IEEE Trans. Cybern. **44**(2), 161–174 (2014)
21. Mattivi, R., Shao, L.: Human action recognition using LBP-TOP as sparse spatio-temporal feature descriptor. In: Computer Analysis of Images and Patterns, pp. 740–747 (2009)
22. Wang, Y., See, J., Phan, R.C.-W., Oh, Y.-H.: Efficient spatio-temporal local binary patterns for spontaneous facial micro-expression recognition. PloS One **10**(5) (2015)
23. Spizhevoy, A.S.: Robust dynamic facial expressions recognition using LBP-TOP descriptors and Bag-of-Words classification model
24. Nurzynska, K., Smolka, B.: Smiling and neutral facial display recognition with the local binary patterns operator. J. Med. Imag. Health Inform. **5**(6), 1374–1382 (2015)
25. Chengeta, K., Viriri, S.: A survey on facial recognition based on local directional and local binary patterns. In: Conference on Information Communications Technology and Society, Durban (2018)
26. Ravi Kumar, Y.B., Ravi Kumar, C.N.: Local binary pattern: an improved LBP to extract nonuniform LBP patterns with Gabor filter to increase the rate of face similarity. In: 2016 Second International Conference on Cognitive Computing and Information Processing (CCIP), Mysore, pp. 1–5 (2016)
27. Viola, P., Jones, M.J.: Robust real-time face detection. Int. J. Comput. Vis. **57** (2004)
28. Aggarwal, C.C.: Data Mining Concepts (2015). ISBN 978-3-319-14141-1, XXIX, 734 p. 180 illus

Identification of Emotional States and Their Potential

Jan Francisti and Zoltan Balogh

Abstract Today's trend is to use IT in every area of our lives. Technology is primarily used to improve the standard of living. Emotions are the basis of human experience, even though it is difficult to define, recognize, and classify them. Nowadays, greater emphasis and attention is placed on the computer's ability to evaluate emotional changes and conditions in humans. Proper assessment and recognition of the human emotions may lead to a better understanding of user behavior. Systems that are able to acquire data, evaluate user status and model them have a broad application in various spheres of human activity (neuro-marketing, automotive control, adaptive learning, mental health, etc.). The cognitive process is carried out at two fundamental levels in the level of sensory perception and intellectual perception. In humans, these two basic levels are progressively developed through age or by their own experience. The chapter describes a research study of individual emotional states that can be captured by various sensors, which can quantify and evaluate emotional states of users and thus adapt their surroundings.

Keywords Emotional states · Physiological functions · Sensory networks · Smart wristbands

1 Introduction

In a field such as didactics, the cognitive process consists of two parts that are sensory and intellectual. Different validations of effectiveness, as well as various psychological and pedagogical research, are shown to indicate that the number of

J. Francisti (✉) · Z. Balogh
Department of Informatics, Faculty of Natural Sciences, Constantine the Philosopher University in Nitra, Tr. A. Hlinku 1, 949 74 Nitra, Slovakia
e-mail: jan.fracisti@ukf.sk

Z. Balogh
e-mail: zbalogh@ukf.sk

© Springer Nature Singapore Pte Ltd. 2019
S. K. Bhatia et al. (eds.), *Advances in Computer Communication and Computational Sciences*, Advances in Intelligent Systems and Computing 924,
https://doi.org/10.1007/978-981-13-6861-5_58

activating sensors is proportional to perception and memory. Therefore, it is necessary to determine what impact they have on the user (whether the teacher or the student himself).

Authors [17] point out that quantitative measurements of human emotions confirm both the psychological and physical management. Based on similar measurements, it is possible to determine the students' emotional reaction to the occurrence of an event, e.g., when students writing examinations or learning. Students' reactions, for simplicity, can be divided into two groups, distraction and attraction, where we can distinguish the positive and negative reactions in the measurement.

At present, a great effort is being made to improve the interaction between humans and computers. The ability of a computer to evaluate the user's emotional state or human is getting more and more attention and interest. For students, a great effort to evaluate emotional state is for determining the perception of a person and also the stress level, in order to minimize the negative state so that students achieve the best possible success in the study [16].

Body sensor networks (BSNs) have become revolutionary technologies in many areas, such as health, smart cities, factories, and other areas of the Internet of things (IoT). Commercially available system solutions offer only a few options for tracking user behavior [11].

Through data mining and designing a convenient model, we try to overcome the barrier between man and non-emotional machine. System capable of data mining, evaluating user status, and creating models has a broad application in a variety of human activities.

The aim of the chapter focuses on the development of the system and the assessment of the emotional state by applying the sensory characteristics of the user himself. The chapter aims to use the obtained data measurement through the sensors as the underlying materials for determining the user's emotional state. Based on the assessed status, after evaluation, it will be possible to customize the learning material for such a student.. The development of the system is the base used for sensory tracing and evaluation of the emotional state. Also, to create feedback from the user that warns that in a given situation it is advisable to silence the voice (to alleviate aggression, calm down) etc.

2   Related Work

In human life, emotions and feelings are very important because they are involved in creating a motivational structure. From human focus, depth, and permanence depends on actions, what he will do. Feelings have different inner and outer manifestations. People can see different changes when they experience some emotion, such as when they are cheerful, they have a smile on their face. The most striking expressions of emotions are crying and laughing [8, 12].

2.1 Emotional Intelligence (EQ)

Emotional intelligence is the ability to identify, evaluate, and control your own emotions and the emotions of others and groups. There are two different kinds of intelligence: emotional and rational (IQ). Success in a person's life depends on both types. Without emotional intelligence, the intellect is not able to exploit all its potential. People's life experiences point to the fact that high intellectual ability is not automatically a prerequisite for successful enrollment at school in learning and in the practical life.

The following table shows the importance of both types of intelligence, not only intellectual but also emotional. In real decision-making and negotiation, the feelings have the same weight as thoughts [6].

2.2 Evaluating Human Emotions

The issue of assessing human feelings not simply utilizing a computer is an exceptionally fascinating point that nowadays has increased considerably. It is significantly noting the intertwining of several non-related areas (automotive) and fields such as computer science and psychology. Therefore, it has great potential to improve interaction between human and computer in different areas—from the educational sphere, through medicine to the commercial area. The examination into the feelings of the human (particularly face-to-face) started in the seventeenth century and turned into the reason for further advancement. For example in a book by John Bulwer titled "Pathomyotomia", where are the issue of facial expression and muscle movement on the face discussed in detail [15].

In 1980, Plutchik described emotions through the shape of a wheel. There were 8 primary different emotional pairs in the wheel. Joy/sadness, anger/fear, trust/resistance, and surprise/expectation. Like colors, primary emotions can be expressed in different shades and can be combined to form different emotions [20].

Another way to distribute emotions is, for example, Löheimheim's emotional cube. This model was made up of three signaling substances (dopamine, noradrenaline, and serotonin) further from the axes of coordinating system and eight basic emotions located in the eight corners. For example, an emotional state of anger is produced as a mix of high dopamine, noradrenaline, and low serotonin. On the other hand, joy is a product of high serotonin and dopamine and low noradrenaline. Because none of the axes is identical to valence 1, it seems that the cube is slightly rotated relative to other models [14].

Ekman and Friesen built up a "Facial Action Coding System" in 1978 to encode outward appearances, where the facial expressions are depicted by an arrangement of activity units [9]. Each AU unit is based on the relationship of the muscles. This mimic coding system is executed manually using a set of rules. Static images of the mimetic, often representing the top of the expression, serve as inputs. This process is time-consuming. Surveying every AU at an enabled time recognizes up to 7 kinds of emotions: sadness, happiness, fear, surprise, disgust, anger, and neutral expression.

Ekman's work roused numerous scientists [18, 21] to dissect outward appearances utilizing the picture and video preparing. By following facial highlights and estimating the measure of development on the face, they attempted to deal with various outward appearances into individual classes. Authors [24] are centered around face demeanor investigation and acknowledgment "fundamental terms" or their subset. Authors Pantic and Rothkrantz created the survey of a considerable lot of the face acknowledgement investigate which is done also nowadays[19].

The method for allocating the emanation by Ekman will also be used in our testing, where we will discover the level of stress due to the presence of certain emotional states.

2.3 Measurement and Detection of Emotional Changes

To test the effect of emotional valency on lower grade students [26] prepared a memory game where students had to answer "yes" or "no" according to whether the word was said earlier or not. Testing took place so that the teacher pronounced the words and the students had to answer, and also some words were spoken with emotional content. The result of the test was that students were more aware and responded correctly to those words that were spoken with emotional content.

Nowadays, stress is another factor that affects our emotions and their distribution. Also, stress is one of the elements that influence our well-being. It is important to understand how people respond to events that can cause stress. Cruz-Albarran et al. [7, 23] developed a model capable of estimating the stress signal (a signal that captured the sensors of the activity of the brain, skin, and thermal camera) of the people during respiration. For testing, they utilized a computational stress signal forecast system created as a result of a help vector machine, a hereditary calculation, and a neural network. Testing took place in the room where cameras took photographs of people faces. The dataset contained physiological and physical sensor signals at the time of meditation activity and used a thermal camera that recorded the temperature changes on the face. Based on the information on the shape of the stress signal, a stress-based predictor based on artificial neurons was created and used as a stress detector for real-time observation and evaluation.

Present high burdens and ongoing technological developments that lead to a constant change and the need to adapt to the technologies in question mean that this problem is increasingly serious for administrative workers. To avoid that the stress is chronic and causes irreversible damage, it needs to be revealed in the early stages. Stress measurements can be carried out along three main techniques, including psychological, physiological, and behavioral procedures, along with contextual measurements to provide advice on the most appropriate techniques that should be used. According to authors' experiments [2, 22], the most precise pressure discovery is created via physiological signals. The results also indicate that electrocardiogram

(ECG), using heart rate variation (HRV) functions, and electrodermal activity (EDA) are the most exact physiological pressure acknowledgment signals.

In emotional state analysis, the authors used [13] the same three basic biometric techniques. They have made a warning that decides the level of emotional state and efficiency at work. By utilizing these three biometric strategies, that can identify four states (push, work efficiency, temperament, enthusiasm for work) and seven fundamental feelings. The system also integrated biometric fingerprints (blood pressure and pulse rate) that were incorporated into a computer mouse with other functions such as body temperature and push-button force. The result of the testing was information from the advisory system, which gave the user their very own ongoing evaluation of production and emotional status.

Authors [25] tested the presence of physical and cognitive stress based on keystroke monitoring designed to detect changes in cognitive and physical functions. Respondents received a random text from a foreign language that contained other characters. Texts had to be overwritten and unknown characters could be done by combining shortcuts on the keyboard. Based on random text writing in a foreign language, they wanted to find out if respondents would be stressed. To evaluate the presence of stress, they used machine learning techniques.

Rates for the right characterization of subjective stress tests were predictable with the rates presently utilized with the utilization of positive computational strategies. The test results were positive in situations where physical and cognitive stress were significant.

For measuring stress, the authors [5] also used cameras, while observing respondents while playing interactive games on smart devices. The camera was used as a secondary stress-sensing sensor, where the primary sensor used the device itself, on which respondents played the game, in which it was necessary to calculate the displayed sum using the least amount of time. Specifically, the authors examined how stress affects acceleration, maximum and medium touch intensity, touch duration, amount of motion, and cognitive performance. The camera was used to record the movement of respondents during the game. When comparing data from the camera and the device, the respondents who measured the stress did not move while playing the game, and also the touch on the device was under greater pressure and faster. Authors [3] used cameras for students' movement monitoring during teaching. The used software was at first calibrated with training data and then deployed in real-time monitoring of the students. In this way, the authors were able to acquire student behavior while viewing e-courses and to modify study materials in which students would not show negative emotions.

Stress also can be measured according to the context. Authors' estimations are divided into two sections where the first was research center estimation and the second was in genuine circumstance. In the research facility, the respondents had wristbands with the volume of blood estimating capacities. Stress volume was assessed by explaining unpleasant circumstances. After the first measurements in the laboratory, the results from the wristband was compared with comes about because of tackling the circumstances. Division and filtering were executed as first. In the event that some piece of information was deficient, authors repeat the test. Earned information was prepared utilizing WEKA machine learning. In this way, information was assessed and a stress detector was made for every respondent. When every respondent has own "stress detector" (a model created by authors for each respondent individually composed of data through biometric sensors during laboratory testing), respondents were given inquiries of stress feeling level amid the day. Since they had a wristband with an implicit stress detector, the respondents' level of pressure was anything but difficult to pursue. Last outcomes correlate in view of the respondents' and the detector reaction [10].

3 Research Methodology

From previously mentioned authors' measurement and evaluation of emotions, we can consider that it is a good solution to use a combination of biometric sensors that can measure the physiological functions of people, which are suitable indicators for capturing the emotional state. We also found that we can use wearable devices when testing, which people could wear it during the test and would not be disturbing them.

To get a better understanding of emotions, it is necessary to focus on three main components: physiological responses, subjective experiences and expressive reactions. The emotions and behavior of most people will make perfect sense if you are at least partly aware of how their emotions can work. Most people are accustomed to mood and emotion, and they do not separate them. Experts and psychologists have different opinions. The main difference between emotions and moodiness is obvious. Emotions are short-lived and usually very intense. Also, they are specific and always have some clearly defined cause. Through emotions, we can feel short-term anger and short-term joy. Based on [1] experiments where Authors identify, that anger reaction can be recognized from the body. As for mood, they are much longer and mild. It may happen that we simply wake up badly and we will not even be able to identify the cause.

3.1 Sensory System Design

The solution of sensory system design would be made utilizing a microcontroller or microcomputer for the need for capturing and assess the emotional state. The point of our examination will be to choose the right device, that can collect information from different sensors. Wristband with previously mentioned functions offers, for example, wristband from the company www.myfeel.co, where they offer wristband with sensors.

The wristband is waterproof, and for connection, Bluetooth and USB port are used. Through the flexible material, it is very easy to put on the hand. Measurement data capturing and evaluation can be processed by support of a microcomputer or a microcontroller, for example, Arduino, Intel Edison, or Raspberry Pi, which also used [4] for their data evaluation. Also, we need a proper application with which we will be able to capture each change and send to the Cloud, where they will be prepared and assessed as a form of the present user's feeling. We expect that claim planned and assemble will be more efficient than other accessible observing; likewise, it will be modular and will have more extensive use than general. Then, we will be able to share the measured data through data networks (e.g., Sigfox) and then use them for further processing and approve their validity for the exploration area. An essential for the automotive working of emotional observing is additionally the system advancement in view of various factual strategies. In view of the system, we will have the chance to grow our exploration job in the field of IoT additionally in different fields, for example, natural knowledge.

Using the system we create, we will be able to customize the learning material, also the learning style and the method of access to specific students, according to their current emotional state. We assume that the adaptive system will be more flexible for learners and more efficient for acquiring new knowledge.

4   Results of Research

Based on a research of the tests performed by previously mentioned authors, we found that most of the sensory network elements were used in assessing emotional states. These wristbands could students wear during exams without been disturbing by them. Devices include sensors for the physiological function measurements of such as brain, cardiac, and activity of the skin.

At the beginning of measurement physiological functions, the authors set individual sensors and defined the valency of the boundary values. For each sensor, they set values to indicate that the emotional state of a person changes during measurement. In some studies, the authors [7, 23] combined sensors that measured physiological functions together with a camera that had the role of capturing facial expression.

Initial measurements were performed under laboratory conditions, due to calibration of devices and sensors. For each sensor, they set a measurement deviation so that the results are as accurate as possible. Also, the software that was used with the camera included a database of tremendous photographs of human facial expressions, able to evaluate the facial expression in real time and assess the emotional state.

After setting and calibration of measuring devices, initial testing was performed. Based on the obtained values, the measured data could be processed, evaluated and the system will be optimized. The results obtained were then displayed through graphs or tables.

We have designed our system as a combination of solutions by other authors. Our aim is to test the evaluation of emotional states of students during classes and examinations. When sensing, we want to use sensor networks and devices that contain sensors for measuring physiological functions, such as cardiac and skin functions. Students would have bracelets installed during testing to record their physiological functions and send information to the computer where they would then be evaluated.

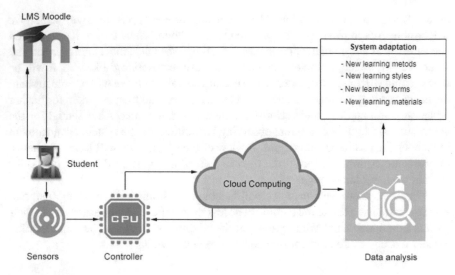

Fig. 1 Scheme of the adaptive learning system

Contrary to the above-mentioned researches by other authors, we would like to deploy wireless devices that would make students more comfortable. As a test result, we would like to propose an adaptive system that can adapt to students to achieve the best possible result during teaching.

We determined the research hypothesis.

H0: Students who showed signs of stress during lessons and examinations have lower results than those who did not show signs of stress.

The aim of the test will be to determine the presence of stress in the students based on changing their emotional state by measuring the physiological functions whose values will be different from the standard values. Based on the data obtained from the devices, it is an attempt to propose courses in the Moodle LMS so that students do not expose the signs of stress during the learning shown in Fig. 1. If there would be stress during learning, we want to focus on changing the style of teaching and also on methods and forms. We want to achieve that students can master the curriculum at a higher level of Bloom's taxonomy and that signs of stress would not be manifested even during examinations.

5 Conclusions

The aim of the chapter was to highlight the possibilities of sensory networks in the context measuring emotional states. Devices or things that are connected to the Internet create the Internet of things network. IoT also includes network that can be used to retrieve different data sensory from connected devices, that is directly sent over the network. Based on previously mentioned experiments through sensory networks, we will be able to capture and record students' emotional states that directly or indirectly related to their effectiveness in learning and acquiring knowledge. Experiments by these authors have shown that the presence of stress in humans has led to an increase in overall productivity, which we would like to test in future experiments on the students during the examinations. Using statistical analysis of the data, we will evaluate and subsequently propose new approaches to e-materials and determine the appropriate teaching style and form and also adapt the teaching material. The next step will be testing different sensors in order to get emotions and compare them with the available datasets.

Acknowledgements This research has been supported by University Grant Agency under the contract No. VII/6/2018

References

1. Abramson, L., Marom, I., Petranker, R., Aviezer, H.: Is fear in your head? A comparison of instructed and real-life expressions of emotion in the face and body. Emotion **17**(3), 557–565 (2017). https://doi.org/10.1037/emo0000252
2. Alberdi, A., Aztiria, A., Basarab, A.: Towards an automatic early stress recognition system for office environments based on multimodal measurements: a review. J. Biomed. Inform. **59**, 49–75 (2016). https://doi.org/10.1016/J.JBI.2015.11.007
3. Bahreini, K., Nadolski, R., Westera, W.: Towards multimodal emotion recognition in e-learning environments. Interact. Learn. Environ **24**(3), 590–605 (2016). https://doi.org/10.1080/10494820.2014.908927
4. Ben Henia Wiem, M., Lachiri, Z.: Emotion recognition system based on physiological signals with Raspberry Pi III implementation. In: 2017 3rd International Conference on Frontiers of Signal Processing (ICFSP), (pp. 20–24), IEEE (2017) https://doi.org/10.1109/ICFSP.2017.8097053
5. Carneiro, D., Castillo, J.C., Novais, P., Fernández-Caballero, A., Neves, J.: Multimodal behavioral analysis for non-invasive stress detection. Expert. Syst. Appl. **39**(18), 13376–13389 (2012). https://doi.org/10.1016/J.ESWA.2012.05.065
6. Caruso, D.: Emoční Inteligence. Grada Publishing, a. s, Praha (2015)
7. Cruz-Albarran, I.A., Benitez-Rangel, J.P., Osornio-Rios, R.A., Morales-Hernandez, L.A.: Human emotions detection based on a smart-thermal system of thermographic images. Infrared Phys. Technol. **81**, 250–261 (2017). https://doi.org/10.1016/J.INFRARED.2017.01.002
8. Czako, M., Seemannova, M., Bratska, M.: Emócie. Slovenské pedagogické nakladateľstvo, Bratislava (1982)
9. Ekman, P., Friesen, W.: Facial Action Coding System: Investigator's Guide. Consulting Psychologists Press, Palo Alto, CA (1978)
10. Gjoreski, M., Luštrek, M., Gams, M., Gjoreski, H.: Monitoring stress with a wrist device using context. J Biomed Inform. (2017)https://doi.org/10.1016/j.jbi.2017.08.006
11. Gravina, R., Alinia, P., Ghasemzadeh, H., Fortino, G.: Multi-sensor fusion in body sensor networks: state-of-the-art and research challenges. Inf. Fusion **35**, 68–80 (2017). https://doi.org/10.1016/J.INFFUS.2016.09.005
12. Hasson, G.: Inteligenční emoce. Praha:: Grada Publishing, a. s (2015)
13. Kaklauskas, A., Zavadskas, E. K., Seniut, M., Dzemyda, G., Stankevic, V., Simkevičius, C., Gribniak, V.: Web-based biometric computer mouse advisory system to analyze a user's emotions and work productivity. Eng. Appl. Artif Intell. **24**(6), 928–945 (2011) https://doi.org/10.1016/J.ENGAPPAI.2011.04.006

14. Lövheim, H.: A new three-dimensional model for emotions and monoamine neurotransmitters. Med. Hypotheses **78**(2), 341–348 (2012). https://doi.org/10.1016/j.mehy.2011.11.016
15. Magdin, M., Turcani, M., Hudec, L.: Evaluating the emotional state of a user using a webcam. Int. J. Interact. Multimedia. Artif. Intell. **4**(1), 61 (2016). https://doi.org/10.9781/ijimai.2016.4112
16. Mattsson, S., Partini, J., Fast-Berglund.: Evaluating four devices that present operator emotions in real-time. Procedia CIRP. **50,** 524–528 (2016) https://doi.org/10.1016/j.procir.2016.05.013
17. Mosciano, F., Mencattini, A., Ringeval, F., Schuller, B., Martinelli, E., Natale, C.Di.: An array of physical sensors and an adaptive regression strategy for emotion recognition in a noisy scenario. Sens. Actuators A **267**, 48–59 (2017). https://doi.org/10.1016/j.sna.2017.09.056
18. Otsuka, T., Ohya, J.: A study of transformation of facial expressions based on expression recognition from temporal image sequences. Inst Electron. Inf. Commun. Eng (IEICE), Technical report. (1997)
19. Pantic, M., Rothkrantz, L.J.: Automatic analysis of facial expressions: the state of art. IEEE. Trans. Pattern. Recogn. Mach. Intell. (2000)
20. Plutchik, R.: The nature of emotions: Clinical implications. In: Clynes, M., Panksepp, J. (eds.) Emotions and Psychopathology, (pp. 1–20). Boston: Springer. (1988) https://doi.org/10.1007/978-1-4757-1987-1
21. Rosenblum, M., Yacoob, Y., Davis, L.: Human expression recognition from motion using a radial basis function network architecture. IEEE Trans Neural Netw. (1996)
22. Shalini, T.B., Vanitha, L.: Emotion detection in human beings using ECG signals. Int. J. Eng. Trends. Technol. (IJETT) **4**(May), 1337–1342 (2013)
23. Sharma, N., Gedeon, T.: Modeling a stress signal. Appl. Soft Comput. **14**, 53–61 (2014). https://doi.org/10.1016/J.ASOC.2013.09.019
24. Tian, Y., Kanade, T., Cohn, J.: Recognizing Action units for facial expression analysis. IEEE Trans. Pattern. Recogn. Mach. Intell. Carnegie_Mellon University (2001)
25. Vizer, L.M., Zhou, L., Sears, A.: Automated stress detection using keystroke and linguistic features: an exploratory study. Int. J. Hum. Comput. Stud. **67**(10), 870–886 (2009). https://doi.org/10.1016/J.IJHCS.2009.07.005
26. Vo, M.L.-H., Jacobs, A.M., Kuchinke, L., Hofmann, M., Conrad, M., Schacht, A., Hutzler, F.: The coupling of emotion and cognition in the eye: introducing the pupil old/new effect. Psychophysiology, **0**(0), 071003012229007–??? (2007) https://doi.org/10.1111/j.1469-8986.2007.00606.x

Effects of a Spectral Window on Frequency Domain HRV Parameters

Jeom Keun Kim and Jae Mok Ahn

Abstract Heart rate variability (HRV) can provide physiological information about the autonomic nervous system. For the HRV spectral analysis, spectral windows are applied as weighting functions to solve the spectral leakage problem. The present study aimed to investigate HRV spectra with respect to spectral windows and to determine the frequency characteristics of an optimal window. Two short-term recordings, comprising twelve 5-minute and twelve 2-minute segments created from the one-hour HRV dataset, were analyzed. The HRV indices with respect to three different windows (Hanning, Hamming, and Blackman) were compared with reference to the Hanning window used widely. Hanning-Blackman window showed significant differences in VLF ($t = -18.341$ with $p < 0.0001$) for the 5-minute dataset. However, the Hanning-Hamming window showed no significant difference in all frequency bands for both 5-minute and 2-minute datasets. We suggest a low sidelobe related to amplitude accuracy and a narrow width of the main lobe as the frequency characteristics.

Keyword Heart rate · Spectrum · Analysis · Frequency · Autonomic nervous system

1 Introduction

Heart rate variability (HRV) is a popular non-invasive marker of both the autonomic nervous system and the physiological phenomenon. Several time and frequency domain HRV techniques have been developed to provide insight into cardiac

J. K. Kim · J. M. Ahn (✉)
Department of Electronics Engineering, College of Engineering, Hallym University, Gangwon-do, Chuncheon-si, Korea
e-mail: ajm@hallym.ac.kr

J. K. Kim
e-mail: jkim@hallym.ac.kr

© Springer Nature Singapore Pte Ltd. 2019
S. K. Bhatia et al. (eds.), *Advances in Computer Communication and Computational Sciences*, Advances in Intelligent Systems and Computing 924,
https://doi.org/10.1007/978-981-13-6861-5_59

697

autonomic regulation in both health and disease, but the exact parameters reflecting the physiological phenomena are controversial.

The strengths and limitations of HRV techniques and the interaction between prevailing heart rate and HRV have been evaluated in the clinic [1, 2]. Numerous studies have been published during the last three decades about frequency domain HRV parameters in many pathological conditions or different physiological conditions including the clinical applications [3–5], but there has been no comprehensive mathematical review for the power spectral density (PSD) analysis with respect to a spectral window. Basically, the PSD analysis is based on the assumption that heart period time series contains only linear and stationary dynamics. It requires resampling of the inherently unevenly sampled heartbeat time series to produce the evenly sampled ones and window function to avoid spectral leakage [6, 7]. It is well known that prior to the PSD estimation, the NN time series (normal-to-normal heart period) is interpolated and then resampled using a sampling frequency of 4 Hz [7, 8], but the effects of window function on HRV parameters are less understood.

In this paper, frequency domain HRV parameters based on discrete Fourier transform (DFT) analysis were investigated with respect to three spectral windows (Hanning, Hamming, and Blackman) to define the characteristics of an optimal window function for clinical use of HRV. The Hanning window has been recommended for HRV analysis, but the mathematical reason for including its frequency characteristics is less understood [7]. Different window functions can have their own distinct effects on evaluating the sympathetic–parasympathetic autonomic balance in terms of the low-frequency (LF, 0.04–0.12 Hz), high-frequency (HF, 0.12–0.4 Hz), and very low-frequency (VLF, 0.0033–0.04 Hz) bands. The frequency domain HRV parameters have been used to assess many cardiovascular and non-cardiovascular diseases such as myocardial infarction, diabetic neuropathy, high blood pressure, and sudden cardiac death [9–11]. In particular, clinical symptoms, such as exhaustion, nervousness, insomnia, and gastrointestinal problems, can occur in workers with frequent or intense subjective symptoms indicative of sympathovagal imbalance [5]. Clinical studies have demonstrated a relationship between HRV fluctuation and the progression of human coronary atherosclerosis as well as the patterns of association between HRV parameters and somatic symptom disorder [12–14]. As HRV methodology has become more significant, the importance of artifact correction, interpolation, normalization, fast Fourier transformation (FFT), spectral window, autoregression, and filtering has been emphasized [15, 16]. For example, the normalization of HRV to heart rate (HR) was presented to differentiate between physiologically and mathematically mediated changes in HRV [17]. Spectral analysis of HRV can be mathematically biased because the NN interval in milliseconds exhibits large changes for slow HR and small changes for fast HR, even for the same fluctuations [18].

Therefore, the effect of all methodologies on the frequency domain of HRV parameters must be thoroughly examined step by step. Among them, we report on the characteristics of a spectral window for the spectral analysis.

2 Methods

2.1 Data Collection

The one-hour HRV dataset recording using a professional pulse analyzer with a finger-type sensor of a photoplethysmogram, TAS9VIEW (or CANOPY9 RSA, IEM-BIO Co. Ltd., Chuncheon-si, South Korea), was used to measure the NN interval time series in milliseconds. Measurements were taken at a sampling rate of 1 kHz while the participant was in the supine position in a quiet room and was not allowed to talk or move. It was not important in this study to confirm the participant's health status regarding blood pressure, arrhythmia, and insomnia because the purpose was to define the optimal characteristics of the window function.

The simulated twelve 5-minute and twelve 2-minute datasets were created from the same one-hour HRV dataset. Frequency domain HRV indices were calculated for twenty-four short-term recordings in total using the research mode program of TAS9VIEW, which is a software package specifically designed for detailed HRV analysis: very low frequency (VLF), low-frequency (LF), and high-frequency (HF) bands.

2.2 Spectral Analysis

The PSD describes how the power of a signal or time series is distributed with frequency. It was calculated by means of discrete Fourier transform (DFT) with 1000 data points resampled by linear interpolation for each 5-minute or 2-minute HRV dataset. Linear interpolation was applied to resample heartbeat periods at frequencies between 0.5 and 3.3 Hz. The sampling frequency with 300 ms was determined to be sufficiently high such that the shortest NN interval in milliseconds for the fastest heart rate was not within the sampling frequency.

Each Hamming, Hanning, and Blackman window was employed in the spectral calculation. The PSD with respect to each window was calculated based on the DFT algorithm. DFT is the most basic transform of a discrete time domain signal. It is defined as follows:

$$X(\omega) = \sum_{0}^{N-1} x(n) e^{\frac{-2i\pi kn}{N}} \tag{1}$$

k as a positive integer represents a frequency for which the Fourier transform is being calculated, N is the number of resampled HRV datasets in the time domain, and n is a time domain sample. As shown in Eq. 1, each HRV signal is multiplied by a complex number. Equation 2 is the transform of a product in a general case of an arbitrary window, $w(n)$, applied to the time series HRV dataset to obtain the final absolute values of the HRV parameters,

$$F_W(\omega) = \sum_{0}^{N-1} w(n)x(n)e^{\frac{-2i\pi kn}{N}}$$ (2)

The following Eq. 3 is equivalent to the convolution of the two corresponding transforms to calculate PSD parameters.

$$F_W(\omega) = X(\omega) * W(\omega)$$ (3)

2.2.1 Hamming Window

The Hamming window can be thought of as a modified Hanning window as described in the following section. The coefficients of the Hamming window are considered to achieve minimum sidelobe levels, -43 dB. A generalized Hamming window with one period of a raised cosine can have an s step discontinuity at the endpoint, but no impulsive points. The Hamming window in the time and frequency domains is shown in Fig. 1 (top). Its sidelobe roll-off is asymptotically -6 dB per octave with equal ripples, and the first sidelobe is decreased by 41 dB. Equation 4 gives the mathematical expression with two degrees of freedom (0.54, -0.46).

$$W(n) = 0.54 - 0.46\cos\left(\frac{2\pi n}{N-1}\right), \quad 0 \le n \le N-1$$ (4)

2.2.2 Hanning Window

The sampled Hanning window can be defined as the sum of the sequences indicated in Eq. 5.

$$W(n) = 0.5 - 0.5\cos\left(\frac{2\pi n}{N-1}\right), \quad 0 \le n \le N-1$$ (5)

The form of the Hanning window in the time and frequency domains is shown in Fig. 1 (middle). The characteristics of the Hanning window include its sharp main lobe but lower sidelobes. Unlike the Hamming window, the Hanning window reaches zero at its endpoints with zero slope. As a result, the sidelobes roll off at approximately -18 dB per octave. In addition to the greatly accelerated roll-off rate, the first sidelobe drops from -13 to -31.5 dB. The Hanning window has been mostly used in frequency domain HRV analysis because it is the variant that places the zero endpoints one-half sample to the left and right of the outmost window samples.

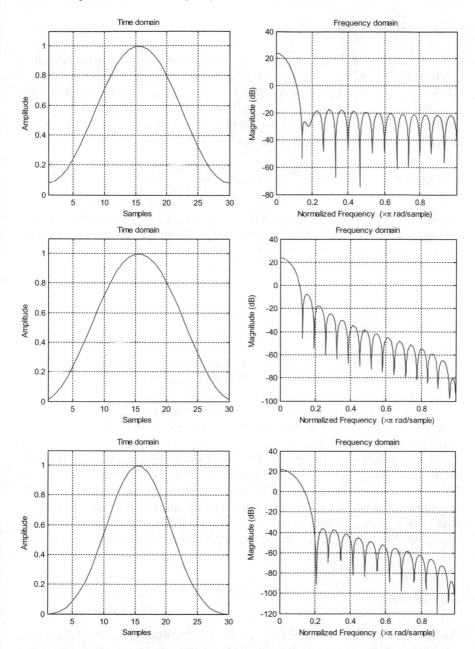

Fig. 1 Time (Left) and frequency (right) characteristics of different smoothing windows: (top) Hamming window, (middle) Hanning window, and (bottom) Blackman window

2.2.3   Blackman Window

Relative to the generalized Hamming window, an additional cosine weighted by 0.08 is added in the Blackman window. It has three degrees of freedom (0.42, −0.5, and 0.08) instead of two degrees of freedom in the Hamming and Hanning windows. One degree of freedom is used to normalize the window amplitude, and then two remaining degrees of freedom are used to maximize the roll-off rate or the sidelobe rejection. The sampled Blackman window can be defined by Eq. 6 as follows.

$$W(n) = 0.42 - 0.5\cos\left(\frac{2\pi n}{N-1}\right) + 0.08\,\cos\left(\frac{4\pi n}{N-1}\right), \quad 0 \le n \le N - 1 \quad (6)$$

Figure 1 (bottom) shows the frequency response and time series of the Blackman window. Its sidelobes roll off at approximately −18 dB per octave, similar to the Hanning window, and its sidelobe level is approximately −58 dB per octave even in the worst case.

2.3   Processing Scheme

The DFT algorithm for spectral information of HRV signals was embedded in the TAS9VIEW pulse analyzer developed with C# development tool (Microsoft Visual Studio Community 2017). In its research mode, all HRV parameters were analyzed by moving 5-minute and 2-minute HRV datasets in time by a specific time interval (5 min) for the same one-hour HRV tachogram. The processing scheme for the DFT analysis is displayed in Fig. 2.

The reason that we used two short-term segments of HRV data was to confirm the periodic continuity because a finite length signal is rarely periodic at its boundaries. Leakage to nearby frequencies due to the finite interval over which the data are sampled takes the form of sidelobes in the periodogram. Two types of short-term data segments were used to investigate how this spectral leakage could be minimized through the same window function according to different individual data lengths.

2.4   Statistics

We used the MedCalc statistical program (MedCalc software, Ostend, Belgium) to conduct the independent t-test statistic. It was run on the data with a 95% confidence interval (CI) for the mean difference. It tests the null hypothesis that the difference between the means of two samples is equal to 0. Bland-Altman plot was used to compare two windows in terms of each HRV index graphically by calculating 95% CI of mean of difference.

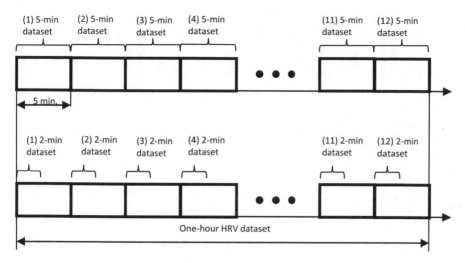

Fig. 2 Processing scheme for DFT analysis: twelve 5-minute segments from the one-hour HRV dataset

3 Results

3.1 The Power of the 5-Minute HRV Dataset

Three spectral parameters (VLF, LF, and HF) with respect to each window were obtained from a spectrum calculated from the 5-minute HRV dataset in Table 1 (top). The statistical results of PSD for each pair of windows (Hanning-Hamming and Hanning-Blackman) are listed in Table 2. It shows that Hanning-Hamming has no significant difference for VLF, LF, and HF ($t = 0.086$ with $p = 0.9321$, $t = -0.473$ with $p = 0.6410$, and $t = -0.813$ with $p = 0.4249$, respectively), but Hanning-Blackman has significant difference for VLF ($t = 18.341$ with $p < 0.0001$). Although the sidelobe roll-off of the Hamming window is three times as high as that of the Hanning window, they showed similar results.

Figure 3 indicates that the difference between the reference window (Hanning) and each Hamming and Blackman window is expressed as percentages of the values on the axis. Increase in the width of mean ± 1.96 SD for the VLF power in Hanning-Blackman was observed with a range of 234.3 to -3054.3, but the difference within mean ± 1.96 SD was not significant in Hanning-Hamming, ranging from 5.6 to -9.1. The Hanning window recommended by the Task Force of The European Society of Cardiology and the North American Society of Pacing and Electrophysiology [7] fell between the Hamming and Blackman windows. It means that minimum sidelobe level should be between -43 dB of the Hamming window and -58 dB of the Blackman window to obtain the frequency characteristics of an optimal window.

Table 1 Absolute values (ms^2) of all HRV parameters for each spectral window: (top) 5-minute HRV dataset and (bottom) 2-minute HRV dataset

No.	VLF			LF			HF		
	Hamming	Hanning	Blackman	Hamming	Hanning	Blackman	Hamming	Hanning	Blackman
1	635.6	655.5	8105.0	535.7	508.9	428.5	336.6	310.7	256.3
2	233.6	234.9	6780.4	340.2	309.1	253.9	537.8	498.0	407.4
3	1246.5	1173.6	5749.2	639.7	604.4	513.5	374.0	336.7	268.6
4	1857.5	1761.9	10,017.4	358.9	320.8	247.2	470.4	431.1	367.5
5	534.7	536.9	7526.4	492.3	455.3	374.5	632.5	580.7	472.4
6	301.5	293.8	6513.6	223.3	196.7	160.9	581.1	527.5	427.1
7	299.4	303.9	6839.3	269.8	243.7	187.9	464.7	422.8	351.3
8	292.9	297.4	7532.0	218.1	191.7	155.3	345.8	310.4	261.7
9	361.3	367.9	7069.4	164.3	143.8	115.1	394.8	361.1	301.7
10	2053.9	2014.7	8260.5	404.9	377.1	330.4	329.0	295.5	259.8
11	714.2	679.5	6697.1	222.0	197.1	148.8	308.0	275.3	227.5
12	612.4	565.9	6897.2	260.3	243.2	199.1	257.5	228.7	185.8
Mean	762.0	740.5	7332.3	344.1	316.0	259.6	419.4	381.6	315.6
SD	623.9	596.8	1092.6	147.8	143.8	125.8	117.5	110.1	88.4

(continued)

Table 1 (continued)

No.	VLF			LF			HF		
	Hamming	Hamming	Blackman	Hamming	Hamming	Blackman	Hamming	Hamming	Blackman
1	451.3	461.8	7152.2	412.7	337.3	246.0	117.2	87.7	67.0
2	168.4	205.9	8073.1	436.3	374.2	266.5	293.7	234.9	174
3	889.9	867.0	10,867.2	606.6	536.4	427.0	355.6	299.5	250.6
4	467.0	481.9	7703.6	415.4	331.4	256.0	277.2	227.1	172.1
5	424.3	417.1	7019.8	675.5	630.7	546.0	323.1	267.1	208.9
6	606.6	582.6	9774.2	1057.5	939.0	766.0	721.9	656.8	538.4
7	468.2	476.7	8005.7	239.7	184.2	151.4	183.6	134.8	107.4
8	512.2	536.4	8264.1	324.1	259.1	195.5	211.7	170.5	134.9
9	206.2	207.6	9361.2	176.8	132.9	95.2	194.8	145.9	116.7
10	281.3	319.7	10,671.0	479.7	382.1	246.6	223.0	178.0	148.0
11	1212.5	1215.3	12,553.8	319.5	253.9	206.6	228.8	185.3	147.7
12	876.4	812.6	4941.3	238.4	174.8	134.4	149.5	109.6	88.7
Mean	547.0	548.7	8698.9	448.5	377.9	294.8	273.3	224.8	179.6
SD	307.6	291.9	2054.5	242.1	228.9	193.9	157.6	149.9	123.9

Table 2 Statistical results of paired windows for the PSD of the 5-minute HRV dataset

	Windows	t-statistic	Degrees of freedom	p-value
VLF	Hanning-Hamming	0.086	22	0.9321
	Hanning-Blackman	18.341	22	<0.0001
LF	Hanning-Hamming	0.473	22	0.6410
	Hanning-Blackman	−1.022	22	0.3177
HF	Hanning-Hamming	0.813	22	0.4249
	Hanning-Blackman	−1.618	22	0.1198

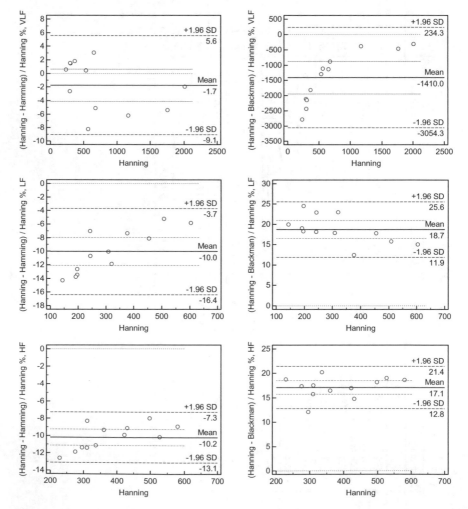

Fig. 3 5-minute HRV parameters (ms^2 on the x-axis) in the frequency domain calculated as the difference between the Hanning window: (left) Hamming and (right) Blackman. Red dotted horizontal line is the reference line of no difference ($y = 0$)

Table 3 Statistical results of paired windows for the PSD of the 2-minute HRV dataset

	Windows	t-statistic	Degrees of freedom	p-value
VLF	Hanning-Hamming	−0.014	22	0.9892
	Hanning-Blackman	13.606	22	<0.0001
LF	Hanning-Hamming	0.734	22	0.4707
	Hanning-Blackman	−0.961	22	0.3472
HF	Hanning-Hamming	0.773	22	0.4475
	Hanning-Blackman	−0.805	22	0.4294

3.2 The Power of the 2-Minute HRV Dataset

The VLF, LF, and HF powers with respect to each window were obtained from the 2-minute HRV dataset as shown in Table 1 (bottom). Their statistical results for each pair of windows (Hanning-Hamming and Hanning-Blackman) are listed in Table 3. For the HF power, the absolute value of the mean was lower in the 2-minute HRV dataset than in the 5-minute HRV dataset for the Hamming (273.31, 419.35), Hanning (224.31, 381.55), and Blackman (179.55, 315.60, respectively) windows. The standard deviations (SD) were higher in the 2-minute HRV dataset than in the 5-minute HRV dataset except for the VLF power.

Table 3 shows that there is no significant difference between the Hanning and Hamming windows ($t = -0.014$ with $p = 0.9892$, $t = 0.734$ with $p = 0.4707$, and $t = 0.773$ with $p = 0.4475$ for VLF, LF, and HF, respectively). However, for the Hanning-Blackman window, it was found that the significance level of VLF was $p < 0.0001$, showing that they would be highly different. Figure 4 plots a relationship between the spectral differences and the averages with respect to three window functions to confirm how they differ from one another as shown in Fig. 3. The LF and HF powers in the Hamming window are slightly higher than those in both the Hanning and Blackman windows, but the VLF power is the highest in the Blackman window.

4 Discussion

In this paper, we performed a PSD analysis for two types of short-term HRV signals (twelve 2-minute and twelve 5-minute datasets) in the frequency domain. The PSD for these HRV datasets was estimated based on the DFT with 1000 evenly spaced resampled data points by linear interpolation. Hamming, Hanning, and Blackman windows were independently implemented for estimation of the HRV spectra to investigate how each window can influence the results of HRV analysis with reference to the Hanning window. Results obtained from the Blackman window differed

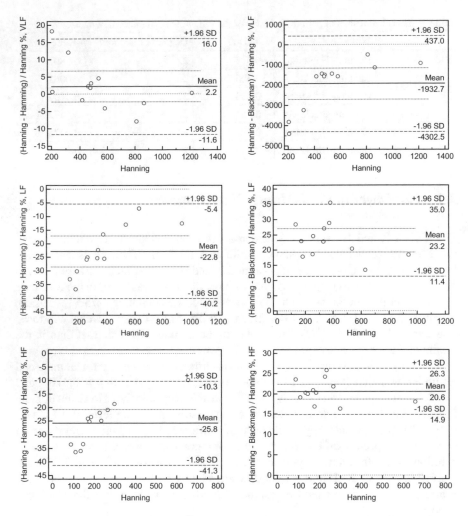

Fig. 4 2-minute HRV parameters (ms² on the x-axis) in the frequency domain calculated as the difference between the Hanning window: (left) Hamming and (right) Blackman. Red dotted horizontal line is the reference line of no difference ($y = 0$)

relatively in VLF, LF, and HF powers with respect to the Hanning window for both the 2-minute and 5-minute HRV datasets.

As a result, the Blackman window is not recommended for clinical use in HRV analysis because the VLF power, which is associated with increased risk of arrhythmic death [19], could show such a large difference. It has very low sidelobe roll-off and sidelobe level in the frequency characteristics compared to the Hanning window. The HRV indices of the Hanning and Hamming windows were mostly similar, but the slight difference between the two windows in the HRV parameters was due to spectral leakage caused by the corresponding window function, as shown in Fig. 1.

Regarding the mathematical characteristics of each window, in the time domain, the Hamming window does not get as close to zero near the edges as the Hanning window. Nearing zero at the boundaries makes the Hanning window to reduce spectral leakage much more than the Hamming window. The Hanning window, in the frequency domain, has a lower sidelobe roll-off rate representing the asymptotic decay rate than the Hamming window, 60 and 20 dB/dec., respectively, as shown in Fig. 1. The widths of the main lobe at −3 dB below the main lobe peak are 1.44 (bins) and 1.30 (bins) for the Hanning and Hamming windows, respectively.

Therefore, for an optimal window function in calculating HRV spectra, the first criterion must be a window that is close to zero at the boundaries, resulting in the continuous periodic extension of the HRV dataset, reflecting a smoothing window to reduce spectral leakage. The second criterion is that the width of the main lobe must be narrow to distinguish two closely spaced frequency components. However, an extremely narrow width of the main lobe leads to the problems of increasing spectral leakage and decreasing amplitude accuracy. In conclusion, an optimal window should be selected considering the various trade-offs between amplitude and spectral resolution from a point of view of both the main lobe and the sidelobes. The basic recommended characteristics of an optimal window function are low sidelobes, which cause cross talk in the estimated spectrum from one frequency to another, and a narrow width of the main lobe to simultaneously improve spectral resolution.

Statement on Ethical Approval: Our study aimed at investigating how different HRV parameters are based on three different spectral windows. Single HRV dataset was required to define the optimal characteristics of the window function. HRV dataset was recorded by the investigator without identifiers. It was not important in this study to confirm the participant's health status and personal medical information as mentioned in Methods (section of Data Collection). I think that this study involves Category 4 in OHRP Exempt Categories 45 CFR 46.101(B) of IRB Institutional Review Board.

Acknowledgements This research was supported by Hallym University Research Fund, 2018 (HRF-201,801-011).

References

1. Billman, G.E., Huikuni, H.V., Sacha, J., Trimmel, K.: An introduction to heart rate variability: methodological considerations and clinical applications. Front. Physiol. **6**(55), 1–3 (2015)
2. Billman, G.E.: The LF/HF ratio does not accurately measure cardiac sympatho-vagal balance. Front. Physiol. **4**(26), 1–5 (2013)
3. Zara, A., Lombardi, F.: Autonomic indexes based on the analysis of heart rate variability: a view from the sinus node. Cardiovasc. Res. **50**(3), 434–442 (2001)
4. Elghozi, J.L., Julien, C.: Sympathetic control of short-term heart rate variability and its pharmacological modulation. Fundam. Clin. Pharmacol. **21**, 337–347 (2007)
5. Benichou, T., Pereira, B., Mermillod, M., Tauveron, I., Pfabigan, D., Magdasy, S., Dutheil, F.: Heart rate variability in type 2 diabetes mellitus: a systematic review and meta-analysis. PLoS One **13**(4), 1–19 (2018)

6. Bartels, R., Neumamm, L., Pecanha, T., Carvalho, S.: SinusCor: an advanced tool for heart rate variability analysis. BioMed. Eng. OnLine. **16**(1), 110–124 (2017)
7. Task Force of The European Society of Cardiology and The North American Society of Pacing and Electrophysiology (Membership of the Task Force listed in the Appendix).: Heart rate variability: standards of measurement, physiological interpretation, and clinical use. Circulation. **93**(5), 1043–1065 (1996)
8. Keselbrener, L., Akselrod, S.: Selective discrete fourier transform algorithm for time-frequency analysis: method and application on simulated and cardiovascular signals. IEEE Trans. Biomed. Eng. **43**(8), 789–802 (1996)
9. Heathers, J.A.: Everything Hertz: methodological issues in short-term frequency-domain HRV. Front Physiol. **5**, 177–200 (2014)
10. Maheshwari, A., Norby, F.L., Soliman, E.Z., Adabag, S., Whitsel, E.A., Alonso, A., Chen, L.Y.: Low heart rate variability in a 2-minute electrocardiogram recording is associated with an increased risk of sudden cardiac death in the general population: the atherosclerosis risk in communities study. PLoS One **11**(8), 1–12 (2016)
11. Malliani, A., Pagani, M., Lombardi, F., Cerutti, S.: Cardiovascular neural regulation explored in the frequency domain. Circulation **84**, 482–492 (1991)
12. Karita, K., Nakao, M., Nishikitani, M., Nomura, K., Yano, E.: Autonomic nervous activity changes in relation to the reporting of subjective symptoms among male workers in an information service company. Int. Arch. Occup. Environ. Health **79**(5), 441–444 (2006)
13. Huikuri, H.V., Jokinen, V., Syvanne, M., Nieminen, M.S., Juhani Airaksien, K.E., Ikaheimo, M.J., Koistinen, J.M., Kauma, H., Kesaniemi, A.Y., Majahalme, S., Niemela, K.O., Heikki Frick, M., the Lopid Coronary Angioplasty (LOCAT) study Group.: Heart rate variability and progression of coronary atherosclerosis. Arterioscler. Thromb. Vasc. Biol. **19**, 1979–1985 (1999)
14. Huang, W.L., Liao, S.C., Yang, C.C., Kuo, T.B., Chen, T.T., Chen, I.M., Gau, S.S.: Measures of heart rate variability in individuals with somatic symptom disorder. Psychosom. Med. **79**(1), 34–42 (2017)
15. Kuss, O., Schumann, B., Kluttiq, A., Greiser, K.H., Haerting, J.: Time domain parameters can be estimated with less statistical error than frequency domain parameters in the analysis of heart rate variability. J. Electrocardiol. **41**(4), 287–291 (2008)
16. Chemla, D., Young, J., Badilini, F., Maison-Blanche, P., Affres, H., Lecarpentier, Y., Chanson, P.: Comparison of fat Fourier transform and autoregressive spectral analysis for the study of heart rate variability in diabetic patients. Int. J. Cardiol. **104**(3), 307–313 (2005)
17. Sacha, J., Barabach, S., Statkiewicz-Barabach, G., Sacha, K., Muller, A., Piskorski, J.: How to strengthen or weaken the HRV dependence on heart rate-description of the method and its perspectives. Int. J. Cardiol. **168**, 1660–1663 (2013)
18. Sacha, J., Pluta, W.: Alterations of an average heart rate change heart rate variability due to mathematical reasons. Int. J. Cardiol. **128**, 444–447 (2013)
19. Biqqer Jr., J.T., Fleiss, J.L., Steinman, R.C., Rolnitzky, L.M., Kleiqer, R.E., Rothman, J.N.: Frequency domain measures of heart period variability and mortality after myocardial infarction. Circulation **85**(1), 164–171 (1992)

Effects of Heart Rate on Results of HRV Analysis

Jae Mok Ahn and Jeom Keun Kim

Abstract Heart rate variability (HRV) expressing variations in the heart period (NN interval) has been widely analyzed to quantify autonomic regulation. However, mathematically biased HRV results are obtained for different average heart rates (HRs) due to the nonlinear relationship between NN interval and HR. The present study aimed to investigate HRV parameters with respect to different means HRs and to evaluate the influence of heart period on HRV. Six simulated 5 min segments from a 10 min HRV dataset were prepared from slow HR to fast HR. The HRV parameters for the six groups were compared to investigate their differences with respect to the HRV segment (average HR of 60 bpm). There was no significant difference in only SDSD, RMSSD, SD1, and pNN50 in terms of relative SD. It was concluded that heart period affected the HRV results of not only the spectrum analysis but also the parameters in the time and geometrical domains.

Keywords Heart rate variability · NN interval · Photoplethysmography · Frequency domain

1 Introduction

Heart rate variability (HRV), which represents the oscillation in an instantaneous heart period over time, is a widely used as noninvasive indicator of autonomic cardiac regulation. HRV techniques have been used in a broad range of clinical applications to predict mortality after a myocardial infarction, sympathetic–parasympathetic autonomic balance and cardiovascular health, the physiological process of sleep, stress and psychiatric condition, and the development of congestive heart failure [1–5].

J. M. Ahn (✉) · J. K. Kim
Department of Electronics Engineering, College of Engineering, Hallym University,
Chuncheon-si, Gangwon-do, South Korea
e-mail: ajm@hallym.ac.kr

J. K. Kim
e-mail: jkim@hallym.ac.kr

© Springer Nature Singapore Pte Ltd. 2019　　　　　　　　　　　711
S. K. Bhatia et al. (eds.), *Advances in Computer Communication
and Computational Sciences*, Advances in Intelligent Systems and Computing 924,
https://doi.org/10.1007/978-981-13-6861-5_60

However, HRV is influenced by the heart period since the normal-to-normal (NN) interval time series and HR are inversely related, and this association is mathematically biased [6]. Therefore, the curvilinear relationship between the NN interval and heartbeat creates a mathematical difference in the HRV parameters of patients with different average HRs. Due to this mathematical problem, fluctuations in the NN interval time series depend on average HR, resulting in deceleration of average HR that increases the oscillations of NNs. Some researchers have shown that the HRV dependence on HR dominates the parasympathetic nervous branch when a patient's slow HR is generated [2, 7]. Several studies have demonstrated an inverse correlation between heartbeat in bpm and time-domain HRV parameters, especially the standard deviation of normal-to-normal (SDNN) beats when the magnitude of HRV increases with slow HR [8, 9].

Frequency-domain HRV parameters are influenced by means of HR as well. It has been demonstrated that the high-frequency (HF) energy of HRV increases with decreases in HR, while the low-frequency (LF) energy of HRV increases with the prevailing HR, as shown in other research [10]. To solve the problem of mathematical bias in the standard HRV analysis, several studies have calculated the variability in NN intervals corrected for the average NN interval [7] and have tried to normalize HRV results for the prevailing HR as HR changes in response to autonomic neural activation or inhibition [11]. HRV techniques have not provided ways to remove both the physiological difference and the mathematical bias in HRV between heart rhythms in patients with different average HRs. That is why it has not been widely appreciated that the nonlinear relationship between HR in bpm and NN interval in millisecond could accept HRV results enough to apply for clinical use.

Therefore, since the HR's commitment to the clinical meaning of HRV was different due to the curvilinear relationship between NN interval and HR, the aim of this paper was to investigate HRV indices in both frequency and time domains, which obtained from six different average HRs and to evaluate the influence of HR on HRV. Thus, the HRV parameters that were less influenced by HR were determined.

2   Methods

2.1   HRV Dataset

The 10 min HRV dataset (Origin) was obtained from finger-type photoplethysmography (PPG) recordings using a commercial pulse analyzer (TAS9VIEW or CANOPY9 RSA, IEMBIO Co. Ltd., Chuncheon-si, Korea) with a time interval of two successive peak waves in milliseconds (ms).

Fig. 1 Curvilinear
relationship between NN
interval and HR

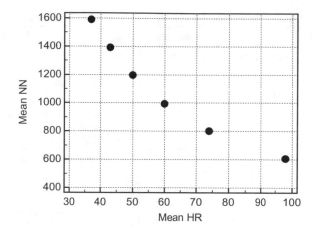

Six simulated heartbeat intervals (Origin—400 ms, Origin—200 ms, Origin, Origin + 200 ms, Origin + 400, Origin + 600 ms) were created for mean heartbeats ranging from 606 to 1588 ms (98–37 bpm) for the 5 min segment in Fig. 1. The standard deviation (SD) and SDNN in each group were ±20.65 beats/min at each mean HR level and 1.81 ms, respectively.

HRV parameters were obtained in the time, frequency, and geometric domains using the CANOPY9 RSA monitor program. The following indices of HRV were described including comment and unit as shown in Table 1.

2.2 Time Domain

HRV measurements in the time domain calculated from the NN intervals between normal heartbeats are as follows. Mean NN represents the average beat interval in milliseconds for all NN intervals, which is equivalent to the heart rate of NN interval. The SDNN in ms reflects the long-term HRV for periods of interest longer than the 5 min dataset.

The root mean square of successive differences (RMSSD) of the NN intervals in ms reflects the parasympathetic activity to control heart rates but is exaggerated by abnormal and irregular heart rhythm. pNN50 is the number of pairs of successive NN intervals with difference of more than 50 ms in the entire recording divided by the total number of all NN intervals; it reflects the parasympathetic activity decreasing heart rate and indicates that the rhythm is normal [12].

Table 1 Descriptions and comments for all HRV parameters

	Parameters	Description	Comment	Unit
Time domain	SDNN	Standard deviation of all interbeat normal-to-normal (NN) intervals	Long-term HRV	ms
	RMSSD	Root mean square of successive NN interval differences	Short-term HRV	ms
	Mean NN	Average of total NN intervals	Long-term HRV	ms
	Mean HR	Average of total heart rates		bpm
	Beat count	Total number of all NN intervals		
	Mode HR	Heartbeat in bpm that appears most often		bpm
	Mode NN	Heart period in ms that appears most often in an 8 ms time interval		ms
	Max. HR	The highest heart rate		bpm
	Min. HR	The lowest heart rate		bpm
	Max_Min HR	Difference between the highest and lowest heart rates		bpm
	Mode count	Number of heart periods in ms that appears most often		
	HRV index	Integral of the density of the NN interval histogram divided by its height	Long-term HRV	
	pNN50	Numbers of pairs of adjacent NN intervals differing by more than 50 divided by beat count	Short-term HRV	%
	CVAA	Coefficient of variation of the NN interval	Short-term HRV	%
	SDSD	Standard deviation of differences between adjacent NN intervals	Short-term HRV	ms

(continued)

Table 1 (continued)

	Parameters	Description	Comment	Unit
Frequency domain	Ln TP	The natural logarithms of the variance of all NN intervals	Long-term HRV	ms^2
	Ln VLF	The natural logarithm of the very low frequency range (0.0033–0.04 Hz)	Parasympathetic activity Long-term HRV	ms^2
	Ln LF	The natural logarithm of the low frequency range (0.04–0.15 Hz)	Baroceptor reflexes Long-term HRV	ms^2
	Ln HF	The natural logarithm of the high frequency range (0.15–0.4 Hz)	Respiratory sinus arrhythmia, parasympathetic activity Short-term HRV	ms^2
	Ln LF/Ln HF	Ln LF/Ln HF ratio	Short-term HRV	
	LF norm	LF power in normalized units (LF/(TP-VLF)*100)	Long-term HRV	
	HF norm	HF power in normalized units (HF/(TP-VLF)*100)	Short-term HRV	
Geometric domain	SD1	Standard deviation of the width of the Poincare plot	Short-term HRV	ms
	SD2	Standard deviation of the length of the Poincare plot	Long-term HRV	ms
	SD2/SD1	SD2/SD1 ratio		
	sArea	Area of the ellipse	Long-term HRV	ms^2
	TINN	Triangular interpolation of RR interval histogram	Long-term HRV	ms

The HRV index is calculated from the density distribution of the number of NN intervals in the individual bin, similar to a discrete scale. Its triangular index is obtained by dividing the total number of NN intervals by the number of modal bin that NN interval appears most often. Short-term HRV is used as a meaningful index in clinical applications to indicate heart periods shorter than the 5 min dataset. Fifteen indices are introduced as time-domain parameters.

2.3 Frequency Domain

Nonparametric spectrum estimation involves the computation of discrete Fourier transform (DFT). Information about spectral parameters can be obtained by decomposing the spectrum of the HRV dataset into quantified frequency components. The intensity of the signal in a given frequency band can be calculated by integrating the frequencies of the band. High-frequency spectrum refers to HF power in the 5 min segment in the range from 0.15 to 0.4 Hz in the HRV power spectrum.

Low-frequency spectrum refers to LF power in the 5 min segment in the range from 0.04 to 0.15 Hz. LF and HF frequency bands are widely used to quantify sympathetic and parasympathetic regulation, respectively, but are not clearly described to mirror sympatho/vagal balance [13].

The very low-frequency spectrum refers to the VLF power in the power spectrum range between 0.0033 and 0.04 Hz. The physiological influence of the VLF index on health is much less understood when interpreting the power spectral density (PSD) of the short-term HRV dataset.

2.4 Geometric Domain

NN interval time series can also be transformed into nonlinear patterns including the density distribution of NN interval durations that of differences between successive NN intervals, and the Poincare plot of NN intervals. The width of the distribution histogram for HRV dataset has been obtained with bins approximately 8 ms long.

The triangular interpolation of the NN interval distribution (TINN) is the baseline width on x-axis of the NN interval distribution and represents the total power of HRV such as variance in time domain as the HRV index does. In the Poincare HRV graph, each NN interval, NN (n), is plotted against the next NN interval, NN $(n+1)$. Many researchers have provided clinical information on how aging and illness might affect the geometry of the Poincare HRV plot [14, 15].

The ellipse fitting method was used to calculate the SD length of points perpendicular to the axis of line-of-identity ($y = -x$) and the SD length of points along the axis of line-of-identity ($y = x$), named SD1 and SD2 in ms, respectively . SD2 corresponds to the ULF (ultralow-frequency band) in the 24-h recording, and SD1 corresponds to the SD of differences between successive NN intervals (SDSD) reflecting the parasympathetic activity.

2.5 Statistical Methods

We used simple statistics to test whether the average and relative SD of each group differ significantly. The mean, SD, and its relative SD were calculated with $n = 6$. The box-and-whisker plot displays a graphical statistical summary to confirm the similarity between meaningful indices and different average HRs. The central box represents the values from the lower to upper quartiles (25–75th percentile). The middle line represents the median. The vertical line extends from the minimum to the maximum value, removing outside values which are displayed as separate points.

3 Results

We used six 5 min HRV segments in the NN interval time series to evaluate how different they are from the averages HRs. The nonlinear relationship between NN intervals and HR can increase or decrease the values of HRV indices in terms of the time, frequency, and geometric domains due to the mathematical rule when average HR is changed.

Table 2 shows the influence of this nonlinear relationship on the results of the HRV analysis. The means, SD, and relative SD with percentage were calculated over a range of 37–98 bpm on average. The values of CVAA, SD2, sArea, and SDNN for slow HR were attenuated, while those for fast HR were boosted; consequently, HRV lost its dependence on HR, as shown in Fig. 2. However, HR has little influence on HRV parameters of SDSD, RMSSD, SD1, and pNN50, as shown in Fig. 3.

These parameters were found to be not significantly different over the range of the six different heart rates, providing a relative SD of less than 2.00%. In particular, SDSD had the lowest difference with a range of 54.59–56.17, a mean (SD) of 55.3817 (0.5589), and a relative SD (percentage) of 0.01009 (1.01%). Figure 4 shows that the total powers (TPs) of HRV spectra were slightly independent on HR with a relative SD (percentage) of 0.01809 (1.81%); however, VLFs, LFs, and HFs were dependent on HR with a relative SD of less than 10%.

This dependence on HR increased from fast HR to slow HR, indicating that the HRV parameters are not easily applied for clinical use without understanding of cycle dependence. In the geometric domain, all indices except for SD1 showed dependence on HR.

Table 2 Descriptions and comments for all HRV parameters

	Parameters	Mean (SD)	Range	Relative SD (%)
Time domain	SDNN	41.8183 (1.8074)	39.52–44.84	0.04322 (4.32)
	RMSSD	55.4833 (0.5813)	54.68–56.32	0.01048 (1.05)
	Mean NN	1096.1667 (368.0241)	606–1588	0.3357 (33.57)
	Mean HR	60.3333 (22.6156)	37–98	0.3748 (37.48)
	Beat count	304.6667 (114.2815)	189–495	0.3751 (37.51)
	Mode HR	61.3333 (20.4418)	39–95	0.3333 (33.33)
	Mode NN	1061 (331.8650)	629–1527	0.3128 (31.28)
	Max. HR	69.8333 (31.3523)	40–124	0.4490 (44.90)
	Min. HR	54.1667 (18.2474)	35–84	0.3369 (33.69)
	Max_Min HR	15.6667 (13.2615)	5–40	0.8465 (84.65)
	Mode count	28.8333 (7.5741)	19–38	0.2627 (26.27)
	HRV index	10.3767 (1.3490)	9.30–13.03	0.1300 (13.00)
	pNN50	39.4867 (0.4474)	38.93–40.07	0.01133 (1.13)
	CVAA	4.2983 (1.8156)	2.49–7.39	0.4224 (42.24)
	SDSD	55.3817 (0.5589)	54.59–56.17	0.01009 (1.01)
Frequency domain	Ln TP	6.4733 (0.1171)	6.25–6.59	0.01809 (1.81)
	Ln VLF	5.7250 (0.1820)	5.4–5.94	0.03178 (3.18)
	Ln LF	4.6867 (0.4291)	4.23–5.30	0.09155 (9.16)
	Ln HF	5.3583 (0.3936)	4.60–5.63	0.07346 (7.35)
	Ln LF/Ln HF	0.8833 (0.1507)	0.75–1.15	0.1706 (17.06)
	LF norm	35.0433 (18.0916)	19.72–66.83	0.5163 (51.63)
	HF norm	64.9567 (18.0916)	33.17–80.28	0.2785 (27.85)
Geometric domain	SD1	39.3050 (0.4273)	38.73–39.93	0.01087 (1.09)
	SD2	44.1583 (3.5280)	39.20–49.64	0.07989 (7.99)
	SD2/SD1	1.1233 (0.09564)	0.98–1.26	0.08514 (8.51)
	sArea	5450.2517 (410.3470)	4917.13–6126.96	0.07529 (7.53)
	TINN	202.5000 (10.5404)	188–220	0.05205 (5.21)

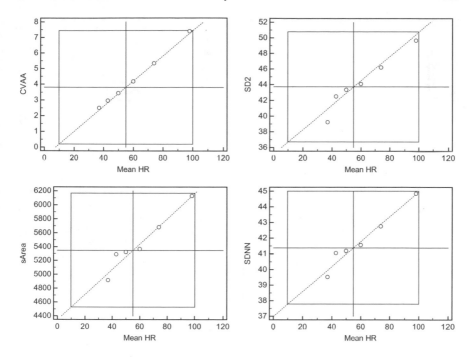

Fig. 2 Dependence of HRV on HR for CVAA, SD2, sArea, and SDNN

4 Discussion

We examined the effects of both slow and fast HR on HRV parameters. The aim of our study was to determine to which extent mathematical rules influence the results of HRV analysis. The HRV measures are more variable in the time domain than in either the frequency or geometric domain. The ratio of Ln LF to Ln HF representing the dominance of sympatho/vagal activity used often should be carefully taken for clinical use because of high relative SD of 17.06%.

The average fast HR is more variable than the slow average HR in the frequency and geometric domains. This result means that the mathematical rules may potentially distort the outcomes of HRV analysis, leading to the false clinical evaluation of the physiological difference when HRV is calculated from instantaneous heart rates. Some research has demonstrated that low HRV in a 2 min HRV segment is associated with an increased risk of sudden cardiac death [16]. HRV analyzed from the NN interval segment with unusually slow or fast HR without consideration of cycle dependence may provide the wrong results for a given patient. If HRV is highly dependent on HR, even slight differences in frequency-domain indices may lead to false outcomes.

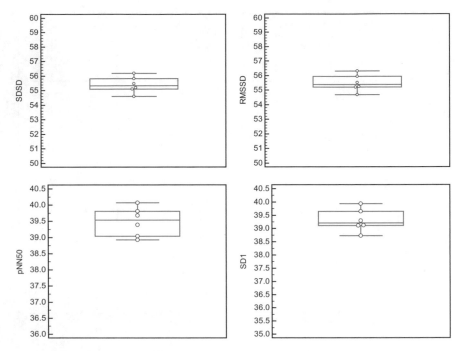

Fig. 3 A graphical statistical summary of SDSD, RMSSD, pNN50, and SD1 using box-and-whisker plots

For decades, HRV biofeedback has been studied to find possible ways to recover a specific body function as a therapy purpose [17, 18]. The patient utilizes biofeedback device to generate the maximal respiratory sinus arrhythmia (RSA) by attempting intentionally one deep breathing per 10 s during inhalation and exhalation, which is the heart rhythm that occurs when heart rate increases and decreases, respectively. This practice influenced time-, frequency-, and geometric-domain HRV indices, but especially, SDNN and Ln LF were highly increased compared to those for the normal breathing during HRV biofeedback [17]. This perfect example of the importance of an understanding HRV dependence on HR is a treatment mechanism for using HRV biofeedback.

One should be careful when indicating how the predictive value of HRV can be applied to the clinical practice if HR is a significant risk factor. Therefore, it is worth mentioning that all HRV parameters discussed in this paper should be carefully reviewed before being used for clinical application if HRV dynamic analysis shows that their indices are significantly associated with HR.

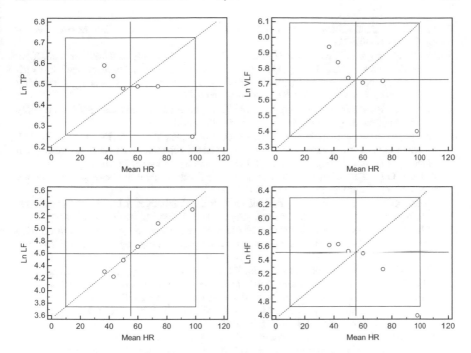

Fig. 4 Dependence of PSD on HR for Ln TP, Ln VLF, Ln LF, and Ln HF

Acknowledgements This research was supported by Hallym University Research Fund, 2017 (HRF-201711-013).

References

1. Buccelletti, E., Gilardi, E., Scanini, E., Galiuto, L., Persiani, R., Biondi, A., Basile, F., Silveri, N.G.: Heart rate variability and myocardial infarction: systematic literature review and metanalysis. Eur. Rev. Med. Pharmacol. Sci. **13**(4), 299–307 (2009)
2. Task Force of The European Society of Cardiology and The North American Society of Pacing and Electrophysiology (Membership of the Task Force listed in the appendix).: Heart rate variability: standards of measurement, physiological interpretation, and clinical use. Circulation. **93**(5), 1043–1065 (1996)
3. Saper, C.B., Cano, G., Scammell, T.E.: Homeostatic, circadian, and emotional regulation of sleep. J. Comp. Neurol. **493**(1), 92–98 (2005)
4. Berntson, G.G., Cacioppo, J.T.: Heart rate variability: stress and psychiatric conditions. In book: Malik, M., Camm, A.J. (eds.) Dynamic Electrocardiography, pp. 56–63, Blackwell Publishing, Oxford (2004)
5. Patel, V.N., Pierce, B.R., Bodapati, R.K., Brown, D.L., Ives, D.G., Stein, P.K.: Association of Holter-derived heart rate variability parameters with the development of congestive heart failure in the cardiovascular health study. J Am Coll Cardiol Heart Fail. **5**(6), 423–431 (2017)
6. Sacha, J., Pluta, W.: Alterations of an average heart rate change heart rate variability due to mathematical reasons. Int. J. Cardiol. **128**(3), 444–447 (2008)

7. Sacha, J.: Interaction between heart rate and heart rate variability. Ann. Noninvasive. Electro-cardiol. **19**(3), 207–216 (2014)
8. Van Hoogenhuyze, D., Weinstein, N., Martin, G.J., Weiss, J.S., Schaad, J.W., Sahyouni, X.N., Fintel, D., Remme, W.J., Singer, D.H.: Reproducibility and relations to mean heart rate vari-ability in normal subjects and patients with congestive heart failure secondary to coronary artery disease. Am. J. Cardiol. **68**(17), 1668–1676 (1991)
9. Fleiss, J.L., Bigger Jr., J.T., Rolinitzky, L.M.: The correlation between heart period variability and mean period length. Stat. Med. **11**(1), 125–129 (1992)
10. Sacha, J., Barabach, S., Statkiewicz-Barabach, G., Sacha, K., Muller, A., Piskorski, J., Barthel, P., Schmidt, G.: How to strengthen or weaken the HRV dependence on heart rate-description of the method and its perspectives. Int. J. Cardiol. **168**(2), 1660–1663 (2013)
11. Billman, G.E.: The effect of heart rate on the heart rate variability response to autonomic interventions. Front Physiol. **4**, 222–230 (2013)
12. Mietus, J.E., Peng, C.-K., Henry, I., Goldsmith, R.L., Goldberger, A.L.: The pNNx files: re-examining a widely used heart rate variability measure. Heart. **88**(4), 378–380 (2002)
13. Milicevic, G.: Low to high frequency ratio of heart rate variability spectra fails to describe sympatho-vagal balance in cardiac patients. Collegium Antropologicum. **29**(1), 295–300 (2005)
14. Hsu, C.H., Tsai, M.Y., Huang, G.S., Lin, T.C., Chen, K.P., Ho, S.T., Shyu, L.Y., Li, C.Y.: Poincare plot indexes of heart rate variability detect dynamic autonomic modulation during general anesthesia induction. Acta Anaesthesiol Taiwan. **50**(1), 12–18 (2012)
15. Singh Yadav, K.P., Saini, B.S.: Study of the aging effects on HRV measures in healthy subjects. Int. J. Comput. Theor. Eng. **4**(3), 346–349 (2012)
16. Maheshwari, A., Norby, F.L., Soliman, E.Z., Adabaq, S., Whitsel, E.A., Alonso, A., Chen, L.Y.: Low heart rate variability in a 2 min electrocardiogram recording is associated with an increased risk of sudden cardiac death in the general population: the atherosclerosis risk in communities study. PLoS One **11**(8), 1–12 (2016)
17. Lehrer, P.M., Gevirtz, R.: Heart rate variability biofeedback: how and why does it work? Front Psychol. **5**, 756–765 (2014)
18. Gevirtz, R.: The promise of heart rate variability biofeedback: evidence-based applications. Biofeedback. **41**, 110–120 (2013)

Review of Camera Calibration Algorithms

Li Long and Shan Dongri

Abstract Camera calibration is used to establish a mathematical model and solve the parameters of the camera through the correspondence between a series of scene points and pixel points. How to establish this mapping relationship is a key issue that needs to be solved in camera calibration. Various algorithms of calibration have been proposed by domestic and foreign scholars, including traditional visual calibration algorithm, camera self-calibration algorithm, and active-vision-based calibration algorithm. This paper focuses on some of the most widely used camera calibration algorithms and compares them.

Keywords Machine vision · Calibration · Algorithm

1 Traditional Visual Calibration Algorithm

The internal parameters of camera are their own attributes. In traditional visual calibration theory, it is generally considered that internal parameters are unchanged, which includes effective focal length, optical center, non-vertical factor, radial distortion, and tangential distortion. The external parameters of camera are the mapping relationship between the world and the image coordinate system, which consists of a rotation matrix and a translation vector. The following paragraphs will show you how to solve these parameters by the traditional calibration algorithm.

1.1 DLT Algorithm

The DLT algorithm [1] was proposed by Abdel-Aziz and Karara. They studied the relationship between images and objects in detail and built a linear model that can be

L. Long · S. Dongri (✉)
School of Mechanical & Automotive Engineering, Qilu University of Technology (Shandong Academy of Sciences), Jinan 250353, China
e-mail: shandongri@126.com

© Springer Nature Singapore Pte Ltd. 2019
S. K. Bhatia et al. (eds.), *Advances in Computer Communication and Computational Sciences*, Advances in Intelligent Systems and Computing 924,
https://doi.org/10.1007/978-981-13-6861-5_61

solved by direct linear transformation. The perspective projection matrix in homogeneous coordinates is shown in the following Eq. (1.1).

$$
\begin{bmatrix} u \\ v \\ 1 \end{bmatrix} = P_{3 \times 4} \begin{bmatrix} X_W \\ Y_W \\ Z_W \\ 1 \end{bmatrix} \tag{1.1}
$$

$(u, v, 1)$ and $(X_w, Y_w, Z_w, 1)$ in Eq. (1.1) represent homogeneous coordinates in the image and world coordinate system, respectively. $P_{3 \times 4}$ is the perspective matrix, which also rotates matrix and translation vector. According to Eq. (1.1), the equation group can be obtained as follows:

$$
\begin{cases} P_{11}X_W + P_{12}Y_W + P_{13}Z_W + P_{14} - P_{31}uX_W - P_{32}uY_W - P_{33}uZ_W - P_{34}u = 0 \\ P_{21}X_W + P_{22}Y_W + P_{23}Z_W + P_{24} - P_{31}vX_W - P_{32}vY_W - P_{33}vZ_W - P_{34}v = 0 \end{cases}
$$
$$\tag{1.2}$$

It can be known from Eq. (1.2) that a point pair formed by the image system and the world coordinate system can construct two equations. When there are n point pairs, $2n$ linear equations can be established. Then, Eq. (1.2) can be written as follows:

$$
MP = 0 \tag{1.3}
$$

where M is a $2n \times 12$ matrix.

Give the constraint: $P_{34} = 1$; $P' = -(C^T C)^{-1} C^T B$.

In the constraint, P' is composed of the first 11 elements in P; C is synthesized by the first 11 columns in M; B is the last column vector in M. Since M is a known matrix. The matrix P is determined by the parameters of the camera, and it can be seen that the matrix P can be obtained by taking six or more known points into formula (1.3).

1.2 Tasi Algorithm

The Tasi algorithm [2] is a classic two-step calibration algorithm proposed by Tasi. The first step is to solve the external parameters by linear transformation equations; the second step is to solve the internal parameters by nonlinear optimization. In The Tasi algorithm, the perspective model of camera and world coordinates is as follows:

$$\begin{bmatrix} x \\ y \\ 1 \end{bmatrix} = K(R,t) \begin{bmatrix} X_w \\ Y_w \\ Z_w \\ 1 \end{bmatrix} \tag{1.5}$$

$(x, y, 1)$ and $(X_w, Y_w, Z_w, 1)$ in Eq. (1.5) are the homogeneous coordinate of the image and world coordinate system; K is the internal parameter matrix; (R, t) are the rotation matrix and the translation vector. They can be expressed as follows:

$$K = \begin{bmatrix} fs & 0 & u_0 \\ 0 & f & v_0 \\ 0 & 0 & 1 \end{bmatrix} R = \begin{bmatrix} r_{11} & r_{12} & r_{13} \\ r_{21} & r_{22} & r_{23} \\ r_{31} & r_{32} & r_{33} \end{bmatrix} t = \begin{bmatrix} t_1 \\ t_2 \\ t_3 \end{bmatrix}$$

In the matrix K, f stands for focal length, (u_0, v_0) for distortion center coordinate, f_s for uncertainty factor. Equation (1.5) can be expanded to Eq. (1.6):

$$\begin{cases} x = u_0 + \dfrac{fs(r_{11}X + r_{12}Y + r_{13}Z + t_1)}{r_{31}X + r_{32}Y + r_{33}Z + t_3} \\ y = v_0 + \dfrac{fs(r_{21}X + r_{22}Y + r_{23}Z + t_2)}{r_{31}X + r_{32}Y + r_{33}Z + t_3} \end{cases} \tag{1.6}$$

As shown in Fig. 1, (x, y) is the coordinate of the P point in the ideal projection point of the camera, and (u, v) is the actual projection point coordinate. If the radial distortion parameter is considered, the following equation exists:

$$\begin{cases} (u_0 - x)\left[k_1(u^2 + v^2) + 1\right] = u_0 - u \\ (v_0 - y)\left[k_1(u^2 + v^2) + 1\right] = v_0 - v \end{cases} \tag{1.7}$$

Fig. 1 Schematic diagram of radial distortion

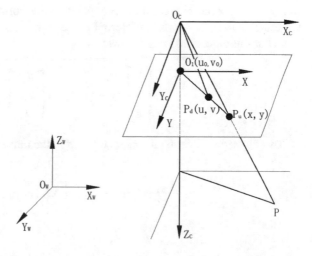

In Eq. (1.7), k_1 is the distortion factor; as shown in Fig. 1,(x, y), (u, v), (u_0, v_0) are collinear, so there exists equation:

$$\frac{(x - u_0)}{(y - v_0)} = \frac{(u - u_0)}{(v - v_0)} \qquad (1.8)$$

Bringing formula (1.6) into formula (1.8) and get formula (1.9)

$$\frac{s(r_{11}X + r_{12}Y + r_{13}Z + t_1)}{r_{21}X + r_{22}Y + r_{23}Z + t_2} = \frac{u - u_0}{v - v_0} \qquad (1.9)$$

According to the relationship between the model point and the image point, it can be linearly obtained $sr_{11}, sr_{12}, sr_{13}, st_1, r_{21}, r_{22}, r_{23}, t_2$; From the equation $r_{11}^2 + r_{12}^2 + r_{13}^2 = 1$, it can calculate $r_{11}, r_{12}, r_{13}, s$, further calculating t_1; Then, calculating r_{31}, r_{32}, r_{33} according to the equation $r_{31}, r_{32}, r_{33} = (r_{11}, r_{12}, r_{13}) \times (r_{21}, r_{22}, r_{23})$. Set the distortion factor $k_1 = 0$ as the initial value, and bring the previously obtained results into Eq. (1.6) to calculate t_3, f; Finally, nonlinear optimization through Eq. (1.10) to obtain a more accurate solution.

$$\begin{cases} \frac{fs(r_{11}X + r_{12}Y + r_{13}Z + t_1)}{r_{31}X + r_{32}Y + r_{33}Z + t_3}\left[1 + k_1(u^2 + v^2)\right] = u - u_0 \\ \frac{f(r_{21}X + r_{22}Y + r_{23}Z + t_2)}{r_{31}X + r_{32}Y + r_{33}Z + t_3}\left[1 + k_1(u^2 + v^2)\right] = v - v_0 \end{cases} \qquad (1.10)$$

1.3 Zhang Zhengyou Planar Pattern Algorithm

Zhang Zhengyou's planar pattern algorithm [3] is a widely used calibration algorithm. It is also a more mature algorithm. First, we use the camera to shoot this checkerboard from all angles, then extract the corners of the checkerboard in the image. Lastly, establish the mapping relationship between the coordinates of the corner points in the image and the coordinates of the corner points in the world coordinate system. So that we can calculate the camera's internal and external parameters. The mathematical model of this algorithm is as follows:

$$\begin{bmatrix} u \\ v \\ 1 \end{bmatrix} = K\begin{pmatrix} r_1 & r_2 & t \end{pmatrix}\begin{bmatrix} X_w \\ Y_w \\ 1 \end{bmatrix} \qquad (1.11)$$

This algorithm defaults the checkerboard in the plane of $Z = 0$.
set:

$$H = \begin{bmatrix} h_1 & h_2 & h_3 \end{bmatrix} = \lambda K(r_1 \ r_2 \ t) \qquad (1.12)$$

Then,

$$\begin{cases} r_1 = \dfrac{1}{\lambda} K^{-1} h_1 \\[2mm] r_2 = \dfrac{1}{\lambda} K^{-1} h_2 \\[2mm] r_3 = r_1 \times r_2 \\[2mm] r_4 = \dfrac{1}{\lambda} K^{-1} h_3 \end{cases} \tag{1.13}$$

According to the nature of the rotation matrix:

$$\begin{cases} r_1^{\mathrm{T}} r_2 = 0 \\ \|r_1\| = \|r_2\| = 1 \end{cases} \tag{1.14}$$

Bringing Eq. (1.13) into Eq. (1.14) can get the two basic constraint equations for each image:

$$\begin{cases} h_1^{\mathrm{T}} K^{-\mathrm{T}} K^{-1} h_2 = 0 \\ h_1^{\mathrm{T}} K^{-\mathrm{T}} K^{-1} h_1 = h_2^{\mathrm{T}} K^{-\mathrm{T}} K^{-1} h_2 \end{cases} \tag{1.15}$$

The camera has five unknown internal parameters. According to formula (1.15), as long as the number of images is greater than or equal to three, the internal parameters can be calculated linearly by constructing six constraint equations, and then Eq. (1.13) is calculated to solve the external parameter r_1, r_2, r_3, t.

1.4   Camera Calibration Method Based on Cross-Ratio Invariance

In the former three methods, nonlinear distortion parameters are as internal parameters of the camera. Its calibration is obtained by pre-setting the initial value and along with other camera internal parameters through the nonlinear optimization. A method of camera calibration based on the constant cross-ratio [4] was proposed by Zhang Guojun et al. It is a method to calibrate the distortion parameters of camera separately. The algorithm implementation principle is as follows.

According to the constant nature of the cross-ratio, we can get:

$$\begin{cases} \dfrac{(X_1 - X_3)(X_2 - X_4)}{(X_2 - X_3)(X_1 - X_4)} = \mathrm{CR} \\[3mm] \dfrac{(Y_1 - Y_3)(Y_2 - Y_4)}{(Y_2 - Y_3)(Y_1 - Y_4)} = \mathrm{CR} \end{cases} \tag{1.16}$$

$(X_1, Y_1), (X_2, Y_2), (X_3, Y_3)$, and (X_4, Y_4) are image point coordinates corresponding to four points of the spatial collinear line, respectively. One-stair radial-distorted camera model is:

$$\begin{cases} X_i = \overline{X_i}(1 + k_1 r_i^2) \\ Y_i = \overline{Y_i}(1 + k_2 r_i^2) \quad (i = 1, 2, 3, 4) \\ r_i^2 = \overline{X_i^2} + \overline{Y_i^2} \end{cases} \tag{1.17}$$

Put Eq. (1.16) into Eq. (1.17) can obtain the distortion parameter k_1, k_2.

2 Camera Self-calibration Algorithm

The self-calibration implementation only needs a set of image series and establishes the constraint equation according to the pole correspondence in each image. It has better flexibility because no specific standard block is needed. The commonly used self-calibration algorithms are based on the projection theory of absolute conic and absolute quadrics.

2.1 Self-calibration Algorithm Based on Absolute Conic

The self-calibration algorithm proposed by Faugeras, Luong, Maybank, etc., which is a method directly solved based on the Kruppa equation [5, 6]. The Kruppa equation is exported by using the theory of absolute curve and epipolar transformation.

As shown in Fig. 2, A_1 and A_2 are the projection image curves of the space quadratic curve A at two viewpoints, respectively. l_{1a}, l_{1b} and l_{2a}, l_{2b} are tangent to quadratic curves A and B, respectively. l_{1a} and l_{1b} intersect at pole e_1, l_{2a} and l_{2b} intersect at pole e_2. π_∞ is the infinity plane. According to the nature of the polar line transformation [7]:

$$[e]_\times^T KK^T [e]_\times \cong F^T KK^T F \tag{2.1}$$

In Eq. (2.1), $[]_\times$ represents the anti-symmetric matrix. From Eq. (2.1), five quadratic equations can be obtained, but only two of them are independent [8]. Equation (1.2) is often referred to the Kruppa equation due to its special form. Two Kruppa equations are available for every two images. If you want to solve the camera's five parameters, you need at least three images.

2.2 Calibration Algorithm Based on Absolute Quadric

The concept of absolute quadrics was introduced into self-calibration research by Triggs [9]. The principle of this method is as follows:

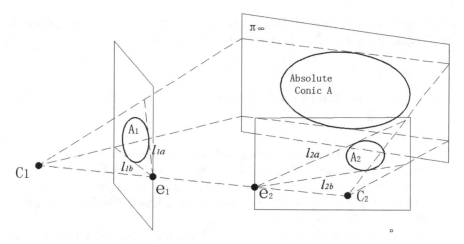

Fig. 2 Schematic diagram of quadratic projection

$$\omega^* \cong KK^T = \lambda P_i \Omega^* P_i^T \qquad (2.2)$$

Cross-multiplying for the corresponding term of the two-sided matrix of (2.2), and eliminating λ can get:

$$[\omega]_{kl}[P_i\Omega^*P_i^T]_{k'l'} - [\omega]_{k'l'}[P_i\Omega^*P_i^T]_{kl} = 0 \qquad (2.3)$$

k, l represents the kth row and the lth column of the matrix. For each image, 15 equations of the form such as (2.3) can be obtained, but only five of them are independent [10]. Triggs proposed using the sequential quadratic programming (SQP) optimization algorithm to directly solve the minimum value of the cost function $\sum_i \|\omega \wedge (P_i\Omega^*P_i^T)\|^2$ under the constraint of det $(\Omega^*) = 0$. Through this method, the camera parameters can be obtained if three images can be given at least, and the initial value needs to be estimated.

3 Active-Vision-Based Calibration Algorithm

The active-vision-based calibration algorithm acquired the image data by the autonomously controlled camera. The camera motion information is known during the calibration process which is the main advantage of this method. In general, the camera parameters can be solved linearly, and the calculation is simple and robust. Ma Songde's triple orthogonal motion method [11] and Hartley's camera-based pure rotation calibration [12] are classic active-vision-based calibration method. The disadvantage of this calibration method is the high cost of the system. It is not applicable for the camera motion unknown and uncontrollable condition.

3.1 Calibration Based on Three Orthogonal Translational Motions

The calibration based on the three orthogonal translational motions was proposed by Ma Songde. It describes a method of active vision camera calibration systematically. And it proves that the calibration of the external parameters of the camera only needs to reconstruct the spatial point and use the pole information. The basic algorithm principle of its calibration is as follows:

By controlling the camera to make a set of three orthogonal motions, calculating the three focus of expansions (FOEs) according to the corresponding points of the acquired images, denoted them as F_1, F_2, F_3, according to the properties of FOE, the constraint equations of f_u, f_v, u_0, v_0 can be obtained:

$$\begin{cases} \left[\begin{array}{ccc} \frac{F_{1u}-u_0}{f_u} & \frac{F_{1v}-v_0}{f_v} & 1 \end{array} \right] \left[\begin{array}{ccc} \frac{F_{2u}-u_0}{f_u} & \frac{F_{2v}-v_0}{f_v} & 1 \end{array} \right]^{\mathrm{T}} = 0 \\ \left[\begin{array}{ccc} \frac{F_{1u}-u_0}{f_u} & \frac{F_{1v}-v_0}{f_v} & 1 \end{array} \right] \left[\begin{array}{ccc} \frac{F_{3u}-u_0}{f_u} & \frac{F_{3v}-v_0}{f_v} & 1 \end{array} \right]^{\mathrm{T}} = 0 \\ \left[\begin{array}{ccc} \frac{F_{2u}-u_0}{f_u} & \frac{F_{2v}-v_0}{f_v} & 1 \end{array} \right] \left[\begin{array}{ccc} \frac{F_{3u}-u_0}{f_u} & \frac{F_{3v}-v_0}{f_v} & 1 \end{array} \right]^{\mathrm{T}} = 0 \end{cases} \tag{3.1}$$

Simplifying Eq. (3.1) and setting $x = u_0$, $y = \frac{v_0 f_u^2}{f_v^2}$, $z = \frac{f_u^2}{f_v^2}$, two linear equations can be obtained:

$$\begin{cases} (F_{1u} - F_{3u})x + (F_{1v} - F_{3v})y - F_{2v}(F_{1v} - F_{3v})z = F_{2u}(F_{1u} - F_{3u}) \\ (F_{2u} - F_{3u})x + (F_{2v} - F_{3v})y - F_{1v}(F_{2v} - F_{3v})z = F_{1u}(F_{2u} - F_{3u}) \end{cases} \tag{3.2}$$

It can be seen that x, y, and z cannot be solved uniquely by Eq. (3.2). It needs to do another triple orthogonal motion. Literature [13] proves that when any four of the six translational motions are not coplanar in two sets of triple orthogonal motions, x, y, and z can be linearly uniquely solved from two constrained equations of the form such as 3.2, and then f_u, f_v, u_0, v_0 can be solved.

3.2 Calibration Based on Camera Pure Rotation

The algorithms of calibrating a camera based on the camera's pure rotational motion around its optical center were proposed by Hartley. The basic principles are as follows:

When the camera makes a pure rotational motion around its optical center, image correspondence before and after motion is:

$$\begin{cases} U_{1i} \approx KX_i \\ U_{2i} \approx KRX_i \end{cases} \tag{3.3}$$

Therefore, $U_{2i} \approx KRK^{-1}U_{1i} \approx HU_{1i}$. The matrix H can be obtained from multiple sets of image corresponding points (≥ 4). If Det $(H) = 1$ is defined, then

$$H = KRK^{-1} \tag{3.4}$$

Transpose and right multiplications the two sides of the original Eq. (3.4), we can get:

$$HKK^T H^T = KK^T \tag{3.5}$$

Equation (3.5) is the basic constraint equation of these calibration algorithms. In Eq. (3.5), $C = KK^T$ is an unknown and H is a known number. If C becomes a known amount, the matrix K can be obtained by Cholesky decomposition, so camera calibration problems will translate into questions about how to solve matrix C.

4 Comparison of Camera Calibration Algorithms

Which calibration method is chosen often depends on a variety of factors. Camera self-calibration has better flexibility and practicality than traditional methods, but its accuracy needs to be further improved. Active-vision-based calibration is simple to calculate and can obtain higher precision, but has higher requirements on calibration system accuracy. The advantages and disadvantages of the three camera calibration algorithms are shown in Table 1.

Table 1 Comparison of camera calibration algorithms

Calibration algorithms	Advantage	Disadvantage	Applications
Traditional visual calibration	High precision	Calibration is complex and requires calibration blocks	High precision, camera parameters are basically unchanged
Camera self-calibration	Not need calibration blocks, good flexibility	Low precision	The camera needs to move often or cannot set a known reference
Active-vision-based calibration	Simple calculation	High system cost	Camera running information is known

5 Conclusion

This paper reviews the camera calibration algorithms that is widely used at home and abroad, and introduces the main algorithm implementation principles. So far, the theoretical problems of camera calibration have been solved basically through decades of unremitting efforts. The current research focuses on how to improve the precision and robustness of camera calibration and how to use these theories to solve problem of actual vision.

References

1. Abdel-Aziz, Y.I., Karara, H.M.: Direct linear transformation into object space coordinates in close-range photo grammetry. In: Urbana proceedings of Symposium. Close-Range Photogrammetry, pp. 1–18 (1971)
2. Tsai, R.Y.: An efficient and accurate camera calibration technique for 3D machine vision. In: Proceedings of IEEE Conference of Computer Vision and Pattern Recognition, pp. 364–374 (1986)
3. Zhang, Z.Y.: A flexible new technique for camera calibration. Microsoft Corporation: Technical Report (1998)
4. Zhang, G., He, J., Yang, X.: Calibrating camera radial distortion with cross-ratio invariability. Opt. Laser Technol. **35**, 457–461 (2003)
5. Ma, S.D.: A self calibration technique for active vision system. IEEE. Trans. Robot. Autom. **12**(1), 114–120 (1996)
6. Hartley, R., Gupta, R., Chang, T.: Stereo from uncalibrated cameras. In: Proceedings of Computer Vision and Pattern Recognition, pp. 761–764 (1992)
7. Faugeras O, Luong, Q.T., Maybank, S.: Camera self calibration: theory and experiments. In: Proceedings of the 2nd European Conference on Computer Vision, pp. 321–334, Italy (1992)
8. Luong, Q.T.: Matrix Fondamentale et Calibration Visuelle sur I'Environnement. [Ph.D thesis]. Centre DcOrsay: Universite Paris-Sud (1992)
9. Triggs, B.: Auto-calibration and the absolute quadric. In: Proceedings of Computer Vision and Pattern Recognition, pp. 609–614 (1997)
10. Quan, L.: Uncalibrated 1D projective camera and 3D affine reconstruction of lines. In: Proceedings of Computer Vision and Pattern Recognition, pp. 60–65, USA (1997)
11. De Ma, S.: A Self-calibration technique for active vision system. IEEE. Trans. Robot. Autom. **12**(1), 114–120 (1996)
12. Hartley, R.: Self-calibration of stationary cameras. Int. J. Comput. Vision. **22**(1), 5–23 (1997)
13. Wu, F., Li, H., Hu, Z.: A new cetive vision based camera self-calibration technique. Acta Automatica Siniea. **27**(6), 736–746 (in Chinese) (2001)

Breast Cancer Prediction: Importance of Feature Selection

Prateek

Abstract In today's world, breast cancer is one of the most widespread causes of death in women. According to an estimation, approximately 40,920 women would die in 2018 just because of breast cancer, which is a highly alarming number. Such alarming numbers could be reduced if the cancer is diagnosed at an early stage. With the advent of technology, making such predictions has become an easier task. Machine learning is one of the latest trends, which enables to make predictions related to diseases based on physical or behavioral characteristics. In this paper, we use various machine learning algorithms like decision trees, k-nearest neighbor (KNN), logistic regression, neural networks (NNs), naïve Bayes, random forest, and support vector machine (SVM). The outcome is then compared based on the precision, recall, and F1 score. Furthermore, we identify the least important features in the dataset, implement all these algorithms again after removing those features, and then compare the outcomes for the two implementation stages in order to understand the importance of feature selection in breast cancer prediction.

Keywords Machine learning · KNN · Feature selection · SVM · Logistic regression · Naïve Bayes · Classification · Prediction algorithms · Breast cancer

1 Introduction

Breast cancer currently is one of the most widespread cancer occurring in women. It is reported that in 2018 alone, more than two millions of new cases of breast cancer has been reported. Due to such alarmingly high numbers, it is also the second most commonly occurring cancer. These statistics show the number of cases that have been diagnosed or treated for breast cancer. There are high chances that there would be many more such cases that have not yet been diagnosed by it but are suffering from breast cancer. Such cases generally pose a threat to the life of the patient

Prateek (✉)
QR No. 1012, SECTOR 4/c, Bokaro Steel City, Jharkhand, India
e-mail: prateek.ps0794@gmail.com

© Springer Nature Singapore Pte Ltd. 2019   733
S. K. Bhatia et al. (eds.), *Advances in Computer Communication
and Computational Sciences*, Advances in Intelligent Systems and Computing 924,
https://doi.org/10.1007/978-981-13-6861-5_62

since the cancer is diagnosed at a much later stage when it is nearly impossible to cure it effectively. Hence, early prediction and treatment of breast cancer are highly important presently. Achieving such feats in the past was a big obstacle but due to the introduction of newer techniques like machine learning, it is becoming highly possible to predict breast cancer based on various factors and symptoms. Algorithms like logistic regression, KNN, SVM, and NN have made it possible for the doctors to identify patients that might have breast cancer. The main motive of this paper is to enable this prediction based on certain observations recorded for each patient.

The dataset used in this paper is Breast Cancer Wisconsin (Diagnostic) Data Set. This data set has 30 features recorded for each patient, and the main motive of this paper is to understand the changes in the prediction accuracy that occur once different sets of features are chosen for the prediction models. Initial implementations involve selecting all 30 features for the prediction models and then compare them with the second stage of implementation in which we chose 15 of the most important features present in the dataset. This paper will be helpful in deciding if removal of features can be done or not which might ultimately decide the speed of the systems implemented. This paper is structured as follows: Sect. 2 contains a literature review for other studies in this field. Section 3 describes the methodology giving a brief overview on the different algorithms used. Section 4 describes the implementation. Section 5 compares the output for the various implementations, and Sect. 6 is the conclusion for this work.

2 Literature Review

Tyrer et al. in [1] state the importance of the roles played by various background factors of a woman that might play a big role in determining if the woman is prone to carrying the breast cancer genes. The factors suggested by the author are records of past breast cancer patients in the family, and in some cases, even the personal medical history has been found to play a role in determining so. Shah and Anjali in [2] used various data analytics algorithms like decision tree, Bayesian algorithm, and k-nearest neighbor, compared them on the bases of some parameters like kappa statistics, relative absolute error, root relative squared error, and concluded by finding that naive Bayes algorithm takes the least response time for the prediction model. In [3], Setiono focuses on the importance of a good algorithm which can be used for extracting rules from a trained NN. The author also emphasizes the importance of cleaning the dataset and removing null values that can result in improving the efficiency of the prediction model by providing faster results, thus enabling the training of more networks that in turn would provide us with even better extraction rules. Sarvestani and Soltani in [4] assert that statistical NN can be of great help in diagnosing breast cancer and to reach to this conclusion the author tests two internationally available datasets across various NN algorithms like multilayer perceptron, self-organizing map, and probabilistic NN. The author then compares the results based on mean-squared error and meant time to find that PNN is the most efficient and accurate way for breast cancer prediction. In [5], the author emphasizes the importance of using machine learning tools in making medical predictions. One such algorithm, which the author uses, is the least square support vector machine (LS-SVM) classifier algorithm. The author examines its efficiency on various factors like confusion matrix, analysis of sensitivity and specificity, classification accuracy, and k-fold cross-validation and found that it is most efficient on a 50–50% training and testing dataset. In [6], the author suggests using decision tree algorithm in top-down approach to categorize the cancer as benign or malignant, which in turn, will help in providing the physician with an extra opinion and improve the ability of ultrasonography in the differential diagnosis of breast cancer. This model turns out to provide an accuracy of 96%. In [7], the importance of association rules along with NN is established. The author states that association rules are used to simplify the input space and then the NN can be applied to obtain faster and more accurate results. The author compares this model with the NN algorithm by threefold cross-validation method to conclude that the combination of two provides a better accuracy (97.4%) than NN (95.2%). In [8], the author predicts the survival rate for breast cancer patients using three classification algorithms, which are artificial NN, decision trees, and logistic regression and then compares them on parameters like sensitivity, classification accuracy, and specificity using tenfold cross-validation method. In addition, the author suggests the importance of different factors in the prediction process with the differential grade being the most important factor followed by the stage of cancer. Gupta et al. in [9] have conducted a detailed review on two of the biggest medical challenges today, which are, diagnosis of the breast cancer category and the recurrence chances on breast cancer in a patient. They studied the various techniques and algorithms that are currently being used and concluded that artificial NNs are most efficient in finding the chances of cancer recurrence in a person where the diagnosis models can be made by extracting the best features from each model to create an efficient model. Hassanien and Ali in [10] suggest the importance of using intelligent systems like rough set theory to find out dependencies in dataset, approximate set classification, reducing data. They select the attributes and generate the rough set dependency rules after normalizing the attributes. These rules are then used to find the redundant data which is the minimal subset of the attributes. The author concludes that this method is quite effective in building intelligent models for predictive analysis. Bellaachia and Erhan predict the survival rate for breast cancer patients using three data mining techniques which are backpropagated NN, Naïve Bayes, and C4.5 decision tree algorithm. After comparing, the three algorithms they state that NN and decision trees perform comparably. In the future, the authors believe that they can include the missing data by taking values from an older dataset, which will immensely increase the dataset size and will enable in performing better predictions. Baker et al. in [12] tries to study if artificial NN can be used to categorize the cancer tumor to be benign or malignant. The network was built on standardized lexicon of Breast Imaging Recording and Data System (BI-RADS) which had 18 inputs of which 10 were from the BI-RADS system and rest 8 of the inputs were taken from the past medical records of the patient. Baker found out that the positive prediction value drastically improved at a particular output network threshold. In addition, the specificity of the network was greater than that of the radiologists. Kharya in [13] has reviewed various data analysis technologies in order to differentiate benign from malignant cancer and to study the cancer prognosis. The author studies numerous methodologies which include decision trees, digital mammography classification using association rule mining and artificial NN, naïve Bayes classifier, support vector machines, logistic regression, and Bayesian networks. The author concludes by finding the decision tree to be the best prediction technique with an accuracy of 93.62%. Xiong in [14] mentions three existing methods to diagnose breast cancer, namely fine needle aspirate (FNA), mammography, and surgical biopsy. The author states that surgical biopsy is the most accurate one, but it is also the most expensive method available and hence suggests building a model using FNA with a data mining algorithm and statistics to find out an even better way. The author uses principal component analysis (PCA) and partial least squares (PLS) linear regression analysis for statistics and combines them with data mining techniques like select attributes, decision trees, and associate rules. The author concludes by encouraging the combined usage of these methodologies with each other in order to achieve the best model. Chaurasia and Saurabh in [15] use RepTree, RBF network, and simple logistic and compare them on tenfold cross-validation to evaluate the efficiency of each system. They found out that the simple logistic was the most accurate of them all with an accuracy of 74.5%. In addition, the authors also concluded that in order to decrease the dimension of the feature space, one could use simple logistics, whereas RepTree and RBF network can be used to get an automatic diagnosis in a much faster way.

3   Methodology

Figure 1 below describes the workflow of the paper diagrammatically. Initially, the dataset is considered and the data headers are carefully studied and examined to get a clear picture of the data mentioned in the set. Various heat maps are plotted to establish a sense of correlation between the various features mentioned in the dataset. Then, this data is also displayed on histograms, pie charts, and bar plots to get a clear understanding of the picture.

The next important step to be taken is the data preprocessing. Preprocessing the data helps in numerous ways. It removes redundant data from the dataset which ultimately provides an unbiased prediction. It also enables to create a model with a faster prediction time by omitting the fields that have null of invalid format of values. The missing values can be taken care of in various ways. One of them is to replace are empty fields by zero, but this method decreases the efficiency of the model drastically. Therefore, the most common method of taking care of missing data is to fill the data field with the average value of that data column. Once the missing data has been handled, the next important task is the features' selection. It might not be necessary that all the features in the dataset might play a role in the model; there might be cases where certain features have no or less effect on the outcome of a model. In such cases, such features shall be omitted in order to achieve a faster model. Once the

Fig. 1 Work flow for the breast cancer prediction model

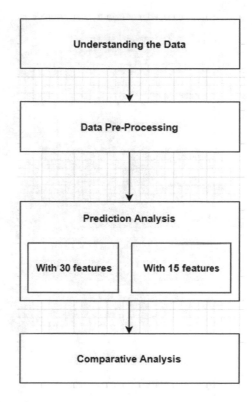

preprocessing part is handled, the next part is to apply the various machine learning algorithms to the datasets. The algorithms that can be used are decision trees, KNN, logistic regression, NN, naïve Bayes, random forest, and SVM. These algorithms shall be used with and without the features in order to compare the efficiency of the two implementations later on to arrive at a final result.

4 Experimentation

Breast Cancer Wisconsin (Diagnostic) Data Set is used for carrying out the experiment. It has a total of 32 attributes that aid in predicting whether the cancer is benign or malignant. First, the dataset is analyzed at depth. The count, mean, standard deviation, twenty-fifth median, and seventy-fifth percentile are calculated. Next, data preprocessing is conducted. The missing values are found, and it is observed that column unnamed has missing values for all entries and hence that field is removed. Also, the first field which has the serial number for entries is removed as it has no relevant contribution to the prediction. Now, there are 30 attributes remaining which are divided into ten attributes related to mean, ten for standard error, and ten for worst.

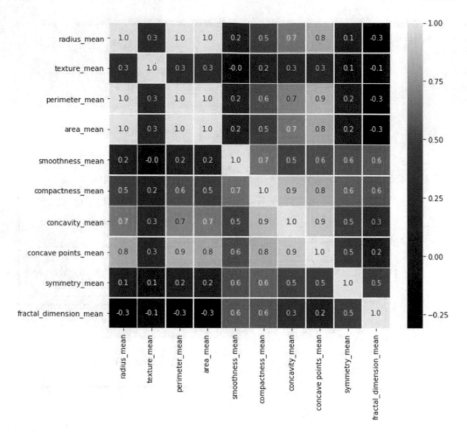

Fig. 2 Heat map for correlation between mean attributes

For analyzing the correlation between attributes, heat maps are used. Correlation is found out between all the attributes, and between the attributes relating to mean, standard error, and worst. Next, the dataset is split into training (70%) and testing (30%) sets. Once all the classification algorithms including logistic regression, decision trees, random forest, SVM, KNN, naïve Bayes, and NN with Keras are applied to the data set with all the thirty attributes. Next, based on the previous correlation study, feature extraction is performed to get the best fifteen attributes with most relevance to the prediction and the algorithms are applied again. Precision, recall, and F1 score are calculated for all to find out the relevancy and accuracy of the tests. At last, a comparative study of the results is done to find out the difference in results with and without feature selection. The dataset has *B* for benign and *M* for malignant which is changed to 0 and 1, respectively, for prediction. I used various algorithms due to their specific advantages. For example, I used decision trees' algorithm since it enables easier feature selection by processing just the top few nodes. I used neural network since it needs much less background statistics and it enables it better feature

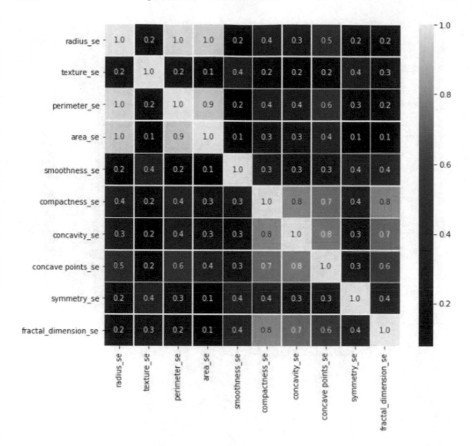

Fig. 3 Heat map for correlation between standard error attributes

selection with the same. The usage of random forest enabled to combine several different independent base classifiers.

5 Results and Discussions

Figures 2, 3, and 4 are the three heat maps for the mean, standard error, and worst attributes. The colors show the degree of correlation among the attributes and hence aid feature selection. Figure 5 shows the histogram plot for precision, recall, and F1 scores for thirty attributes for all algorithms, and Fig. 6 shows the same for fifteen attributes after feature selection. Figure 7 is a comparative plot of the precision score for thirty attributes versus fifteen attributes.

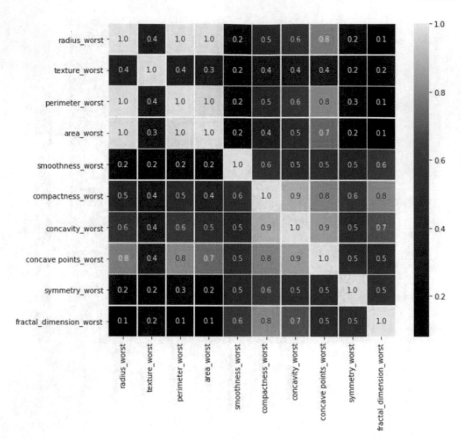

Fig. 4 Heat map for correlation between worst attributes

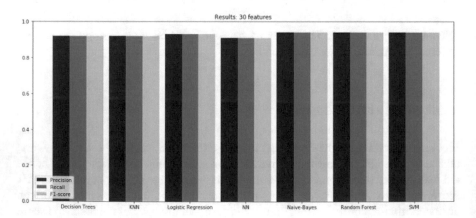

Fig. 5 Histogram for 30 features

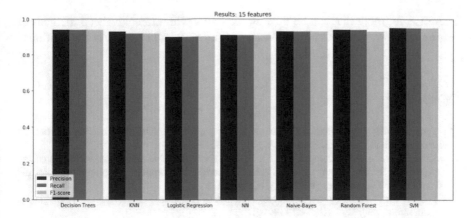

Fig. 6 Histogram for 15 features

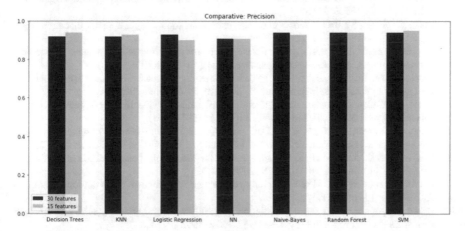

Fig. 7 Histogram for comparative precision scores

6 Conclusion

With the application of machine learning algorithms, the research makes an effort to predict breast cancer in its early stages. With thirty features, naïve Bayes, random forest, and SVM yield the best result with a precision score of 0.94, while SVM performs the best with fifteen features with a precision score of 0.95. This study shows that prediction algorithms with feature selection outperform those without feature selection in most cases. Decision trees, KNN, and SVM clearly show better results while random forest and NN show a similar result. Only logistic regression and naïve Bayes show better results for thirty attributes. It is also advantageous to exclude less relevant features as it is beneficial in the face of computational cost.

There is a lot more that can be done in the future with more advanced methods of feature selection and in-depth application of different and better-suited algorithms.

References

1. Tyrer, J., Duffy, S.W., Cuzick, J.: A breast cancer prediction model incorporating familial and personal risk factors. Stat. Med. **23**(7), 1111–1130 (2004)
2. Shah, C., Jivani, A.G: Comparison of data mining classification algorithms for breast cancer prediction. In: 2013 Fourth International Conference on Computing, Communications and Networking Technologies (ICCCNT), IEEE (2013)
3. Setiono, R.: Generating concise and accurate classification rules for breast cancer diagnosis. Artif. Intell. Med. **18**(3), 205–219 (2000)
4. Sarvestani, A., Soltani, et al.: Predicting breast cancer survivability using data mining techniques. In: 2010 2nd International Conference on Software technology and Engineering (ICSTE), vol. 2, IEEE (2010)
5. Polat, K., Güneş, S.: Breast cancer diagnosis using least square support vector machine. Digit. Signal Proc. **17**(4), 694–701 (2007)
6. Kuo, W-J., et al.: Data mining with decision trees for diagnosis of breast tumor in medical ultrasonic images. Breast. Cancer. Res. Treat. **66**(1), 51–57 (2001)
7. Karabatak, M., Ince, M.C.: An expert system for detection of breast cancer based on association rules and neural network. Expert. Syst. Appl. **36**(2), 3465–3469 (2009)
8. Delen, D., Walker, G., Kadam, A.: Predicting breast cancer survivability: a comparison of three data mining methods. Artif. Intell. Med. **34**(2), 113–127 (2005)
9. Gupta, S., Kumar, D., Sharma, A.: Data mining classification techniques applied for breast cancer diagnosis and prognosis. Indian. J. Comput. Sci. Eng. (IJCSE). **2**(2), 188–195 (2011)
10. Hassanien, A E., Ali, J.M.H.: Rough set approach for generation of classification rules of breast cancer data. Informatica. **15**(1), 23–38 (2004)
11. Bellaachia, A., Guven, E.: Predicting breast cancer survivability using data mining techniques. Age. **58**(13), 10–110 (2006)
12. Baker, J.A., et al.: Breast cancer: prediction with artificial neural network based on BI-RADS standardized lexicon. Radiology. **196**(3), 817–822 (1995)
13. Kharya, S.: Using data mining techniques for diagnosis and prognosis of cancer disease. arXiv preprint arXiv:1205.1923 (2012)
14. Xiong, X., et al.: Analysis of breast cancer using data mining and statistical techniques. In: Sixth International Conference on Software Engineering, Artificial Intelligence, Networking and Parallel/Distributed Computing, 2005 and First ACIS International Workshop on Self-Assembling Wireless Networks. SNPD/SAWN 2005, IEEE (2005)
15. Chaurasia, V., Pal, S.: Data mining techniques: to predict and resolve breast cancer survivabilitys (2017)

Enactment of LDPC Code Over DVB-S2 Link System for BER Analysis Using MATLAB

Shariq Siraj Uddin Ghouri, Sajid Saleem and Syed Sajjad Haider Zaidi

Abstract In this paper, Digital Video Broadcast—Satellite Second Generation (DVB-S2) Simulink model on MATLAB is evaluated. To analyze the performance especially while working on 8PSK modulation, FEC rates as 3/4, 3/5, and 9/10 along with low-density parity-check (LDPC) codes chained with Bose–Chaudhuri–Hochquenghem (BCH) codes for effective transmission. To examine this application performance, several entries of energy per symbol noise ratio (Es/No) were changed for model behavior detection. LDPC decoder iteration relationship analysis and LDPC bit error rate analysis have been carried out. After examining this has been evaluated at which FEC rate this model response finest.

Keywords 8PSK · LDPC · BCH · DVB-S2 · MATLAB · BER

1 Introduction

In 2003, Digital Video Broadcasting (DVB) project extended the second-generation design for applications related to satellite broadband services as DVB-S2 standard [ETSI EN 302 307-1] [1]. This classification allows satellite operations such as Internet access, digital satellite news gathering (DSNG), and TV and Radio broadcasting. DVB-S2 has been stated to fulfill three characteristics reasonable receiver complexity, total flexibility, and best transmission performance reaching the Shannon limit. Low-density parity-check (LDPC) codes as channel coding techniques

S. S. U. Ghouri (✉)
Faculty of Electrical Engineering (Communication Systems), Pakistan Navy Engineering College—National University of Sciences and Technology (NUST), Karachi, Pakistan
e-mail: shariq.ghouri2015@pnec.nust.edu.pk

S. Saleem · S. S. H. Zaidi
National University of Sciences & Technology (NUST), Islamabad, Pakistan
e-mail: ssaleem@pnec.nust.edu.pk

S. S. H. Zaidi
e-mail: sajjadzaidi@pnec.nust.edu.pk

© Springer Nature Singapore Pte Ltd. 2019
S. K. Bhatia et al. (eds.), *Advances in Computer Communication and Computational Sciences*, Advances in Intelligent Systems and Computing 924, https://doi.org/10.1007/978-981-13-6861-5_63

and BPSK, QPSK, 8PSK, 16APSK, and 32APSK as latest ModCods have been adopted in combination for better working and performance of the system on nonlinear satellite channel. Maximum flexibility and synchronization specified by framing structure are considered for the worst low SNR configuration cases. Adaptive Coding Modulation (ACM) allows effective communication parameters in one-to-one links communication. Modes with the backward-compatibility allow integrated receivers and decoders of DVB-S2 which works during the period in transition [2].

The objective of this paper is to execute the DVB-S2 Link Simulink Model in MATLAB and study the performance especially while working on 8PSK modulation, FEC rates as 3/4, 3/5, and 9/10 along with low-density parity-check (LDPC) codes chained with Bose–Chaudhuri–Hochquenghem (BCH) codes for effective transmission. To analyze bit error rate and occurrence related to iteration which substantiates an adequate system performance with reduced system complexity and poor channel conditions. After analysis, this has been evaluated at which FEC rate this model response best and finally validate the results whether they are according to the standards of ESTI which states that lower the modulation coding rate the better will be the communication performance.

In this paper, 8PSK modulation along with LDPC and BCH channel coding techniques is used structure of this paper follows: Sect. 2 covers basics related to this research such as Shannon–Hartley Theorem, 8PSK bit mapping constellation and LDPC codes with DVB-S2 standard. Section 3 describes physical layer packet and frame structure of DVB-S2 along with the technical description of MATLAB Simulation Model. Section 4 reveals the simulation process leading to the results. In Sect. 5, conclusions are depicted.

2   Basics of this Research

2.1   Shannon–Hartley Theorem

The Shannon–Hartley theorem depicts noisy channel capacity keeping random noise as assumption. Theorem states that

$$''C = B \log 2[1 + (S/N)]''$$

"(C = channel capacity in bps, B = channel bandwidth in Hz, S/N = signal-to-noise ratio at the channel output or receiver input)."

This theorem highlights signal-to-noise ratio importance and bandwidth in the application of communication. In order to provide channel capacity, increased bandwidth comprehends decreased signal power in provided channel capacity. "Increase in the channel bandwidth increases the channel capacity of a noisy channel", as would actually be suggested by the Shannon–Hartley theorem. "Increasing the bandwidth also increases noise resulting in decreasing the signal-to-noise" [3].

Fig. 1 8PSK constellation

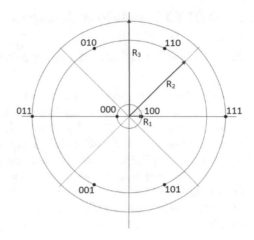

2.2 Bit Mapping into 8PSK Constellation

Constellations of 8PSK and 8APSK are of eight points, distributed over three rings. The first ring consists of two points, second consist of four points, and on the third ring, two points are located $(2 + 4+2 = 8)$ [1]. Constellation is illustrated in Fig. 1.

2.3 LDPC Coding Over DVB-S2 Link

The ETSI EN 302 307 standard has been developed to broadcast, provision of interactive services, to gather news and supports other services related to broadband satellite (DVB-S2) [4]. Coding schemes have increased the channel capacity. In 1960, Gallager while working over his seminal doctoral thesis invented LDPC code which can attain exceptionally low error rates close to channel capacity. Combination of low-density parity-check and Bose–Chaudhuri–Hochquenghem codes is the basic requirement of this scheme of code [5].

BCH codes outer layer is needed to rectify the sporadic type of errors generated through decoder of LDPC. DVB-S.2 offers quasi-error-free process regardless of any type of transmission medium (error rate of packet is below 10^{-7}) from 0.7 dB till 1 dB according to "Shannon limit" [6].

3 DVB-S2 Standard with Application

3.1 DVB-S2 Packet and Frame Structure

DVB-S2 standard comprises of two frames, one is the physical layer frame while other is an FEC frame. FEC frame includes the transmitted data in a structured form of the transport stream. Data for transmission contained within the frame of FEC known as generic data. Baseband header is a part of data field comprises of 80 bit in a data field layer [7]. Error protection code rate is then padded to baseband header and data field. Selected code rate to protect from error is expanded to the data block with the baseband header and afterward Bose–Chaudhuri–Hochquenghem code along with low-density parity-check code appended. FEC frame of DVB-S2 having frame length 16,200 or 64,800 bits is depicted in Fig. 2 [6].

3.2 Technical Description of Simulation Model

The DVB-S2 simulation is realized in Simulink, MATLAB is shown in Fig. 3.
  Its main components are the following signal processing blocks:

- BBFRAME Buffering block—used to prepare BB (Base Band) frames to serve as input frames for the BCH encoder. All frames are arranged according to the BCH

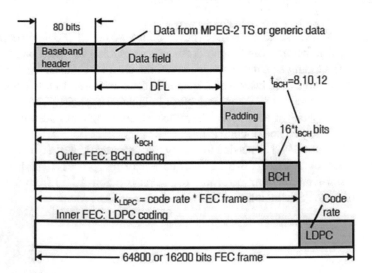

Fig. 2 FEC frame of DVB-S2

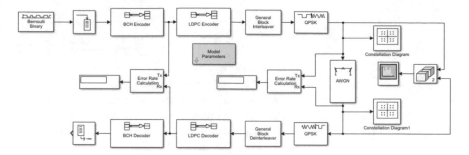

Fig. 3 Transceiver Simulink diagram of DVB-S2 system

encoder input data size. Input data frames (188 bytes or 1504 bit) are stacked up
to the size determined by the number of information bits transferred within one
BCH code word. Where necessary, the input BCH frame is stuffed with zeroes to
ensure the fixed size of all encoder input frames.

- BCH encoder block—performs forward error correction encoding. BB frames
  prepared in the BBFRAME Buffering block are processed by the BCH encoder.
  BCH encoder adds redundant bits that are used for correction of errors caused by
  transmission over error-prone wireless channel.
- LDPC encoder block—performs internal error correction encoding based on par-
  ity bit calculation and their insertion into the information bit sequence. In this
  simulation, the output FEC frame (after BCH and LDPC encoding) will always
  retain a fixed size of 64,800 bits. LDPC encoding is the last block of the error
  correction processing.
- Block Interleaver—performs interleaving of bits from received FEC frames in
  order to distribute energy and reduce burst errors. In the simulation, bit interleaving
  is performed by writing the frame data into columns and reading three consecutive
  columns as rows.
- Modulator block—performs signal modulation. The simulation offers two modulation scheme options: QPSK with any of the eleven code ratio values, and 8PSK
  with the 3/5, 2/3, 3/4, 5/6, 8/9, or 9/10 code ratio.

The Simulink-designed simulation covers the basic mechanisms of signal processing and transmission during signal broadcast in the DVB-S2 system [8].

4 Results

This research has been carried out on Simulink MATLAB over which the designed
transceiver DVB-S2 system has been re-evaluated. The transmitted signal (TX) was
compared with the received signal (RX) after that qualitative and quantitative com-

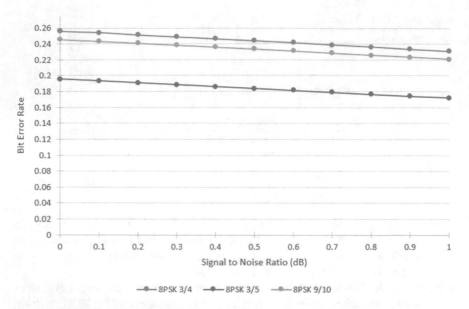

Fig. 4 LDPC BER result using DVB-S2 model—SNR ranges 0.1–1.0 dB

parison of the received signal was conducted in the form of occurrence of iteration and bit error rate analysis. Performance of this system and correction of the errors made by the LDPC codes was studied under the simulation. Modulated system was changed thrice (8PSK 3/4, 8PSK 3/5, 8PSK 9/10). Values of SNR ranging from 0.1 to 1 dB were changed and the system was put under test to analyze the results. LDPC codes were especially noted and evaluated at multiple values to test this model under the light of E_b/N_o's vector and LDPC curves were generated.

Figures 4 and 5 depict the SNR relationship at 8PSK 3/4, 8PSK 3/5 and 8PSK 9/10. These figures palpably noticed that 8PSK 3/5 converges over least BER values while 8PSK 3/4 and 8PSK 9/10 falls behind which is due to positive effects of the LDPC decoder.

Figure 6 shows no. of iterations distribution executed by decoder of LDPC coding.

Author has analyzed the performance of 8PSK modulation scheme along with the FEC rates which are 3/4, 3/5, and 9/10 over DVB-S2 Link Simulink Model in MATLAB. Detailed analysis has been done which results as 8PSK 3/5 performs best as its BER converges at 6 dB of signal-to-noise ratio and this model obeys ETSI standard for DVB-S2 Link.

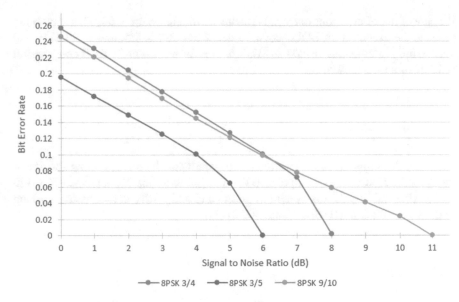

Fig. 5 LDPC BER result using DVB-S2 model—SNR ranges 0–11 dB

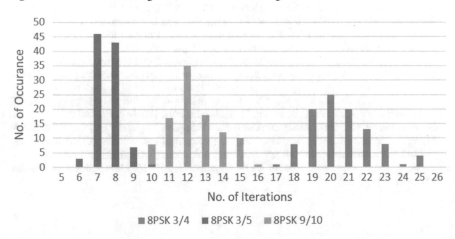

Fig. 6 No. of iterations distribution executed by the decoder of LDPC

5 Conclusions

Flourishing results has been simulated by the use of DVB-S2 Link Simulink Model in MATLAB where LDPC and BCH coders as well as decoders were incorporated. This model obeys the standard of ESTI according to performance regarding bit error rates. Overall demonstration of least ModCods produced improved bit error rates. Performance of the system is well appreciated as convergence of maximum limit in the case of no. of iterations occurred remains within the count of 50. This simulation resolute the execution of LDPC codes at 8PSK 3/4, 8PSK 3/5, and 8PSK 9/10 and successfully attained the results as 8PSK 3/5 performs best as its BER converges at 6 dB of signal-to-noise ratio.

6   Future Work

Future work in this scope of research can be conducted by incorporating sound and video recorded stream as an input to DVB-S2 Link Simulink Model in MATLAB to analyze the stability of this system to understand and justify the scenarios working under ETSI standard [ETSI EN 302 307-2] [1].

References

1. ETSI, EN: 302 307–2 V1.1.1 (2014–10); citation_journal_title = Digital Video Broadcasting (DVB). In: Second Generation Framing Structure, Channel Coding and Modulation Systems For Broadcasting, Interactive Services, News Gathering and Other Broadband Satellite Applications
2. Morello, A., Mignone, V.: DVB-S2: the second generation standard for satellite broad-band services. In: Proceedings of the IEEE 2006, vol. 94(1), pp. 210–227
3. Maini, A.K., Agrawal, V.: Satellite technology: principles and applications. 2011. Wiley
4. ETSI, EN: 302 307 V1.1.1 (2004–06). European Standard (Telecommunications series) (2004)
5. Gallager, R.: Low-density parity-check codes. IRE Trans Inf Theor **8**(1), 21–28 (1962)
6. Abusedra, L.F., Daeri, A.M., Zerek, A.R.: Implementation and performance study of the LDPC coding in the DVB-S2 link system using MATLAB. In: 2016 17th International Conference on Sciences and Techniques of Automatic Control and Computer Engineering (STA), IEEE (2016)
7. Méric, H., Piquer, J.M.: DVB-S2 spectrum efficiency improvement with hierarchical modulation. In: 2014 IEEE International Conference on Communications (ICC), IEEE (2014)
8. Baotic, P., et al.: Simulation model of DVB-S2 system. In: 2013 55th International Symposium ELMAR, IEEE (2013)

Correction to: User Behaviour-Based Mobile Authentication System

**Adnan Bin Amanat Ali, Vasaki Ponnusamy
and Anbuselvan Sangodiah**

Correction to:
**Chapter "User Behaviour-Based Mobile Authentication
System" in: S. K. Bhatia et al. (eds.),** *Advances in Computer
Communication and Computational Sciences,*
Advances in Intelligent Systems and Computing 924,
https://doi.org/10.1007/978-981-13-6861-5_40

The original version of the book was inadvertently published with the co-author's name as "Vasaki Ponnusamay", which has been now corrected as "Vasaki Ponnusamy" in Chapter 40. The erratum chapter and the book have been updated with the changes.

The updated version of this chapter can be found at
https://doi.org/10.1007/978-981-13-6861-5_40

Author Index

Printed in the United States
By Bookmasters